W9-CRV-090

Light Scattering by Nonspherical Particles

Theory, Measurements, and Applications

Light Scattering by Nonspherical Particles

Theory, Measurements, and Applications

Edited by

Michael I. Mishchenko

NASA Goddard Institute for Space Studies
New York, New York

Joop W. Hovenier

Free University and University of Amsterdam
Amsterdam, The Netherlands

Larry D. Travis

NASA Goddard Institute for Space Studies
New York, New York

ACADEMIC PRESS

A Harcourt Science and Technology Company

San Diego   San Francisco   New York   Boston   London   Sydney   Tokyo

Cover art description and credit: Scattering patterns for particles of a single size are usually burdened by the so-called interference structure. This effect is demonstrated by the figure on the cover, which shows T-matrix computations of the degree of linear polarization versus scattering angle and surface-equivalent-sphere size parameter for an oblate spheroid with an aspect ratio of 1.7 and a refractive index of $1.53 + 0.008i$. The spheroid has a fixed orientation with respect to the incident beam, and the scattering plane is defined as a plane through the incident beam and the spheroid axis. The figure displays a characteristic "butterfly wing" pattern composed of sharp alternating maxima and minima. These strong oscillations of polarization within a small range of scattering angle and/or size parameter are a typical manifestation of the interference structure and make comparisons of scattering characteristics of different monodisperse particles particularly difficult. From a modified version of Plate 2.1. Refer to Chapter 2 for more details. Cover art courtesy of the authors.

Academic Press
A division of Harcourt Brace & Company
525 B Street, Suite 1900, San Diego, California 92101-4495, USA
http://www.apnet.com

Academic Press
24–28 Oval Road, London NW1 7DX, UK
http://www.hbuk.co.uk/ap/

Library of Congress Catalog Card Number: 99-61962

International Standard Book Number: 0-12-498660-9

PRINTED IN THE UNITED STATES OF AMERICA
99 00 01 02 03 04 MM 9 8 7 6 5 4 3 2 1

Contents

Part I
Introduction

Chapter 1
Concepts, Terms, Notation
Michael I. Mishchenko, Joop W. Hovenier, and Larry D. Travis

Chapter 2
Overview of Scattering by Nonspherical Particles
Michael I. Mishchenko, Warren J. Wiscombe, Joop W. Hovenier, and Larry D. Travis

Chapter 3
Basic Relationships for Matrices Describing Scattering by Small Particles
Joop W. Hovenier and Cornelis V. M. van der Mee

Part II
Theoretical and Numerical Techniques

Chapter 4
Separation of Variables for Electromagnetic Scattering by Spheroidal Particles
Ioan R. Ciric and Francis R. Cooray

Chapter 5

The Discrete Dipole Approximation for Light Scattering by Irregular Targets

Bruce T. Draine

Chapter 6

T-Matrix Method and Its Applications

Michael I. Mishchenko, Larry D. Travis, and Andreas Macke

Chapter 7

Finite Difference Time Domain Method for Light Scattering by Nonspherical and Inhomogeneous Particles

Ping Yang and K. N. Liou

Part III

Compounded, Heterogeneous, and Irregular Particles

Chapter 8

Electromagnetic Scattering by Compounded Spherical Particles

Kirk A. Fuller and Daniel W. Mackowski

Part IV
Laboratory Measurements

Part V
Applications

Chapter 15

Light Scattering and Radiative Transfer in Ice Crystal Clouds: Applications to Climate Research

K. N. Liou, Yoshihide Takano, and Ping Yang

Chapter 16

Centimeter and Millimeter Wave Scattering from Nonspherical Hydrometeors

Kültegin Aydın

Chapter 17

Microwave Scattering by Precipitation

Jeffrey L. Haferman

Chapter 18

Polarized Light Scattering in the Marine Environment

Mary S. Quinby-Hunt, Patricia G. Hull, and Arlon J. Hunt

Chapter 19

Scattering Properties of Interplanetary Dust Particles

Kari Lumme

Chapter 20

Biophysical and Biomedical Applications of Nonspherical Scattering

Alfons G. Hoekstra and Peter M. A. Sloot

Contributors

Numbers in parentheses indicate the pages on which the authors' contributions begin

Kültegin Aydın (451), Department of Electrical Engineering, Pennsylvania State University, University Park, Pennsylvania 16802

Petr Chýlek (273), Atmospheric Science Program, Departments of Physics and Oceanography, Dalhousie University, Halifax, Nova Scotia B3H 3J5, Canada

Ioan R. Ciric (89), Department of Electrical and Computer Engineering, University of Manitoba, Winnipeg, Manitoba R3T 5V6, Canada

Francis R. Cooray (89), CSIRO Telecommunications and Industrial Physics, Epping, New South Wales 1710, Australia

J. Steven Dobbie (273), Atmospheric Science Program, Department of Physics, Dalhousie University, Halifax, Nova Scotia B3H 3J5, Canada

Bruce T. Draine (131), Princeton University Observatory, Princeton, New Jersey 08544

Kirk A. Fuller (225), Department of Atmospheric Science, Colorado State University, Fort Collins, Colorado 80523

D. J. Wally Geldart (273), Atmospheric Science Program, Department of Physics, Dalhousie University, Halifax, Nova Scotia B3H 3J5, Canada

Bo Å. S. Gustafson (367), Department of Astronomy, University of Florida, Gainesville, Florida 32611

Jeffrey L. Haferman (481), Fleet Numerical Meteorology and Oceanography Center, Satellite Division, Monterey, California 93943

Alfons G. Hoekstra (585), Faculty of Mathematics, Computer Science, Physics, and Astronomy, University of Amsterdam, Kruislaan 403, 1098 SJ Amsterdam, The Netherlands

Joop W. Hovenier (xix, 3, 29, 61, 355), Department of Physics and Astronomy, Free University, De Boelelaan 1081, 1081 HV Amsterdam, The Netherlands; and Astronomical Institute "Anton Pannekoek," University of Amsterdam, Kruislaan 403, 1098 SJ Amsterdam, The Netherlands

Patricia G. Hull (525), Department of Physics, Tennessee State University, Nashville, Tennessee 37209

Arlon J. Hunt (525), Lawrence Berkeley National Laboratory, University of California, Berkeley, California 94720

K. N. Liou (173, 417), Department of Atmospheric Sciences, University of California, Los Angeles, Los Angeles, California 90095

Kari Lumme (555), Observatory, University of Helsinki, FIN-00014 Helsinki, Finland

Andreas Macke (147, 309), Institut für Meereskunde, Abt. Maritime Meteorologie, Düsternbrooker Weg 20, 24105 Kiel, Germany

Daniel W. Mackowski (225), Department of Mechanical Engineering, Auburn University, Auburn, Alabama 36849

Michael I. Mishchenko (xix, 3, 29, 147), NASA Goddard Institute for Space Studies, New York, New York 10025

Karri Muinonen (323), Observatory, Kopernikuksentie 1, University of Helsinki, FIN-00014 Helsinki, Finland

Mary S. Quinby-Hunt (525), Lawrence Berkeley National Laboratory, University of California, Berkeley, California 94720

Kenneth Sassen (393), Department of Meteorology, University of Utah, Salt Lake City, Utah 84112

Peter M. A. Sloot (585), Faculty of Mathematics, Computer Science, Physics, and Astronomy, University of Amsterdam, Kruislaan 403, 1098 SJ Amsterdam, The Netherlands

Yoshihide Takano (417), Department of Atmospheric Sciences, University of California, Los Angeles, Los Angeles, California 90095

Larry D. Travis (xix, 3, 29, 147), NASA Goddard Institute for Space Studies, New York, New York 10025

H. C. William Tso (273), Atmospheric Science Program, Department of Physics, Dalhousie University, Halifax, Nova Scotia B3H 3J5, Canada

Hendrik Christoffel van de Hulst (xxv), Leiden Observatory, 2300 RA Leiden, The Netherlands

Cornelis V. M. van der Mee (61), Dipartimento di Matematica, Università di Cagliari, Via Ospedale 72, 09124 Cagliari, Italy

Gorden Videen (273), U.S. Army Research Laboratory, Adelphi, Maryland 20783

Warren J. Wiscombe (29), NASA Goddard Space Flight Center, Greenbelt, Maryland 20771

Ping Yang (173, 417), Department of Atmospheric Sciences, University of California, Los Angeles, Los Angeles, California 90095

Preface

A great variety of science and engineering disciplines have significant interest in scattering of light and other electromagnetic radiation by small particles. For example, this subject is important to climatology because the Earth's radiation budget is strongly affected by scattering of solar radiation by cloud and aerosol particles. Another example is remote sensing of the Earth and planetary atmospheres, which relies largely on analysis of the parameters of radiation scattered by aerosols, clouds, and precipitation. The scattering of light by homogeneous and layered spherical particles composed of isotropic materials can be computed readily using the conventional Lorenz–Mie theory or its modifications. However, many natural and artificial small particles have nonspherical overall shapes or lack a spherically symmetric internal structure. Examples of such nonspherical particles are mineral and soot aerosols, cirrus cloud and contrail particles, liquid cloud particles with asymmetrically located inclusions, hydrometeors, snow and frost crystals, particles composed of anisotropic materials, ocean hydrosols, interplanetary and cometary dust grains, planetary ring particles, particles forming planetary and asteroid surfaces, and biological microorganisms. It is now well recognized that the scattering properties of nonspherical particles can differ dramatically from those of "equivalent" Mie spheres. Therefore, the ability to accurately compute or measure light scattering by nonspherical particles in order to clearly understand the effects of particle nonsphericity on scattering patterns is very important.

Electromagnetic scattering by nonspherical particles was considered in some detail in the classical monographs by van de Hulst (1957), Kerker (1969), and Bohren and Huffman (1983), as well as in a collection of papers edited by Schuerman (1980a). However, the rapid advancement in computers and experimental techniques as well as the development of improved analytical and numerical methods over the past two decades has resulted in a much better understanding and knowledge of scattering by nonspherical particles that has not been systematically summarized. Furthermore, papers on different aspects of this subject are contained in dozens of various scientific and engineering journals, a fact that often leads to an inefficient use of the accumulated knowledge and unnecessary redundancy in research activities. Therefore, the primary aim of this collective treatise is to provide a systematic state-of-the-art summary of the field, including analytical and numerical methods for computing electromagnetic scattering by nonspherical particles, measurement approaches, knowledge of typical features in scattering patterns, retrieval and remote sensing techniques, nonspherical particle characterization and practical applications. Considering the widespread need for this information in optics, geophysics, remote sensing, astrophysics, engineering, medicine, and biology, we hope that the book will be useful to many graduate students, scientists, and engineers working on various aspects of electromagnetic scattering and its applications.

The framework for this book was conceived during the organizing of an international conference, "Light Scattering by Nonspherical Particles: Theory, Measurements, and Applications." This conference was held 29 September to 1 October 1998 at the NASA Goddard Institute for Space Studies in New York and was sponsored by the National Aeronautics and Space Administration, the American Meteorological Society, and the American Geophysical Union in cooperation with the Optical Society of America. Leading experts in respective areas were invited to give designated review talks and write review chapters that formed this volume. Although the book is the product of a number of authors, a concerted effort was made to present the material with a unified notation and consistent terminology in order to create a coherent work.

The volume opens with a foreword written by Professor H. C. van de Hulst and is divided into five parts with 20 chapters. Part I (Chapters 1–3) provides a general introduction to the book. Chapter 1 formulates the subject of the book and introduces basic definitions and a unified system of terminology and notation, which are systematically and consistently used throughout most of the text. Chapter 2 gives a concise overview of electromagnetic scattering by nonspherical particles, describes the extent to which various aspects of this subject are covered by the book, and provides essential references in a few areas not reviewed in a dedicated chapter. Chapter 3 uses first principles to derive general relationships for matrices describing electromagnetic scattering by small particles. These relationships can be employed for theoretical purposes as well as for testing scattering matrices that have been obtained numerically or experimentally.

Part II (Chapters 4–7) describes several widely used techniques for computing electromagnetic scattering by nonspherical particles based on numerically solving Maxwell's equations. Chapter 4 reviews the method of separation of variables as applied to the spheroidal wave equations and its implementation in scattering of a uniform plane wave with an arbitrary incidence direction by a single spheroid. Appropriate rotational–translational addition theorems are then used to formulate the solution for systems of spheroids. Numerical results computed with a specified accuracy are useful for evaluating simpler, approximate analysis methods. Chapter 5 introduces the discrete dipole approximation, provides a clear statement of the precise nature of the approximation, explains how calculations based on this approximation may be performed, describes the public-domain DDSCAT package of Draine and Flatau, and briefly mentions practical applications. Chapter 6 presents a general formulation of the T-matrix approach, describes an efficient analytical method for computing orientation-averaged scattering characteristics, discusses methods for calculating the T matrix for single and composite/aggregated particles, and reviews practical applications. Chapter 7 outlines the finite difference time domain method, describes current efforts aimed at optimizing the computer CPU and memory requirements and at improving the numerical accuracy, and discusses applications of the method to computations of scattering properties of aerosol and cloud particles.

Part III (Chapters 8–11) deals with light scattering by aggregated, heterogeneous, and randomly shaped particles. Chapter 8 presents a review of theoretical and applied studies of electromagnetic scattering and absorption by compound systems of spheres. Both aggregates of two or more spherical particles and spherical particles with one or more arbitrarily located spherical inclusions are considered. Chapter 9 reviews plausible derivations of classical effective medium approximations for heterogeneous particles as well as approximations extended to inclusion sizes beyond the limit of Rayleigh scattering and establishes their ranges of applicability by comparing approximate results with exact theoretical computations and laboratory measurements. Chapter 10 describes a hybrid approach for computing electromagnetic scattering by a large particle with multiple internal inclusions. This approach uses the Lorenz–Mie theory to compute single scattering on an inclusion and a Monte Carlo solution of the radiative transfer equation to describe multiple scattering inside the host particle. Chapter 11 describes how the shapes of small irregular particles can be modeled using multivariate lognormal statistics resulting in Gaussian random spheres and cylinders. It then uses the Rayleigh, Rayleigh–Gans, and geometric optics approximations as well as volume integral equation, discrete dipole approximation, and perturbation series methods to compute scattering by such particles in different size parameter ranges.

Part IV (Chapters 12 and 13) deals with laboratory measurements of electromagnetic scattering by nonspherical particles. Chapter 12 describes a successful experimental setup developed and built at the Free University, Amsterdam, for measuring all 16 elements of the scattering matrix in the visible part of the spectrum and discusses results of measurements for a representative selection of natural particle samples. Chapter 13 introduces the microwave analog technique and briefly reviews previous microwave analog measurements. It then describes a modern, fully automated, broadband microwave scattering facility developed and built at the University of Florida and discusses a large collection of measurements obtained with this facility.

Applied aspects of electromagnetic scattering by nonspherical particles are discussed in Part V (Chapters 14–20). Chapter 14 explains the basic physical principle of the lidar backscatter depolarization technique, evaluates its current cloud and aerosol capabilities and limitations, and describes how this technique has contributed to our knowledge of the composition and structure of a variety of cloud and aerosol types. Chapter 15 describes theoretical computations and measurements of single and multiple scattering of visible and thermal radiation by ice cloud crystals and discusses effects of ice particle nonsphericity on results of remote sensing retrievals and studies of the cloud direct radiative forcing of climate. Chapter 16 reviews centimeter and millimeter wave scattering character-istics of nonspherical hydrometeors, such as hailstones, graupel, snowflakes, rain-drops, and ice crystals, and describes the use of this information for distinguishing habits of hydrometeors and estimating their mass and precipitation rate with radar. Chapter 17 reviews the radiative transfer theory necessary for modeling precipita-tion at microwave frequencies and examines the effect of partially and randomly oriented nonspherical hydrometeors on results of remote sensing retrievals based on measurements of microwave brightness temperature of upwelling radiation by satellite radiometers. Chapter 18 examines the scattering from irregularly shaped particles in the ocean (microorganisms and mineral matter) and in the marine atmosphere (resulting from the desiccation of sea spray and contributions from land), as well as the highly complex scattering associated with sea ice. Chapter 19 discusses scattering properties of interplanetary dust grains and particles forming cometary comas and surfaces of atmosphereless solar system bodies. The chap-ter reviews results of observational studies as well as modeling results based on state-of-the-art techniques for computing scattering by nonspherical and aggre-gated particles. Chapter 20 reviews biophysical and biomedical applications of electromagnetic scattering, with some emphasis on the polarization state of the scattered light, discusses two techniques (flowcytometry and ektacytometry) that are in routine use in biomedicine, and introduces a theoretical framework for in-terpreting experimental scattering data.

Although the individual authors were primarily responsible for their chapters, the chapters are not completely self contained. The form of each chapter has been standardized to the extent possible, redundancy has been minimized, and a unified list of references appears at the back of the book. In each chapter, equations are numbered in one sequence throughout and referred to as Eq. (1), Eqs. (3)–(5), for example. Similar rules apply to numbering of figures and tables.

We thank the authors for their efforts to make the book a complete volume rather than simply a collection of individual contributions. We appreciate the following chapter reviewers for their constructive criticism on earlier versions of the text: Kültegin Aydın, Franklin Evans, Martha Hanner, Nikolai Khlebtsov, Alexei Kouzoubov, Valeri Maltsev, Toshiyuki Murayama, Timo Nousiainen, Ronald Pinnick, Jussi Rahola, Michael Schulz, Cornelis van der Mee, and J. Vivekanandan. We thankfully acknowledge the continuing support of this project by the NASA Radiation Science Program managed by Robert Curran. We are grateful to the Academic Press staff, especially to Frank Cynar, Cathleen Ryan, and Cheryl Uppling for their patience, encouragement, and assistance, and to the Technical Typesetting Inc. staff for the excellent copy editing and typesetting of the manuscript. Special thanks are due to Nadia Zakharova of Science Systems and Applications, Inc., who provided valuable assistance at all stages of preparing this book.

MICHAEL I. MISHCHENKO
New York

JOOP W. HOVENIER
Amsterdam

LARRY D. TRAVIS
New York

April 1999

Hints from History: A Foreword

Hendrik Christoffel van de Hulst

Leiden Observatory
Leiden
The Netherlands

I. LORENZ–MIE THEORY

Nearly 20 years ago a book fully devoted to scattering by nonspherical particles appeared (Schuerman, 1980a). I have no clear memory of the conference preceding it, but I cherished the book. At that time, the microwave analog method seemed to stand out in clarity and reliability over the courageous attempts of various numerical methods.

The balance has now shifted. Taking advantage of the rising tide of computer development, authors from many countries have developed numerical methods with great ingenuity and admirable perseverance. I shall not say much about this subject, for it will be reviewed in great detail later in this monograph.

Instead, my intention is to look back to over half a century ago, when Mie scattering was getting ripe for a similar review of analytical and numerical results. I am aware that many readers of this book were not even born at that time. Yet, a look at this ancient history may yield certain hints for choosing a wise strategy in dealing with light scattering by nonspherical particles.

With nostalgia I think back on the year before my thesis, 53 years ago, when I never tired of entering yet a different library or making a glossary of the notations used by yet a different author. A pleasant surprise was that the Utrecht University library, after some insistence on my part, managed to dig up from the basement the thesis of Peter Debye, written almost 40 years earlier in 1908. The Mie paper of the same year impressed me less and the still earlier Lorenz solution had not yet been rediscovered.

What got me started on this subject was some advice from Jan Oort. One day, I told him what I was doing for my then intended thesis, which aimed at explaining the extinction curve of interstellar dust. He said, "You *cite* the results which Schalén and Greenstein have computed by means of the Mie theory. But have you ever looked at that theory yourself?" I had not, so I started right away. This subject had so many intriguing aspects that the one chapter on Mie theory I had planned for my thesis developed into many. Finally, under some pressure of time to accept a postdoctoral position, I decided to make this the sole topic of my thesis and to postpone the astrophysical applications until later.

In the years that followed, I had many opportunities to give talks about Mie scattering at various departments, physics, astronomy, meteorology, chemistry, and engineering, throughout the United States and elsewhere. In each place I learned something new about the details that could be misunderstood. Gradually, this experience prepared me for writing a book, which appeared in 1957 and was reprinted in 1981 (van de Hulst, 1957). Of course, the development did not stop at that time, but this is all I need in the way of introduction.

II. LIMITS OF APPLICABILITY

In the majority of applications, the assumption of homogeneous spherical particles is highly unrealistic. That much was clear from the start to all serious research workers. Meteorologists studying raindrops and chemists working with latex spheres were free from this criticism. Even so, meteorologists should be aware that hailstones sometimes have eccentric layers of different composition, and I remember at least one paper discussing the distortion of the rainbow because a falling raindrop is not exactly round. More outrageous was the assumption of spherical shape for interstellar dust. One early author computed Mie scattering by many kinds of metal particles (e.g., zinc spheres). I read those papers with a smile, suspecting that the only reason for making this assumption was that metal spheres require a smaller size, and hence a shorter computation, to come anywhere near a curve resembling the observed extinction curve.

Another assumption, which is crucial for particles that are large compared to a wavelength, is the smoothness of the surface. The many rapidly shifting peaks and valleys in the Mie curves for large particles clearly arise from interference effects. Yet anyone who has tried to arrange a demonstration experiment with light showing interference effects knows how much cleaning and adjusting is necessary to make the demonstration successful. How would interstellar dust particles grown over millions of years by accretion ever acquire a polished surface? In fact since 1949, the year of the accidental discovery of interstellar polarization, we have known that they must be nonspherical, while the galactic magnetic field prevents fully random orientation.

The upshot of these objections is that the efforts of most scientists applying Mie computations to a problem in nature were unwarranted. They introduced the assumption of a smooth sphere merely to make the problem manageable and hoped that the answer would not be too far from the truth. Perhaps now, half a century later, we have reached a time in which more realistic assumptions about the shape can be effectively introduced in the calculations.

III. MIE THEORY: DETAILED RESULTS

Let me return briefly to the Mie theory. About 1956 I noticed that authors at conferences started to be apologetic, saying: "I won't bother you with details, for of course you can compute those yourself." This gave me about 10 years (1946–1956) to collect from many sides a great deal of useful details and interim results that had been published in articles and even in thick volumes of tables. These included:

(a) The final results of Mie computations, in particular, the extinction curve and the scattering pattern. The extinction is the real part of a complex function. It is often helpful to plot that complex function, rather than the extinction alone. This is a typical example of the use of an interim result.
(b) The values of the coefficients a_n and b_n, which can (for nonabsorbing spheres) also be represented by two real angles, α_n and β_n.
(c) The complex Riccati–Bessel functions (see Fig. 1), which were used in the calculations of items (a) and (b).

From this abundant collection of data, I could freely select those that fit in the didactic concept of my book. In fact, that book contains very few calculations of my own.

I would like to show one illustration, which I made some 10 years ago, plotting numbers from an old book of tables of Riccati–Bessel functions. The function

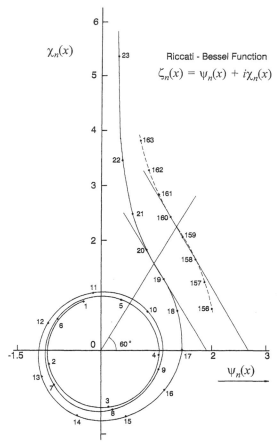

Figure 1   Complex Riccati–Bessel function $\zeta_n(x)$ plotted for $x = 20$ (solid curve) and partially for $x = 160$ (dashed curve).

plotted in the complex plane is $\zeta_n(x) = \psi_n(x) + i\chi_n(x)$. In both curves, the size parameter x is fixed, and the order n is an integer, running in principle from 1 to infinity. The full curve, for $x = 20$, shows a spiral around the unit circle, widening with increasing n. At $n + 1/2 = x$ it reaches a transition point, which is also an inflection point. Beyond this point the imaginary part goes to infinity and the real part to zero. Of the dashed curve, plotted for the eight times higher value, $x = 160$, only four points at both sides of the transition point are shown. The pattern is entirely similar, on a scale proportional to $x^{1/6}$, that is, $2^{1/2}$ larger than the other curve.

The interpretation is found in the conversion of the Mie solution for large x into ray optics. The transition point stands for rays tangential to the sphere; for smaller n the rays hit the sphere; and for larger n they miss it. Of course, Debye knew all this 90 years ago, but to see the transition so clearly displayed was new to me. It is aesthetically pleasing to see in Fig. 1 that the transition point lies at the top of an equilateral triangle with one corner at the origin. Once seen in the figure, all these properties can also be derived (with some effort!) from the exact formulas.

In more recent work on Mie theory two developments please me most:

(a) The many papers, mainly by Gouesbet and co-workers, on scattering of a focused (laser) beam that illuminates the sphere eccentrically
(b) The glare points (in some papers wrongly called rainbows) showing under which angles the most intense radiation exits from a sphere fully illuminated by a distant source

Items (a) and (b) are in a sense reciprocal, but to my knowledge a clear analysis of their interrelation has not yet been given.

IV. STRATEGY FOR NONSPHERICAL PARTICLES

The aim of the preceding summary of Mie scattering is to extract possible hints about what to do with nonspherical particles. Such hints may help to increase both the ease and the practical use of the computations. Experience shows that such lengthy computations have multiple uses and users with different needs. First, there may be the immediate "client," who may wish to have just one number or one curve. Second, later analysts may wish to interpolate between various assumptions, and for that purpose need more detailed numbers. Here are the hints that I have collected, but more might be apparent after reading this book.

A. AGREE ON TERMINOLOGY AND NOTATION AND, WHERE USEFUL, ON METHOD

This is obvious and to a large extent is being done already. It may be advisable to define special guidelines for cases with narrowed-down assumptions, such as:

- Particles with axial symmetry
- Particles of any shape in random orientation
- Particles with sharp edges
- Porous particles

B. Aim at Effective Mutual Verification

In Mie theory this happened by itself, for many authors selected the refractive indices 1.5 (glass), or 1.33 (water), or $1 + \epsilon$ (particles in biological tissues). The numerous parameters of irregular particles make such a chance coincidence very unlikely. Yet, mutual verification of the results obtained by different persons, and particularly by different methods, remains strongly advisable. Selecting some standard problems for this purpose might be wise.

C. Read between the Lines

An impressive body of computational results is gradually being built. With determination and good luck, this may yield extra information by interpolation, that is, by "reading between the lines." A well-known example from Mie scattering is that the extinction curves for different refractive indices, when plotted versus x times (refractive index -1) instead of x, reveal a striking common pattern. Looking for similar extras, however hidden, in the scattering by nonspherical particles may be equally rewarding.

D. Save Interim Results and Make Them Available

Several examples of useful "interim results" in Mie computations were mentioned in the preceding section. Presumably, the trick of plotting the complex function, of which the extinction is the real part, can be applied directly to nonspherical particles. In the angular distribution of scattered light, the first "customer" may be content with a plot of intensity versus angle. Future research, however, may profit far more from the complex amplitudes or from the development in generalized spherical functions. And such key quantities as albedo and asymmetry parameter should never be forgotten. To keep options of future use open, for example, for interpolation between different assumptions on size, composition, or shape, authors are advised to save key interim data and to make them available either in print or in electronic file.

I hope these few hints may prove useful.

Part I

Introduction

Chapter 1

Concepts, Terms, Notation

Michael I. Mishchenko
NASA Goddard Institute
for Space Studies
New York
New York 10025

Joop W. Hovenier
Department of Physics
and Astronomy
Free University
1081 HV Amsterdam
The Netherlands

Larry D. Travis
NASA Goddard Institute
for Space Studies
New York
New York 10025

I. INTRODUCTION

The literature on electromagnetic scattering by small particles is extremely diverse and often contains discrepancies and inconsistencies in the definition and usage of terms and notation. This situation is not unexpected because alternative terminologies have been developed somewhat independently by physical scientists, antenna engineers, astrophysicists, and geophysicists. Some of the differences are trivial, but others can cause serious confusion. Among typical differences are the use of right-handed versus left-handed coordinate systems, the use of different definitions of Stokes parameters, and the use of the time factor $\exp(-i\omega t)$ versus $\exp(i\omega t)$, where $i = (-1)^{1/2}$, ω is the angular frequency, and t is time.

This book is written by experts with diverse scientific backgrounds, and it is essentially impossible to maintain the use of a universal and comprehensive convention on terms and notation throughout all individual chapters. However, it is also clear that publishing all chapters under the same cover and making sure that the entire book is useful to the reader requires a substantial degree of uniformity and unification. Thus, the primary aim of this chapter is to introduce a minimal system of basic terminology and notation that will be consistently used in the remainder of the book as a "reference frame." All deviations from this system in individual chapters will be explicitly indicated, the relation of newly introduced and alternative quantities and definitions to the basic ones will be explained, and the reader will be warned when different conventions may lead to confusion.

II. INDEPENDENT SCATTERING

Because all aspects of light scattering by small particles and its applications cannot be entirely covered in one book, we will limit our treatment to scattering by individual particles within the framework of Maxwell's equations and linear optics. As a first restriction, we will consider only time-harmonic quasi-monochromatic light by assuming that the (complex) amplitude of the electric field fluctuates with time much slower than the time factor $\exp(-i\omega t)$. Second, we will assume that light scattering occurs without frequency redistribution; that is, the scattered light has the same frequency as the incident light. This excludes phenomena such as Raman scattering or fluorescence. Third, we will consider only scattering in the far-field zone, where the propagation of the scattered wave is away from the particle, and the scattered field is polarized transverse to the propagation direction and decays inversely with distance from the particle. Fourth, and most important, we will consider only independently scattering, randomly positioned particles. This means that particles are separated widely enough, so that each particle scatters light in exactly the same way as if all other particles did not exist. Furthermore, there are no systematic phase relations between partial electromagnetic waves scattered by different particles, so that the intensities (or, more generally, Stokes parameters) of the partial waves can be added without regard to phase. In other words, we will assume that each particle is in the far-field zones of all other particles, and that scattering by different particles is incoherent.

The assumption of independent scattering greatly simplifies the problem of computing multiple light scattering by a collection of particles (e.g., by a cloud of water droplets or ice crystals) because it allows the use of the radiative transfer equation (Ishimaru, 1978, Chapter 7; Tsang *et al.*, 1985, Chapter 3; Ulaby *et al.*, 1986, Chapter 13; Fung, 1994, Chapter 1). It should be noted, however, that even for tenuous collections of randomly positioned particles, scattering in the exact forward direction is always coherent and causes attenuation of the incident wave as it propagates through the medium (e.g., van de Hulst, 1957, Chapter 4; Bohren and Huffman, 1983, Chapter 3). Furthermore, so-called self-avoiding reciprocal multiple-scattering paths always interfere constructively at the exact backscattering direction and cause a coherent intensity peak (Tsang *et al.*, 1985, Chapter 5; Barabanenkov *et al.*, 1991). The latter phenomenon is called coherent backscattering (or weak photon localization) and is not accounted for explicitly by the classical radiative transfer theory. Fortunately, the angular width of the backscattering intensity peak caused by weak photon localization is proportional to the ratio wavelength/(photon transport mean free path) and thus is negligibly small for tenuous particle collections such as clouds, aerosols, and precipitation.

It is not always easy to determine what minimal separation makes scattering by a collection of particles independent. Exact scattering calculations for randomly oriented two-sphere clusters composed of identical wavelength-sized spheres suggest that particles can scatter independently when the distance between their centers is as small as four times their radius (Mishchenko *et al.*, 1995). Even though this result is not necessarily a universal rule and may be expected to be inapplicable for decidedly subwavelength-sized particles, it can be considered a simple approximate condition for independent scattering by particles comparable to and larger than a wavelength.

Although the subject of this book is incoherent far-field scattering, the computation of single-scattering properties of some particles may require an explicit treatment of coherent and near-field effects. Indeed, scattering particles are not always single homogeneous bodies such as spheres or spheroids and in many cases exist in the form of clusters of small monomers. Because cluster components form a specific configuration and are in close proximity to each other, each component affects the fields scattered by all other components. Therefore, electromagnetic scattering by a cluster must be computed using special techniques such as those described in Chapters 4–8. At a large distance from a cluster, however, it acts as a single scatterer and, as such, is also a subject of this book.

III. REFERENCE FRAMES AND PARTICLE ORIENTATION

To describe the scattering of a plane electromagnetic wave by a nonspherical particle in an arbitrary orientation, we must first specify the directions of the incident and scattered waves and the orientation of the particle with respect

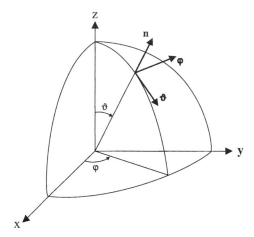

Figure 1 Laboratory coordinate system.

to a reference frame. Let this reference frame be a right-handed Cartesian co-ordinate system L with orientation fixed in space, having its origin inside the particle. In what follows, this coordinate system will be referred to as the labo-ratory reference frame. The direction of propagation of a transverse electromag-netic wave is specified by a unit vector **n** or, equivalently, by a couple (ϑ, φ), where $\vartheta \in [0, \pi]$ is the polar (zenith) angle measured from the positive z axis, and $\varphi \in [0, 2\pi]$ is the azimuth angle measured from the positive x axis in the clockwise sense, when looking in the direction of the positive z axis (Fig. 1). The ϑ and φ components of the electric field are denoted as $\mathbf{E}_\vartheta$ and $\mathbf{E}_\varphi$, respec-tively. The component $\mathbf{E}_\vartheta = E_\vartheta \boldsymbol{\vartheta}$ lies in the meridional plane (plane through the beam and the z axis), whereas the component $\mathbf{E}_\varphi = E_\varphi \boldsymbol{\varphi}$ is perpendicular to this plane. Here, $\boldsymbol{\vartheta}$ and $\boldsymbol{\varphi}$ are the corresponding unit vectors such that $\mathbf{n} = \boldsymbol{\vartheta} \times \boldsymbol{\varphi}$. Note that $\mathbf{E}_\vartheta$ and $\mathbf{E}_\varphi$ are often denoted as $\mathbf{E}_v$ and $\mathbf{E}_h$ and called the vertical and horizontal electric field vector components, respectively (e.g., Ulaby *et al.*, 1986, Chapter 13; Tsang *et al.*, 1985, Chapter 3; Ulaby and Elachi, 1990, Chap-ter 1).

To specify the orientation of the particle with respect to the laboratory refer-ence frame, we introduce a right-handed coordinate system P fixed to the particle and having the same origin as L. This coordinate system will be referred to as the particle or body reference frame. The choice of the particle frame is, in prin-ciple, arbitrary, although in many cases it is naturally suggested by the particle shape. For example, for axially symmetric particles it is useful to choose the body frame such that its z axis coincides with the axis of particle symmetry. Now it is convenient to specify the orientation of the particle with respect to the labo-ratory frame L by three Euler angles of rotation α, β, and γ that transform the

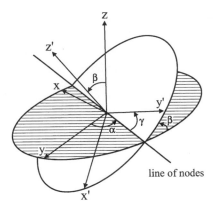

Figure 2   Euler angles of rotation α, β, and γ transforming the laboratory reference frame $L\{x, y, z\}$ into the body frame of the scattering particle $P\{x', y', z'\}$.

coordinate system $L\{x, y, z\}$ into the coordinate system $P\{x', y', z'\}$ as shown in Fig. 2 (Arfken and Weber, 1995, Chapter 3). The Euler rotations are performed as follows:

   a. Rotation about the z axis through an angle $\alpha \in [0, 2\pi]$ reorienting the y axis such that it coincides with the line of nodes (line formed by the intersection of the xy and $x'y'$ planes).

   b. Rotation about the new y axis through an angle $\beta \in [0, \pi]$.

   c. Rotation about the z' axis through an angle $\gamma \in [0, 2\pi]$.

An angle of rotation is positive if the rotation is in the clockwise direction, when looking in the positive direction of the rotation axis.

IV. AMPLITUDE MATRIX

Consider a monochromatic plane electromagnetic wave with electric field vector

$$\mathbf{E}^{\text{inc}}(\mathbf{R}) = \left(E_{\vartheta 0}^{\text{inc}} \boldsymbol{\vartheta}^{\text{inc}} + E_{\varphi 0}^{\text{inc}} \boldsymbol{\varphi}^{\text{inc}}\right) \exp\left(ik\mathbf{n}^{\text{inc}}\mathbf{R}\right) = \mathbf{E}_0^{\text{inc}} \exp(ik\mathbf{n}^{\text{inc}}\mathbf{R})$$

$$= E_{\vartheta}^{\text{inc}} \boldsymbol{\vartheta}^{\text{inc}} + E_{\varphi}^{\text{inc}} \boldsymbol{\varphi}^{\text{inc}} \tag{1}$$

incident upon a nonspherical particle in a direction $\mathbf{n}^{\text{inc}}$; here, $k = 2\pi/\lambda$ is the free-space wavenumber, λ is the free-space wavelength, and $\mathbf{R}$ is the radius vector with its origin at the origin of the laboratory coordinate system. Throughout most of the chapter, we assume and suppress the time factor $\exp(-i\omega t)$. Because of the linearity of Maxwell's equations and boundary conditions, it must always be possible to express the scattered electric field linearly in the incident electric field. In the far-field region ($kR \gg 1$, $R = |\mathbf{R}|$), the scattered wave becomes spherical and is given by (Tsang *et al.*, 1985, Chapter 3)

$$\mathbf{E}^{\mathrm{sca}}(\mathbf{R}) = E_{\vartheta}^{\mathrm{sca}}(R, \mathbf{n}^{\mathrm{sca}}) \boldsymbol{\vartheta}^{\mathrm{sca}} + E_{\varphi}^{\mathrm{sca}}(R, \mathbf{n}^{\mathrm{sca}}) \boldsymbol{\varphi}^{\mathrm{sca}}, \qquad \mathbf{n}^{\mathrm{sca}} = \frac{\mathbf{R}}{R}, \qquad (2)$$

$$\mathbf{R}\mathbf{E}^{\mathrm{sca}}(\mathbf{R}) = 0, \qquad (3)$$

$$\begin{bmatrix} E_{\vartheta}^{\mathrm{sca}} \\ E_{\varphi}^{\mathrm{sca}} \end{bmatrix} = \frac{\exp(ikR)}{R} \mathbf{S}(\mathbf{n}^{\mathrm{sca}}; \mathbf{n}^{\mathrm{inc}}; \alpha, \beta, \gamma) \begin{bmatrix} E_{\vartheta 0}^{\mathrm{inc}} \\ E_{\varphi 0}^{\mathrm{inc}} \end{bmatrix}, \qquad (4)$$

where $\mathbf{S}$ is a 2×2 amplitude matrix that linearly transforms the electric field vector components of the incident wave into the electric field vector components of the scattered wave. The amplitude matrix depends on the directions of incidence and scattering as well as on the size, morphology, and composition of the scattering particle and on its orientation with respect to the laboratory reference frame as specified by the Euler angles of rotation α, β and γ. The amplitude matrix is the primary quantity that defines the single-scattering law. If known, it enables one to compute any other light-scattering characteristic of the particle.

Note that both van de Hulst (1957) and Bohren and Huffman (1983) assume that the incident light always propagates along the z axis of the laboratory co-ordinate system and introduce the amplitude matrix as the one describing the transformation of the electric field components defined with respect to the scattering plane, that is, the plane through the directions of incidence and scattering. Their amplitude matrix can be considered a particular case of the $\mathbf{S}$ matrix defined in Eq. (4) and corresponds, apart from a scalar, to the choice $\vartheta^{\mathrm{inc}} = 0$ and $\varphi^{\mathrm{inc}} = \varphi^{\mathrm{sca}}$. One should also take into account that van de Hulst decomposes the electric vectors of the incident and scattered light using unit vectors $\mathbf{l} = \boldsymbol{\vartheta}$ and $\mathbf{r} = -\boldsymbol{\varphi}$, whereas Bohren and Huffman (1983) use the same unit vectors as van de Hulst but denote them by $\mathbf{e}_{\parallel}$ and $\mathbf{e}_{\perp}$, respectively. Therefore, our electric field components are translated into those of Bohren and Huffman (1983) by the formulas $E_{\parallel} = E_{\vartheta}$ and $E_{\perp} = -E_{\varphi}$. This also leads to a sign difference in the off-diagonal elements of the amplitude matrix.

It should be pointed out that the choice of the laboratory reference frame with the z axis along the direction of the incident light can sometimes be inconvenient in computations for nonspherical particles because any change in the direction of light incidence also changes the orientation of the particles with respect to the reference frame. Therefore, it is often preferable to use a fixed (laboratory) co-ordinate system to specify *both* the directions of propagation and the states of polarization of the incident and scattered light *and* the particle orientation (Tsang *et al.*, 1985, Chapter 3). An important advantage of this approach is that the laboratory coordinate system can be chosen such that it most adequately represents the physical mechanism of particle orientation and is especially convenient in solving the radiative transfer equation.

Our choice of the exponent sign in the time factor $\exp(-i\omega t)$ causes the imaginary part of the refractive index to be nonnegative. This convention is used in many books on optics and light scattering (e.g., Born and Wolf, 1970; Bohren and Huffman, 1983; Barber and Hill, 1990) and is a nearly universal choice in electromagnetics (e.g., Jackson, 1975; Tsang *et al.*, 1985) and solid-state physics. However, van de Hulst (1957) and Kerker (1969) use the time factor $\exp(i\omega t)$, which implies a nonpositive imaginary part of the refractive index. It is important to remember that once a choice of the exponent sign has been made, it must be used consistently in all derivations. To distinguish the use of the $\exp(-i\omega t)$ and $\exp(i\omega t)$ time factor conventions in different chapters of this book, the latter will be written in the form $\exp(j\omega t)$ and, accordingly, j will be used to denote the imaginary unity.

V. STOKES PARAMETERS

Photometric and polarimetric optical instruments cannot measure the electric field associated with a beam of light but rather measure some quantities that are quadratic combinations of the electric field components. Therefore, it is convenient to introduce the so-called Stokes parameters of a monochromatic transverse electromagnetic wave as

$$I = E_\vartheta E_\vartheta^* + E_\varphi E_\varphi^*, \tag{5}$$

$$Q = E_\vartheta E_\vartheta^* - E_\varphi E_\varphi^*, \tag{6}$$

$$U = -E_\vartheta E_\varphi^* - E_\varphi E_\vartheta^*, \tag{7}$$

$$V = i\left(E_\varphi E_\vartheta^* - E_\vartheta E_\varphi^*\right), \tag{8}$$

where the asterisk denotes the complex conjugate value. Note that despite some difference in notation, our definition of the Stokes parameters is identical to that of van de Hulst (1957, Chapter 5) and Bohren and Huffman (1983, Chapter 3), whereas the definition of the Stokes parameter U by Tsang *et al.* (1985, Chapter 3) has the opposite sign. All Stokes parameters have the same dimension: energy per unit area per unit time per unit wavelength. The first Stokes parameter is equal, up to a factor common to all four parameters, to the net monochromatic energy flux. The Stokes parameters Q and U describe the state of linear polarization and V describes the state of circular polarization of the wave. A detailed discussion of the polarimetric definitions can be found in Bohren and Huffman (1983) and Hovenier and van der Mee (1983) (see also the Appendix at the end of this chapter). For further use, it is convenient to define the Stokes vector as a 4×1 column having the Stokes parameters as its components as follows:

$$\mathbf{I} = \begin{bmatrix} I \\ Q \\ U \\ V \end{bmatrix}. \tag{9}$$

The Stokes parameters are always defined with respect to a reference plane, in our case the meridional plane of the beam. If the reference frame is rotated through an angle η in the anticlockwise direction when looking in the direction of propagation, the transformation of the old Stokes vector $\mathbf{I}$ into the new Stokes vector $\mathbf{I}'$ is given by a 4×4 rotation matrix $\mathbf{L}$ (see the Appendix):

$$\mathbf{I}' = \mathbf{L}(\eta)\mathbf{I} = \begin{bmatrix} 1 & 0 & 0 & 0 \\ 0 & \cos 2\eta & \sin 2\eta & 0 \\ 0 & -\sin 2\eta & \cos 2\eta & 0 \\ 0 & 0 & 0 & 1 \end{bmatrix} \mathbf{I}. \tag{10}$$

As follows from the definition of the Stokes parameters, for an elementary monochromatic plane or spherical electromagnetic wave

$$I^2 \equiv Q^2 + U^2 + V^2. \tag{11}$$

However, light beams that are commonly measured are incoherent mixtures of many elementary plane waves of the type given by Eq. (1). Because there are no permanent phase relations between these elementary waves, the Stokes parameters obey the fundamental property of additivity: The Stokes parameters for the mixture are sums of the respective Stokes parameters of the elementary waves (Chandrasekhar, 1950, Chapter 1; Tsang *et al.*, 1985, Chapter 3). In this case the identity of Eq. (11) is replaced by the inequality

$$I^2 \geq Q^2 + U^2 + V^2. \tag{12}$$

The ratio $p = [Q^2 + U^2 + V^2]^{1/2}/I \leq 1$ is called the degree of polarization and is equal to unity for fully polarized light, whereas the ratios $[Q^2 + U^2]^{1/2}/I$ and V/I are called the degrees of linear and circular polarization, respectively. If U vanishes, the ratio $-Q/I$ is also often used as a measure of the degree of linear polarization. For unpolarized (natural) light, $Q = U = V = 0$.

The Stokes vector defined by Eqs. (5)–(9) is one of many possible representations of polarization, and alternative definitions of the Stokes parameters can be found to be more convenient for specific applications. For example, radar engineers often use so-called modified Stokes parameters, whereas physicists widely employ the so-called circular polarization representation and the density matrix. The relationship between different representations of polarization is discussed, for example, by Hovenier and van der Mee (1983), Ulaby and Elachi (1990), and Dolginov *et al.* (1995).

VI. PHASE MATRIX

As follows from the definitions of the amplitude matrix and the Stokes parameters, the transformation of the Stokes parameters of the incident plane wave into those of the scattered spherical wave owing to light scattering by a single particle is given by

$$\mathbf{I}^{\text{sca}} = \frac{1}{R^2}\mathbf{Z}(\mathbf{n}^{\text{sca}}; \mathbf{n}^{\text{inc}}; \alpha, \beta, \gamma)\mathbf{I}^{\text{inc}}$$

$$= \frac{1}{R^2}\mathbf{Z}(\vartheta^{\text{sca}}, \varphi^{\text{sca}}; \vartheta^{\text{inc}}, \varphi^{\text{inc}}; \alpha, \beta, \gamma)\mathbf{I}^{\text{inc}}, \tag{13}$$

where $\mathbf{Z}$ is the 4×4 phase matrix with elements

$$Z_{11} = \frac{1}{2}\left(|S_{11}|^2 + |S_{12}|^2 + |S_{21}|^2 + |S_{22}|^2\right), \tag{14}$$

$$Z_{12} = \frac{1}{2}\left(|S_{11}|^2 - |S_{12}|^2 + |S_{21}|^2 - |S_{22}|^2\right), \tag{15}$$

$$Z_{13} = -\operatorname{Re}\left(S_{11}S_{12}^* + S_{22}S_{21}^*\right), \tag{16}$$

$$Z_{14} = -\operatorname{Im}\left(S_{11}S_{12}^* - S_{22}S_{21}^*\right), \tag{17}$$

$$Z_{21} = \frac{1}{2}\left(|S_{11}|^2 + |S_{12}|^2 - |S_{21}|^2 - |S_{22}|^2\right), \tag{18}$$

$$Z_{22} = \frac{1}{2}\left(|S_{11}|^2 - |S_{12}|^2 - |S_{21}|^2 + |S_{22}|^2\right), \tag{19}$$

$$Z_{23} = -\operatorname{Re}\left(S_{11}S_{12}^* - S_{22}S_{21}^*\right), \tag{20}$$

$$Z_{24} = -\operatorname{Im}\left(S_{11}S_{12}^* + S_{22}S_{21}^*\right), \tag{21}$$

$$Z_{31} = -\operatorname{Re}\left(S_{11}S_{21}^* + S_{22}S_{12}^*\right), \tag{22}$$

$$Z_{32} = -\operatorname{Re}\left(S_{11}S_{21}^* - S_{22}S_{12}^*\right), \tag{23}$$

$$Z_{33} = \operatorname{Re}\left(S_{11}S_{22}^* + S_{12}S_{21}^*\right), \tag{24}$$

$$Z_{34} = \operatorname{Im}\left(S_{11}S_{22}^* + S_{21}S_{12}^*\right), \tag{25}$$

$$Z_{41} = -\operatorname{Im}\left(S_{21}S_{11}^* + S_{22}S_{12}^*\right), \tag{26}$$

$$Z_{42} = -\operatorname{Im}\left(S_{21}S_{11}^* - S_{22}S_{12}^*\right), \tag{27}$$

$$Z_{43} = \operatorname{Im}\left(S_{22}S_{11}^* - S_{12}S_{21}^*\right), \tag{28}$$

$$Z_{44} = \operatorname{Re}\left(S_{22}S_{11}^* - S_{12}S_{21}^*\right). \tag{29}$$

In general, all 16 elements of the phase matrix are nonzero. However, the phase matrix elements of a single particle are expressed in terms of only seven independent real constants resulting from the four complex elements of the amplitude matrix minus an irrelevant phase (van de Hulst, 1957, Chapter 5). Therefore, only 7 of the phase matrix elements are independent, and there are several relations among the 16 quantities Z_{ij}, $i, j = 1, \ldots, 4$ (Chapter 3).

Note that van de Hulst (1957, Chapter 5) and Bohren and Huffman (1983, Chapter 3) always direct the incident light along the z axis of the laboratory reference frame and introduce the scattering matrix to describe the Stokes vector transformation provided that both $\mathbf{I}^{\mathrm{inc}}$ and $\mathbf{I}^{\mathrm{sca}}$ are defined with respect to the scattering plane. Therefore, the matrix denoted by $\mathbf{F}$ in Section 5.14 of van de Hulst (1957) and the matrix denoted by $\mathbf{S}$ in Section 3.3 of Bohren and Huffman (1983) are, apart from a scalar, a particular case of the phase matrix $\mathbf{Z}$ corresponding to the choice $\vartheta^{\mathrm{inc}} = 0$ and $\varphi^{\mathrm{inc}} = \varphi^{\mathrm{sca}}$. Note also that the use of the $\exp(i\omega t)$ versus the $\exp(-i\omega t)$ time factor convention in actual computations of the phase matrix causes an opposite sign in numerical values of the elements in the fourth row and the fourth column except the Z_{44} element.

For a small volume element dv comprising randomly positioned particles, the waves scattered by different particles are random in phase, and the Stokes parameters of these incoherent waves add up. Therefore, the phase matrix for the volume element is the sum of the phase matrices of the individual particles:

$$\mathbf{Z} = \sum_n \mathbf{Z}_n = n_0 \langle \mathbf{Z} \rangle dv, \tag{30}$$

where the index n numbers the particles, n_0 is the particle number density, and $\langle \mathbf{Z} \rangle$ is the ensemble-averaged phase matrix per particle.

VII. TOTAL OPTICAL CROSS SECTIONS

The phase matrix describes the angular distribution of the light *scattered* by a particle. Scattering and absorption of light by the particle also change the characteristics of the incident beam after it passes the particle. For a nonspherical particle in a fixed orientation, the extinction cross section C_{ext} describes the total attenuation of the incident beam resulting from the combined effect of scattering and absorption by the particle and is defined as follows: The product of the extinction cross section and the incident monochromatic energy flux is equal to the total monochromatic power removed by the particle from the incident beam. The extinction cross section has the dimension of area, depends on the particle orientation and the direction and the polarization state of the incident beam, and is expressed in the elements of the extinction matrix given by Eqs. (35)–(41) as follows:

$$C_{\mathrm{ext}} = \frac{1}{I^{\mathrm{inc}}} \left[K_{11} I^{\mathrm{inc}} + K_{12} Q^{\mathrm{inc}} + K_{13} U^{\mathrm{inc}} + K_{14} V^{\mathrm{inc}} \right]. \tag{31}$$

Similarly, the product of the scattering cross section and the monochromatic incident energy flux gives the total monochromatic power removed from the incident beam resulting solely from light scattering by the particle. The scattering cross section has the same dimension as the extinction cross section and also depends on the particle orientation and the direction of incidence and the state of polarization of the illuminating beam. As follows from Eq. (13), the total scattering cross section is given by

$$C_{\text{sca}} = \frac{1}{I^{\text{inc}}} \int_{4\pi} d\mathbf{n}^{\text{sca}} I^{\text{sca}}(\mathbf{n}^{\text{sca}}), \tag{32}$$

with

$$I^{\text{sca}} = Z_{11}I^{\text{inc}} + Z_{12}Q^{\text{inc}} + Z_{13}U^{\text{inc}} + Z_{14}V^{\text{inc}}. \tag{33}$$

Finally, the product of the absorption cross section C_{abs} and the incident monochromatic energy flux is equal to the total monochromatic power removed from the incident beam as a result of absorption of light by the particle and is equal to the difference between the extinction and scattering cross sections:

$$C_{\text{abs}} = C_{\text{ext}} - C_{\text{sca}}. \tag{34}$$

VIII. DICHROISM AND EXTINCTION MATRIX

A nonspherical particle in a fixed orientation can change not only the energy characteristics of the incident beam after it passes the particle, but also its state of polarization. This phenomenon is called dichroism of the scattering medium and results from different values of extinction for different polarization components of the incident light. Two well-known manifestations of dichroism are interstellar polarization (change of polarization of starlight after it passes a cloud of interstellar dust grains preferentially oriented by cosmic magnetic fields) (Martin, 1978) and cross polarization of radio waves propagating through partially aligned nonspherical hydrometeors (Oguchi, 1983; Chapter 16). Because of dichroism, the full description of the light extinction process requires the introduction of the so-called 4×4 extinction matrix $\mathbf{K}$. According to the optical theorem, the elements of the extinction matrix are expressed in the elements of the forward-scattering amplitude matrix as follows (Tsang *et al.*, 1985, Chapter 3):

$$K_{jj}(\mathbf{n}) = \frac{2\pi}{k} \text{Im}\big[S_{11}(\mathbf{n}, \mathbf{n}) + S_{22}(\mathbf{n}, \mathbf{n})\big], \qquad j = 1, \ldots, 4, \tag{35}$$

$$K_{12}(\mathbf{n}) = K_{21}(\mathbf{n}) = \frac{2\pi}{k} \text{Im}\big[S_{11}(\mathbf{n}, \mathbf{n}) - S_{22}(\mathbf{n}, \mathbf{n})\big], \tag{36}$$

$$K_{13}(\mathbf{n}) = K_{31}(\mathbf{n}) = -\frac{2\pi}{k} \text{Im}\big[S_{12}(\mathbf{n}, \mathbf{n}) + S_{21}(\mathbf{n}, \mathbf{n})\big], \tag{37}$$

$$K_{14}(\mathbf{n}) = K_{41}(\mathbf{n}) = \frac{2\pi}{k} \text{Re}\big[S_{21}(\mathbf{n}, \mathbf{n}) - S_{12}(\mathbf{n}, \mathbf{n})\big], \tag{38}$$

$$K_{23}(\mathbf{n}) = -K_{32}(\mathbf{n}) = \frac{2\pi}{k} \, \mathrm{Im}\big[S_{21}(\mathbf{n}, \mathbf{n}) - S_{12}(\mathbf{n}, \mathbf{n})\big], \tag{39}$$

$$K_{24}(\mathbf{n}) = -K_{42}(\mathbf{n}) = -\frac{2\pi}{k} \, \mathrm{Re}\big[S_{12}(\mathbf{n}, \mathbf{n}) + S_{21}(\mathbf{n}, \mathbf{n})\big], \tag{40}$$

$$K_{34}(\mathbf{n}) = -K_{43}(\mathbf{n}) = \frac{2\pi}{k} \, \mathrm{Re}\big[S_{22}(\mathbf{n}, \mathbf{n}) - S_{11}(\mathbf{n}, \mathbf{n})\big], \tag{41}$$

where we have omitted for brevity the arguments α, β, and γ specifying particle orientation. All extinction matrix elements have the dimension of area. For a small volume element dv comprising randomly positioned particles, the extinction matrix is the sum of the extinction matrices of individual particles:

$$\mathbf{K} = \sum_n \mathbf{K}_n = n_0 \langle \mathbf{K} \rangle \, dv, \tag{42}$$

where $\langle \mathbf{K} \rangle$ is the ensemble-averaged extinction matrix per particle.

The physical meaning of the extinction matrix is as follows. Consider a parallel beam of light propagating in a direction $\mathbf{n}$ through a medium filled with randomly positioned, sparsely distributed particles. Then the change of the Stokes vector of the beam along the light path is described by the equation

$$\frac{d\mathbf{I}(\mathbf{n})}{ds} = -n_0 \langle \mathbf{K}(\mathbf{n}) \rangle \mathbf{I}(\mathbf{n}), \tag{43}$$

where the pathlength element ds is measured along the direction of propagation $\mathbf{n}$.

IX. RECIPROCITY

A fundamental property of the amplitude matrix is the reciprocity relation (van de Hulst, 1980, Section 3.2.1; Tsang, 1991, Appendix A)

$$\mathbf{S}(-\mathbf{n}'; -\mathbf{n}; \alpha, \beta, \gamma) = \begin{bmatrix} S_{11}(\mathbf{n}; \mathbf{n}'; \alpha, \beta, \gamma) & -S_{21}(\mathbf{n}; \mathbf{n}'; \alpha, \beta, \gamma) \\ -S_{12}(\mathbf{n}; \mathbf{n}'; \alpha, \beta, \gamma) & S_{22}(\mathbf{n}; \mathbf{n}'; \alpha, \beta, \gamma) \end{bmatrix}. \tag{44}$$

Reciprocity is a manifestation of the symmetry of the light-scattering process with respect to an inversion of time, directly follows from Maxwell's equations, and holds for any particle in an arbitrary orientation. The only requirement is that the dielectric, permeability, and conductive tensors be symmetric. Using Eqs. (13)–(29), (35)–(41), and (44), we easily derive the reciprocity relations for the phase and extinction matrices:

$$\mathbf{Z}(-\mathbf{n}'; -\mathbf{n}; \alpha, \beta, \gamma) = \mathbf{Q}\mathbf{Z}^{\mathrm{T}}(\mathbf{n}; \mathbf{n}'; \alpha, \beta, \gamma)\mathbf{Q}, \tag{45}$$

$$\mathbf{K}(-\mathbf{n}; \alpha, \beta, \gamma) = \mathbf{Q}\mathbf{K}^{\mathrm{T}}(\mathbf{n}; \alpha, \beta, \gamma)\mathbf{Q}, \tag{46}$$

where

$$\mathbf{Q} = \mathbf{Q}^{-1} = \mathbf{Q}^{\mathrm{T}} = \mathrm{diag}[1, 1, -1, 1], \tag{47}$$

and T denotes matrix transposition. Another consequence of reciprocity is the so-called backscattering theorem (van de Hulst, 1957, Section 5.32):

$$S_{21}(-\mathbf{n}; \mathbf{n}; \alpha, \beta, \gamma) = -S_{12}(-\mathbf{n}; \mathbf{n}; \alpha, \beta, \gamma). \tag{48}$$

This theorem along with Eqs. (14), (19), (24), and (29) leads to the following general property of the backscattering phase matrix:

$$Z_{11}(-\mathbf{n}; \mathbf{n}; \alpha, \beta, \gamma) - Z_{22}(-\mathbf{n}; \mathbf{n}; \alpha, \beta, \gamma)$$
$$+ Z_{33}(-\mathbf{n}; \mathbf{n}; \alpha, \beta, \gamma) - Z_{44}(-\mathbf{n}; \mathbf{n}; \alpha, \beta, \gamma) = 0. \tag{49}$$

Owing to the property of additivity, Eqs. (45), (46), and (49) hold not only for a single particle in an arbitrary orientation, but also for ensemble-averaged phase and extinction matrices.

Because of the fundamental nature of reciprocity, Eqs. (44)–(46), (48), and (49) are important tests in computations or measurements of light scattering by non-spherical particles: Violation of reciprocity means that the computations or measurements are incorrect or inaccurate. Alternatively, the use of reciprocity in computer calculations can substantially shorten the central processing unit (CPU) time consumption because one may calculate light scattering for only half of all scattering geometries and then employ Eqs. (44)–(46) for the reciprocal geometries.

X. ENSEMBLE AVERAGING

Natural scattering media are usually mixtures of particles with different sizes, shapes, orientations, and refractive indices. Therefore, in practice the phase and extinction matrices must be averaged over the particle ensemble. Computation of ensemble averages is, in principle, rather straightforward. For example, for homogeneous ellipsoids with axes $a \in [a_1, a_2]$, $b \in [b_1, b_2]$, and $c \in [c_1, c_2]$ and the same composition, the ensemble-averaged phase matrix per particle is

$$\langle \mathbf{Z}(\mathbf{n}; \mathbf{n}') \rangle = \int_{a_1}^{a_2} da \int_{b_1}^{b_2} db \int_{c_1}^{c_2} dc \int_0^{2\pi} d\alpha \int_0^\pi d\beta \sin\beta \int_0^{2\pi} d\gamma$$
$$\times\, p(\alpha, \beta, \gamma; a, b, c) \mathbf{Z}(\mathbf{n}; \mathbf{n}'; \alpha, \beta, \gamma), \tag{50}$$

where Euler angles α, β, and γ specify particle orientations with respect to the laboratory reference frame, and $p(a, b, c; \alpha, \beta, \gamma)$ is a probability density function satisfying the normalization condition

$$\int_{a_1}^{a_2} da \int_{b_1}^{b_2} db \int_{c_1}^{c_2} dc \int_0^{2\pi} d\alpha \int_0^{\pi} d\beta \sin \beta$$

$$\times \int_0^{2\pi} d\gamma\, p(a, b, c; \alpha, \beta, \gamma) = 1. \tag{51}$$

It is often assumed that the size/shape and orientation distributions are statistically independent. Then the total probability function is simplified by representing it as a product of two functions, one of which, $p_s(a, b, c)$, describes the particle size/shape distribution and the other one, $p_o(\alpha, \beta, \gamma)$, describes the distribution of particle orientations:

$$p(a, b, c; \alpha, \beta, \gamma) = p_s(a, b, c) p_o(\alpha, \beta, \gamma), \tag{52}$$

each normalized to unity. As a consequence, the problems of computing size/shape and orientational averages are separated. Similarly, it is often convenient to separate averaging over sizes and shapes by assuming that particle sizes and shapes are statistically independent. For example, the shape of a spheroidal particle can be specified by its aspect ratio ξ (along with the designation of either prolate or oblate), whereas the particle size can be specified by the radius of the equal-volume sphere r. Then the probability density function p_s can be represented as a product

$$p_s(a, b, c) = p(\xi) n(r), \tag{53}$$

where $p(\xi)$ describes the distribution of spheroid aspect ratios and $n(r)$ is the distribution of equivalent-sphere radii. Again, both $p(\xi)$ and $n(r)$ are normalized to unity.

In the absence of external forces such as magnetic, electrostatic, or aerodynamic forces, all orientations of a nonspherical particle are equiprobable. In this practically important case of randomly oriented particles, the orientation distribution function is uniform with respect to the Euler angles of rotation, and we have

$$p_{o,\text{random}}(\alpha, \beta, \gamma) \equiv \frac{1}{8\pi^2}. \tag{54}$$

An external force can make the orientation distribution axially symmetric with the axis of symmetry given by the direction of the force. For example, interstellar dust grains can be axially oriented by a cosmic magnetic field (Martin, 1978), whereas nonspherical hydrometeors can be axially oriented by the aerodynamic force resulting from their finite falling velocity (Liou, 1992). In this case it is convenient to choose the laboratory reference frame with the z axis along the external force direction so that the orientation distribution is uniform with respect to the Euler angles α and γ:

$$p_{0,\text{axial}}(\alpha, \beta, \gamma) = \frac{1}{4\pi^2} p(\beta). \qquad (55)$$

Particular details of the particle shape can also simplify the orientation distribution function. For example, for rotationally symmetric bodies it is convenient to direct the z axis of the particle reference frame along the axis of rotation, in which case the orientation distribution function in the laboratory reference frame becomes independent of the Euler angle γ:

$$p_0(\alpha, \beta, \gamma) \equiv \frac{1}{2\pi} p_0(\alpha, \beta). \qquad (56)$$

XI. SCATTERING MATRIX AND MACROSCOPICALLY ISOTROPIC AND SYMMETRIC MEDIA

By definition, the phase matrix relates the Stokes parameters of the incident and scattered beams defined relative to their respective meridional planes. Unlike the phase matrix, the scattering matrix $\mathbf{F}$ relates the Stokes parameters of the incident and scattered beams defined with respect to the scattering plane, that is, the plane through the unit vectors $\mathbf{n}^{\text{inc}}$ and $\mathbf{n}^{\text{sca}}$ (van de Hulst, 1957, Chapter 5; Bohren and Huffman, 1983, Chapter 3). The simplest way to introduce the scattering matrix is to direct the z axis of the laboratory reference frame along the incident beam ($\vartheta^{\text{inc}} = 0$) and superpose the principal azimuthal plane with $\varphi^{\text{inc}} = \varphi^{\text{sca}} = 0$ with the scattering plane. Then the scattering matrix $\mathbf{F}$ can be defined as a matrix proportional to the phase matrix $\mathbf{Z}(\vartheta^{\text{sca}}, \varphi^{\text{sca}} = 0; \vartheta^{\text{inc}} = 0, \varphi^{\text{inc}} = 0)$. In general, all 16 elements of the scattering matrix are nonzero and depend on the particle orientation distribution with respect to the incident and scattered beams.

The concept of the scattering matrix is especially useful when external forces are absent or do not cause scattering particles to become partially or perfectly aligned. In this case all orientations of the particles are equiprobable and the particles are randomly oriented in three-dimensional space. If, furthermore, each particle has a plane of symmetry and/or particles and their mirror particles are present in equal numbers, then such a scattering medium is called macroscopically isotropic and symmetric. Because of symmetry, the scattering matrix $\mathbf{F}$ for macroscopically isotropic and symmetric media is invariant with respect to the choice of the scattering plane and depends only on the angle between the incident and scattered beams; that is, the scattering angle $\Theta = \vartheta^{\text{sca}} \in [0, \pi]$ (van de Hulst, 1957). We thus have

$$\mathbf{F}(\Theta) = \frac{4\pi}{C_{\text{sca}}} \langle \mathbf{Z}(\Theta, 0; 0, 0) \rangle, \tag{57}$$

where the average scattering cross section per particle is given by

$$C_{\text{sca}} = 2\pi \int_0^\pi d\Theta \sin \Theta \langle Z_{11}(\Theta, 0; 0, 0) \rangle, \tag{58}$$

and the proportionality constant $4\pi/C_{\text{sca}}$ in Eq. (57) follows from the traditional normalization condition on the phase function, Eq. (65). Note that for macroscopically isotropic and symmetric media, the average scattering cross section is independent of the direction and polarization of the incident beam. The extinction matrix is also considerably simplified: It becomes independent of the direction and polarization of the incident beam and is diagonal with diagonal elements being equal to the average extinction cross section per particle:

$$\langle \mathbf{K}(\mathbf{n}) \rangle = \text{diag}[C_{\text{ext}}, C_{\text{ext}}, C_{\text{ext}}, C_{\text{ext}}]. \tag{59}$$

The average absorption cross section $C_{\text{abs}} = C_{\text{ext}} - C_{\text{sca}}$ is also independent of the direction and polarization state of the incident light. The probability that a photon incident on a small volume element will survive is equal to the ratio of the scattering and extinction cross sections and is called the albedo for single scattering ϖ:

$$\varpi = \frac{C_{\text{sca}}}{C_{\text{ext}}}. \tag{60}$$

The scattering matrix for macroscopically isotropic and symmetric media has the well-known block-diagonal structure (van de Hulst, 1957)

$$\mathbf{F}(\Theta) = \begin{bmatrix} a_1(\Theta) & b_1(\Theta) & 0 & 0 \\ b_1(\Theta) & a_2(\Theta) & 0 & 0 \\ 0 & 0 & a_3(\Theta) & b_2(\Theta) \\ 0 & 0 & -b_2(\Theta) & a_4(\Theta) \end{bmatrix}, \tag{61}$$

so that, in general, only eight elements of the scattering matrix are nonzero and only six of them are independent. Furthermore, there are special relations for the scattering angles 0 and π (van de Hulst, 1957; Mishchenko and Hovenier, 1995):

$$a_2(0) = a_3(0), \qquad a_2(\pi) = -a_3(\pi), \tag{62}$$
$$b_1(0) = b_2(0) = b_1(\pi) = b_2(\pi) = 0, \tag{63}$$
$$a_4(\pi) = a_1(\pi) - 2a_2(\pi). \tag{64}$$

The $(1, 1)$ element of the scattering matrix $a_1(\Theta)$ is called the scattering function or phase function and satisfies the normalization condition

$$\frac{1}{4\pi} \int_{4\pi} d\mathbf{n}^{\text{sca}}\, a_1(\Theta) = \frac{1}{2} \int_0^\pi d\Theta \sin\Theta\, a_1(\Theta) = 1. \tag{65}$$

The quantity

$$\langle \cos\Theta \rangle = \frac{1}{2} \int_{-1}^1 d(\cos\Theta)\, a_1(\Theta) \cos\Theta \tag{66}$$

is called the asymmetry parameter of the phase function and is positive for particles that scatter predominantly in the forward direction, negative for backscattering particles, and zero for symmetric phase functions with $a_1(\pi - \Theta) = a_1(\Theta)$.

Rotationally symmetric particles have an additional property (Mishchenko and Travis, 1994c; Hovenier and Mackowski, 1998)

$$a_4(0) = 2a_2(0) - a_1(0). \tag{67}$$

The structure of the scattering matrix further simplifies for spherically symmetric particles (i.e., homogeneous or radially inhomogeneous spherical particles composed of isotropic materials with refractive index depending only on the distance from the particle center) because in this case $a_1(\Theta) \equiv a_2(\Theta)$ and $a_4(\Theta) \equiv a_3(\Theta)$.

Equation (10) can be used to express the phase matrix for macroscopically isotropic and symmetric media in terms of the scattering and rotation matrices as follows (Hovenier and van der Mee, 1983):

$$\langle \mathbf{Z}(\mathbf{n}^{\text{sca}}, \mathbf{n}^{\text{inc}}) \rangle = \langle \mathbf{Z}(\vartheta^{\text{sca}}, \vartheta^{\text{inc}}, \varphi^{\text{sca}} - \varphi^{\text{inc}}) \rangle$$
$$= \frac{C_{\text{sca}}}{4\pi} \mathbf{L}(\pi - \sigma_2) \mathbf{F}(\Theta) \mathbf{L}(-\sigma_1), \tag{68}$$

where the scattering angle Θ and the angles of rotation σ_1 and σ_2 are shown in Fig. 3 and can be calculated from ϑ^{inc}, ϑ^{sca}, and $\varphi^{\text{inc}} - \varphi^{\text{sca}}$ using spherical trigonometry:

$$\cos\Theta = \cos\vartheta^{\text{sca}} \cos\vartheta^{\text{inc}} + \sin\vartheta^{\text{sca}} \sin\vartheta^{\text{inc}} \cos\left(\varphi^{\text{inc}} - \varphi^{\text{sca}}\right), \tag{69}$$

$$\cos\sigma_1 = \frac{\cos\vartheta^{\text{sca}} - \cos\vartheta^{\text{inc}} \cos\Theta}{\sin\vartheta^{\text{inc}} \sin\Theta}, \tag{70}$$

$$\cos\sigma_2 = \frac{\cos\vartheta^{\text{inc}} - \cos\vartheta^{\text{sca}} \cos\Theta}{\sin\vartheta^{\text{sca}} \sin\Theta}. \tag{71}$$

The first equality of Eq. (68) emphasizes the fact that for macroscopically isotropic and symmetric media the phase matrix depends only on the difference of the azimuthal angles of the scattered and incident beams rather than on their specific values.

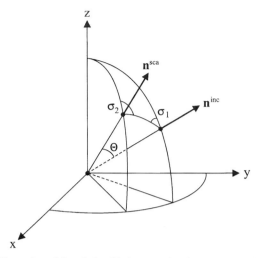

Figure 3   Illustration of the relationship between the phase and scattering matrices.

In computations of light scattering by a macroscopically isotropic and symmetric medium, an efficient approach is to expand the elements of the scattering matrix as follows (de Haan *et al.*, 1987):

$$a_1(\Theta) = \sum_{s=0}^{s_{\max}} \alpha_1^s P_{00}^s(\cos\Theta), \tag{72}$$

$$a_2(\Theta) + a_3(\Theta) = \sum_{s=2}^{s_{\max}} \left(\alpha_2^s + \alpha_3^s\right) P_{22}^s(\cos\Theta), \tag{73}$$

$$a_2(\Theta) - a_3(\Theta) = \sum_{s=2}^{s_{\max}} \left(\alpha_2^s - \alpha_3^s\right) P_{2,-2}^s(\cos\Theta), \tag{74}$$

$$a_4(\Theta) = \sum_{s=0}^{s_{\max}} \alpha_4^s P_{00}^s(\cos\Theta), \tag{75}$$

$$b_1(\Theta) = \sum_{s=2}^{s_{\max}} \beta_1^s P_{02}^s(\cos\Theta), \tag{76}$$

$$b_2(\Theta) = \sum_{s=2}^{s_{\max}} \beta_2^s P_{02}^s(\cos\Theta), \tag{77}$$

where $P_{mn}^s(x)$ are generalized spherical functions (Gelfand *et al.*, 1963), and the upper summation limit, $s_{\max}$, depends on the desired numerical accuracy. For given m and n, the generalized spherical functions $P_{mn}^s(x)$ with $s \geqslant \max(|m|, |n|)$ form a complete orthonormal set of functions on the interval $x \in [-1, 1]$. Taking into account the orthogonality relation

$$\int_{-1}^{+1} dx\, P_{mn}^s(x) P_{mn}^{s'}(x) = \frac{2}{2s+1}(-1)^{m+n}\delta_{ss'}, \tag{78}$$

where $\delta_{ss'}$ is the Kronecker delta, we have from Eqs. (72)–(77)

$$\alpha_1^s = \left[s + \frac{1}{2}\right]\int_{-1}^{+1} d(\cos\Theta)\, P_{00}^s(\cos\Theta)a_1(\Theta), \tag{79}$$

$$\alpha_2^s + \alpha_3^s = \left[s + \frac{1}{2}\right]\int_{-1}^{+1} d(\cos\Theta)\, P_{22}^s(\cos\Theta)\big[a_2(\Theta) + a_3(\Theta)\big], \tag{80}$$

$$\alpha_2^s - \alpha_3^s = \left[s + \frac{1}{2}\right]\int_{-1}^{+1} d(\cos\Theta)\, P_{2,-2}^s(\cos\Theta)\big[a_2(\Theta) - a_3(\Theta)\big], \tag{81}$$

$$\alpha_4^s = \left[s + \frac{1}{2}\right]\int_{-1}^{+1} d(\cos\Theta)\, P_{00}^s(\cos\Theta)a_4(\Theta), \tag{82}$$

$$\beta_1^s = \left[s + \frac{1}{2}\right]\int_{-1}^{+1} d(\cos\Theta)\, P_{02}^s(\cos\Theta)b_1(\Theta), \tag{83}$$

$$\beta_2^s = \left[s + \frac{1}{2}\right]\int_{-1}^{+1} d(\cos\Theta)\, P_{02}^s(\cos\Theta)b_2(\Theta). \tag{84}$$

Note that generalized spherical functions are closely related to Wigner's d functions used in the quantum theory of angular momentum (e.g., Hovenier and van der Mee, 1983; Varshalovich *et al.*, 1988, Chapter 4). General properties of the expansion coefficients in Eqs. (72)–(77) are discussed by van der Mee and Hovenier (1990).

If the expansion coefficients appearing in Eqs. (72)–(77) are known, then the elements of the scattering matrix can be easily calculated for practically any number of scattering angles with a minimum expense of computer time. Thus instead of specifying the elements of the scattering matrix for a large number of scattering angles and using interpolation to calculate the scattering matrix in intermediate points, one can use a limited (and usually small) number of numerically significant expansion coefficients. This also makes the expansion coefficients especially convenient in ensemble averaging: Instead of computing ensemble-averaged scattering matrix elements, one can average a (much) smaller number of expansion coefficients. An additional advantage of expansions (72)–(77) is that they enable one to employ the addition theorem for generalized spherical functions and directly calculate the Fourier components of the phase matrix appearing in the Fourier decomposition of the radiative transfer equation for plane-parallel scattering media (Kuščer and Ribarič, 1959; de Haan *et al.*, 1987). The Fourier decomposition is used to handle analytically the azimuthal dependence of multiply scattered radiation.

Because, for each s, $P_{00}^s(x)$ is a Legendre polynomial $P_s(x)$, Eq. (72) is the well-known expansion of the phase function in Legendre polynomials, which is widely employed in the scalar radiative transfer theory, that is, when the polarization of light is ignored and only the radiance of multiply scattered light is considered (Chandrasekhar, 1950; van de Hulst, 1980; Liou, 1980; Lenoble, 1993). Note that

$$\langle \cos \Theta \rangle = \frac{\alpha_1^1}{3}. \tag{85}$$

XII. MULTIPLE SCATTERING AND RADIATIVE TRANSFER EQUATION

The quantities introduced previously can be used to describe not only single scattering of light by independently scattering particles, but also multiple light scattering by a large collection of such particles. The general radiative transfer equation for a nonemitting medium comprising sparsely and randomly distributed, spherical or arbitrarily oriented nonspherical particles is as follows (Tsang *et al.*, 1985, Chapter 3):

$$\frac{d\mathbf{I}(\mathbf{n})}{ds} = -n_0 \langle \mathbf{K}(\mathbf{n}) \rangle \mathbf{I}(\mathbf{n}) + n_0 \int_{4\pi} d\mathbf{n}' \langle \mathbf{Z}(\mathbf{n}, \mathbf{n}') \rangle \mathbf{I}(\mathbf{n}'), \tag{86}$$

where the four-component column $\mathbf{I}$ is the specific intensity vector (radiance vector) of multiply scattered light propagating in the direction $\mathbf{n}$, and the pathlength element ds is measured along the unit vector $\mathbf{n}$. The first term on the right-hand side of this equation describes the change of the specific intensity vector caused by extinction, whereas the second term describes the contribution of light illuminating a small volume element from all directions $\mathbf{n}'$ and scattered in the direction $\mathbf{n}$. It is important to realize that although we use the same symbol $\mathbf{I}$ to denote the Stokes vector of a transverse electromagnetic wave in Eq. (9) and the specific intensity vector in Eq. (86), their dimensions are different. Whereas the elements of the Stokes vector have the dimension of monochromatic energy flux, those of the specific intensity vector have the dimension of monochromatic radiance: energy per unit area per unit time per unit wavelength per unit solid angle (Hansen and Travis, 1974). The radiative transfer equation must be supplemented by boundary conditions appropriate for a particular physical problem. For example, in the case of light scattering by planetary atmospheres, a standard model is a plane-parallel system illuminated from above by solar radiation. General solutions of Eq. (86) for plane-parallel scattering media are discussed by Mishchenko (1990).

For macroscopically isotropic and symmetric scattering media, the radiative transfer equation can be substantially simplified as follows:

$$\frac{d\mathbf{I}(\mathbf{n})}{d\tau} = -\mathbf{I}(\mathbf{n}) + \frac{\varpi}{4\pi} \int_{-1}^{+1} d(\cos\vartheta') \int_0^{2\pi} d\varphi\, \mathbf{Z}_n(\vartheta', \vartheta, \varphi' - \varphi)\mathbf{I}(\vartheta', \varphi'), \quad (87)$$

where $d\tau = n_0 C_{ext}\, ds$ is the optical pathlength element, and

$$\mathbf{Z}_n(\vartheta', \vartheta, \varphi' - \varphi) = \frac{4\pi}{C_{sca}} \langle \mathbf{Z}(\vartheta', \vartheta, \varphi' - \varphi) \rangle \qquad (88)$$

is the normalized phase matrix [cf. Eq. (68)]. The dependence of the normalized phase matrix on the difference $\varphi' - \varphi$ rather than on φ' and φ enables efficient analytical treatment of the azimuthal dependence of the multiply scattered light using Fourier decomposition of the radiative transfer equation in azimuth. Numerical methods for solving Eq. (87) for the plane-parallel geometry are reviewed, for example, by Hansen and Travis (1974) and Lenoble (1985).

Equation (87) can be further simplified by neglecting polarization and replacing the specific intensity vector $\mathbf{I}$ by its first element (i.e., radiance I) and the normalized phase matrix $\mathbf{Z}_n$ by its $(1, 1)$ element (i.e., the phase function). Although ignoring the vector nature of light and replacing the rigorous vector radiative transfer equation by its approximate scalar counterpart has no rigorous physical justification, this simplification is widely used in atmospheric optics and astrophysics when only the radiance of multiply scattered light needs to be computed. Analytical and numerical solutions of this scalar radiative transfer equation are discussed by Sobolev (1974), van de Hulst (1980), Lenoble (1985), Goody and Yung (1989), and Yanovitskij (1997).

For media with thermal emission, the right-hand side of Eq. (86) contains an additional term $\mathbf{K}_a(\mathbf{n}) I_b(T)$, where T is the temperature of the particles, $I_b(T)$ is the Planck blackbody radiance, and $\mathbf{K}_a(\mathbf{n})$ is an absorption coefficient vector with components

$$K_{a1}(\mathbf{n}) = n_0 \langle K_{11}(\mathbf{n}) \rangle - n_0 \int_{4\pi} d\mathbf{n}' \langle Z_{11}(\mathbf{n}, \mathbf{n}') \rangle, \qquad (89)$$

$$K_{a2}(\mathbf{n}) = n_0 \langle K_{21}(\mathbf{n}) \rangle - n_0 \int_{4\pi} d\mathbf{n}' \langle Z_{21}(\mathbf{n}, \mathbf{n}') \rangle, \qquad (90)$$

$$K_{a3}(\mathbf{n}) = n_0 \langle K_{31}(\mathbf{n}) \rangle - n_0 \int_{4\pi} d\mathbf{n}' \langle Z_{31}(\mathbf{n}, \mathbf{n}') \rangle, \qquad (91)$$

$$K_{a4}(\mathbf{n}) = n_0 \langle K_{41}(\mathbf{n}) \rangle - n_0 \int_{4\pi} d\mathbf{n}' \langle Z_{41}(\mathbf{n}, \mathbf{n}') \rangle \qquad (92)$$

(Tsang, 1991). Numerical solutions of the vector radiative transfer equation for media with thermal emission are discussed in Chapter 17.

XIII. APPENDIX: GEOMETRICAL INTERPRETATION OF STOKES PARAMETERS AND THE ROTATION TRANSFORMATION LAW FOR I, Q, U, AND V

The main purpose of this appendix is to show how the ellipsometric characteristics of a monochromatic transverse electromagnetic wave can be derived from the Stokes parameters defined by Eqs. (5)–(8). Our discussion closely follows Chandrasekhar (1950), van de Hulst (1957), and Hovenier and van der Mee (1983).

Consider a monochromatic plane electromagnetic wave with electric field vector [cf. Eq. (1)]

$$\mathbf{E(R)} = \begin{bmatrix} E_\vartheta \\ E_\varphi \end{bmatrix} = \begin{bmatrix} E_{\vartheta 0} \\ E_{\varphi 0} \end{bmatrix} \exp(ik\mathbf{nR} - i\omega t) \tag{A1}$$

propagating in the direction $\mathbf{n} = \boldsymbol{\vartheta} \times \boldsymbol{\varphi}$. Introducing a real electric vector $\boldsymbol{\xi} = \mathrm{Re}(\mathbf{E})$ and writing $E_{\vartheta 0} = a_{\vartheta 0} \exp(i\Delta_{\vartheta 0})$ and $E_{\varphi 0} = a_{\varphi 0} \exp(i\Delta_{\varphi 0})$ with real positive $a_{\vartheta 0}$ and $a_{\varphi 0}$, we have

$$\xi_\vartheta = a_{\vartheta 0} \cos(\Delta_{\vartheta 0} + k\mathbf{nR} - \omega t), \tag{A2}$$

$$\xi_\varphi = a_{\varphi 0} \cos(\Delta_{\varphi 0} + k\mathbf{nR} - \omega t). \tag{A3}$$

We now consider a fixed point, O, in the beam and rewrite Eqs. (A2) and (A3) in the form

$$\xi_\vartheta = a_{\vartheta 0} \sin(\omega t - \Delta_1), \tag{A4}$$

$$\xi_\varphi = a_{\varphi 0} \sin(\omega t - \Delta_2), \tag{A5}$$

with $\Delta_1 = \Delta_{\vartheta 0} + k\mathbf{nR} - \pi/2$ and $\Delta_2 = \Delta_{\varphi 0} + k\mathbf{nR} - \pi/2$. Using Eqs. (5)–(8), we find for the Stokes parameters

$$I = a_{\vartheta 0}^2 + a_{\varphi 0}^2, \tag{A6}$$

$$Q = a_{\vartheta 0}^2 - a_{\varphi 0}^2, \tag{A7}$$

$$U = -2a_{\vartheta 0}a_{\varphi 0} \cos(\Delta_1 - \Delta_2), \tag{A8}$$

$$V = 2a_{\vartheta 0}a_{\varphi 0} \sin(\Delta_1 - \Delta_2). \tag{A9}$$

At the location O in the beam, the endpoint of the real electric vector given by Eqs. (A4) and (A5) describes an ellipse (Fig. 4). The major axis of this ellipse makes an angle ζ with the positive φ axis so that $0 \leqslant \zeta < \pi$. This angle is obtained by rotating φ in the anticlockwise direction when looking in the direction of propagation, until φ is directed along the major ellipse axis. The ellipticity, that is, the ratio of the minor to major axis of the ellipse, is given by $|\tan \beta|$,

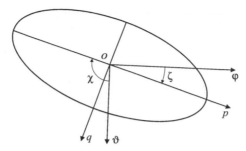

Figure 4   Ellipse described by the endpoint of the real electric vector at a point O in a beam propagating in the direction $\vartheta \times \varphi$.

where $-\pi/4 \leqslant \beta \leqslant \pi/4$. The sign of β and thus of $\tan \beta$ is positive when the real electric vector at O rotates clockwise, as viewed by an observer looking in the direction of propagation. This polarization is called right-handed, as opposed to the left-handed polarization corresponding to the anticlockwise rotation of the electric vector.

To derive the orientation of the ellipse and the ellipticity from the Stokes parameters, we first note that the equations representing the rotation of the real electric vector can also be written in the form

$$\xi_p = a_0 \cos \beta \sin \omega t, \tag{A10}$$
$$\xi_q = a_0 \sin \beta \cos \omega t. \tag{A11}$$

Here, ξ_p and ξ_q are the electric vector components along the major and minor axis of the ellipse, respectively (see Fig. 4), and $a_0^2 = a_{\vartheta 0}^2 + a_{\varphi 0}^2 = I$. The connection between Eqs. (A4)–(A5) and (A10)–(A11) can be established by using the well-known transformation rule for rotation of a coordinate system:

$$\xi_\vartheta = a_0(\sin \beta \cos \omega t \cos \zeta + \cos \beta \sin \omega t \sin \zeta), \tag{A12}$$
$$\xi_\varphi = a_0(- \sin \beta \cos \omega t \sin \zeta + \cos \beta \sin \omega t \cos \zeta). \tag{A13}$$

It is now straightforward to show that the Stokes parameters can also be written as

$$I = a_0^2, \tag{A14}$$
$$Q = -a_0^2 \cos 2\beta \cos 2\zeta, \tag{A15}$$
$$U = -a_0^2 \cos 2\beta \sin 2\zeta, \tag{A16}$$
$$V = -a_0^2 \sin 2\beta. \tag{A17}$$

These equations yield the orientation of the ellipse as follows. Equations (A15) and (A16) give

$$\tan 2\zeta = \frac{U}{Q}.$$ (A18)

Because $|\beta| \leqslant \pi/4$, we have $\cos 2\beta \geqslant 0$ so that $\cos 2\zeta$ has the same sign as $-Q$. Therefore, from the different values of ζ that satisfy Eq. (A18) but differ by $\pi/2$, we must choose the value that makes the sign of $\cos 2\zeta$ the same as that of $-Q$. The ellipticity and handedness now follow from

$$\tan 2\beta = -\frac{V}{(Q^2 + U^2)^{1/2}}.$$ (A19)

Equations (A18) and (A19) also hold for quasi-monochromatic light. The sign of V specifies the handedness of polarization: The polarization is left-handed for positive V and right-handed for negative V. Note that the sign of the Stokes parameter V given by Eq. (A17) is opposite to that on page 42 of van de Hulst (1957). This is a direct consequence of our use of the $\exp(-i\omega t)$ time factor convention instead of the $\exp(j\omega t)$ convention adopted by van de Hulst.

For comparison with other authors (e.g., van de Hulst, 1957), we note that in terms of the angle $\chi = \zeta + 90°$ (see Fig. 4) we have

$$Q = a_0^2 \cos 2\beta \cos 2\chi,$$ (A20)
$$U = a_0^2 \cos 2\beta \sin 2\chi,$$ (A21)

yielding

$$\tan 2\chi = \frac{U}{Q},$$ (A22)

where $\cos 2\chi$ must have the same sign as Q.

Consider now a rotation of the coordinate axes ϑ and φ through an angle $\eta \geqslant 0$ in the anticlockwise direction when looking in the direction of propagation. We thus make the transformation $\zeta \to \zeta - \eta$. Equations (A14)–(A17) show that I and V remain unaltered after such a transformation, whereas Q and U do change. Using primes to denote the Stokes parameters in the new system, we derive from Eqs. (A15) and (A16)

$$Q' = -a_0^2 \cos 2\beta(\cos 2\zeta \cos 2\eta + \sin 2\zeta \sin 2\eta),$$ (A23)
$$U' = -a_0^2 \cos 2\beta(\sin 2\zeta \cos 2\eta - \cos 2\zeta \sin 2\eta),$$ (A24)

which can also be written as

$$\begin{pmatrix} Q' \\ U' \end{pmatrix} = \begin{pmatrix} \cos 2\eta & \sin 2\eta \\ -\sin 2\eta & \cos 2\eta \end{pmatrix} \begin{pmatrix} Q \\ U \end{pmatrix}.$$ (A25)

This equation is the rotation transformation law for the Stokes parameters and agrees with Eq. (10).

ACKNOWLEDGMENT

We thank C. V. M. van der Mee for many valuable discussions.

Chapter 2

Overview of Scattering by Nonspherical Particles

Michael I. Mishchenko
NASA Goddard Institute for Space Studies
New York, New York 10025

Joop W. Hovenier
Department of Physics and Astronomy
Free University
1081 HV Amsterdam, The Netherlands

Warren J. Wiscombe
NASA Goddard Space Flight Center
Greenbelt, Maryland 20771

Larry D. Travis
NASA Goddard Institute for Space Studies
New York, New York 10025

Light Scattering by Nonspherical Particles: Theory, Measurements, and Applications
Copyright © 2000 by Academic Press. All rights of reproduction in any form reserved.

29

I. INTRODUCTION

The scattering characteristics introduced in the preceding chapter are inti-
mately related to the physical characteristics of particles such as size, shape, and
refractive index. Accurate quantitative knowledge of the scattering interaction is
required to understand natural optical phenomena as well as to develop remote
sensing and laboratory techniques that exploit the interaction in order to determine all or some of the particle characteristics.

The scattering properties of homogeneous or layered spheres can be easily
computed via the conventional Lorenz–Mie theory using one of the efficient and
well-documented computer algorithms (e.g., Wiscombe, 1980; Bohren and Huffman, 1983, Appendixes A and B). The convenient availability of the Lorenz–Mie
solution has resulted in the widespread practice of treating nonspherical particles as if they were spheres to which Lorenz–Mie results are applicable. The
assumption of sphericity, however, is rarely made after first studying the effect
of nonsphericity and then discounting it. Furthermore, there is the overwhelming
evidence that scattering properties of nonspherical particles can differ quantitatively and even qualitatively from those of volume- or surface-equivalent spheres.
For example, spherically symmetric particles are incapable of producing such
remarkable optical phenomena as halos, arcs, and pillars and zenith-enhanced
lidar backscatter observed for ice crystals (Greenler, 1990; Platt, 1977); interstellar polarization caused by dust grains oriented by cosmic magnetic fields (Martin, 1978); and single-scattering lidar and radar depolarization observed for cirrus
clouds and precipitation (Chapters 14 and 16). Hence, the last two decades have
demonstrated a major effort aimed at a much better understanding of the effect of
nonsphericity on light scattering.

The scattering properties of nonspherical particles can be either computed theoretically or measured experimentally, and both approaches have their strengths
and weaknesses. Theoretical modeling does not involve expensive instrumentation, can be used to compute any scattering characteristic, and often allows switching to another shape, size, refractive index, or orientation by changing a few lines
in a code. However, accurate computations for realistic polydispersions of irregular particles can be very costly, if at all possible, and often have to be replaced
by computations for simplified model shapes. Experimental measurements using
visible or infrared light can deal with real particles, either natural or artificial.
However, these experiments use complex and expensive hardware, are often incapable of simultaneously and accurately measuring all scattering characteristics,
and traditionally suffer from the inability to characterize the sample precisely. The
microwave analog technique allows a much greater degree of sample characterization and enables true controlled laboratory measurements, but involves even
costlier equipment and cannot be readily applied to realistic particle polydispersions. It is thus clear that only a combination of various theoretical and experi-

mental approaches can lead to a much improved knowledge of electromagnetic scattering by nonspherical particles.

This chapter presents a brief outline of the subject of electromagnetic scattering by nonspherical particles and, along with Chapter 1, is intended to serve as a general introduction to the entire book. It also provides essential references to a few subject areas not covered in detail by specific book chapters. We review and compare numerical and experimental techniques, summarize the current knowledge of typical manifestations of nonsphericity in light scattering, and briefly discuss practical applications. We also mention important problems that are yet to be solved.

In what follows, particle characteristics will often be specified by (i) the size parameter $x = 2\pi a/\lambda$, where a is a characteristic particle size (e.g., semi-major dimension or radius of surface- or volume-equivalent sphere) and λ is the wavelength of light in the surrounding medium; (ii) the aspect ratio, which is the ratio of the maximum to the minimum particle dimensions; and (iii) the index of refraction m relative to the surrounding medium. The efficiency of a numerical technique will often be characterized by its computational complexity, that is, the dependence of the number of computer operations on the particle size parameter. It should be remembered, however, that although the computational complexity of two different techniques can be proportional to the same power of size parameter, the respective proportionality factors can be vastly different, thereby making one technique much slower than the other. Because our main interest is in three-dimensional scattering by finite particles, we will not discuss specifically measurements and computations for such peculiar two-dimensional scatterers as infinite cylinders. For the reader's convenience, all abbreviations used throughout this chapter are summarized in Section VI.

II. EXACT THEORIES AND NUMERICAL TECHNIQUES

All exact theories and numerical techniques for computing the scattered electromagnetic field are based on solving Maxwell's equations either analytically or numerically. The search for an exact analytical solution has been traditionally reduced to solving the vector Helmholtz equation for the time-harmonic electric field using the separation of variables technique in one of the few coordinate systems in which this equation is separable. The incident electric field and the field inside the scatterer are expanded in eigenfunctions that are regular inside the scatterer, whereas the scattered field outside the scatterer is expanded in eigenfunctions that reduce to outgoing waves at infinity. These series are double series in general; degeneration to single series occurs only for spheres and infinite cylinders. Subject to the requirement of continuity of the tangential electric field component at the particle boundary, the unknown expansion coefficients of the internal and scattered fields are determined from the known expansion coefficients of the incident field.

Unfortunately, the separation of variables technique results in an analytical solution only for the few simplest cases. Lorenz in 1890 and, independently, Love (1899), Mie (1908), and Debye (1909) derived the solution for an isotropic homogeneous sphere [see the historical remarks in Section 3.4 of Kerker (1969)]. This solution has been extended to concentric core–mantle spheres (Aden and Kerker, 1951), concentric multilayered spheres (Wait, 1963; Mikulski and Murphy, 1963; Bhandari, 1985), radially inhomogeneous spheres (Wyatt, 1962), and optically active (chiral) spheres (Bohren, 1974). Wait (1955) gave a full solution for electromagnetic scattering by a homogeneous, isotropic, infinite circular cylinder. This solution was extended to optically active cylinders by Bohren (1978), while Kim and Yeh (1991) solved the general problem of scattering by an infinite, isotropic elliptical cylinder. Finally, Oguchi (1973) and Asano and Yamamoto (1975) derived a general solution for homogeneous, isotropic spheroids.

It is unlikely that this list of exact analytical solutions will be significantly extended in the future. Indeed, the solution for the simplest finite nonspherical particles, spheroids, is already so complex that it behaves like a numerical solution and offers little practical advantage. Some exact numerical solutions, in turn, often behave like an analytical solution by means of expanding the incident and scattered fields in a set of orthogonal eigenfunctions with well-known and convenient properties. As a result, the formerly rigid distinction between analytical and numerical solutions for nonspherical particles has become semantic rather than practical.

Although the numerical techniques for computing electromagnetic scattering by nonspherical particles may seem to be innumerable, some of them have been rederived under different names several times, and most of them fall into two broad categories. Differential equation methods compute the scattered field by solving the vector wave equation in the frequency or in the time domain, whereas integral equation methods are based on the volume or surface integral counterparts of Maxwell's equations. Exceptions are hybrid techniques or methods that can be derived using different approaches.

In the rest of this section we will describe several methods that have found extensive practical applications and will compare their relative performance and ranges of applicability. Most numerical techniques compute the scattered electric field for a single particle in a fixed orientation, whereas practical applications often require knowledge of ensemble-averaged quantities such as the optical cross sections and scattering matrix elements. Therefore, we will specifically indicate how ensemble averaging affects the performance of a technique. Because traditional versions of many techniques are applicable only to homogeneous, isotropic, optically inactive particles, we will explicitly mention possible extensions to anisotropic, inhomogeneous, and/or chiral scatterers. More detailed information about the available numerical techniques can be found either in dedicated chapters in this book or in the literature cited. In most cases we cite a recent review paper or a monograph where further references are provided.

A. SEPARATION OF VARIABLES METHOD FOR SPHEROIDS

The separation of variables method (SVM) for single homogeneous, isotropic spheroids was pioneered by Oguchi (1973) and Asano and Yamamoto (1975) and then significantly improved by Voshchinnikov and Farafonov (1993). The method solves the electromagnetic scattering problem for a prolate or an oblate spheroid in the respective spheroidal coordinate system and is based on expanding the incident, internal, and scattered fields in vector spheroidal wave functions. The expansion coefficients of the incident field are computed analytically, whereas the unknown expansion coefficients of the internal and scattered fields are determined through the requirement of the boundary condition. Because the vector spheroidal wave functions are not orthogonal on the spheroidal surface, this procedure results in an infinite set of linear algebraic equations for the unknown expansion coefficients which has to be truncated and solved numerically. For spheroids significantly larger than a wavelength and/or for large refractive indices, the system of linear equations becomes large and ill conditioned. Furthermore, the computation of vector spheroidal wave functions is a difficult mathematical and numerical problem, especially for absorbing particles. These factors have limited the applicability of SVM to semi-major-axis size parameters less than about 40. The obvious limitation of the method is that it is applicable only to spheroidal particles. The main advantage of SVM is that it can produce very accurate results. Furthermore, the improved version of SVM (Voshchinnikov and Farafonov, 1993) is applicable to spheroids with large aspect ratios. The computational complexity of SVM is $O(x^3)$–$O(x^4)$.

SVM was extended to core–mantle spheroids by Onaka (1980), Cooray and Ciric (1992), and Farafonov *et al.* (1996) and to optically active spheroids by Cooray and Ciric (1993). Schulz *et al.* (1998a) developed an analytical technique for computing electromagnetic scattering by an ensemble of randomly oriented spheroids. Extensive SVM computations for spheroids were reported by Asano (1979, 1983), Rogers and Martin (1979), Asano and Sato (1980), Schaefer (1980), de Haan (1987), Stammes (1989), Kurtz and Salib (1993), Kim and Martin (1995), Voshchinnikov (1996), Somsikov (1996), Li *et al.* (1998), and Schulz *et al.* (1998b). Available SVM computer codes are listed by Flatau (1998). Detailed information about SVM can be found in Chapter 4.

B. FINITE ELEMENT METHOD

The finite element method (FEM) is a differential equation method that computes the scattered time-harmonic electric field by solving numerically the vector Helmholtz equation subject to boundary conditions at the particle surface (Morgan and Mei, 1979; Silvester and Ferrari, 1996). The scatterer is embedded in a finite computational domain that is discretized into many small-volume cells called elements with about 10 to 20 elements per wavelength. The unknown field values are specified at the nodes of these elements. Through the requirement of the boundary conditions, the differential equation is converted into a matrix equation for the unknown node field values that can be solved using, for example, Gaussian elimination (GE) or one of the preconditioned iterative methods, for example, the conjugate gradient method (CGM). Because of the local nature of the differential equation, electric fields at the nodes interact only with their neighbors, and the resultant matrix equation is sparse and banded, thereby significantly reducing the computational effort. The computational complexity of FEM with sparse GE is $O(x^7)$ and that of FEM with CGM is only $O(x^4)$. However, the disadvantage of FEM with CGM is that computations must be repeated for each new direction of incidence. It is thus clear that symmetries of the scattering problem should be used as much as possible to reduce the number of these directions.

Although electromagnetic scattering in the far-field zone is an open-space problem, FEM is always implemented in a finite computational domain in order to limit the number of unknowns to a manageable size that can be accommodated by the computer memory and central processing unit (CPU). Therefore, approximate absorbing boundary conditions have to be imposed at the outer boundary of the computational domain in order to suppress wave reflections back into the domain and permit all outward-propagating numerical wave analogs to exit the domain almost as if the domain were infinite (Mittra and Ramahi, 1990). An alternative approach (e.g., Volakis *et al.*, 1998; Sheng *et al.*, 1998) is to couple FEM with a surface integral equation in order to accurately satisfy the radiation condition at infinity (i.e., to ensure the $1/R$ decay of the transverse component and a faster than $1/R$ decay of the radial component of the scattered electric field with $R \to \infty$, where R is the distance from the particle). A drawback of the latter approach is that it can destroy the sparsity of the finite element matrix.

Another way of enforcing the radiation condition is the so-called unimoment method (Mei, 1974; Morgan and Mei, 1979; Morgan, 1980). This modification of FEM uses a spherical computational domain and an expansion of the scattered field outside the computational domain in outgoing spherical wave functions with unknown coefficients. The total external field is the sum of this expansion and the known spherical harmonic expansion of the incident field. The unknown expansion coefficients are found by applying FEM inside the computational domain and matching the nodal field solution and the spherical wave function expansions at the computational domain boundary. The scattered field in the far-field zone is calculated by evaluating the spherical wave function expansion and automatically satisfies the radiation condition. Because the unimoment method always uses a spherical computational domain, it can be inefficient for objects with high aspect ratios.

Some important advantages of FEM are that it permits the modeling of arbitrarily shaped and inhomogeneous particles, is simple in concept and execution, and avoids the singular-kernel problem typical of the integral equation methods (see Section II.E). However, FEM computations are spread over the entire computational domain rather than confined to the scatterer itself, as with integral equation methods. This tends to make FEM computations rather time consuming and limited to size parameters less than about 10. The finite spatial discretization and the approximate absorbing boundary condition make FEM poorly suitable for achieving high and controllable numerical accuracy. Further information about FEM and the closely related finite difference method (FDM) can be found in Morgan (1990), Silvester and Ferrari (1996), and Volakis *et al.* (1998). Several FEM computer codes are listed by Wriedt (1998).

C. FINITE DIFFERENCE TIME DOMAIN METHOD

Unlike FEM, the finite difference time domain method (FDTDM) calculates electromagnetic scattering in the time domain by directly solving Maxwell's time-dependent curl equations (Yee, 1966; Kunz and Luebbers, 1993; Taflove, 1995, 1998). The space and time derivatives of the electric and magnetic fields are approximated using a finite difference scheme with space and time discretizations selected so that they bound computational errors and ensure numerical stability of the algorithm. Hence, time is approximated by discrete time steps, and a marching-in-time procedure is used to track the evolution of the fields from their initial values at some initial time. As in FEM, the scattering particle is embedded in a finite computational domain, and absorbing boundary conditions are employed to model scattering in the open space (e.g., Berenger, 1996; Grote and Keller, 1998; Yang and Liou, 1998b). The fields are specified at spatial grid points with discretization density similar to that needed for FEM. The grid values at the previous and current time steps are used to calculate the values at the next time step, thereby making the system of equations to update the fields fully explicit. Hence, there is no need to solve a large system of linear equations, and the memory storage requirement is proportional to the total number of grid points. The operation count grows approximately as the fourth power of the particle size parameter. Because FDTDM computes the near field in the time domain, a special near zone to far zone transformation must be invoked in order to compute the scattered far field in the frequency domain.

FDTDM has become rather popular recently because of its conceptual simplicity and ease of implementation. It shares the advantages of FEM with CGM as well as its limitations in terms of accuracy, computational complexity, size parameter range, and the need to repeat all computations with changing direction of illumination. Applications of FDTDM to far-field scattering computations are described by Tang and Aydin (1995), Yang and Liou (1996a), Aydin and Tang (1997a, b), and Yang *et al.* (1997). Additional information about FDTDM is given in Chapter 7. Available FDTDM computer codes are listed by Wriedt (1998).

D. POINT MATCHING METHOD

In this differential equation technique, the incident and internal fields are expanded in vector spherical wave functions (VSWFs) regular at the origin, and the scattered field outside the scatterer is expanded in outgoing VSWFs. The expansion coefficients of the incident field are known, whereas the expansion coefficients of the internal and scattered fields are found by truncating the expansions to a finite size and matching the fields at the surface of the scatterer via the application of boundary conditions. In the simple point matching method (PMM), the fields are matched at as many points on the surface as there exist unknown expansion coefficients (Oguchi, 1973). However, the validity of the simple PMM is questionable and depends on the applicability of the Rayleigh hypothesis, that is, the assumption that the scattered field can be accurately expanded in the outgoing spherical waves in the region enclosed between the particle surface and the smallest circumscribing sphere (e.g., Millar, 1973; Lewin, 1970; Bates, 1975). This problem appears to be ameliorated in the generalized PMM (GPMM) by forming an overdetermined system of equations for the unknown expansion coefficients. This is done by matching the fields in the least-squares sense at a significantly greater number of surface points than the number of unknowns (Morrison and Cross, 1974; Oguchi and Hosoya, 1974; Al-Rizzo and Tranquilla, 1995a). A further improvement of the GPMM has been achieved by utilizing multiple spherical expansions to describe the fields both inside and outside the scattering object (Joo and Iskander, 1990; Al-Rizzo and Tranquilla, 1995b). This multiple-expansion GPMM (ME-GPMM) is otherwise known as the generalized multipole technique (Hafner, 1990). It is claimed that ME-GPMM for rotationally symmetric scatterers is numerically stable, sufficiently accurate, and applicable to very large size parameters (Al-Rizzo and Tranquilla, 1995b). Piller and Martin (1998a) extended ME-GPMM to anisotropic scatterers.

E. INTEGRAL EQUATION METHODS

The scattering of a plane electromagnetic wave by an object of volume V can be described by an integral equation (Shifrin, 1951; Van Bladel, 1961),

$$\mathbf{E}(\mathbf{r}) = \mathbf{E}^{\text{inc}}(\mathbf{r}) + k^2 \int_V d^3\mathbf{r}' \left[\mathbf{1} + \frac{1}{k^2} \nabla\nabla \right] \frac{\exp(ik|\mathbf{r} - \mathbf{r}'|)}{4\pi|\mathbf{r} - \mathbf{r}'|} [m^2(\mathbf{r}') - 1]\mathbf{E}(\mathbf{r}'), \quad (1)$$

where $\mathbf{E}(\mathbf{r})$ is the total electric field, $\mathbf{E}^{\text{inc}}(\mathbf{r})$ is the incident field, k is the free-space wavenumber, $m(\mathbf{r})$ is the refractive index, and $\mathbf{1}$ is the identity dyad. The factor $[m^2(\mathbf{r}') - 1]\mathbf{E}(\mathbf{r}')$ is the interior polarization current, which can be interpreted as the source of the scattered field. The calculation of the scattered field via Eq. (1) would be straightforward except that the internal electric field is unknown. Therefore, Eq. (1) must first be solved for the internal field. Physically, this is equivalent to representing the internal field at each point of the volume V as a sum of the incident field and the field induced by sources at all interior points, including the self point. The integral in Eq. (1) is calculated by discretizing the interior region into N small cubic cells with about 10 to 20 cells per wavelength and assuming that the field within each cell is constant. This procedure is repeated for each of the N unknown internal field values and results in a matrix equation that must be solved numerically. Because the source elements interact with each other all throughout the scatterer, the resultant matrix is full. Once the internal field is obtained, the external field is found from Eq. (1). Finally, the scattered field is computed by subtracting the incident field from the external field. This version of the volume integral equation method (VIEM) is known as the method of moments (MOM) (Harrington, 1968).

Several modifications of MOM have been developed under different names: the discrete dipole approximation (DDA), also known as the coupled dipole method (Purcell and Pennypacker, 1973; Draine, 1988); the digitized Green's function algorithm (Goedecke and O'Brien, 1988); the volume integral equation formulation (Iskander *et al.*, 1989a; Hage *et al.*, 1991; Lumme and Rahola, 1998); and the variational volume integral equation method (Peltoniemi, 1996). The main difference between these techniques is the way in which they treat the self-interaction term.

The straightforward approach to solving the MOM matrix equation using GE entails a computational complexity of $O(N^3) \propto O(x^9)$ and is not practical for size parameters exceeding unity. The conjugate gradient method together with the fast Fourier transform (CGM-FFT) (Peterson *et al.*, 1998, Chapter 4) has the computational complexity $O(N^{1+\alpha} \log N) \propto O(x^{3+3\alpha} \log x)$, where N^α with $0 < \alpha < 1$ is the total number of iterations required to achieve a certain accuracy, and can be applied to significantly larger size parameters. Furthermore, CGM-FFT and related techniques can significantly reduce computer memory requirements. The standard drawback of using CGM and other preconditioned iterative methods is that the computations must be fully repeated for each new illumination direction.

The major advantages of VIEMs are that they automatically satisfy the radiation condition at infinity; are confined to the scatterer itself, thereby resulting in fewer unknowns than the differential equation methods; and can be applied to inhomogeneous, anisotropic, and optically active scatterers (e.g., Su, 1989). Their main drawbacks are a low computational accuracy and slow improvement with increasing N and the fast CPU time growth with increasing size parameter. Further information on MOM can be found in Miller *et al.* (1992) and Wang (1991).

Electromagnetic scattering by homogeneous or layered dielectric bodies can be computed using the surface integral counterpart of Eq. (1) (Poggio and Miller, 1973; Umashankar *et al.*, 1986; Medgyesi-Mitschang *et al.*, 1994). Although surface integral equation methods (SIEMs) cannot be applied to highly inhomogeneous scatterers, their important advantages are that the dimensionality of the problem is reduced by one and the number of unknowns N is proportional to x^2 rather than to x^3, as in VIEMs, thereby resulting in a computational complexity of $O(x^6)$ for SIEMs with GE and $O(x^{4+2\alpha})$ for SIEMs with CGM.

F. DISCRETE DIPOLE APPROXIMATION

The original derivation of DDA (otherwise known as the coupled dipole method) by Purcell and Pennypacker (1973) is based on partitioning a particle into a number N of elementary polarizable units called dipoles. The electromagnetic response of the dipoles to the local electric field is assumed to be known. The field exciting a dipole is a superposition of the external field and the fields scattered by all other dipoles. This allows one to write a system of N linear equations for N fields exciting the N dipoles. The numerical solution of this system is used to compute N partial fields scattered by the dipoles and, thus, the total scattered field.

Since the original paper by Purcell and Pennypacker, DDA has been improved by modifying the treatment of dipole polarizability (Draine, 1988; Dungey and Bohren, 1991; Draine and Goodman, 1993; Lumme and Rahola, 1994), including magnetic in addition to electric dipole terms (Mulholland *et al.*, 1994), applying CGM-FFT to solve the DDA matrix equation with an $O(x^{3+3\alpha} \log x)$ computational complexity (Goodman *et al.*, 1991), and employing concepts of the sampling theory (Piller and Martin, 1998b). Varadan *et al.* (1989) and Lakhtakia (1992) extended DDA to anisotropic and bianisotropic materials, respectively. Lakhtakia and Mulholland (1993) gave a general derivation of DDA from the volume integral equation formulation and discussed its close relation to MOM. Ku (1993) compared the numerical performance of MOM (Iskander *et al.*, 1989a) and DDA (Dungey and Bohren, 1991). McClain and Ghoul (1986) and Singham *et al.* (1986) developed analytical DDA procedures for computing scattering by randomly oriented particles based on reexpanding the Cartesian tensor products in terms of spherical tensor products and exploiting the analytical properties of Wigner D functions. Unfortunately, this approach assumes time-consuming matrix inversions [$O(x^9)$ computational complexity] and is applicable only to particles smaller than a wavelength. Chiappetta (1980a) and Singham and Bohren (1987, 1988) developed a scattering-order formulation of DDA.

The most important advantage of DDA is its applicability to arbitrarily shaped, inhomogeneous, and anisotropic particles. The disadvantages of the technique are limited numerical accuracy, especially for scattering matrix elements; slow convergence of results with increasing N; and the need to repeat the entire calculation for each new direction of incidence (for DDA with CGM-FFT) (Singham, 1989; Draine and Flatau, 1994; Okamoto *et al.*, 1995; Liu and Illingworth, 1997). These factors have made DDA computations time consuming, especially for particle size and/or orientation distributions, and have limited the particle size parameter to relatively small values. The attractiveness and simplicity of the physical idea of DDA and the public availability of a well-documented DDA code by Draine and Flatau (1997) have led to widespread applications of this technique during the last decade. A detailed review of DDA and its applications is given in Chapter 5.

G. FREDHOLM INTEGRAL EQUATION METHOD

Equation (1) is a Fredholm-type integral equation with a singular kernel at $\mathbf{r}' = \mathbf{r}$. Holt *et al.* (1978) have removed the singularity by applying the Fourier transform to the internal field and converting the volume integral to one in the wave number coordinate space. Discretization of the latter integral results in a matrix equation that must be solved numerically and gives the scattered field. The scattered field obtained with the Fredholm integral equation method (FIEM) satisfies a variational principle and is claimed to be numerically stable and convergent to the exact result even for particles with large aspect ratios, although the actual size parameter in reported computations for highly aspherical scatterers has been small. The numerical implementation of the technique becomes much simpler for homogeneous, rotationally symmetric bodies.

The major limitation of FIEM is that the matrix elements must be evaluated analytically, thereby leading to different programs for each scatterer and restricting computations to only a few model shapes such as spheroids, triaxial ellipsoids, and finite circular cylinders (Evans and Holt, 1977; Holt *et al.*, 1978; Holt and Shepherd, 1979; Shepherd and Holt, 1983; Matsumura and Seki, 1991, 1996; Stamatakos *et al.*, 1997). Most published FIEM calculations pertain to size parameters smaller than 5 and tend to be rather time consuming (Holt, 1982). Larger particles (equal-volume-sphere size parameters up to 36.7) were considered by Stamatakos *et al.* (1997). However, the refractive index was limited to 1.04, and a comparison of FIEM results for a sphere with exact Mie computations showed poor agreement. An important advantage of FIEM is that a significant part of the calculation, the integrals, depends only on the size parameter and the shape. Hence, changing the refractive index and/or the direction and polarization state of the incident light does not require a complete new calculation. A similar saving of CPU time is achieved in performing convergence checks. Recently, Papadakis *et al.* (1990) extended FIEM to anisotropic dielectric ellipsoids, whereas Stamatakos and Uzunoglu (1997) applied FIEM to scattering by a linear chain of triaxial dielectric ellipsoids.

H. *T*-MATRIX APPROACH

The *T*-matrix method (TMM) is based on expanding the incident field in VSWFs regular at the origin and expanding the scattered field outside a circumscribing sphere of the scatterer in VSWFs regular at infinity. The *T* matrix transforms the expansion coefficients of the incident field into those of the scattered field and, if known, can be used to compute any scattering characteristic of a nonspherical particle. TMM was first introduced by Waterman (1971) for single homogeneous scatterers and was then generalized to multilayered scatterers and arbitrary clusters of nonspherical particles by Peterson and Ström (1973, 1974) and to nonspherical chiral scatterers by Lakhtakia *et al.* (1985b). For spheres, all TMM formulas reduce to those of the standard Mie theory. Furthermore, in the case of clusters composed of spherical components, the *T*-matrix method is equivalent to the multisphere superposition method (Borghese *et al.*, 1984; Mackowski, 1994; Mishchenko *et al.*, 1996b).

The *T* matrix for single homogeneous and multilayered scatterers is usually computed using the Huygens principle or, equivalently, the extended boundary condition method (Waterman, 1971; Peterson and Ström, 1974; Barber and Yeh, 1975). A more recent and simpler derivation of TMM equations by Waterman (1979) clearly avoids the use of the Rayleigh hypothesis and shows that scattering objects need not be convex and close to spherical in order to ensure the applicability of TMM. This result appears to be very important because recent computations by Videen *et al.* (1996), Ngo *et al.* (1997) and Doicu *et al.* (1999) seem to indicate that the Rayleigh hypothesis may be wrong. In general, TMM can be applied to any particle shape, although TMM computations are much simpler and more efficient for bodies of revolution. Accordingly, almost all existing computer codes assume rotationally symmetric shapes both smooth, for example, spheroids and so-called Chebyshev particles, and sharp edged, for example, finite circular cylinders (Wiscombe and Mugnai, 1986; Barber and Hill, 1990;

Kuik *et al.*, 1994; Mishchenko and Travis, 1998). Special procedures have been developed to improve the numerical stability of TMM computations for large size parameters and/or extreme aspect ratios (Lakhtakia *et al.*, 1983; Mishchenko and Travis, 1994a; Wielaard *et al.*, 1997). Recent work has demonstrated the practical applicability of TMM to particles without axial symmetry, for example, general ellipsoids, cubes, and clusters of spheres (Schneider and Peden, 1988; Mackowski and Mishchenko, 1996; Wriedt and Doicu, 1998; Laitinen and Lumme, 1998), although the computational complexity of these calculations is significantly greater than that for rotationally symmetric scatterers.

The computation of the T matrix for a cluster assumes that the T matrices of all components are known and is based on the use of the translation addition theorem for VSWFs (Peterson and Ström, 1973). The computation becomes much simpler for clusters composed of spherical particles and, especially, for linear sphere configurations.

The loss of efficiency for particles with large aspect ratios or with shapes lacking axial symmetry is the main drawback of TMM. The main advantages of TMM are that it is highly accurate, fast [computational complexity $O(x^3)$–$O(x^4)$ for rotationally symmetric scatterers], and applicable to particles with equivalent-sphere size parameters exceeding 100 (Mishchenko and Travis, 1998). The elements of the T matrix are independent of the incident and scattered fields and depend only on the shape, size parameter, and refractive index of the scattering particle and on its orientation with respect to the reference frame, so that the T matrix need be computed only once and then can be used in computations for any directions of light incidence and scattering. Mishchenko (1991a) developed an analytical orientation-averaging T-matrix procedure that makes computations for randomly oriented, rotationally symmetric particles as fast as those for a particle in a fixed orientation. Recently, this procedure has been extended to arbitrary clusters of spheres (Mackowski and Mishchenko, 1996). A similar analytical procedure was developed for computing the extinction matrix for nonspherical particles axially oriented by magnetic, electric, or aerodynamic forces (Mishchenko, 1991b). Further information about TMM can be found in Chapter 6.

I. SUPERPOSITION METHOD FOR COMPOUNDED SPHERES AND SPHEROIDS

The separation of variables solution for a single sphere can be extended to clusters of spheres by using the translation addition theorem for VSWFs (Bruning and Lo, 1971; Borghese *et al.*, 1979; Hamid *et al.*, 1990; Fuller, 1991, 1994a; Mackowski, 1991). The total scattered field from a multisphere cluster can be represented as a superposition of individual fields scattered from each sphere. These individual fields are interdependent because of electromagnetic interactions between the spheres. The external electric field illuminating the cluster and the individual fields scattered by the component spheres are expanded in VSWFs with origins at the individual sphere centers. To exploit the orthogonality of the VSWFs in the sphere boundary conditions, the translation addition theorem is used in which a VSWF centered at one sphere origin is reexpanded about a second origin. This procedure ultimately leads to a matrix equation for the scattered-field expansion coefficients of each sphere. The numerical solution of this equation for the specific direction and polarization state of the incident field gives the individual scattered fields and thus the total scattered field.

Alternatively, inversion of the cluster matrix equation gives sphere-centered transition matrices that transform the expansion coefficients of the incident field into the expansion coefficients of the individual scattered fields. The advantage of this approach is that the individual-sphere transition matrices are independent of the direction and polarization state of the incident field. In the far-field region, the individual scattered-field expansions can be transformed into a single expansion centered at a single origin inside the cluster. This procedure gives a matrix that transforms the incident-field expansion coefficients into the single-origin expansion coefficients of the total scattered field. This matrix is completely equivalent to the cluster T matrix (Borghese *et al.*, 1984; Mackowski, 1994) and can be used in analytical averaging of scattering characteristics over cluster orientations (Fucile *et al.*, 1993, 1995; Mackowski and Mishchenko, 1996). In this regard, the superposition method (SM) for spheres can be considered a particular case of TMM in which the latter is applied to an aggregate of spherical scatterers (Peterson and Ström, 1973; Mishchenko *et al.*, 1996b).

SM has been extended to clusters of concentrically layered spheres (Hamid *et al.*, 1992), spheres with one or more spherical inclusions (Fikioris and Uzunoglu, 1979; Borghese *et al.*, 1992, 1994; Fuller, 1995b; Mackowski and Jones, 1995), and to pairs of osculating spheres (Videen *et al.*, 1996). Also, Cooray and Ciric (1991a) have developed SM for systems of two dielectric spheroids by combining the SVM solution for individual spheroids with the use of rotational–translational addition theorems for vector spheroidal wave functions (Cooray and Ciric, 1989a).

The computational complexity of multisphere SM calculations strongly depends on the number of components and their size parameter and configuration. In practice, obtaining converged results for a larger number of components often necessitates a smaller component size parameter, and vice versa. SM computations become especially efficient for linear configurations of spheres. Because of the analyticity of the SM mathematical formulation, the method is capable of producing very accurate results. Chapter 8 provides a review of SM for compounded spheres, while Chapter 4 reviews SM for systems of spheroids.

J. Other Exact Techniques

Kattawar *et al.* (1987) found the solution of Eq. (1) by solving first a simpler equation for a resolvent kernel matrix. An advantage of this approach is that the resolvent kernel matrix is computed only once for the entire range of refractive indices. Rother and Schmidt (1996) and Rother (1998) developed a differential equation method, called the discretized Mie formalism, that solves the vector Helmholtz equation for homogeneous scatterers using a method of lines. The main advantage of this technique is an analytic incorporation of the radiation condition at infinity. Another version of the differential equation method was developed by Tosun (1994). Eremin and Orlov (1998) suggested the so-called discrete sources method, which is based on principles similar to those of the multiple-expansion GPMM and closely resembles the so-called Yasuura method (Kawano *et al.*, 1996). Vechinski *et al.* (1994) developed a time domain SIEM to compute transient scattering from arbitrarily shaped homogeneous dielectric bodies. The advantage of their approach over FDTDM is that the radiation condition at infinity is automatically satisfied and the memory requirement is considerably reduced (see also Pocock *et al.*, 1998).

Further information on numerical scattering techniques can be found in the special journal issues edited by Shafai (1991), Barber *et al.* (1994), Hovenier (1996), Lumme (1998) and Mishchenko *et al.* (1999).

K. Comparison of Methods, Benchmark Results, and Computer Codes

The very existence and use of many numerical techniques for computing electromagnetic scattering by nonspherical particles indicate that there is no single technique that provides the best results in all cases. Depending on the application at hand, each technique may prove to be more appropriate in terms of efficiency, accuracy, and applicability to the particle physical parameters. Furthermore, it is very difficult to develop and apply simple and objective criteria in order to examine the relative performance of different numerical techniques in a wide range of applications. In principle, one should use the same computer and treat the same scattering problems using codes written by authors with comparable levels of programming skills. Even in this idealistic case, however, the specific characteristics of the computer used can favorably enhance the performance of one technique and degrade the efficiency of another. For example, one technique may especially benefit from the use of a parallel computer, whereas the performance of another technique may strongly depend on the availability and efficient organization of double or extended precision computations. Furthermore, direct comparisons of different techniques can face almost insuperable organizational problems (e.g.,

Hovenier *et al.*, 1996) and have always been limited to a few techniques and a few scattering problems (Cooper *et al.*, 1996; Wriedt and Comberg, 1998). It may thus appear in many cases that the only practical way of deciding which technique is most suitable for a specific problem has been an indirect comparison of techniques based on semiquantitative evidence scattered over many publications (e.g., Oguchi, 1981; Holt, 1982). The availability of a well-documented public-domain computer code is also an important factor to take into account.

Other than Mie theory, the only methods capable of providing very accurate results (five and more correct decimals) for particles comparable to and larger than a wavelength are SVM, TMM, and SM. Each of these techniques incorporates an internal convergence test, which is a good measure of the absolute convergence (Kuik *et al.*, 1992; Hovenier *et al.*, 1996). Benchmark results for spheroids, finite circular cylinders, Chebyshev particles, and two-sphere clusters in fixed and random orientations were reported by Mishchenko (1991a), Kuik *et al.* (1992), Mishchenko *et al.* (1996a), Mishchenko and Mackowski (1996), Hovenier *et al.* (1996), and Wielaard *et al.* (1997). These results cover a wide range of equivalent-sphere size parameters from a few units to 60 (Wielaard *et al.*, 1997) and are given with up to nine correct decimals. Because the accuracy of these numbers is guaranteed up to a few units in the last digits, they are an important tool for testing the accuracy of other theoretical approaches. Other valuable means of checking the physical correctness of the theoretical computations include the reciprocity relation (Chapter 1) and analytical properties of the phase matrix elements (Chapter 3).

SVM, TMM, SM, and GPMM are the only methods that have been extensively used in computations for particles significantly larger than a wavelength. SVM, TMM, and SM also seem to be the most efficient techniques for computing electromagnetic scattering by homogeneous and composite particles of revolution. The availability of the analytical orientation-averaging procedure makes TMM an efficient technique for randomly oriented rotationally symmetric particles with moderate aspect ratios. Scattering by particles with larger aspect ratios can be computed with an improved version of SVM (Voshchinnikov and Farafonov, 1993), a modification of TMM called the iterative extended boundary condition method (Lakhtakia *et al.*, 1983), and the multiple-expansion GPMM. SIEM (Zuffada and Crisp, 1997) and FIEM can also be applied to homogeneous, rotationally symmetric particles with large aspect ratios, although they seem to be significantly slower and less accurate than the other techniques.

Computations for particles that cannot be treated with SVM, GPMM, and the commonly used version of TMM (e.g., anisotropic scatterers and homogeneous and inhomogeneous particles lacking rotational symmetry) must rely on more flexible techniques such as FEM, FDM, FDTDM, and different versions of VIEM, including DDA. All these techniques are simple in concept and software implementation and seem to have comparable performance characteristics (e.g., Wriedt and Comberg, 1998). Unfortunately, the simplicity and flexibility of these techniques are often accompanied by a substantial loss in efficiency and accuracy and by stronger practical limitations on the maximal particle size parameter. Further work is obviously required in order to develop a method that is efficient, flexible, and applicable to a wide range of size parameters.

A number of software implementations of the previously mentioned techniques are currently available, and many of them are in the public domain. Lists of available computer codes were compiled by Flatau (1998) and Wriedt (1998).

III. APPROXIMATIONS

A. RAYLEIGH, RAYLEIGH–GANS, AND ANOMALOUS DIFFRACTION APPROXIMATIONS

Rayleigh (1897) derived an approximation for scattering in the small-particle limit ($x \ll 1$) by assuming that the incident field inside and near the particle behaves almost as an electrostatic field and the internal field is homogeneous. A detailed account of the Rayleigh approximation (RA) was given by Kleinman and Senior (1986). A completely analytical solution has been found for only a few simple shapes, including ellipsoids. For general shapes, one must solve numerically a simple integral equation for the polarizability tensor. Bohren and Huffman (1983, Chapter 5) gave analytical formulas for the optical cross sections and the scattering matrix elements of randomly oriented spheroids. Note that the direct use of the optical theorem (Section VIII of Chapter 1) in the framework of RA gives only the absorption component of the extinction cross section and must be supplemented by the computation of the scattering cross section via Eqs. (32) and (33) of Chapter 1 (Bohren and Huffman, 1983, Section 5.5). Mishchenko (1991b) studied the range of applicability of RA in computations of the extinction matrix for partially and randomly oriented spheroids. Jones (1979) extended RA to clusters of small spheres (see also Mackowski, 1995, and references therein).

Stevenson (1953) generalized RA by expanding the fields in powers of the size parameter x. The first term $O(x^2)$ gives RA, while the second term $O(x^4)$ gives the so-called Rayleigh–Gans–Stevenson approximation. This approach was further improved by von Ross (1971) and applied to various scattering problems by Stevenson (1968) and Khlebtsov (1979).

The conditions of validity of the Rayleigh–Gans approximation (RGA) (otherwise known as the Rayleigh–Debye or Born approximation; e.g., Ishimaru, 1978, Chapter 2) are $x|m-1| \ll 1$ and $|m-1| \ll 1$. In other words, the particles must be not too large (although they may be larger than in the case of Rayleigh scattering) and optically "soft." The fundamental RGA assumption is that each small-volume element in the scatterer is excited only by the incident field. The trade-off is that significant analytical progress can be made in this case. Also, like many asymptotic approximations, RGA is useful outside its strictly defined range of validity (e.g., Barber and Wang, 1978). Acquista (1976) generalized RGA by applying the method of successive iterations (Shifrin, 1951) to Eq. (1). This approach was used in computations for spheroids and finite circular cylinders and is valid for $x|m - 1| \lesssim 1$ (Haracz *et al.*, 1984, 1985, 1986). Khlebtsov (1984) derived an exact integral equation of Lippman–Schwinger type by taking the Fourier transform of Eq. (1). Successive iterations of this equation give RGA, the Acquista result, and higher order approximations. This approach was improved and applied to suspensions of aligned anisotropic particles by Khlebtsov and Melnikov (1991) and Khlebtsov *et al.* (1991). Muinonen (1996b) applied RA and RGA to particles with Gaussian random surfaces.

The anomalous diffraction approximation (ADA) was initially developed by van de Hulst (1957, Chapter 11) as a means for computing the extinction cross section for large, optically soft spheres: $x \gg 1$ and $|m - 1| \ll 1$. Because the second condition means that rays are negligibly deviated as they cross the particle boundary and are negligibly reflected, ADA assumes that extinction is caused by the absorption of light passing through the particle and by the interference of light passing through the particle and passing around the particle. This allows a general representation of the extinction and absorption cross sections as simple integrals over the particle projection on the plane perpendicular to the incident beam. The integrals can be evaluated numerically or, in some special cases, analytically. ADA was applied to prismatic columns (Chýlek and Klett, 1991a, b), hexagonal ice columns (Chýlek and Videen, 1994), spheroids (Evans and Fournier, 1994; Baran *et al.*, 1998), cubes (Masłowska *et al.*, 1994), ellipsoids (Streekstra *et al.*, 1994), and finite circular cylinders (Liu *et al.*, 1998). Comparisons of ADA and exact *T*-matrix computations (Liu *et al.*, 1998) suggest that ADA extinction is more accurate for randomly oriented nonspherical particles than for spheres, and ADA absorption errors decrease with increasing imaginary part of the refractive index. Meeten (1982) and Khlebtsov (1993) extended ADA to scattering by anisotropic particles and fractal clusters, respectively. ADA and the closely related Wentzel–Kramers–Brillouin and eikonal approximations belong to the family of high-energy approximations (e.g., Perrin and Lamy, 1986; Bourrely *et al.*, 1989; Klett and Sutherland, 1992).

B. Geometric Optics Approximation

The geometric optics approximation (GOA) (otherwise known as the ray-tracing or ray optics approximation) is a universal approximate method for computing light scattering by particles much larger than a wavelength. GOA is based on the assumption that the incident plane wave can be represented as a collection of independent parallel rays. The history of each ray impinging on the particle surface is traced using the Snell law and Fresnel's equations. The sampling of all escaping light rays into predefined narrow angular bins supplemented by the computation of Fraunhofer diffraction of the incident wave on the particle projection yields a quantitative representation of the particle scattering properties.

GOA is particularly simple for spheres (Liou and Hansen, 1971) because the ray paths remain in a plane. For other particles, ray tracing is usually performed using a Monte Carlo approach. Wendling *et al.* (1979), Cai and Liou (1982), Takano and Jayaweera (1985), Rockwitz (1989), Takano and Liou (1989a), Masuda and Takashima (1992), and Xu *et al.* (1997) applied GOA to finite hexagonal cylinders in random and horizontal orientations, whereas Yang and Cai (1991) and Macke and Mishchenko (1996) computed scattering by randomly oriented spheroids and finite circular cylinders. Light scattering by other polyhedral shapes was computed by Liou *et al.* (1983), Muinonen *et al.* (1989), Macke (1993), Iaquinta *et al.* (1995), Takano and Liou (1995), Liu *et al.* (1996a), Macke *et al.* (1996b), and Yang and Liou (1998a). GOA was also applied to distorted raindrops (Macke and Großklaus, 1998) and large stochastic particles, both randomly shaped (Peltoniemi *et al.*, 1989; Muinonen *et al.*, 1996; Macke *et al.*, 1996b; Yang and Liou, 1998a, Chapter 11) and containing multiple randomly positioned inclusions (Macke *et al.*, 1996a, Chapter 10).

The main advantage of GOA is that it can be applied to essentially any shape. However, GOA is approximate by definition, and its range of applicability in terms of the smallest size parameter x_{min} must be checked by comparing GOA results with exact numerical solutions of Maxwell's equations. Comparisons of GOA and Mie results (Hansen and Travis, 1974) show that GOA phase function calculations for spheres are reasonably accurate only for x_{min} exceeding several hundred. Analogous comparisons of GOA and TMM phase function results for randomly oriented spheroids and finite circular cylinders (Macke *et al.*, 1995; Wielaard *et al.*, 1997) show that for moderately absorbing nonspherical particles x_{min} can be as small as 80. Unfortunately, decreasing absorption increases x_{min} even for nonspherical particles, and accurate GOA calculations of the full scattering matrix require much larger x_{min} values than analogous GOA calculations of the phase function (Hansen and Travis, 1974; Wielaard *et al.*, 1997).

Absorbing particles (imaginary part of the refractive index not equal to zero) require special treatment because in this case the refracted wave becomes inhomogeneous so that the surface of constant amplitude does not coincide with the surface of constant phase. The Snell law can still be formally used, but needs to be modified as derived by Stratton (1941, Section 9.8) (see also Ulaby *et al.*, 1981, Section 2-8). The effect of this modification on the scattering properties of absorbing ice crystals is discussed by Yang and Liou (1995) and Zhang and Xu (1995).

A modification of GOA, the so-called physical optics or Kirchhoff approximation, was developed by Ravey and Mazeron (1982, 1983). This method implies the computation of the surface field using ray tracing and the computation of the corresponding far field via the vector Kirchhoff integral. The method was used, with some variations, by Muinonen (1989), Yang and Liou (1995, 1996b), and Mazeron and Muller (1996) and was found to be rather time consuming. Because this technique is still an approximation, its accuracy as a function of size parameter should be extensively tested versus exact solutions, especially when the full scattering matrix is computed. Ways of generalizing GOA and incorporating wave effects into GOA computations are discussed by Muinonen *et al.* (1989), Arnott and Marston (1991), Marston (1992), Lock (1996), Mishchenko *et al.* (1997b), Mishchenko and Macke (1998), and Kravtsov and Orlov (1998).

C. Perturbation Theory

The idea of perturbation theory (PT) is to define the surface of an irregular particle in spherical coordinates by $r(\vartheta, \varphi) = r_0[1 + \xi f(\vartheta, \varphi)]$, where r_0 is the radius of the "unperturbed" sphere, ξ is a "smallness parameter," and $f(\vartheta, \varphi)$ obeys the condition $|\xi f(\vartheta, \varphi)| < 1$. The fields inside and outside the particle are expanded in VSWFs, and the expansion coefficients, which are determined through the boundary condition, are expressed as power series in ξ (Oguchi, 1960; Yeh, 1964; Erma, 1969). A similar approach was developed by Ogura and Takahashi (1990). Note that the application of the boundary condition explicitly relies on the (unproven) validity of the Rayleigh hypothesis. Schiffer (1990) combined PT with an analytical orientation-averaging procedure to compute the scattering properties of randomly oriented particles. He also gave many numerical results obtained with the second-order PT and compared them with exact T-matrix computations by Mugnai and Wiscombe (1980) and Wiscombe and Mugnai (1988) for Chebyshev particles. The second-order PT showed good accuracy only for $2\pi r_0/\lambda \lesssim 7$ and only if the surface deviations from the unperturbed sphere were much smaller than the wavelength. Similar conclusions were reached by Kiehl *et al.* (1980) on the basis of first-order PT computations.

Lacoste *et al.* (1998) considered light scattering by a Faraday-active dielectric sphere imbedded in an isotropic medium with no magnetooptical properties and subject to a homogeneous external magnetic field. They computed the amplitude matrix by using a perturbation approach and keeping only terms proportional to the first order of the magnetic field.

D. OTHER APPROXIMATIONS

If the thickness of a particle in one of its dimensions is much smaller than a wavelength, it is often possible to approximate the integral equations describing scattering. This approach was applied to thin finite cylinders by Uzunoglu *et al.* (1978), to thin cylinders and disks by Schiffer and Thielheim (1979) and Fung (1994, Section 11.2), to thin disks by Weil and Chu (1980), and to thin-walled cylinders by Senior and Weil (1977).

Equation (1) can be used to compute the scattered field if the internal field is known. Le Vine *et al.* (1985) calculated scattering from a homogeneous dielectric disk with a radius much larger than the thickness by approximating the internal field by the field that would exist inside an infinite homogeneous slab of the same thickness, orientation, and dielectric properties. Similarly, Karam and Fung (1988) computed scattering from long cylinders by approximating the internal field using the exact solution for an infinitely long cylinder with the same radius, orientation, and refractive index.

A similar approach was applied by Kuzmin and Babenko (1981) to the problem of scattering by spherical particles composed of a weakly anisotropic material. They computed the scattered field via Eq. (1) by approximating the internal field by that of an "equivalent" isotropic sphere.

Pollack and Cuzzi (1980) developed a semiempirical theory based on the microwave analog measurements of Zerull (1976). They used equal-volume Mie computations for particles with $x \lesssim x_0$, where x_0 is a tunable parameter typically close to 5. The absorption cross section for larger particles is still gotten from Mie theory, while the phase function is represented as a sum of the Fraunhofer diffraction, reflected rays from a sphere, and transmitted rays fitted to mimic Zerull's measurements by use of another tunable parameter. Coletti (1984) developed another semiempirical theory based on his own optical measurements and similar in some respects to that of Pollack and Cuzzi.

Latimer (1975) proposed several hybrid approximations for spheroids based on using the Lorenz–Mie theory and assigning an effective sphere radius and refractive index depending on the spheroid orientation and axial ratio. Latimer and Barber (1978) examined the accuracy of this approach by comparing approximate and exact T-matrix results. A discussion of approximate scattering theories can be found in Jones (1999) and Kokhanovsky (1999).

IV. MEASUREMENTS

We will survey here only measurements specifically designed to study the effects of nonsphericity on electromagnetic scattering. There are commercially available instruments that use light scattering, arguably, to measure the size of particles whatever their shape, but they are of no use for our purposes because all most of them do is to make trivial empirical adjustments to account for non-sphericity (if they account for it at all).

Existing measurements of electromagnetic scattering by nonspherical particles are constrained by the state of source and detector technology and the lack of windows in the spectrum of Earth's atmosphere, and traditionally fall into two categories:

- Scattering of visible and infrared light by particles with sizes ranging from several hundredths of a micrometer to several hundred micrometers
- Microwave scattering by millimeter- and centimeter-sized objects

Visible and infrared measurements involve relatively simpler, cheaper, and more portable instrumentation and can be performed in the field as well as in the laboratory. Microwave scattering experiments, by contrast, require expensive instrumentation and large measurement facilities.

A. VISIBLE AND INFRARED MEASUREMENTS

In an advanced experimental setup for measuring the full scattering matrix using visible or infrared light, the beam produced by a light source (usually a laser) passes a linear polarizer and a polarization modulator and then illuminates particles contained in a jet stream or a scattering chamber. Light scattered by the particles at an angle Θ passes a quarter-wave plate (optionally) and a polarization analyzer before its intensity is measured by a detector. The Stokes vector of the scattered light reaching the detector, $\mathbf{I}'$, is given by

$$\mathbf{I}'(\Theta) \propto \mathbf{AQF}(\Theta)\mathbf{MPI}, \tag{2}$$

where $\mathbf{I}$ is the Stokes vector of the beam leaving the light source; $\mathbf{A}$, $\mathbf{Q}$, $\mathbf{M}$, and $\mathbf{P}$ are 4×4 Mueller matrices of the analyzer, quarter-wave plate, modulator, and polarizer, respectively; and $\mathbf{F}(\Theta)$ is the scattering matrix of the particles. It is assumed that the scattering plane acts as the plane of reference for defining the Stokes parameters. The Mueller matrices of the polarizer, modulator, quarter-wave plate, and analyzer depend on their orientation with respect to the scattering plane and can be precisely varied. Several measurements with different orientations of the optical components with respect to the scattering plane are required for the full determination of the scattering matrix. In the case of macroscopically isotropic and symmetric scattering media, the six independent scattering matrix elements [Eq. (61) of Chapter 1] can be determined using four different orientation combinations (Kuik *et al.*, 1991). This procedure is repeated at different scattering angles in order to determine the angular profile of the scattering matrix.

Visible and infrared measurements often suffer from the inability to accurately characterize the size and shape of scattering particles. Another serious problem is that the arrangement of the light source and the detector usually precludes measurements at scattering angles close to 0° and 180°. For example, the range of scattering angles in the measurements by Kuik (1992) is (5°, 175°). This makes difficult an accurate estimate of the scattering cross section (and hence of the phase function) from integrating the scattered intensities over angle. As a result, experimental phase functions are often normalized to the value at a fixed scattering angle or to equivalent-sphere phase functions at 5° or 10°, which may introduce biases into the conclusions drawn. Measurements of the scattering cross section using integrating nephelometers (Heintzenberg and Charlson, 1996) are also known to produce significant errors reaching 20 to 50% for particles larger than a wavelength (Anderson *et al.*, 1996; Rosen *et al.*, 1997a). Extinction cross-sectional measurements have traditionally suffered from the problem that a detector with a finite aperture picks up some of the light scattered by particles in the forward direction. Depending on the average particle size and thus the magnitude and angular width of the diffraction component of the phase function, the extinction can be underestimated by as much as a factor of 2. Correction for the diffraction is possible if the average particle projection is known and is large, but this is not always the case. And with large errors in the extinction and scattering cross sections, little can be said about the absorption cross section and hence the single scattering albedo.

Ashkin and Dziedzic (1980) have obtained direct backscatter measurements by using the optical levitation technique, which involves suspension of particles by light pressure from the source laser beam alone. An instrument specifically designed for measurements at the exact backscattering direction is a lidar (Chapter 14). Because lidars usually measure backscattering from aerosol and cloud particles located at large distances (hundreds and thousands of meters) from the instrument, the scattering angle can be made arbitrarily close to 180°. Important quantities measured by a polarization lidar are so-called linear and circular depolarization ratios. Because both ratios are always equal to zero for spherically symmetric scatterers, nonzero ratios directly indicate the presence of nonspherical particles (Sassen, 1991; Mishchenko and Hovenier, 1995).

Early scattering experiments used unpolarized incident light and were limited to measurements of the scattered intensity and the degree of linear polarization (Hodkinson, 1963; Napper and Ottewill, 1963). The first measurements of other elements of the scattering matrix were performed by Pritchard and Elliott (1960) using a simple subtraction method. The measurement accuracy was significantly improved by implementing polarization modulation techniques (Hunt and Huffman, 1973; Thompson, 1978; Thompson *et al.*, 1980; Anderson, 1992). A sophisticated, fully computerized setup is described by Stammes (1989), Kuik *et al.* (1991), and Kuik (1992) (see also Chapter 12).

Despite the availability of advanced experimental techniques, the number of measurements of the complete scattering matrix remains relatively small. Beardsley (1968), Kadyshevich *et al.* (1976), Thompson *et al.* (1978), Voss and Fry (1984), Fry and Voss (1985), Lofftus *et al.* (1992) and Witkowski *et al.* (1998) measured the scattering matrix for marine particulates. Holland and Gagne (1970) reported measurements for randomly oriented 0.1- to 1-μm quartz crystals with flat platelike shapes and irregular edges. Dugin and Mirumyants (1976) published results for artificially grown platelike and star-shaped ice crystals with sizes ranging from 20 to 150 μm. Bickel and Stafford (1980) measured the scattering matrix for a variety of biological particles. Perry *et al.* (1978) measured all scattering matrix elements for irregular, nearly cubical salt crystals and rounded ammonium sulfate particles of 0.1 to 2 μm. A much greater degree of particle shape and size control than in previous experiments was achieved by Bottiger *et al.* (1980), who measured the scattering matrix for individual, electrostatically levitated aggregates of two, three, and four polystyrene latex spheres. Kuik *et al.* (1991) and Kuik (1992) reported scattering matrix measurements for four samples of polydisperse, irregularly shaped SiO_2 particles with effective radii from 8.6 to 25.8 μm.

A variety of less complete laboratory and *in situ* measurements of visible and infrared light scattering were aimed at specific practical applications. For example, Huffman and Thursby (1969), Liou and Lahore (1974), Dugin *et al.* (1977), Nikiforova *et al.* (1977), Sassen and Liou (1979a, b), Volkovitskiy *et al.* (1980), Pluchino (1987), Arnott *et al.* (1995), Gayet *et al.* (1997, 1998), Crépel *et al.* (1997), Rimmer and Saunders (1997, 1998), and Lawson *et al.* (1998) measured the phase function and extinction, polarization, and depolarization characteristics of artificially grown and natural cirrus cloud crystals. Kadyshevich and Lyubovtseva (1973), Quinby-Hunt *et al.* (1989), and Volten *et al.* (1998) measured several scattering matrix elements for ocean hydrosols and marine *Chlorella*. Bickel *et al.* (1976) reported measurements of several scattering matrix elements for two strains of spores. Pinnick *et al.* (1976), Jaggard *et al.* (1981), Nakajima *et al.* (1989), and West *et al.* (1997) measured phase functions and the degree of linear polarization for natural and artificial soil and mineral aerosol samples. Ben-David (1998) measured the backscattering Mueller matrix for atmospheric aerosols at CO_2 laser wavelengths ranging from 9.2 to 10.7 μm. Holler *et al.* (1998) detected the interference structure (see Section V.A) in two-dimensional patterns of scattered intensity for single airborne microparticles. Weiss-Wrana (1983) reported phase function and linear polarization measurements for meteoritic grains. Pope *et al.* (1992) measured the phase function and the degree of linear polarization for artificially grown ammonia ice crystals as prototypes of particulates making up the visible clouds of Jupiter and Saturn. Multiple applications of the polarization lidar technique are reviewed in Chapter 14.

Additional sources of experimental information include Donn and Powell (1963), Chýlek *et al.* (1976), Kirmaci and Ward (1979), Saunders (1980), Coletti (1984), Arnott and Marston (1991), Spinrad and Brown (1993), Quirantes and Delgado (1995, 1998), McGuire and Hapke (1995), Videen *et al.* (1997), Rosen *et al.* (1997b), Barkey *et al.* (1999) and Shvalov *et al.* (1999).

B. MICROWAVE MEASUREMENTS

Information about the scattering properties of millimeter- and centimeter-sized objects at microwave frequencies is important for such applications as remote sensing of precipitation (Chapters 16 and 17) and communication technology. In addition, it has long been realized that particle size in the theoretical formulation of the electromagnetic scattering problem is only encountered as a ratio to the wavelength. Therefore, the main idea of microwave analog measurements is to manufacture a centimeter-sized scattering object with the desired shape and refractive index, study the scattering of a microwave beam by this object, and finally extrapolate the results to other wavelengths (e.g., visible or infrared) by keeping the ratio size/wavelength fixed (Greenberg *et al.*, 1961; Lind *et al.*, 1965).

In a typical microwave scattering setup, radiation from a transmitting antenna passes a polarizer, is scattered by the target in question, passes another polarizer, and is measured by a receiving antenna. Microwave measurements allow wide coverage of scattering angles including the exact forward direction and a much greater degree of control over the target size, shape, and orientation than optical/infrared measurements. By special techniques, even the extinction cross section (or, more generally, extinction matrix) could be measured. Because the target size is on the order of centimeters, high-precision target manufacturing is easy and can involve computer-controlled milling or stereolithography. As a result, controlled laboratory measurements at microwave frequencies can often be used to check scattering theories and computer codes. A disadvantage of microwave measurements is that they can be performed only for one particle size, shape, and orientation at a time, thereby making ensemble averaging a time-consuming and tedious procedure.

Although the microwave analog technique was introduced more than 35 years ago, the complexity and the high cost of the equipment involved have limited the number of actual hardware implementations of the technique to only a few. Apparently the most sophisticated, fully automated microwave scattering setup was described by Gustafson (1996) (see also Chapter 13). The number of published microwave analog measurements is also relatively small. Waterman (1971) measured scattered microwave intensities for metal spheroids, regular cylinders, cylinders with spherical and hemispherical end caps, and smoothly joined cone spheres, primarily in order to test his T-matrix computations. Zerull and Giese (1974), Zerull (1976), and Zerull *et al.* (1977) published microwave analog measurements of the phase function and the degree of linear polarization for roughened spheres, cubes, octahedrons, and fluffy and irregular convex and concave particles with effective size parameters up to 20. Allan and McCormick (1978, 1980) measured the backscattering matrix for dielectric spheroids with different refractive indices and varying orientations relative to the direction of illumination. Greenberg and Gustafson (1981) and Gustafson (1985) used the microwave analog technique to simulate visible light scattering by zodiacal light and cometary particles. Schuerman *et al.* (1981) reported measurements of scattering matrix elements for 28 targets in the form of circular cylinders and disks and prolate and oblate spheroids with surface-equivalent-sphere size parameters up to 6 and a common refractive index of $1.61 - 0.004i$. Wang *et al.* (1981) and Fuller *et al.* (1986) published measurements of microwave scattering by systems of two interacting spheres and compared them with theoretical computations. Chýlek *et al.* (1988) used microwave measurements to check the accuracy of effective medium approximations in scattering and absorption computations for nonabsorbing particles containing highly absorbing inclusions. Hage *et al.* (1991) measured scattering matrix elements for an irregular porous cube in order to verify their volume integral equation formulation computations. Zerull *et al.* (1993) reported measurements for loose aggregates of 250 to 500 spheres with and without an absorbing mantle. Fuller *et al.* (1994a) used the microwave analog technique to model scattering by ice prisms. Finally, Wang and van de Hulst (1995) used microwave measurements for finite circular cylinders to check the accuracy of approximate computations based on the solution for infinitely long cylinders.

Radars are a special class of instruments providing active polarization measurements for remote targets at microwave and radiowave frequencies. Monostatic radars use the same antenna to transmit and receive electromagnetic radiation and are limited to measurements at the exact backscattering direction. Bistatic radars use one or more additional receiving antennas, which provide supplementary information for other scattering angles. Applications of the radar technique are reviewed in Chapter 16.

V. MANIFESTATIONS OF NONSPHERICITY IN ELECTROMAGNETIC SCATTERING

A. INTERFERENCE STRUCTURE, ENSEMBLE AVERAGING, AND STATISTICAL APPROACH

It is well known that scattering patterns for particles of a single size are usually burdened with what is called the interference structure (van de Hulst, 1957). This effect is demonstrated in Plate 2.1a (see color Plates 2.1–2.4), which depicts the degree of linear polarization for unpolarized incident light (i.e., the ratio

$-F_{21}/F_{11}$ of the elements of the scattering matrix; see Section XI of Chapter 1) versus the size parameter and scattering angle for monodisperse spheres with a refractive index of $1.53 + 0.008i$. This refractive index is typical of dustlike tropospheric aerosols in Earth's atmosphere at visible wavelengths (d'Almeida *et al.*, 1991). Plate 2.1a shows the characteristic "butterfly" pattern (Hansen and Travis, 1974) composed of red and blue spots that represent alternating sharp maxima and minima. These strong oscillations of polarization within a small range of scattering angle and/or size parameter result from the interference of light diffracted and reflected/transmitted by a particle. Plate 2.2a shows the interference structure for the same monodisperse spheres using a surface rather than a contour diagram and makes more visible the sharpness of the local spikes and holes.

Although the interference structure is an intrinsic property of electromagnetic scattering, it is rarely observed in practice because even a narrow distribution of particle sizes is sufficient to smooth out most of the interference effects (Hansen and Travis, 1974). Plates 2.1b and 2.2b depict polarization computations for a narrow power law size distribution of spheres (Mishchenko *et al.*, 1996b) and show that size averaging tends to level off the spikes and fill in the holes, thereby making the polarization pattern smoother and less featured. This effect is easy to understand qualitatively in terms of taking weighted averages along the vertical lines in Plate 2.1a.

The interference structure for monodisperse nonspherical particles is even more complicated than that for spheres because now scattering patterns depend on the orientation of a particle with respect to the incident and scattered beams. This is demonstrated in Plates 2.1c and 2.1d (see also Plates 2.2c and 2.2d), which show T-matrix computations of the degree of linear polarization versus the scattering angle and equal-surface-area-sphere size parameter for monodisperse oblate spheroids with aspect ratio $\epsilon = 1.7$ and two orientations of the spheroid axis with respect to the scattering plane. Indeed, it is seen that the polarization patterns for the two spheroid orientations are totally different. In particular, the lack of axial symmetry for the light-scattering geometry in Plate 2.1d results in nonzero polarization at $0°$ and $180°$ scattering angles.

Plates 2.1e and 2.2e show that the polarization pattern computed for monodisperse spheroids in random orientation is much smoother and less complicated than that for spheroids in a fixed orientation. This smoothing effect of averaging over orientations is obviously reinforced by averaging over sizes (Plates 2.1f and 2.2f), which totally removes the residual interference structure still seen in Plates 2.1e and 2.2e.

The most obvious reason for performing polydisperse rather than monodisperse computations and measurements of light scattering is better modeling of natural particle ensembles in which particles are most often distributed over a range of sizes and orientations. The second reason is the presence of the complicated and highly variable interference structure for monodisperse particles in a fixed orientation, which makes it essentially impossible to use computations and/or measurements for monodisperse particles in order to derive useful conclusions about the effect of nonsphericity on light scattering. Contrasting the individual panels in Plates 2.1 and 2.2 shows that averaging over sizes for spheres and averaging over orientations and sizes for nonspherical particles largely removes the interference structure and enables meaningful comparisons of the scattering properties of different particles. For example, an interesting feature of the polarization pattern for polydisperse, randomly oriented spheroids (Plates 2.1f and 2.2f) is the bridge of positive polarization at side-scattering angles that extends from the region of Rayleigh scattering (size parameters less than about 1) toward larger size parameters and separates two regions of negative polarization at small and large scattering angles. This bridge is absent in Plates 2.1b and 2.2b for surface-equivalent spheres and may be a typical scattering characteristic of some nonspherical particles (cf. Perry *et al.*, 1978).

In addition to size and orientation averaging, averaging over shapes may prove to be necessary in many cases. More often than not, natural and artificial particle samples exhibit a great variety of shapes, thereby making questionable the ability of a single nonspherical shape to represent scattering properties of a shape mixture. Indeed, it can be demonstrated that even after size and orientation averaging, essentially any particle shape produces a unique, shape-specific scattering pattern, whereas experimental measurements for real nonspherical particles usually show smooth, featureless patterns (e.g., Perry *et al.*, 1978). As an example, Plate 2.3a depicts the phase function [i.e., the (1, 1) element of the scattering matrix given by Eq. (61) of Chapter 1] for a monodisperse sphere with a radius of 1.163 μm and surface-equivalent, monodisperse, randomly oriented prolate spheroids with aspect ratios increasing from 1.2 to 2.4. The wavelength is 0.443 μm, and the refractive index is $1.53 + 0.008i$. Whereas the monodisperse curves form a tangle of lines with no clear message, averaging over sizes (Plate 2.3b) makes the phase functions much smoother and reveals a highly systematic change with increasing aspect ratio, rendering each phase function curve unique and dissimilar to all other curves. However, this uniqueness is suppressed and ultimately removed by averaging over an increasingly wide aspect ratio distribution of prolate spheroids (Plate 2.3c) and by subsequent mixing of prolate and oblate spheroids (Plate 2.3d). The resulting phase function (red curve in Plate 2.3d) is very smooth and featureless and, in fact, almost perfectly coincides with the phase function experimentally measured by Jaggard *et al.* (1981) for micrometer-sized, irregularly shaped soil particles (Mishchenko *et al.*, 1997a).

This example may have two important implications. First, it may indicate that the often observed smooth scattering-angle dependence of the elements of the scattering matrix for samples of natural and artificial nonspherical particles is caused by the diversity of particle shapes in the samples. Second, it may suggest that at least some scattering properties of many kinds of irregular particles can be rather adequately modeled using a polydisperse shape mixture of simple particles such as spheroids. The assumption that particles chosen for the purposes of ensemble averaging need not be in one-to-one correspondence with the ensemble of irregular particles of interest and may have relatively simple shapes is central to the so-called statistical approach (Shifrin and Mikulinsky, 1987; Mugnai and Wiscombe, 1989; Bohren and Singham, 1991; Goncharenko *et al.*, 1999). The need for this kind of approach stems from the fact that it is often impossible to exactly specify the shapes and sizes of all particles forming a natural or artificial sample. Even if it were possible, the low efficiency of numerical scattering techniques applicable to arbitrarily shaped particles would entail an enormous computational effort. On the other hand, the availability of techniques such as the T-matrix method, which is very fast for randomly oriented, rotationally symmetric particles and is applicable to large size parameters, makes the statistical approach quite practical. Successful applications of the statistical approach by Bohren and Huffman (1983, Chapter 12), Nevitt and Bohren (1984), Hill *et al.* (1984), and Mishchenko *et al.* (1997a) suggest that it may indeed be a valuable practical tool in many cases.

Another promising approach is to assume that the scattering properties of an ensemble of irregular particles can be well reproduced by mixing only a few "statistically representative" artificial particles created by a numerical random shape generator. This approach was pursued, among others, by Peltoniemi *et al.* (1989), Muinonen *et al.* (1996), and Macke *et al.* (1996b), who used GOA for particles much larger than a wavelength, and by Lumme and Rahola (1998), who used VIEM for particles with equivalent-sphere size parameters smaller than 6 (see also Chapters 11 and 19). Although this approach tends to be rather time consuming and has a limited size parameter range, it will undoubtedly find more applications as computers become more powerful and methods such as FEM, FDTDM, and VIEM become more efficient. By using more sophisticated shape generator programs and forming richer and more representative particle mixtures, one may hope to provide an increasingly accurate theoretical description of the scattering properties of real ensembles of irregular particles.

B. Effects of Nonsphericity on Scattering Patterns

The most fundamental effects of particle nonsphericity on light scattering are that the 4×4 extinction matrix does not, in general, degenerate to a direction- and polarization-independent scalar; the extinction, scattering, and absorption cross sections depend on the direction and polarization state of the incident beam; and the scattering matrix **F** (Section XI of Chapter 1) does not have the well-known

Lorenz–Mie structure given by

$$
\mathbf{F} = \begin{bmatrix}
F_{11} & F_{12} & 0 & 0 \\
F_{12} & F_{11} & 0 & 0 \\
0 & 0 & F_{33} & F_{34} \\
0 & 0 & -F_{34} & F_{33}
\end{bmatrix}.
\tag{3}
$$

In general, all 16 elements of the scattering matrix for nonspherical particles can be nonzero and depend on the orientation of the scattering plane. All these effects can directly indicate the presence of nonspherical particles. For example, measurements of interstellar polarization are used in astrophysics to detect preferentially oriented dust grains that cause different values of extinction for different polarization components of the transmitted starlight (Martin, 1978). Similarly, depolarization of radiowave signals propagating through Earth's atmosphere may indicate the presence of partially aligned nonspherical hydrometeors (Oguchi, 1983).

When nonspherical particles are randomly oriented and form a microscopically isotropic and symmetric scattering medium (Section XI of Chapter 1), the extinction matrix does reduce to the scalar extinction cross section, and all optical cross sections become orientation and polarization independent. Furthermore, the scattering matrix becomes block diagonal [Eq. (61) of Chapter 1] and has almost the same structure as the Lorenz–Mie scattering matrix. However, the remaining fundamental difference is that the Lorenz–Mie identities $a_2(\Theta) \equiv a_1(\Theta)$ and $a_4(\Theta) \equiv a_3(\Theta)$ do not, in general, hold for nonspherical particles (cf. Plates 2.4a–2.4c). This factor makes measurements of the linear depolarization ratio $\delta_L = (a_1 - a_2)/(a_1 + a_2)$ and the closely related circular depolarization ratio the most reliable indicators of particle nonsphericity (Sassen, 1991; Mishchenko and Hovenier, 1995; Gobbi, 1998). It is widely accepted that strong depolarization by nonabsorbing or weakly absorbing nonspherical particles much larger than a wavelength can be explained in ray optics terms as the result of rotations of the vibration plane of the electric vector as a ray is multiply internally reflected inside the particle (Liou and Lahore, 1974). However, another explanation, perhaps referring to some kind of resonance effect, is needed for wavelength-sized particles that can produce even stronger depolarization (Mishchenko and Sassen, 1998).

In addition to this qualitative difference, which unequivocally distinguishes randomly oriented nonspherical particles from spheres, there can be significant quantitative differences in specific scattering patterns. The most obvious examples are halos, arcs, pillars, and other pronounced phase function features exhibited by regularly shaped nonspherical crystals such as hexagonal columns and plates (Greenler, 1990). In general, nonspherical particles tend to exhibit enhanced scattering at intermediate scattering angles and reduced backscattering, whereas phase function differences at forward-scattering angles are often negligibly small (Plate 2.3d; Zerull, 1976; Perry *et al.*, 1978). The degree of linear polarization $-b_1(\Theta)/a_1(\Theta)$ for nonspherical particles tends to be positive at side-scattering angles, and the element $a_4(\Theta)$ tends to be larger than $a_3(\Theta)$ (Perry *et al.*, 1978; Kuik, 1992; Mishchenko and Travis, 1994b; West *et al.*, 1997; Plates 2.1f, 2.4b, and 2.4c). The overall scattering-angle behavior of the ratio $b_2(\Theta)/a_1(\Theta)$ is similar for nonspherical and surface- or volume-equivalent spherical particles (Perry *et al.*, 1978; Mishchenko *et al.*, 1996b; Plates 2.4d and 2.4f). All elements of the scattering matrix tend to be (much) less shape dependent at scattering angles less than 60° than at side- and backscattering angles (Perry *et al.*, 1978; Mishchenko *et al.*, 1996b; Plate 2.4). Spherical/nonspherical differences in the elements of the scattering matrix are maximal for nonabsorbing particles and diminish with increasing absorption (Mishchenko and Travis, 1994b; Mishchenko *et al.*, 1997a). All these effects of nonsphericity may have very serious implications for remote-sensing studies of small particles (e.g., Coffeen, 1979; Liou and Takano, 1994; Francis, 1995; Mishchenko *et al.*, 1996c; Chepfer *et al.*, 1998). Although differences in the optical cross sections, single scattering albedo, and asymmetry parameter of the phase function can also be noticeable, they are usually much smaller than the differences in the elements of the scattering matrix (Mugnai and Wiscombe, 1986; Mishchenko *et al.*, 1996b). Further information can be found in systematic theoretical surveys of light scattering by ensembles of Chebyshev particles (Mugnai and Wiscombe, 1980, 1986, 1989; Wiscombe and Mugnai, 1988), spheroids (Mishchenko and Travis, 1994b; Mishchenko *et al.*, 1996b), finite circular cylinders (Mishchenko *et al.*, 1996a), Gaussian random particles (Muinonen, 1996b; Chapter 11), and polycrystals (Takano and Liou, 1995; Macke *et al.*, 1996b).

Clusters of small monomers form a special class of nonspherical particles. Although the scattering properties of randomly oriented two-sphere clusters closely resemble those of a single sphere (Mishchenko *et al.*, 1995), the effect of cooperative phenomena in many-particle clusters can be very strong (e.g., Mackowski and Mishchenko, 1996). Sometimes the scattering properties of clusters are a complex combination of those for a single monomer and those for a solid particle circumscribing the cluster and having the same average projected area (e.g., West, 1991; Lumme *et al.*, 1997). Scattering by clusters is reviewed in detail in Chapters 8, 13, and 19.

VI. ABBREVIATIONS

ADA   Anomalous diffraction approximation
CGM   Conjugate gradient method
CPU   Central processing unit

DDA   Discrete dipole approximation
FDM   Finite difference method
FDTDM   Finite difference time domain method
FEM   Finite element method
FFT   Fast Fourier transform
FIEM   Fredholm integral equation method
GE   Gaussian elimination
GOA   Geometric optics approximation
GPMM   Generalized point matching method
ME-GPMM   Multiple-expansion generalized point matching method
MOM   Method of moments
PMM   Point matching method
PT   Perturbation theory
RA   Rayleigh approximation
RGA   Rayleigh–Gans approximation
SIEM   Surface integral equation method
SM   Superposition method
SVM   Separation of variables method
TMM T-matrix method
VIEM   Volume integral equation method
VSWF   Vector spherical wave function

ACKNOWLEDGMENTS

We thank Nadia Zakharova for help with the graphics and Zoe Wai for bibliographical assistance.

Chapter 3

Basic Relationships for Matrices Describing Scattering by Small Particles

Joop W. Hovenier

Department of Physics and Astronomy
Free University
1081 HV Amsterdam, The Netherlands

Cornelis V. M. van der Mee

Dipartimento di Matematica
Università di Cagliari
09124 Cagliari, Italy

I. INTRODUCTION

As discussed in Chapters 1 and 2, polarization of light (electromagnetic radiation) plays an important role in studies of light scattering by small particles. A convenient way to treat the polarization of a beam of light is to use the four

Stokes parameters and to make these the elements of a column vector, called the Stokes vector. Scattering by a particle in a fixed orientation can then be described by means of a real 4 × 4 matrix that transforms the Stokes vector of the incident beam into that of the scattered beam. Such a matrix is a pure phase (or scattering) matrix (Hovenier, 1994), because its elements follow directly from the corresponding 2×2 amplitude matrix that transforms the two electric field components. A large number of scalar and matrix properties of pure phase matrices has been reported and the same is true for sums of such matrices, which are needed to describe independent single scattering by collections of particles.

An important goal of this chapter is to present in a systematic way the main properties of matrices describing single scattering by small particles in atmospheres and water bodies (Sections II and III). The emphasis is on the basic relationships from which others can be derived and on simple relationships. In principle, all relationships can be used for theoretical purposes or to test whether an experimentally or numerically determined matrix can be a pure phase matrix or a sum of pure phase matrices. Some strong and convenient tests are presented in Section IV. Our analysis provides the most general and objective criteria for testing phase and scattering matrices. Section V is devoted to a discussion and outlook.

II. RELATIONSHIPS FOR SCATTERING BY ONE PARTICLE IN A FIXED ORIENTATION

A. Relationships between Amplitude Matrix and Pure Phase Matrix

Consider the laboratory reference frame used in Chapter 1 with its origin inside an arbitrary particle in a fixed orientation. Scattering of electromagnetic radiation by this particle is fully characterized by the 2 × 2 amplitude matrix $\mathbf{S}(\mathbf{n}^{\text{sca}}; \mathbf{n}^{\text{inc}}; \alpha, \beta, \gamma)$, which linearly transforms the electric field vector components of the incident wave into the electric field vector components of the scattered wave (see Section IV of Chapter 1). The four elements of the amplitude matrix are, in general, four different complex functions. The element in the ith row and the jth column will be denoted as S_{ij}.

Using Stokes parameters, as defined in Section V of Chapter 1, the scattering by one particle in a fixed orientation can also be described by means of a 4 × 4 phase matrix $\mathbf{Z}(\vartheta^{\text{sca}}, \varphi^{\text{sca}}, \vartheta^{\text{inc}}, \varphi^{\text{inc}}; \alpha, \beta, \gamma)$, which in the most general case has 16 different real nonvanishing elements [see Eq. (13) of Chapter 1]. Each element of such a phase matrix can be completely expressed in the elements of the amplitude matrix pertaining to the same scattering problem. We will call such a phase matrix a pure phase matrix, because this is merely a special case of the general concept of a pure Mueller matrix (Hovenier, 1994). Explicit expressions for the elements of a pure phase matrix were first given by van de Hulst (1957). In our terminology and notation they were presented in Chapter 1 as Eqs. (14)–(29). However, the relationship between **S** and **Z** can also be expressed by the matrix relation (see, e.g., O'Neill, 1963)

$$\mathbf{Z} = \mathbf{\Gamma}_s(\mathbf{S} \otimes \mathbf{S}^*)\mathbf{\Gamma}_s^{-1}, \tag{1}$$

where

$$\mathbf{\Gamma}_s = \frac{1}{\sqrt{2}} \begin{pmatrix} 1 & 0 & 0 & 1 \\ 1 & 0 & 0 & -1 \\ 0 & -1 & -1 & 0 \\ 0 & -i & i & 0 \end{pmatrix} \tag{2}$$

is a unitary matrix with inverse

$$\mathbf{\Gamma}_s^{-1} = \frac{1}{\sqrt{2}} \begin{pmatrix} 1 & 1 & 0 & 0 \\ 0 & 0 & -1 & i \\ 0 & 0 & -1 & -i \\ 1 & -1 & 0 & 0 \end{pmatrix}, \tag{3}$$

while the Kronecker product is defined by

$$\mathbf{S} \otimes \mathbf{S}^* = \begin{pmatrix} S_{11}\mathbf{S}^* & S_{12}\mathbf{S}^* \\ S_{21}\mathbf{S}^* & S_{22}\mathbf{S}^* \end{pmatrix} \tag{4}$$

and an asterisk denotes the complex conjugate. Both recipes for obtaining a pure phase matrix from the corresponding amplitude matrix have their specific advantages and disadvantages. Equation (1) is particularly useful for formula manipulations if one is familiar with the properties of Kronecker products (see, e.g., Horn and Johnson, 1991).

Employing one of the preceding recipes, one can readily verify the following relations between a pure phase matrix **Z** and its corresponding amplitude matrix **S**.

a. If

$$d = |\det \mathbf{S}|, \tag{5}$$

where det stands for the determinant, we have

$$d^2 = Z_{11}^2 - Z_{21}^2 - Z_{31}^2 - Z_{41}^2. \tag{6}$$

The right-hand side of this equation may be replaced by similar four-term expressions, as will be explained later (see Section II.B).

b.

$$|\text{Tr}\,\mathbf{S}|^2 = \text{Tr}\,\mathbf{Z}, \tag{7}$$

where Tr stands for the trace, that is, the sum of the diagonal elements. Apparently, Tr $\mathbf{Z}$ is always nonnegative.

c.

$$d^4 = \det \mathbf{Z}, \tag{8}$$

which implies that det $\mathbf{Z}$ can never be negative.

d. If $d \neq 0$, the inverse matrix

$$\mathbf{S}^{-1} \sim \mathbf{Z}^{-1}, \tag{9}$$

where the symbol $\sim$ stands for "corresponds to" in the sense of Eqs. (14)–(29) of Chapter 1.

e. The product $\mathbf{S}_1\mathbf{S}_2$ of two amplitude matrices corresponds to the product $\mathbf{Z}_1\mathbf{Z}_2$ of the corresponding pure phase matrices; that is, $\mathbf{Z}_1\mathbf{Z}_2$ is a pure phase matrix and

$$\mathbf{S}_1\mathbf{S}_2 \sim \mathbf{Z}_1\mathbf{Z}_2. \tag{10}$$

Another type of relationship can be obtained by investigating the changes experienced by a pure phase matrix if the corresponding amplitude matrix is subjected to an elementary algebraic operation. Suppose

$$\mathbf{S} = \begin{pmatrix} S_{11} & S_{12} \\ S_{21} & S_{22} \end{pmatrix} \sim \mathbf{Z}. \tag{11}$$

Then

i.

$$\alpha\mathbf{S} \sim |\alpha|^2\mathbf{Z}, \tag{12}$$

where α is an arbitrary real or complex constant;

ii.

$$\tilde{\mathbf{S}} \sim \boldsymbol{\Delta}_4\tilde{\mathbf{Z}}\boldsymbol{\Delta}_4, \tag{13}$$

where a tilde above a matrix means its transpose and $\boldsymbol{\Delta}_4 = \text{diag}(1, 1, 1, -1)$;

iii.

$$\tilde{\mathbf{S}}^* \sim \tilde{\mathbf{Z}}; \tag{14}$$

iv.

$$\begin{pmatrix} S_{11} & -S_{12} \\ -S_{21} & S_{22} \end{pmatrix} \sim \mathbf{\Delta}_{3,4}\mathbf{Z}\mathbf{\Delta}_{3,4}, \tag{15}$$

where $\mathbf{\Delta}_{3,4} = \mathrm{diag}(1, 1, -1, -1)$;

v.

$$\begin{pmatrix} S_{22} & S_{12} \\ S_{21} & S_{11} \end{pmatrix} \sim \mathbf{\Delta}_2\tilde{\mathbf{Z}}\mathbf{\Delta}_2, \tag{16}$$

where $\mathbf{\Delta}_2 = \mathrm{diag}(1, -1, 1, 1)$.

Several of the previous relations are directly clear for physical reasons. For instance, Eq. (15) originates from mirror symmetry (see van de Hulst, 1957; Hovenier, 1969). Other relations may be obtained by successive application of two or more relations. For instance, Eqs. (13) and (15) yield the reciprocity relation

$$\begin{pmatrix} S_{11} & -S_{21} \\ -S_{12} & S_{22} \end{pmatrix} \sim \mathbf{\Delta}_3\tilde{\mathbf{Z}}\mathbf{\Delta}_3, \tag{17}$$

where $\mathbf{\Delta}_3 = \mathrm{diag}(1, 1, -1, 1)$ is the same matrix as $\mathbf{Q}$ in Eq. (47) of Chapter 1. Furthermore, the relation

$$\mathbf{S}^* \sim \mathbf{\Delta}_4\mathbf{Z}\mathbf{\Delta}_4 \tag{18}$$

may be obtained by combining Eqs. (13) and (14).

It should be noted that Eq. (12) is especially useful when dealing with an amplitude matrix with a different normalization than that of $\mathbf{S}$. It also shows that multiplication of $\mathbf{S}$ by a factor $e^{i\varepsilon}$ with $i = \sqrt{-1}$ and arbitrary real ε does not affect $\mathbf{Z}$. Conversely, if $\mathbf{Z}$ is known then $\mathbf{S}$ can be reconstructed up to a factor $e^{i\varepsilon}$, as follows from Eqs. (1) and (4). As another corollary of the preceding expressions, we observe that in view of Eqs. (12), (15), and (16) we have, for $d \neq 0$,

$$\mathbf{S}^{-1} = \frac{1}{\det \mathbf{S}}\begin{pmatrix} S_{22} & -S_{12} \\ -S_{21} & S_{11} \end{pmatrix} \sim \frac{1}{d^2}\mathbf{G}\tilde{\mathbf{Z}}\mathbf{G}, \tag{19}$$

with $\mathbf{G} = \mathbf{G}^{-1} = \mathrm{diag}(1, -1, -1, -1)$. Employing Eq. (9), we thus find a simple expression for the inverse of $\mathbf{Z}$, namely,

$$\mathbf{Z}^{-1} = \frac{1}{d^2}\mathbf{G}\tilde{\mathbf{Z}}\mathbf{G}. \tag{20}$$

Taking determinants on both sides corroborates Eq. (8). When we premultiply both sides of Eq. (20) by $\mathbf{Z}$ we find

$$\mathbf{Z}\mathbf{G}\tilde{\mathbf{Z}} = d^2\mathbf{G}, \tag{21}$$

whereas postmultiplication of both sides of Eq. (20) by $\mathbf{Z}$ gives

$$\tilde{\mathbf{Z}}\mathbf{G}\mathbf{Z} = d^2\mathbf{G}. \tag{22}$$

By taking the trace on both sides of Eqs. (21) and (22) we obtain

$$\mathrm{Tr}(\mathbf{Z}\mathbf{G}\tilde{\mathbf{Z}}) = -2d^2 \tag{23}$$

and

$$\mathrm{Tr}(\tilde{\mathbf{Z}}\mathbf{G}\mathbf{Z}) = -2d^2. \tag{24}$$

Equations (22) and (24) were first reported by Barakat (1981) and Simon (1982).

As noted in Chapter 1, a variety of conventions are used in publications on light scattering. In this connection Eqs. (12)–(18) as well as the following observation are useful. All relations in this chapter remain valid when $\mathbf{S}$ is replaced by

$$\overline{\mathbf{S}} = \begin{pmatrix} \cos\eta_2 & \sin\eta_2 \\ -\sin\eta_2 & \cos\eta_2 \end{pmatrix} \mathbf{S} \begin{pmatrix} \cos\eta_1 & \sin\eta_1 \\ -\sin\eta_1 & \cos\eta_1 \end{pmatrix} \tag{25}$$

for arbitrary angles η_1 and η_2 and simultaneously $\mathbf{Z}$ is replaced by [cf. Eq. (10) of Chapter 1]

$$\overline{\mathbf{Z}} = \mathbf{L}(\eta_2)\mathbf{Z}\mathbf{L}(\eta_1). \tag{26}$$

This follows directly from Eq. (10). Consequently, no essential difference occurs when instead of $\mathbf{S}$ and $\mathbf{Z}$ use is made of a 2×2 amplitude matrix, which describes the transformation of the electric field components defined with respect to the scattering plane, and the corresponding 4×4 scattering matrix (see Section XI of Chapter 1 and van de Hulst, 1957). This should be kept in mind when consulting the literature, in particular when using published relationships for the scattering matrix.

B. INTERNAL STRUCTURE OF A PURE PHASE MATRIX

The phase matrix of a particle in a fixed orientation may contain 16 real, different, nonvanishing elements. On the other hand, the corresponding amplitude matrix is essentially determined by no more than seven real numbers, because only phase differences occur in Eqs. (14)–(29) of Chapter 1. Consequently, interrelations for the elements of a pure phase matrix must exist or, in other words, a pure phase matrix has a certain internal structure. As mentioned in Section I, many investigators have studied such interrelations. Using simple trigonometric relations, Hovenier *et al.* (1986) first derived equations that involve the real and imaginary parts of products of the type $S_{ij} S_{kl}^*$ and then translated these into relations for the elements of the corresponding pure phase matrix. This approach is very simple and yields a plethora of properties.

On seeking the internal structure of a pure phase matrix we are, of course, interested in simple relations that involve its elements. From the work of Hovenier *et al.* (1986) one obtains the following two sets of simple interrelations for the elements of an arbitrary pure phase matrix **Z**.

1. Seven relations for the squares of the elements of **Z**. These equations can be written in the form

$$
\begin{aligned}
Z_{11}^2 - Z_{21}^2 - Z_{31}^2 - Z_{41}^2 &= -Z_{12}^2 + Z_{22}^2 + Z_{32}^2 + Z_{42}^2 \\
&= -Z_{13}^2 + Z_{23}^2 + Z_{33}^2 + Z_{43}^2 \\
&= -Z_{14}^2 + Z_{24}^2 + Z_{34}^2 + Z_{44}^2 \\
&= Z_{11}^2 - Z_{12}^2 - Z_{13}^2 - Z_{14}^2 \\
&= -Z_{21}^2 + Z_{22}^2 + Z_{23}^2 + Z_{24}^2 \\
&= -Z_{31}^2 + Z_{32}^2 + Z_{33}^2 + Z_{34}^2 \\
&= -Z_{41}^2 + Z_{42}^2 + Z_{43}^2 + Z_{44}^2.
\end{aligned}
\tag{27}
$$

In view of Eq. (6) each four-term expression in Eq. (27) equals d^2. A convenient way to describe the relations for the squares of the elements of **Z** is to consider the matrix

$$
\mathbf{Z}^s =
\begin{pmatrix}
Z_{11}^2 & -Z_{12}^2 & -Z_{13}^2 & -Z_{14}^2 \\
-Z_{21}^2 & Z_{22}^2 & Z_{23}^2 & Z_{24}^2 \\
-Z_{31}^2 & Z_{32}^2 & Z_{33}^2 & Z_{34}^2 \\
-Z_{41}^2 & Z_{42}^2 & Z_{43}^2 & Z_{44}^2
\end{pmatrix}
\tag{28}
$$

and require that all sums of the four elements of a row or column of $\mathbf{Z}^s$ are the same.

2. Thirty relations that involve products of different elements of **Z**. A convenient overview of these equations may be obtained by means of a graphical code. Let a 4×4 array of dots in a pictogram represent the elements of a pure phase matrix, a solid curve or line connecting two elements represent a positive product, and a dotted curve or line represent a negative product. Let us further adopt the convention that all positive and negative products must be added to get zero. The result is shown in parts a and b of Fig. 1. For example, the pictogram in the upper left corner of Fig. 1a means

$$
Z_{11} Z_{12} - Z_{21} Z_{22} - Z_{31} Z_{32} - Z_{41} Z_{42} = 0,
\tag{29}
$$

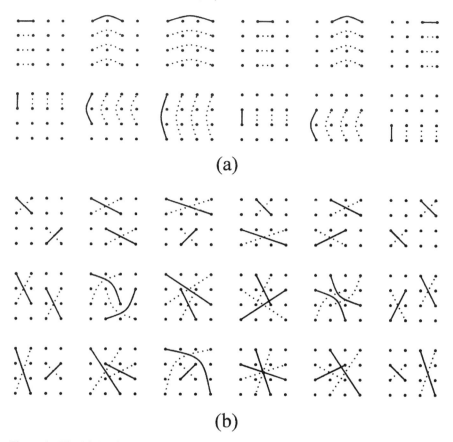

(a)

(b)

Figure 1 The 16 dots in each pictogram represent the elements of a pure phase matrix. A solid line or curve connecting two elements stands for a positive product and a dotted curve or line for a negative product. In each pictogram the sum of all positive and negative products vanishes. (a) Twelve pictograms that represent equations that carry corresponding products of any two chosen rows and columns. (b) Eighteen pictograms that demonstrate that the sum or difference of any chosen pair of complementary subdeterminants vanishes.

and the pictogram in the upper left corner of Fig. 1b stands for

$$Z_{11}Z_{22} - Z_{12}Z_{21} - Z_{33}Z_{44} + Z_{34}Z_{43} = 0. \tag{30}$$

Together all 120 possible products of two distinct elements appear in the 30 relations, and each such product occurs only once. The 30 relations subdivide into the following two types. The 12 equations shown in Fig. 1a carry corresponding products of any two chosen rows and columns. The 18 equations shown in Fig. 1b demonstrate that the sum or difference of any chosen pair of complementary sub-determinants vanishes. Here, the term "complementary" refers to the remaining rows and columns. Sums and differences of subdeterminants alternate in each column and row of the logical arrangement of pictograms shown in Fig. 1b. Keeping the signs in mind for the first pictograms in parts a and b of Fig. 1, one should have little trouble reproducing all pictograms, and thus all 30 equations, from memory.

We have thus shown that every pure phase matrix has a simple and elegant internal structure that is embodied by interrelations that involve either only squares of the elements or only products of different elements. These interrelations may be clearly visualized by means of Eq. (28) and Fig. 1. It is readily verified that all interrelations remain true if the rows and columns of $\mathbf{Z}$ are interchanged. This reflects the fact that if $\mathbf{Z}$ is a pure phase matrix, then $\tilde{\mathbf{Z}}$ can also be a pure phase matrix [cf. Eq. (14)]. Similarly, if we first switch the signs of the elements in the second row and then those in the second column (so that Z_{22} is unaltered), all interrelations remain true [cf. Eqs. (14) and (16)], and this also holds if we apply such operations on the third or fourth row and column [cf. Eqs. (13), (14), and (17)] or even if we combine a number of those sign-switching operations. Consequently, by considering not merely one pure phase matrix but also related ones, several features of the internal structure can easily be explained. An important corollary is that all interrelations are invariant on using Eq. (13) of Chapter 1 with polarization parameters that differ from our Stokes parameters in having a different sign for Q, U, or V, or any combination of them.

Evidently, each interrelation for the elements of $\mathbf{Z}$ also holds for the elements of the scattering matrix (as used in Section XI of Chapter 1 and by van de Hulst, 1957) and for those of $c\mathbf{Z}$, where c is an arbitrary real scalar. In particular, the normalization of $\mathbf{Z}$ does not influence its internal structure. It should be noted, however, that if $\mathbf{Z}$ is a pure phase matrix, $c\mathbf{Z}$ cannot be a pure phase matrix for $c < 0$, as, according to Eq. (14) of Chapter 1,

$$Z_{11} \geq 0 \tag{31}$$

for every pure phase matrix. Clearly the case where Z_{11} vanishes is very exceptional and implies that $\mathbf{S}$ and $\mathbf{Z}$ are null matrices. We may call this the trivial case.

The internal structure described previously is not only simple and elegant, but also fundamental, because all interrelations for the elements of $\mathbf{Z}$ can be derived from this structure. To prove this theorem, we first make the assumption

$$Z_{11} + Z_{22} - Z_{12} - Z_{21} \neq 0. \tag{32}$$

As shown by Hovenier *et al.* (1986) there are in this case nine relations, each involving products and squares of sums and differences of elements, from which all interrelations can be derived. These relations are

$$(Z_{11} + Z_{22})^2 - (Z_{12} + Z_{21})^2 = (Z_{33} + Z_{44})^2 + (Z_{34} - Z_{43})^2, \quad (33)$$

$$(Z_{11} - Z_{12})^2 - (Z_{21} - Z_{22})^2 = (Z_{31} - Z_{32})^2 + (Z_{41} - Z_{42})^2, \quad (34)$$

$$(Z_{11} - Z_{21})^2 - (Z_{12} - Z_{22})^2 = (Z_{13} - Z_{23})^2 + (Z_{14} - Z_{24})^2, \quad (35)$$

$$(Z_{11} + Z_{22} - Z_{12} - Z_{21})(Z_{13} + Z_{23})$$
$$= (Z_{31} - Z_{32})(Z_{33} + Z_{44}) - (Z_{41} - Z_{42})(Z_{34} - Z_{43}), \quad (36)$$

$$(Z_{11} + Z_{22} - Z_{12} - Z_{21})(Z_{34} + Z_{43})$$
$$= (Z_{31} - Z_{32})(Z_{14} - Z_{24}) + (Z_{41} - Z_{42})(Z_{13} - Z_{23}), \quad (37)$$

$$(Z_{11} + Z_{22} - Z_{12} - Z_{21})(Z_{33} - Z_{44})$$
$$= (Z_{31} - Z_{32})(Z_{13} - Z_{23}) - (Z_{41} - Z_{42})(Z_{14} - Z_{24}), \quad (38)$$

$$(Z_{11} + Z_{22} - Z_{12} - Z_{21})(Z_{14} + Z_{24})$$
$$= (Z_{31} - Z_{32})(Z_{34} - Z_{43}) + (Z_{41} - Z_{42})(Z_{33} + Z_{44}), \quad (39)$$

$$(Z_{11} + Z_{22} - Z_{12} - Z_{21})(Z_{31} + Z_{32})$$
$$= (Z_{33} + Z_{44})(Z_{13} - Z_{23}) + (Z_{34} - Z_{43})(Z_{14} - Z_{24}), \quad (40)$$

$$(Z_{11} + Z_{22} - Z_{12} - Z_{21})(Z_{41} + Z_{42})$$
$$= (Z_{33} + Z_{44})(Z_{14} - Z_{24}) - (Z_{34} - Z_{43})(Z_{13} - Z_{23}). \quad (41)$$

By rewriting these nine relations so that only the squares and products of elements appear, we can readily verify that they follow from Eq. (27) and Fig. 1. If Eq. (32) does not hold, then we either have the trivial case or at least one of the following inequalities must hold:

$$Z_{11} + Z_{22} + Z_{12} + Z_{21} \neq 0, \quad (42)$$

$$Z_{11} - Z_{22} - Z_{12} + Z_{21} \neq 0, \quad (43)$$

$$Z_{11} - Z_{22} + Z_{12} - Z_{21} \neq 0. \quad (44)$$

If one of Eqs. (42)–(44) holds, we have a set of nine relations with which to deal that differs from Eqs. (33)–(41), but we can follow a similar procedure. This completes the proof of our theorem.

To illustrate the preceding theorem, let us give three examples. First, the well-known relation

$$\sum_{i=1}^{4} \sum_{j=1}^{4} Z_{ij}^2 = 4Z_{11}^2 \quad (45)$$

given by Fry and Kattawar (1981) is easily obtained from Eq. (27) by successive application of the following operations on $\mathbf{Z}^s$ [cf. Eq. (28)]:

1. Add the elements of the second, third, and fourth columns.
2. Subtract the elements of the first column.
3. Equate the result to twice the sum of the elements of the first row.

Thus, Eq. (45) is a composite of five simple interrelations. Note that it is obeyed by the elements of diag(1, 1, 1, -1), for example, though this is not a pure phase matrix [cf. Eq. (30)].

Second, as shown by Barakat (1981) and Simon (1982), we have the matrix equation [cf. Eqs. (22) and (24)]

$$\tilde{\mathbf{Z}}\mathbf{G}\mathbf{Z} = -\frac{1}{2}[\mathrm{Tr}\,(\tilde{\mathbf{Z}}\mathbf{G}\mathbf{Z})]\mathbf{G}. \tag{46}$$

Evidently, a matrix equation of the type given by Eq. (46) is equivalent to a set of 16 scalar equations for the elements of $\mathbf{Z}$. The nondiagonal elements yield 12 equations, but the elements (i, j) and (j, i) yield the same equation if $i \neq j$. Thus six equations arise for products of different elements of $\mathbf{Z}$. These are exactly the same equations as shown by the top six pictograms of Fig. 1a. Equating the diagonal elements on both sides of Eq. (46) yields four equations. If one of these is used to eliminate $\mathrm{Tr}(\tilde{\mathbf{Z}}\mathbf{G}\mathbf{Z})$, we obtain three equations that involve only squares of elements of $\mathbf{Z}$. These are precisely the first three equations contained in Eq. (27). However, not all interrelations for the elements of $\mathbf{Z}$ follow from Eq. (46). Indeed, if this were the case Eq. (30), for example, should follow from Eq. (46). However, the matrix diag(1, 1, 1, -1) obeys Eq. (46) but does not satisfy Eq. (30).

Third, using the internal structure of $\mathbf{Z}$ and Eq. (6), it can be shown (Hovenier *et al.*, 1986) that

$$\left(I^{\mathrm{sca}}\right)^2\left(1 - p_{\mathrm{sca}}^2\right) = \frac{d^2}{R^4}\left(1 - p_{\mathrm{inc}}^2\right)\left(I^{\mathrm{inc}}\right)^2, \tag{47}$$

where p_{sca} and p_{inc} are the degrees of polarization of the scattered and incident light, respectively, as defined in Section V of Chapter 1 and R is the distance to the origin located inside the particle [see Eq. (13) of Chapter 1]. Consequently, if the incident light is fully polarized, so is the scattered light and if $d = 0$ the scattered light is always completely polarized. However, when the incident light is only partially polarized, p_{sca} may be either larger or smaller than p_{inc} (see Hovenier and van der Mee, 1995), which shows that adjectives such as "nondepolarizing" and "totally polarizing" instead of "pure" are less desirable.

C. SYMMETRY

The elements of the amplitude matrix of a single particle in a fixed orientation are, in general, four different complex functions, or, in other words, they are specified by eight real functions of $\mathbf{n}^{\mathrm{sca}}$ and $\mathbf{n}^{\mathrm{inc}}$. Symmetry properties may reduce this number. Particles can have a large variety of symmetry shapes, as is well known from crystallography and molecular physics. Group theory is helpful for a systematic treatment of these symmetry shapes (see, e.g., Hamermesh, 1962; Heine, 1960). Hu *et al.* (1987) presented a comprehensive study of strict forward ($\Theta = 0$) and strict backward ($\Theta = \pi$) scattering by an individual particle in a fixed orientation. For strict forward scattering they distinguished 16 different symmetry shapes, which were classified into five symmetry classes, and for backward scattering four different symmetry shapes, which were classified into two symmetry classes. A large number of relations for the amplitude matrix and the corresponding pure phase matrix were derived in this way.

A comprehensive treatment of all symmetry properties of the amplitude matrix and the corresponding pure phase matrix for arbitrary $\mathbf{n}^{\mathrm{sca}}$ and $\mathbf{n}^{\mathrm{inc}}$ is beyond the scope of this chapter. An important case, however, is the following. Consider a particle located in the origin of the coordinate system shown in Fig. 1 of Chapter 1. The particle has a plane of symmetry coinciding with the x–z plane. Suppose the incident light propagates along the positive z axis and let us consider light scattering in a direction in the x–z plane. Because the particle is its own mirror image we must have (see van de Hulst, 1957)

$$\begin{pmatrix} S_{11} & S_{12} \\ S_{21} & S_{22} \end{pmatrix} = \begin{pmatrix} S_{11} & -S_{12} \\ -S_{21} & S_{22} \end{pmatrix}, \tag{48}$$

so that

$$S_{12} = S_{21} = 0. \tag{49}$$

Using Eqs. (14)–(29) of Chapter 1, we find that the corresponding phase matrix in this case obtains the simple form

$$\mathbf{Z} = \begin{pmatrix} Z_{11} & Z_{12} & 0 & 0 \\ Z_{12} & Z_{11} & 0 & 0 \\ 0 & 0 & Z_{33} & Z_{34} \\ 0 & 0 & -Z_{34} & Z_{33} \end{pmatrix}, \tag{50}$$

with

$$Z_{11} = \left[Z_{12}^2 + Z_{33}^2 + Z_{34}^2 \right]^{1/2}. \tag{51}$$

A simple example of this case occurs for a spherically symmetric particle composed of an isotropic substance. Another example is a homogeneous body of revolution with its rotation axis in the x–z plane. This was numerically established by Hovenier *et al.* (1996) for scattering of light by four homogeneous bodies of revolution, namely, an oblate spheroid, a prolate spheroid, a finite circular cylinder, and a bisphere with equal touching components, where in each case the incident light propagated along the positive z axis and the scattered light in the x–z plane.

The second kind of symmetry we wish to consider is reciprocity. This was already mentioned in Section IX of Chapter 1. The main results for arbitrary directions of incidence and scattering are embodied by Eqs. (44) and (45) of that section. When time inversion yields the same scattering problem, we have

$$S_{21} = -S_{12} \tag{52}$$

and the corresponding pure phase matrix has the form

$$\mathbf{Z} = \begin{pmatrix} Z_{11} & Z_{12} & Z_{13} & Z_{14} \\ Z_{12} & Z_{22} & Z_{23} & Z_{24} \\ -Z_{13} & -Z_{23} & Z_{33} & Z_{34} \\ Z_{14} & Z_{24} & -Z_{34} & Z_{44} \end{pmatrix}, \tag{53}$$

with

$$Z_{11} - Z_{22} + Z_{33} - Z_{44} = 0, \tag{54}$$

as follows from Eq. (52) together with Eqs. (14)–(29) of Chapter 1. This case occurs, for example, for strict backscattering by an arbitrary particle [cf. Eq. (49) of Chapter 1].

D. INEQUALITIES

Many inequalities may be derived from the internal structure of a pure phase matrix. We do not aim here at a comprehensive list of inequalities, but in addition to Eq. (31) we mention the following:

$$|Z_{ij}| \leq Z_{11}, \qquad i, j = 1, 2, 3, 4, \tag{55}$$

$$Z_{11} + Z_{22} + Z_{12} + Z_{21} \geq 0, \tag{56}$$

$$Z_{11} + Z_{22} - Z_{12} - Z_{21} \geq 0, \tag{57}$$

$$Z_{11} - Z_{22} + Z_{12} - Z_{21} \geq 0, \tag{58}$$

$$Z_{11} - Z_{22} - Z_{12} + Z_{21} \geq 0, \tag{59}$$

$$Z_{11} + Z_{22} + Z_{33} + Z_{44} \geq 0, \tag{60}$$

$$Z_{11} + Z_{22} - Z_{33} - Z_{44} \geq 0, \tag{61}$$

$$Z_{11} - Z_{22} + Z_{33} - Z_{44} \geq 0, \tag{62}$$

$$Z_{11} - Z_{22} - Z_{33} + Z_{44} \geq 0. \tag{63}$$

We refer to Hovenier *et al.* (1986) for proofs of these and other inequalities.

III. RELATIONSHIPS FOR SINGLE SCATTERING BY A COLLECTION OF PARTICLES

A. THE GENERAL CASE

In this section we discuss relationships for the phase matrix of a collection of independently scattering particles, each of them characterized by an individual amplitude matrix. Because the waves scattered by each particle are essentially incoherent, the Stokes vectors of the scattered waves of the constituent particles are to be added to get the Stokes vector of the wave scattered by the collection. If we indicate the individual particles in the collection by a superscript g, then the phase matrix $\mathbf{Z}^c$ of the collection is the sum of the pure phase matrices $\mathbf{Z}^g$ of the individual particles, that is,

$$\mathbf{Z}^c = \sum_g \mathbf{Z}^g = n_0 \langle \mathbf{Z} \rangle \, dv, \tag{64}$$

where n_0 is the particle number density, $\langle \mathbf{Z} \rangle$ is the collection-averaged phase matrix per particle, and dv is a small volume element containing all particles of the collection [cf. Eq. (30) of Chapter 1]. Instead of a sum of pure phase matrices we may have an integral of a pure phase matrix with respect to size or orientation. The properties of such matrices are the same as for a sum of pure phase matrices. A special case of this occurs in light-scattering experiments for one particle that involve averaging over orientations.

Linear inequalities for the elements of a pure phase matrix are also valid for the phase matrix of a collection of particles, because these are obtained by adding the corresponding elements of the phase matrices of the constituent particles. In particular, we find the following linear inequalities:

$$Z_{11}^c \geq 0, \tag{65}$$

$$\left| Z_{ij}^c \right| \leq Z_{11}^c, \tag{66}$$

$$Z_{11}^c + Z_{22}^c + Z_{12}^c + Z_{21}^c \geq 0, \tag{67}$$

$$Z_{11}^c + Z_{22}^c - Z_{12}^c - Z_{21}^c \geq 0, \tag{68}$$

$$Z_{11}^c - Z_{22}^c + Z_{12}^c - Z_{21}^c \geq 0, \tag{69}$$

$$Z_{11}^c - Z_{22}^c - Z_{12}^c + Z_{21}^c \geq 0. \tag{70}$$

Quadratic relations between the elements of a pure phase matrix such as Eqs. (33)–(41) are generally lost when the phase matrix of a collection of particles is formed by adding the pure phase matrices of the individual particles. However, the following six quadratic inequalities, first obtained by Fry and Kattawar (1981), are always valid:

$$\left(Z^c_{11} + Z^c_{12}\right)^2 - \left(Z^c_{21} + Z^c_{22}\right)^2 \geq \left(Z^c_{31} + Z^c_{32}\right)^2 + \left(Z^c_{41} + Z^c_{42}\right)^2, \quad (71)$$

$$\left(Z^c_{11} - Z^c_{12}\right)^2 - \left(Z^c_{21} - Z^c_{22}\right)^2 \geq \left(Z^c_{31} - Z^c_{32}\right)^2 + \left(Z^c_{41} - Z^c_{42}\right)^2, \quad (72)$$

$$\left(Z^c_{11} + Z^c_{21}\right)^2 - \left(Z^c_{12} + Z^c_{22}\right)^2 \geq \left(Z^c_{13} + Z^c_{23}\right)^2 + \left(Z^c_{14} + Z^c_{24}\right)^2, \quad (73)$$

$$\left(Z^c_{11} - Z^c_{21}\right)^2 - \left(Z^c_{12} - Z^c_{22}\right)^2 \geq \left(Z^c_{13} - Z^c_{23}\right)^2 + \left(Z^c_{14} - Z^c_{24}\right)^2, \quad (74)$$

$$\left(Z^c_{11} + Z^c_{22}\right)^2 - \left(Z^c_{12} + Z^c_{21}\right)^2 \geq \left(Z^c_{33} + Z^c_{44}\right)^2 + \left(Z^c_{34} - Z^c_{43}\right)^2, \quad (75)$$

$$\left(Z^c_{11} - Z^c_{22}\right)^2 - \left(Z^c_{12} - Z^c_{21}\right)^2 \geq \left(Z^c_{33} - Z^c_{44}\right)^2 + \left(Z^c_{34} + Z^c_{43}\right)^2. \quad (76)$$

Indeed, to derive Eq. (72), we start from Eq. (34), where each term carries the superscript g to denote the individual particles. Because Eqs. (68) and (70) also hold for the elements of each $\mathbf{Z}^g$, we can find nonnegative quantities N^g_1 and N^g_2 and angles θ^g such that

$$\begin{cases} N^g_1 = \sqrt{Z^g_{11} - Z^g_{12} - Z^g_{21} + Z^g_{22}}, \\ N^g_2 = \sqrt{Z^g_{11} - Z^g_{12} + Z^g_{21} - Z^g_{22}}, \\ N^g_1 N^g_2 \cos\theta^g = Z^g_{31} - Z^g_{32}, \\ N^g_1 N^g_2 \sin\theta^g = Z^g_{41} - Z^g_{42}. \end{cases} \quad (77)$$

Consequently,

$$\begin{aligned}
&\left(Z^c_{11} - Z^c_{12}\right)^2 - \left(Z^c_{21} - Z^c_{22}\right)^2 - \left(Z^c_{31} - Z^c_{32}\right)^2 - \left(Z^c_{41} - Z^c_{42}\right)^2 \\
&= \left(Z^c_{11} - Z^c_{12} - Z^c_{21} + Z^c_{22}\right)\left(Z^c_{11} - Z^c_{12} + Z^c_{21} - Z^c_{22}\right) \\
&\quad - \left(Z^c_{31} - Z^c_{32}\right)^2 - \left(Z^c_{41} - Z^c_{42}\right)^2 \\
&= \sum_g \left(N^g_1\right)^2 \sum_h \left(N^h_2\right)^2 - \sum_{g,h} N^g_1 N^g_2 N^h_1 N^h_2 \cos\left(\theta^g - \theta^h\right) \\
&\geq \sum_g \left(N^g_1\right)^2 \sum_h \left(N^h_2\right)^2 - \sum_{g,h} N^g_1 N^g_2 N^h_1 N^h_2 \\
&= \sum_{g \neq h} \left\{\left(N^g_1\right)^2 \left(N^h_2\right)^2 - N^g_1 N^g_2 N^h_1 N^h_2\right\} \\
&= \sum_{g < h} \left(N^g_1 N^h_2 - N^h_1 N^g_2\right)^2 \geq 0, \quad (78)
\end{aligned}$$

which implies Eq. (72). Equations (71) and (73)–(76) are proved analogously.

It is clear from the preceding discussion that for a collection of particles with proportional amplitude matrices (with real or complex proportionality constants) the inequalities (71)–(76) reduce to equalities, as is the case for a pure phase matrix. This occurs, in particular, for a collection of identical particles with the same orientation in space or for a collection of identical spherically symmetric particles.

Many other inequalities can be found from Eqs. (65)–(76). For instance, by adding Eqs. (71)–(76), observing that the double products cancel each other, and rearranging terms, one obtains the inequality [cf. Fry and Kattawar (1981)]

$$\sum_{i=1}^{4}\sum_{j=1}^{4}\left(Z_{ij}^{c}\right)^{2} \leq 4\left(Z_{11}^{c}\right)^{2}. \tag{79}$$

Note that Eq. (79) becomes an equality for a pure phase matrix [cf. Eq. (45)].

Evidently, all interrelations for Z_{ij}^{c} keep their validity for a sum of matrices of the type given by Eq. (26) and in particular for a sum of pure scattering matrices as considered by van de Hulst (1957).

B. SYMMETRY

The description of light scattering by a cloud of particles simplifies when the particles themselves or their orientations in space possess certain symmetry properties. For an extensive treatment of this subject we must refer to the literature (see, e.g., Perrin, 1942; van de Hulst, 1957), but a few remarks here are in order.

As shown by Eq. (64), the phase matrix of a collection of identical particles all having the same orientation is a pure phase matrix [cf. Eq. (12)] with the internal structure discussed in Section II. Another extreme situation is rendered by a collection of particles in (three-dimensional) random orientation. Then the scattered light depends on the scattering angle, but there is rotational symmetry about the direction of incidence. Assuming reciprocity (see Section IX of Chapter 1), we find that a collection of particles in random orientation has a scattering matrix (see Section XI of Chapter 1) of the form

$$\mathbf{F}(\Theta) = \begin{pmatrix} a_1(\Theta) & b_1(\Theta) & b_3(\Theta) & b_5(\Theta) \\ b_1(\Theta) & a_2(\Theta) & b_4(\Theta) & b_6(\Theta) \\ -b_3(\Theta) & -b_4(\Theta) & a_3(\Theta) & b_2(\Theta) \\ b_5(\Theta) & b_6(\Theta) & -b_2(\Theta) & a_4(\Theta) \end{pmatrix}. \tag{80}$$

If we also assume that all particles have a plane of symmetry or, equivalently, that particles and their mirror particles are present in equal numbers, we obtain the block-diagonal structure [cf. Eq. (61) of Chapter 1]

$$\mathbf{F}(\Theta) = \begin{pmatrix} a_1(\Theta) & b_1(\Theta) & 0 & 0 \\ b_1(\Theta) & a_2(\Theta) & 0 & 0 \\ 0 & 0 & a_3(\Theta) & b_2(\Theta) \\ 0 & 0 & -b_2(\Theta) & a_4(\Theta) \end{pmatrix}. \tag{81}$$

Equations (67)–(76) now reduce to the four simple inequalities [cf. Eqs. (65) and (66)]

$$(a_3 + a_4)^2 + 4b_2^2 \leq (a_1 + a_2)^2 - 4b_1^2, \tag{82}$$

$$|a_3 - a_4| \leq a_1 - a_2, \tag{83}$$

$$|a_2 - b_1| \leq a_1 - b_1, \tag{84}$$

$$|a_2 + b_1| \leq a_1 + b_1. \tag{85}$$

Consequently, all available information is contained in Eqs. (82) and (83) plus the fact that no element of $\mathbf{F}(\Theta)$ in Eq. (81) is larger in absolute value than a_1. The properties of the corresponding (normalized) phase matrix were studied by Hovenier and van der Mee (1988).

Special cases arise for strict forward ($\Theta = 0$) and backward ($\Theta = \pi$) scattering (see Sections IX and XI of Chapter 1 and Hovenier and Mackowski, 1998). Some important results are summarized in Tables I and II. In the case of backscattering, consequences for the linear and circular depolarization ratios have been reported by Mishchenko and Hovenier (1995), whereas bounds for p_{sca} in terms of p_{inc} have been derived by Hovenier and van der Mee (1995).

IV. TESTING MATRICES DESCRIBING SCATTERING BY SMALL PARTICLES

This section is devoted to the following problem. Suppose we have a real 4×4 matrix $\mathbf{M}$ with elements M_{ij}, which may have been obtained from experiments or numerical calculations. If we wish to know if $\mathbf{M}$ can be a pure phase matrix or a phase matrix of a collection of particles, what tests can be applied? In either case, there exist tests providing necessary and sufficient conditions for a real 4×4 matrix to have all of the mathematical requirements of a pure phase matrix or of the phase matrix of a collection of particles. These tests can only be performed if one knows all 16 elements of the matrix $\mathbf{M}$, which is not always the case. There

Table I

Properties of the Scattering Matrix for Exact Forward Scattering by a Collection of Randomly Oriented Identical Particles Each Having a Plane of Symmetry or by a Mixture of Such Collections

Scattering matrix

$$\mathbf{F} = \begin{pmatrix} a_1 & 0 & 0 & 0 \\ 0 & a_2 & 0 & 0 \\ 0 & 0 & a_2 & 0 \\ 0 & 0 & 0 & a_4 \end{pmatrix}$$

- In general:

$$|a_2| \leq a_1$$
$$|a_4| \leq a_1$$
$$a_4 \geq 2|a_2| - a_1$$

- Special case, each particle is rotationally symmetric:

$$0 \leq a_2 \leq a_1$$
$$a_4 = 2a_2 - a_1$$

- Special case, each particle is homogeneous, optically inactive, and spherical:

$$a_1 \geq 0$$
$$a_1 = a_2 = a_4$$

also exist tests providing only necessary conditions. These tests are particularly useful if not all 16 elements of the given matrix $\mathbf{M}$ are available or if $\mathbf{M}$ has a property that allows one to exclude it directly on the basis of a simple test. Once a given matrix has been shown to have the mathematical properties of a pure phase matrix or the phase matrix of a collection of particles, the matrix can, in principle, describe certain scattering situations but not necessarily the scattering problem intended. This is particularly true if scaling or symmetry errors have been made. Thus the tests are useful to verify if a given matrix can describe certain scattering events, but they are not sufficient to be certain of its "physical correctness." We refer the reader to Hovenier and van der Mee (1996) for a systematic study of tests for scattering matrices, which are completely analogous to those for phase matrices.

To test if a given real 4×4 matrix can be a pure phase matrix, one can distinguish between five types of tests:

  a. *Visual tests*, where one checks a simple property of the given matrix. For instance, one checks if the sum of the rows and the columns of the matrix

Table II

Properties of the Scattering Matrix for Exact Backward Scattering by a Collection of Randomly Oriented Identical Particles or by a Mixture of Such Collections

---

Scattering matrix

$$\mathbf{F} = \begin{pmatrix} a_1 & 0 & 0 & b_5 \\ 0 & a_2 & 0 & 0 \\ 0 & 0 & -a_2 & 0 \\ b_5 & 0 & 0 & a_4 \end{pmatrix}$$

• In general:

$$0 \leq a_2 \leq a_1$$
$$a_4 = a_1 - 2a_2$$
$$a_2 - a_1 \leq b_5 \leq a_1 - a_2$$

• Special case, each particle has a plane of symmetry:

$$0 \leq a_2 \leq a_1$$
$$a_4 = a_1 - 2a_2$$
$$b_5 = 0$$

• Special case, each particle is homogeneous, optically inactive, and spherical:

$$a_1 \geq 0$$
$$a_1 = a_2 = -a_4$$
$$b_5 = 0$$

---

in Eq. (28) are all equal to the same nonnegative number. Other examples of visual tests are to verify Eq. (45), some of the identities represented by the pictograms in Fig. 1, or some of the inequalities (55)–(63).

b. *Tests consisting of nine relations.* For instance, when Eq. (32) holds, Eqs. (33)–(41) form one such set. Other sets can be pointed out if one of Eqs. (42)–(44) is fulfilled. The advantage of such a test is that the nine relations are complete in the sense that $\mathbf{M}$ can be written in the form of Eq. (1) for a suitable amplitude matrix $\mathbf{S}$ that is unique apart from a phase factor of the form $e^{i\varepsilon}$ (cf. Hovenier *et al.*, 1986).

c. *Tests based on analogy with the Lorentz group,* such as verifying Eq. (46). However, this test is incomplete, because the matrix diag$(1, 1, 1, -1)$, for example, satisfies Eq. (46) but is not a pure phase matrix.

d. *Tests based on reconstructing the underlying amplitude matrix.* Starting from $\mathbf{M}$, one computes $\boldsymbol{\Gamma}_s^{-1}\mathbf{M}\boldsymbol{\Gamma}_s$, where $\boldsymbol{\Gamma}_s$ and $\boldsymbol{\Gamma}_s^{-1}$ are given by Eqs. (2) and (3), and checks if it has the form of the right-hand side of Eq. (4) (cf. November, 1993; Anderson and Barakat, 1994).

e. *Tests based on the coherency matrix.* In this test one computes from the given real 4×4 matrix $\mathbf{M}$, a complex Hermitian 4×4 matrix $\mathbf{T}$ (i.e., $T_{ij} = T_{ji}^*$) in a linear one-to-one way. Then $\mathbf{M}$ can be a pure nontrivial phase matrix if and only if $\mathbf{T}$ has one positive and three zero eigenvalues. If so desired, the underlying amplitude matrix can then be computed from the eigenvector corresponding to the positive eigenvalue. Tests of this type, with different coherency matrices that are unitarily equivalent, have been developed by Cloude (1986) and Simon (1982, 1987).

We now discuss the coherency matrix in more detail. This matrix $\mathbf{T}$ is easily derived from a given 4×4 matrix $\mathbf{M}$ and is defined as follows:

$$
\left.
\begin{aligned}
T_{11} &= \tfrac{1}{2}(M_{11} + M_{22} + M_{33} + M_{44}) \\
T_{22} &= \tfrac{1}{2}(M_{11} + M_{22} - M_{33} - M_{44}) \\
T_{33} &= \tfrac{1}{2}(M_{11} - M_{22} + M_{33} - M_{44}) \\
T_{44} &= \tfrac{1}{2}(M_{11} - M_{22} - M_{33} + M_{44})
\end{aligned}
\right\} ,
\tag{86}
$$

$$
\left.
\begin{aligned}
T_{14} &= \tfrac{1}{2}(M_{14} - i M_{23} + i M_{32} + M_{41}) \\
T_{23} &= \tfrac{1}{2}(i M_{14} + M_{23} + M_{32} - i M_{41}) \\
T_{32} &= \tfrac{1}{2}(-i M_{14} + M_{23} + M_{32} + i M_{41}) \\
T_{41} &= \tfrac{1}{2}(M_{14} + i M_{23} - i M_{32} + M_{41})
\end{aligned}
\right\} ,
\tag{87}
$$

$$
\left.
\begin{aligned}
T_{12} &= \tfrac{1}{2}(M_{12} + M_{21} - i M_{34} + i M_{43}) \\
T_{21} &= \tfrac{1}{2}(M_{12} + M_{21} + i M_{34} - i M_{43}) \\
T_{34} &= \tfrac{1}{2}(i M_{12} - i M_{21} + M_{34} + M_{43}) \\
T_{43} &= \tfrac{1}{2}(-i M_{12} + i M_{21} + M_{34} + M_{43})
\end{aligned}
\right\} ,
\tag{88}
$$

$$
\left.
\begin{aligned}
T_{13} &= \tfrac{1}{2}(M_{13} + M_{31} + i M_{24} - i M_{42}) \\
T_{31} &= \tfrac{1}{2}(M_{13} + M_{31} - i M_{24} + i M_{42}) \\
T_{24} &= \tfrac{1}{2}(-i M_{13} + i M_{31} + M_{24} + M_{42}) \\
T_{42} &= \tfrac{1}{2}(i M_{13} - i M_{31} + M_{24} + M_{42})
\end{aligned}
\right\} .
\tag{89}
$$

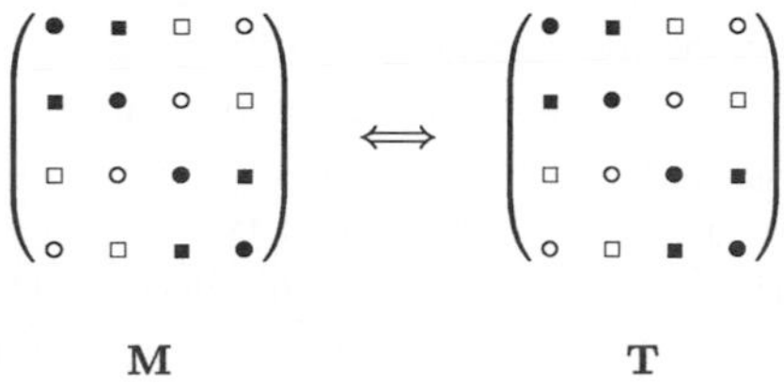

$$\begin{pmatrix} \bullet & \blacksquare & \square & \circ \\ \blacksquare & \bullet & \circ & \square \\ \square & \circ & \bullet & \blacksquare \\ \circ & \square & \blacksquare & \bullet \end{pmatrix} \Longleftrightarrow \begin{bmatrix} \bullet & \blacksquare & \square & \circ \\ \blacksquare & \bullet & \circ & \square \\ \square & \circ & \bullet & \blacksquare \\ \circ & \square & \blacksquare & \bullet \end{bmatrix}$$

$$\mathbf{M} \qquad\qquad\qquad \mathbf{T}$$

Figure 2   Transformation of the 4×4 matrix $\mathbf{M}$ to the coherency matrix $\mathbf{T}$. Four basic groups of elements are distinguished by four different symbols.

In fact, $\mathbf{T}$ depends linearly on $\mathbf{M}$ and the linear relation between them is given by four sets of linear transformations between corresponding elements of $\mathbf{M}$ and $\mathbf{T}$ (see Fig. 2). Moreover, $\mathbf{T}$ is always Hermitian, so that it has four real eigenvalues. If three of the eigenvalues vanish and one is positive, $\mathbf{M}$ can be a pure nontrivial phase matrix. This is a simple and complete test. It was discovered in the theory of radar polarization [see Cloude (1986), where $\mathbf{T}$ is defined with factors $\frac{1}{4}$ in Eqs. (86)–(89) instead of factors $\frac{1}{2}$]. Another complete test using the coherency matrix, namely, verifying

$$\operatorname{Tr}\mathbf{T} \geq 0, \qquad \mathbf{T}^2 = (\operatorname{Tr}\mathbf{T})\mathbf{T}, \tag{90}$$

is mostly due to Simon (1982, 1987), where, instead of $\mathbf{T}$, a Hermitian matrix $\mathbf{N}$ was used that is unitarily equivalent to the coherency matrix, namely,

$$\mathbf{N} = \mathbf{\Gamma}^{-1}\mathbf{T}\mathbf{\Gamma}, \tag{91}$$

where $\mathbf{\Gamma} = \operatorname{diag}(1, 1, -1, -1)\mathbf{\Gamma}_s$ and $\mathbf{\Gamma}_s$ is given by Eq. (2). The transformation from $\mathbf{M}$ to $\mathbf{N}$ is displayed in Fig. 3.

To test if a given real 4×4 matrix $\mathbf{M}$ can be the phase matrix of a collection of particles, one may employ two types of tests, specifically visual tests and tests based on the coherency matrix. The comparatively simple visual tests can often

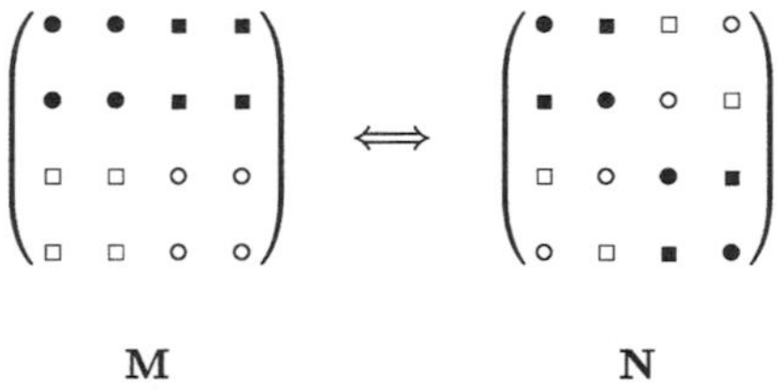

$$\begin{pmatrix} \bullet & \bullet & \blacksquare & \blacksquare \\ \bullet & \bullet & \blacksquare & \blacksquare \\ \square & \square & \circ & \circ \\ \square & \square & \circ & \circ \end{pmatrix} \Longleftrightarrow \begin{bmatrix} \bullet & \blacksquare & \square & \circ \\ \blacksquare & \bullet & \circ & \square \\ \square & \circ & \bullet & \blacksquare \\ \circ & \square & \blacksquare & \bullet \end{bmatrix}$$

$$\mathbf{M} \qquad\qquad\qquad \mathbf{N}$$

Figure 3   As in Fig. 2, but for the transformation from $\mathbf{M}$ to $\mathbf{N}$.

be applied if one has incomplete knowledge of the matrix **M**. Examples abound. For instance, one can verify any of the inequalities of Eqs. (65)–(76) and (79). The inequalities of Eqs. (65)–(70) are useful eyeball tests that often allow one to quickly dismiss a given matrix as a phase matrix of a collection of particles. The six inequalities of Eqs. (71)–(76) are commonly used to test matrices, especially in the form of Eqs. (82)–(85) for matrices **M** of the form of the right-hand side of Eq. (81).

Using the coherency matrix, one obtains a most effective method to verify if a given real 4×4 matrix **M** can be the phase matrix of a collection of particles. It was developed in radar polarimetry by Huynen (1970) for matrices with one special symmetry and by Cloude (1986) for general real 4×4 matrices. As before, one constructs the complex Hermitian matrix **T** from the given matrix **M** by using Eqs. (86)–(89) and computes the four eigenvalues of **T**, which must necessarily be real. Then **M** can only be a nontrivial phase matrix of a collection of particles if and only if all four eigenvalues of **T** are nonnegative and at least one of them is positive.

The coherency matrix test allows some fine tuning. First of all, recalling that **M** can be a pure nontrivial phase matrix whenever **T** has one positive and three zero eigenvalues, the ratio of the second largest to the largest positive eigenvalue of **T** may be viewed as a measure of the degree to which a phase matrix is pure (Cloude, 1989, 1992a, b; Anderson and Barakat, 1994). Second, because a complex Hermitian matrix can always be diagonalized by a unitary matrix whose columns form an orthonormal basis of its eigenvectors, one can write any phase matrix as a sum of four pure phase matrices. This result may come as a big surprise in the light-scattering community, but it is well known in radar polarimetry where it is called target decomposition (cf. Cloude, 1989).

In the coherency matrix test described previously, the matrix **T** may be replaced by the matrix **N**. This is obvious, because **T** and **N** are unitarily equivalent and therefore have the same eigenvalues. As a test for phase matrices of a collection of particles, this was clearly understood by Cloude (1992a, b) and by Anderson and Barakat (1994). The details of the "target decomposition," but not its principle, are different but can easily be transformed into each other. The testing procedures described in this section have been used in practice in a number of publications, including Kuik *et al.* (1991), Mishchenko *et al.* (1996a), Lumme *et al.* (1997), and Hess *et al.* (1998).

V. DISCUSSION AND OUTLOOK

The phase matrices studied so far all transform a beam of light with degree of polarization not exceeding 1 into a beam of light having the same property; that is, they satisfy the Stokes criterion. The latter is defined as follows. If a real four-vector $\mathbf{I}^{\text{inc}}$ whose components I^{inc}, Q^{inc}, U^{inc}, and V^{inc} satisfy the inequality

$$I^{\text{inc}} \geq \left[\left(Q^{\text{inc}} \right)^2 + \left(U^{\text{inc}} \right)^2 + \left(V^{\text{inc}} \right)^2 \right]^{1/2} \qquad (92)$$

is transformed by $\mathbf{M}$ into the vector $\mathbf{I}^{\text{sca}} = \mathbf{M}\mathbf{I}^{\text{inc}}$ with components I^{sca}, Q^{sca}, U^{sca}, and V^{sca}, and the latter satisfy the inequality

$$I^{\text{sca}} \geq \left[\left(Q^{\text{sca}} \right)^2 + \left(U^{\text{sca}} \right)^2 + \left(V^{\text{sca}} \right)^2 \right]^{1/2}, \qquad (93)$$

then $\mathbf{M}$ is said to satisfy the Stokes criterion. The real 4×4 matrices satisfying the Stokes criterion have been studied in detail. Konovalov (1985), van der Mee and Hovenier (1992), and Nagirner (1993) have indicated which matrices $\mathbf{M}$ of the form of the right-hand side of Eq. (81) satisfy the Stokes criterion. Givens and Kostinski (1993) and van der Mee (1993) have given necessary and sufficient conditions for a general real 4×4 matrix $\mathbf{M}$ to satisfy the Stokes criterion. These conditions involve the eigenvalues and eigenvectors of the matrix $\mathbf{G\tilde{M}GM}$, where $\mathbf{G} = \text{diag}(1, -1, -1, -1)$. Givens and Kostinski (1993) assumed diagonalizability of the matrix $\mathbf{G\tilde{M}GM}$, whereas no such constraint appeared in van der Mee (1993). Unfortunately, all of these studies are of limited value for describing scattering by particles, because the class of matrices satisfying the Stokes criterion is too large, as exemplified by the matrices $\text{diag}(1, 1, 1, -1)$ and $\mathbf{G} = \text{diag}(1, -1, -1, -1)$, which satisfy the Stokes criterion but fail to satisfy the coherency matrix test discussed in Section IV [see also Eqs. (82) and (83)]. Moreover, the coherency matrix test is more easily implemented than any known general test to verify the Stokes criterion.

Hitherto we have given tests to verify if a given real 4×4 matrix $\mathbf{M}$ can be a pure phase matrix or the phase matrix of a collection of particles, as if this matrix consisted of exact data. However, if $\mathbf{M}$ has been numerically or experimentally determined, a test might cause one to reject $\mathbf{M}$ as a (pure) phase matrix, whereas there exists a small perturbation of $\mathbf{M}$ within the numerical or experimental error that leads to a positive test result. In such a case, $\mathbf{M}$ should not have been rejected.

One way of dealing with experimental or numerical error is to treat a deviation from a positive test result as an indication of numerical or experimental errors. Assuming that the given matrix $\mathbf{M}$ is the sum of a perturbation $\Delta\mathbf{M}$ and an "exact" matrix $\mathbf{M}^e$, which can be a (pure) phase matrix, an error bound formula is derived in terms of the given matrix $\mathbf{M}$ such that $\mathbf{M}$ passes the test whenever the error bound is less than a given threshold value. Such a procedure has been implemented for the coherency matrix test by Anderson and Barakat (1994) and by Hovenier and van der Mee (1996). In either paper, a "corrected" (pure) phase matrix is sought that minimizes the error bound. Procedures to correct given matrices go back as far as Konovalov (1985), who formulated such a method for the Stokes criterion.

Table III

Eigenvalues λ_i of the Coherency Matrix T if M Is One of the Three
Matrices Given in Table II of Cariou *et al.* (1990). These Matrices Describe
Underwater Scattering for Different Scatterer Amounts and Therefore
Different Approximate Values of the Optical Extinction Coefficient k_{ext}

k_{ext} (m^{-1})	λ_1	λ_2	λ_3	λ_4
0.5	1.9878	0.0444	−0.0273	−0.0048
1.0	1.5333	0.0776	0.2166	0.1725
2.0	1.2395	0.3795	0.1571	0.2239

The application of error bound tests to a given real 4×4 matrix can lead
to conclusions that primarily depend on the choice of the error bound formula.
Moreover, no information on known numerical or experimental errors is taken
into account. One possible way out is to test three matrices $\mathbf{M}^0$, $\mathbf{M}^+$, and $\mathbf{M}^-$
such that $\mathbf{M}^0$ is the given real 4×4 matrix and

$$\delta_{ij} = M_{ij}^0 - M_{ij}^- = M_{ij}^+ - M_{ij}^0, \qquad i, j = 1, 2, 3, 4, \qquad (94)$$

are the errors in the elements of $\mathbf{M}^0$. Then the matrix $\mathbf{M}^0$ is accepted as a (pure)
phase matrix if all of these three matrices satisfy the appropriate "exact" test.

By way of example we will now apply the coherency matrix test to three real
4×4 matrices describing forward scattering by kaolinite particles suspended in
water as measured using pulsed laser radiation (see Table II of Cariou *et al.*, 1990).
The corresponding eigenvalues of the coherency matrix $\mathbf{T}$ are then given by Table III and are all nonnegative, except for the two smallest eigenvalues pertaining
to the first matrix. Hence the second and third matrices can be scattering matrices
of a collection of particles. The first matrix has one large positive eigenvalue and
three eigenvalues that are very small in absolute value. One may expect that this
matrix coincides with a pure scattering matrix within experimental errors.

When the scattering matrix of a collection of particles has the form of the right-
hand side of Eq. (81), its six different nontrivial elements can be expanded into
a series involving generalized spherical functions [see Eqs. (72)–(77) of Chapter 1]. With the help of Eqs. (82)–(85) and the orthogonality property given by
Eq. (78) of Chapter 1, a plethora of equalities and inequalities for the expansion
coefficients can be derived (see van der Mee and Hovenier, 1990). Some of these
relations are very convenient for testing purposes, because sometimes the expansion coefficients rather than the elements of the scattering matrix are given (see,
e.g., Mishchenko and Mackowski, 1994; Mishchenko and Travis, 1998). Unfortunately, the problem of finding necessary and sufficient conditions on the matrix

in Eq. (81) to be a scattering matrix of a collection of particles in terms of the expansion coefficients has not yet been solved.

Multiple scattering of polarized light by small particles in atmospheres and oceans can also be described by matrices that transform the Stokes parameters. Examples are provided by the reflection and transmission matrices for plane-parallel media. For macroscopically isotropic and symmetric scattering media (Section XI of Chapter 1) above a Lambert or Fresnel reflecting surface, the elements of such multiple-scattering matrices obey the same relationships as the elements of a sum of pure phase (scattering) matrices considered in preceding sections (Hovenier and van der Mee, 1997). Further work in this field is in progress.

Part II

Theoretical and Numerical Techniques

Chapter 4

Separation of Variables for Electromagnetic Scattering by Spheroidal Particles

Ioan R. Ciric

Department of Electrical and Computer Engineering
University of Manitoba
Winnipeg, Manitoba
Canada R3T 5V6

Francis R. Cooray

CSIRO Telecommunications and Industrial Physics
Epping, New South Wales 1710
Australia

I. INTRODUCTION

The method of separation of variables can be used to obtain exact analytic solutions to problems involving scattering of electromagnetic waves by objects whose surfaces can be made to coincide with coordinate surfaces in certain curvilinear orthogonal systems. Spheroidal coordinate systems are such coordinate systems and, because spheroidal surfaces are appropriate for approximating the surfaces of a large variety of real-world, nonspherical objects, in this chapter we derive exact expressions for the electromagnetic fields scattered by single spheroids and also by systems of spheroids in arbitrary orientation. These expressions represent benchmark solutions that are useful for evaluating the accuracy and efficiency of various simpler, approximate methods of solution.

Exact solutions for the scalar Helmholtz equation in spheroidal coordinates, obtained by using the method of separation of variables, are expressed in terms of scalar spheroidal wave functions. These functions and their vector counterparts, both of which are defined later, are the key to obtaining exact analytic solutions to scattering problems involving single or multiple spheroids. For a historical survey on the application of spheroidal wave functions, the reader is referred to Flammer's monograph (Flammer, 1957).

Even though various applications of spheroidal wave functions have been presented since 1880, it is in the work of Schultz (1950) that a formulation was given for the first time for obtaining an exact solution to the problem of scattering of electromagnetic waves by a perfectly conducting prolate spheroid at axial incidence. Based on Schultz's formulation, Siegel *et al.* (1956) carried out quantitative calculations of the backscattering from a prolate spheroid and plotted the variation of the backscattering cross section with the size of the spheroid, for a prolate spheroid of axial ratio 10. An exact solution for the more general case of the scattering of electromagnetic waves by a perfectly conducting prolate spheroid for arbitrary polarization and angle of incidence was given by Reitilinger (1957), but no numerical results were published. Numerical results for this problem, in the form of scattering cross sections, were presented by Sinha and Mac-Phie (1977), whose formulation also eliminated some drawbacks in the analysis given by Reitilinger (1957). An exact solution to a similar problem involving a dielectric spheroid was obtained by Asano and Yamamoto (1975) using a different type of vector wave functions than that used in the work of Sinha and MacPhie (1977). Analytic solutions for scattering by a dielectric spheroid with a confocal lossy dielectric coating (Cooray and Ciric, 1992) and by a chiral spheroid (Cooray and Ciric, 1993) were first given by the authors. A similar solution for a prolate spheroid with a lossy dielectric coating was also presented by Sebak and Sinha (1992), but only for axial incidence. Recent calculations using the method of separation of variables for the scattering by single homogeneous spheroids and by coated spheroids were also reported by Voshchinnikov (1996). On the other hand, exact analytical results for a single spheroidal particle have been used to derive its T matrix in spheroidal coordinates (Schulz *et al.*, 1998a). The T matrix expressed in spherical coordinates is applied to evaluate optical properties of ensembles of randomly oriented spheroidal particles by performing the averaging analytically (Mishchenko, 1991a).

Research on the application of the method of separation of variables to the electromagnetic scattering by systems of two spheroids is not as extensive as in the case of a single spheroid because of the fact that the analysis is now more complicated by the necessity of implementing translational and rotational–translational addition theorems for spheroidal wave functions. An exact solution for electromagnetic scattering by two parallel conducting spheroids was first obtained by Sinha and MacPhie (1983). A similar solution, but using again a different type of vector wave function was given by Dalmas and Deleuil (1985). The scattering from two parallel dielectric spheroids was analyzed by Cooray *et al.* (1990). Analytic solutions to the problems of electromagnetic scattering by two conducting and two dielectric spheroids of arbitrary orientation were first obtained by the authors (Cooray and Ciric, 1989b; Cooray and Ciric, 1991a) using the rotational–translational addition theorems for vector spheroidal wave functions previously derived by them (Cooray and Ciric, 1989a) and independently by Dalmas *et al.* (1989). The authors also presented a solution for the case of an arbitrary number of dielectric spheroids in arbitrary orientation (Cooray and Ciric, 1991b), with numerical results given for the case of two spheroids. The case of scattering by two homogeneous lossy dielectric spheroids of arbitrary orientation was recently considered by Nag and Sinha (1995).

In the following sections of this chapter, we present the spheroidal coordinate systems, define different spheroidal wave functions used in the analysis of scattering problems, and present exact solutions to the problems of electromagnetic scattering by a single spheroid (coated dielectric and chiral) and by systems with an arbitrary number of dielectric spheroids in arbitrary orientation, with illustrative numerical results computed for spheroids of different sizes. A time-harmonic dependence $\exp(j\omega t)$, where ω is the angular frequency and $j^2 = -1$, is assumed throughout in the formulas and derivations given in this chapter (cf. Section IV of Chapter 1).

II. SPHEROIDAL COORDINATE SYSTEMS

The prolate and oblate spheroidal coordinate systems are constructed by rotating a two-dimensional elliptic coordinate system, consisting of confocal ellipses and hyperbolas, about the major and minor axes of the ellipses, respectively. In the following definition of the spheroidal coordinates, the z axis of the Cartesian coordinate system attached to the center of the confocal spheroids has been considered as the axis of revolution. Let F be the corresponding semi-interfocal distance. The prolate spheroidal coordinates (η, ξ, φ) are then related to the Cartesian coordinates by the transformation

$$x = F\big[(1 - \eta^2)(\xi^2 - 1)\big]^{1/2} \cos\varphi, \qquad y = F\big[(1 - \eta^2)(\xi^2 - 1)\big]^{1/2} \sin\varphi,$$

$$z = F\eta\xi,$$
(1)

with $-1 \leq \eta \leq 1, 1 \leq \xi < \infty, 0 \leq \varphi \leq 2\pi$. The oblate spheroidal coordinates are related to the Cartesian coordinates by the transformation

$$x = F\big[(1 - \eta^2)(\xi^2 + 1)\big]^{1/2} \cos\varphi, \qquad y = F\big[(1 - \eta^2)(\xi^2 + 1)\big]^{1/2} \sin\varphi,$$

$$z = F\eta\xi,$$
(2)

with either $-1 \leq \eta \leq 1, 0 \leq \xi < \infty, 0 \leq \varphi \leq 2\pi$ or $0 \leq \eta \leq 1, -\infty < \xi < \infty$, $0 \leq \varphi \leq 2\pi$.

III. SPHEROIDAL WAVE FUNCTIONS

The scalar Helmholtz equation can be written in prolate spheroidal coordinates as

$$\left[\frac{\partial}{\partial\eta}(1 - \eta^2)\frac{\partial}{\partial\eta} + \frac{\partial}{\partial\xi}(\xi^2 - 1)\frac{\partial}{\partial\xi} + \frac{\xi^2 - \eta^2}{(\xi^2 - 1)(1 - \eta^2)}\frac{\partial^2}{\partial\varphi^2} + h^2(\xi^2 - \eta^2)\right]\psi$$
$$= 0,$$
(3)

and in oblate spheroidal coordinates as

$$\left[\frac{\partial}{\partial\eta}(1 - \eta^2)\frac{\partial}{\partial\eta} + \frac{\partial}{\partial\xi}(\xi^2 + 1)\frac{\partial}{\partial\xi} + \frac{\xi^2 + \eta^2}{(\xi^2 + 1)(1 - \eta^2)}\frac{\partial^2}{\partial\varphi^2} + h^2(\xi^2 + \eta^2)\right]\psi$$
$$= 0,$$
(4)

where ψ is the scalar spheroidal wave function and $h = kF$, with k being the wavenumber. It can be seen that Eq. (4) can be obtained from Eq. (3) using the transformation $\xi \to \pm j\xi$, $h \to \mp jh$. Thus this transformation can be used to pass from one system to the other.

By applying the method of separation of variables, elementary solutions to Eqs. (3) and (4) are obtained in the form

$$\psi_{\substack{e\\o}mn} = S_{mn}(c, \eta)R_{mn}(c, \upsilon)\frac{\cos m\varphi}{\sin m\varphi},$$
(5)

where $c = h$, $\upsilon = \xi$ correspond to Eq. (3) and $c = -jh$, $\upsilon = j\xi$ to Eq. (4), and e and o indicate the even and the odd functions, respectively. The functions

$S_{mn}(h, \eta)$ and $R_{mn}(h, \xi)$ satisfy the differential equations

$$\frac{d}{d\eta}\left[(1 - \eta^2)\frac{d}{d\eta}S_{mn}(h, \eta)\right] + \left[\lambda_{mn} - h^2\eta^2 - \frac{m^2}{1 - \eta^2}\right]S_{mn}(h, \eta) = 0 \quad (6)$$

and

$$\frac{d}{d\xi}\left[(\xi^2 - 1)\frac{d}{d\xi}R_{mn}(h, \xi)\right] - \left[\lambda_{mn} - h^2\xi^2 + \frac{m^2}{\xi^2 - 1}\right]R_{mn}(h, \xi) = 0, \quad (7)$$

where λ_{mn} and m are separation constants, with λ_{mn} being a function of h. The discrete values of λ_{mn} $(n = m, m + 1, m + 2, \ldots)$ for which the differential equation (6) gives solutions that are finite at $\eta = \pm 1$ are the desired eigenvalues, with m being an integer including 0 (Flammer, 1957). For a discussion on how to evaluate the eigenvalues the reader is referred to Flammer (1957) and Sinha *et al.* (1973). The equations satisfied by the corresponding oblate functions $S_{mn}(-jh, \eta)$ and $R_{mn}(-jh, j\xi)$ are obtained from Eqs. (6) and (7), respectively, by replacing h by $-jh$ and ξ by $j\xi$.

A. SPHEROIDAL ANGLE FUNCTIONS

The prolate spheroidal angle functions are the eigenfunctions $S_{mn}(h, \eta)$ corresponding to the eigenvalues λ_{mn} of Eq. (6). The solution to Eq. (6) yields two kinds of angle functions $S_{mn}^{(1)}(h, \eta)$ and $S_{mn}^{(2)}(h, \eta)$. It is only $S_{mn}^{(1)}(h, \eta)$ that is used in physical problems, because of its regularity throughout the interval $-1 \leq \eta \leq 1$. Thus we simplify the notation by writing $S_{mn}(h, \eta)$ to mean the angle functions of the first kind. This remark is also valid for the oblate spheroidal angle functions $S_{mn}(-jh, \eta)$.

The spheroidal angle functions can be expressed in the form of an infinite series of associated Legendre functions of the first kind as (Flammer, 1957)

$$S_{mn}(c, \eta) = \sum_{r=0,1}^{\infty}{}' d_r^{mn}(c)P_{m+r}^m(\eta), \quad (8)$$

in which c is either h or $-jh$, depending on whether we consider the prolate or the oblate cases, and the prime indicates that the summation is over even values of r when $n - m$ is even and over odd values of r when $n - m$ is odd. The evaluation of the spheroidal expansion coefficients $d_r^{mn}(c)$ is described in detail in the previous reference.

A power series expansion can also be used to evaluate the angle functions, which for prolate functions is given by (Flammer, 1957)

$$S_{mn}(h, \eta) = (1 - \eta^2)^{m/2} \sum_{k=0}^{\infty} c_{2k}^{mn}(h)(1 - \eta^2)^k, \qquad (n - m) \text{ even}, \qquad (9)$$

$$S_{mn}(h, \eta) = \eta(1 - \eta^2)^{m/2} \sum_{k=0}^{\infty} c_{2k}^{mn}(h)(1 - \eta^2)^k, \qquad (n - m) \text{ odd}, \qquad (10)$$

where

$$c_{2k}^{mn}(h) = \frac{1}{2^m k!(m + k)!} \sum_{r=k}^{\infty} \frac{(2m + 2r)!}{(2r)!}(-r)_k \left(m + r + \frac{1}{2}\right)_k d_{2r}^{mn}(h),$$

$$(n - m) \text{ even}, \qquad (11)$$

$$c_{2k}^{mn}(h) = \frac{1}{2^m k!(m + k)!} \sum_{r=k}^{\infty} \frac{(2m + 2r + 1)!}{(2r + 1)!}(-r)_k \left(m + r + \frac{3}{2}\right)_k d_{2r+1}^{mn}(h),$$

$$(n - m) \text{ odd}, \qquad (12)$$

in which

$$(\alpha)_k = \alpha(\alpha + 1)(\alpha + 2) \cdots (\alpha + k - 1), \qquad (\alpha)_0 = 1. \qquad (13)$$

A recursion formula for calculating $c_{2k}^{mn}(h)$ is also given in this reference. The power series expansion for the oblate case is obtained from Eqs. (9) and (10) by replacing h by $-jh$.

An important property of spheroidal angle functions is the orthogonality in the interval $-1 \leq \eta \leq 1$ that results from the Sturm–Liouville theory for the differential equations, which can be expressed as

$$\int_{-1}^{1} S_{mn}(c, \eta) S_{mn'}(c, \eta) \, d\eta = N_{mn}(c)\delta_{nn'}, \qquad (14)$$

where $\delta_{nn'}$ is the Kronecker delta and

$$N_{mn}(c) = 2 \sum_{r=0,1}^{\infty}{}' \frac{(2m + r)!\{d_r^{mn}(c)\}^2}{(2m + 2r + 1)r!} \qquad (15)$$

is the normalization constant, with $c = h$ or $c = -jh$ depending on whether we consider prolate or oblate functions.

B. Spheroidal Radial Functions

The prolate spheroidal radial functions are solutions of the differential equation (7), and the range of the coordinate ξ is $1 \leq \xi < \infty$.

In physical problems one usually requires the spheroidal radial functions of both the first kind $R_{mn}^{(1)}(h, \xi)$ and the second kind $R_{mn}^{(2)}(h, \xi)$, which are independent solutions of Eq. (7). The functions of the third and fourth kind, $R_{mn}^{(3)}(h, \xi)$ and $R_{mn}^{(4)}(h, \xi)$, are linear combinations of $R_{mn}^{(1)}(h, \xi)$ and $R_{mn}^{(2)}(h, \xi)$.

Like the spheroidal angle functions, the spheroidal radial functions $R_{mn}^{(1)}(h, \xi)$ and $R_{mn}^{(2)}(h, \xi)$ can also be expanded as an infinite series in the form (Morse and Feshbach, 1953; Flammer, 1957; Sinha, 1974)

$$R_{mn}^{(1)}(h, \xi) = \left(\frac{\xi^2 - 1}{\xi^2} \right)^{m/2} \sum_{r=0,1}^{\infty}{}' a_r(h|mn) j_{m+r}(h\xi), \tag{16}$$

$$R_{mn}^{(2)}(h, \xi) = \left(\frac{\xi^2 - 1}{\xi^2} \right)^{m/2} \sum_{r=0,1}^{\infty}{}' a_r(h|mn) y_{m+r}(h\xi), \tag{17}$$

where $j_{m+r}(h\xi)$ and $y_{m+r}(h\xi)$ are spherical Bessel and Neumann functions, respectively, of order $m + r$ and argument $h\xi$, and $a_r(h|mn)$ are the expansion coefficients, whose calculation is given in the three references mentioned previously. The series representation of $R_{mn}^{(1)}(h, \xi)$ has good convergence, whereas the one of $R_{mn}^{(2)}(h, \xi)$ is an asymptotic series that is very slowly convergent for small values of $h\xi$. In such cases, $R_{mn}^{(2)}(h, \xi)$ can be calculated using an integral method that is presented in Sinha (1974) and Sinha and MacPhie (1975).

Prolate spheroidal radial functions can also be expressed as a series of powers of $\xi^2 - 1$ (Flammer, 1957). Series of this type are particularly advantageous and provide an alternate method for the calculation of $R_{mn}^{(2)}(h, \xi)$ for small values of $h\xi$. The series expansions of $R_{mn}^{(1)}(h, \xi)$ are of the form

$$R_{mn}^{(1)}(h, \xi) = \left[\kappa_{mn}^{(1)}(h) \right]^{-1} (\xi^2 - 1)^{m/2} \sum_{k=0}^{\infty} (-1)^k c_{2k}^{mn}(h) (\xi^2 - 1)^k,$$

$$(n - m) \text{ even}, \tag{18}$$

$$R_{mn}^{(1)}(h, \xi) = \left[\kappa_{mn}^{(1)}(h) \right]^{-1} \xi (\xi^2 - 1)^{m/2} \sum_{k=0}^{\infty} (-1)^k c_{2k}^{mn}(h) (\xi^2 - 1)^k,$$

$$(n - m) \text{ odd}, \tag{19}$$

where

$$\kappa_{mn}^{(1)}(h) = \frac{(2m+1)(n+m)! \sum\limits_{r=0}^{\infty}{}' d_r^{mn}(h) \dfrac{(2m+r)!}{r!}}{2^{n+m} d_0^{mn}(h) h^m m! \left(\dfrac{n-m}{2}\right)! \left(\dfrac{n+m}{2}\right)!},$$

$(n-m)$ even, $\qquad\qquad$ (20)

$$\kappa_{mn}^{(1)}(h) = \frac{(2m+3)(n+m+1)! \sum\limits_{r=1}^{\infty}{}' d_r^{mn}(h) \dfrac{(2m+r)!}{r!}}{2^{n+m} d_1^{mn}(h) h^{m+1} m! \left(\dfrac{n-m-1}{2}\right)! \left(\dfrac{n+m+1}{2}\right)!},$$

$(n-m)$ odd, $\qquad\qquad$ (21)

and $c_{2k}^{mn}(h)$ are given in Eqs. (11) and (12). The series expansions for $R_{mn}^{(2)}(h, \xi)$ are of the form

$$R_{mn}^{(2)}(h, \xi) = \frac{1}{2} Q_{mn} R_{mn}^{(1)}(h, \xi) \log \frac{\xi+1}{\xi-1} + g_{mn}(h, \xi), \qquad (22)$$

where

$$Q_{mn} = \frac{[\kappa_{mn}^{(1)}(h)]^2}{h} \sum_{r=0}^{m} \alpha_r^{mn}(h) \frac{(-1)^{m-r+1}(2m-2r)!}{r![2^{m-r}(m-r)!]^2},$$

$(n-m)$ even, $\qquad\qquad$ (23)

$$Q_{mn} = \frac{[\kappa_{mn}^{(1)}(h)]^2}{h} \sum_{r=0}^{m} \alpha_r^{mn}(h) \frac{(-1)^{m-r+1}(2m-2r+1)!}{r![2^{m-r}(m-r)!]^2},$$

$(n-m)$ odd, $\qquad\qquad$ (24)

in which

$$\alpha_r^{mn}(h) = \left\{ \frac{d^r}{dx^r} \left[\sum_{k=0}^{\infty} (-1)^k c_{2k}^{mn}(h) x^k \right]^{-2} \right\}_{x=0}, \qquad (25)$$

and

$$g_{mn}(h, \xi) = \xi (\xi^2 - 1)^{-m/2} \sum_{r=0}^{\infty} b_r^{mn} (\xi^2 - 1)^r, \qquad (n-m) \text{ even,} \quad (26)$$

$$g_{mn}(h, \xi) = (\xi^2 - 1)^{-m/2} \sum_{r=0}^{\infty} b_r^{mn} (\xi^2 - 1)^r, \qquad (n-m) \text{ odd.} \quad (27)$$

Recurrence formulas for the evaluation of the coefficients b_r^{mn} and the explicit forms of $\alpha_r^{mn}(h)$ for $r = 0, 1, 2, 3, 4$ are also given in Flammer (1957). The accuracy of the calculated prolate spheroidal radial functions can be checked by using the Wronskian relation

$$R_{mn}^{(1)}(h, \xi) \frac{d}{d\xi} R_{mn}^{(2)}(h, \xi) - R_{mn}^{(2)}(h, \xi) \frac{d}{d\xi} R_{mn}^{(1)}(h, \xi) = \frac{1}{h(\xi^2 - 1)}. \tag{28}$$

The oblate spheroidal wave functions can be accurately calculated using a power series expansion in terms of $\xi^2 + 1$ (Flammer, 1957). For radial functions of the first kind, this expansion is given by

$$R_{mn}^{(1)}(-jh, j\xi) = \left[j^{-m}\kappa_{mn}^{(1)}(-jh)\right]^{-1}(\xi^2 + 1)^{m/2} \sum_{k=0}^{\infty} c_{2k}^{mn}(-jh)(\xi^2 + 1)^k,$$

$$(n - m) \text{ even}, \tag{29}$$

$$R_{mn}^{(1)}(-jh, j\xi) = \left[j^{-m-1}\kappa_{mn}^{(1)}(-jh)\right]^{-1}\xi(\xi^2 + 1)^{m/2} \sum_{k=0}^{\infty} c_{2k}^{mn}(-jh)(\xi^2 + 1)^k,$$

$$(n - m) \text{ odd}, \tag{30}$$

where $\kappa_{mn}^{(1)}(-jh)$ and $c_{2k}^{mn}(-jh)$ are obtained from Eqs. (20), (21) and Eqs. (11), (12), respectively, by changing h in these equations to $-jh$. The power series expansion for radial functions of the second kind is given by

$$R_{mn}^{(2)}(-jh, j\xi) = \bar{Q}_{mn} R_{mn}^{(1)}(-jh, j\xi)\left[\tan^{-1}\xi - \frac{\pi}{2}\right] + g_{mn}(-jh, j\xi), \tag{31}$$

where

$$\bar{Q}_{mn} = \frac{[j^{-m}\kappa_{mn}^{(1)}(-jh)]^2}{h} \sum_{r=0}^{m} \bar{\alpha}_r^{mn}(-jh)\frac{(2m - 2r)!}{r![2^{m-r}(m - r)!]^2},$$

$$(n - m) \text{ even}, \tag{32}$$

$$\bar{Q}_{mn} = -\frac{[j^{-m-1}\kappa_{mn}^{(1)}(-jh)]^2}{h} \sum_{r=0}^{m} \bar{\alpha}_r^{mn}(-jh)\frac{(2m - 2r + 1)!}{r![2^{m-r}(m - r)!]^2},$$

$$(n - m) \text{ odd}, \tag{33}$$

in which

$$\bar{\alpha}_r^{mn}(-jh) = \left\{\frac{d^r}{dx^r}\left[\sum_{k=0}^{\infty} c_{2k}^{mn}(-jh)x^k\right]^{-2}\right\}_{x=0}, \tag{34}$$

and

$$g_{mn}(-jh, j\xi) = \left(\xi^2 + 1\right)^{-m/2} \sum_{r=0}^{\infty} B_{2r}^{mn} \xi^{2r+1}, \qquad (n - m) \text{ even}, \quad (35)$$

$$g_{mn}(-jh, j\xi) = \left(\xi^2 + 1\right)^{-m/2} \sum_{r=0}^{\infty} B_{2r}^{mn} \xi^{2r}, \qquad (n - m) \text{ odd}. \quad (36)$$

Recursion relations for calculating the coefficients B_{2r}^{mn} are given in Flammer (1957). The Wronskian relation in this case is

$$R_{mn}^{(1)}(-jh, j\xi) \frac{d}{d\xi} R_{mn}^{(2)}(-jh, j\xi) - R_{mn}^{(2)}(-jh, j\xi) \frac{d}{d\xi} R_{mn}^{(1)}(-jh, j\xi)$$

$$= \frac{1}{h(\xi^2 + 1)}. \qquad (37)$$

IV. SPHEROIDAL VECTOR WAVE FUNCTIONS

By the application of vector differential operators to the scalar spheroidal wave function given in Eq. (5), the vector spheroidal wave functions **M** and **N** are defined as (Flammer, 1957)

$$\mathbf{M}_{mn} = \nabla \psi_{mn} \times \mathbf{a}, \qquad (38)$$

$$\mathbf{N}_{mn} = k^{-1}(\nabla \times \mathbf{M}_{mn}), \qquad (39)$$

where **a** is either an arbitrary constant unit vector or the position vector **r**. None of the coordinate unit vectors $\boldsymbol{\eta}$, $\boldsymbol{\xi}$, or $\boldsymbol{\varphi}$ in the spheroidal coordinate systems has the properties required for **a**. Instead, the Cartesian unit vectors can efficiently be used, because the transformation from the Cartesian system to the spheroidal systems is relatively simpler.

The three Cartesian unit vectors **x**, **y**, **z** and the radial vector **r** generate the following prolate spheroidal vector wave functions $\mathbf{M}(h; \eta, \xi, \varphi)$ and $\mathbf{N}(h; \eta, \xi, \varphi)$:

$$\mathbf{M}_{\substack{e \\ o}mn}^{p(i)}(h; \eta, \xi, \varphi) = \nabla \psi_{\substack{e \\ o}mn}^{(i)}(h; \eta, \xi, \varphi) \times \mathbf{p},$$

$$\mathbf{p} = \mathbf{x}, \mathbf{y}, \mathbf{z}, \qquad p = x, y, z, \qquad (40)$$

$$\mathbf{M}_{\substack{e \\ o}mn}^{r(i)}(h; \eta, \xi, \varphi) = \nabla \psi_{\substack{e \\ o}mn}^{(i)}(h; \eta, \xi, \varphi) \times \mathbf{r}, \qquad (41)$$

$$\mathbf{N}_{\substack{e \\ o}mn}^{p(i)}(h; \eta, \xi, \varphi) = k^{-1}\left[\nabla \times \mathbf{M}_{\substack{e \\ o}mn}^{p(i)}(h; \eta, \xi, \varphi)\right], \qquad (42)$$

$$\mathbf{N}_{\substack{e \\ o}mn}^{r(i)}(h; \eta, \xi, \varphi) = k^{-1}\left[\nabla \times \mathbf{M}_{\substack{e \\ o}mn}^{r(i)}(h; \eta, \xi, \varphi)\right], \qquad (43)$$

in which e and o refer to the even and odd functions, respectively, and i indicates the kind of function, $i = 1, 2, 3, 4$. The oblate spheroidal vector wave functions

$\mathbf{M}(-jh; \eta, j\xi, \varphi)$ and $\mathbf{N}(-jh; \eta, j\xi, \varphi)$ can be obtained from the corresponding prolate spheroidal ones by changing h to $-jh$ and ξ to $j\xi$. Explicit expressions for these vector spheroidal wave functions are available in Flammer (1957). In the functions $\mathbf{M}_{\substack{e\\o}mn}^{x(i)}$, $\mathbf{M}_{\substack{e\\o}mn}^{y(i)}$, $\mathbf{N}_{\substack{e\\o}mn}^{x(i)}$, and $\mathbf{N}_{\substack{e\\o}mn}^{y(i)}$, the φ dependence of various components is simply given by the product of $\cos \varphi$ or $\sin \varphi$ with either $\cos m\varphi$ or $\sin m\varphi$. It is convenient therefore to define the following additional vector wave functions:

$$\mathbf{M}_{\substack{e\\o}m+1,n}^{+(i)}(h; \eta, \xi, \varphi) = \frac{1}{2}\left[\mathbf{M}_{\substack{e\\o}mn}^{x(i)}(h; \eta, \xi, \varphi) \mp \mathbf{M}_{\substack{o\\e}mn}^{y(i)}(h; \eta, \xi, \varphi)\right], \quad (44)$$

$$\mathbf{M}_{\substack{e\\o}m-1,n}^{-(i)}(h; \eta, \xi, \varphi) = \frac{1}{2}\left[\mathbf{M}_{\substack{e\\o}mn}^{x(i)}(h; \eta, \xi, \varphi) \pm \mathbf{M}_{\substack{o\\e}mn}^{y(i)}(h; \eta, \xi, \varphi)\right], \quad (45)$$

$$\mathbf{N}_{\substack{e\\o}m+1,n}^{+(i)}(h; \eta, \xi, \varphi) = \frac{1}{2}\left[\mathbf{N}_{\substack{e\\o}mn}^{x(i)}(h; \eta, \xi, \varphi) \mp \mathbf{N}_{\substack{o\\e}mn}^{y(i)}(h; \eta, \xi, \varphi)\right], \quad (46)$$

$$\mathbf{N}_{\substack{e\\o}m-1,n}^{-(i)}(h; \eta, \xi, \varphi) = \frac{1}{2}\left[\mathbf{N}_{\substack{e\\o}mn}^{x(i)}(h; \eta, \xi, \varphi) \pm \mathbf{N}_{\substack{o\\e}mn}^{y(i)}(h; \eta, \xi, \varphi)\right], \quad (47)$$

where the components with an index $m+1$ have a φ dependence of either $\cos(m+1)\varphi$ or $\sin(m+1)\varphi$, whereas those with an index $m-1$ have a φ dependence of either $\cos(m-1)\varphi$ or $\sin(m-1)\varphi$. The $-$ and $+$ signs in Eqs. (44) and (46), and the $+$ and $-$ signs in Eqs. (45) and (47), on the right-hand sides, are associated with the even and odd vector wave functions, respectively. Explicit expressions for $\mathbf{M}_{\substack{e\\o}m+1,n}^{+(i)}$, $\mathbf{M}_{\substack{e\\o}m-1,n}^{-(i)}$, $\mathbf{N}_{\substack{e\\o}m+1,n}^{+(i)}$, and $\mathbf{N}_{\substack{e\\o}m-1,n}^{-(i)}$ are also given in Flammer (1957). In Sinha and MacPhie (1983), these latter even and odd functions were combined correspondingly to have an exponential variation of φ and the modified functions have been denoted, respectively, by $\mathbf{M}_{mn}^{+(i)}$, $\mathbf{M}_{mn}^{-(i)}$, $\mathbf{N}_{mn}^{+(i)}$, and $\mathbf{N}_{mn}^{-(i)}$, so that the φ dependence in $\mathbf{M}_{mn}^{\pm(i)}$ and $\mathbf{N}_{mn}^{\pm(i)}$ is $\exp[j(m \pm 1)]\varphi$. The even and odd prolate spheroidal vector wave functions $\mathbf{M}_{\substack{e\\o}mn}^{z(i)}$ and $\mathbf{N}_{\substack{e\\o}mn}^{z(i)}$ can also be combined to form two new functions $\mathbf{M}_{mn}^{z(i)}$ and $\mathbf{N}_{mn}^{z(i)}$ given by

$$\mathbf{M}_{mn}^{z(i)} = \mathbf{M}_{emn}^{z(i)} + j\mathbf{M}_{omn}^{z(i)}, \qquad \mathbf{N}_{mn}^{z(i)} = \mathbf{N}_{emn}^{z(i)} + j\mathbf{N}_{omn}^{z(i)}, \quad (48)$$

with an $\exp(jm\varphi)$ φ dependence. Similarly, the even and odd prolate spheroidal vector wave functions $\mathbf{M}_{\substack{e\\o}mn}^{r(i)}$ and $\mathbf{N}_{\substack{e\\o}mn}^{r(i)}$ can be combined to form the functions

$$\mathbf{M}_{mn}^{r(i)} = \mathbf{M}_{emn}^{r(i)} + j\mathbf{M}_{omn}^{r(i)}, \qquad \mathbf{N}_{mn}^{r(i)} = \mathbf{N}_{emn}^{r(i)} + j\mathbf{N}_{omn}^{r(i)}, \quad (49)$$

which also have an $\exp(jm\varphi)$ φ dependence. The corresponding oblate spheroidal wave functions can be obtained from the prolate spheroidal ones by using the transformation $\xi \to j\xi$ and $h \to -jh$. In the following sections of this chapter we use the vector spheroidal wave functions $\mathbf{M}_{mn}^{\pm(i)}$, $\mathbf{M}_{mn}^{z(i)}$, $\mathbf{M}_{mn}^{r(i)}$, $\mathbf{N}_{mn}^{\pm(i)}$, $\mathbf{N}_{mn}^{z(i)}$, and $\mathbf{N}_{mn}^{r(i)}$ to obtain exact solutions to the problems of scattering of electromagnetic waves by a single spheroid or by an ensemble of spheroids in arbitrary orientation.

V. ELECTROMAGNETIC SCATTERING BY A COATED LOSSY SPHEROID

To illustrate the application of the method of separation of variables, we derive in this section an exact solution to the problem of scattering of electromagnetic waves by a homogeneous dielectric spheroid coated with a confocal lossy dielectric material of arbitrary thickness. The electric and magnetic fields inside the spheroid, within the coating, and outside the coating are expressed in terms of a set of vector spheroidal wave functions, and the solution is obtained by imposing the appropriate boundary conditions at each spheroidal surface.

A. FIELD EXPRESSIONS

Consider a linearly polarized monochromatic plane electromagnetic wave with an electric field of unit amplitude, incident on a dielectric spheroid with a confocal lossy dielectric coating of arbitrary thickness, as shown in Fig. 1. The materials of the spheroid and the coating layer are assumed to be linear, homogeneous, isotropic, and lossy, with permittivities ε_2 and ε_1 and permeabilities μ_2 and μ_1, respectively. The medium outside the coating layer is assumed to be linear, homogeneous, and isotropic, with permittivity ε and permeability μ. The semiaxial lengths of the spheroidal core are denoted by a_2 and b_2, and those of the outer surface of the confocal layer by a_1 and b_1. The inner and outer spheroidal surfaces

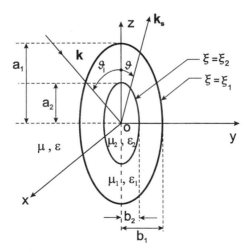

Figure 1   Geometry of the scattering system.

are defined by $\xi = \xi_2$ and $\xi = \xi_1$, respectively (see Fig. 1), with ξ being the radial coordinate of a spheroidal coordinate system whose origin is at the center O of the spheroid. The major axes of the spheroidal surfaces are along the z axis of the Cartesian system $Oxyz$.

Without any loss of generality, the incident plane can be considered to be the xz plane ($\varphi_i = 0$). The incident propagation vector $\mathbf{k}$ makes an angle ϑ_i with the z axis. A linearly polarized incident wave can in general be resolved into transverse electric (TE) and transverse magnetic (TM) components. These components are usually defined in terms of the polarization angle ζ, which is the angle between the direction of the incident electric field vector and the direction of the normal to the plane of incidence. For TE polarization $\zeta = 0$ and for TM polarization $\zeta = \pi/2$.

The incident electric field $\mathbf{E}_i$ and magnetic field $\mathbf{H}_i$ can be expressed in terms of the vector spheroidal wave functions $\mathbf{M}$ and $\mathbf{N}$ defined in Section IV as (Sinha and MacPhie, 1983; Cooray and Ciric, 1989b)

$$\mathbf{E}_i = \sum_{m=-\infty}^{\infty} \sum_{n=|m|}^{\infty} \left(p_{mn}^{+} \mathbf{M}_{mn}^{+(1)} + p_{mn}^{-} \mathbf{M}_{mn}^{-(1)} \right), \tag{50}$$

$$\mathbf{H}_i = j(\varepsilon/\mu)^{1/2} \sum_{m=-\infty}^{\infty} \sum_{n=|m|}^{\infty} \left(p_{mn}^{+} \mathbf{N}_{mn}^{+(1)} + p_{mn}^{-} \mathbf{N}_{mn}^{-(1)} \right), \tag{51}$$

where

$$p_{mn}^{\pm} = \frac{2j^{n-1}}{kN_{mn}(h)} S_{mn}(h, \cos\vartheta_i) \left(\frac{\cos\zeta}{\cos\vartheta_i} \mp j\sin\zeta \right), \tag{52}$$

in which $N_{mn}(h)$ is the normalization constant of the spheroidal angle function $S_{mn}(h, \cos\vartheta_i)$ defined in Section III.A and $h = kF$, with k being the wavenumber in the medium outside the coating layer and F the semi-interfocal distance of the spheroid. Equation (50) can now be rewritten in matrix form

$$\mathbf{E}_i = \bar{\mathbf{M}}_i^{(1)\mathrm{T}} \bar{I}, \tag{53}$$

with the overbar denoting a column matrix and T denoting the transpose of a matrix, where

$$\bar{\mathbf{M}}_i^{(1)\mathrm{T}} = \begin{bmatrix} \bar{\mathbf{M}}_{i0}^{\mathrm{T}} & \bar{\mathbf{M}}_{i1}^{\mathrm{T}} & \bar{\mathbf{M}}_{i2}^{\mathrm{T}} & \cdots \end{bmatrix}, \qquad \bar{I}^{\mathrm{T}} = \begin{bmatrix} \bar{p}_0^{\mathrm{T}} & \bar{p}_1^{\mathrm{T}} & \bar{p}_2^{\mathrm{T}} & \cdots \end{bmatrix} \tag{54}$$

and

$$\bar{\mathbf{M}}_{i0}^{\mathrm{T}} = \begin{bmatrix} \bar{\mathbf{M}}_{-1}^{+(1)\mathrm{T}} & \bar{\mathbf{M}}_{1}^{-(1)\mathrm{T}} \end{bmatrix},$$

$$\bar{\mathbf{M}}_{i\sigma}^{\mathrm{T}} = \begin{bmatrix} \bar{\mathbf{M}}_{\sigma-1}^{+(1)\mathrm{T}} & \bar{\mathbf{M}}_{\sigma+1}^{-(1)\mathrm{T}} & \bar{\mathbf{M}}_{-(\sigma+1)}^{+(1)\mathrm{T}} & \bar{\mathbf{M}}_{-(\sigma-1)}^{-(1)\mathrm{T}} \end{bmatrix} \qquad \text{for } \sigma \geqslant 1, \tag{55}$$

with

$$\bar{\mathbf{M}}_{\tau}^{\pm(1)\mathrm{T}}$$
$$= \left[\mathbf{M}_{\tau,|\tau|}^{\pm(1)}(h;\eta,\xi,\varphi) \quad \mathbf{M}_{\tau,|\tau|+1}^{\pm(1)}(h;\eta,\xi,\varphi) \quad \mathbf{M}_{\tau,|\tau|+2}^{\pm(1)}(h;\eta,\xi,\varphi) \quad \cdots \right],$$

(56)

and

$$\bar{p}_0^{\mathrm{T}} = \left[\bar{p}_{-1}^{+\mathrm{T}} \quad \bar{p}_1^{-\mathrm{T}} \right],$$

$$\bar{p}_\sigma^{\mathrm{T}} = \left[\bar{p}_{\sigma-1}^{+\mathrm{T}} \quad \bar{p}_{\sigma+1}^{-\mathrm{T}} \quad \bar{p}_{-(\sigma+1)}^{+\mathrm{T}} \quad \bar{p}_{-(\sigma-1)}^{-\mathrm{T}} \right] \qquad \text{for } \sigma \geqslant 1,$$

(57)

with

$$\bar{p}_\tau^{\pm\mathrm{T}} = \left[p_{\tau,|\tau|}^{\pm} \quad p_{\tau,|\tau|+1}^{\pm} \quad p_{\tau,|\tau|+2}^{\pm} \quad \cdots \right].$$

(58)

Equation (51) can similarly be written in matrix form (Cooray *et al.*, 1990)

$$\mathbf{H}_{\mathrm{i}} = j \left(\frac{\varepsilon}{\mu} \right)^{1/2} \bar{\mathbf{N}}_{\mathrm{i}}^{(1)\mathrm{T}} \bar{I},$$

(59)

with the elements of $\bar{\mathbf{N}}_{\mathrm{i}}^{(1)\mathrm{T}}$ obtained from the corresponding elements of $\bar{\mathbf{M}}_{\mathrm{i}}^{(1)\mathrm{T}}$ by replacing the vector wave functions **M** by **N**.

The scattered field $\mathbf{E}_{\mathrm{s}}$ for $\xi > \xi_1$ can be expanded in terms of vector spheroidal wave functions of the fourth kind as (Cooray and Ciric, 1989b)

$$\mathbf{E}_{\mathrm{s}} = \sum_{m=0}^{\infty} \sum_{n=m}^{\infty} \left(\tilde{\alpha}_{mn}^{+} \mathbf{M}_{mn}^{+(4)} + \tilde{\alpha}_{m+1,n+1}^{z} \mathbf{M}_{m+1,n+1}^{z(4)} \right)$$
$$+ \sum_{n=0}^{\infty} \left(\tilde{\alpha}_{-1,n+1}^{+} \mathbf{M}_{-1,n+1}^{+(4)} + \tilde{\alpha}_{0n}^{z} \mathbf{M}_{0n}^{z(4)} \right)$$
$$+ \sum_{m=0}^{\infty} \sum_{n=m}^{\infty} \left(\tilde{\alpha}_{-mn}^{-} \mathbf{M}_{-mn}^{-(4)} + \tilde{\alpha}_{-(m+1),n+1}^{z} \mathbf{M}_{-(m+1),n+1}^{z(4)} \right),$$

(60)

where $\tilde{\alpha}_{uv}$ are the unknown expansion coefficients. This equation can now be written in matrix form

$$\mathbf{E}_{\mathrm{s}} = \bar{\mathbf{M}}_{\mathrm{s}}^{(4)\mathrm{T}} \tilde{\bar{\alpha}},$$

(61)

where

$$\bar{\mathbf{M}}_{\mathrm{s}}^{(4)\mathrm{T}} = \left[\bar{\mathbf{M}}_{\mathrm{s}0}^{\mathrm{T}} \quad \bar{\mathbf{M}}_{\mathrm{s}1}^{\mathrm{T}} \quad \bar{\mathbf{M}}_{\mathrm{s}2}^{\mathrm{T}} \quad \cdots \right],$$

$$\tilde{\bar{\alpha}}^{\mathrm{T}} = \left[\tilde{\bar{\alpha}}_0^{\mathrm{T}} \quad \tilde{\bar{\alpha}}_1^{\mathrm{T}} \quad \tilde{\bar{\alpha}}_2^{\mathrm{T}} \quad \cdots \right],$$

(62)

in which

$$\bar{\mathbf{M}}_{s0}^{\mathrm{T}} = \begin{bmatrix} \bar{\mathbf{M}}_{-1}^{+(4)\mathrm{T}} & \bar{\mathbf{M}}_0^{z(4)\mathrm{T}} \end{bmatrix},$$

$$\bar{\mathbf{M}}_{s\sigma}^{\mathrm{T}} = \begin{bmatrix} \bar{\mathbf{M}}_{\sigma-1}^{+(4)\mathrm{T}} & \bar{\mathbf{M}}_\sigma^{z(4)\mathrm{T}} & \bar{\mathbf{M}}_{-(\sigma-1)}^{-(4)\mathrm{T}} & \bar{\mathbf{M}}_{-\sigma}^{z(4)\mathrm{T}} \end{bmatrix} \qquad \text{for } \sigma \geqslant 1,$$

(63)

with

$$\bar{\mathbf{M}}_\tau^{\pm(4)\mathrm{T}} = \begin{bmatrix} \mathbf{M}_{\tau,|\tau|}^{\pm(4)}(h; \eta, \xi, \varphi) & \mathbf{M}_{\tau,|\tau|+1}^{\pm(4)}(h; \eta, \xi, \varphi) & \mathbf{M}_{\tau,|\tau|+2}^{\pm(4)}(h; \eta, \xi, \varphi) & \cdots \end{bmatrix},$$

(64)

$$\bar{\mathbf{M}}_\tau^{z(4)\mathrm{T}} = \begin{bmatrix} \mathbf{M}_{\tau,|\tau|}^{z(4)}(h; \eta, \xi, \varphi) & \mathbf{M}_{\tau,|\tau|+1}^{z(4)}(h; \eta, \xi, \varphi) & \mathbf{M}_{\tau,|\tau|+2}^{z(4)}(h; \eta, \xi, \varphi) & \cdots \end{bmatrix},$$

(65)

and

$$\tilde{\bar{\alpha}}_0^{\mathrm{T}} = \begin{bmatrix} \tilde{\bar{\alpha}}_{-1}^{+\mathrm{T}} & \tilde{\bar{\alpha}}_0^{z\mathrm{T}} \end{bmatrix},$$

$$\tilde{\bar{\alpha}}_\sigma^{\mathrm{T}} = \begin{bmatrix} \tilde{\bar{\alpha}}_{\sigma-1}^{+\mathrm{T}} & \tilde{\bar{\alpha}}_\sigma^{z\mathrm{T}} & \tilde{\bar{\alpha}}_{-(\sigma-1)}^{-\mathrm{T}} & \tilde{\bar{\alpha}}_{-\sigma}^{z\mathrm{T}} \end{bmatrix} \qquad \text{for } \sigma \geqslant 1,$$

(66)

with

$$\tilde{\bar{\alpha}}_\tau^{\pm\mathrm{T}} = \begin{bmatrix} \tilde{\alpha}_{\tau,|\tau|}^\pm & \tilde{\alpha}_{\tau,|\tau|+1}^\pm & \tilde{\alpha}_{\tau,|\tau|+2}^\pm & \cdots \end{bmatrix},$$

$$\tilde{\bar{\alpha}}_\tau^{z\mathrm{T}} = \begin{bmatrix} \tilde{\alpha}_{\tau,|\tau|}^z & \tilde{\alpha}_{\tau,|\tau|+1}^z & \tilde{\alpha}_{\tau,|\tau|+2}^z & \cdots \end{bmatrix}.$$

(67)

Using the Maxwell equation $\nabla \times \mathbf{E} = -j\omega\mu\mathbf{H}$, the corresponding expansion for the scattered magnetic field for $\xi > \xi_1$ can be derived from Eq. (61) as

$$\mathbf{H}_s = j\left(\frac{\varepsilon}{\mu}\right)^{1/2} \bar{\mathbf{N}}_s^{(4)\mathrm{T}}\tilde{\bar{\alpha}},$$

(68)

with the elements of $\bar{\mathbf{N}}_s^{(4)\mathrm{T}}$ obtained from those of $\bar{\mathbf{M}}_s^{(4)\mathrm{T}}$ by replacing the vector wave functions $\mathbf{M}$ by $\mathbf{N}$.

The electric field $^{(1)}\mathbf{E}_t$ transmitted in the region $\xi_2 < \xi < \xi_1$ contains both the first and the second kinds of vector spheroidal wave functions, whereas the field $^{(2)}\mathbf{E}_t$ transmitted in the region $\xi < \xi_2$ only contains the vector wave functions of the first kind. Thus, their expansions in terms of vector spheroidal wave functions can be written in matrix form

$$^{(1)}\mathbf{E}_t = {^{(1)}\bar{\mathbf{M}}_t^{(1)\mathrm{T}}}\tilde{\bar{\beta}} + {^{(1)}\bar{\mathbf{M}}_t^{(2)\mathrm{T}}}\tilde{\bar{\gamma}},$$

(69)

$$^{(2)}\mathbf{E}_t = {^{(2)}\bar{\mathbf{M}}_t^{(1)\mathrm{T}}}\tilde{\bar{\delta}},$$

(70)

where the elements of the matrices $^{(1)}\bar{\mathbf{M}}_t^{(1)\mathrm{T}}$ and $^{(1)}\bar{\mathbf{M}}_t^{(2)\mathrm{T}}$ are obtained from the corresponding elements of the matrix $\bar{\mathbf{M}}_s^{(4)\mathrm{T}}$ by replacing the spheroidal vector wave functions of the fourth kind by those of the first kind and the second kind, respectively, and h by $h_1 = k_1 F$, with $k_1 = (\sqrt{\mu_1/\mu}\sqrt{\varepsilon_1/\varepsilon})k$. The elements of the matrix ${}^{(2)}\bar{\mathbf{M}}_{\mathrm{t}}^{(1)\mathrm{T}}$ are then obtained from the corresponding elements of the matrix ${}^{(1)}\bar{\mathbf{M}}_{\mathrm{t}}^{(1)\mathrm{T}}$ by replacing h_1 by $h_2 = k_2 F$, where $k_2 = (\sqrt{\mu_2/\mu}\sqrt{\varepsilon_2/\varepsilon})k$, and those of the unknown coefficient matrices $\bar{\bar{\beta}}$, $\bar{\bar{\gamma}}$, and $\bar{\bar{\delta}}$ are obtained from the corresponding elements of $\bar{\bar{\alpha}}$ by replacing $\tilde{\alpha}$ by $\tilde{\beta}$, $\tilde{\gamma}$, and $\tilde{\delta}$, respectively.

Referring to the derivation of $\mathbf{H}_s$ in Eq. (68) from $\mathbf{E}_s$ in Eq. (61), the expansions of the corresponding magnetic fields can be written as

$$
{}^{(1)}\mathbf{H}_{\mathrm{t}} = j\left(\frac{\varepsilon_1}{\mu_1}\right)^{1/2}\left[{}^{(1)}\bar{\mathbf{N}}_{\mathrm{t}}^{(1)\mathrm{T}}\bar{\bar{\beta}} + {}^{(1)}\bar{\mathbf{N}}_{\mathrm{t}}^{(2)\mathrm{T}}\bar{\bar{\gamma}}\right], \tag{71}
$$

$$
{}^{(2)}\mathbf{H}_{\mathrm{t}} = j\left(\frac{\varepsilon_2}{\mu_2}\right)^{1/2}{}^{(2)}\bar{\mathbf{N}}_{\mathrm{t}}^{(1)\mathrm{T}}\bar{\bar{\delta}}, \tag{72}
$$

where the matrices ${}^{(1)}\bar{\mathbf{N}}_{\mathrm{t}}^{(1)\mathrm{T}}$, ${}^{(1)}\bar{\mathbf{N}}_{\mathrm{t}}^{(2)\mathrm{T}}$, and ${}^{(2)}\bar{\mathbf{N}}_{\mathrm{t}}^{(1)\mathrm{T}}$ are obtained from ${}^{(1)}\bar{\mathbf{M}}_{\mathrm{t}}^{(1)\mathrm{T}}$, ${}^{(1)}\bar{\mathbf{M}}_{\mathrm{t}}^{(2)\mathrm{T}}$, and ${}^{(2)}\bar{\mathbf{M}}_{\mathrm{t}}^{(1)\mathrm{T}}$, respectively, by replacing the vector wave functions $\mathbf{M}$ by $\mathbf{N}$.

B. BOUNDARY VALUE PROBLEM SOLUTION

The boundary conditions require that the tangential components of the total electric and magnetic fields across each of the spheroidal surfaces $\xi = \xi_1$ and $\xi = \xi_2$ be continuous, that is,

$$
(\mathbf{E}_i + \mathbf{E}_s) \times \boldsymbol{\xi}|_{\xi=\xi_1} = {}^{(1)}\mathbf{E}_{\mathrm{t}} \times \boldsymbol{\xi}|_{\xi=\xi_1}, \qquad (\mathbf{H}_i + \mathbf{H}_s) \times \boldsymbol{\xi}|_{\xi=\xi_1} = {}^{(1)}\mathbf{H}_{\mathrm{t}} \times \boldsymbol{\xi}|_{\xi=\xi_1} \tag{73}
$$

and

$$
{}^{(1)}\mathbf{E}_{\mathrm{t}} \times \boldsymbol{\xi}|_{\xi=\xi_2} = {}^{(2)}\mathbf{E}_{\mathrm{t}} \times \boldsymbol{\xi}|_{\xi=\xi_2}, \qquad {}^{(1)}\mathbf{H}_{\mathrm{t}} \times \boldsymbol{\xi}|_{\xi=\xi_2} = {}^{(2)}\mathbf{H}_{\mathrm{t}} \times \boldsymbol{\xi}|_{\xi=\xi_2}, \tag{74}
$$

with $\boldsymbol{\xi}$ being the unit vector normal to the respective spheroidal surface. After substituting the different electric and magnetic fields in Eqs. (73) and (74) with the respective expressions in Eqs. (53), (59), (61), and (68)–(72), applying the orthogonality properties of the trigonometric functions and the spheroidal angle functions, and integrating correspondingly over each spheroidal surface, a matrix equation is finally obtained in the form

$$
[G_d]\bar{S} = [R_d]\bar{I}, \tag{75}
$$

where $\bar{S} = [\bar{\beta}^{\mathrm{T}} \;\; \bar{\gamma}^{\mathrm{T}} \;\; \bar{\alpha}^{\mathrm{T}} \;\; \bar{\delta}^{\mathrm{T}}]^{\mathrm{T}}$ is the column matrix of the unknown coefficients. The structure and elements of the matrices $[G_d]$ and $[R_d]$ are similar to those in Cooray and Ciric (1991b). The details regarding the calculation of the integrals that appear as a result of applying the orthogonality of the spheroidal wave functions are given in Sinha and MacPhie (1977) and Cooray and Ciric (1991b). The solution of Eq. (75) can now be written in the form

$$\bar{S} = [G]\bar{I}, \tag{76}$$

where $[G] = [G_d]^{-1}[R_d]$ is the system matrix, whose elements are independent of the direction and polarization of the incident wave. Once the expansion coefficients $\tilde{\alpha}$, $\tilde{\beta}$, $\tilde{\gamma}$, and $\tilde{\delta}$ are known, it is possible to calculate the fields inside the spheroid, inside the coating layer, and outside the coating layer, by substituting back in the appropriate series expansions of the fields.

C. SCATTERED FAR FIELD

The scattered electric field in the far zone can be written by using the asymptotic forms of the vector spheroidal wave functions **M** as

$$\mathbf{E}_s(r, \vartheta, \varphi) = \frac{e^{-jkr}}{kr}\left[F_\vartheta(\vartheta, \varphi)\boldsymbol{\vartheta} + F_\varphi(\vartheta, \varphi)\boldsymbol{\varphi}\right], \tag{77}$$

where

$$
\begin{aligned}
F_\vartheta(\vartheta, \varphi) = {}& -k \sum_{m=0}^{\infty} \sum_{n=m}^{\infty} j^{n+1} \frac{S_{mn}(h, \cos\vartheta)}{2} \\
& \times \left\{ \left(\tilde{\alpha}_{mn}^+ - \tilde{\alpha}_{-mn}^-\right)\cos(m+1)\varphi + j\left(\tilde{\alpha}_{mn}^+ + \tilde{\alpha}_{-mn}^-\right)\sin(m+1)\varphi \right\} \\
& - k \sum_{n=1}^{\infty} j^{n+1} \frac{S_{1n}(h, \cos\vartheta)}{2}\tilde{\alpha}_{-1n}^+,
\end{aligned} \tag{78}
$$

$$
\begin{aligned}
F_\varphi(\vartheta, \varphi) = {}& k \sum_{m=0}^{\infty} \sum_{n=m}^{\infty} j^n \left[\cos\vartheta \frac{S_{mn}(h, \cos\vartheta)}{2} \right. \\
& \times \left\{ \left(\tilde{\alpha}_{mn}^+ + \tilde{\alpha}_{-mn}^-\right)\cos(m+1)\varphi + j\left(\tilde{\alpha}_{mn}^+ - \tilde{\alpha}_{-mn}^-\right)\sin(m+1)\varphi \right\} \\
& - j\sin\vartheta\, S_{m+1,n+1}(h, \cos\vartheta)\left\{ \left(\tilde{\alpha}_{m+1,n+1}^z + \tilde{\alpha}_{-(m+1),n+1}^z\right) \right. \\
& \left. \times \cos(m+1)\varphi + j\left(\tilde{\alpha}_{m+1,n+1}^z - \tilde{\alpha}_{-(m+1),n+1}^z\right)\sin(m+1)\varphi \right\} \Bigg] \\
& + k\cos\vartheta \sum_{n=1}^{\infty} j^n \frac{S_{1n}(h, \cos\vartheta)}{2}\tilde{\alpha}_{-1n}^+ \\
& - k\sin\vartheta \sum_{n=0}^{\infty} j^n S_{0n}(h, \cos\vartheta)\tilde{\alpha}_{0n}^z,
\end{aligned} \tag{79}
$$

with r, ϑ, φ being the spherical coordinates of the point of observation and $\boldsymbol{\vartheta}$, $\boldsymbol{\varphi}$ being the unit vectors in the ϑ and φ directions, respectively.

The normalized bistatic cross section is given by

$$F_0 \equiv k^2 r^2 |\mathbf{E}_s(r, \vartheta, \varphi)|^2 = |F_\vartheta(\vartheta, \varphi)|^2 + |F_\varphi(\vartheta, \varphi)|^2. \tag{80}$$

The normalized backscattering cross section F_1 is obtained from Eq. (80) with $\vartheta = \vartheta_i$ and $\varphi = \varphi_i = 0$.

Of most practical interest are the normalized bistatic cross sections in the E and H planes (i.e., the planes determined by $\varphi = \pi/2$ and $\varphi = 0$, respectively) for different values of the scattering angle ϑ and the normalized backscattering cross section for different values of the angle of incidence ϑ_i.

D. QUANTITATIVE RESULTS

Numerical results are presented in this section as plots of the far-field normalized bistatic and backscattering cross sections, for a selected set of spheroids with various coatings. Because the series expansions of all the electric and magnetic fields in terms of vector spheroidal wave functions consist of an infinite number of terms, all the matrices involved in the field expressions have infinite dimensions. To obtain numerical results, these infinite series and matrices are to be appropriately truncated. The number of terms in the series expansion of the fields (or the number of matrix elements) required to obtain a given accuracy in the computed scattering cross sections depends on the frequency, size, spheroid material, and material used for coating. For the illustrative examples considered in this section, it has been found sufficient to consider only the φ harmonics e^{j0}, $e^{\pm j\varphi}$, and $e^{\pm 2j\varphi}$ in the vector spheroidal wave functions $\mathbf{M}_{mn}$ and $\mathbf{N}_{mn}$, and $n = |m|, |m| + 1, \ldots, |m| + 5$ for each value of m, in order to obtain a two-significant-digit accuracy in the computed scattering cross sections.

Numerical experiments performed to validate the software used in the calculation of the scattering cross sections are given in Cooray and Ciric (1992). All normalized bistatic cross sections in the following examples are calculated for $\mu_1 = \mu_2 = \mu$ and for the case of an axial incidence ($\vartheta_i = 0°$).

Figure 2 shows plots of the normalized bistatic cross section versus the scattering angle and of the normalized backscattering cross section versus the angle of incidence of the plane wave, for a dielectric spheroid of semi-major axis length $\lambda/4$, relative permittivity 3, confocally coated with a lossy material of relative permittivity $2.13 - j0.055$ and thickness $a_1 - a_2 = 0.02\lambda$, with λ being the free-space wavelength. The axial ratio of the spheroid is 2 in one case and 10 in the other. The behavior of the scattering cross section in the two cases is very similar. However, the magnitude of the scattering cross section is lower when the axial ratio is 10, because the area available for scattering is less in this case. For both

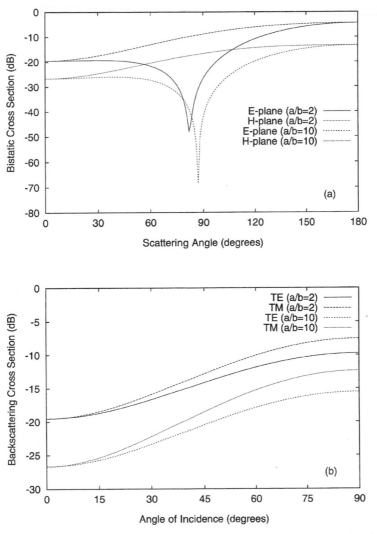

Figure 2   Normalized scattering cross sections of a dielectric prolate spheroid of semi-major axis length $\lambda/4$ and relative permittivity 3, coated with a material of relative permittivity $2.13 - j0.055$ and thickness 0.02λ, for two axial ratios.

spheroids, the magnitude of the backscattering cross section in the TM case is larger than that of the corresponding TE case.

In Fig. 3 we show the variation of the normalized backscattering cross section with the angle of incidence for a lossless dielectric spheroid and for a perfectly

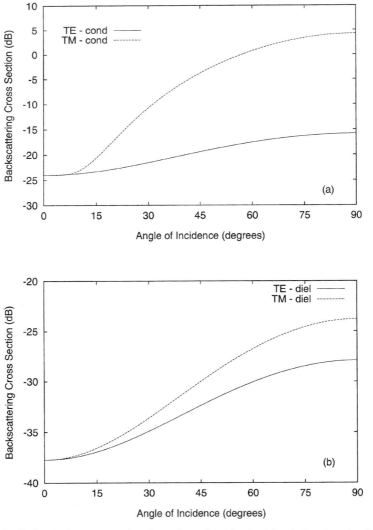

Figure 3 Backscattering cross sections for prolate spheroids of axial ratio 5 and semi-major axis length $\lambda/4$. (a) Conducting spheroid and (b) dielectric spheroid of relative permittivity 3.

conducting spheroid, for both TE and TM polarizations of the incident uniform plane wave. The relative permittivity of the lossless material of the dielectric spheroid considered is 3, and the plots are given for both spheroids having a semi-major axis length $\lambda/4$. The axial ratio of the spheroids in both cases is 5. The results for the case of the dielectric spheroid have been obtained from those corresponding to a coated dielectric spheroid, as a special case when the confocal layer has the material constants reduced to those of a free space. On the other hand, the conducting spheroid has been simulated from the dielectric spheroid by using a very high value of relative permittivity. The variation of the backscattering cross section in the TE case is almost the same for both types of spheroids. However, in the TM case the magnitude of the scattering cross section for the conducting spheroid increases quite rapidly as compared to that for the dielectric spheroid.

VI. SCATTERING OF ELECTROMAGNETIC WAVES BY A CHIRAL SPHEROID

In this section, an exact solution to the problem of plane wave scattering by an optically active (or chiral) spheroid is derived using the same method of separation of variables. Fields outside as well as inside the spheroid are expanded in terms of vector spheroidal wave functions, and a set of simultaneous linear equations is obtained by imposing the boundary conditions on the surface of the spheroid. The solution to this system of equations yields the unknown coefficients in the series expansions of the corresponding fields.

A. FIELD EXPANSIONS

Consider a homogeneous, isotropic, and chiral spheroid, with chirality admittance ξ_c, permittivity ε_s, and permeability μ_s, located in a medium assumed to be linear, homogeneous, and isotropic, with permittivity ε and permeability μ. An arbitrarily polarized monochromatic uniform plane electromagnetic wave with an electric field intensity of unit amplitude is incident at an arbitrary angle on the spheroid whose center is at the origin O of a Cartesian coordinate system and whose axis of symmetry is along the z axis. The same time dependence $\exp(j\omega t)$ is used in the derivations and formulas given later. The plane of incidence is assumed to be the xz plane. Then, the expansion of the incident electric field in terms of the radial vector spheroidal wave functions defined in Section IV is given by (Cooray and Ciric, 1993)

$$\mathbf{E_i} = \sum_{m=-\infty}^{\infty} \sum_{n=|m|}^{\infty} \left[p_{mn}(h, \vartheta_i)\mathbf{M}_{mn}^{r(1)}(h; \mathbf{r}) + q_{mn}(h, \vartheta_i)\mathbf{N}_{mn}^{r(1)}(h; \mathbf{r}) \right], \qquad (81)$$

where $\mathbf{r}$ denotes the coordinate triad η, ξ, φ,

$$p_{mn}(h, \vartheta_i) = f_{mn}(h, \vartheta_i) \cos \zeta - j g_{mn}(h, \vartheta_i) \sin \zeta, \qquad (82)$$

$$q_{mn}(h, \vartheta_i) = g_{mn}(h, \vartheta_i) \cos \zeta - j f_{mn}(h, \vartheta_i) \sin \zeta, \qquad (83)$$

in which $h = kF$, with k being the wavenumber in the region outside the spheroid and F the semi-interfocal distance of the spheroid, and ζ is the polarization angle defined in Section V. The coefficients $f_{mn}(h, \vartheta_{\mathrm{i}})$ and $g_{mn}(h, \vartheta_{\mathrm{i}})$ in Eqs. (82) and (83) are given by (Cooray and Ciric, 1993)

$$f_{mn}(h, \vartheta_{\mathrm{i}}) = -\frac{2j^n}{N_{mn}(h)} \sum_{r=0,1}^{\infty}{}' \frac{d_r^{mn}(h)}{(r+m)(r+m+1)} \frac{d}{d\vartheta_{\mathrm{i}}} P_{m+r}^m(\cos\vartheta_{\mathrm{i}}), \quad (84)$$

$$g_{mn}(h, \vartheta_{\mathrm{i}}) = -\frac{2mj^n}{N_{mn}(h)} \sum_{r=0,1}^{\infty}{}' \frac{d_r^{mn}(h)}{(r+m)(r+m+1)} \frac{P_{m+r}^m(\cos\vartheta_{\mathrm{i}})}{\sin\vartheta_{\mathrm{i}}}. \quad (85)$$

When $\vartheta_{\mathrm{i}} = 0$, only $f_{1n}(h, 0)$ and $g_{1n}(h, 0)$ remain nonzero, and are given by

$$f_{1n}(h, 0) = g_{1n}(h, 0) = -\frac{j^n}{N_{1n}(h)} \sum_{r=0,1}^{\infty}{}' d_r^{1n}(h). \quad (86)$$

Equation (81) can now be written in a matrix form similar to that of Eq. (53) as

$$\mathbf{E}_{\mathrm{i}} = {}_r\bar{\mathbf{M}}_{\mathrm{i}}^{(1)\mathrm{T}}{}_r\bar{I}_p + {}_r\bar{\mathbf{N}}_{\mathrm{i}}^{(1)\mathrm{T}}{}_r\bar{I}_q, \quad (87)$$

with the overbar denoting a column matrix and T denoting the transpose of a matrix, where

$$_r\bar{\mathbf{M}}_{\mathrm{i}}^{(1)\mathrm{T}} = \begin{bmatrix} \bar{\mathbf{M}}_0^{(1)\mathrm{T}} & \bar{\mathbf{M}}_1^{(1)\mathrm{T}} & \bar{\mathbf{M}}_{-1}^{(1)\mathrm{T}} & \bar{\mathbf{M}}_2^{(1)\mathrm{T}} & \bar{\mathbf{M}}_{-2}^{(1)\mathrm{T}} & \cdots \end{bmatrix}, \quad (88)$$

with

$$\bar{\mathbf{M}}_w^{(1)\mathrm{T}} = \begin{bmatrix} \mathbf{M}_{w,|w|}^{r(1)}(h; \mathbf{r}) & \mathbf{M}_{w,|w|+1}^{r(1)}(h; \mathbf{r}) & \mathbf{M}_{w,|w|+2}^{r(1)}(h; \mathbf{r}) & \cdots \end{bmatrix}, \quad (89)$$

and

$$_r\bar{I}_p^{\mathrm{T}} = \begin{bmatrix} \bar{I}_0^{\mathrm{T}} & \bar{I}_1^{\mathrm{T}} & \bar{I}_{-1}^{\mathrm{T}} & \bar{I}_2^{\mathrm{T}} & \bar{I}_{-2}^{\mathrm{T}} & \cdots \end{bmatrix}, \quad (90)$$

with

$$\bar{I}_w^{\mathrm{T}} = \begin{bmatrix} p_{w,|w|} & p_{w,|w|+1} & p_{w,|w|+2} & \cdots \end{bmatrix}. \quad (91)$$

The elements of $_r\bar{\mathbf{N}}_{\mathrm{i}}^{(1)\mathrm{T}}$ and $_r\bar{I}_q$ are obtained from those of $_r\bar{\mathbf{M}}_{\mathrm{i}}^{(1)\mathrm{T}}$ and $_r\bar{I}_p$, respectively, by replacing $\mathbf{M}$ by $\mathbf{N}$ and p by q.

The electric field scattered by the spheroid is expanded in terms of radial vector spheroidal wave functions of the fourth kind in the form

$$\mathbf{E}_{\mathrm{s}} = \sum_{m=-\infty}^{\infty} \sum_{n=|m|}^{\infty} [\alpha_{mn}\mathbf{M}_{mn}^{r(4)}(h; \mathbf{r}) + \beta_{mn}\mathbf{N}_{mn}^{r(4)}(h; \mathbf{r})], \quad (92)$$

where α_{mn} and β_{mn} are the unknown expansion coefficients to be determined. This can be written in matrix form

$$\mathbf{E}_{\mathrm{s}} = {}_r\bar{\mathbf{M}}_{\mathrm{s}}^{(4)\mathrm{T}}\bar{\alpha} + {}_r\bar{\mathbf{N}}_{\mathrm{s}}^{(4)\mathrm{T}}\bar{\beta}, \qquad (93)$$

where the elements of the vector matrices are obtained from those of the matrices in Eq. (87) by replacing $\mathbf{M}_{mn}^{r(1)}(h; \mathbf{r})$ by $\mathbf{M}_{mn}^{r(4)}(h; \mathbf{r})$ and $\mathbf{N}_{mn}^{r(1)}(h; \mathbf{r})$ by $\mathbf{N}_{mn}^{r(4)}(h; \mathbf{r})$. The elements α_{mn} and β_{mn} of $\bar{\alpha}$ and $\bar{\beta}$ are obtained from those of ${}_r\bar{I}_p$ and ${}_r\bar{I}_q$, respectively, by replacing p by α and q by β.

The electric field $\mathbf{E}_{\mathrm{c}}$ inside the chiral spheroid can be expressed as (Bohren, 1974; Bohren, 1978; Cooray and Ciric, 1993)

$$\mathbf{E}_{\mathrm{c}} = \mathbf{E}_{\mathrm{R}} + \mathbf{E}_{\mathrm{L}}, \qquad (94)$$

where $\mathbf{E}_{\mathrm{R}}$ and $\mathbf{E}_{\mathrm{L}}$ are the electric fields corresponding to the right-handed and left-handed circularly polarized waves. These electric fields can be expanded in the form

$$\mathbf{E}_{\mathrm{R}} = \sum_{m=-\infty}^{\infty} \sum_{n=|m|}^{\infty} \gamma_{mn}\left[\mathbf{M}_{mn}^{r(1)}(h_{\mathrm{R}}; \mathbf{r}) + \mathbf{N}_{mn}^{r(1)}(h_{\mathrm{R}}; \mathbf{r})\right], \qquad (95)$$

$$\mathbf{E}_{\mathrm{L}} = \sum_{m=-\infty}^{\infty} \sum_{n=|m|}^{\infty} \delta_{mn}\left[\mathbf{M}_{mn}^{r(1)}(h_{\mathrm{L}}; \mathbf{r}) - \mathbf{N}_{mn}^{r(1)}(h_{\mathrm{L}}; \mathbf{r})\right], \qquad (96)$$

in which γ_{mn}, δ_{mn} are the unknown coefficients that have to be determined, $h_{\mathrm{R}} = k_{\mathrm{R}}F$, $h_{\mathrm{L}} = k_{\mathrm{L}}F$, where $k_{\mathrm{R,L}} = \omega\sqrt{\mu_{\mathrm{s}}\varepsilon_{\mathrm{c}}} \pm \omega\mu_{\mathrm{s}}\xi_{\mathrm{c}}$ with $\varepsilon_{\mathrm{c}} = \varepsilon_{\mathrm{s}} + \mu_{\mathrm{s}}\xi_{\mathrm{c}}^2$ being the effective permittivity and ξ_{c} the chirality admittance. Equation (94) can now be written in the form

$$\mathbf{E}_{\mathrm{c}} = \bar{\mathbf{M}}_{\mathrm{c}}^{(1)\mathrm{T}}\bar{\gamma} + \bar{\mathbf{N}}_{\mathrm{c}}^{(1)\mathrm{T}}\bar{\delta}, \qquad (97)$$

where

$$\bar{\mathbf{M}}_{\mathrm{c}}^{(1)\mathrm{T}} = \left[\bar{\mathbf{M}}_0^{(1)\mathrm{T}'} \quad \bar{\mathbf{M}}_1^{(1)\mathrm{T}'} \quad \bar{\mathbf{M}}_{-1}^{(1)\mathrm{T}'} \quad \bar{\mathbf{M}}_2^{(1)\mathrm{T}'} \quad \bar{\mathbf{M}}_{-2}^{(1)\mathrm{T}'} \quad \cdots\right], \qquad (98)$$

$$\bar{\mathbf{N}}_{\mathrm{c}}^{(1)\mathrm{T}} = \left[\bar{\mathbf{N}}_0^{(1)\mathrm{T}'} \quad \bar{\mathbf{N}}_1^{(1)\mathrm{T}'} \quad \bar{\mathbf{N}}_{-1}^{(1)\mathrm{T}'} \quad \bar{\mathbf{N}}_2^{(1)\mathrm{T}'} \quad \bar{\mathbf{N}}_{-2}^{(1)\mathrm{T}'} \quad \cdots\right], \qquad (99)$$

with

$$\bar{\mathbf{M}}_w^{(1)\mathrm{T}'} = \left[\mathbf{M}_{w,|w|}^{r(1)}(h_{\mathrm{R}}; \mathbf{r}) + \mathbf{N}_{w,|w|}^{r(1)}(h_{\mathrm{R}}; \mathbf{r})\right.$$
$$\left.\mathbf{M}_{w,|w|+1}^{r(1)}(h_{\mathrm{R}}; \mathbf{r}) + \mathbf{N}_{w,|w|+1}^{r(1)}(h_{\mathrm{R}}; \mathbf{r}) \quad \cdots\right], \qquad (100)$$

$$\bar{\mathbf{N}}_w^{(1)\mathrm{T}'} = \left[\mathbf{M}_{w,|w|}^{r(1)}(h_{\mathrm{L}}; \mathbf{r}) - \mathbf{N}_{w,|w|}^{r(1)}(h_{\mathrm{L}}; \mathbf{r})\right.$$
$$\left.\mathbf{M}_{w,|w|+1}^{r(1)}(h_{\mathrm{L}}; \mathbf{r}) - \mathbf{N}_{w,|w|+1}^{r(1)}(h_{\mathrm{L}}; \mathbf{r}) \quad \cdots\right]. \qquad (101)$$

The elements of $\bar{\gamma}$ and $\bar{\delta}$ are obtained from those of $\bar{\alpha}$ and $\bar{\beta}$ by replacing α and β by γ and δ, respectively.

The expressions for the incident and scattered magnetic fields can be obtained from those of the corresponding electric fields as in the previous section:

$$\mathbf{H}_i = j\left(\frac{\varepsilon}{\mu}\right)^{1/2} \left[{}_r\bar{\mathbf{N}}_i^{(1)\mathrm{T}} {}_r\bar{I}_p + {}_r\bar{\mathbf{M}}_i^{(1)\mathrm{T}} {}_r\bar{I}_q \right], \tag{102}$$

$$\mathbf{H}_s = j\left(\frac{\varepsilon}{\mu}\right)^{1/2} \left[{}_r\bar{\mathbf{N}}_s^{(4)\mathrm{T}} \bar{\alpha} + {}_r\bar{\mathbf{M}}_s^{(4)\mathrm{T}} \bar{\beta} \right]. \tag{103}$$

Similarly, the magnetic field inside the chiral spheroid is obtained from $\mathbf{E}_c$, in terms of $\mathbf{E}_R$ and $\mathbf{E}_L$, as

$$\mathbf{H}_c = j\left(\frac{\varepsilon_c}{\mu_s}\right)^{1/2} (\mathbf{E}_R - \mathbf{E}_L). \tag{104}$$

Using Eqs. (95) and (96), $\mathbf{H}_c$ can now be expressed in matrix form as

$$\mathbf{H}_c = j\left(\frac{\varepsilon_c}{\mu_s}\right)^{1/2} \left[-\bar{\mathbf{N}}_c^{(1)\mathrm{T}} \bar{\delta} + \bar{\mathbf{M}}_c^{(1)\mathrm{T}} \bar{\gamma} \right]. \tag{105}$$

B. BOUNDARY CONDITIONS

The tangential electric and magnetic fields are continuous across the surface of the chiral spheroid $\xi = \xi_0$, that is,

$$(\mathbf{E}_s + \mathbf{E}_i) \times \boldsymbol{\xi}|_{\xi=\xi_0} = \mathbf{E}_c \times \boldsymbol{\xi}|_{\xi=\xi_0}, \tag{106}$$

$$(\mathbf{H}_s + \mathbf{H}_i) \times \boldsymbol{\xi}|_{\xi=\xi_0} = \mathbf{H}_c \times \boldsymbol{\xi}|_{\xi=\xi_0}, \tag{107}$$

where $\boldsymbol{\xi}$ is the unit vector normal to the spheroid surface. To obtain a set of simultaneous equations, the η and φ components of Eqs. (106) and (107) are first multiplied by $l_\eta = (\xi_0^2 - \eta^2)^{5/2}$ and $l_\varphi = (\xi_0^2 - \eta^2)^2 (\xi_0^2 - 1)^{-1/2}$, respectively, and then each of the equations corresponding to these components is multiplied by $S_{|m|-1,|m|-1+\kappa}(h, \eta) \exp(\pm jm\varphi)$ for $m \neq 0$ and by $S_{1,1+\kappa}(h, \eta)$ for $m = 0$ and integrated over the surface of the spheroid. After applying the orthogonality properties of the trigonometric functions and of the spheroidal angle functions, the resultant set of simultaneous linear equations can be written in matrix form as

$$[G_c]\bar{S}_c = [R_c]\bar{I}_c, \tag{108}$$

where $\bar{S}_c$ is the column matrix of the unknown coefficients and $\bar{I}_c$ is the column matrix of the known incident field coefficients. The elements of the matrices $[G_c]$ and $[R_c]$ are given in Cooray and Ciric (1993). The solution to the previous matrix equation yields the unknown expansion coefficients, and thus the electromagnetic field inside and outside the spheroid.

C. SCATTERING CROSS SECTIONS AND NUMERICAL RESULTS

The scattered electric field in the far zone can be written as in Eq. (77):

$$\mathbf{E}_s(r, \vartheta, \varphi) = \frac{e^{-jkr}}{kr} \left[F_\vartheta(\vartheta, \varphi)\boldsymbol{\vartheta} + F_\varphi(\vartheta, \varphi)\boldsymbol{\varphi} \right], \tag{109}$$

where, using the asymptotic forms of the radial vector spheroidal wave functions $\mathbf{M}_{mn}^{r(4)}$ and $\mathbf{N}_{mn}^{r(4)}$,

$$F_\vartheta(\vartheta, \varphi) = \sum_{m=1}^{\infty} \sum_{n=m}^{\infty} \left\{ -j^n \frac{m S_{mn}(h, \cos\vartheta)}{\sin\vartheta} \right.$$

$$\times \left[(\alpha_{mn} - \alpha_{-mn}) \cos m\varphi + j(\alpha_{mn} + \alpha_{-mn}) \sin m\varphi \right]$$

$$+ j^n \frac{d}{d\vartheta} S_{mn}(h, \cos\vartheta)$$

$$\times \left[(\beta_{mn} + \beta_{-mn}) \cos m\varphi + j(\beta_{mn} - \beta_{-mn}) \sin m\varphi \right] \Big\}$$

$$+ \sum_{n=0}^{\infty} j^n \beta_{0n} \frac{d}{d\vartheta} S_{0n}(h, \cos\vartheta), \tag{110}$$

$$F_\varphi(\vartheta, \varphi) = \sum_{m=1}^{\infty} \sum_{n=m}^{\infty} \left\{ j^{n+1} \frac{m S_{mn}(h, \cos\vartheta)}{\sin\vartheta} \right.$$

$$\times \left[(\beta_{mn} - \beta_{-mn}) \cos m\varphi + j(\beta_{mn} + \beta_{-mn}) \sin m\varphi \right]$$

$$- j^{n+1} \frac{d}{d\vartheta} S_{mn}(h, \cos\vartheta)$$

$$\times \left[(\alpha_{mn} + \alpha_{-mn}) \cos m\varphi + j(\alpha_{mn} - \alpha_{-mn}) \sin m\varphi \right] \Big\}$$

$$- \sum_{n=0}^{\infty} j^{n+1} \alpha_{0n} \frac{d}{d\vartheta} S_{0n}(h, \cos\vartheta), \tag{111}$$

with r, ϑ, φ being the spherical coordinates of the point of observation.

The normalized bistatic and backscattering cross sections are given, respectively, by

$$F_0 \equiv k^2 r^2 |\mathbf{E}_s(r, \vartheta, \varphi)|^2 = |F_\vartheta(\vartheta, \varphi)|^2 + |F_\varphi(\vartheta, \varphi)|^2 \qquad (112)$$

and

$$F_1 \equiv k^2 r^2 |\mathbf{E}_s(r, \vartheta_i, 0)|^2 = |F_\vartheta(\vartheta_i, 0)|^2 + |F_\varphi(\vartheta_i, 0)|^2. \qquad (113)$$

Numerical results are again presented in the form of normalized bistatic and backscattering cross sections in the far zone, for a few chiral spheroids. The bistatic cross section is calculated for the case of an axial incidence. To obtain numerical results with a required accuracy, the series and the corresponding matrices involved in the field expressions need to be truncated appropriately. For the examples considered, a two-digit accuracy in the computed bistatic cross section is obtained by only retaining the terms corresponding to $m = 1$ and $n = |m|$, $|m| + 1, \ldots, |m| + 12$. However, to obtain the same accuracy for the backscattering cross section, it is necessary to use terms with $m = 0, 1, 2, 3$, and the same n as before. All the results presented are for the case $\mu_s = \mu$, but the formulation and the software used for the calculations are valid for any $\mu_s \neq \mu$, as long as the material is linear. Numerical experiments performed to validate the software and the accuracy of the results obtained are described in Cooray and Ciric (1993).

Figure 4 shows the variation of the normalized bistatic cross section with the scattering angle and the variation of the normalized backscattering cross section with the angle of incidence for both prolate and oblate chiral spheroids, under a uniform plane wave illumination. Each spheroid has a relative permittivity given by $\sqrt{\varepsilon_r} = 1.33$, a semi-major axis length of $\lambda/4$, and a chirality admittance $\xi_c = 0.001S$. The axial ratio of the spheroids is 2. It can be seen that the bistatic cross section patterns are similar for the prolate and the oblate spheroids. However, the magnitude of the cross section at a given scattering angle is lower for the prolate spheroid than for the oblate spheroid. The behavior of the backscattering cross section is completely different for the two spheroids. In the case of the prolate spheroid the magnitude of the cross section increases with the angle of incidence, whereas in the case of the oblate spheroid it decreases.

VII. SCATTERING BY SYSTEMS OF ARBITRARILY ORIENTED SPHEROIDS

An exact analytic solution to the problem of scattering of a monochromatic uniform plane electromagnetic wave of arbitrary polarization and angle of incidence by a system of dielectric prolate spheroids of arbitrary orientation can be obtained by expanding the incident, the scattered, and the transmitted elec-

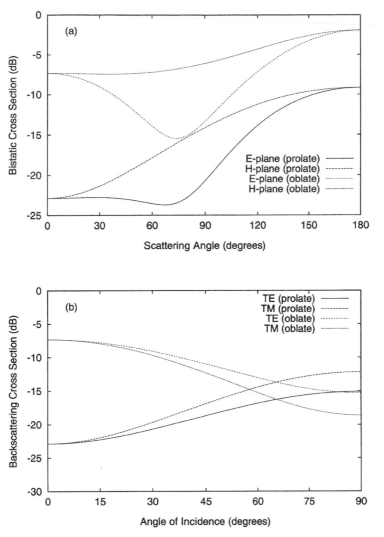

Figure 4 Scattering cross sections of a chiral prolate spheroid and of a chiral oblate spheroid, with a refractive index of 1.33 and a chirality admittance of 0.001S, illuminated by a uniform plane wave. The axial ratio is 2 and the semi-major axis length is $\lambda/4$ for both spheroids.

tromagnetic fields in terms of appropriate vector spheroidal wave functions. The boundary conditions at the surface of a given spheroid are imposed by express-ing the electromagnetic field scattered by all the other spheroids in terms of the spheroidal coordinates attached to the spheroid under consideration, employing the rotational–translational addition theorems for vector spheroidal wave functions. The solution of the resultant set of algebraic equations yields the unknown scattered and transmitted field expansion coefficients, in terms of which the electromagnetic field can be evaluated at any given point.

A. PROBLEM FORMULATION

Consider N arbitrarily oriented prolate spheroids, with their centers located at the origins O_r of the local Cartesian coordinate systems $O_r x_r y_r z_r$, $r = 1$, $2, \ldots, N$, attached to the N spheroids as shown in Fig. 5. The major axes of these spheroids are along the z axes of the respective local Cartesian systems. The position of each of the origins O_r with respect to the global Cartesian coordinate system $Oxyz$ is given by the spherical coordinates d_r, ϑ_{0_r}, φ_{0_r}, and each system $O_r x_r y_r z_r$ is rotated with respect to $Oxyz$ through the Euler angles α_r, β_r, γ_r (Chapter 1).

Suppose a linearly polarized uniform plane electromagnetic wave with an electric field intensity of unit amplitude is incident on the system of spheroids, at an angle ϑ_i with respect to the z axis of the global system $Oxyz$, the direction of propagation being in the xz plane ($\varphi_i = 0$). This linearly polarized incident wave can be decomposed into its TE and TM components using the polarization

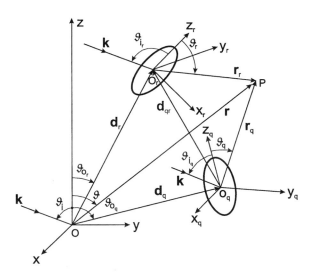

Figure 5 Geometry of the qth and rth prolate spheroids in arbitrary orientation and the associated coordinate systems.

angle ζ, with respect to the xz plane, as described in Sections V and VI. The medium in which the spheroids are embedded is assumed to be homogeneous and nonconducting, of permittivity ε and permeability μ.

The direction of the incident wave vector **k** with respect to the global system $Oxyz$ is specified by the angular spherical coordinates ϑ_i and $\varphi_i = 0$ (see Fig. 5). Thus,

$$\mathbf{k} = -k(\sin \vartheta_i \mathbf{x} + \cos \vartheta_i \mathbf{z}), \tag{114}$$

where **x** and **z** are the unit vectors along the x and z axes of the system $Oxyz$ and k is the wavenumber of the medium outside the spheroids. If the direction of **k** with respect to the local system $O_r x_r y_r z_r$ is specified by the angular spherical coordinates $\vartheta_{i_r}, \varphi_{i_r}$, then

$$\mathbf{k} = -k(\sin \vartheta_{i_r} \cos \varphi_{i_r} \mathbf{x}_r + \sin \vartheta_{i_r} \sin \varphi_{i_r} \mathbf{y}_r + \cos \vartheta_{i_r} \mathbf{z}_r), \tag{115}$$

where $\mathbf{x}_r$, $\mathbf{y}_r$, $\mathbf{z}_r$ are the unit vectors along the x_r, y_r, and z_r axes of the system $O_r x_r y_r z_r$. The unit vectors **x**, **y**, **z** are related to $\mathbf{x}_r, \mathbf{y}_r, \mathbf{z}_r$ by

$$\mathbf{a} = {}^r c_{ax}\mathbf{x}_r + {}^r c_{ay}\mathbf{y}_r + {}^r c_{az}\mathbf{z}_r, \qquad \mathbf{a} = \mathbf{x}, \mathbf{y}, \mathbf{z}, \qquad a = x, y, z, \tag{116}$$

where

$$
\begin{aligned}
{}^r c_{xx} &= \cos \alpha_r \cos \beta_r \cos \gamma_r - \sin \alpha_r \sin \gamma_r, \\
{}^r c_{xy} &= -(\cos \alpha_r \cos \beta_r \sin \gamma_r + \sin \alpha_r \cos \gamma_r), \\
{}^r c_{xz} &= \cos \alpha_r \sin \beta_r, \\
{}^r c_{yx} &= \sin \alpha_r \cos \beta_r \cos \gamma_r + \cos \alpha_r \sin \gamma_r, \\
{}^r c_{yy} &= -\sin \alpha_r \cos \beta_r \sin \gamma_r + \cos \alpha_r \cos \gamma_r, \\
{}^r c_{yz} &= \sin \alpha_r \sin \beta_r, \\
{}^r c_{zx} &= -\sin \beta_r \cos \gamma_r, \qquad {}^r c_{zy} = \sin \beta_r \sin \gamma_r, \qquad {}^r c_{zz} = \cos \beta_r,
\end{aligned}
\tag{117}
$$

with the Euler angles $\alpha_r, \beta_r, \gamma_r$. Substituting **x** and **z** from Eq. (116) into Eq. (114) and identifying the corresponding coefficients of $\mathbf{x}_r, \mathbf{y}_r, \mathbf{z}_r$ with those in Eq. (115) yield the relations

$$
\begin{aligned}
\sin \vartheta_{i_r} \cos \varphi_{i_r} &= {}^r c_{xx} \sin \vartheta_i + {}^r c_{zx} \cos \vartheta_i, \\
\sin \vartheta_{i_r} \sin \varphi_{i_r} &= {}^r c_{xy} \sin \vartheta_i + {}^r c_{zy} \cos \vartheta_i, \\
\cos \vartheta_{i_r} &= {}^r c_{xz} \sin \vartheta_i + {}^r c_{zz} \cos \vartheta_i,
\end{aligned}
\tag{118}
$$

which are used to evaluate ϑ_{i_r} and φ_{i_r}.

The incident electric field ${}^r\mathbf{E}_i$ in the system $O_r x_r y_r z_r$ can be expressed as a linear combination of its TE and TM components ${}^r\mathbf{E}_i^{\mathrm{TE}}$ and ${}^r\mathbf{E}_i^{\mathrm{TM}}$ in the form

$$
{}^r\mathbf{E}_i = {}^r\mathbf{E}_i^{\mathrm{TE}} \cos \zeta + {}^r\mathbf{E}_i^{\mathrm{TM}} \sin \zeta, \tag{119}
$$

where ζ is the polarization angle. ${}^r\mathbf{E}_i$ can now be expanded in terms of vector spheroidal wave functions ${}^r\mathbf{M}$ associated with the system $O_r x_r y_r z_r$ as

$$
{}^r\mathbf{E}_i = \exp(-j\mathbf{k}\cdot\mathbf{d}_r) \sum_{m=-\infty}^{\infty} \sum_{n=|m|}^{\infty} \left({}^r p_{mn}^+ \, {}^r\mathbf{M}_{mn}^{+(1)} + {}^r p_{mn}^- \, {}^r\mathbf{M}_{mn}^{-(1)} + {}^r p_{mn}^z \, {}^r\mathbf{M}_{mn}^{z(1)} \right),
$$

(120)

where

$$
{}^r p_{mn}^{\pm} = \frac{2 j^{n-1}}{k N_{mn}(h_r)} S_{mn}(h_r, \cos\vartheta_{i_r}) \exp(-jm\varphi_{i_r}) \left[\left({}^r c_{yx} \mp j \, {}^r c_{yy} \right) \sin\zeta \right.
$$

$$
\left. + \begin{cases} \left({}^r c_{xx} \mp j \, {}^r c_{xy} \right) \dfrac{\cos\zeta}{\cos\vartheta_i} \right] & \text{for } \vartheta_i \neq \pi/2, \\[2ex] -\left({}^r c_{zx} \mp j \, {}^r c_{zy} \right) \dfrac{\cos\zeta}{\sin\vartheta_i} \right] & \text{for } \vartheta_i \neq 0, \pi, \end{cases}
$$

(121)

$$
{}^r p_{mn}^z = \frac{2 j^{n-1}}{k N_{mn}(h_r)} S_{mn}(h_r, \cos\vartheta_{i_r}) \exp(-jm\varphi_{i_r})
$$

$$
\times \left[{}^r c_{yz} \sin\zeta + \begin{cases} {}^r c_{xz} \dfrac{\cos\zeta}{\cos\vartheta_i} \right] & \text{for } \vartheta_i \neq \pi/2, \\[2ex] -{}^r c_{zz} \dfrac{\cos\zeta}{\sin\vartheta_i} \right] & \text{for } \vartheta_i \neq 0, \pi. \end{cases}
$$

(122)

$S_{mn}(h_r, \cos\vartheta_{i_r})$ and $N_{mn}(h_r)$ are the spheroidal angle function and the normalization constant associated with it, respectively, as defined in Section III.A, $h_r = k F_r$ with F_r being the semi-interfocal distance of the rth spheroid, and $\mathbf{d}_r$ is the position vector of O_r relative to O. In Eq. (120), the argument $(h_r; \eta_r, \xi_r, \varphi_r)$ of each ${}^r\mathbf{M}$, where η_r, ξ_r, φ_r are the spheroidal coordinates of the rth system, has been suppressed for convenience. Arranging the terms in the series expansion of ${}^r\mathbf{E}_i$ in the sequence $e^{j0}, e^{\pm j\varphi_r}, e^{\pm 2j\varphi_r}, \ldots$ yields the matrix form (Cooray, 1990; Cooray and Ciric, 1991b)

$$
{}^r\mathbf{E}_i = {}^r\bar{\mathbf{M}}_i^{(1)T} \, {}^r\bar{I},
$$

(123)

where ${}^r\bar{\mathbf{M}}_i^{(1)}$ and ${}^r\bar{I}$ are column matrices whose elements are the prolate spheroidal vector wave functions of the first kind expressed in terms of the coordinates in the rth spheroidal system and the corresponding incident field expansion coefficients, respectively. The incident electric field on any of the spheroids can similarly be expanded in terms of the spheroidal coordinates attached to that particular spheroid.

The electric field ${}^q\mathbf{E}_s$ scattered by the qth spheroid can be expanded in terms of a set of vector spheroidal wave functions associated with the system $O_q x_q y_q z_q$ as (Cooray, 1990; Cooray and Ciric, 1991b)

$$^q\mathbf{E}_s = \sum_{m=0}^{\infty} \sum_{n=m}^{\infty} (^q\beta_{mn}^+ {}^q\mathbf{M}_{mn}^{+(4)} + {}^q\beta_{m+1,n+1}^z {}^q\mathbf{M}_{m+1,n+1}^{z(4)})$$

$$+ \sum_{n=0}^{\infty} (^q\beta_{-1,n+1}^+ {}^q\mathbf{M}_{-1,n+1}^{+(4)} + {}^q\beta_{0n}^z {}^q\mathbf{M}_{0n}^{z(4)})$$

$$+ \sum_{m=0}^{\infty} \sum_{n=m}^{\infty} (^q\beta_{-mn}^- {}^q\mathbf{M}_{-mn}^{-(4)} + {}^q\beta_{-(m+1),n+1}^z {}^q\mathbf{M}_{-(m+1),n+1}^{z(4)}), \quad (124)$$

where $^q\beta_{\mu\nu}$ are the unknown expansion coefficients. Arranging the terms in this expansion in the sequence $e^{j0}, e^{\pm j\varphi_q}, e^{\pm 2j\varphi_q}, \ldots$ yields

$$^q\mathbf{E}_s = {}^q\bar{\mathbf{M}}_s^{(4)\mathrm{T}} {}^q\bar{\beta}, \quad (125)$$

where $^q\bar{\mathbf{M}}_s^{(4)}$ and $^q\bar{\beta}$ are column matrices whose elements are prolate spheroidal vector wave functions of the fourth kind expressed in terms of the spheroidal coordinates associated with the system $O_q x_q y_q z_q$ and the corresponding unknown expansion coefficients, respectively. These matrices are similar to the corresponding ones defined in Section V.A.

To impose the boundary conditions at the surface of the rth spheroid, the electromagnetic fields scattered by all the other $N - 1$ spheroids have to be expressed as incoming fields with respect to the rth spheroid. This can be achieved by employing rotational–translational addition theorems. Consider the qth spheroid whose scattered field is denoted by $^q\mathbf{E}_s$. To express this field as an incoming field with respect to the rth spheroid, the vector spheroidal wave functions of the fourth kind associated with the system $O_q x_q y_q z_q$ have to be expanded in terms of the vector spheroidal wave functions of the first kind associated with the system $O_r x_r y_r z_r$, using the appropriate rotational–translational addition theorems for vector spheroidal wave functions (Cooray and Ciric, 1989a; Dalmas *et al.*, 1989),

$$^q\mathbf{M}_{mn}^{+(4)}(h_q; \mathbf{r}_q) = \sum_{\nu=0}^{\infty} \sum_{\mu=-\nu}^{\nu} {}_{qr}^{(4)}Q_{\mu\nu}^{mn}(\alpha_{qr}, \beta_{qr}, \gamma_{qr}; \mathbf{d}_{qr})$$

$$\times [^{qr}C_1 {}^r\mathbf{M}_{\mu\nu}^{+(1)}(h_r; \mathbf{r}_r) + {}^{qr}C_2 {}^r\mathbf{M}_{\mu\nu}^{-(1)}(h_r; \mathbf{r}_r)$$

$$+ {}^{qr}C_3 {}^r\mathbf{M}_{\mu\nu}^{z(1)}(h_r; \mathbf{r}_r)], \quad (126)$$

$$^q\mathbf{M}_{mn}^{-(4)}(h_q; \mathbf{r}_q) = \sum_{\nu=0}^{\infty} \sum_{\mu=-\nu}^{\nu} {}_{qr}^{(4)}Q_{\mu\nu}^{mn}(\alpha_{qr}, \beta_{qr}, \gamma_{qr}; \mathbf{d}_{qr})$$

$$\times [^{qr}C_2^* {}^r\mathbf{M}_{\mu\nu}^{+(1)}(h_r; \mathbf{r}_r) + {}^{qr}C_1^* {}^r\mathbf{M}_{\mu\nu}^{-(1)}(h_r; \mathbf{r}_r)$$

$$+ {}^{qr}C_3^* {}^r\mathbf{M}_{\mu\nu}^{z(1)}(h_r; \mathbf{r}_r)], \quad (127)$$

$$
{}^{q}\mathbf{M}_{mn}^{z(4)}(h_q;\mathbf{r}_q) = \sum_{\nu=0}^{\infty}\sum_{\mu=-\nu}^{\nu} {}^{(4)}_{qr}Q_{\mu\nu}^{mn}(\alpha_{qr},\beta_{qr},\gamma_{qr};\mathbf{d}_{qr})
$$

$$
\times\big[{}^{qr}C_4\,{}^{r}\mathbf{M}_{\mu\nu}^{+(1)}(h_r;\mathbf{r}_r) + {}^{qr}C_4^{*}\,{}^{r}\mathbf{M}_{\mu\nu}^{-(1)}(h_r;\mathbf{r}_r)
$$

$$
+ {}^{qr}C_5\,{}^{r}\mathbf{M}_{\mu\nu}^{z(1)}(h_r;\mathbf{r}_r)\big], \tag{128}
$$

which are valid for $r_r \le d_{qr}$, where the asterisk denotes the complex conjugate, $\mathbf{r}_q$ and $\mathbf{r}_r$ denote the coordinate triads (η_q,ξ_q,φ_q) and (η_r,ξ_r,φ_r), respectively, $\mathbf{d}_{qr}$ is the position vector of O_r relative to O_q (see Fig. 5), $\alpha_{qr},\beta_{qr},\gamma_{qr}$ are the Euler angles that describe the rotation of the system $O_r x_r y_r z_r$ relative to $O_q x_q y_q z_q$, and

$$
{}^{qr}C_1 = \tfrac{1}{2}\big[\big({}^{qr}c_{xx}+{}^{qr}c_{yy}\big) - j\big({}^{qr}c_{xy}-{}^{qr}c_{yx}\big)\big],
$$

$$
{}^{qr}C_2 = \tfrac{1}{2}\big[\big({}^{qr}c_{xx}-{}^{qr}c_{yy}\big) + j\big({}^{qr}c_{xy}+{}^{qr}c_{yx}\big)\big], \tag{129}
$$

$$
{}^{qr}C_3 = \tfrac{1}{2}\big({}^{qr}c_{xz}+j\,{}^{qr}c_{yz}\big), \qquad {}^{qr}C_4 = {}^{qr}c_{zx} - j\,{}^{qr}c_{zy}, \qquad {}^{qr}C_5 = {}^{qr}c_{zz}.
$$

The coefficients ${}^{qr}c_{xx}, {}^{qr}c_{xy}, \ldots$ are obtained, respectively, from ${}^{r}c_{xx}, {}^{r}c_{xy}, \ldots$ defined in Eq. (117), by replacing α_r, β_r, and γ_r by α_{qr}, β_{qr}, and γ_{qr}, respectively. ${}^{(4)}_{qr}Q_{\mu\nu}^{mn}$ are the rotational–translational coefficients in the expansion of scalar spheroidal wave functions of the fourth kind associated with $O_q x_q y_q z_q$ in terms of the same functions of the first kind associated with $O_r x_r y_r z_r$, for $r_r \le d_{qr}$, and are given by MacPhie *et al.* (1987). After arranging the terms in the series expansions of Eqs. (126)–(128) in the sequence $e^{j0}, e^{\pm j\varphi_r}, e^{\pm 2j\varphi_r}, \ldots$, the outgoing vector wave functions ${}^{q}\mathbf{M}_s^{(4)}$ can be expressed in terms of the incoming vector wave functions ${}^{qr}\mathbf{M}^{(1)}$ in the form

$$
{}^{q}\bar{\mathbf{M}}_s^{(4)} = [\Gamma_{qr}]\,{}^{qr}\bar{\mathbf{M}}^{(1)}, \tag{130}
$$

where the matrix $[\Gamma_{qr}]$ is defined in Cooray (1990) and

$$
{}^{qr}\bar{\mathbf{M}}^{(1)\mathrm{T}} = \big[{}^{qr}\bar{\mathbf{M}}_0^{(1)\mathrm{T}} \quad {}^{qr}\bar{\mathbf{M}}_1^{(1)\mathrm{T}} \quad {}^{qr}\bar{\mathbf{M}}_2^{(1)\mathrm{T}} \quad \cdots\big], \tag{131}
$$

with

$$
{}^{qr}\bar{\mathbf{M}}_0^{(1)\mathrm{T}} = \big[{}^{r}\bar{\mathbf{M}}_{-1}^{+(1)\mathrm{T}} \quad {}^{r}\bar{\mathbf{M}}_{1}^{-(1)\mathrm{T}} \quad {}^{r}\bar{\mathbf{M}}_{0}^{z(1)\mathrm{T}}\big], \tag{132}
$$

$$
{}^{qr}\bar{\mathbf{M}}_\sigma^{(1)\mathrm{T}} = \big[{}^{r}\bar{\mathbf{M}}_{\sigma-1}^{+(1)\mathrm{T}} \quad {}^{r}\bar{\mathbf{M}}_{\sigma+1}^{-(1)\mathrm{T}} \quad {}^{r}\bar{\mathbf{M}}_{\sigma}^{z(1)\mathrm{T}} \quad {}^{r}\bar{\mathbf{M}}_{-(\sigma+1)}^{+(1)\mathrm{T}} \quad {}^{r}\bar{\mathbf{M}}_{-(\sigma-1)}^{-(1)\mathrm{T}} \quad {}^{r}\bar{\mathbf{M}}_{-\sigma}^{z(1)\mathrm{T}}\big]
$$

$$
\text{for } \sigma \ge 1. \tag{133}
$$

Let us denote the secondary incident field on the rth spheroid caused by the electric field ${}^{q}\mathbf{E}_s$ scattered by the qth spheroid by ${}^{qr}\mathbf{E}_s$. Then, taking the transpose of both sides of Eq. (130) and substituting for ${}^q\bar{\mathbf{M}}_s^{(4)\mathrm{T}}$ in Eq. (125) yield

$$
{}^{qr}\mathbf{E}_s = {}^{qr}\bar{\mathbf{M}}^{(1)\mathrm{T}}[\Gamma_{qr}]^\mathrm{T}\,{}^q\bar{\beta} \qquad \text{for } q = 1, 2, \ldots, r-1, r+1, \ldots, N. \quad (134)
$$

These $N-1$ secondary incident fields and the primary incident plane wave field determine the electric field ${}^r\mathbf{E}_s$ scattered by the rth spheroid, which is expanded as in Eq. (125), in the form

$$
{}^r\mathbf{E}_s = {}^r\bar{\mathbf{M}}_s^{(4)\mathrm{T}}\,{}^r\bar{\beta}. \quad (135)
$$

The column matrices ${}^r\bar{\mathbf{M}}_s^{(4)}$ and ${}^r\bar{\beta}$ have the same form as those of ${}^q\bar{\mathbf{M}}_s^{(4)}$ and ${}^q\bar{\beta}$, respectively, with the vector wave functions evaluated with respect to the spheroidal coordinate system attached to the rth spheroid.

The electric field ${}^r\mathbf{E}_t$ of the electromagnetic field transmitted inside the rth spheroid can be expanded in terms of a set of vector spheroidal wave functions as (Cooray *et al.*, 1990)

$$
{}^r\mathbf{E}_t = {}^r\bar{\mathbf{M}}_t^{(1)\mathrm{T}}\,{}^r\bar{\alpha}, \quad (136)
$$

where the unknown coefficients in the series expansion are contained in the column matrix ${}^r\bar{\alpha}$ and the elements of the column matrix ${}^r\bar{\mathbf{M}}_t^{(1)}$ are prolate spheroidal wave functions of the first kind expressed in terms of the spheroidal coordinates associated with $O_r x_r y_r z_r$, taking into account the permittivity of the material inside the rth spheroid, respectively.

The expansions of the different magnetic fields in terms of vector spheroidal wave functions can be obtained from those of the corresponding electric fields as in the previous sections, by replacing $\mathbf{M}$ by $\mathbf{N}$, where $\mathbf{N} = k^{-1}(\nabla \times \mathbf{M})$, and by multiplying each expansion by the appropriate value of $j(\varepsilon/\mu)^{1/2}$. This yields

$$
{}^r\mathbf{H}_i = j\left(\frac{\varepsilon}{\mu}\right)^{1/2}\,{}^r\bar{\mathbf{N}}_i^{(1)\mathrm{T}}\,{}^r\bar{I}, \quad (137)
$$

$$
{}^{qr}\mathbf{H}_s = j\left(\frac{\varepsilon}{\mu}\right)^{1/2}\,{}^{qr}\bar{\mathbf{N}}^{(1)\mathrm{T}}[\Gamma_{qr}]^\mathrm{T}\,{}^q\bar{\beta}, \quad (138)
$$

$$
{}^r\mathbf{H}_s = j\left(\frac{\varepsilon}{\mu}\right)^{1/2}\,{}^r\bar{\mathbf{N}}_s^{(4)\mathrm{T}}\,{}^r\bar{\beta}, \quad (139)
$$

$$
{}^r\mathbf{H}_t = j\left(\frac{\varepsilon_r}{\mu_r}\right)^{1/2}\,{}^r\bar{\mathbf{N}}_t^{(1)\mathrm{T}}\,{}^r\bar{\alpha}, \quad (140)
$$

with ε, ε_r being the permittivities and μ, μ_r the permeabilities of the media outside and inside the rth spheroid, respectively. The elements of the matrices ${}^r\bar{\mathbf{N}}_i^{(1)\mathrm{T}}$, ${}^{qr}\bar{\mathbf{N}}^{(1)\mathrm{T}}$, ${}^r\bar{\mathbf{N}}_s^{(4)\mathrm{T}}$, and ${}^r\bar{\mathbf{N}}_t^{(1)\mathrm{T}}$ are obtained from the elements of the matrices $^r\bar{\mathbf{M}}_i^{(1)\mathrm{T}}$, $^{qr}\bar{\mathbf{M}}^{(1)\mathrm{T}}$, $^r\bar{\mathbf{M}}_s^{(4)\mathrm{T}}$, and $^r\bar{\mathbf{M}}_t^{(1)\mathrm{T}}$, respectively, by replacing the vector spheroidal wave functions $\mathbf{M}$ by the corresponding $\mathbf{N}$ functions.

B. IMPOSING THE BOUNDARY CONDITIONS

On the surface of each dielectric spheroid $\xi_r = \xi_{r_0}$, $r = 1, 2, \ldots, N$, the tangential components of both $\mathbf{E}$ and $\mathbf{H}$ fields must be continuous across the boundary. With Eqs. (123) and (134)–(140), these conditions can be expressed in the form

$$\left({}^r\bar{\mathbf{M}}_i^{(1)\mathrm{T}}\,{}^r\bar{I} + \sum_{q=1}^{N} [(1 - \delta_{qr})\,{}^{qr}\bar{\mathbf{M}}^{(1)\mathrm{T}}[\Gamma_{qr}]^{\mathrm{T}} + \delta_{qr}\,{}^q\bar{\mathbf{M}}_s^{(4)\mathrm{T}}]\,{}^q\bar{\beta} \right) \times \boldsymbol{\xi}_r|_{\xi_r=\xi_{r_0}}$$

$$= ({}^r\bar{\mathbf{M}}_t^{(1)\mathrm{T}}\,{}^r\bar{\alpha}) \times \boldsymbol{\xi}_r|_{\xi_r=\xi_{r_0}}, \tag{141}$$

$$\left({}^r\bar{\mathbf{N}}_i^{(1)\mathrm{T}}\,{}^r\bar{I} + \sum_{q=1}^{N} [(1 - \delta_{qr})\,{}^{qr}\bar{\mathbf{N}}^{(1)\mathrm{T}}[\Gamma_{qr}]^{\mathrm{T}} + \delta_{qr}\,{}^q\bar{\mathbf{N}}_s^{(4)\mathrm{T}}]\,{}^q\bar{\beta} \right) \times \boldsymbol{\xi}_r|_{\xi_r=\xi_{r_0}}$$

$$= \left(\frac{\varepsilon_r \mu}{\varepsilon \mu_r} \right)^{1/2} ({}^r\bar{\mathbf{N}}_t^{(1)\mathrm{T}}\,{}^r\bar{\alpha}) \times \boldsymbol{\xi}_r|_{\xi_r=\xi_{r_0}} \tag{142}$$

for $r = 1, 2, \ldots, N$, where δ_{qr} is the Kronecker delta and $\boldsymbol{\xi}_r$ is the unit vector normal to the surface of the rth spheroid. Taking the scalar product of both sides of Eqs. (141) and (142) by $\eta_r\,{}^r l_\eta S_{m,|m|+\kappa}(h_r, \eta_r)e^{\pm j(m\pm 1)\varphi_r}$ and $\boldsymbol{\varphi}_r\,{}^r l_\varphi S_{m,|m|+\kappa}(h_r, \eta_r)e^{\pm j(m\pm 1)\varphi_r}$, respectively, for $r = 1, 2, \ldots, N$, $m = \ldots, -2, -1, 0, 1, 2, \ldots$, $\kappa = 0, 1, 2, \ldots$, integrating correspondingly over the surfaces of the N spheroids, and using the orthogonality properties of the trigonometric and spheroidal angle functions, yield finally (Cooray, 1990)

$$\bar{S}_N^d = [G_N]\bar{I}_N, \tag{143}$$

where $[G_N]$ is the spheroid system matrix, which is independent of the direction and polarization of the incident wave, $\bar{I}_N$ is the column matrix of the known incident field coefficients, and $\bar{S}_N^d$ is the column matrix of the unknown expansion coefficients. The coefficients $^r l_\eta$ and $^r l_\varphi$ used for Eq. (141) are given by $^r l_\eta = 2jF_r(\xi_{r_0}^2 - \eta_r^2)^{1/2}$, $^r l_\varphi = 2F_r(\xi_{r_0}^2 - \eta_r^2)$ and those used for Eq. (142) by $^r l_\eta = 2F_r^2(\xi_{r_0}^2 - \eta_r^2)^{5/2}/(\xi_{r_0}^2 - 1)^{1/2}$, $^r l_\varphi = 2jF_r^2(\xi_{r_0}^2 - \eta_r^2)/(\xi_{r_0}^2 - 1)$. The elements of the matrices in Eq. (143) are given in Cooray (1990) and Cooray and Ciric (1991b). Equation (143) provides the unknown coefficients in the expansion of the fields scattered and transmitted by the N arbitrarily oriented spheroids.

The solution for the case of N arbitrarily oriented perfectly conducting spheroids can be derived from the previous solution by letting the permittivity

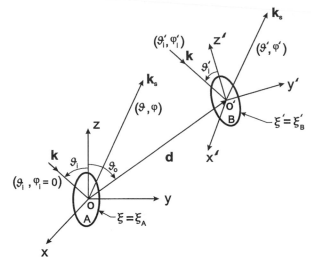

Figure 6   Geometry of two spheroids in arbitrary orientation and the associated coordinate systems.

of each of the spheroids become very high (theoretically infinite). Because there is no field inside the spheroids, the boundary condition in Eq. (141) now becomes

$$\left({}^r\bar{\mathbf{M}}_i^{(1)\mathrm{T}} \, {}^r\bar{I} + \sum_{q=1}^{N} [(1 - \delta_{qr}) \, {}^{qr}\bar{\mathbf{M}}^{(1)\mathrm{T}}[\Gamma_{qr}]^{\mathrm{T}} + \delta_{qr} \, {}^q\bar{\mathbf{M}}_s^{(4)\mathrm{T}}] \, {}^q\bar{\beta} \right) \times \boldsymbol{\xi}_r |_{\xi_r = \xi_{r_0}} = 0$$

(144)

for $r = 1, 2, \ldots, N$. Following a similar procedure as in the case of dielectric spheroids, a system of algebraic equations is obtained in a matrix form as

$$\bar{S}_N^c = [G_N'] \bar{I}_N,$$

(145)

where $[G_N']$ is the system matrix, which has the same properties as those of $[G_N]$ in Eq. (143), and $\bar{S}_N^c$ is the column matrix of the unknown coefficients. The structure and the elements of $[G_N']$ are given in Cooray (1990).

C. SCATTERING CROSS SECTIONS

To illustrate the derivation of the normalized scattering cross section expressions, we consider a system of two spheroids A and B in an arbitrary configuration. The Cartesian system $Oxyz$ attached to the spheroid A and the system $O'x'y'z'$ attached to the spheroid B are shown in Fig. 6. The observation point has spherical coordinates r, ϑ, φ and r', ϑ', φ' and spheroidal coordinates η, ξ, φ and η', ξ', φ', in $Oxyz$ and $O'x'y'z'$, respectively. Using the asymptotic expressions for different vector spheroidal wave functions of the fourth kind, the electric field in the far zone can be written in the form

$$
\begin{aligned}
\mathbf{E}_s(r, \vartheta, \varphi) &= \mathbf{E}_{sA}(r, \vartheta, \varphi) + \mathbf{E}_{sB}(r', \vartheta', \varphi') \\
&= \frac{e^{-jkr}}{kr}\big[F_{\vartheta A}(\vartheta, \varphi)\boldsymbol{\vartheta} + F_{\varphi A}(\vartheta, \varphi)\boldsymbol{\varphi} \\
&\qquad\qquad + F_{\vartheta' B}(\vartheta', \varphi')\boldsymbol{\vartheta}' + F_{\varphi' B}(\vartheta', \varphi')\boldsymbol{\varphi}'\big] \\
&= \frac{e^{-jkr}}{kr}\big[F_{\vartheta A}(\vartheta, \varphi)\boldsymbol{\vartheta} + F_{\varphi A}(\vartheta, \varphi)\boldsymbol{\varphi} + F_{\vartheta' B}(\vartheta', \varphi')\{g_1\boldsymbol{\vartheta} + g_2\boldsymbol{\varphi}\} \\
&\qquad\qquad + F_{\varphi' B}(\vartheta', \varphi')\{g_3\boldsymbol{\vartheta} + g_4\boldsymbol{\varphi}\}\big] \\
&= \frac{e^{-jkr}}{kr}\big[F_{\vartheta}(\vartheta, \varphi)\boldsymbol{\vartheta} + F_{\varphi}(\vartheta, \varphi)\boldsymbol{\varphi}\big],
\end{aligned}
\tag{146}
$$

where

$$
[g_1 \quad g_2]^{\mathrm{T}} = [\Omega][C][\cos\vartheta'\cos\varphi' \quad \cos\vartheta'\sin\varphi' \quad -\sin\vartheta']^{\mathrm{T}}, \tag{147}
$$

$$
[g_3 \quad g_4]^{\mathrm{T}} = [\Omega][C][-\sin\varphi' \quad \cos\varphi' \quad 0]^{\mathrm{T}}, \tag{148}
$$

with

$$
[\Omega] = \begin{bmatrix} \cos\vartheta\cos\varphi & \cos\vartheta\sin\varphi & -\sin\vartheta \\ -\sin\varphi & \cos\varphi & 0 \end{bmatrix},
$$

$$
[C] = \begin{bmatrix} c_{xx'} & c_{xy'} & c_{xz'} \\ c_{yx'} & c_{yy'} & c_{yz'} \\ c_{zx'} & c_{zy'} & c_{zz'} \end{bmatrix}.
\tag{149}
$$

The expressions of $F_{\vartheta A}$ and $F_{\varphi A}$ are exactly the same as those of F_{ϑ} and F_{φ} in Eqs. (78) and (79). The expansion coefficients in the expressions of $F_{\vartheta A}$ and $F_{\varphi A}$ are obtained from the appropriate set of algebraic equations of the form $\bar{S}_2 = [G_2]\bar{I}_2$, which is obtained from Eq. (143) or (145) by particularizing it for the case of two spheroids. Explicit forms of these equations are given in Cooray (1990) and Cooray and Ciric (1991b). Expressions for the coefficients $c_{ax'}$, $c_{ay'}$, $c_{az'}$ for $a = x, y, z$ are obtained from the corresponding ones for $^r c_{ax'}$, $^r c_{ay'}$, $^r c_{az'}$, respectively, by replacing α_r, β_r, γ_r by α_0, β_0, γ_0, which are the Euler angles characterizing the rotation of the spheroid B with respect to the spheroid A. The explicit expressions of $F_{\vartheta' B}(\vartheta', \varphi')$ and $F_{\varphi' B}(\vartheta', \varphi')$ are obtained from $F_{\vartheta A}(\vartheta, \varphi)$ and $F_{\varphi A}(\vartheta, \varphi)$, respectively, by replacing the scattered field expansion coefficients of the spheroid A by those of the spheroid B, ϑ and φ by ϑ' and φ', respectively, and by multiplying each expression by an overall phase factor $\exp(j\mathbf{k}_s \cdot \mathbf{d})$ with $\mathbf{k}_s$ given by

$$\mathbf{k}_s = k(\sin\vartheta\,\cos\varphi\,\mathbf{x} + \sin\vartheta\,\sin\varphi\,\mathbf{y} + \cos\vartheta\,\mathbf{z}). \tag{150}$$

The scattering cross sections are computed from Eq. (80).

D. NUMERICAL RESULTS FOR TWO ARBITRARILY ORIENTED SPHEROIDS

Normalized bistatic and backscattering cross sections in the far field are presented for a system of two nonmagnetic prolate spheroids in arbitrary orientation. The bistatic cross sections are calculated for an axial incidence ($\vartheta_i = 0°$) and for a TE polarization of the incident wave with respect to the system $Oxyz$. Numerical results of a given accuracy can be obtained by truncating appropriately the series and the matrices involved. To obtain results with an accuracy of two significant digits for the examples considered in this section, it has been found sufficient to consider only the φ harmonics e^{j0}, $e^{\pm j\varphi}$, and $e^{\pm 2j\varphi}$, so that the index m in Section VII.B is limited to $m = -2, -1, 0, 1, 2$, and to associate with m only $\kappa = 0, 1, \ldots, 5$. Details regarding how this truncation affects the size of the different submatrices in the system matrix are given explicitly in Cooray (1990). It should be noticed that, owing to the axial symmetry of the spheroids, the results presented next are independent of the Euler angle γ_0.

Results are presented in Fig. 7 for normalized bistatic and backscattering cross sections for a system of two identical dielectric prolate spheroids A and B, of semi-major axis length $\lambda/4$, relative permittivity 2, and axial ratio 2, with the orientation of B relative to A given by the Euler angles $\alpha_0 = 30°$, $\beta_0 = 45°$, $\gamma_0 = 90°$ and the center of B displaced relative to that of A along the z axis of the coordinate system attached to A. As the distance between the centers of the spheroids increases from $\lambda/2$ to λ, more oscillations appear in both the bistatic and the backscattering cross sections due to the phenomenon of multiple scattering in a more resonant system. The maxima and minima of the backscattering cross section patterns corresponding to TE and TM polarizations occur at the same angle of incidence in both cases.

Figure 8 shows the normalized scattering cross sections for two different dielectric spheroids A and B, each of semi-major axis length $\lambda/4$, but with axial ratios 2 and 10, and relative permittivities 2 and 3, respectively. The orientation and displacements of B relative to A are the same as those in Fig. 7. The E- and H-plane patterns in this case remain practically the same when the distance between the centers of the spheroids changes from $\lambda/2$ to λ because of the fact that the field scattered by B is quite small as compared to that of A since B has a high axial ratio. The backscattering cross section patterns corresponding to TE

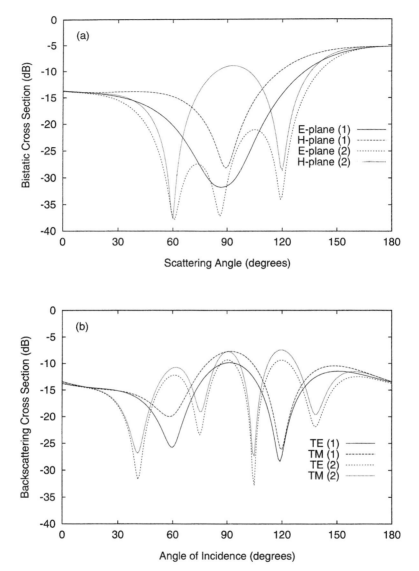

Figure 7 Normalized scattering cross sections for two identical dielectric prolate spheroids of semi-major axis length $\lambda/4$, axial ratio 2, and relative permittivity 2, with one spheroid rotated by the Euler angles $30°$, $45°$, $90°$ and displaced by a distance (1) $\lambda/2$ and (2) λ along the major axis of the other spheroid.

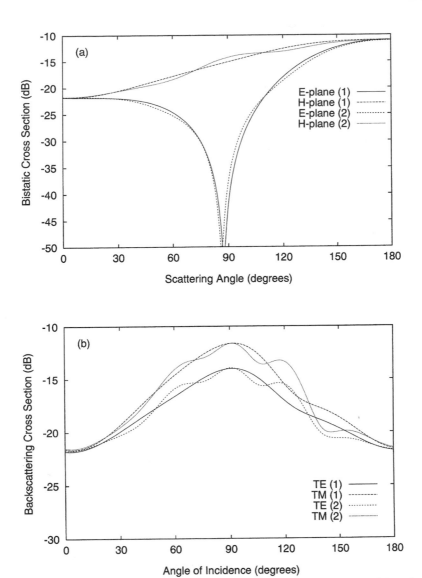

Figure 8   Scattering cross sections of two different dielectric prolate spheroids A and B, each of semi-major axis length $\lambda/4$, but with axial ratios 2 and 10, and relative permittivities 2 and 3, respectively. The rotation and the displacements of B relative to A are the same as those in Fig. 7.

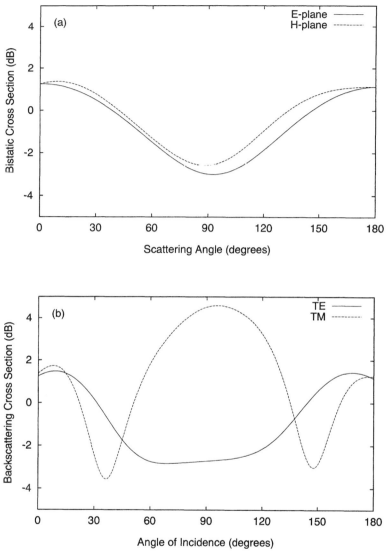

Figure 9 Scattering cross sections for two identical conducting prolate spheroids A and B, each of semi-major axis length $\lambda/4$ and axial ratio 5, with B rotated by the Euler angles $45°$, $90°$, $45°$ relative to A and displaced by a distance $\lambda/2$ along the x axis of the coordinate system attached to A.

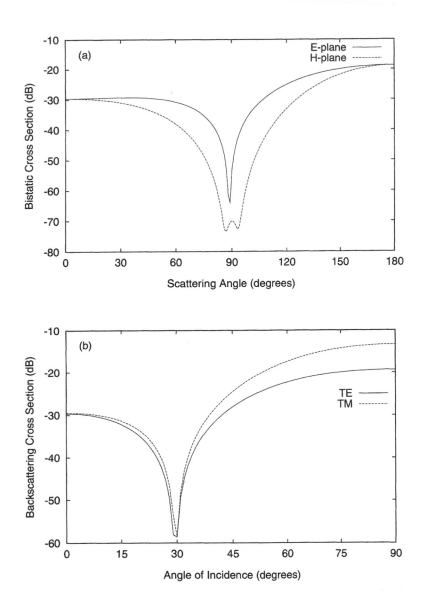

Figure 10   Scattering cross sections for two identical dielectric prolate spheroids of semi-major axis length $\lambda/4$, axial ratio 5, and relative permittivity 3, in a parallel configuration, displaced by a distance $\lambda/2$ along the x axis.

and TM polarizations have the same behavior for both cases, but the magnitude for the TM case is higher than the magnitude for the TE case.

In Fig. 9 the normalized bistatic and backscattering cross sections are plotted for two identical conducting spheroids A and B of axial ratio 5 and semi-major axis length $\lambda/4$, separated by a center-to-center distance of $\lambda/2$ along the x axis of A, with the rotation of B relative to A specified by the Euler angles $45°$, $90°$, $45°$. The behavior of both the E-plane and the H-plane patterns is quite similar. However, the backscattering cross section pattern for TM polarization has more oscillations as compared to the pattern for TE polarization, with two minima around $\vartheta_i = 35°$, $145°$ and a maximum around $\vartheta_i = 90°$.

Numerical results for two identical dielectric spheroids A and B of axial ratio 5, semi-major axis length $\lambda/4$, and relative permitivity 3, in a parallel configuration, are given in Fig. 10. The two spheroids are separated by a center-to-center distance of $\lambda/2$ along the x axis. In this case the E-plane pattern has a sharp minimum around $\vartheta = 90°$, whereas the H-plane pattern presents some oscillation about the same minimum point. The backscattering cross section has a minimum around $\vartheta_i = 30°$ for both polarizations and exhibits a similar behavior, with a higher magnitude for the TM case than for the TE case at higher angles of incidence, as expected. Because of symmetry about $\vartheta_i = 90°$, the variation of the backscattering cross section is shown only for the range from $\vartheta_i = 0°$ to $\vartheta_i = 90°$.

Chapter 5

The Discrete Dipole Approximation for Light Scattering by Irregular Targets

Bruce T. Draine

Princeton University Observatory
Princeton, New Jersey 08544

I. INTRODUCTION

The discrete dipole approximation—also referred to as the coupled dipole approximation—is a flexible technique for studying scattering and absorption of electromagnetic radiation by targets with sizes comparable to the wavelength. Advances in numerical techniques coupled with the increasing speed and memory of scientific workstations now make it possible for calculations to be carried out for targets with dimensions as large as several times the wavelength of the incident

Light Scattering by Nonspherical Particles: Theory, Measurements, and Applications

radiation. Although the technique is not well suited for targets with very large complex refractive index m, it works well for materials with $|m - 1| \lesssim 3$ and target dimension $D \lesssim 5\lambda$, where λ is the wavelength in the surrounding medium. The discrete dipole approximation (DDA) has been applied to compute scattering and absorption by targets of size comparable to the wavelength in a broad range of problems, including interstellar dust grains (Draine and Malhotra, 1993; Draine and Weingartner, 1996, 1997; Wolff et al., 1998), ice crystals in the atmosphere of the Earth (Okamoto et al., 1995; Lemke et al., 1998) and other planets (West and Smith, 1991), interplanetary dust (Mann et al., 1994; Kimura and Mann, 1998), cometary dust (Okamoto et al., 1994; Xing and Hanner, 1997; Yanamandra-Fisher and Hanner, 1999), soot produced in flames (Ivezić and Mengüc, 1996; Ivezić et al., 1997), surface features on semiconductor devices (Schmehl et al., 1997; Nebeker et al., 1998), and optical characteristics of human blood cells (Hoekstra et al., 1998).

The discrete dipole approximation was introduced by Purcell and Pennypacker (1973) and has undergone a number of theoretical developments since then, including the introduction of radiative reaction corrections (Draine, 1988), application of fast Fourier transform techniques (Goodman et al., 1991), and a prescription for dipole polarizabilities based on the lattice dispersion relation (Draine and Goodman, 1993). The discrete dipole approximation was reviewed recently by Draine and Flatau (1994).

II. WHAT IS THE DISCRETE DIPOLE APPROXIMATION?

There has been some confusion about exactly *what* is being approximated in the "discrete dipole approximation." The actual approximation can be simply stated:

> The discrete dipole approximation consists of approximating the actual target by an array of polarizable points (the "dipoles").

This is the *only* essential approximation—once the location and polarizability of the points are specified, calculation of the scattering and absorption of light by the array of polarizable points can be carried out to whatever accuracy is required (within the practical limits imposed by the computational hardware).

Suppose we have an array of points $\mathbf{x}_j$, $j = 1, \ldots, N$, each with complex polarizability tensor $\boldsymbol{\alpha}_j$, and with a monochromatic incident wave $\mathbf{E}_{\text{inc},j} e^{-i\omega t}$ at each location, where ω is the angular frequency and t is time. Each of the dipoles is subject to an electric field that is the sum of the incident wave plus the electric fields resulting from all of the other dipoles. The self-consistent solution for the dipole moments $\mathbf{P}_j e^{-i\omega t}$ satisfies a system of $3N$ linear equations, which can be written as

$$\sum_{j=1}^{N} \mathbf{A}_{ij}\mathbf{P}_j = \mathbf{E}_{\text{inc},i}, \qquad i = 1, \ldots, N, \tag{1}$$

where the elements $\mathbf{A}_{ij}$ are 3×3 matrices (see Draine and Flatau, 1994). The diagonal elements $\mathbf{A}_{ii} = \boldsymbol{\alpha}_i^{-1}$, where $\boldsymbol{\alpha}$ is the 3×3 complex polarizability tensor (see Section VI). The off-diagonal elements $\mathbf{A}_{i \neq j}$ depend only on $k = \omega/c$, where c is the speed of light, and the vector displacement $\mathbf{r}_{ij} \equiv \mathbf{x}_i - \mathbf{x}_j$. Equation (1) is a system of $3N$ complex linear equations; the computational challenge is to find the solution $\mathbf{P}_j$ satisfying this equation.

Once the solution $\mathbf{P}_j$ has been found, it is straightforward to calculate the complete scattering matrix for the target, as well as other quantities, such as the absorption and extinction cross sections, the intensity and polarization of scattered radiation, and the force and torque exerted on the target by the electromagnetic field (Draine and Weingartner, 1996).

There are some obvious issues surrounding the discrete dipole approximation:

1. How many dipoles are required for the dipole array to adequately approximate the target?
2. If a given lattice is to approximate a given continuum target, what choice of dipole polarizabilities will result in the most accurate approximation?
3. For a given target geometry, size, complex refractive index m, and incident wavelength λ, how accurate is the discrete dipole approximation?
4. What are the computational requirements of the discrete dipole approximation?

These are addressed next.

III. THE DDSCAT SCATTERING CODE

DDSCAT is a portable f77 code developed by B. T. Draine and P. J. Flatau to carry out calculations using the discrete dipole aproximation. The DDSCAT code is publicly available[1] (Draine and Flatau, 1994), and a comprehensive user guide is now available (Draine and Flatau, 1997). The calculations reported in the following discussion were carried out using the current "release," DDSCAT.5a8.

---

[1] http://astro.princeton.edu/~draine/, or anonymous ftp to astro.princeton.edu, directory draine/scat/ddscat.

IV. DIPOLE ARRAY GEOMETRY

Equation (1) applies for any array geometry. For general geometry, there are $(3N)^2$ distinct complex elements of the matrix $\mathbf{A}$; when considering $N \gtrsim 10^4$ dipoles, it is apparent that computing this matrix would be very central processing unit (CPU) intensive, and storing the elements for reuse would require large amounts of random access memory (RAM). For example, in the following discussion we show results computed for a target with $N \approx 200,000$ dipoles; storing the complete $\mathbf{A}$ matrix, with 8 bytes per complex number, would require 2.8 TB! There are great advantages to be gained if the dipoles are located on a lattice, because now many different pairs i, j have identical $\mathbf{r}_{ij}$, and hence identical $\mathbf{A}_{ij}$.

Because of this, DDSCAT requires the target array to reside on a cubic lattice. It is then possible to use fast Fourier transform (FFT) techniques to evaluate matrix–vector products $\mathbf{A} \cdot \mathbf{v}$ (Goodman *et al.*, 1991); because they allow evaluation of the product in $O[(3N)\ln(3N)]$ rather than $O[(3N)^2]$ operations, FFT techniques allow use of much larger values of N.

DDSCAT includes routines to generate dipole arrays for a variety of target geometries, including ellipsoids, rectangular prisms, hexagonal prisms, and tetrahedra; it can also accept a user-supplied list of occupied lattice sites. The target material can be anisotropic and the target can be inhomogeneous.

V. TARGET GENERATION

There is some arbitrariness in the construction of the array of point dipoles intended to represent a solid target of specified geometry. DDSCAT uses a straightforward procedure: In the "target frame," construct a target of volume V. Let the target centroid define the origin of coordinates. Choose a "trial" lattice spacing d and construct a lattice $(x, y, z) = (n_x, n_y, n_z)d + (o_x, o_y, o_z)d$, where the n_j are integers and the "offset" vector (o_x, o_y, o_z) allows the target centroid to be located at a lattice point or between lattice points, as appropriate.

Having chosen d and (o_x, o_y, o_z), the target array is now taken to consist of the lattice points located within the target volume; let N be the number of such points. With these N lattice points now determined, we make a small adjustment to the lattice spacing and set $d = (V/N)^{1/3}$ (when N is large, the d so obtained is nearly the same as the original "trial" d).[2]

In Fig. 1 we show an $N = 59,728$ dipole representation of a sphere. The array fits within a $48 \times 48 \times 48$ region on the lattice. In Fig. 2 we show an $N = 61,432$

---

[2]Because each dipole "represents" a volume d^3 of material, we require the array volume $Nd^3 = V$.

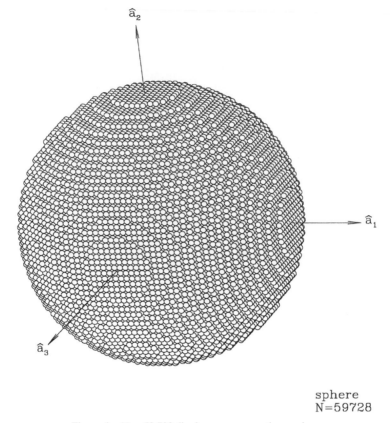

sphere
N=59728

Figure 1 $N = 59,728$ dipole array representing a sphere.

dipole representation for a regular tetrahedron. In this case the dipole array fits within a $78 \times 82 \times 96$ region on the lattice.

It is convenient to characterize the target size by the "effective radius"

$$a_{\text{eff}} \equiv \left(\frac{3V}{4\pi}\right)^{1/3} = \left(\frac{3N}{4\pi}\right)^{1/3} d; \tag{2}$$

a_{eff} is simply the radius of a sphere of equal volume. DDSCAT reports dimensionless scattering, absorption, and extinction efficiency factors, Q_{sca}, Q_{abs}, and Q_{ext}—these are simply the corresponding cross sections divided by πa_{eff}^2.

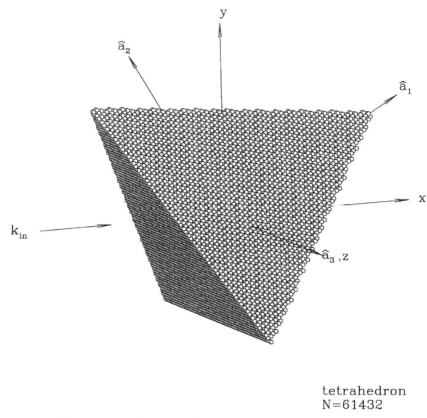

Figure 2   Array of $N = 61,432$ dipoles representing a regular tetrahedron.

VI. DIPOLE POLARIZABILITIES

After constructing a target array with lattice spacing d, it remains to specify the dipole polarizabilities α_j. In the limit $d/\lambda \to 0$, the Clausius–Mossotti prescription would apply: $\alpha^{(CM)} = (3d^3/4\pi)(\varepsilon - 1)/(\varepsilon + 2)$, where $\varepsilon = m^2$ is the complex dielectric constant. For finite d/λ, however, this is not the best choice. It is important to include radiative-reaction corrections, which are of $O[(kd)^3]$ (Draine, 1988). However, there are also corrections of $O[(kd)^2]$, and there has been some controversy about what these should be taken to be (Goedecke and O'Brien, 1988; Iskander *et al.*, 1989a; Hage and Greenberg, 1990; Dungey and Bohren, 1991; Lumme and Rahola, 1994; Okamoto, 1995). Because it is intended that the dipole array mimic a continuum material of dielectric constant ε, a natural way to specify the polarizabilities α is to require that an infinite lattice of points with this polarizability should have the same dispersion relation $k(\omega)$ as the continuum material it is intended to mimic. Using this "lattice dispersion relation" (LDR) approach, Draine and Goodman (1993) obtained a prescription for assigning dipole polarizabilities, including $O[(kd)^2]$ and $O[(kd)^3]$ corrections to the Clausius–Mossotti estimate. DDSCAT.5a8 uses these LDR polarizabilities.

For targets with large size parameters, Okamoto (1995) advocates approximating the target as a cluster of spherical monomers, with each monomer then approximated by a point dipole, but with a polarizability obtained from Mie theory. This approach (Okamoto and Xu, 1998) is similar in spirit, though different in detail, from that of Dungey and Bohren (1991).

VII. ACCURACY AND VALIDITY CRITERIA

It is intuitively clear that one validity criterion should be

$$|m|kd \lesssim 1 \tag{3}$$

because we would like to have the "phase" vary by less than approximately 1 rad between dipoles. It is also clear that we would like to have N as large as feasible, in order that the dipole array accurately mimic the target geometry.

What accuracies are obtained using the discrete dipole approximation? One way to answer this question is to apply the discrete dipole approximation to compute scattering and absorption by a sphere, for which exact results are readily available.

In Fig. 3 we show fractional errors in the absorption and scattering cross sections C_{abs} and C_{sca} for a sphere with refractive index $m = 1.7 + 0.1i$. For $N = 17,904$, accuracies are better than 2% up to $x = 9.6$, and accuracies of better than 1% are attainable for size parameter $x = ka$ as large as 21.4 using $N \approx 2 \times 10^5$ dipoles, where a is the radius.

The accuracy does depend on the refractive index m; for smaller values of $|m - 1|$ the accuracy for a given N and $|m|kd$ is generally better. Draine and Flatau (1994) show accuracies for $m = 1.33 + 0.01i, 1.7 + 0.1i, 2 + i$, and $3 + 4i$, and comparison to exact results for touching spheres has been made by Flatau *et al.* (1993).

VIII. SOLUTION METHOD

When considering $N \gtrsim 10^4$, it is apparent that direct solution of Eq. (1) by standard techniques, requiring approximately $O[(3N)^3]$ operations, is utterly infeasible. However, iterative techniques exist that find excellent approximations

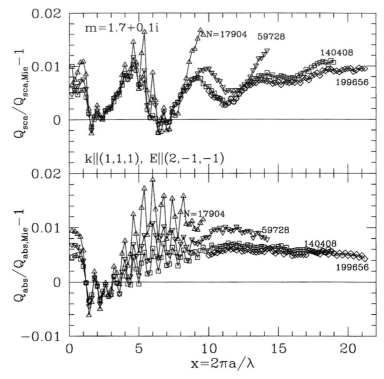

Figure 3 Fractional errors for scattering and absorption cross sections for an $m = 1.7 + 0.1i$ pseudo-sphere for different numbers N of dipoles. For each N, results are shown up to the value of x for which $|m|kd \approx 1$ [i.e., $x = (3N/4\pi)^{1/3}/|m|$]. In the "target frame," the incident radiation is propagating along the $(1, 1, 1)$ direction and polarized along the $(2, -1, -1)$ direction.

to the true solution **P** in a modest number (often only 10–100) iterations (Draine, 1988; Flatau, 1997). Each iteration involves computation of a matrix–vector product $\sum_{j=1}^{N} \mathbf{A}_{ij} \mathbf{v}_j$ or $\sum_{j=1}^{N} \mathbf{A}_{ij}^{\dagger} \mathbf{v}_j$, where $\mathbf{v}_j$ is a $3N$-dimensional vector and $\mathbf{A}^{\dagger}$ is the Hermitian conjugate of **A**. The products are evaluated using FFT techniques, as described previously. Hoekstra *et al.* (1998) have demonstrated that the FFT calculations can be parallelized. A number of different iterative procedures of the complex conjugate gradient type are available (Flatau, 1997). Draine (1988) originally employed the method of Petravic and Kuo-Petravic (1979), which DDSCAT retains as an option. Lumme and Rahola (1994) and Nebeker *et al.* (1998) recommend the "quasi-minimum residual" method (Freund, 1992) as the most computationally efficient. Flatau (1997) compared a number of different methods and recommends the "stabilized bi-conjugate gradient" method (van der Vorst, 1992) with preconditioning, available as an option within DDSCAT.

For an approximation **P**, we define the fractional error to be

$$\text{err} \equiv \frac{|\mathbf{AP} - \mathbf{E}_{\text{inc}}|}{|\mathbf{E}_{\text{inc}}|}, \qquad (4)$$

where **P** and $\mathbf{E}_{\text{inc}}$ are $3N$-dimensional vectors and **A** is the $3N \times 3N$ matrix from Eq. (1). We routinely iterate until err $< 10^{-5}$.

IX. COMPUTATIONAL REQUIREMENTS

Numerical techniques for computing scattering by irregular bodies are computationally intensive, and their utility may be limited by computational requirements.

A. MEMORY

Because of the use of FFT techniques, the memory requirements of DDSCAT are proportional to the number $N_{\text{FFT}} = N_x N_y N_z$ of sites in the "computational volume"—an $N_x \times N_y \times N_z$ region of the lattice containing the "occupied" lattice sites.[3] For a rectangular target, $N_{\text{FFT}} = N$, but for other targets $N_{\text{FFT}} > N$. For spherical targets, $N_{\text{FFT}} \approx (6/\pi)N$; for a tetrahedron, $N_{\text{FFT}} \approx 6N$.

Using 8 bytes per complex number, DDSCAT requires approximately $1.0 + 0.61(N_{\text{FFT}}/1000)$ MB. Thus a 32^3 computational volume requires only 21 MB, but a 64^3 volume would require 161 MB.

B. CENTRAL PROCESSING UNIT TIME

Most of the computing time is spent iterating until the solution vector **P** satisfies Eq. (1) to the required accuracy. The time spent per iteration scales approximately as N_{FFT}. For a given scattering problem (target geometry, refractive index m, and $x = ka_{\text{eff}}$), the number of iterations required is essentially independent of N_{FFT}, so the overall CPU time per scattering problem scales approximately linearly with N_{FFT}.

For example, for an $m = 1.7 + 0.1i$ sphere and $x = 9$, solving for two incident polarizations using a 167-MHz Sun Ultrasparc required 1400 CPU-s for the $N = 17,904$ sphere and 14,800 CPU-s for $N = 140,408$.

---

[3] When Temperton's (1992) "generalized prime factor algorithm" is used, N_x, N_y, and N_z must each be of the form $2^p 3^q 5^r$, with p, q, r integers.

X. BENCHMARK CALCULATIONS: SCATTERING BY TETRAHEDRA

In addition to the discrete dipole approximation, there are other approaches that can be used to calculate scattering and absorption by irregular targets, including the "extended boundary condition method," often referred to as the "T-matrix method" (Barber and Yeh, 1975; Mishchenko, 1991a; Chapter 6); the "finite difference time domain method" (e.g., Yang and Liou, 1996a; Chapter 7); and the "volume-integral method" (e.g., Eremin and Ivakhnenko, 1998; Chapter 2; and references therein).

The DDA has been compared to some of these other methods for various target shapes, including spheroids, cylinders, and bispheres (Hovenier *et al.*, 1996) and cubes (Wriedt and Comberg, 1998).

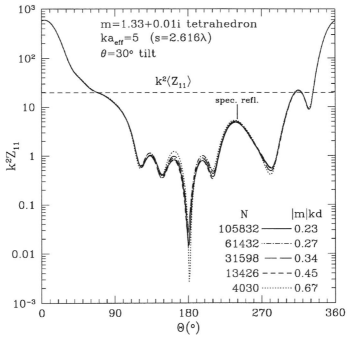

Figure 4   Scattering of unpolarized incident light by a tetrahedron with $m = 1.33 + 0.01i$ and $ka_{\rm eff} = 5$. The incident radiation is propagating in the x direction, and $k^2 Z_{11}$ is shown for scattering in the x–y plane. Axis $\hat{a}_1$ of the tetrahedron (see Fig. 2) is in the x–y plane at an angle $\theta = 30°$ from the x axis. Axis $\hat{a}_2$ of the tetrahedron is in the x–y plane. The broken line labeled $k^2 \langle Z_{11} \rangle$ is $k^2 Z_{11}$ averaged over all scattering directions for this orientation. The peak at $\Theta = 240°$ is from "specular" reflection. Accurate results for Z_{11} are obtained for $|m|kd \approx 0.5$, which appears to be sufficient for good accuracy.

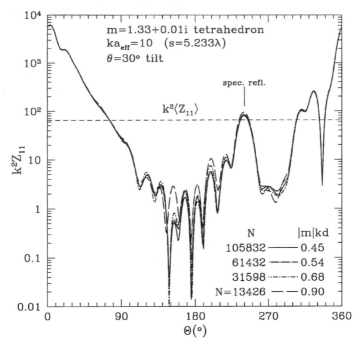

Figure 5   Same as Fig. 4, but for $ka_{\mathrm{eff}} = 10$. The specular reflection peak at $\Theta = 240°$ has become more pronounced. Once again, good accuracy is obtained for $|m|kd \approx 0.5$.

Here I present additional "benchmark" calculations that can be used to compare different techniques. From the standpoint of the discrete dipole approximation, there are no "special" shapes—a target is just a list of occupied lattice sites.

A regular tetrahedron is a simple target shape with "edges" that can be used to test different approaches to computing light scattering. The orientation is as shown in Fig. 2, with the tetrahedron axis tilted $\theta = 30°$ away from the direction of propagation of the incident radiation.

Scattering of incident unpolarized light is measured by Z_{11}, one element of the 4×4 phase matrix (see Chapter 1). Figure 4 shows $k^2 Z_{11}$ for an $m = 1.33 + 0.01i$ tetrahedron with $ka_{\mathrm{eff}} = 5$, for scattering in the x–y plane. Z_{11} measures the scattered intensity for incident unpolarized light. Results are shown for different numbers N of dipoles. For $ka_{\mathrm{eff}} = 5$, the tetrahedron side $s = 2.616\lambda$.

Increasing the size of the target by a factor of 2 in linear extent, we obtain the results shown in Fig. 5. A further increase to $s = 7.849\lambda$ gives the scattering properties shown in Fig. 6. Figures 7 and 8 show the results for a different refractive index.

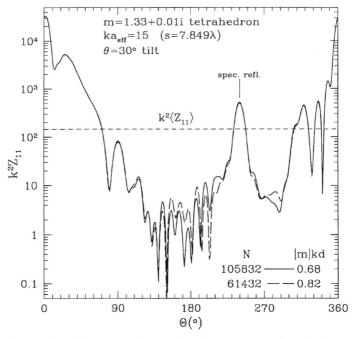

Figure 6 Same as Fig. 4, but for $x = ka_{eff} = 15$. Notice that the specular reflection peak at $\Theta = 240°$ has become more pronounced. The results are well converged where $Z_{11} \gtrsim 0.1\langle Z_{11}\rangle$, but do not appear to be fully converged for directions where the scattering is weak.

Each plot shows the value of $k^2 Z_{11}$ averaged over all scattering directions for this orientation: $k^2\langle Z_{11}\rangle = k^2 C_{sca}/4\pi$, where C_{sca} is the scattering cross section. We see that for $|m|kd \lesssim 1$ the fractional errors in Z_{11} are large only for scattering directions where the scattering is relatively weak to begin with.

In each of these plots it is interesting to note the peak in Z_{11} at a scattering angle $\Theta = 240°$, corresponding to "specular reflection" off the face of the tetrahedron upon which the radiation is incident. This specular reflection peak becomes narrower and stronger as the target size is increased, as expected from diffraction theory.

The CPU time required for these calculations is given in Table I. It will be of interest to compare these scattering results for tetrahedra with the results of other computational techniques.[4]

---

[4]For those wishing to repeat these benchmark calculations on their system, the ddscat.par files can be obtained by anonymous ftp from astro.princeton.edu, directory draine/scat/ddscat/benchmarks.

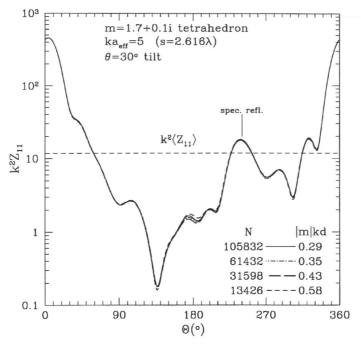

Figure 7   Same as Fig. 4, but for refractive index $m = 1.7 + 0.1i$. The results appear to be well converged for $|m|kd \lesssim 0.5$.

Table I

Timings on Sun Ultrasparc 170 (167 MHz)

N	$ka_{eff} = 5$	$ka_{eff} = 10$	$ka_{eff} = 15$
	$m = 1.33 + 0.01i$ tetrahedron: CPU time (s)		
4,030	78	—	—
13,426	311	565	—
31,598	776	1,390	2,280
61,432	1,580	2,850	4,650
105,832	3,600	6,320	11,300

N	$ka_{eff} = 5$	$ka_{eff} = 10$
	$m = 1.7 + 0.1i$ tetrahedron: CPU time (s)	
13,426	767	—
31,598	1,920	4,180
61,432	4,080	8,420
105,832	9,590	19,800

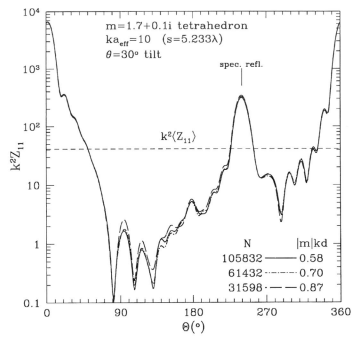

Figure 8 Same as Fig. 7, but for $x = ka_{\text{eff}} = 10$. Notice the strong specular reflection peak at $\Theta = 240°$. The good agreement between the results for $N = 61,432$ and $105,832$ indicates that $N = 105,832$ gives a good approximation to the exact results.

XI. SUMMARY

The discrete dipole approximation is an effective technique for computing scattering and absorption by irregular targets, provided the target dimension $D \lesssim 5\lambda$ and $|m − 1| \lesssim 3$, where m is the complex refractive index. The portable f77 code DDSCAT can be used to apply the discrete dipole approximation to a broad range of scattering problems. It is quite easy to apply DDSCAT to study new target geometries.

A set of benchmark calculations has been proposed, using tetrahedral targets with $m = 1.33 + 0.01i$ and $m = 1.7 + 0.1i$. It is hoped that other methods for computing scattering from irregular targets will be applied to these test problems, as this will make possible direct comparison of the accuracy and computational demands of the different numerical approaches.

ACKNOWLEDGMENTS

This research was supported in part by NSF grant AST-9619429. The author wishes to acknowledge the gracious hospitality of Osservatorio Astrofisico di Arcetri, where part of this work was completed.

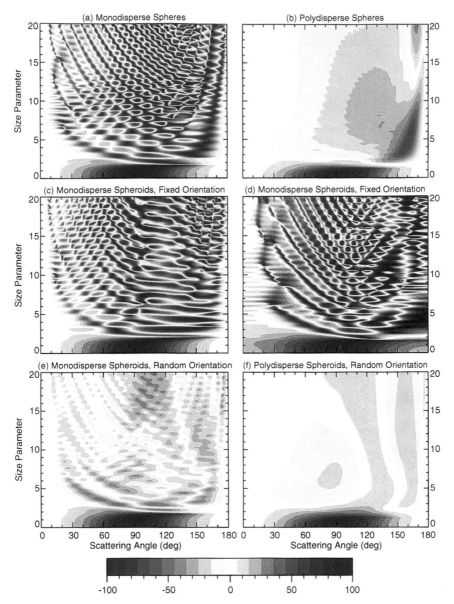

Plate 2.1 Color contour plots of the degree of linear polarization $-F_{21}/F_{11}$ (%) vs the scattering angle and size parameter for monodisperse and polydisperse spheres and surface-equivalent oblate spheroids in fixed and random orientations. External light is incident along the z axis (see Fig. 3 of Chapter 1), and the scattered beam lies in the xz plane. The scattering angle is equal to the polar angle of the scattered beam. In panel (c) the spheroid axis is parallel to the z axis, while in panel (d) the spheroid axis is parallel to the x axis. The vertical axes in panels (b) and (f) show the values of the effective size parameter for a narrow power law size distribution. The refractive index is $1.53 + 0.008i$ and the spheroid aspect ratio is 1.7.

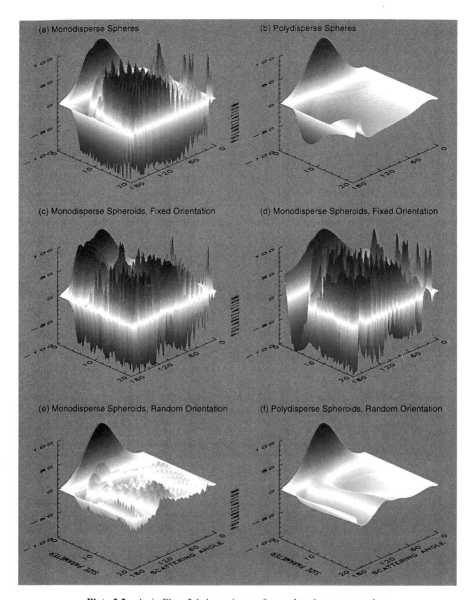

Plate 2.2 As in Plate 2.1, but using surface rather than contour plots.

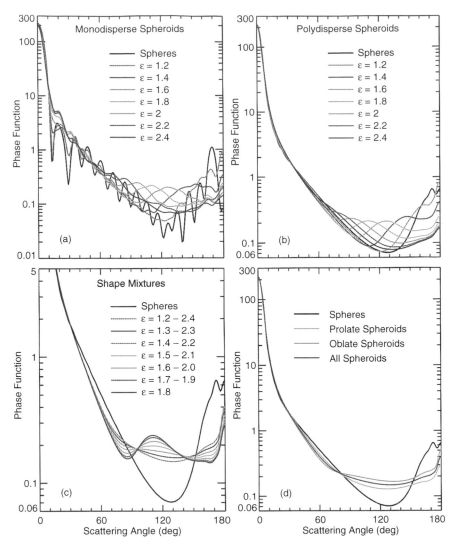

Plate 2.3 *T*-matrix computations of the phase function vs the scattering angle for monodisperse and polydisperse spheres and randomly oriented spheroids with a refractive index of 1.53 + 0.008*i*. Panel (a) shows results for monodisperse spheres with a radius of 1.163 μm and surface-equivalent prolate spheroids with aspect ratios increasing from 1.2 to 2.4. Panel (b) shows similar computations but for a log-normal size distribution with an effective radius of 1.163 μm and an effective variance of 0.168 (Mishchenko *et al.*, 1997a). Panel (c) demonstrates the effect of varying width of the spheroid aspect ratio distribution and shows ensemble-averaged phase functions for equiprobable shape mixtures of polydisperse prolate spheroids with different aspect ratio ranges. For all shape distributions the aspect ratio step size is equal to 0.1. Panel (d) shows the phase functions for polydisperse spheres and ensemble-averaged phase functions for equiprobable shape mixtures of prolate spheroids (green curve), oblate spheroids (blue curve), and prolate and oblate spheroids (red curve) with aspect ratios ranging from 1.2 to 2.4 in steps of 0.1. The wavelength is 0.443 μm.

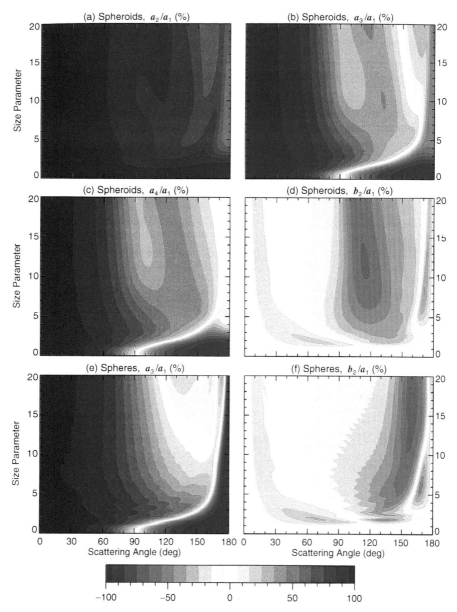

Plate 2.4 Color contour plots of normalized elements of the scattering matrix vs the scattering angle and effective size parameter for a narrow power law distribution of spheres and surface-equivalent oblate spheroids in random orientation. The refractive index is $1.53 + 0.008i$ and the spheroid aspect ratio is 1.7.

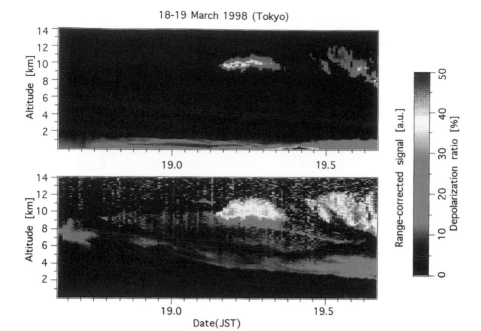

Plate 14.1 Altitude vs time (in Japan Standard Time) displays of linear depolarization ratios (see δ-value scale) and attenuated backscattering (in arbitrary units) obtained on March 18, 1998, from a lidar station in Tokyo, showing returns from cirrus clouds, an Asian dust plume, and aerosols in the moist boundary layer (courtesy of T. Murayama).

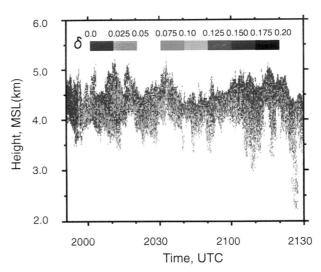

Plate 14.2 Height vs time (in UTC) displays of FARS ruby lidar displays of linear depolarization ratios (see inserted key) and attenuated backscattering (in relative units based on a logarithmic gray scale) collected on July 3, 1998, from a dense elevated smoke layer generated by a local forest fire. In this desert environment, strong scattering from moistened aerosols in the boundary layer is absent (compare with Plate 14.1).

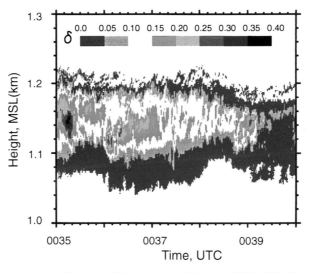

Plate 14.3 Height vs time displays of high-resolution (1.5 m by 10 Hz) PDL displays of depolarization and attenuated backscattering of a thin continental stratus cloud layer studied on September 22, 1997, at the U.S. Department of Energy Southern Great Plains CART site. The gradual decrease in δ values over the 5-min period corresponds to the narrowing of the receiver FOV from 3.8 to 0.28 mrad.

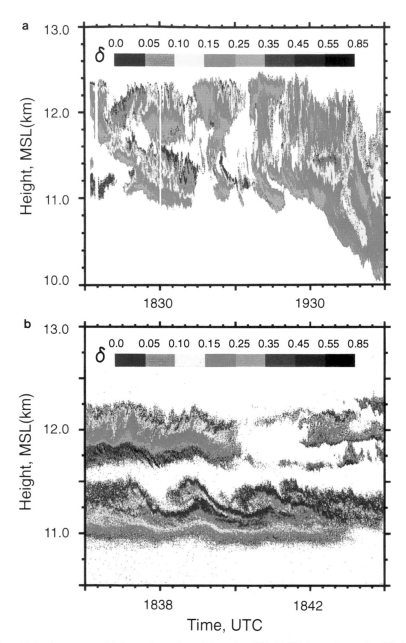

Plate 14.4 A montage of information collected from the SPG CART site on September 26, 1997, from the cirrus cloud shield produced by hurricane Nora, including PDL displays of (a) the very low depolarizing cirrus layer, (b) an expanded view of breaking Kelvin–Helmholtz waves, (c) a wide-angle photograph of some of the brilliant optical displays generated by the cirrus (courtesy of G. G. Mace), and (d) a number of replicas of the ice crystal shapes obtained from a supporting aircraft (courtesy of W. P. Arnott). *(Continues)*

Flight a092697a 190528 Temp(C) -43 Lat(deg) 36.632 Long(deg) 97.497
Press(mb) 238.455 Alt(km) 10.6565

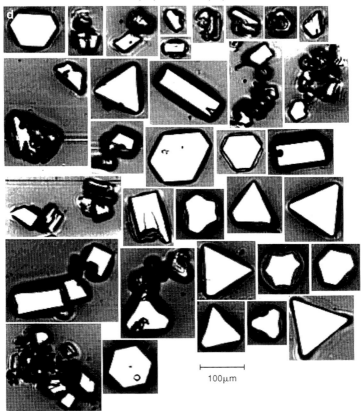

100μm

Plate 14.4 *(Continued)*

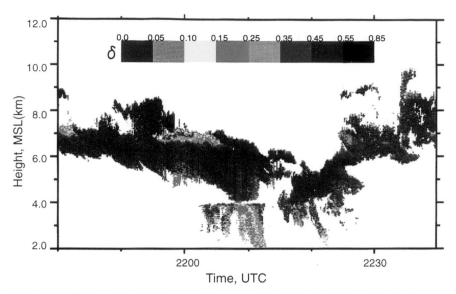

Plate 14.5   FARS zenith ruby lidar returned energy and depolarization displays depicting the passage of a summer monsoon thunderstorm that produced light rain at approximately 2212 UTC on August 7, 1992. This cross section of a propagating convective shower reveals that the lidar-observed backscattering behavior in the melting region can be quite variable, depending, presumably, on the type of ice particles producing the rain.

Chapter 6

T-Matrix Method and Its Applications

Michael I. Mishchenko
NASA Goddard Institute for Space Studies
New York, New York 10025

Larry D. Travis
NASA Goddard Institute for Space Studies
New York, New York 10025

Andreas Macke
Institut für Meereskunde
24105 Kiel, Germany

I. INTRODUCTION

The *T*-matrix method was initially introduced by Waterman (1965, 1971) as a technique for computing electromagnetic scattering by single, homogeneous nonspherical particles based on the Huygens principle (otherwise known as the extended boundary condition, Schelkunoff equivalent current method, Ewald–Oseen extinction theorem, and null-field method). However, the meant-to-be auxiliary concept of expanding the incident and the scattered waves in vector spherical wave functions (VSWFs) and relating these expansions by means of a *T* matrix has proved to be extremely powerful by itself and has dramatically expanded the realm of the *T*-matrix approach. The latter now includes electromagnetic, acoustic, and elastodynamic wave scattering by single and compounded scatterers, multiple scattering in discrete random media, and scattering by gratings and periodically rough surfaces (Varadan and Varadan, 1980; Tsang *et al.*, 1985; Varadan *et al.*, 1988).

At present, the T-matrix approach is one of the most powerful and widely used tools for rigorously computing electromagnetic scattering by single and compounded nonspherical particles. In many applications it compares favorably with other frequently used techniques in terms of efficiency, accuracy, and size parameter range and is the only method that has been used in systematic surveys of nonspherical scattering based on calculations for thousands of particles in random orientation. Recent improvements have made this method applicable to size parameters exceeding 100 and, therefore, suitable for checking the accuracy of the geometric optics approximation and its modifications at lower frequencies.

In this chapter, we review the current status of the T-matrix approach and its various applications. The chapter is composed of seven sections. The following section introduces the general concept of the T-matrix approach in application to an arbitrary nonspherical particle, either single or composite. Section III describes an efficient analytical method for computing orientation-averaged scattering characteristics for ensembles of nonspherical particles based on exploiting the rotational properties of VSWFs. In Section IV we consider the standard scheme for computing the T matrix for single scatterers, either homogeneous or layered, and discuss special techniques for improving the numerical stability of T-matrix computations for particles that are much larger than a wavelength and/or have large aspect ratios. Section V introduces the superposition T-matrix method for computing electromagnetic scattering by aggregated particles based on the translation addition theorem for VSWFs. Section VI briefly describes public-domain T-matrix codes available on the World Wide Web and discusses their ranges of applicability. The concluding section reviews multiple practical applications of the T-matrix approach.

II. THE T-MATRIX APPROACH

Consider scattering of a plane electromagnetic wave by a single particle, as discussed in Section IV of Chapter 1, and expand the incident and scattered fields in VSWFs as follows:

$$\mathbf{E}^{\text{inc}}(\mathbf{R}) = \sum_{n=1}^{\infty} \sum_{m=-n}^{n} \left[a_{mn} \operatorname{Rg} \mathbf{M}_{mn}(k\mathbf{R}) + b_{mn} \operatorname{Rg} \mathbf{N}_{mn}(k\mathbf{R}) \right], \tag{1}$$

$$\mathbf{E}^{\text{sca}}(\mathbf{R}) = \sum_{n=1}^{\infty} \sum_{m=-n}^{n} \left[p_{mn} \mathbf{M}_{mn}(k\mathbf{R}) + q_{mn} \mathbf{N}_{mn}(k\mathbf{R}) \right], \qquad R > R_>, \tag{2}$$

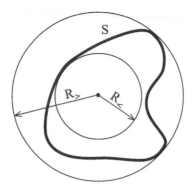

Figure 1   Cross section of a general scattering object bounded by a closed surface S. $R_>$ is the radius of the smallest circumscribed sphere and $R_<$ is the radius of a concentric inscribed sphere.

where

$$\mathbf{M}_{mn}(k\mathbf{R}) = (-1)^m d_n h_n^{(1)}(kR)\mathbf{C}_{mn}(\vartheta)\exp(im\varphi), \tag{3}$$

$$\mathbf{N}_{mn}(k\mathbf{R}) = (-1)^m d_n \left\{ \frac{n(n+1)}{kR} h_n^{(1)}(kR)\mathbf{P}_{mn}(\vartheta) \right.$$
$$\left. + \frac{1}{kR}\left[kR h_n^{(1)}(kR)\right]' \mathbf{B}_{mn}(\vartheta) \right\} \exp(im\varphi), \tag{4}$$

$$\mathbf{B}_{mn}(\vartheta) = \hat{\vartheta}\frac{d}{d\vartheta}d_{0m}^n(\vartheta) + \hat{\varphi}\frac{im}{\sin\vartheta}d_{0m}^n(\vartheta), \tag{5}$$

$$\mathbf{C}_{mn}(\vartheta) = \hat{\vartheta}\frac{im}{\sin\vartheta}d_{0m}^n(\vartheta) - \hat{\varphi}\frac{d}{d\vartheta}d_{0m}^n(\vartheta), \tag{6}$$

$$\mathbf{P}_{mn}(\vartheta) = \hat{\mathbf{R}}\frac{d_{0m}^n(\vartheta)}{R}, \tag{7}$$

$$d_n = \left[\frac{2n+1}{4\pi n(n+1)}\right]^{1/2}, \tag{8}$$

$k = 2\pi/\lambda$ is the wavenumber, λ is the wavelength in the surrounding medium, $R_>$ is the radius of the smallest circumscribing sphere of the scattering particle centered at the origin of the coordinate system (Fig. 1), and $d_{lm}^n(\vartheta)$ are Wigner d functions (Varshalovich *et al.*, 1988) given by

$$d_{lm}^n(\vartheta) = A_{lm}^n(1-\cos\vartheta)^{(l-m)/2}(1+\cos\vartheta)^{-(l+m)/2}$$
$$\times \frac{d^{n-m}}{(d\cos\vartheta)^{n-m}}\left[(1-\cos\vartheta)^{n-l}(1+\cos\vartheta)^{n+l}\right] \tag{9}$$

for $n \geq n_* = \max(|l|, |m|)$ and by $d_{lm}^n(\vartheta) = 0$ for $n < n_*$. In Eq. (9),

$$A_{lm}^n = \frac{(-1)^{n-m}}{2^n} \left[\frac{(n+m)!}{(n-l)!(n+l)!(n-m)!} \right]^{1/2}. \tag{10}$$

The d functions can be expressed in terms of generalized spherical functions as follows (Hovenier and van der Mee, 1983):

$$d_{lm}^n(\vartheta) = i^{m-l} P_{lm}^n(\cos\vartheta). \tag{11}$$

The expressions for the functions $\mathrm{Rg}\,\mathbf{M}_{mn}$ and $\mathrm{Rg}\,\mathbf{N}_{mn}$ can be obtained from Eqs. (3) and (4) by replacing spherical Hankel functions $h_n^{(1)}$ by spherical Bessel functions j_n. Note that the functions $\mathrm{Rg}\,\mathbf{M}_{mn}$ and $\mathrm{Rg}\,\mathbf{N}_{mn}$ are regular at the origin, while the use of the outgoing functions $\mathbf{M}_{mn}$ and $\mathbf{N}_{mn}$ in the expansion of Eq. (2) ensures that the scattered field satisfies the radiation condition at infinity (i.e., the transverse component of the scattered electric field decays as $1/R$, whereas the radial component decays faster than $1/R$ with $R \to \infty$). The requirement $R > R_>$ in Eq. (2) means that the scattered field is considered only outside the smallest circumscribing sphere of the scatterer (Fig. 1). The so-called Rayleigh hypothesis (e.g., Bates, 1975; Paulick, 1990) assumes that the scattered field can be expanded in outgoing waves not only in the outside region, but also in the region between the particle surface and the circumscribing sphere (Fig. 1). Because the range of applicability of this hypothesis is still unknown and is in fact questionable, as recent results by Videen *et al.* (1996) and Ngo *et al.* (1997) seem to indicate, the requirement $R > R_>$ in Eq. (2) is important in order to make sure that the Rayleigh hypothesis is not implicitly used (Lewin, 1970).

The expansion coefficients of the plane incident wave are given by the following simple analytical formulas (Tsang *et al.*, 1985, Chapter 3; Eq. (1) of Chapter 1):

$$a_{mn} = 4\pi(-1)^m i^n d_n \mathbf{C}_{mn}^*(\vartheta^{\mathrm{inc}}) \mathbf{E}_0^{\mathrm{inc}} \exp(-im\varphi^{\mathrm{inc}}), \tag{12}$$

$$b_{mn} = 4\pi(-1)^m i^{n-1} d_n \mathbf{B}_{mn}^*(\vartheta^{\mathrm{inc}}) \mathbf{E}_0^{\mathrm{inc}} \exp(-im\varphi^{\mathrm{inc}}), \tag{13}$$

where an asterisk indicates complex conjugation. Owing to the linearity of Maxwell's equations and boundary conditions, the relation between the scattered field coefficients p_{mn} and q_{mn} on the one hand and the incident field coefficients a_{mn} and b_{mn} on the other hand must be linear and is given by a transition (or T matrix) $\mathbf{T}$ as follows (Waterman, 1971; Tsang *et al.*, 1985, Chapter 3):

$$p_{mn} = \sum_{n'=1}^{\infty} \sum_{m'=-n'}^{n'} \left[T_{mnm'n'}^{11} a_{m'n'} + T_{mnm'n'}^{12} b_{m'n'} \right], \tag{14}$$

$$q_{mn} = \sum_{n'=1}^{\infty} \sum_{m'=-n'}^{n'} \left[T_{mnm'n'}^{21} a_{m'n'} + T_{mnm'n'}^{22} b_{m'n'} \right]. \tag{15}$$

In compact matrix notation, Eqs. (14) and (15) can be rewritten as

$$\begin{bmatrix} \mathbf{p} \\ \mathbf{q} \end{bmatrix} = \mathbf{T} \begin{bmatrix} \mathbf{a} \\ \mathbf{b} \end{bmatrix} = \begin{bmatrix} \mathbf{T}^{11} & \mathbf{T}^{12} \\ \mathbf{T}^{21} & \mathbf{T}^{22} \end{bmatrix} \begin{bmatrix} \mathbf{a} \\ \mathbf{b} \end{bmatrix}. \tag{16}$$

Equation (16) forms the basis of the T-matrix approach. Indeed, if the T matrix for a given scatterer is known, Eqs. (14), (15), (12), (13), and (2) give the scattered field and, thus, the amplitude matrix appearing in Eq. (4) of Chapter 1. Specifically, making use of the large argument asymptotic for spherical Hankel functions,

$$h_n^{(1)}(kR) \simeq \frac{(-1)^{n+1} \exp(ikR)}{kR}, \qquad kR \gg n^2, \tag{17}$$

we easily derive in dyadic notation

$$\mathbf{S}(\mathbf{n}^{\text{sca}}, \mathbf{n}^{\text{inc}})$$
$$= \frac{4\pi}{k} \sum_{nmn'm'} i^{n'-n-1} (-1)^{m+m'} d_n d_{n'} \exp\left[i\left(m\varphi^{\text{sca}} - m'\varphi^{\text{inc}}\right)\right]$$
$$\times \left\{ \left[T_{mnm'n'}^{11} \mathbf{C}_{mn}\left(\vartheta^{\text{sca}}\right) + T_{mnm'n'}^{21} i\mathbf{B}_{mn}\left(\vartheta^{\text{sca}}\right) \right] \mathbf{C}_{m'n'}^*\left(\vartheta^{\text{inc}}\right) \right.$$
$$\left. + \left[T_{mnm'n'}^{12} \mathbf{C}_{mn}\left(\vartheta^{\text{sca}}\right) + T_{mnm'n'}^{22} i\mathbf{B}_{mn}\left(\vartheta^{\text{sca}}\right) \right] \mathbf{B}_{m'n'}^*\left(\vartheta^{\text{inc}}\right) / i \right\}. \tag{18}$$

Knowledge of the amplitude matrix allows one to compute any scattering characteristic introduced in Chapter 1.

A fundamental feature of the T-matrix approach is that the T matrix depends only on the physical and geometrical characteristics of the scattering particle (refractive index, size, shape, and orientation with respect to the reference frame) and is completely independent of the incident and scattered fields. This means that the T matrix need be computed only once and then can be used in calculations for any directions of incidence and scattering and for any polarization state of the incident field.

Using Eq. (18) and the reciprocity relation for the amplitude matrix [Eq. (44) of Chapter 1], we derive the following general symmetry relation for the T-matrix elements:

$$T_{mnm'n'}^{kl} = (-1)^{m+m'} T_{-m'n'-mn}^{lk}, \qquad k, l = 1, 2. \tag{19}$$

Energy conservation leads to another important property of the T matrix:

$$\sum_{ln_1m_1} \left(T_{m_1n_1mn}^{lp}\right)^* T_{m_1n_1m'n'}^{lq} \leq -\frac{1}{2}\left[\left(T_{m'n'mn}^{qp}\right)^* + T_{mnm'n'}^{pq}\right]. \tag{20}$$

The equality holds only for nonabsorbing particles and is called the unitarity property (Tsang *et al.*, 1985, Chapter 3). These relations are helpful in practice for checking the numerical accuracy of computing the T matrix.

It is useful to note that the VSWFs defined by Eqs. (3) and (4) are not the only possible class of expansion functions. Other classes of expansion functions resulting in somewhat different T matrices have been used (e.g., Waterman, 1971; Barber and Yeh, 1975; Ström and Zheng, 1987). We use the VSWFs defined by Eqs. (3) and (4) because their convenient analytical properties greatly simplify mathematical derivations. Note also that Eqs. (5)–(7) are often written in terms of associated Legendre functions $P_n^m(\cos \vartheta) = (-1)^m[(n+m)!/(n-m)!]^{1/2}d_{0m}^n(\vartheta)$ (e.g., Tsang *et al.*, 1985, Chapter 3). It is well known, however, that the numerical computation of associated Legendre functions via recurrence formulas becomes highly unstable for large n and m, whereas recurrence formulas for Wigner d functions (e.g., Varshalovich *et al.*, 1988) remain quite stable and provide accurate results.

III. ANALYTICAL AVERAGING OVER ORIENTATIONS

An essential feature of the T-matrix approach is analyticity of its mathematical formulation. Indeed, the analytical properties of the special functions involved are well known and can be used to derive general properties of the T matrix and also to analytically average scattering characteristics over particle orientations. The latter feature is particularly important because, in most practical circumstances, particles are distributed over a range of orientations rather than being perfectly aligned.

To derive the rotation transformation rule for the T matrix, consider a laboratory (L) and a particle (P) coordinate system having a common origin inside the scattering particle. Let α, β, and γ be Euler angles of rotation that transform the laboratory coordinate system into the particle coordinate system (Section III of Chapter 1) and let (kR, ϑ_L, φ_L) and (kR, ϑ_P, φ_P) be the spherical coordinates of the same radius vector $k\mathbf{R}$ in the two coordinate systems, respectively. We then have

$$\mathbf{M}_{mn}(kR, \vartheta_P, \varphi_P) = \sum_{m'=-n}^{n} D_{m'm}^n(\alpha, \beta, \gamma)\mathbf{M}_{m'n}(kR, \vartheta_L, \varphi_L), \qquad (21)$$

$$\mathbf{M}_{mn}(kR, \vartheta_L, \varphi_L) = \sum_{m'=-n}^{n} \left[D_{mm'}^n(\alpha, \beta, \gamma)\right]^*\mathbf{M}_{m'n}(kR, \vartheta_P, \varphi_P), \qquad (22)$$

where

$$D_{m'm}^n(\alpha, \beta, \gamma) = \exp(-im'\alpha)d_{m'm}^n(\beta)\exp(-im\gamma) \tag{23}$$

are Wigner D functions (Varshalovich *et al.*, 1988). Analogous expansions hold for the functions $\mathbf{N}_{mn}$, $\mathrm{Rg}\,\mathbf{M}_{mn}$, and $\mathrm{Rg}\,\mathbf{N}_{mn}$. Let $\mathbf{T}(P)$ and $\mathbf{T}(L)$ be the T matrices of the particle with respect to the coordinate systems P and L, respectively. Taking into account Eqs. (1), (2), (14), (15), (21), and (22), we derive (Varadan, 1980)

$$T_{mnm'n'}^{kl}(L) = \sum_{m_1=-n}^{n} \sum_{m_2=-n'}^{n'} [D_{m'm_2}^{n'}(\alpha, \beta, \gamma)]^* T_{m_1nm_2n'}^{kl}(P)D_{mm_1}^{n}(\alpha, \beta, \gamma),$$

$$k, l = 1, 2. \tag{24}$$

If we now assume that the T matrix $\mathbf{T}(P)$ is already known and use the Euler angles of rotation α, β, and γ to specify the orientation of the particle with respect to the laboratory coordinate system, then Eq. (24) gives the particle T matrix in the laboratory coordinate system. Therefore, Eqs. (18) and (24) are ideally suited for computing orientationally averaged scattering characteristics using a single precalculated $\mathbf{T}(P)$ matrix (Tsang *et al.*, 1985, Chapter 3).

Note that for particles with special symmetries, a proper choice of the particle coordinate system can substantially simplify the T-matrix calculations. For example, for rotationally symmetric particles, it is convenient to direct the z axis of the particle coordinate system along the axis of symmetry. In this case, the $\mathbf{T}(P)$ matrix becomes diagonal with respect to azimuthal indices m and m',

$$T_{mnm'n'}^{kl}(P) = \delta_{mm'} T_{mnmn'}^{kl}(P), \tag{25}$$

where $\delta_{mm'}$ is the Kronecker delta, and also has the property

$$T_{mnmn'}^{kl}(P) = (-1)^{k+l} T_{-mn-mn'}^{kl}(P). \tag{26}$$

Other possible symmetries of the T matrix are discussed by Schulz *et al.* (1999a). The T matrix becomes especially simple for spherically symmetric particles, in which case we have for any coordinate system:

$$T_{mnm'n'}^{11} = -\delta_{nn'}b_n, \tag{27}$$

$$T_{mnm'n'}^{22} = -\delta_{nn'}a_n, \tag{28}$$

$$T_{mnm'n'}^{12} = T_{mnm'n'}^{21} \equiv 0, \tag{29}$$

where a_n and b_n are the well-known Lorenz–Mie coefficients for homogeneous spheres or their analogs for radially inhomogeneous spheres (Bohren and Huffman, 1983, Chapters 4 and 8). Moreover, in the case of spherically symmetric particles all formulas of the T-matrix approach become identical to the corresponding Lorenz–Mie formulas. Therefore, the T-matrix approach can be considered an extension of the Lorenz–Mie theory to particles without spherical symmetry.

Equation (24) can be used to develop analytical procedures for averaging scattering characteristics over particle orientations. In the practically important case of randomly oriented particles, all particle orientations are equiprobable, and the orientation distribution function $p_0(\alpha, \beta, \gamma)$ is equal to $(8\pi^2)^{-1}$ [Eq. (54) of Chapter 1]. Therefore, using Eq. (24) and the orthogonality property of Wigner D functions (Varshalovich *et al.*, 1988),

$$
\int_0^{2\pi} d\alpha \int_0^{\pi} d\beta \sin\beta \int_0^{2\pi} d\gamma \, D_{mm'}^n(\alpha, \beta, \gamma) \left[D_{m_1 m_1'}^{n'}(\alpha, \beta, \gamma) \right]^*
$$
$$
= \frac{8\pi^2}{2n+1} \delta_{nn'} \delta_{mm_1} \delta_{m'm_1'}, \tag{30}
$$

we derive for the orientation-averaged T matrix

$$
\langle T_{mnm'n'}^{kl} \rangle = \frac{1}{8\pi^2} \int_0^{2\pi} d\alpha \int_0^{\pi} d\beta \sin\beta \int_0^{2\pi} d\gamma \, T_{mnm'n'}^{kl}(L)
$$
$$
= \frac{1}{2n+1} \delta_{mm'} \delta_{nn'} \sum_{m_1=-n}^{n} T_{m_1 n m_1 n}^{kl}(P), \qquad k, l = 1, 2. \tag{31}
$$

As a result, we obtain the following general formula for the extinction cross section of randomly oriented particles (Mishchenko, 1991b):

$$
\langle C_{\text{ext}} \rangle = \frac{2\pi}{k} \, \text{Im} \left[\langle S_{11}(\mathbf{n}, \mathbf{n}) \rangle + \langle S_{22}(\mathbf{n}, \mathbf{n}) \rangle \right]
$$
$$
= -\frac{2\pi}{k^2} \, \text{Re} \sum_{n=1}^{\infty} \sum_{m=-n}^{n} \sum_{k=1}^{2} T_{mnmn}^{kk}(P). \tag{32}
$$

A similar but less simple formula was derived by Borghese *et al.* (1984). Because the choice of the particle coordinate system is arbitrary, we can conclude that the orientation-averaged extinction cross section is proportional to the real part of the trace of the T matrix computed in an arbitrary reference frame. An equally simple formula can be derived for the scattering cross section of randomly oriented particles (Mishchenko, 1991a; Khlebtsov, 1992):

$$
\langle C_{\text{sca}} \rangle = \frac{2\pi}{k^2} \sum_{n=1}^{\infty} \sum_{n'=1}^{\infty} \sum_{m=-n}^{n} \sum_{m'=-n'}^{n'} \sum_{k=1}^{2} \sum_{l=1}^{2} |T_{mnm'n'}^{kl}(P)|^2. \tag{33}
$$

The orientation-averaged extinction and scattering cross sections must be invariant with respect to the choice of the coordinate system. And indeed, using

Eq. (24) and the unitarity property of Wigner D functions (Varshalovich *et al.*, 1988),

$$\sum_{m''=-n}^{n} \left[D^n_{m''m}(\alpha, \beta, \gamma) \right]^* D^n_{m''m'}(\alpha, \beta, \gamma) = \delta_{mm'}, \tag{34}$$

we derive the following two general invariants:

$$\sum_{m} T^{kl}_{mnmn}(L) = \sum_{m} T^{kl}_{mnmn}(P), \tag{35}$$

$$\sum_{mm'} |T^{kl}_{mnm'n'}(L)|^2 = \sum_{mm'} |T^{kl}_{mnm'n'}(P)|^2, \qquad k, l = 1, 2. \tag{36}$$

Energy conservation requires that the orientation-averaged extinction cross section always be larger than or equal to the orientation-averaged scattering cross section. Therefore, the elements of the T matrix computed in an arbitrary reference frame must satisfy the partial inequality [cf. Eqs. (32), (33), (35), and (36)]

$$\sum_{nn'mm'kl} |T^{kl}_{mnm'n'}(P)|^2 \leq -\text{Re} \sum_{nmk} T^{kk}_{mnmn}(P), \tag{37}$$

where the equality holds only for lossless particles. It is easy to show that this partial inequality is consistent with Eq. (20).

Computation of the elements of the scattering matrix given by Eq. (61) of Chapter 1 requires orientation averaging of products of amplitude matrix elements. This problem was addressed by Mishchenko (1991a) for the case of rotationally symmetric particles, that is, when Eqs. (25) and (26) apply. His approach is based on exploiting the Clebsch–Gordan expansion

$$d^n_{mm'}(\beta)d^{n'}_{m_1 m_1'}(\beta) = \sum_{n_1=|n-n'|}^{n+n'} C^{n_1 m+m_1}_{nmn'm_1} C^{n_1 m'+m_1'}_{nm n'm_1'} d^{n_1}_{m+m_1, \, m'+m_1'}(\beta) \tag{38}$$

and the orthogonality relation

$$\int_0^\pi d\beta \sin\beta \, d^n_{mm'}(\beta)d^{n'}_{mm'}(\beta) = \delta_{nn'} \frac{2}{2n+1}, \tag{39}$$

where C^{ij}_{klmn} are the well-known Clebsch–Gordan coefficients, which can be efficiently computed using the recurrence formulas listed by Varshalovich *et al.* (1988). Furthermore, instead of directly computing the scattering matrix elements, Mishchenko (1991a) first computed the expansion coefficients appearing in Eqs. (72)–(77) of Chapter 1. Because both the expansion coefficients and the $\mathbf{T}(P)$ matrix elements are independent of the directions and polarization states of the incident and scattered beams, one may expect a direct relationship between these two sets of quantities that does not involve any angular variables. And indeed,

Mishchenko (1991a) derived simple analytical formulas directly expressing the expansion coefficients of Eqs. (72)–(77) of Chapter 1 in the elements of the $\mathbf{T}(P)$ matrix. As a result, the computation of the highly complicated angular structure of light scattered by a nonspherical particle in a fixed orientation (Section V of Chapter 2) with further numerical integration over orientations is avoided, thereby making the analytical averaging method very accurate and fast. The most time-consuming part in any computations based on the T-matrix method is evaluation of multiple nested summations, and an important advantage of the analytical approach is that the maximal order of nested summations involved is only three. This makes the analytical approach ideally suited to developing an efficient computer code (Section VI). Direct comparisons of the analytical method and the straightforward averaging procedure using numerical angular integrations over particle orientations (Wiscombe and Mugnai, 1986; Barber and Hill, 1990; Sid'ko *et al.*, 1990) have shown that the former is faster by a factor of several tens (Mishchenko, 1991a; W. M. F. Wauben, personal communication).

Mackowski and Mishchenko (1996) extended the analytical approach to randomly oriented particles lacking rotational symmetry. Khlebtsov (1992) and Fucile *et al.* (1993) studied the same problem but did not use the idea of expanding the scattering matrix elements in generalized spherical functions.

Mishchenko (1991b, 1992a) considered the problem of computing the extinction matrix for nonspherical particles axially oriented by an external force. An orientation distribution function symmetric with respect to the z axis of the laboratory reference frame is given by Eq. (55) of Chapter 1, which, along with Eqs. (24) and (38), leads to a simple formula for the orientationally averaged T matrix in the laboratory frame:

$$
\langle T^{kl}_{mnm'n'}(L) \rangle
$$

$$
= \int_0^{2\pi} d\alpha \int_0^{\pi} d\beta \, \sin\beta \int_0^{2\pi} d\gamma \, T^{kl}_{mnm'n'}(L; \alpha, \beta, \gamma) p_0(\alpha, \beta, \gamma)
$$

$$
= \delta_{mm'} \sum_{m_1=-M}^{M} \sum_{n_1=|n-n'|}^{n+n'} (-1)^{m+m_1} p_{n_1} C^{n_1 0}_{nmn'-m} C^{n_1 0}_{nm_1n'-m_1} T^{kl}_{m_1nm_1n'}(P), \quad (40)
$$

where $M = \min(n, n')$ and

$$
p_n = \int_0^{\pi} d\beta \, \sin\beta \, p(\beta) d^n_{00}(\beta) \tag{41}
$$

are coefficients in the expansion of the function $p(\beta)$ in Legendre polynomials:

$$
p(\beta) = \sum_{n=0}^{\infty} \frac{2n+1}{2} p_n P_n(\cos\beta). \tag{42}
$$

Equations (18) and (40) along with the optical theorem, Eqs. (35)–(41) of Chapter 1, provide a fast and accurate method for computing the ensemble-averaged extinction matrix with respect to the laboratory frame. Equation (40) was later rederived by Fucile *et al.* (1995).

The analytical orientation-averaging approach for randomly and axially oriented nonspherical particles was straightforwardly extended by Paramonov (1995) to arbitrary quadratically integrable orientation distribution functions. Unfortunately, the resulting formulas involve highly nested summations, and their efficient numerical implementation may be problematic. In this case, the standard averaging approach employing numerical integrations over orientation angles may prove to be more efficient. This approach was described by Wiscombe and Mugnai (1986), Barber and Hill (1990), and Vivekanandan *et al.* (1991) and is based on the equivalence of averaging over particle orientations and averaging over directions of light incidence and scattering and the fact that knowledge of the $\mathbf{T}(P)$ matrix enables computations of the amplitude matrix for any direction of light incidence and scattering with respect to the particle coordinate system, Eq. (18).

IV. COMPUTATION OF THE *T* MATRIX FOR SINGLE PARTICLES

The standard scheme for computing the *T* matrix for single homogeneous scatterers in the particle reference frame is called the extended boundary condition method (EBCM) and is based on the vector Huygens principle (Waterman, 1971). The general problem is to find the field scattered by an object bounded by a closed surface *S* (Fig. 1). The Huygens principle establishes the following relationship between the incident field $\mathbf{E}^{\mathrm{inc}}(\mathbf{R})$, the total external field $\mathbf{E}(\mathbf{R})$ (i.e., the sum of the incident and the scattered fields), and the surface field on the exterior of *S*:

$$\left.\begin{matrix} \mathbf{E}(\mathbf{R}) \\ 0 \end{matrix}\right\} = \mathbf{E}^{\mathrm{inc}}(\mathbf{R}) + \text{ integral over } S, \qquad \mathbf{R}\left\{\begin{matrix} \text{outside } S \\ \text{inside } S, \end{matrix}\right. \tag{43}$$

where the integral term involves the unknown surface field on the exterior of *S*. The gist of the numerical procedure is to find the surface field on the exterior of *S* by applying Eq. (43) to points inside *S* and then to use this surface field to compute the integral term on the right-hand side of Eq. (43) for points outside *S*, that is, the scattered field.

In more technical terms, the incident and the scattered waves are expanded in regular and outgoing VSWFs, respectively, according to Eqs. (1) and (2). The convergence of the expansion of Eq. (1) is guaranteed inside an inscribed sphere with radius $R_<$, whereas Eq. (2) is strictly valid only for points outside the circumscribing sphere. The internal field can also be expanded in VSWFs regular at the origin:

$$\mathbf{E}^{\text{int}}(\mathbf{R}) = \sum_{n=1}^{\infty} \sum_{m=-n}^{n} \left[c_{mn} \operatorname{Rg} \mathbf{M}_{mn}(mk\mathbf{R}) + d_{mn} \operatorname{Rg} \mathbf{N}_{mn}(mk\mathbf{R}) \right],$$
$$\mathbf{R} \text{ inside } S, \tag{44}$$

where m is the refractive index of the particle relative to that of the surrounding medium. Via boundary conditions, the surface field on the exterior of S can be expressed in the surface field on the interior of S. The latter is given by Eq. (44). As a result, the application of Eqs. (1), (43), and (44) to points with $R < R_<$ gives a matrix equation

$$\begin{bmatrix} \mathbf{a} \\ \mathbf{b} \end{bmatrix} = \begin{bmatrix} \mathbf{Q}^{11} & \mathbf{Q}^{12} \\ \mathbf{Q}^{21} & \mathbf{Q}^{22} \end{bmatrix} \begin{bmatrix} \mathbf{c} \\ \mathbf{d} \end{bmatrix}, \tag{45}$$

in which the elements of the $\mathbf{Q}$ matrix are simple surface integrals of products of VSWFs that depend only on the particle size, shape, and refractive index. Inversion of this matrix equation expresses the unknown expansion coefficients of the internal field $\mathbf{c}$ and $\mathbf{d}$ in the known expansion coefficients of the incident field $\mathbf{a}$ and $\mathbf{b}$. Analogously, the application of boundary conditions and Eq. (44) to the integral term on the right-hand side of Eq. (43) for points with $R > R_>$ and using Eq. (2) gives the following matrix expression:

$$\begin{bmatrix} \mathbf{p} \\ \mathbf{q} \end{bmatrix} = - \begin{bmatrix} \operatorname{Rg} \mathbf{Q}^{11} & \operatorname{Rg} \mathbf{Q}^{12} \\ \operatorname{Rg} \mathbf{Q}^{21} & \operatorname{Rg} \mathbf{Q}^{22} \end{bmatrix} \begin{bmatrix} \mathbf{c} \\ \mathbf{d} \end{bmatrix}, \tag{46}$$

where the elements of the $\operatorname{Rg} \mathbf{Q}$ matrix are also given by simple integrals over the particle surface and depend only on the particle characteristics. By comparing Eqs. (16), (45), and (46), we obtain

$$\mathbf{T} = -\operatorname{Rg} \mathbf{Q} \mathbf{Q}^{-1}. \tag{47}$$

Finally, Eq. (16) gives the expansion coefficients of the scattered field and, thus, the scattered field itself.

General formulas for computing the matrices $\mathbf{Q}$ and $\operatorname{Rg} \mathbf{Q}$ for particles of an arbitrary shape are given by Tsang *et al.* (1985, Chapter 3). These formulas become much simpler for rotationally symmetric particles provided that the z axis of the particle coordinate system coincides with the axis of particle symmetry [pages 187 and 188 of Tsang *et al.* (1985); cf. Eqs. (25) and (26)]. This simplicity explains why nearly all numerical results computed with EBCM pertain to bodies of revolution. However, several successful attempts have been made to apply EBCM to scatterers lacking rotational symmetry such as triaxial ellipsoids (Schneider and Peden, 1988; Schneider *et al.*, 1991) and cubes (Wriedt and Doicu, 1998; Wriedt and Comberg, 1998; Laitinen and Lumme, 1998). Peterson and Ström (1974) (see also Bringi and Seliga, 1977a; Wang and Barber, 1979) extended EBCM to layered scatterers, while Lakhtakia *et al.* (1985b) and Lakhtakia (1991) applied EBCM to light scattering by chiral particles embedded in an achiral isotropic or chiral host medium.

Alternative derivations and formulations of EBCM are discussed by Ström (1975), Barber and Yeh (1975), Agarwal (1976), Bates and Wall (1977), and Morita (1979). The derivation given by Waterman (1979) is especially simple and makes it quite clear that EBCM is not based on the Rayleigh hypothesis, and that scattering objects need not be convex and close to spherical in order to ensure the validity of EBCM. It is interesting that EBCM can in fact be derived from the Rayleigh hypothesis (Burrows, 1969; Bates, 1975; Chew, 1990, Section 8.5; Schmidt *et al.*, 1998). This does not mean, however, that EBCM is equivalent to the Rayleigh hypothesis or requires it to be valid (Lewin, 1970). The equivalence of the two approaches would follow from a reciprocal derivation of the Rayleigh hypothesis from EBCM, but this has not been done so far.

A serious practical difficulty with EBCM is the poor numerical stability of calculations for particles with very large real and/or imaginary parts of the refractive index, large sizes compared with a wavelength, and/or extreme geometries such as spheroids with large axial ratios. The origin of this problem can be explained as follows. Although the expansions of Eqs. (1) and (2) are, in general, infinite, in practical computer calculations they must be truncated to a finite maximum size. This size depends on the required accuracy of computations and is found by increasing the size of the $\mathbf{Q}$ and Rg $\mathbf{Q}$ matrices in unit steps until an accuracy criterion is satisfied. Unfortunately, different elements of the $\mathbf{Q}$ matrix can differ by many orders of magnitude, thus making the numerical calculation of the inverse matrix $\mathbf{Q}^{-1}$ an ill-conditioned process strongly influenced by round-off errors. The ill-conditionality means that even small numerical errors in the computed elements of the $\mathbf{Q}$ matrix can result in large errors in the elements of the inverse matrix $\mathbf{Q}^{-1}$. The round-off errors become increasingly significant with increasing particle size parameter and/or aspect ratio and rapidly accumulate with increasing size of the $\mathbf{Q}$ matrix. As a result, T-matrix computations for large and/or highly aspherical particles can be slowly convergent or even divergent (Barber, 1977; Varadan and Varadan, 1980; Wiscombe and Mugnai, 1986).

Efficient approaches for overcoming the numerical instability problem in computing the T matrix for highly elongated particles are the so-called iterative EBCM (IEBCM) and a closely related multiple multipole EBCM (Iskander *et al.*, 1983; Lakhtakia *et al.*, 1983; Iskander and Lakhtakia, 1984; Iskander *et al.*, 1989b; Doicu and Wriedt, 1997a, b; Wriedt and Doicu, 1997, 1998). The main idea of IEBCM is to represent the internal field by several subdomain spherical function expansions centered on the major axis of an elongated scatterer. These subdomain expansions are linked to each other by being explicitly matched in the appropriate overlapping zones. IEBCM has been used to compute light scattering and absorption by highly elongated lossy and low-loss dielectric scatterers with aspect ratios as large as 17. In some cases the use of IEBCM instead of the regular EBCM allows one to more than quadruple the maximal convergent size parameter. The disadvantage of IEBCM is that its numerical stability is achieved at the expense of a considerable increase in computer code complexity and required central processing unit (CPU) time.

Another approach to deal with the numerical instability of the regular EBCM exploits the unitarity property of the T matrix for nonabsorbing particles (Waterman, 1973; Lakhtakia *et al.*, 1984, 1985a). This technique is based on iterative orthogonalization of the T matrix, is simple and computationally efficient, and results in numerically stable T matrices for elongated and flattened spheroids with aspect ratios as large as 20. The obvious disadvantage of this approach is that it is applicable only to perfectly conducting or lossless dielectric scatterers. Wielaard *et al.* (1997) demonstrated that a better approach is to invert the $\mathbf{Q}$ matrix using a special form of the lower triangular–upper triangular (LU) factorization method. This technique is applicable not only to nonabsorbing but also to lossy particles and increases the maximum convergent size parameter for lossless and low-loss particles by a factor of several units.

Mishchenko and Travis (1994a) showed that an efficient general method for ameliorating the numerical instability of inverting the $\mathbf{Q}$ matrix is to improve the accuracy with which this matrix is calculated and inverted. Specifically, they calculated the elements of the $\mathbf{Q}$ matrix and performed the matrix inversion using extended-precision (REAL*16 and COMPLEX*32) instead of double-precision (REAL*8 and COMPLEX*16) floating-point variables. Extensive checks have shown that this approach more than doubles the maximum size parameter for which convergence of T-matrix computations can be achieved. Timing tests performed on IBM RISC workstations show that the use of extended-precision arithmetic slows computations down by a factor of only 5 to 6. Other key features of this approach are its simplicity and the fact that little additional programming effort and negligibly small extra memory are required.

An interesting method for computing the T matrix for spheroids was developed by Schulz *et al.* (1998a), who used the separation of variables method to derive the T matrix in spheroidal coordinates and then converted it into the regular T matrix in spherical coordinates.

V. AGGREGATED AND COMPOSITE PARTICLES

According to Eqs. (21) and (22), VSWFs in a rotated reference frame can be expanded in VSWFs in the original reference frame, thereby leading to a simple rotation transformation rule for the T matrix, Eq. (24). Analogously, VSWFs in a translated coordinate system can be expressed in VSWFs in the original coordinate system via the translation addition theorem, resulting in a translation transformation rule for the T matrix. The latter can be used to develop a T-matrix scheme to compute light scattering by aggregated particles. This superposition T-matrix approach was developed by Peterson and Ström (1973) (see also Peterson, 1977) for the general case of a cluster composed of an arbitrary number of nonspherical components.

Consider a cluster consisting of N arbitrarily shaped and arbitrarily oriented particles illuminated by a plane external electromagnetic wave and assume that the T matrices of each of the particles are known with respect to their local coordinate systems with origins inside the particles. Assume also that all these local coordinate systems have the same spatial orientation as the laboratory reference frame and that the smallest circumscribing spheres of the component particles centered at the origins of their respective local coordinate systems do not overlap. [Note that Peterson (1977) discusses weaker restrictions on possible particle configurations.] The total electric field scattered by the entire cluster can be represented as a superposition of individual scattering contributions from each particle:

$$\mathbf{E}^{\text{sca}}(\mathbf{R}) = \sum_{j=1}^{N} \mathbf{E}_j^{\text{sca}}(\mathbf{R}), \tag{48}$$

where $\mathbf{R}$ connects the origin of the laboratory coordinate system and the observation point. Because of electromagnetic interactions between the component particles, the individual scattered fields are interdependent and the total electric field illuminating each particle is the superposition of the external incident field $\mathbf{E}_0^{\text{inc}}$ and the sum of the individual fields scattered by all other component particles:

$$\mathbf{E}_j^{\text{inc}}(\mathbf{R}) = \mathbf{E}_0^{\text{inc}}(\mathbf{R}) + \sum_{l \neq j} \mathbf{E}_l^{\text{sca}}(\mathbf{R}), \qquad j = 1, \ldots, N. \tag{49}$$

To make use of the information contained in the jth particle T matrix, we must expand the fields incident on and scattered by this particle in VSWFs centered at the origin of the particle's local coordinate system:

$$\mathbf{E}_j^{\text{inc}}(\mathbf{R}) = \sum_{mn} [a_{mn}^j \, \text{Rg} \, \mathbf{M}_{mn}(k\mathbf{R}_j) + b_{mn}^j \, \text{Rg} \, \mathbf{N}_{mn}(k\mathbf{R}_j)]$$

$$= \sum_{mn} \left[\left(a_{mn}^{j0} + \sum_{l \neq j} a_{mn}^{jl} \right) \text{Rg} \, \mathbf{M}_{mn}(k\mathbf{R}_j) \right.$$

$$\left. + \left(b_{mn}^{j0} + \sum_{l \neq j} b_{mn}^{jl} \right) \text{Rg} \, \mathbf{N}_{mn}(k\mathbf{R}_j) \right], \qquad j = 1, \ldots, N,$$

$$\tag{50}$$

$$\mathbf{E}_j^{\text{sca}}(\mathbf{R}) = \sum_{nm} \left[p_{mn}^j \mathbf{M}_{mn}(k\mathbf{R}_j) + q_{mn}^j \mathbf{N}_{mn}(k\mathbf{R}_j) \right], \qquad R_j > R_{>j},$$

$$j = 1, \ldots, N, \tag{51}$$

where the $\mathbf{R}_j$ connects the origin of the jth particle local coordinate system and the observation point $\mathbf{R}$, $R_{>j}$ is the radius of the smallest circumscribing sphere of the jth particle, the expansion coefficients a_{mn}^{j0} and b_{mn}^{j0} describe the external incident field, and the expansion coefficients a_{mn}^{jl} and b_{mn}^{jl} describe the contribution of the lth particle to the field illuminating the jth particle:

$$\mathbf{E}_0^{\text{inc}}(\mathbf{R}) = \sum_{nm} \left[a_{mn}^{j0} \operatorname{Rg} \mathbf{M}_{mn}(k\mathbf{R}_j) + b_{mn}^{j0} \operatorname{Rg} \mathbf{N}_{mn}(k\mathbf{R}_j) \right],$$

$$j = 1, \ldots, N, \tag{52}$$

$$\mathbf{E}_l^{\text{sca}}(\mathbf{R}) = \sum_{nm} \left[a_{mn}^{jl} \operatorname{Rg} \mathbf{M}_{mn}(k\mathbf{R}_j) + b_{mn}^{jl} \operatorname{Rg} \mathbf{N}_{mn}(k\mathbf{R}_j) \right],$$

$$j, l = 1, \ldots, N. \tag{53}$$

The expansion coefficients of the illuminating and scattered fields are related via the jth particle T matrix $\mathbf{T}^j$:

$$\begin{bmatrix} \mathbf{p}^j \\ \mathbf{q}^j \end{bmatrix} = \mathbf{T}^j \left(\begin{bmatrix} \mathbf{a}^{j0} \\ \mathbf{b}^{j0} \end{bmatrix} + \sum_{l \neq j} \begin{bmatrix} \mathbf{a}^{jl} \\ \mathbf{b}^{jl} \end{bmatrix} \right), \qquad j = 1, \ldots, N. \tag{54}$$

The field scattered by the lth particle can also be expanded in VSWFs centered at the origin of the lth local coordinate system:

$$\mathbf{E}_l^{\text{sca}}(\mathbf{R}) = \sum_{\nu\mu} \left[p_{\mu\nu}^l \mathbf{M}_{\mu\nu}(k\mathbf{R}_l) + q_{\mu\nu}^l \mathbf{N}_{\mu\nu}(k\mathbf{R}_l) \right], \qquad R_l > R_{>l}, \tag{55}$$

where $\mathbf{R}_l$ connects the origin of the lth particle coordinate system and the observation point $\mathbf{R}$. Using the translation addition theorem (Tsang *et al.*, 1985, Chapter 6), the VSWFs in Eq. (55) can be expanded in regular VSWFs originating inside the jth particle:

$$\mathbf{M}_{\mu\nu}(k\mathbf{R}_l) = \sum_{nm} \left[A_{mn\mu\nu}(k\mathbf{R}_{lj}) \operatorname{Rg} \mathbf{M}_{mn}(k\mathbf{R}_j) + B_{mn\mu\nu}(k\mathbf{R}_{lj}) \operatorname{Rg} \mathbf{N}_{mn}(k\mathbf{R}_j) \right],$$

$$R_j < R_{lj}, \tag{56}$$

$$\mathbf{N}_{\mu\nu}(k\mathbf{R}_l) = \sum_{nm} \left[B_{mn\mu\nu}(k\mathbf{R}_{lj}) \operatorname{Rg} \mathbf{M}_{mn}(k\mathbf{R}_j) + A_{mn\mu\nu}(k\mathbf{R}_{lj}) \operatorname{Rg} \mathbf{N}_{mn}(k\mathbf{R}_j) \right],$$

$$R_j < R_{lj}, \tag{57}$$

where the vector $\mathbf{R}_{lj} = \mathbf{R}_l - \mathbf{R}_j$ connects the origins of the local coordinate systems of the lth and the jth particles, and the translation coefficients $A_{mn\mu\nu}(k\mathbf{R}_{lj})$

and $B_{mn\mu\nu}(k\mathbf{R}_{lj})$ are given by the analytical expressions listed on page 449 of Tsang *et al.* (1985). Comparing Eqs. (53)–(57), we finally derive in matrix notation

$$
\begin{bmatrix} \mathbf{p}^j \\ \mathbf{q}^j \end{bmatrix} = \mathbf{T}^j \left(\begin{bmatrix} \mathbf{a}^{j0} \\ \mathbf{b}^{j0} \end{bmatrix} + \sum_{l \neq j} \begin{bmatrix} \mathbf{A}(k\mathbf{R}_{lj}) & \mathbf{B}(k\mathbf{R}_{lj}) \\ \mathbf{B}(k\mathbf{R}_{lj}) & \mathbf{A}(k\mathbf{R}_{lj}) \end{bmatrix} \begin{bmatrix} \mathbf{p}^l \\ \mathbf{q}^l \end{bmatrix} \right),
$$
$$
j = 1, \ldots, N. \tag{58}
$$

Because the expansion coefficients of the external plane electromagnetic wave a_{mn}^{j0} and b_{mn}^{j0} and the translation coefficients $A_{mn\mu\nu}(k\mathbf{R}_{lj})$ and $B_{mn\mu\nu}(k\mathbf{R}_{lj})$ can be computed via closed-form analytical formulas, Eq. (58) can be considered a system of linear algebraic equations that can be solved numerically and yields the expansion coefficients of the individual scattered fields p_{mn}^j and q_{mn}^j for each of the cluster components. When these coefficients are known, Eqs. (51) and (48) give the total field scattered by the cluster.

Equation (58) becomes especially simple for a cluster composed of spherical particles because in this case the individual particle T matrices are diagonal with standard Lorenz–Mie coefficients standing along their main diagonal [Eqs. (27)–(29)]. The resulting equation is identical to that derived using the so-called multisphere superposition formulation or multisphere separation of variables technique (Bruning and Lo, 1971; Borghese *et al.*, 1979, 1984; Hamid *et al.*, 1990; Fuller, 1991; Mackowski, 1991; Ioannidou *et al.*, 1995). In this regard, the latter can be considered a particular case of the superposition T-matrix method. Solutions of Eq. (58) for clusters of spheres have been obtained using different numerical techniques (direct matrix inversion, method of successive orders of scattering, conjugate gradients method, method of iterations, recursive method) and have been extensively reported in the literature (Hamid *et al.*, 1991; Fuller, 1994a, 1995a; de Daran *et al.*, 1995; Xu, 1995; Tishkovets and Litvinov, 1996; Rannou *et al.*, 1997; Videen *et al.*, 1998). Fikioris and Uzunoglu (1979), Borghese *et al.* (1992, 1994), Fuller (1995b), Mackowski and Jones (1995), and Skaropoulos *et al.* (1994, 1996) extended the superposition approach to the case of internal aggregation by solving the problem of light scattering by spherical particles with eccentric spherical inclusions, whereas Videen *et al.* (1995b) considered a more general case of a sphere with an irregular inclusion. It should be noted that particles with single inclusions can also be treated using the standard EBCM for multilayered scatterers (Peterson and Ström, 1974).

Inversion of Eq. (58) gives (Mackowski, 1994)

$$
\begin{bmatrix} \mathbf{p}^j \\ \mathbf{q}^j \end{bmatrix} = \sum_{l=1}^{N} \mathbf{T}^{jl} \begin{bmatrix} \mathbf{a}^{l0} \\ \mathbf{b}^{l0} \end{bmatrix}, \qquad j = 1, \ldots, N, \tag{59}
$$

where the matrix $\mathbf{T}^{jl}$ transforms the expansion coefficients of the incident field centered at the lth particle into the jth-particle-centered expansion coefficients of the field scattered by the jth particle. The calculation of the $\mathbf{T}^{jl}$ matrices implies numerical inversion of a large matrix and can be a time-consuming process. However, these matrices are independent of the incident field and depend only on the cluster configuration and shapes and orientations of the component particles. Therefore, they need be computed only once and then can be used in computations for any direction and polarization state of the incident field.

Furthermore, in the far-field region the scattered-field expansions from the individual particles can be transformed into a single expansion centered at the origin of the laboratory reference frame. This single origin can represent the average of the component particle positions but in general can be arbitrary. The first step is to expand the incident and total scattered fields in VSWFs centered at the origin of the laboratory reference frame according to Eqs. (1) and (2). We again employ the translation addition theorem given by

$$\mathrm{Rg}\,\mathbf{M}_{mn}(k\mathbf{R}) = \sum_{\nu\mu}\left[\mathrm{Rg}\,A_{\mu\nu mn}(k\mathbf{R}_{0l})\mathrm{Rg}\,\mathbf{M}_{\mu\nu}(k\mathbf{R}_l)\right.$$
$$\left. + \mathrm{Rg}\,B_{\mu\nu mn}(k\mathbf{R}_{0l})\mathrm{Rg}\,\mathbf{N}_{\mu\nu}(k\mathbf{R}_l)\right], \tag{60}$$

$$\mathrm{Rg}\,\mathbf{N}_{mn}(k\mathbf{R}) = \sum_{\nu\mu}\left[\mathrm{Rg}\,B_{\mu\nu mn}(k\mathbf{R}_{0l})\mathrm{Rg}\,\mathbf{M}_{\mu\nu}(k\mathbf{R}_l)\right.$$
$$\left. + \mathrm{Rg}\,A_{\mu\nu mn}(k\mathbf{R}_{0l})\mathrm{Rg}\,\mathbf{N}_{\mu\nu}(k\mathbf{R}_l)\right] \tag{61}$$

and by reciprocal formulas

$$\mathbf{M}_{mn}(k\mathbf{R}_j) = \sum_{\nu\mu}\left[\mathrm{Rg}\,A_{\mu\nu mn}(k\mathbf{R}_{j0})\mathbf{M}_{\mu\nu}(k\mathbf{R}) + \mathrm{Rg}\,B_{\mu\nu mn}(k\mathbf{R}_{j0})\mathbf{N}_{\mu\nu}(k\mathbf{R})\right],$$
$$R > R_{j0}, \tag{62}$$

$$\mathbf{N}_{mn}(k\mathbf{R}_j) = \sum_{\nu\mu}\left[\mathrm{Rg}\,B_{\mu\nu mn}(k\mathbf{R}_{j0})\mathbf{M}_{\mu\nu}(k\mathbf{R}) + \mathrm{Rg}\,A_{\mu\nu mn}(k\mathbf{R}_{j0})\mathbf{N}_{\mu\nu}(k\mathbf{R})\right],$$
$$R > R_{j0}, \tag{63}$$

where $\mathbf{R}_{0l} = \mathbf{R} - \mathbf{R}_l$, $\mathbf{R}_{j0} = \mathbf{R}_j - \mathbf{R}$, and the translation coefficients $\mathrm{Rg}\,A_{\mu\nu mn}(k\mathbf{R}_{0l})$ and $\mathrm{Rg}\,B_{\mu\nu mn}(k\mathbf{R}_{0l})$ differ from $A_{\mu\nu mn}(k\mathbf{R}_{0l})$ and $B_{\mu\nu mn}(k\mathbf{R}_{0l})$ in that they are based on spherical Bessel functions rather than on spherical Hankel functions. We then easily derive

$$\begin{bmatrix} \mathbf{a}^{l0} \\ \mathbf{b}^{l0} \end{bmatrix} = \begin{bmatrix} \mathrm{Rg}\,\mathbf{A}(k\mathbf{R}_{0l}) & \mathrm{Rg}\,\mathbf{B}(k\mathbf{R}_{0l}) \\ \mathrm{Rg}\,\mathbf{B}(k\mathbf{R}_{0l}) & \mathrm{Rg}\,\mathbf{A}(k\mathbf{R}_{0l}) \end{bmatrix}\begin{bmatrix} \mathbf{a} \\ \mathbf{b} \end{bmatrix}, \qquad l = 1, \ldots, N, \tag{64}$$

$$\begin{bmatrix} \mathbf{p} \\ \mathbf{q} \end{bmatrix} = \sum_{j=1}^{N} \begin{bmatrix} \text{Rg A}(k\mathbf{R}_{j0}) & \text{Rg B}(k\mathbf{R}_{j0}) \\ \text{Rg B}(k\mathbf{R}_{j0}) & \text{Rg A}(k\mathbf{R}_{j0}) \end{bmatrix} \begin{bmatrix} \mathbf{p}^j \\ \mathbf{q}^j \end{bmatrix}. \tag{65}$$

Finally, using Eqs. (1), (2), (59), (64), and (65), we obtain Eq. (16), in which the cluster T matrix is given by

$$\mathbf{T} = \sum_{j,l=1}^{N} \begin{bmatrix} \text{Rg A}(k\mathbf{R}_{j0}) & \text{Rg B}(k\mathbf{R}_{j0}) \\ \text{Rg B}(k\mathbf{R}_{j0}) & \text{Rg A}(k\mathbf{R}_{j0}) \end{bmatrix} \mathbf{T}^{jl} \begin{bmatrix} \text{Rg A}(k\mathbf{R}_{0l}) & \text{Rg B}(k\mathbf{R}_{0l}) \\ \text{Rg B}(k\mathbf{R}_{0l}) & \text{Rg A}(k\mathbf{R}_{0l}) \end{bmatrix} \tag{66}$$

(Peterson and Ström, 1973; Mackowski, 1994). This cluster T matrix can be used in Eq. (18) to compute the amplitude matrix and in the analytical procedure for averaging over orientations described in Section III (Mishchenko and Mackowski, 1994; Mackowski and Mishchenko, 1996).

It is rather straightforward to derive a translation transformation law for the T matrix analogous to the rotation transformation law given by Eq. (24). Suppose that the T matrix of an arbitrary (single or clustered) nonspherical particle is known in coordinate system 1 and we seek the T matrix in a translated coordinate system 2 having the same spatial orientation. After simple manipulations, we obtain

$$\mathbf{T}(2) = \begin{bmatrix} \text{Rg A}(-k\mathbf{R}_{21}) & \text{Rg B}(-k\mathbf{R}_{21}) \\ \text{Rg B}(-k\mathbf{R}_{21}) & \text{Rg A}(-k\mathbf{R}_{21}) \end{bmatrix} \mathbf{T}(1) \begin{bmatrix} \text{Rg A}(k\mathbf{R}_{21}) & \text{Rg B}(k\mathbf{R}_{21}) \\ \text{Rg B}(k\mathbf{R}_{21}) & \text{Rg A}(k\mathbf{R}_{21}) \end{bmatrix},$$

$$\tag{67}$$

where the vector $\mathbf{R}_{21}$ originates at the origin of coordinate system 2 and connects it with the origin of coordinate system 1. Because the extinction and scattering cross sections averaged over a uniform orientation distribution must be independent of the choice of the coordinate system, Eqs. (32) and (33) lead to the following invariants with respect to translations of the coordinate system:

$$\sum_{nmk} T^{kk}_{mnmn}(2) = \sum_{nmk} T^{kk}_{mnmn}(1), \tag{68}$$

$$\sum_{nmn'm'kl} \left| T^{kl}_{mnm'n'}(2) \right|^2 = \sum_{nmn'm'kl} \left| T^{kl}_{mnm'n'}(1) \right|^2. \tag{69}$$

Different versions of the superposition T-matrix approach were derived by Chew *et al.* (1994), Tseng and Fung (1994), and Şahin and Miller (1998). An important modification of the T-matrix superposition method was developed by Ström and Zheng (1988), Zheng (1988), and Zheng and Ström (1989, 1991). Several alternative expressions for the T matrix of a composite object were derived, which enabled the authors to avoid the geometrical constraints inherent in the standard approach. As a result, this technique can be applied to composite particles with concavo-convex components and can also be used in computations for particles with extreme geometries, for example, highly elongated or flattened spheroids. In this regard, the technique can be considered a supplement to the methods for suppressing the numerical instability of the regular T-matrix approach described in the previous section.

VI. PUBLIC-DOMAIN T-MATRIX CODES

Several Fortran T-matrix codes for computing electromagnetic scattering by rotationally symmetric particles in fixed and random orientations are available on the World Wide Web at http://www.giss.nasa.gov/~crmim. The codes incorporate all the latest developments, including the analytical orientation averaging procedure for randomly oriented scatterers (Mishchenko, 1991a) and an automatic convergence procedure (Mishchenko, 1993), are extensively documented, have been thoroughly tested, and provide a reliable and efficient practical instrument. The codes compute the complete set of scattering characteristics, that is, the amplitude matrix for particles in a fixed orientation (Section IV of Chapter 1) and the optical cross sections, expansion coefficients, and scattering matrix for randomly oriented particles (Section XI of Chapter 1).

The code for two-sphere clusters with touching and separated components is based on the superposition T-matrix technique (Mackowski, 1994; Mishchenko and Mackowski, 1994, 1996). The codes for homogeneous nonspherical particles are based on EBCM and are provided in two versions. One version utilizes only double-precision floating-point variables, whereas the other one computes the T-matrix elements using extended-precision variables. The extended-precision code is slower than the double-precision code, especially on supercomputers, but allows computations for significantly larger particles. The EBCM codes have an option for inverting the **Q** matrix using either standard Gaussian elimination with partial pivoting or a special form of the LU factorization (Wielaard *et al.*, 1997). The latter approach is especially beneficial for nonabsorbing or weakly absorbing scatterers. In the present setting, the EBCM codes are directly applicable to spheroids, finite circular cylinders, and even-order Chebyshev particles (Fig. 2). Note that Chebyshev particles are rotationally symmetric bodies obtained by continuously deforming a sphere by means of a Chebyshev polynomial of degree n (Wiscombe and Mugnai, 1986). Their shape in the particle coordinate system with the z axis along the axis of symmetry is given by

$$r(\vartheta, \varphi) = r_0 \left[1 + \xi T_n (\cos \vartheta) \right], \qquad |\xi| < 1, \tag{70}$$

where r_0 is the radius of the unperturbed sphere, ξ is the deformation parameter, and $T_n(\cos \vartheta) = \cos n\vartheta$ is the Chebyshev polynomial of degree n. The codes can be easily modified to accommodate any rotationally symmetric particle having a plane of symmetry perpendicular to the axis of rotation. Mishchenko and Travis (1998) provide a detailed user guide to the EBCM codes.

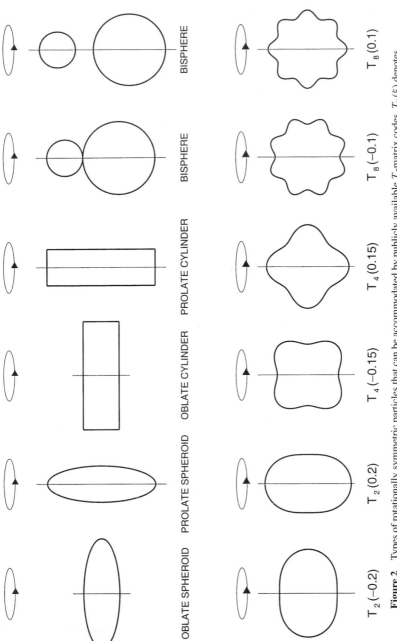

Figure 2 Types of rotationally symmetric particles that can be accommodated by publicly available T-matrix codes. $T_n(\xi)$ denotes the nth-degree Chebyshev particle with deformation parameter ξ.

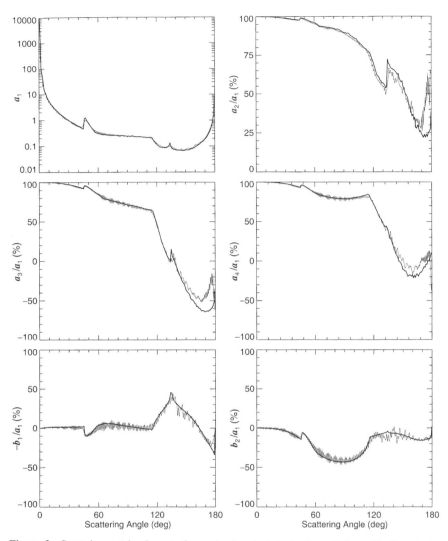

Figure 3 Scattering matrix elements for randomly oriented circular cylinders with diameter-to-length ratio 1, surface-equivalent sphere size parameter 180, and refractive index 1.311. Thin curves show T-matrix computations; thick curves represent ray-tracing results.

As for all exact numerical techniques for computing electromagnetic scattering by nonspherical particles, the performance of the T-matrix codes in terms of convergence and memory and CPU time requirements strongly depends on the options used and such particle characteristics as shape, size parameter (defined

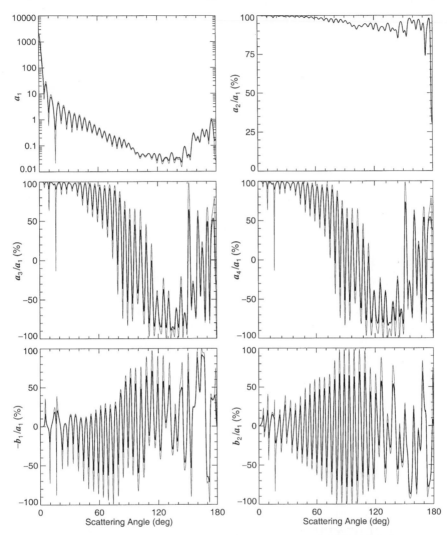

Figure 4 Scattering matrix elements for a two-sphere cluster in random orientation (thick curves) and a single sphere (thin curves). The component spheres and the single sphere have the same size parameter 40 and the same refractive index $1.5 + 0.005i$.

here as the wavenumber times the surface-equivalent sphere radius), and refractive index. For example, the maximal convergent size parameter increases from 12 for oblate spheroids with an aspect ratio of 20 and a refractive index of 1.311 to more than 160 for composition-equivalent oblate spheroids with an aspect ratio of 1.5. The sensitivity to refractive index is weaker but is also significant. The use of extended-precision variables more than doubles the maximal convergent size parameter, but makes computations slower. The use of the special LU-factorization scheme in place of standard Gaussian elimination to compute the $\mathbf{Q}^{-1}$ matrix can more than triple the maximal convergent size parameter for nonabsorbing or weakly absorbing particles. All these factors should be carefully taken into account, especially in planning massive computer calculations for large particle ensembles (Mishchenko and Travis, 1998).

Figures 3 and 4 exemplify the capabilities of the T-matrix codes. Figure 3 compares T-matrix and geometric optics computations for randomly oriented circular cylinders with diameter-to-length ratio 1, surface-equivalent sphere size parameter 180, and refractive index 1.311. The small-amplitude oscillations in the T-matrix curves are a manifestation of the interference structure typical of monodisperse particles (Section V.A of Chapter 2). For this large size parameter, the T-matrix computations closely reproduce the asymptotic geometric optics behavior, in particular, such pronounced phase function features as the $46°$ halo caused by minimum deviation at $90°$ prisms and the strong backscattering peak caused by double internal reflections from mutually perpendicular facets (Macke and Mishchenko, 1996). Figure 4 compares scattering matrix elements for a randomly oriented two-sphere cluster with identical touching components and a single sphere with size parameter equal to that of the cluster component spheres. It is obvious that the dominant feature in the cluster scattering is the single scattering from the component spheres, although this feature is somewhat reduced by cooperative scattering effects and orientation averaging (Mishchenko et al., 1995). The only distinct manifestations of the cluster nonsphericity are the departure of the ratio a_2/a_1 from unity and the inequality of the ratios a_3/a_1 and a_4/a_1 (cf. Section V.B of Chapter 2).

An older collection of EBCM codes developed by Barber and Hill (1990) is also available on the World Wide Web (Flatau, 1998). The codes use single-precision floating-point variables and do not incorporate the most recent developments.

VII. APPLICATIONS

Because of its high numerical accuracy, the T-matrix method is ideally suited for producing benchmark results. Benchmark numbers for particles in fixed and random orientations were reported by Mishchenko (1991a), Kuik et al. (1992), Mishchenko et al. (1996a), Mishchenko and Mackowski (1996), Hovenier et al. (1996), and Wielaard et al. (1997). They cover a range of equivalent-sphere size parameters from a few units to 60 and are given with up to nine correct decimals.

The great computational efficiency of the T-matrix approach has been employed by many authors to study electromagnetic scattering by representative ensembles of nonspherical particles with various shapes and sizes. Systematic computations for homogeneous and layered spheroids, finite circular cylinders, Chebyshev particles, and two-sphere clusters in random orientation were reported and analyzed by Mugnai and Wiscombe (1980, 1986, 1989), Wiscombe and Mugnai (1986, 1988), Kuik *et al.* (1994), Mishchenko and Travis (1994b, c), Mishchenko and Hovenier (1995), Mishchenko *et al.* (1995, 1996a, b), and Quirantes (1999).

The T-matrix approach has been used in many practical applications. Warner and Hizal (1976), Bringi and Seliga (1977b), Yeh *et al.* (1982), Aydin and Seliga (1984), Kummerow and Weinman (1988), Vivekanandan *et al.* (1991), Sturniolo *et al.* (1995), Haferman *et al.* (1997), Bringi *et al.* (1998), Aydin *et al.* (1998), Seow *et al.* (1998), Czekala (1998), Czekala and Simmer (1998), Prodi *et al.* (1998), and Roberti and Kummerow (1999) used T-matrix computations in remote-sensing studies of precipitation, whereas Toon *et al.* (1990), Flesia *et al.* (1994), Mannoni *et al.* (1996), and Mishchenko and Sassen (1998) analyzed depolarization measurements of stratospheric aerosols and contrail particles. Bantges *et al.* (1998) computed cirrus cloud radiance spectra in the thermal infrared wavelength region. Mishchenko *et al.* (1997b) and Mishchenko and Macke (1998) studied zenith-enhanced lidar backscatter and δ-function transmission by ice plates. Hill *et al.* (1984), Iskander *et al.* (1986), Lacis and Mishchenko (1995), Khlebtsov and Mel'nikov (1995), Mishchenko *et al.* (1997a), Kahn *et al.* (1997), Liang and Mishchenko (1997), Krotkov *et al.* (1997, 1999), Pilinis and Li (1998), and von Hoyningen-Huene (1998) modeled scattering properties of soil particles and mineral and soot aerosols using size/shape mixtures of randomly oriented spheroids. Carslaw *et al.* (1998), Tsias *et al.* (1998), and Trautman *et al.* (1998) applied the T-matrix technique to remote sensing of polar stratospheric clouds. Kouzoubov *et al.* (1998) computed the scattering matrix for nonspherical ocean water particulates. Kolokolova *et al.* (1997) used T-matrix computations to model the photometric and polarization properties of nonspherical cometary dust grains. Khlebtsov *et al.* (1996) calculated the extinction properties of colloidal gold sols. Quirantes and Delgado (1995, 1998) and Jalava *et al.* (1998) applied the T-matrix method to particle size/shape determination. Nilsson *et al.* (1998) analyzed near and far fields originating from light interaction with a spheroidal red blood cell. Latimer and Barber (1978), Barber and Wang (1978), Wang *et al.* (1979), Goedecke and O'Brien (1988), Iskander *et al.* (1989a), Evans and Fournier (1994), Streekstra *et al.* (1994), Macke *et al.* (1995), Peltoniemi (1996), Wielaard *et al.* (1997), Mishchenko *et al.* (1997b), Baran *et al.* (1998), Mishchenko and Macke (1998), and Liu *et al.* (1998) used numerically exact T-matrix computations to check the accuracy of various approximate and numerical approaches. Lai *et al.* (1991), Mazumder *et al.* (1992), Ngo and Pinnick (1994), and Borghese *et al.* (1998) analyzed the effect of nonsphericity and inhomogeneity on morphology-dependent resonances in small particles.

Mishchenko (1996) studied coherent effects in two-sphere clusters. Pitter *et al.* (1998) analyzed second-order fluctuations of the polarization state of light scattered by ensembles of randomly positioned spheroidal particles. Ruppin (1998) studied polariton modes of spheroidal microcrystals of dispersive materials over a wide range of spheroid sizes and eccentricities. Ho and Allen (1994) and Liu *et al.* (1999) analyzed the effect of nonsphericity on numerical solutions of inverse problems. Other applications of the T-matrix method were reported by Geller *et al.* (1985), Hofer and Glatter (1989), Ruppin (1990), Ryde and Matijević (1994), Xing and Greenberg (1994), Lumme and Rahola (1998), Balzer *et al.* (1998), Mishchenko and Macke (1999), Evans *et al.* (1999), and Petrova (1999).

ACKNOWLEDGMENTS

We thank Nadia Zakharova for help with the graphics and Zoe Wai for bibliographical assistance.

Chapter 7

Finite Difference Time Domain Method for Light Scattering by Nonspherical and Inhomogeneous Particles

Ping Yang and K. N. Liou
Department of Atmospheric Sciences
University of California, Los Angeles
Los Angeles, California 90095

I. Introduction
II. Conceptual Basis of the Finite Difference Time Domain Method
III. Finite Difference Equations for the Near Field
  A. Scheme 1
  B. Scheme 2
  C. Scheme 3
  D. Schemes 4, 5, and 6
IV. Absorbing Boundary Condition
  A. Mur's Absorbing Boundary Condition
  B. Liao's Transmitting Boundary Condition

C. Perfectly Matched Layer Absorbing Boundary Condition
V. Field in Frequency Domain
VI. Transformation of Near Field to Far Field
  A. Scattered Far Field
  B. Extinction and Absorption Cross Sections
VII. Scattering Properties of Aerosols and Ice Crystals
  A. Aerosols
  B. Small Ice Crystals
VIII. Conclusions

Light Scattering by Nonspherical Particles: Theory, Measurements, and Applications

I. INTRODUCTION

The finite difference time domain (FDTD) technique has been demonstrated to be one of the most robust and efficient computational methods to solve for the interaction of electromagnetic waves with scatterers, particularly those with complicated geometries and inhomogeneous compositions. In this method, the space containing a scattering particle is discretized by using a grid mesh and the existence of the particle is represented by assigning suitable electromagnetic constants in terms of permittivity, permeability, and conductivity over the grid points. Because it is not necessary to impose the electromagnetic boundary conditions at the particle surface, the FDTD approach with appropriate and minor modifications can be applied to the solution of light scattering by various nonspherical and inhomogeneous particles such as irregular ice crystals and aerosols with inclusions.

Conventional numerical methods for light scattering solve the electromagnetic wave equations in the frequency domain. However, the FDTD approach directly seeks numerical solutions of Maxwell's equations in the time domain. Mathematically, Maxwell's equations in the frequency domain are elliptic and the solution of the scattering problem for an incident electromagnetic wave is carried out as a boundary value problem. On the other hand, Maxwell's equations are hyperbolic if they are expressed in the time domain and the scattering process is described as an initial value problem whose solution is relatively simpler, particularly when a complicated particle geometry is involved. Moreover, it has been recognized that the time domain approach can be more efficient in the numerical modeling of electromagnetic interactions (Holland *et al.*, 1991).

The FDTD method was developed and pioneered by Yee (1966), but it did not receive significant recognition until high-quality absorbing boundary conditions were derived in the 1980s. Through the persistent efforts of a number of electrical engineers and computational physicists (Taflove, 1980, 1995; Kunz and Luebbers, 1993; Holland *et al.*, 1980), several advantages of the FDTD method have become widely recognized. In recent years, the FDTD technique has been applied to solve for the interactions between targets and electromagnetic waves involving such problems as antenna scattering, numerical modeling of microstrip structures, and electromagnetic absorption by human tissues (Andrew *et al.*, 1997; Sheen *et al.*, 1990; Sullivan *et al.*, 1987). Applications of this method to the solutions of the scattering and polarization properties of atmospheric nonspherical particles have also been carried out recently by Yang and Liou (1995, 1996a, 1998b) and Tang and Aydin (1995).

This chapter is organized as follows. In Section II, the physical basis of the FDTD technique is reviewed. In Section III, we recapitulate the FDTD algorithm involving the computation of the near field. Six numerical schemes for the discretization of Maxwell's equations in time and space are presented. In Section IV, we review the three algorithms for the absorbing boundary condition that have been used to suppress the artificial reflection from the boundary of the computational domain. Presented in Section V is the transformation of the near field from the time domain to the frequency domain. In Section VI, we present the fundamental integral equations for the mapping of the near field to the far field. The amplitude matrix is explicitly formulated with respect to the two incident polarization configurations parallel and perpendicular to the scattering plane. In Section VII, we discuss the scattering and polarization properties of nonspherical ice crystals and aerosols that are computed using the FDTD technique. Finally, conclusions are given in Section VIII.

II. CONCEPTUAL BASIS OF THE FINITE DIFFERENCE TIME DOMAIN METHOD

The FDTD technique is a direct implementation of Maxwell's time-dependent curl equations to solve for the temporal variation of electromagnetic waves within a finite space that contains the scattering object. In practice, this space is discretized by a number of rectangular cells of which a grid mesh is composed. Variations of the electromagnetic properties as functions of the spatial location are specified by defining the permittivity, permeability, and conductivity at each grid point, as shown in the conceptual diagram in Fig.1a. The time-dependent Maxwell's curl equations are subsequently discretized by using the finite difference approximation in both time and space. At the initial time $t = 0$, a plane wave source, not necessarily harmonic, is turned on. The excited wave then propagates toward the particle and eventually interacts with it, thereby causing a scattering event. The spatial and temporal variations of the electromagnetic field are simulated by directly applying the discretized Maxwell's equations in a manner of time-marching iterations over the entire computational domain. Information on the convergent scattered field can be obtained when a steady-state field is established at each grid point if a sinusoidal source is used or when the electric and magnetic fields in the computational domain have reduced to significantly small values if a pulse source is implemented. The second approach is more popular in practical computations because a time domain pulse can provide a wide frequency range.

The conventional FDTD numerical algorithm is based on Cartesian grid meshes. When a scattering particle with a nonrectangular surface is discretized over a Cartesian grid mesh, a staircasing effect is inherent because of the step-by-step approximation of the particle shape. In recent years, significant efforts have been focused on various FDTD algorithms associated with global curvilinear and obliquely Cartesian grids (Fusco, 1990; Fusco *et al.*, 1991; Jurgens *et al.*, 1992; Lee, 1993) and local target-conforming grids (Holland *et al.*, 1991;

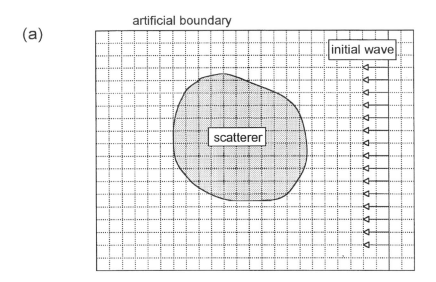

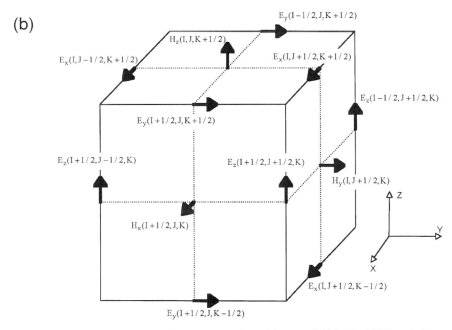

Figure 1 (a) Conceptual diagram for the computation of the near field by the FDTD technique. (b) Locations of various field components on a cubic cell.

Yee *et al.*, 1992). These endeavors are employed to avoid the staircasing approximation of an oblique or curved surface in a rectangular Cartesian mesh. In addition, numerical schemes based on computational fluid dynamics have received considerable attention of late for some special electromagnetic problems, such as the propagation of a pulse in which a steep gradient appears (Vinh *et al.*, 1992; Omick and Castillo, 1993). To economize the computer memory and central processing unit (CPU) time demands, FDTD algorithms that allow a coarse grid size and the subgridding technique are also subjects of active research (Kunz and Simpson, 1981; Cole, 1995). Although the curvilinear grid and target-conforming schemes are more accurate, they are usually derived for some special geometries and are relatively inflexible when the scatterers have various sizes and shapes. In addition, the cells in a globally irregular mesh usually differ in size so much that one must use a small time increment in order to obtain a stable solution. Further, irregular schemes are inherently more complicated and tedious than rectangular Cartesian schemes. In particular, the implementation of absorbing boundary conditions for the truncation of the computational domain is not as straightforward for a curvilinear grid mesh as for a Cartesian one. It has been shown that the staircasing effect is not a serious problem for the FDTD technique when it is applied to the computation of light scattering by nonferromagnetic and nonconducting ice crystals and aerosols, once a proper method is developed to evaluate the dielectric constants over the grid points (Yang and Liou, 1995, 1996a).

Although the actual process of scattering of an electromagnetic wave by a particle occurs in unbounded space, it must be truncated by imposing artificial boundaries in practical applications of the FDTD technique. In order for the simulated field within the truncated region to be the same as in the unbounded case, an artificial boundary must be imposed with a property known as the absorbing or transmitting boundary condition. Otherwise, the spurious reflections off the boundary would contaminate the near field within the truncated domain. The construction of an efficient absorbing boundary condition is an important aspect of the FDTD technique. An inappropriate boundary condition may lead to numerical instability. In addition, an absorbing boundary condition with poor performance may require a substantially large free space between a modeled scatterer and the boundary, thereby wasting computer memory and CPU time.

Values of the near field computed by the finite difference analog of Maxwell's equations are in the time domain. To obtain the frequency response of the scattering particle, one must transform the field from the time domain to the frequency domain. If we use a Gaussian pulse as an initial excitation, the discrete Fourier transform technique can be employed to obtain the frequency spectrum of the time-dependent signal. This procedure, however, is not so straightforward. In order to avoid numerical aliasing and dispersion, one must correctly select the width of the pulse in the time domain and properly consider the available frequency spectrum provided by the pulse.

To obtain the particle scattering and polarization properties involving the phase and extinction matrices, one must make the transformation of the frequency response from the near field to the far field. To do that, a common approach has been used that invokes a surface-integration technique on the basis of the electromagnetic equivalence principle (Umashankar and Taflove, 1982; Britt, 1989; Yang and Liou, 1995) associated with the tangential components of the electric and magnetic fields on a surface enclosing the particle. Because it is equivalent in electrodynamics to define either the field everywhere on the particle surface or the field everywhere inside the particle in the case of nonconducting object, a volume-integration technique can also be used to obtain the far field solution, as formulated by Yang and Liou (1996a).

On the basis of the preceding discussions, the major steps required in the application of the FDTD technique to the solution of light scattering by a particle can be summarized as follows:

1. Discretize the finite space containing the particle by a grid mesh and simulate the field in this region by the finite difference analog of Maxwell's time-dependent curl equations.
2. Apply an absorbing boundary condition to suppress the spurious reflection from the boundary of the computational domain.
3. Transform the near field from the time domain to the frequency domain.
4. Transform the near field in the frequency domain to the corresponding far field based on a rigorous electromagnetic integral method.

III. FINITE DIFFERENCE EQUATIONS FOR THE NEAR FIELD

As stated in the preceding section, the advantage of the FDTD technique is that the electromagnetic wave is simulated in the time domain so that its interaction with a target is formulated as an initial value problem. The well-known time-dependent Maxwell's curl equations are given by

$$\nabla \times \mathbf{H}(\mathbf{r}, t) = \frac{\varepsilon(\mathbf{r})}{c} \frac{\partial \mathbf{E}(\mathbf{r}, t)}{\partial t}, \tag{1a}$$

$$\nabla \times \mathbf{E}(\mathbf{r}, t) = -\frac{1}{c} \frac{\partial \mathbf{H}(\mathbf{r}, t)}{\partial t}, \tag{1b}$$

where ε is the permittivity of the dielectric medium, usually a complex variable, and c is the speed of light in vacuum. In Eq. (1b), the permeability has been assumed to be unity because cloud and aerosol particles in the atmosphere and many other scattering targets are mostly nonferromagnetic materials. When a particle is absorptive, the imaginary part of the permittivity is nonzero. In this case, effective values can be introduced (Yang and Liou, 1996a) to circumvent the complex calculations required in Eq. (1a). The two effective parameters are defined by ε_r and $kc\varepsilon_i/4\pi$, respectively, where ε_i and ε_r are the imaginary and real parts of the permittivity, $k = 2\pi/\lambda$ is the wavenumber of the incident radiation, and λ is the wavelength. Based on the effective dielectric constants, Eq. (1a) can be expressed equivalently as follows:

$$\nabla \times \mathbf{H}(\mathbf{r}, t) = \frac{\varepsilon_r(\mathbf{r})}{c} \frac{\partial \mathbf{E}(\mathbf{r}, t)}{\partial t} + k\varepsilon_i(\mathbf{r})\mathbf{E}(\mathbf{r}, t). \tag{2}$$

In the formulation, we select a harmonic time-dependent factor of $\exp(-ikct)$ for the electromagnetic wave in the frequency domain. The permittivity can then be related to the refractive index m as

$$\varepsilon_r = m_r^2 - m_i^2, \tag{3a}$$

$$\varepsilon_i = 2m_r m_i, \tag{3b}$$

where m_r and m_i are the real and imaginary parts of the refractive index, respectively. Note that the imaginary part of the refractive index is negative if the harmonic time-dependent factor is selected as $\exp(jkct)$ (see Section IV of Chapter 1).

We can now use Eqs. (1b) and (2) to construct the finite difference analog of Maxwell's equations. We first discretize the computational space that contains the particle by using a number of small rectangular cells. A spatial location in the discretized space is denoted by the indices $(I, J, K) = (I\Delta x, J\Delta y, K\Delta z)$ and any variable as a function of space and time is defined as

$$F^n(I, J, K) = F(I\Delta x, J\Delta y, K\Delta z, n\Delta t),$$

in which Δx, Δy, and Δz are the cell dimensions along the x, y, and z axes, respectively, and Δt is the time increment. The permittivity must be homogeneous within each cell. For a given cell with its center located at a lattice index (I, J, K), the mean permittivity can be evaluated on the basis of the Maxwell-Garnett (1904) rule via

$$\frac{\overline{\varepsilon}(I, J, K) - 1}{\overline{\varepsilon}(I, J, K) + 2} = \frac{1}{\Delta x \Delta y \Delta z} \iiint_{\text{cell}(I,J,K)} \frac{\varepsilon(x, y, z) - 1}{\varepsilon(x, y, z) + 2} \, dx \, dy \, dz. \tag{4}$$

Using the mean permittivity produces smaller staircasing errors than using a sharp step-by-step approximation for a nonspherical geometry (Yang and Liou, 1996a).

Following Yee (1966), we select components of the magnetic field at the center of cell faces and the electric field counterparts at the cell edges, as shown in Fig. 1b. Such an arrangement ensures that the tangential components of the E field and the normal components of the H field are continuous at the cell interfaces. In reference to Fig. 1b, the general form of the finite difference analog of

Eqs. (1b) and (2) can be written for each Cartesian component as follows:

$$E_x^{n+1}\left(I, J + \frac{1}{2}, K + \frac{1}{2}\right)$$

$$= a\left(I, J + \frac{1}{2}, K + \frac{1}{2}\right) E_x^n\left(I, J + \frac{1}{2}, K + \frac{1}{2}\right) + b\left(I, J + \frac{1}{2}, K + \frac{1}{2}\right)$$

$$\times \left\{ \frac{c\Delta t}{\Delta y}\left[H_z^{n+1/2}\left(I, J + 1, K + \frac{1}{2}\right) - H_z^{n+1/2}\left(I, J, K + \frac{1}{2}\right)\right]\right.$$

$$\left. + \frac{c\Delta t}{\Delta z}\left[H_y^{n+1/2}\left(I, J + \frac{1}{2}, K\right) - H_y^{n+1/2}\left(I, J + \frac{1}{2}, K + 1\right)\right]\right\}, \quad (5a)$$

$$E_y^{n+1}\left(I + \frac{1}{2}, J, K + \frac{1}{2}\right)$$

$$= a\left(I + \frac{1}{2}, J, K + \frac{1}{2}\right) E_y^n\left(I + \frac{1}{2}, J, K + \frac{1}{2}\right) + b\left(I + \frac{1}{2}, J, K + \frac{1}{2}\right)$$

$$\times \left\{ \frac{c\Delta t}{\Delta x}\left[H_z^{n+1/2}\left(I, J, K + \frac{1}{2}\right) - H_z^{n+1/2}\left(I + 1, J, K + \frac{1}{2}\right)\right]\right.$$

$$\left. + \frac{c\Delta t}{\Delta z}\left[H_x^{n+1/2}\left(I + \frac{1}{2}, J, K + 1\right) - H_x^{n+1/2}\left(I + \frac{1}{2}, J, K\right)\right]\right\}, \quad (5b)$$

$$E_z^{n+1}\left(I + \frac{1}{2}, J + \frac{1}{2}, K\right)$$

$$= a\left(I + \frac{1}{2}, J + \frac{1}{2}, K\right) E_z^n\left(I + \frac{1}{2}, J + \frac{1}{2}, K\right) + b\left(I + \frac{1}{2}, J + \frac{1}{2}, K\right)$$

$$\times \left\{ \frac{c\Delta t}{\Delta y}\left[H_x^{n+1/2}\left(I + \frac{1}{2}, J, K\right) - H_x^{n+1/2}\left(I + \frac{1}{2}, J + 1, K\right)\right]\right.$$

$$\left. + \frac{c\Delta t}{\Delta x}\left[H_y^{n+1/2}\left(I + 1, J + \frac{1}{2}, K\right) - H_y^{n+1/2}\left(I, J + \frac{1}{2}, K\right)\right]\right\}, \quad (5c)$$

$$H_x^{n+1/2}\left(I + \frac{1}{2}, J, K\right)$$

$$= H_x^{n-1/2}\left(I + \frac{1}{2}, J, K\right)$$

$$+ \left\{ \frac{c\Delta t}{\Delta y}\left[E_z^n\left(I + \frac{1}{2}, J - \frac{1}{2}, K\right) - E_z^n\left(I + \frac{1}{2}, J + \frac{1}{2}, K\right)\right]\right.$$

$$\left. + \frac{c\Delta t}{\Delta z}\left[E_y^n\left(I + \frac{1}{2}, J, K + \frac{1}{2}\right) - E_y^n\left(I + \frac{1}{2}, J, K - \frac{1}{2}\right)\right]\right\}, \quad (5d)$$

$$H_y^{n+1/2}\left(I, J + \frac{1}{2}, K\right)$$

$$= H_y^{n-1/2}\left(I, J + \frac{1}{2}, K\right)$$

$$+ \left\{\frac{c\Delta t}{\Delta z}\left[E_x^n\left(I, J + \frac{1}{2}, K - \frac{1}{2}\right) - E_x^n\left(I, J + \frac{1}{2}, K + \frac{1}{2}\right)\right]\right.$$

$$\left. + \frac{c\Delta t}{\Delta x}\left[E_z^n\left(I + \frac{1}{2}, J + \frac{1}{2}, K\right) - E_z^n\left(I - \frac{1}{2}, J + \frac{1}{2}, K\right)\right]\right\}, \quad (5e)$$

$$H_z^{n+1/2}\left(I, J, K + \frac{1}{2}\right)$$

$$= H_z^{n-1/2}\left(I, J, K + \frac{1}{2}\right)$$

$$+ \left\{\frac{c\Delta t}{\Delta x}\left[E_y^n\left(I - \frac{1}{2}, J, K + \frac{1}{2}\right) - E_y^n\left(I + \frac{1}{2}, J, K + \frac{1}{2}\right)\right]\right.$$

$$\left. + \frac{c\Delta t}{\Delta y}\left[E_x^n\left(I, J + \frac{1}{2}, K + \frac{1}{2}\right) - E_x^n\left(I, J - \frac{1}{2}, K + \frac{1}{2}\right)\right]\right\}. \quad (5f)$$

It can be proven that the truncation errors of this finite difference analog of Maxwell's curl equations are of second order both in time and in space. Other schemes with truncation errors of high order have been suggested (Shlager *et al.*, 1993), but they are less practical. From Eqs. (5a)–(5f) we see that the E and H fields are interlaced both in time and in space. These equations are in explicit forms that can be applied to the time-marching iteration directly, provided that the initial values of the electric and magnetic fields are given. The propagation of the wave can then be simulated by updating the E and H fields in a straightforward manner without imposing the electromagnetic boundary condition at the particle surface. Because this finite difference iterative scheme is completely explicit without the requirement of revision of the coefficient matrix of a set of linear equations, the FDTD technique is simple in concept and is also efficient in numerical computations. It should be pointed out that the location of the spatial and temporal increments Δx, Δy, Δz, and Δt cannot be specified arbitrarily. To circumvent numerical instability, the cell dimensions and time increments must satisfy the Courant–Friedrichs–Levy (CFL) condition (Taflove and Brodwin, 1975) in the form

$$c\Delta t \leq \frac{1}{\sqrt{1/\Delta x^2 + 1/\Delta y^2 + 1/\Delta z^2}}. \quad (6)$$

In addition to the preceding CFL condition, the spatial increments, Δx, Δy, and Δz, should also be smaller than approximately $1/20$ of the incident wavelength so that the phase variation of the electromagnetic wave is negligible over the distance of the cell dimensions. Determination of the coefficients a and b in Eqs. (5a)–(5c) depends on the scheme that is used to discretize the temporal and spatial derivatives. Based on various integral approximations, six schemes are presented in the following.

A. SCHEME 1

Consider the equation for the E_z component as an example. Integration of the z component on the right-hand side of Eq. (2) over a rectangular region enclosed by four apices with grid indices (I, J, K), $(I, J + 1, K)$, $(I + 1, J, K)$, and $(I + 1, J + 1, K)$ leads to the following:

$$
\int_{I\Delta x}^{(I+1)\Delta x} \int_{J\Delta y}^{(J+1)\Delta y} \mathbf{z} \cdot \left[\frac{\varepsilon_r(\mathbf{r})}{c} \frac{\partial \mathbf{E}(\mathbf{r}, t)}{\partial t} + k\varepsilon_i(\mathbf{r})\mathbf{E}(\mathbf{r}, t) \right]_{z=K\Delta z} dx\, dy
$$

$$
\approx \left[\frac{\bar{\varepsilon}_r(I + 1/2, J + 1/2, K)}{c} \frac{\partial E_z(I + 1/2, J + 1/2, K)}{\partial t} \right.
$$

$$
\left. + k\bar{\varepsilon}_i\left(I + \frac{1}{2}, J + \frac{1}{2}, K\right) E_z\left(I + \frac{1}{2}, J + \frac{1}{2}, K\right) \right] \Delta x\, \Delta y, \quad (7)
$$

where $\mathbf{z}$ is the unit vector along the z axis of the Cartesian coordinate system, and the mean values of the real and imaginary parts of the permittivity at the location indicated by lattice index $(I + 1/2, J + 1/2, K)$ are determined by the averages of those associated with four adjacent cells as follows:

$$
\bar{\varepsilon}_r\left(I + \frac{1}{2}, J + \frac{1}{2}, K\right) = \frac{1}{4}\left[\bar{\varepsilon}_r(I, J, K) + \bar{\varepsilon}_r(I, J + 1, K) + \bar{\varepsilon}_r(I + 1, J, K)\right.
$$

$$
\left. + \bar{\varepsilon}_r(I + 1, J + 1, K)\right], \quad (8a)
$$

$$
\bar{\varepsilon}_i\left(I + \frac{1}{2}, J + \frac{1}{2}, K\right) = \frac{1}{4}\left[\bar{\varepsilon}_i(I, J, K) + \bar{\varepsilon}_i(I, J + 1, K) + \bar{\varepsilon}_i(I + 1, J, K)\right.
$$

$$
\left. + \bar{\varepsilon}_i(I + 1, J + 1, K)\right]. \quad (8b)
$$

By applying the Stokes theorem to the integration of the z component on the left-hand side of Eq. (2) over the same integral domain as in Eq. (7), we obtain

$$
\int_{I\Delta x}^{(I+1)\Delta x} \int_{J\Delta y}^{(J+1)\Delta y} \left[\nabla \times \mathbf{H}(\mathbf{r}, t) \cdot \mathbf{z}\right]_{z=K\Delta z} dx\, dy
$$

$$
\approx \Delta x \left[H_x^{n+1/2}\left(I + \frac{1}{2}, J, K\right) - H_x^{n+1/2}\left(I + \frac{1}{2}, J + 1, K\right) \right]
$$

$$+ \Delta y \left[H_y^{n+1/2}\left(I+1, J+\frac{1}{2}, K \right) - H_y^{n+1/2}\left(I, J+\frac{1}{2}, K \right) \right]. \quad (9)$$

It follows from Eqs. (2), (7), and (9) that

$$\frac{\bar{\varepsilon}_r(I+1/2, J+1/2, K)}{c} \frac{\partial E_z(I+1/2, J+1/2, K)}{\partial t}$$

$$+ k\bar{\varepsilon}_i\left(I+\frac{1}{2}, J+\frac{1}{2}, K \right) E_z\left(I+\frac{1}{2}, J+\frac{1}{2}, K \right)$$

$$\approx \left[H_x^{n+1/2}\left(I+\frac{1}{2}, J, K \right) - H_x^{n+1/2}\left(I+\frac{1}{2}, J+1, K \right) \right] \Big/ \Delta y$$

$$+ \left[H_y^{n+1/2}\left(I+1, J+\frac{1}{2}, K \right) - H_y^{n+1/2}\left(I, J+\frac{1}{2}, K \right) \right] \Big/ \Delta x. \quad (10a)$$

Equation (10a) can also be written in a more compact form as follows:

$$\frac{\bar{\varepsilon}_r(I+1/2, J+1/2, K)}{c}$$

$$\times \frac{\partial}{\partial t} \left\{ \exp\left[\bar{\tau}\left(I+\frac{1}{2}, J+\frac{1}{2}, K \right) t \right] E_z\left(I+\frac{1}{2}, J+\frac{1}{2}, K \right) \right\}$$

$$\approx \exp\left[\bar{\tau}\left(I+\frac{1}{2}, J+\frac{1}{2}, K \right) t \right]$$

$$\times \left\{ \left[H_x^{n+1/2}\left(I+\frac{1}{2}, J, K \right) - H_x^{n+1/2}\left(I+\frac{1}{2}, J+1, K \right) \right] \Big/ \Delta y \right.$$

$$\left. + \left[H_y^{n+1/2}\left(I+1, J+\frac{1}{2}, K \right) - H_y^{n+1/2}\left(I, J+\frac{1}{2}, K \right) \right] \Big/ \Delta x \right\}, \quad (10b)$$

where $\bar{\tau}(I+1/2, J+1/2, K) = kc\bar{\varepsilon}_i(I+1/2, J+1/2, K)/\bar{\varepsilon}_r(I+1/2, J+1/2, K)$. For the present scheme, the temporal derivatives in Eq. (10a) are discretized according to the following expressions:

$$\int_{n\Delta t}^{(n+1)\Delta t} \frac{\partial E_z}{\partial t} \, dt = E_z^{n+1} - E_z^n, \quad (11a)$$

$$\int_{n\Delta t}^{(n+1)\Delta t} E_z \, dt \approx \Delta t E_z^{n+1/2} \approx \frac{\Delta t}{2}\left(E_z^{n+1} + E_z^n \right), \quad (11b)$$

$$\int_{n\Delta t}^{(n+1)\Delta t} H_{x,y} \, dt \approx \Delta t H_{x,y}^{n+1/2}. \quad (11c)$$

Based on Eqs. (10a) and (11a)–(11c), it can be shown that the coefficients of the finite difference analog of Maxwell's equations are given by

$$a\left(I + \frac{1}{2}, J + \frac{1}{2}, K\right) = \frac{1 - \overline{\tau}(I + 1/2, J + 1/2, K)\Delta t/2}{1 + \overline{\tau}(I + 1/2, J + 1/2, K)\Delta t/2}, \tag{12a}$$

$$b\left(I + \frac{1}{2}, J + \frac{1}{2}, K\right)$$

$$= \frac{1}{[1 + \overline{\tau}(I + 1/2, J + 1/2, K)\Delta t/2]\overline{\varepsilon}_r(I + 1/2, J + 1/2, K)}. \tag{12b}$$

B. SCHEME 2

In this scheme, discretizations of two exponential integrals are applied to Eq. (10b) for the difference approximation of the temporal derivatives given by

$$\int_{n\Delta t}^{(n+1)\Delta t} \frac{\partial[\exp(\overline{\tau}t)E_z]}{\partial t} dt = \exp[(n+1)\overline{\tau}\Delta t]E_z^{n+1} - \exp(n\overline{\tau}\Delta t)E_z^n, \tag{13a}$$

$$\int_{n\Delta t}^{(n+1)\Delta t} \exp(\overline{\tau}t)H_{x,y} dt \approx H_{x,y}^{n+1/2} \int_{n\Delta t}^{(n+1)\Delta t} \exp(\overline{\tau}t) dt$$

$$= H_{x,y}^{n+1/2} \frac{\exp[(n+1)\overline{\tau}\Delta t][1 - \exp(-\overline{\tau}\Delta t)]}{\overline{\tau}}. \tag{13b}$$

Thus, after some algebraic manipulations we can obtain the coefficients a and b in Eqs. (5a)–(5c) as follows:

$$a\left(I + \frac{1}{2}, J + \frac{1}{2}, K\right) = \exp\left[-\overline{\tau}\left(I + \frac{1}{2}, J + \frac{1}{2}, K\right)\Delta t\right], \tag{14a}$$

$$b\left(I + \frac{1}{2}, J + \frac{1}{2}, K\right) = \frac{1 - \exp[-\overline{\tau}(I + 1/2, J + 1/2, K)\Delta t]}{\overline{\tau}(I + 1/2, J + 1/2, K)\Delta t\overline{\varepsilon}_r(I + 1/2, J + 1/2, K)}. \tag{14b}$$

C. SCHEME 3

This scheme is similar to scheme 2, except that a different approximation is used to replace Eq. (13b) for the temporal discretization associated with the magnetic field in the form

$$\int_{n\Delta t}^{(n+1)\Delta t} \exp(\overline{\tau}t)H_{x,y} dt \approx \Delta t \exp\left[\left(n + \frac{1}{2}\right)\overline{\tau}\Delta t\right]H_{x,y}^{n+1/2}. \tag{15}$$

The coefficients a and b in Eqs. (5a)–(5c) can then be obtained from

$$a\left(I + \frac{1}{2}, J + \frac{1}{2}, K\right) = \exp\left[-\overline{\tau}\left(I + \frac{1}{2}, J + \frac{1}{2}, K\right)\Delta t\right], \qquad (16a)$$

$$b\left(I + \frac{1}{2}, J + \frac{1}{2}, K\right) = \frac{\exp[-\overline{\tau}(I + 1/2, J + 1/2, K)\Delta t/2]}{\overline{\varepsilon}_r(I + 1/2, J + 1/2, K)}. \qquad (16b)$$

D. Schemes 4, 5, and 6

Different from schemes 1–3, the present three schemes first discretize the temporal derivatives. Using the algorithms of schemes 1–3 for the integration over the time increment $n\Delta t$ to $(n + 1)\Delta t$, we obtain the following three time-difference/space-differential equations for schemes 4, 5, and 6, respectively:

$$\mathbf{E}^{n+1} = \frac{1 - \tau\Delta t/2}{1 + \tau\Delta t/2}\mathbf{E}^n + \frac{c\Delta t}{(1 + \tau\Delta t/2)\varepsilon_r}\nabla \times \mathbf{H}^{n+1/2}, \qquad (17a)$$

$$\mathbf{E}^{n+1} = \exp(-\tau\Delta t)\mathbf{E}^n + \frac{[1 - \exp(-\tau\Delta t)]c\Delta t}{\tau\Delta t\varepsilon_r}\nabla \times \mathbf{H}^{n+1/2}, \qquad (17b)$$

$$\mathbf{E}^{n+1} = \exp(-\tau\Delta t)\mathbf{E}^n + \frac{\exp(-\tau\Delta t/2)c\Delta t}{\varepsilon_r}\nabla \times \mathbf{H}^{n+1/2}. \qquad (17c)$$

Further, we use one of the preceding equations to carry out the spatial discretization based on the same procedure described in Eqs. (7) and (9). For example, for scheme 6 represented by Eq. (17c), the coefficients a and b in Eqs. (5a)–(5c) are obtained based on the following averaging procedure:

$$a\left(I + \frac{1}{2}, J + \frac{1}{2}, K\right)$$

$$= \int_{I\Delta x}^{(I+1)\Delta x}\int_{J\Delta y}^{(J+1)\Delta y} \exp\left[-\tau(x, y, z)\Delta t\right]\Big|_{z=K\Delta z}dx\,dy$$

$$= \frac{1}{4}\Big\{\exp\left[-\overline{\tau}(I, J, K)\Delta t\right] + \exp\left[-\overline{\tau}(I, J + 1, K)\Delta t\right]$$

$$+ \exp\left[-\overline{\tau}(I + 1, J, K)\Delta t\right] + \exp\left[-\tau(I + 1, J + 1, K)\Delta t\right]\Big\}, \qquad (18a)$$

$$b\left(I + \frac{1}{2}, J + \frac{1}{2}, K\right)$$

$$= \int_{I\Delta x}^{(I+1)\Delta x}\int_{J\Delta y}^{(J+1)\Delta y} \frac{\exp[-\tau(x, y, z)\Delta t/2]}{\varepsilon_r(x, y, z)}\Big|_{z=K\Delta z}dx\,dy$$

$$
= \frac{1}{4} \left\{ \frac{\exp[-\bar{\tau}(I, J, K)\Delta t/2]}{\bar{\varepsilon}_r(I, J, K)} + \frac{\exp[-\bar{\tau}(I, J+1, K)\Delta t/2]}{\bar{\varepsilon}_r(I, J+1, K)} \right.
$$

$$
\left. + \frac{\exp[-\bar{\tau}(I+1, J, K)\Delta t/2]}{\bar{\varepsilon}_r(I+1, J, K)} + \frac{\exp[-\bar{\tau}(I+2, J+1, K)\Delta t/2]}{\bar{\varepsilon}_r(I+1, J+1, K)} \right\}.
$$

(18b)

For the nonabsorptive case, that is, $\varepsilon_i = 0$, schemes 1–3 and 4–6 reduce to two schemes (hereafter referred to as schemes A and B). The only difference between schemes A and B is that the spatial discretization is applied first in the former, whereas the temporal discretization is carried out first in the latter. In other words, for the former scheme a mean permittivity is evaluated first based on four adjacent homogeneous grid cells and is then used to calculate the coefficients a and b in Eqs. (5a) and (5b). For scheme B, the coefficients a and b are calculated first for the four cells based on their homogeneous permittivities. Then the averages of the coefficient values are taken in the time-marching iteration of the electromagnetic field using the finite difference analog of Maxwell's equations. After the coefficients are determined, the updating iterations of electromagnetic waves are straightforward.

To compare the accuracy of the six schemes, we have carried out the phase function computations for an ice sphere with size parameters $x = kR$ of 5 at 0.5- and 10-μm wavelengths, where R is the sphere radius. In the computations, the perfectly matched layer (PML) absorbing boundary condition (Berenger, 1994) is used and the cell size is selected as 1/25 of the incident wavelength. The relative errors are determined from a comparison with the Lorenz–Mie solution. As shown in the left panel of Fig. 2, schemes 1–3 produce essentially the same results and the error patterns of the three schemes are indistinguishable even for the case of strong absorption ($\lambda = 10$ μm). Schemes 4–6 produce the same accuracy for the computed phase functions (not shown in the diagram). However, the accuracy of schemes 1–3 differs from that of schemes 4–6. As stated previously, the six schemes reduce to two schemes for nonabsorptive cases. From the computational perspective, the six schemes also reduce to two schemes for strong absorption cases. The middle and right panels in Fig. 2 show the phase functions computed using schemes A and B for the same 0.5- and 10-μm wavelengths. Scheme A is more accurate than scheme B, particularly for side scattering. The difference between these two schemes increases if the grid size increases. Thus, in order to discretize Maxwell's equations, permittivity should first be averaged over space. The calculation of the coefficients can then be performed by using the discretized electromagnetic difference equations.

The electromagnetic fields involved in Eqs. (5a)–(5f) represent the total (incident + scattered) field. However, the absorbing boundary condition at the artificial boundary, as discussed in Section IV, is applicable only to the induced or scattered field produced by the existence of the particle. To overcome this

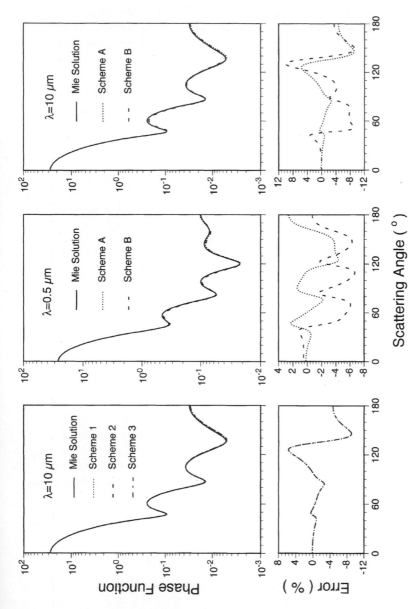

Figure 2 Comparisons of the phase functions (i.e., F_{11} elements of the scattering matrix; see Section XI of Chapter 1) for an ice sphere computed based on various discretization schemes. The refractive index is $1.313 + 1.91 \times 10^{-9}i$ at $\lambda = 0.5$ μm and $1.1991 + 5.1 \times 10^{-2}i$ at $\lambda = 10$ μm.

difficulty, two approaches can be used. First, a connecting surface (also called the Huygens surface) located between the scatterer and the boundary is introduced in the computational domain. Inside and on the connecting surface the total field is computed, but outside the surface only the scattered field is evaluated. Because the fields computed in these two regions are not consistent, a connecting condition must be imposed at the surface. Let the cells enclosed by the connecting surface be defined by the indices $I \in [IA, IB]$ and $J \in [JA, JB]$. The connecting conditions can be derived for each electric and magnetic field component. For the $E_x(I, J + 1/2, K + 1/2)$ connecting condition, $I \in [IA, IB]$, we have

$$
\left.
\begin{aligned}
\tilde{E}_x^{n+1}\left(I, JA - \frac{1}{2}, K + \frac{1}{2}\right) &= E_x^{n+1}\left(I, JA - \frac{1}{2}, K + \frac{1}{2}\right) \\
&\quad - \frac{c\Delta t}{\Delta y} H_{o,z}^{n+1/2}\left(I, JA - 1, K + \frac{1}{2}\right) \\
\tilde{E}_x^{n+1}\left(I, JB + \frac{1}{2}, K + \frac{1}{2}\right) &= E_x^{n+1}\left(I, JB + \frac{1}{2}, K + \frac{1}{2}\right) \\
&\quad + \frac{c\Delta t}{\Delta y} H_{o,z}^{n+1/2}\left(I, JB + 1, K + \frac{1}{2}\right)
\end{aligned}
\right\},
$$

$$
K \in [KA - 1, KB], \tag{19a}
$$

$$
\left.
\begin{aligned}
\tilde{E}_x^{n+1}\left(I, J + \frac{1}{2}, KA - \frac{1}{2}\right) &= E_x^{n+1}\left(I, J + \frac{1}{2}, KA - \frac{1}{2}\right) \\
&\quad + \frac{c\Delta t}{\Delta z} H_{o,y}^{n+1/2}\left(I, J + \frac{1}{2}, KA - 1\right) \\
\tilde{E}_x^{n+1}\left(I, J + \frac{1}{2}, KB + \frac{1}{2}\right) &= E_x^{n+1}\left(I, J + \frac{1}{2}, KB + \frac{1}{2}\right) \\
&\quad - \frac{c\Delta t}{\Delta z} H_{o,y}^{n+1/2}\left(I, J + \frac{1}{2}, KB + 1\right)
\end{aligned}
\right\},
$$

$$
J \in [JA - 1, JB]. \tag{19b}
$$

For the $E_y(I + 1/2, J, K + 1/2)$ connecting condition, $J \in [JA, JB]$, we have

$$
\left.
\begin{aligned}
\tilde{E}_y^{n+1}\left(IA - \frac{1}{2}, J, K + \frac{1}{2}\right) &= E_y^{n+1}\left(IA - \frac{1}{2}, J, K + \frac{1}{2}\right) \\
&\quad + \frac{c\Delta t}{\Delta x} H_{o,z}^{n+1/2}\left(IA - 1, J, K + \frac{1}{2}\right) \\
\tilde{E}_y^{n+1}\left(IB + \frac{1}{2}, J, K + \frac{1}{2}\right) &= E_y^{n+1}\left(IB + \frac{1}{2}, J, K + \frac{1}{2}\right) \\
&\quad - \frac{c\Delta t}{\Delta x} H_{o,z}^{n+1/2}\left(IB + 1, J, K + \frac{1}{2}\right)
\end{aligned}
\right\},
$$

$$K \in [KA - 1, KB],\tag{19c}$$

$$
\left.
\begin{aligned}
\tilde{E}_y^{n+1}\left(I + \frac{1}{2}, J, KA - \frac{1}{2}\right) &= E_y^{n+1}\left(I + \frac{1}{2}, J, KA - \frac{1}{2}\right) \\
&\quad - \frac{c\Delta t}{\Delta z} H_{o,x}^{n+1/2}\left(I + \frac{1}{2}, J, KA - 1\right) \\
\tilde{E}_y^{n+1}\left(I + \frac{1}{2}, J, KB + \frac{1}{2}\right) &= E_y^{n+1}\left(I + \frac{1}{2}, J, KB + \frac{1}{2}\right) \\
&\quad + \frac{c\Delta t}{\Delta z} H_{o,x}^{n+1/2}\left(I + \frac{1}{2}, J, KB + 1\right)
\end{aligned}
\right\},
$$
$$I \in [IA - 1, IB].\tag{19d}$$

For the $E_z(I + 1/2, J + 1/2, K)$ connecting condition, $K \in [KA, KB]$, we have

$$
\left.
\begin{aligned}
\tilde{E}_z^{n+1}\left(I + \frac{1}{2}, JA - \frac{1}{2}, K\right) &= E_z^{n+1}\left(I + \frac{1}{2}, JA - \frac{1}{2}, K\right) \\
&\quad + \frac{c\Delta t}{\Delta y} H_{o,x}^{n+1/2}\left(I + \frac{1}{2}, JA - 1, K\right) \\
\tilde{E}_z^{n+1}\left(I + \frac{1}{2}, JB + \frac{1}{2}, K\right) &= E_z^{n+1}\left(I + \frac{1}{2}, JB + \frac{1}{2}, K\right) \\
&\quad - \frac{c\Delta t}{\Delta y} H_{o,x}^{n+1/2}\left(I + \frac{1}{2}, JB + 1, K\right)
\end{aligned}
\right\},
$$
$$I \in [IA - 1, IB],\tag{19e}$$

$$
\left.
\begin{aligned}
\tilde{E}_z^{n+1}\left(IA - \frac{1}{2}, J + \frac{1}{2}, K\right) &= E_z^{n+1}\left(IA - \frac{1}{2}, J + \frac{1}{2}, K\right) \\
&\quad - \frac{c\Delta t}{\Delta x} H_{o,y}^{n+1/2}\left(IA - 1, J + \frac{1}{2}, K\right) \\
\tilde{E}_z^{n+1}\left(IB + \frac{1}{2}, J + \frac{1}{2}, K\right) &= E_z^{n+1}\left(IB + \frac{1}{2}, J + \frac{1}{2}, K\right) \\
&\quad + \frac{c\Delta t}{\Delta x} H_{o,y}^{n+1/2}\left(IB + 1, J + \frac{1}{2}, K\right)
\end{aligned}
\right\},
$$
$$J \in [JA - 1, JB].\tag{19f}$$

For the $H_x(I + 1/2, J, K)$ connecting condition, $I \in [IA - 1, IB]$, we have

$$
\left.
\begin{aligned}
\tilde{H}_x^{n+1/2}\left(I + \frac{1}{2}, JA - 1, K\right) &= H_x^{n+1/2}\left(I + \frac{1}{2}, JA - 1, K\right) \\
&\quad + \frac{c\Delta t}{\Delta y} E_{o,z}^n\left(I + \frac{1}{2}, JA - \frac{1}{2}, K\right) \\
\tilde{H}_x^{n+1/2}\left(I + \frac{1}{2}, JB + 1, K\right) &= H_x^{n+1/2}\left(I + \frac{1}{2}, JB + 1, K\right) \\
&\quad - \frac{c\Delta t}{\Delta y} E_{o,z}^{n+1/2}\left(I + \frac{1}{2}, JB + \frac{1}{2}, K\right)
\end{aligned}
\right\},
$$

$$K \in [KA, KB], \tag{19g}$$

$$
\left.
\begin{aligned}
\tilde{H}_x^{n+1/2}\left(I + \frac{1}{2}, J, KA - 1\right) &= H_x^{n+1/2}\left(I + \frac{1}{2}, J, KA - 1\right) \\
&\quad - \frac{c\Delta t}{\Delta z} E_{o,y}^n\left(I + \frac{1}{2}, J, KA - \frac{1}{2}\right) \\
\tilde{H}_x^{n+1/2}\left(I + \frac{1}{2}, J, KB + 1\right) &= H_x^{n+1/2}\left(I + \frac{1}{2}, J, KB + 1\right) \\
&\quad + \frac{c\Delta t}{\Delta z} E_{o,y}^{n+1/2}\left(I + \frac{1}{2}, J, KB + \frac{1}{2}\right)
\end{aligned}
\right\},
$$

$$J \in [JA, JB]. \tag{19h}$$

For the $H_y(I, J + 1/2, K)$ connecting condition, $J \in [JA - 1, JB]$, we have

$$
\left.
\begin{aligned}
\tilde{H}_y^{n+1/2}\left(IA - 1, J + \frac{1}{2}, K\right) &= H_y^{n+1/2}\left(IA - 1, J + \frac{1}{2}, K\right) \\
&\quad - \frac{c\Delta t}{\Delta x} E_{o,z}^n\left(IA - \frac{1}{2}, J + \frac{1}{2}, K\right) \\
\tilde{H}_y^{n+1/2}\left(IB + 1, J + \frac{1}{2}, K\right) &= H_y^{n+1/2}\left(IB + 1, J + \frac{1}{2}, K\right) \\
&\quad + \frac{c\Delta t}{\Delta x} E_{o,z}^{n+1/2}\left(IB + \frac{1}{2}, J + \frac{1}{2}, K\right)
\end{aligned}
\right\},
$$

$$K \in [KA, KB], \tag{19i}$$

$$\tilde{H}_y^{n+1/2}\left(I, J + \frac{1}{2}, KA - 1\right) = H_y^{n+1/2}\left(I, J + \frac{1}{2}, KA - 1\right)$$

$$+ \frac{c\Delta t}{\Delta z} E_{o,x}^n\left(I, J + \frac{1}{2}, KA - \frac{1}{2}\right)$$

$$\tilde{H}_y^{n+1/2}\left(I, J + \frac{1}{2}, KB + 1\right) = H_y^{n+1/2}\left(I, J + \frac{1}{2}, KB + 1\right)$$

$$- \frac{c\Delta t}{\Delta z} E_{o,x}^n\left(I, J + \frac{1}{2}, KB + \frac{1}{2}\right)$$

$$I \in [IA, IB]. \tag{19j}$$

For the $H_z(I, J, K + 1/2)$ connecting condition, $K \in [KA - 1, KB]$, we have

$$\tilde{H}_z^{n+1/2}\left(I, JA - 1, K + \frac{1}{2}\right) = H_z^{n+1/2}\left(I, JA - 1, K + \frac{1}{2}\right)$$

$$- \frac{c\Delta t}{\Delta y} E_{o,x}^n\left(I, JA - \frac{1}{2}, K + \frac{1}{2}\right)$$

$$\tilde{H}_z^{n+1/2}\left(I, JB + 1, K + \frac{1}{2}\right) = H_z^{n+1/2}\left(I, JB + 1, K + \frac{1}{2}\right)$$

$$+ \frac{c\Delta t}{\Delta y} E_{o,x}^n\left(I, JB + 1, K + \frac{1}{2}\right)$$

$$I \in [IA, IB], \tag{19k}$$

$$\tilde{H}_z^{n+1/2}\left(IA - 1, J, K + \frac{1}{2}\right) = H_z^{n+1/2}\left(IA - 1, J, K + \frac{1}{2}\right)$$

$$+ \frac{c\Delta t}{\Delta x} E_{o,y}^n\left(IA - \frac{1}{2}, J, K + \frac{1}{2}\right)$$

$$\tilde{H}_z^{n+1/2}\left(IB + 1, J, K + \frac{1}{2}\right) = H_z^{n+1/2}\left(IB + 1, J, K + \frac{1}{2}\right)$$

$$- \frac{c\Delta t}{\Delta x} E_{o,y}^n\left(IB + \frac{1}{2}, J, K + \frac{1}{2}\right)$$

$$J \in [JA, JB]. \tag{19l}$$

On the right-hand sides of Eqs. (19a)–(19l), the second terms with the subscript o are the incident fields, whereas the first terms are evaluated by finite difference equations (5a)–(5f). In this way (hereafter referred to as the total-field FDTD algorithm), the governing equations are the same for both the scattered- and the total-field regions, except that the connecting conditions are imposed at the surface. The previous connecting conditions, in principle, are an application of Schelkunoff's electromagnetic equivalence theorem (Schelkunoff, 1943). As pointed out by Merewether *et al.* (1980), for the region inside the connecting surface the existence of the incident field can be substituted by specifying the equivalent electric and magnetic currents on the surface.

Unlike the total-field algorithm, other approaches construct a global scattered-field formulation within the entire computational domain. Because the electric properties of the medium in our consideration are linear, the total field is the superposition of the incident and scattered fields. Therefore, the pure scattered field is given by

$$\mathbf{E}_s(\mathbf{r}, t) = \mathbf{E}_t(\mathbf{r}, t) - \mathbf{E}_o(\mathbf{r}, t), \tag{20a}$$

$$\mathbf{H}_s(\mathbf{r}, t) = \mathbf{H}_t(\mathbf{r}, t) - \mathbf{H}_o(\mathbf{r}, t), \tag{20b}$$

where the subscripts s, t and o denote the scattered, total, and incident fields, respectively. Note that Eqs. (1b) and (2) can also be applied to the incident field except that the permittivity is set at unity. A set of equations similar to Eqs. (5a)–(5f) can then be derived for the scattered-field algorithm. It should be pointed out that the total-field algorithm is more accurate than the pure scattered-field algorithm for metal objects or heavily shielded cavities (Mur, 1981; Umashankar and Taflove, 1982). The total field algorithm is also more efficient in terms of numerical computations because specification of the incident field is only required at the layer associated with the connecting conditions.

To demonstrate the difference between the near fields computed by the total- and scattered-field algorithms, we have simulated the scattering of a sinusoidal wave propagating along the z direction with the x-polarized E field by a sphere with a refractive index of $(m_r, m_i) = (3, 0)$ and a size parameter of 3.35. The cell dimensions along the three coordinate axes are selected to be equal, that is, $\Delta x = \Delta y = \Delta z = \Delta s$. The wavelength of the sinusoidal wave and the radius of the sphere are selected to be $30\Delta s$ and $16\Delta s$, respectively. The second-order modified Liao transmitting boundary condition (Yang and Liou, 1998b) is used with a "white space" of 15 cells between the target and the boundary. The left panel in Fig. 3 shows the snapshot of E_x contours on the xy plane through the center of the sphere, which is computed by the scattered-field algorithm at the time step $n = 600$. The existence of the scattering particle is clear as the contour gradient is much larger inside than outside the particle because the wavelength inside the sphere decreases by a factor of m_r. The right panel in Fig. 3 is the result computed by the total-field algorithm, where the square in the right diagram is the connecting surface. It can be seen that the results by the two algorithms have the same patterns outside the connecting surface, but some differences are noted for the region inside the boxes.

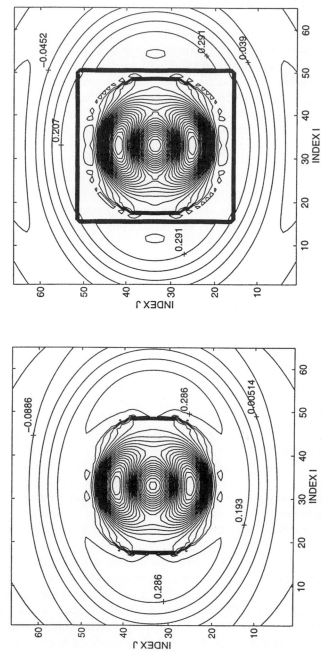

Figure 3 E_x contour plots computed by the scattered- (left panel) and total-field (right panel) algorithms for the scattering by a sphere. The results are observed in the xy plane through the center of the sphere at the time step $n = 600$.

IV. ABSORBING BOUNDARY CONDITION

The numerical implementation of the FDTD technique requires the imposition of an appropriate absorbing boundary condition, which is critical for the stability of numerical computations and the reliability of results. In addition, the "white space" between the boundary and the scatterer required by a specific boundary condition is an important factor determining the computational effort.

The earliest implementation of the absorbing boundary condition in the application of the FDTD technique to electromagnetic scattering problems used the average space–time extrapolating method (Taflove and Brodwin, 1975). Other approaches such as the mode-annihilating operator (Bayliss and Turkel, 1980) and the extrapolating scheme based on the Poynting vector of the scattered wave (Britt, 1989) have also been developed to suppress the reflectivity of the artificial boundary. In the 1980s, the absorbing boundary conditions derived from the one-way wave equation (OWWE) were extensively applied in FDTD implementations. As reviewed by Moore *et al.* (1988) and Blaschak and Kriegsmann (1988), various approximations of the pseudo-differential operator in OWWE can be used to derive numerical schemes for the boundary conditions. Among them, the algorithm developed by Mur (1981) has been widely used. The second-order or higher Mur's absorbing boundary condition involves the wave values at the intersections of boundary faces. However, the corresponding boundary equations cannot be posed in a self-closing form; that is, a less accurate first-order boundary equation or an extrapolating scheme must be used at the intersections. Moreover, Mur's algorithm is rather tedious, especially in the three-dimensional (3D) higher order formulation. The field values at the intersections are not required for updating the field values at interior grid points in the computation of the scattering of electromagnetic waves. This disadvantage of Mur's absorbing boundary condition can be avoided by using the transmitting boundary condition developed by Liao *et al.* (1984) because only the wave values at the interior grid points along the direction normal to the boundary are involved.

Most recently, Berenger (1994, 1996) has developed a novel numerical technique called the perfectly matched layer (PML) boundary condition for the absorption of outgoing waves. With this technique, an absorbing medium is assigned to the outermost layers of the computational domain backed by a perfectly conducting surface. The absorbing medium is specified such that it absorbs the outgoing wave impinging on it without reflecting it back. Theoretically, the PML medium has a null reflection factor for a plane wave striking at the interface between the free space and the PML layers at any frequency and at any incident angle, as shown by Berenger (1994). Numerical experiments have shown that the spurious reflection produced by the PML boundary condition is about 1/3000 of that generated by the analytical absorbing boundary condition derived from the wave equation (Katz *et al.*, 1994). In this section, we recapitulate the physical basis and numerical implementation for three commonly used absorbing boundary conditions.

A. Mur's Absorbing Boundary Condition

To review the conceptual basis of this boundary condition, we begin with the governing equation for a scalar wave or any Cartesian component of the vector electromagnetic field given by

$$\frac{1}{c^2}\frac{\partial^2}{\partial t^2}U - \nabla^2 U = 0, \tag{21}$$

where U denotes the scalar wave displacement or the component of an electromagnetic field vector. This wave equation can be expressed in an operator form as follows (Moore *et al.*, 1988):

$$L_x^+ L_x^- U = 0, \tag{22}$$

where L_x^+ and L_x^- are the OWWE operators for the wave propagation along positive and negative directions of the x axis, respectively, given by

$$L_x^+ = D_x + \frac{D_t}{c}\sqrt{1 - \frac{c^2(D_y^2 + D_z^2)}{D_t^2}}, \tag{23a}$$

$$L_x^- = D_x - \frac{D_t}{c}\sqrt{1 - \frac{c^2(D_y^2 + D_z^2)}{D_t^2}}. \tag{23b}$$

In Eqs. (23a) and (23b), D_x, D_y, and D_t stand for $\partial/\partial x$, $\partial/\partial y$, and $\partial/\partial t$, respectively. For a boundary at, say, $x = 0$, it is completely reflectionless if the following OWWE is satisfied (Blaschak and Kriegsmann, 1988):

$$L_x^- U|_{x=0} = 0. \tag{24}$$

The operator L_x^-, however, is a pseudo-differential operator resulting from the existence of the radical. Thus, Eq. (24) cannot be discretized as a finite difference equation. To obtain the discrete form of OWWE, a rational function should be used to approximate the OWWE operator. The most common approach is the expansion of the radical in terms of the Taylor series, as presented by Engquist and Majda (1977) and Mur (1981). Keeping the first or the first two terms in the Taylor expansion will lead to the first- and second-order Mur absorbing boundary conditions, respectively. For the second-order Mur absorbing boundary condition, the pseudo-differential operator is expanded in the form

$$\frac{D_t}{c}\sqrt{1-\frac{c^2(D_y^2+D_z^2)}{D_t^2}} \approx \frac{D_t}{c} - \frac{1}{2}\frac{c(D_y^2+D_z^2)}{D_t}. \tag{25}$$

Based on Eqs. (24) and (25), a second-order approximation of OWWE can be defined explicitly as follows:

$$\left[\frac{1}{c}\frac{\partial^2}{\partial t\partial x} - \frac{1}{c^2}\frac{\partial^2}{\partial t^2} + \frac{1}{2}\left(\frac{\partial^2}{\partial y^2} + \frac{\partial^2}{\partial z^2}\right)\right]U = 0. \tag{26}$$

As suggested by Mur (1981), Eq. (26) can be discretized by using a central difference scheme for the differentials in time and space such that

$$\begin{aligned}
U^{n+1}(0, J, K) = &-U^{n-1}(1, J, K) \\
&- \frac{\Delta s - c\Delta t}{\Delta s + c\Delta t}\left[U^{n-1}(0, J, K) + U^{n+1}(1, J, K)\right] \\
&+ \frac{2\Delta s}{c\Delta t + \Delta s}\left[U^n(0, J, K) + U^n(1, J, K)\right] \\
&+ \frac{(c\Delta t)^2}{2\Delta s(c\Delta t + \Delta s)}\left[U^n(0, J+1, K) - 4U^n(0, J, K)\right. \\
&+ U^n(0, J-1, K) + U^n(1, J+1, K) - 4U^n(1, J, K) \\
&+ U^n(1, J-1, K) + U^n(0, J, K+1) + U^n(0, J, K-1) \\
&+ \left. U^n(1, J, K+1) + U^n(1, J, K-1)\right],
\end{aligned} \tag{27}$$

where equal cell dimensions along the three coordinate axes are used, that is, $\Delta x = \Delta y = \Delta z = \Delta s$. Equation (27) cannot be applied to the corners and edges of the computational boundary. For these locations, an extrapolation scheme or the first-order Mur absorbing boundary condition must be used. The continuous form of the first-order Mur absorbing boundary condition equation is given by

$$\left[\frac{1}{c}\frac{\partial}{\partial t} - \frac{\partial}{\partial x}\right]U = 0. \tag{28}$$

The discrete form of this equation can also be obtained by using a central differencing scheme in both time and space given by

$$U^{n+1}(0, J, K) = U^n(1, J, K) - \frac{\Delta s - c\Delta t}{\Delta s + c\Delta t}\left[U^{n-1}(0, J, K) + U^{n+1}(1, J, K)\right]. \tag{29}$$

B. LIAO'S TRANSMITTING BOUNDARY CONDITION

This numerical scheme, in principle, is based on the propagation of a wave in the time domain; that is, the wave values at the boundary are the arrivals of those located at interior grid points at earlier time steps. In the construction of this transmitting boundary condition, it is assumed that the outgoing scattered wave can be locally approximated as a plane wave (not necessarily a time-harmonic plane wave) in the vicinity of the boundary. Under such an approximation, the boundary values for normal incidence or the one-dimensional (1D) case can be easily obtained by using an extrapolation scheme in time or space, as noted by Taflove and Brodwin (1975). However, in the two-dimensional (2D) or 3D case with oblique incidence, the interior points cannot be located because of the unknown incident angle of outgoing waves. To overcome this difficulty, Liao *et al.* (1984) have developed a multitransmitting method to define the boundary values in terms of the interior values equally spaced along the directions normal to the boundary faces.

The fundamental postulation of the multitransmitting method is that the original outgoing or scattered wave can be transmitted through the boundary along the direction normal to the boundary face in an artificial transmitting speed with a remaining error wave that can also be transmitted in the same manner. Consequently, a second-order error wave is produced. After this procedure is carried out sequentially, the outgoing wave can be eventually transmitted through the boundary regardless of the incident angle. Based on this principle and the plane wave condition for the outgoing wave, the wave values at a boundary, say, the right-side boundary ($x = x_b$), can be expressed as follows:

$$U(t + \Delta t, x_b) = \sum_{L=1}^{N} (-1)^{L+1} \frac{N!}{(N-L)!L!} U\left[t - (L-1)\Delta t, x_b - Lc_\alpha \Delta t\right], \quad (30)$$

where U is the wave value, c_α is an artificial transmitting speed, which may differ from that of the corresponding real physical wave, and Δt is the temporal increment. Because the ratio of the temporal increment to the spatial increment in the finite difference computation is subject to the CFL condition given by Eq. (6), the wave values on the right-hand side of Eq. (30) are usually not located at grid points. To circumvent this shortcoming, Liao *et al.* (1984) used a quadratic interpolation to obtain the wave values and developed the following algorithm:

$$U(t + \Delta t, x_b) = \sum_{L=1}^{N} (-1)^{L+1} \frac{N!}{(N-L)!L!} \mathbf{T}_L \mathbf{U}_L, \quad (31a)$$

$$\mathbf{T}_L = [T_{L,1} \quad T_{L,2} \quad \cdots \quad T_{L,2L+1}], \quad (31b)$$

$$\mathbf{U}_L = [U_{1,L} \quad U_{2,L} \quad \cdots \quad U_{2L+1,L}]^T, \quad (31c)$$

where the superscript T denotes the transpose of the matrix and $U_{i,j} = U[t - (j-1)\Delta t, x_b - (i-1)\Delta s]$ in which Δs is the spatial increment. The matrix $\mathbf{T}_L$ can be calculated from a recursive equation given by

$$
\mathbf{T}_L = \mathbf{T}_1
$$
$$
\times \begin{bmatrix} T_{L-1,1} & T_{L-1,2} & \cdots & \cdots & T_{L-1,2L-1} & 0 & 0 \\ 0 & T_{L-1,1} & T_{L-1,2} & \cdots & \cdots & T_{L-1,2L-1} & 0 \\ 0 & 0 & T_{L-1,1} & T_{L-1,2} & \cdots & \cdots & T_{L-1,2L-1} \end{bmatrix}
$$
$$
\text{for } L \geq 2, \tag{31d}
$$

in which the three elements of $\mathbf{T}_1$ are $T_{1,1} = (2-\beta)(1-\beta)/2$, $T_{1,2} = \beta(2-\beta)$, and $T_{1,3} = \beta(\beta-1)/2$, where $\beta = c_\alpha \Delta t / \Lambda s$. The preceding algorithm is not stable, and double-precision arithmetic must be used in numerical computations. To stabilize the transmitting boundary condition algorithm, one can introduce artificial diffusive coefficients to suppress the amplification of the wave magnitude in the FDTD time-marching iteration calculations (Moghaddam and Chew, 1991; Yang and Liou, 1998b).

The original explanation of the transmitting boundary equation given by the multitransmitting theory is somewhat misleading because fictitious waves, which may propagate faster than real physical waves, are assumed. It has been shown that Eq. (30) can be directly derived from the extrapolation of boundary values in terms of wave values located at interior grid points at earlier steps using the coefficients that minimize the extrapolation errors (Yang and Liou, 1998b). Modified versions of the transmitting boundary condition equations have been suggested to produce multiple reflection minima so that the transparency of the boundary of the computational domain is enhanced for large incident angles (Chew and Wagner, 1992; Yang and Liou, 1998b). Steich *et al.* (1993) have compared the performance of Liao's boundary condition with that of Mur's absorbing boundary algorithm and noted that the latter approach requires a larger "white space" between a modeled scatterer and the boundary to achieve a convergent scattering solution.

C. Perfectly Matched Layer Absorbing Boundary Condition

Absorption of the outgoing wave by the PML method is based on the absorption by a medium located at the outermost layers in the computational domain. The conventional technique based on an absorbing medium is to specifically define the wave impedance of the medium so that it matches that of the free space. Such a simple matching approach produces substantial nonzero reflections when a scattered wave impinges on the absorbing medium obliquely. To overcome the disadvantage of the conventional method, Berenger (1994) has developed a perfectly matched layer method, in which the absorbing medium is selected such that the wave decay due to absorption is imposed on the field components parallel to boundary layers. To achieve this goal, each Cartesian component of the electromagnetic field is split into two parts as follows:

$$(E_x, E_y, E_z) = \left[(E_{x2} + E_{x3}), (E_{y1} + E_{y3}), (E_{z1} + E_{z2})\right], \qquad (32a)$$

$$(H_x, H_y, H_z) = \left[(H_{x2} + H_{x3}), (H_{y1} + H_{y3}), (H_{z1} + H_{z2})\right], \qquad (32b)$$

where the subscripts 1, 2, or 3 denote the component of the electric (or magnetic) field that is associated with the spatial differential of the magnetic (or electric) field component along the x, y, and z directions, respectively. With the split field components, the six scalar equations expressed in a discrete form in Eqs. (5a)–(5f) that govern the propagation of electromagnetic waves are replaced by 12 equations. The exponential wave decay factors for these equations can be expressed by

$$\frac{\exp[-\tau_1(x)t]}{c} \frac{\partial}{\partial t}\left\{\exp[\tau_1(x)t]E_{y1}\right\} = -\frac{\partial(H_{z1} + H_{z2})}{\partial x}, \qquad (33a)$$

$$\frac{\exp[-\tau_1(x)t]}{c} \frac{\partial}{\partial t}\left\{\exp[\tau_1(x)t]E_{z1}\right\} = \frac{\partial(H_{y1} + H_{y2})}{\partial x}, \qquad (33b)$$

$$\frac{\exp[-\tau_2(y)t]}{c} \frac{\partial}{\partial t}\left\{\exp[\tau_2(y)t]E_{x2}\right\} = \frac{\partial(H_{z1} + H_{z2})}{\partial y}, \qquad (33c)$$

$$\frac{\exp[-\tau_2(y)t]}{c} \frac{\partial}{\partial t}\left\{\exp[\tau_2(y)t]E_{z2}\right\} = -\frac{\partial(H_{x2} + H_{x3})}{\partial y}, \qquad (33d)$$

$$\frac{\exp[-\tau_3(z)t]}{c} \frac{\partial}{\partial t}\left\{\exp[\tau_3(z)t]E_{x3}\right\} = -\frac{\partial(H_{y1} + H_{y3})}{\partial z}, \qquad (33e)$$

$$\frac{\exp[-\tau_3(z)t]}{c} \frac{\partial}{\partial t}\left\{\exp[\tau_3(z)t]E_{y3}\right\} = \frac{\partial(H_{x2} + H_{x3})}{\partial z}, \qquad (33f)$$

$$\frac{\exp[-\tau_1(x)t]}{c} \frac{\partial}{\partial t}\left\{\exp[\tau_1(x)t]H_{y1}\right\} = \frac{\partial(E_{z1} + E_{z2})}{\partial x}, \qquad (33g)$$

$$\frac{\exp[-\tau_1(x)t]}{c} \frac{\partial}{\partial t}\left\{\exp[\tau_1(x)t]H_{z1}\right\} = -\frac{\partial(E_{y1} + E_{y3})}{\partial x}, \qquad (33h)$$

$$\frac{\exp[-\tau_2(y)t]}{c} \frac{\partial}{\partial t}\left\{\exp[\tau_2(y)t]H_{x2}\right\} = -\frac{\partial(E_{z1} + E_{z2})}{\partial y}, \qquad (33i)$$

$$\frac{\exp[-\tau_2(y)t]}{c} \frac{\partial}{\partial t}\left\{\exp[\tau_2(y)t]H_{z2}\right\} = \frac{\partial(E_{x2} + E_{x3})}{\partial y}, \qquad (33j)$$

$$\frac{\exp[-\tau_3(z)t]}{c} \frac{\partial}{\partial t}\left\{\exp[\tau_3(z)t]H_{x3}\right\} = \frac{\partial(E_{y1} + E_{y3})}{\partial z}, \qquad (33k)$$

$$\frac{\exp[-\tau_3(z)t]}{c} \frac{\partial}{\partial t}\left\{\exp[\tau_3(z)t]H_{y3}\right\} = -\frac{\partial(E_{x2} + E_{x3})}{\partial z}, \qquad (33l)$$

where $\tau_1(x)$, $\tau_2(y)$, and $\tau_3(z)$ are zero, except in boundary layers perpendicular to the x, y, and z axes. It can be shown that these 12 equations are equivalent to those given by Berenger (1996) and Chew and Weedon (1994). In practical computations, the parameters $\tau_1(x)$, $\tau_2(y)$, and $\tau_3(z)$ can be specified from zero at the interface of the free space and PML medium to their maximum values at the outermost layer. For example, $\tau_1(x)$ can be defined as

$$\tau_1(x) = \tau_{1,\max} \left(\frac{x - x_o}{D} \right)^p, \tag{34}$$

where $(x - x_o)$ is the distance of a grid point from the interface of the free space and PML medium, $D = L\Delta x$ is the thickness of the PML medium for the boundary perpendicular to the x axis, and p is usually selected between 2 and 2.5. The parameter $\tau_{1,\max}$ can be specified by the reflectance of the boundary with normal incidence as follows:

$$\tau_{1,\max} = -\frac{p+1}{2D} \ln\left[R(0°) \right] c, \tag{35}$$

where $R(0°)$ is the boundary reflection factor. The mean absorption must be taken into account for each cell distance in discrete computations. Thus, the following two mean values for the electric and magnetic fields can be used:

$$\begin{aligned}
\overline{\tau}_1(I) &= \frac{1}{\Delta x} \int_{(I-1/2)\Delta x}^{(I+1/2)\Delta x} \tau_1(x)\, dx \\
&= \frac{\tau_{1,\max}}{n+1} \frac{(I+1/2)^{p+1} - (I-1/2)^{p+1}}{L^{P+1}} \quad \text{for the } E \text{ field,} \quad (36a)
\end{aligned}$$

$$\overline{\tau}_1\left(I + \frac{1}{2} \right) = \frac{1}{\Delta x} \int_{I\Delta x}^{(I+1)\Delta x} \tau_1(x)\, dx = \frac{\tau_{1,\max}}{n+1} \frac{(I+1)^{p+1} - (I)^{p+1}}{L^{P+1}}$$
$$\text{for the } H \text{ field.} \tag{36b}$$

The discretization of Eqs. (33a)–(33l) can be carried out in a manner similar to that described in Section III. To economize on computer memory usage, one uses the preceding 12 equations, Eqs. (33a)–(33l), for the boundary layers, while the six conventional governing equations given by Eqs. (5a)–(5f) are employed in the interior domain inside the boundary layers. A number of comparison studies of the boundary reflection (Berenger, 1994; Katz et al., 1994; Lazzi and Gandhi, 1996) have been made between the PML method and the analytical boundary condition such as Mur's absorbing condition (Mur, 1981) and the retarded time boundary condition (Berntsen and Hornsleth, 1994). The reflection produced by the PML boundary condition is three orders smaller in magnitude than that obtained by using the analytical boundary equations.

V. FIELD IN FREQUENCY DOMAIN

The values of the near field computed by the preceding FDTD algorithm are in the time domain. A transformation of the time-dependent field values to their corresponding counterparts in the frequency domain is required to obtain the single-scattering properties. The transformation algorithm depends on what kind of initial wave is used. If the input is a continuous sinusoidal wave, the magnitude and phase information of the final steady-state field can be obtained by determining the peak positive- and negative-going excursions of the field over a complete cycle of the incident wave (Umashankar and Taflove, 1982). Three successive data sets in the time sequence of the field can be compared to determine if a peak has been reached. When the peak is detected, these data sets can then be used to determine the amplitude and phase of the steady-state field. The positive- and negative-going peak transitions, however, usually do not occur at the exact peak of the wave. For this reason, the magnitude and phase obtained by this algorithm may produce numerical errors. A longer time is also required to obtain a convergent solution by using a sinusoidal wave as the initial excitation, especially for lower frequencies (Furse *et al.*, 1990). Further, if a continuous sinusoidal wave is used, each individual run of the FDTD code can also provide just one frequency response. However, with a pulse excitation each individual run of the FDTD code will provide various frequency responses.

In FDTD simulations of scattering phenomena, the shape and size of the particle are fixed for a given execution and the dielectric constants are independent of the wave frequency if the particle is nonabsorptive. Thus, the FDTD method with an incident pulse can provide the results for a number of size parameters simultaneously. For this reason, a Gaussian pulse will be used as the initial excitation in the computations presented in this chapter. The width of the pulse must be properly selected to avoid numerical dispersion caused by the finite difference approximation. To illustrate the dispersion problem, let us consider the x'-polarized wave propagating along the z' direction in the incident coordinate system $ox'y'z'$. The finite difference equations governing the field variation can be expressed by

$$E_{x'}^{n+1}(I) = E_{x'}^{n}(I) + \frac{c\Delta t}{\Delta s}\left[H_{y'}^{n+1/2}\left(I - \frac{1}{2}\right) - H_{y'}^{n+1/2}\left(I + \frac{1}{2}\right)\right], \quad (37a)$$

$$H_{y'}^{n+1/2}\left(I + \frac{1}{2}\right) = H_{y'}^{n-1/2}\left(I + \frac{1}{2}\right) + \frac{c\Delta t}{\Delta s}\left[E_{x'}^{n}(I) - E_{x'}^{n}(I + 1)\right]. \quad (37b)$$

Consider a harmonic solution given by

$$E_{x'}^{n}(I) = E_o \exp\left[ik(I\Delta s - c^* n\Delta t)\right], \quad (38a)$$

$$H_{y'}^{n+1/2}\left(I + \frac{1}{2}\right) = H_o \exp\left\{ik\left[\left(I + \frac{1}{2}\right)\Delta s - c^*\left(n + \frac{1}{2}\right)\Delta t\right]\right\}, \quad (38b)$$

where c and c^* are the physical and computational phase speeds, respectively, which are different because of the numerical dispersion. From the preceding equations, we obtain

$$\frac{c^*}{c} = \frac{2 \sin^{-1}[c\Delta t \sin(k\Delta s/2)/\Delta s]}{kc\Delta t}. \tag{39}$$

Equation (39) implies that the waves with higher frequencies (shorter wavelengths) suffer a larger numerical dispersion. The dispersion relationships in the 2D and 3D cases have been discussed by Taflove and Umashankar (1990). In these cases, the numerical dispersion depends not only on frequency, but also on the propagation direction of the wave in the grid mesh. Because various frequencies are contained in a pulse, progressive pulse distortion can be produced as higher frequency components propagate slower than lower frequency components. The frequency spectrum of a pulse is determined by the pulse width in the time domain. Thus, the width of an input pulse should be properly selected to reduce numerical dispersion. The input Gaussian pulse at the time step n can be represented in a discrete form as follows:

$$G_n = A \exp\left[-\left(\frac{n}{w} - 5\right)^2\right], \tag{40}$$

where A is a constant and w is a parameter controlling the width of the pulse. The center of a Gaussian pulse is shifted by $5w$ so that the pulse can start with a very small value ($\sim 10^{-11}$) at the initial time.

As stated in Section III, the incident wave is required at the connecting surface when using the total-field algorithm or over the global grid mesh when using the scattered-field algorithm. The incident pulse at these locations at the time step n should not be specified analytically in terms of the exact pulse values given by Eq. (40), because the numerical solution of Eqs. (5a)–(5f) coupled with an analytical specification of the incident wave may cause inconsistent dispersion and aliasing, leading to numerical instability. The 1D FDTD scheme given by Eqs. (37a) and (37b) can be applied to the simulation of the propagation of the incident pulse, which is subsequently interpolated to the required locations by using a natural spline or a linear algorithm. These two interpolation algorithms produce similar results because the grid sizes smaller than approximately $1/20$ of the incident wavelength are usually required.

The frequency or wavenumber spectrum of the simulated field can be obtained by the discrete Fourier transform if a pulse is employed as the initial excitation. Let f be a component of the field and its value at the time step n be f_n. Then, the time variation of f can be written as

$$f(t) = \sum_{n=0}^{N} f_n \delta(t - n\Delta t), \tag{41}$$

where δ is the Dirac delta function and the maximum time step N is chosen such that the field in the time domain is reduced to a small value. The corresponding spectrum in the wavenumber domain is given by

$$F(k) = \int_{-\infty}^{\infty} \left[\sum_{n=0}^{N} f_n \delta(t - n\Delta t) \right] \exp(ikct)\, dt = \sum_{n=0}^{N} f_n \exp(ikcn\Delta t), \tag{42}$$

where k is the wavenumber in vacuum. To avoid aliasing and numerical dispersion and to obtain a correct frequency spectrum, one must band the maximum wavenumber or the minimum wavelength for the region within which the frequency response of the scattering is evaluated. In any finite difference equation, it is required that the wavelength of a simulated wave be larger than the grid size. Therefore, if we let $k_{\text{grid}} = 2\pi/\Delta d$ where Δd is the minimum among Δx, Δy, and Δz, the permitted wavenumber is

$$k = q k_{\text{grid}}, \qquad q \in [0, 1). \tag{43}$$

In practice, the frequency response obtained by the discrete Fourier transform technique would be inaccurate if the selected parameter q in Eq. (43) were larger than 0.1 because of significant computational wave dispersion and aliasing. For light scattering by a nonspherical particle, the effective permittivity and conductivity described in Section III can be specified to be independent of the wavenumber used in the Fourier transform. Thus, by selecting various q values for the frequency spectrum given by Eq. (42), we can obtain the scattering properties for various size parameters by carrying out near field computations. This procedure has been discussed in more detail in Yang and Liou (1995).

The field values in the frequency domain obtained by this procedure must be normalized by the Fourier transform of the incident wave at the center of the grid mesh so that the frequency response of the scattering particle will return to a unit incident harmonic wave. The discrete Fourier transform given by Eq. (42) is different from that developed by other researchers (e.g., Furse *et al.*, 1990) by a constant. However, this constant will eventually be canceled in the procedure of normalization.

VI. TRANSFORMATION OF NEAR FIELD TO FAR FIELD

To obtain the scattered far field, either a surface- or a volume-integration approach can be used. In the former, a regular enclosing surface that contains the particle is selected. The far field is then given by the integration of the near field over the surface. In the latter, the integration of the near field is carried out over the entire domain inside the particle surface. In the following we review the basic electromagnetic relationship between the near field and the far field.

A. SCATTERED FAR FIELD

Owing to the electromagnetic equivalence theorem, the field detected by an observer outside the surface would be the same if the scatterer were removed and replaced by the equivalent electric and magnetic currents given by

$$\mathbf{J} = \mathbf{n}_S \times \mathbf{H}, \tag{44a}$$

$$\mathbf{M} = \mathbf{E} \times \mathbf{n}_S, \tag{44b}$$

where the electric field $\mathbf{E}$ and the magnetic field $\mathbf{H}$ are the total fields that include the incident and the scattered fields produced by the scatterer and $\mathbf{n}_S$ is the outward unit vector normal to the surface. The Hertz vectors or potentials given by the equivalent currents are

$$\mathbf{j}_m(\mathbf{r}) = \iint_S \mathbf{M}(\mathbf{r}')G(\mathbf{r}, \mathbf{r}') \, d^2\mathbf{r}', \tag{45a}$$

$$\mathbf{j}_e(\mathbf{r}) = \iint_S \mathbf{J}(\mathbf{r}')G(\mathbf{r}, \mathbf{r}') \, d^2\mathbf{r}', \tag{45b}$$

where $G(\mathbf{r}, \mathbf{r}')$ is the Green's function in free space, which is defined by

$$G(\mathbf{r}, \mathbf{r}') = \frac{\exp(ik|\mathbf{r} - \mathbf{r}'|)}{4\pi|\mathbf{r} - \mathbf{r}'|}. \tag{46}$$

In the preceding equations, $\mathbf{r}$ is the position vector of the observation point; $\mathbf{r}'$ is the position vector of the source point. The electric field induced by the Hertz vectors can be written in the form

$$\mathbf{E}_s(\mathbf{r}) = -\nabla \times \mathbf{j}_m(\mathbf{r}) + \frac{i}{k}\nabla \times \nabla \times \mathbf{j}_e(\mathbf{r}). \tag{47}$$

For the radiation zone or far field region, that is, $kr \to \infty$, Eq. (47) reduces to

$$\mathbf{E}_s(\mathbf{r})|_{kr \to \infty} = \frac{\exp(ikr)}{-ikr}\frac{k^2}{4\pi}\mathbf{n} \times \iint_S \{\mathbf{n}_S \times \mathbf{E}(\mathbf{r}') - \mathbf{n} \times [\mathbf{n}_S \times \mathbf{H}(\mathbf{r}')]\}$$
$$\times \exp(-ik\mathbf{n} \cdot \mathbf{r}') \, d^2\mathbf{r}', \tag{48}$$

where $\mathbf{n} = \mathbf{r}/r$ is a unit vector in the scattering direction. It is evident that the far field can be obtained exactly if the tangential components of the electric and magnetic fields on the surface S are precisely known.

Equation (48) involves both electric and magnetic fields. An equivalent counterpart of Eq. (46), which involves only the electric field, can also be derived. Based on the vector algebra, it can be proven that the following relationships hold for two arbitrary vectors $\mathbf{P}$ and $\mathbf{Q}$ and a scalar function ϕ:

$$\iiint_V (\mathbf{Q} \cdot \nabla \times \nabla \times \mathbf{P} - \mathbf{P} \cdot \nabla \times \nabla \times \mathbf{Q})\, dV$$

$$= \iiint_V (\mathbf{P} \cdot \nabla^2 \mathbf{Q} - \mathbf{Q} \cdot \nabla^2 \mathbf{P})\, dV + \iint_S \mathbf{n}_S \cdot (\mathbf{Q}\nabla \cdot \mathbf{P} - \mathbf{P}\nabla \cdot \mathbf{Q})\, dS, \quad (49a)$$

$$\iiint_V (\phi \nabla^2 \mathbf{P} - \mathbf{P}\nabla^2 \phi)\, dV = \iint_S \left(\phi \frac{\partial \mathbf{P}}{\partial n_S} - \mathbf{P} \frac{\partial \phi}{\partial n_S} \right) dS, \quad (49b)$$

where S is an arbitrary surface enclosing the volume domain V. Further, we let

$$\mathbf{P} = \mathbf{a} \cdot \mathbf{G}, \qquad \mathbf{Q} = \mathbf{E}, \qquad \phi = G, \qquad (50)$$

where $\mathbf{a}$ is an arbitrary constant vector, G is the Green's function, and $\mathbf{G}$ is the dyadic Green's function given by

$$\mathbf{G}(\mathbf{r}, \mathbf{r}') = \left(\mathbf{I} + \frac{1}{k^2} \nabla_r \nabla_r \right) G(\mathbf{r}, \mathbf{r}'), \qquad (51)$$

where $\mathbf{I}$ is a unit dyad (Tai, 1971). The volume domain V is selected to be the region outside S and S_o but bounded by S_∞, where S_o encloses the source that generates the incident wave (active source), S encloses the scatterer (passive source), and S_∞ denotes a surface infinitely far away. The distance between S_o and S must be large enough so that the impact of the scattered field on the source inside S_o can be neglected. Using Eqs. (49)–(51), we obtain the electric field inside the region of V as follows:

$$\mathbf{E}(\mathbf{r}) = \frac{1}{k^2} \nabla \times \nabla \times \left[\iint_{S_\infty} + \iint_S + \iint_{S_o} \right]$$

$$\times \left[\mathbf{E}(\mathbf{r}') \frac{\partial G(\mathbf{r}, \mathbf{r}')}{\partial n_S} - G(\mathbf{r}, \mathbf{r}') \frac{\partial \mathbf{E}(\mathbf{r}')}{\partial n_S} \right] d^2 \mathbf{r}', \qquad \mathbf{r} \in V, \quad (52)$$

where the integral over S_o is associated with the incident or initial wave. There will be no contribution from the integral over S_∞ if the following Sommerfeld's radiation condition (Sommerfeld, 1952) is applied:

$$\lim_{r \to \infty} r [\nabla \times \mathbf{G} - i k \mathbf{n} \times \mathbf{G}] = 0. \qquad (53)$$

It follows that the scattered or induced field resulting from the presence of a scatterer is

$$\mathbf{E}_s(\mathbf{r}) = \frac{1}{k^2} \nabla \times \nabla \times \int\int_S \left[\mathbf{E}(\mathbf{r}') \frac{\partial G(\mathbf{r}, \mathbf{r}')}{\partial n_S} - G(\mathbf{r}, \mathbf{r}') \frac{\partial \mathbf{E}(\mathbf{r}')}{\partial n_S} \right] d^2\mathbf{r}', \qquad \mathbf{r} \in V. \tag{54}$$

Instead of using the macroelectrodynamics, the preceding expression can also be obtained from the molecular optics (Oseen, 1915). For the far field region, Eq. (54) reduces to

$$\mathbf{E}_s(\mathbf{r})|_{kr \to \infty} = \frac{\exp(ikr)}{-ikr} \frac{k^2}{4\pi} \mathbf{n} \times \left\{ \mathbf{n} \times \int\int_S \left[\mathbf{n}_S \cdot \mathbf{n}\mathbf{E}(\mathbf{r}') + \frac{1}{ik} \frac{\partial \mathbf{E}(\mathbf{r}')}{\partial n_S} \right] \right.$$
$$\left. \times \exp(-ik\mathbf{n} \cdot \mathbf{r}') \, d^2\mathbf{r}' \right\}. \tag{55}$$

Equation (55) is equivalent to Eq. (48), but it contains only the E field and is also simpler for numerical computations.

To derive the far field given by the integration of the near field over the particle volume, we begin with the electromagnetic wave equation in the frequency domain written for a dielectric medium in the source-dependent form (Goedecke and O'Brien, 1988) as follows:

$$(\nabla^2 + k^2)\mathbf{E}(\mathbf{r}) = -4\pi (k^2\mathbf{I} + \nabla\nabla) \cdot \mathbf{P}(\mathbf{r}), \tag{56}$$

where $\mathbf{P}(\mathbf{r})$ is the polarization vector given by

$$\mathbf{P}(\mathbf{r}) = \frac{\varepsilon(\mathbf{r}) - 1}{4\pi} \mathbf{E}(\mathbf{r}). \tag{57}$$

The material medium here is the scattering particle, thereby making the polarization vector nonzero only within the finite region inside the particle. The solution for Eq. (56) is given by an integral equation as follows:

$$\mathbf{E}(\mathbf{r}) = \mathbf{E}_o(\mathbf{r}) + 4\pi \int\int\int_V G(\mathbf{r}, \mathbf{r}')(k^2\mathbf{I} + \nabla_{r'}\nabla_{r'}) \cdot \mathbf{P}(\mathbf{r}') \, d^3\mathbf{r}', \tag{58}$$

where the first term on the right-hand side is the incident wave. The domain of the integration, V, is the region inside the dielectric particle. For the far field, $k(|\mathbf{r} - \mathbf{r}'|) \to \infty$, it can be proven by using Eq. (58) that the scattered or induced far field caused by the presence of the particle is

$$\mathbf{E}_s(\mathbf{r}) = \frac{k^2 \exp(ikr)}{4\pi r} \int\int\int_V [\varepsilon(\mathbf{r}') - 1]\{\mathbf{E}(\mathbf{r}') - \mathbf{n}[\mathbf{n} \cdot \mathbf{E}(\mathbf{r}')]\} \exp(-ik\mathbf{n} \cdot \mathbf{r}') \, d^3\mathbf{r}'. \tag{59}$$

Equation (59) is not applicable if a conducting scatterer is involved. In this case, either Eq. (49) or Eq. (55) can be employed to obtain the scattered far field by carrying out the involved integration over a regular surface enclosing the scatterer.

To compute the scattering matrix, the scattered field given by Eq. (48), (55), or (59) must be expressed in terms of the amplitude matrix (Chapter 1). We will present the required formulation based on the volume integration technique given by Eq. (59). Similar expressions can be derived for the surface-integration techniques based on Eqs. (48) and (55). Because the scattered field is a transverse wave with respect to the scattering direction, it can be decomposed into the components parallel and perpendicular to the scattering plane in the form

$$\mathbf{E}_s(\mathbf{r}) = \alpha E_{s,\alpha}(\mathbf{r}) + \beta E_{s,\beta}(\mathbf{r}), \tag{60}$$

where α and β are the unit vectors parallel and perpendicular to the scattering plane, respectively, and satisfy

$$\mathbf{n} = \beta \times \alpha. \tag{61}$$

Writing Eq. (60) in matrix form, we obtain

$$
\begin{pmatrix} E_{s,\alpha}(\mathbf{r}) \\ E_{s,\beta}(\mathbf{r}) \end{pmatrix} = \frac{k^2 \exp(ikr)}{4\pi r} \int\!\!\int\!\!\int_V [\varepsilon(\mathbf{r}') - 1] \begin{pmatrix} \alpha \cdot \mathbf{E}(\mathbf{r}) \\ \beta \cdot \mathbf{E}(\mathbf{r}) \end{pmatrix} \exp(-ik\mathbf{n}\cdot\mathbf{r}')\, d^3\mathbf{r}'
$$
$$
= \frac{\exp(ikr)}{r} \mathbf{S} \begin{pmatrix} E_{o,\alpha} \\ E_{o,\beta} \end{pmatrix}, \tag{62}
$$

where $\mathbf{S}$ is a 2×2 amplitude scattering matrix and $E_{o,\alpha}$ and $E_{o,\beta}$ are the incident E-field components defined with respect to the scattering plane. In the FDTD method, the incident wave is defined with respect to the incident coordinate system given by $E_{o,x}$ and $E_{o,y}$. Based on the geometry implied by Eqs. (60) and (61), we have

$$
\begin{pmatrix} E_{o,\alpha} \\ E_{o,\beta} \end{pmatrix} = \begin{pmatrix} \beta \cdot \mathbf{x} & -\beta \cdot \mathbf{y} \\ \beta \cdot \mathbf{y} & \beta \cdot \mathbf{x} \end{pmatrix} \begin{pmatrix} E_{o,y} \\ E_{o,x} \end{pmatrix}, \tag{63}
$$

where $\mathbf{x}$ and $\mathbf{y}$ are the unit vectors along the x and y axes, respectively. To obtain the scattering properties of the particle with complete polarization information, we can select two incident cases: (a) $E_{o,x} = 1$ and $E_{o,y} = 0$ and (b) $E_{o,x} = 0$ and $E_{o,y} = 1$, and define the following quantities:

$$
\begin{pmatrix} F_{\alpha,x} \\ F_{\beta,x} \end{pmatrix} = \frac{k^2}{4\pi} \int\!\!\int\!\!\int_V [\varepsilon(\mathbf{r}') - 1] \begin{pmatrix} \alpha \cdot \mathbf{E}(\mathbf{r}') \\ \beta \cdot \mathbf{E}(\mathbf{r}') \end{pmatrix}
$$
$$
\times \exp(-ik\mathbf{n}\cdot\mathbf{r}')\, d^3\mathbf{r}' \Big|_{E_{o,x}=1,\ E_{o,y}=0}, \tag{64a}
$$

$$
\begin{pmatrix} F_{\alpha,y} \\ F_{\beta,y} \end{pmatrix} = \frac{k^2}{4\pi} \iiint_V [\varepsilon(\mathbf{r}') - 1] \begin{pmatrix} \boldsymbol{\alpha} \cdot \mathbf{E}(\mathbf{r}') \\ \boldsymbol{\beta} \cdot \mathbf{E}(\mathbf{r}') \end{pmatrix}
$$

$$
\times \exp(-ik\mathbf{n} \cdot \mathbf{r}') \, d^3\mathbf{r}' \bigg|_{E_{o,x}=0,\ E_{o,y}=1} . \qquad (64b)
$$

Using Eqs. (62)–(64) along with some algebraic manipulations, it can be proven that

$$
\mathbf{S} = \begin{pmatrix} F_{\alpha,y} & F_{\alpha,x} \\ F_{\beta,y} & F_{\beta,x} \end{pmatrix} \begin{pmatrix} \boldsymbol{\beta} \cdot \mathbf{x} & \boldsymbol{\beta} \cdot \mathbf{y} \\ -\boldsymbol{\beta} \cdot \mathbf{y} & \boldsymbol{\beta} \cdot \mathbf{x} \end{pmatrix}. \qquad (65)
$$

The amplitude matrix defined in Chapter 1 is obtained by changing the sign of the off-diagonal elements of the matrix $\mathbf{S}$ given by Eq. (65). After defining the amplitude matrix, the scattering matrix $\mathbf{F}$ can be determined and numerically computed based on the formulas given in Chapter 1. For nonspherical ice crystals and aerosols oriented randomly in space, the scattering matrix normally has a block-diagonal structure with eight nonzero elements among which only six are independent (Section XI of Chapter 1; van de Hulst, 1957).

B. Extinction and Absorption Cross Sections

To derive the integral equations for the absorption and extinction cross sections, we start from Maxwell's equations. For a nonferromagnetic dielectric medium with an incident harmonic wave whose time dependence is given by $\exp(-ikct)$, Maxwell's curl equations in the frequency domain can be written as

$$
c\nabla \times \mathbf{H} = -i\omega(\varepsilon_r + i\varepsilon_i)\mathbf{E}, \qquad (66a)
$$

$$
c\nabla \times \mathbf{E} = i\omega\mathbf{H}, \qquad (66b)
$$

where $\omega = kc$. Using the preceding equations along with vector algebra, we have

$$
-\nabla \cdot \mathbf{s} = \frac{i\omega}{4\pi}\left(\varepsilon_r \mathbf{E} \cdot \mathbf{E}^* - \mathbf{H} \cdot \mathbf{H}^*\right) + \frac{\omega\varepsilon_i}{4\pi}\mathbf{E} \cdot \mathbf{E}^*, \qquad (67a)
$$

$$
\mathbf{s} = \frac{c}{4\pi}\mathbf{E} \times \mathbf{H}^*, \qquad (67b)
$$

where $\mathbf{s}$ is the complex Poynting vector and the asterisk denotes the complex conjugate. Taking the real part of Eq. (67a) and integrating it over the region inside the scattering particle lead to

$$
-\mathrm{Re}\left[\iiint_V \nabla \cdot \mathbf{s}(\mathbf{r}') \, d^3\mathbf{r}'\right] = -\mathrm{Re}\left[\iint_S \mathbf{n}_S \cdot \mathbf{s}(\mathbf{r}') \, d^2\mathbf{r}'\right]
$$

$$
= \frac{\omega}{4\pi}\iiint_V \varepsilon_i(\mathbf{r}')\mathbf{E}(\mathbf{r}') \cdot \mathbf{E}^*(\mathbf{r}') \, d^3\mathbf{r}', \qquad (68)
$$

where $\mathbf{n}_S$ is the outward-pointing unit vector normal to the particle surface. Based on the physical meaning of the Poynting vector (Jackson,1975), the surface integration term in Eq. (68) is the net rate at which electromagnetic energy intersects with the particle surface, that is, the energy absorbed by the particle. Further, the incident electromagnetic flux is given by

$$F_o = \frac{c}{4\pi}\mathbf{E}_o \cdot \mathbf{E}_o^* = \frac{c}{4\pi}|\mathbf{E}_o|^2. \tag{69}$$

It follows that the absorption cross section of the particle is given by

$$
\begin{aligned}
C_{\text{abs}} &= \frac{-\text{Re}\left[\iint_S \mathbf{n}_S \cdot \mathbf{s}(\mathbf{r}')\, d^2\mathbf{r}'\right]}{F_o} \\
&= \frac{k}{|\mathbf{E}_o|^2} \iiint_V \varepsilon_{\text{i}}(\mathbf{r}')\mathbf{E}(\mathbf{r}') \cdot \mathbf{E}^*(\mathbf{r}')\, d^3\mathbf{r}'.
\end{aligned} \tag{70}
$$

In conjunction with the derivation of the extinction cross section, we note that the Poynting vector can be decomposed into the incident, scattered, and extinction components as follows:

$$\mathbf{s} = \mathbf{s}_o + \mathbf{s}_s + \mathbf{s}_e. \tag{71}$$

The complex extinction component of the Poynting vector is given by

$$\mathbf{s}_e = \frac{c}{4\pi}\left(\mathbf{E}_o \times \mathbf{H}^* + \mathbf{E}^* \times \mathbf{H}_o\right). \tag{72}$$

Using Eqs. (71) and (72), we can prove that the electromagnetic energy associated with extinction is defined by

$$-\text{Re}\left[\iint_S \mathbf{n}_S \cdot \mathbf{s}_e(\mathbf{r}')\, d^2 r'\right] = \frac{\omega}{4\pi}\,\text{Im}\left[\iiint_V [\varepsilon(\mathbf{r}')-1]\mathbf{E}(\mathbf{r}')\cdot\mathbf{E}_o^*(\mathbf{r}')\, d^3\mathbf{r}'\right]. \tag{73}$$

Consequently, the extinction cross section is given by

$$
\begin{aligned}
C_{\text{ext}} &= \frac{\omega}{4\pi}\frac{\text{Im}\left\{\iiint_V [\varepsilon(\mathbf{r}') - 1]\mathbf{E}(\mathbf{r}') \cdot \mathbf{E}_o^*(\mathbf{r}')\, d^3\mathbf{r}'\right\}}{F_o} \\
&= \text{Im}\left\{\frac{k}{|\mathbf{E}_o|^2} \iiint_V [\varepsilon(\mathbf{r}') - 1]\mathbf{E}(\mathbf{r}') \cdot \mathbf{E}_o^*(\mathbf{r}')\, d^3\mathbf{r}'\right\}.
\end{aligned} \tag{74}
$$

For scattering by a nonspherical particle, the absorption and extinction cross sections depend on the polarization of the incident wave (see Section VII of Chapter 1). However, if the mean values of a cross section (the average of the cross sections computed with respect to two perpendicularly polarized incident waves) are considered, they are independent of the plane on which the polarization of the incident wave is defined. Using Eqs. (62) and (74) along with integration by parts, it can be proven that the mean extinction cross section is

$$\overline{C}_{ext} = \frac{C_{ext,\parallel} + C_{ext,\perp}}{2} = \frac{2\pi}{4} \operatorname{Im}\left[S_{11}(\mathbf{n}^{inc}) + S_{22}(\mathbf{n}^{inc})\right], \qquad (75)$$

where $\mathbf{n}^{inc}$ is a unit vector indicating the incident direction (Chapter 1). The previous equation is actually a particular form of the optical or extinction theorem. The mean absorption cross section can be computed from Eq. (70) using the preceding two incident cases.

The amplitude matrix and the absorption and extinction cross sections given by Eqs. (64a), (64b), (65), (70), and (74) are presented in continuous integral form. In practical computations, these equations must be discretized so that the near field values at the grid points can be summed. Consider the computation of the extinction cross section as an example. We first normalize the near field values obtained by the discrete Fourier transform with respect to the Fourier spectrum of the incident wave calculated at the center of the computational grid lattice. The Cartesian component of the electric field at a cell center is given by the average of the field component at four cell edges. Thus, we obtain

$$
\begin{aligned}
C_{ext} = {} & \frac{k}{4} \operatorname{Im} \sum_I \sum_J \sum_K [\varepsilon(I, J, K) - 1] \\
& \times \left\{ \left[E_x\left(I, J - \frac{1}{2}, K - \frac{1}{2}\right) + E_x\left(I, J - \frac{1}{2}, K + \frac{1}{2}\right) \right. \right. \\
& \qquad \left. + E_x\left(I, J + \frac{1}{2}, K - \frac{1}{2}\right) + E_x\left(I, J + \frac{1}{2}, K + \frac{1}{2}\right) \right] e_x \\
& \quad + \left[E_y\left(I - \frac{1}{2}, J, K - \frac{1}{2}\right) + E_y\left(I - \frac{1}{2}, J, K + \frac{1}{2}\right) \right. \\
& \qquad \left. + E_y\left(I + \frac{1}{2}, J, K - \frac{1}{2}\right) + E_y\left(I + \frac{1}{2}, J, K + \frac{1}{2}\right) \right] e_y \\
& \quad + \left[E_z\left(I - \frac{1}{2}, J - \frac{1}{2}, K\right) + E_z\left(I - \frac{1}{2}, J + \frac{1}{2}, K\right) \right. \\
& \qquad \left. \left. + E_z\left(I + \frac{1}{2}, J - \frac{1}{2}, K\right) + E_z\left(I + \frac{1}{2}, J + \frac{1}{2}, K\right) \right] e_z \right\} \\
& \times \exp(-ik_x I \Delta x - ik_y I \Delta y - ik_z I \Delta z), \qquad (76)
\end{aligned}
$$

where e_x, e_y, and e_z are the three coordinate components of a unit vector pointing along the polarization direction of the incident electric field and k_x, k_y, and k_z are the components of the incident wavenumber vector projected on the three coordinate axes. The discrete expressions can also be obtained for the amplitude matrix and the absorption cross section.

VII. SCATTERING PROPERTIES OF AEROSOLS AND ICE CRYSTALS

In this section we apply the FDTD technique to compute the scattering properties of ice crystals and aerosols with various geometries and compositions. The numerical results shown in this chapter are intended to be representative rather than extensive. It should be pointed out that we have carried out comprehensive validations regarding the accuracy of the FDTD method using infinite circular cylinders and spheres for which the exact solutions are available (Yang and Liou, 1995, 1996a). Because the FDTD technique does not pose a preferential treatment to any specific geometry (with a possible exception of rectangular targets in a Cartesian grid mesh), these canonical comparison studies constitute a representative test of the accuracy of the FDTD method. In general, when the size of the grid cells is on the order of $1/20$ of the incident wavelength, the FDTD solutions are in good agreement with their corresponding analytical counterparts. The relative errors of the scattered energy are smaller than 3%. The accuracy of the FDTD solution is improved when the ratio of the grid size to the incident wavelength decreases. For size parameters larger than about 15, its accuracy in terms of the relative errors for the phase matrix elements in some scattering directions, for example, at backscattering, can reach 40%, although the errors in total scattered energy are small. The time-marching iterative steps should be sufficiently small in order to obtain a convergent solution in the near field computation when size parameters are larger than 10–20. Through numerical experiments we have found that errors in the FDTD solution can be reduced to less than 1% if the grid size used is on the order of $1/40$ of the incident wavelength and sufficiently small time steps are employed in the time-marching iteration. Our previous validation efforts demonstrated that the FDTD method can achieve reliable results for size parameters smaller than about 15–20.

It is well recognized that the approximation of nonspherical particles by using spheres is physically inadequate and often misleading (Liou and Takano, 1994; Mishchenko *et al.*, 1996c). Figure 4 shows the FDTD solution for the extinction efficiency (the ratio of the extinction cross section to the projected area of particle) of hexagonal ice columns randomly oriented in space, along with results for equivalent-volume and equivalent-surface spheres for comparison. Both spherical solutions overestimate the extinction efficiency. This overestimation increases with increasing size parameter. The equivalent-volume spherical approximation produces smaller overestimation because the induced dipoles inside the particle, whose number is proportional to the particle volume, contribute significantly to the attenuation of incident radiation.

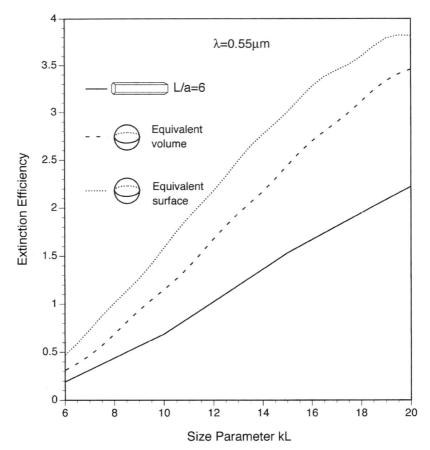

Figure 4   Extinction efficiencies of randomly oriented hexagonal ice crystals computed by the FDTD method and the results for equivalent-volume and -surface spheres computed by the Lorenz–Mie theory. L and a are crystal length and hexagonal diameter, respectively. The refractive index is $1.311 + 3.11 \times 10^{-9}i$.

A. AEROSOLS

Aerosols in the atmosphere exhibit a variety of shapes ranging from quasi-spheres to highly irregular geometries (e.g., Hill *et al.*, 1984; Nakajima *et al.*, 1989; Okada *et al.*, 1987). In addition, aerosols usually appear as a mixed product of different compositions involving dustlike, water-soluble, soot, oceanic, sulfate, mineral, water, and organic materials. The refractive indices for these components have been compiled by d'Almeida *et al.* (1991). To understand the scattering characteristics of aerosols, we have defined various representative aerosol geometries and inhomogeneous compositions for light-scattering computations based on the FDTD method.

The left panels of Fig. 5 show the phase function (i.e., the F_{11} element of the scattering matrix; see Section XI of Chapter 1) and the degree of linear polarization (DLP) $-F_{12}/F_{11}$ at $\lambda = 0.5$ μm for two randomly oriented dustlike aerosol shapes with 10 and 6 faces. The size parameters of these irregular aerosols are specified in terms of the dimensions of their peripheral spheres. Although the two polyhedrons have the same size parameters, the particle with 10 faces scatters more energy in the forward direction than its 6-face counterpart. This is because the volume of the former is larger. The ratio of the extinction cross sections for these two aerosol shapes is 3.92. Dustlike aerosols are absorptive in the visible wavelength as indicated by the single-scattering albedos of 0.9656 and 0.9626 for the two polyhedral geometries with 10 and 6 faces, respectively. The middle panels of Fig. 5 show the other scattering matrix elements associated with the polarization state of the scattered wave. The detailed structures of aerosol geometry show a substantial impact on the polarization configuration. From the results, it appears inadequate to characterize irregular aerosols in terms of peripheral spheres.

Black carbon or soot aerosols generated from the incomplete combustion of fossil fuel and biomass burning can serve as condensation nuclei or become outside attachments to water droplets, a potential possibility perhaps relevant for anomalous cloud absorption (Chýlek *et al.*, 1984a). The right panels of Fig. 5 show the phase function and DLP values for water droplets containing irregular soot inclusions as compared with the Lorenz–Mie result for a homogeneous water sphere. The water droplets with inclusions scatter more light in the side directions between 40° and 100° than the corresponding homogeneous spheres. Furthermore, the single-scattering properties of these aerosols are also dependent on the detailed structure of inclusions. The single-scattering albedos of water droplets with black carbon inclusions are substantially less than 1 for the visible wavelength. They are 0.9510 and 0.8852 for the shapes with 10 and 6 faces, respectively.

Figure 6 shows the phase functions and PDLs at a visible wavelength for four aerosol models. In the diagram the prime and double prime denote that the associated parameters are for mineral/dustlike and soot components, respectively, whereas the corresponding unprimed parameters are for water parts of the compounded particles. Polyhedral particles and sphere clusters produce smoother angular scattering patterns in comparison with the cases involving spheres with inclusions and/or attachments. For the latter, the spherical parts of the compounded aerosols dominate the scattering properties. From Figs. 5 and 6 it is evident that the phase functions of polyhedral and cluster aerosols are substantially different from those of homogeneous spheres.

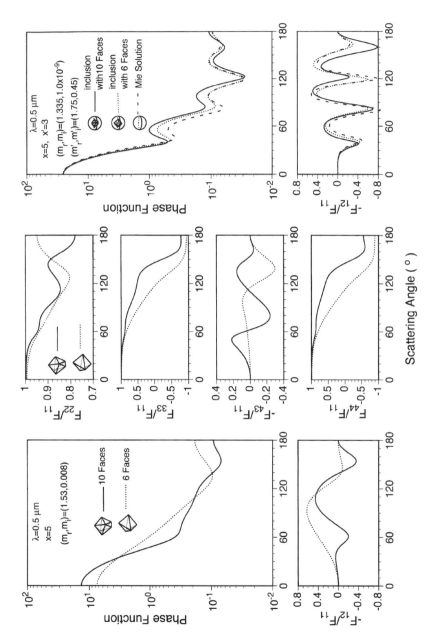

Figure 5 Phase function and polarization properties for randomly oriented nonspherical and inhomogeneous aerosols.

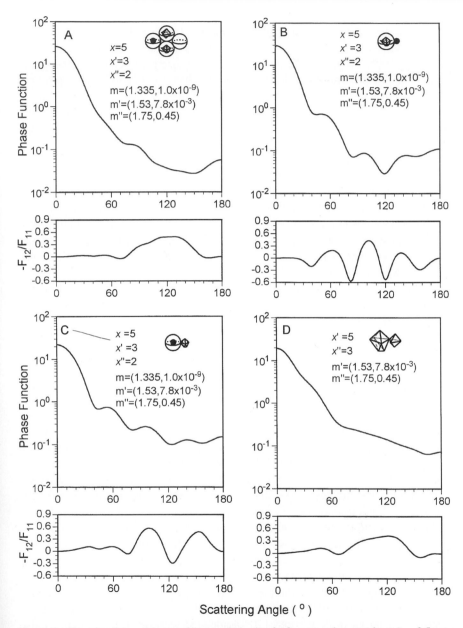

Figure 6   Phase function and degree of linear polarization for four aerosol geometries at $\lambda = 0.5$ μm.

B. SMALL ICE CRYSTALS

The scattering characteristics of nonspherical ice crystals with small size parameters have been investigated previously based on the FDTD technique (Yang and Liou, 1995, 1996a). We demonstrated that the size parameter and aspect ratio of ice crystals are critical to their scattering behaviors. For example, the phase functions for ice plates and long columns are distinctly different, particularly in the scattering angle region larger than approximately 120°. Long columns produce a broad scattering maximum at about 150° and a weak backscattering, but both are absent in the plate case.

Figure 7 illustrates the effects of air bubble and soot inclusions and a cavity on the phase function and DLP for ice crystals at the 0.5- and 10-μm wavelengths. The refractive index of ice for the two wavelengths is $(m_r, m_i) = (1.313, 1.91 \times 10^{-9})$ and $(1.1991, 5.1 \times 10^{-2})$. For $\lambda = 0.5$ μm, the effect of the inclusion does not appear to be significant for the phase function but it substantially affects the DLP patterns. The air bubble and soot inclusions as well as the cavity structures have a substantial impact on the extinction cross sections. For $\lambda = 0.5$ μm the extinction efficiencies are 3.966, 3.289, 2.768, and 3.782 for ice crystals with an air bubble, a soot inclusion, a cavity, and for a solid hexagon (note that the projected areas are the same for these shapes with the geometry parameters specified in Fig. 7), respectively. Moreover, the soot inclusion significantly affects absorption for ice crystals at $\lambda = 0.5$ μm. The single-scattering albedo in this case is 0.7904. For $\lambda = 10$ μm the extinction efficiencies are 2.357, 2.310, 1.892, and 2.729 for the four cases, whereas the corresponding single-scattering albedo values are 0.6833, 0.5710, 0.6802, and 0.6931. The inclusion of soot clearly enhances the absorption of ice crystals at $\lambda = 10$ μm. For the ice crystal size parameters presented in Fig. 7, the distinct scattering peaks associated with halos are absent. The phase interference of the scattered waves produces fluctuations in the phase functions, which are more pronounced for ice crystals with cavities at $\lambda = 10$ μm. The fluctuations and scattering maxima in the ice crystal phase functions are due to the interference pattern of the scattered wave associated with the specific geometries that cannot be completely smoothed out by the random orientation average.

To understand the size parameter values required for the production of halo peaks, we perform computations for infinitely long hexagons with size parameters (with reference to their cross-sectional dimension) of 10 and 60 for two specific incidence configurations, as shown in Fig. 8. For $ka = 10$, fluctuations due to phase interference dominate the phase function pattern. When $ka = 60$, a pronounced peak around 20° is noted for both incidence configurations. In Fig. 8a, a strong scattering peak at about 120° associated with the 120° parhelia is also evident. For hexagonal ice crystals to produce halo patterns, we find that the size parameters must be greater than about 50.

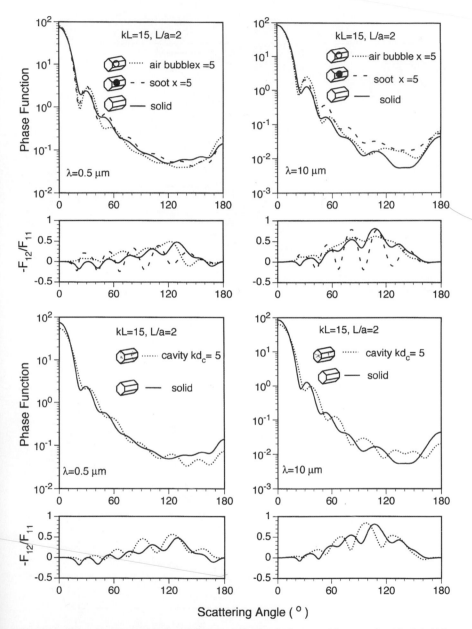

Figure 7   Phase function and degree of linear polarization for hexagonal ice crystals with air bubble and soot inclusions and cavity in comparison with the results for solid columns at $\lambda = 0.5$ μm. The cavity depth of a hollow column is indicated by d_c.

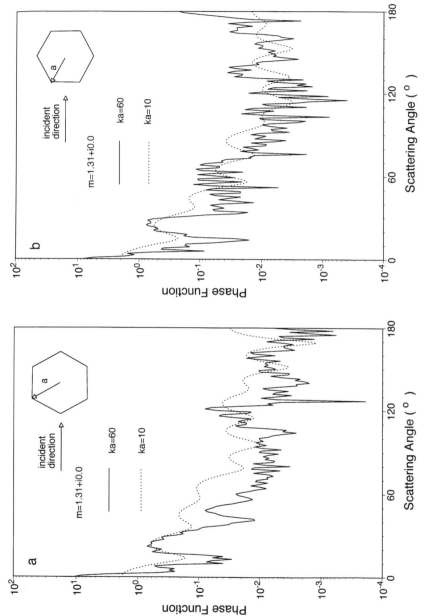

Figure 8 Phase function for infinitely long hexagonal cylinders with $ka = 10$ and 60 for two incident directions.

218

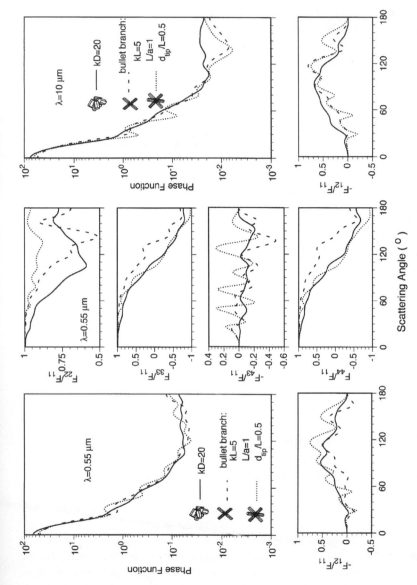

Figure 9 Comparison of the phase function and polarization properties for ice crystal bullet rosettes and aggregates at $\lambda = 0.5$ and 10 μm. The length of the pyramidal tips of bullet branches is indicated by d_{tip}.

Figure 9 illustrates the scattering features for ice bullet rosettes and aggregates at λ = 0.5 and 10 μm. The aggregates consist of eight hexagonal elements whose geometries and relative spatial positions are defined following Yang and Liou (1998a). The definition of bullet rosettes follows Takano and Liou (1995). The lengths of the pyramidal tips of bullet branches are assumed to be equal in the present computation. It is evident from Fig. 9 that bullet rosettes produce similar phase functions as aggregates at λ = 0.5 μm, except that the phase functions for the former display some fluctuations. In addition, aggregates generate weaker backscattering. For λ = 10 μm, substantial differences between the phase functions for bullet rosettes and aggregates are noted in the scattering region around 140°. From Fig. 9 it is also evident that the polarization properties depend on the detailed particle geometry. Because the present results are for ice crystals with small size parameters, the phase functions shown in Fig. 9 differ from those obtained for large ice crystals based on the geometric ray-tracing technique. For a large ice crystal, the scattered wave with respect to different substructures is essentially incoherent. Thus, the phase function for aggregates with hexagonal elements is similar to that associated with a single ice hexagon, as illustrated in Yang and Liou (1998a).

From the computed scattering matrix elements shown in Figs. 5–7 and 9 for complicated ice crystals and compounded aerosols including ice crystal aggregates, aerosols with irregular inclusions, and aerosol clusters, it is clear that the detailed particle structures are important in the determination of their scattering and polarization properties for size parameters in the resonant regime. The scattering properties of nonspherical ice crystals for small size parameters are substantially different from those for large size parameters in the applicable regime of geometric optics. The effect of small ice crystals in remote-sensing applications and radiative transfer calculation deserves future study.

VIII. CONCLUSIONS

In this chapter, we have reviewed the physical basis and numerical implementation of the FDTD technique for light-scattering calculations involving dielectric particles. We discuss four aspects of the methodology including (1) the time-marching iteration for the near field, (2) the absorbing boundary condition for the truncation of the computational domain, (3) the field transformation from the time domain to the frequency domain, and (4) mapping the near field to the far field. For the discretization of Maxwell's equations in both space and time, we show that the best approach is to carry out the spatial discretization first by averaging the permittivity based on the values of four adjacent cells and then performing the temporal discretization. Further, the mathematical formulations of the FDTD method for specific applications to the solutions of phase matrix and extinction and absorption cross sections are presented for computational purposes.

To demonstrate the capability and flexibility of the FDTD technique in dealing with nonspherical and inhomogeneous particles, the single-scattering and polarization properties of aerosols and ice crystals commonly occurring in the atmosphere are presented. New results are illustrated for complicated ice crystals and compounded aerosols including ice crystal aggregates, aerosols with irregular inclusions, and aerosol clusters. It is shown that the detailed particle structures are important in the determination of their scattering and polarization properties for size parameters in the resonant regime.

ACKNOWLEDGMENTS

During the course of this research, we were supported by National Science Foundation Grant ATM-97-96277, Department of Energy Grant DE-FG03-95ER61991, and NASA Grants NAG-5-6160 and NAG-1-1966.

Part III

Compounded, Heterogeneous, and Irregular Particles

Chapter 8

Electromagnetic Scattering by Compounded Spherical Particles

Kirk A. Fuller

Department of Atmospheric Science
Colorado State University
Fort Collins, Colorado 80523

Daniel W. Mackowski

Department of Mechanical Engineering
Auburn University
Auburn, Alabama 36849

Light Scattering by Nonspherical Particles: Theory, Measurements, and Applications
Copyright © 2000 by Academic Press. All rights of reproduction in any form reserved.

I. INTRODUCTION

The scattering of electromagnetic radiation by a collection of objects with sizes comparable to the wavelength of incident radiation, and which are in turn in close enough proximity to each other to allow any one constituent of this assembly to affect the total coherent field scattered by the others is, today, a problem of considerable interest to a broad range of scientific applications. Among these are climatology, visibility modeling, atmospheric remote sensing, planetary astronomy, stellar evolution, fractal optics, nonlinear optics, photonics, sensors, quality control, and process diagnostics.

Beginning with Lorenz–Mie (LM) theory for light scattering by single spheres and the addition theorem for vector spherical harmonics, we review advances made in the study of the optical properties of systems of compound spheres. For organizational purposes, these systems will be divided into two categories: (1) aggregates of two or more spherical particles and (2) one or more spherical inclusions located arbitrarily within a spherical host. As will be discussed, these two categories differ only by certain details at the level of the addition theorem and are essentially described by a single theory that encompasses clusters of included spheres. We discuss comparisons of theory and measurement, as well as applications in fields ranging from astrophysics to immunoassays.

II. HISTORICAL OVERVIEW

The study of scattering and absorption by spheres shares a common ancestry with studies of rainbows, fogbows, glories, and the coloration of minerals and glass—the latter being Gustav Mie's motivation for taking up the subject (Mie, 1908). Much has been written on this history and articles appearing in a special issue on optical particle sizing in *Applied Optics* (cf. Kragh, 1991; Lilienfeld, 1991; Kerker, 1991), the collection of papers assembled by Kerker (1988), and surveys by Logan (1965, 1990) provide abundant resources for its pursuit.

Historically, exact theories of dependent scattering of vector fields began with the use of Fresnel coefficients to study the optical properties of interacting planar surfaces—interferometry. Even earlier, Fresnel (1866a, pp. 639–653) used arguments of energy and momentum conservation, rather than differential equations and boundary conditions to arrive at his expressions for the amplitudes of reflected and transmitted electric vibration. Stokes (1966a, pp. 89–103) confirmed these laws by applying the principle of reciprocity to wave propagation. Both Fresnel (1866b, pp. 787–792) and Stokes (1966b, pp. 155–196; 1966c, pp. 145–156) derived expressions for reflection and transmission coefficients of two or more parallel planar surfaces. Fresnel and Stokes were by no means the only ones to make great contributions to this subject; Newton, Arago, Poisson, and Airy are just a few others, but the works referenced here make fascinating study and lay the foundations for interferometry and, to a degree, plane-parallel radiative transfer. Twersky (1952) provides a review of the history of multiple scattering involving very small objects. Scattering by arrays of parallel cylinders were also studied near the turn of the century, the work of Rayleigh (1903) being the most noteworthy. (These arrays were of primary interest in the study of diffraction gratings, but electromagnetic coupling between the constituents was neglected.) In this context, there is an interesting parallel between Rayleigh–Gans–Debye theory and Newton's rings and between the exact theories for two interacting particles and the Fabry–Perot cavity.

A. Scattering by Interacting Cylinders

A theory for dependent scattering by cylinders evolved somewhat earlier than that for spheres because the addition theorem for cylindrical wave functions had existed since near the turn of the century (Watson, 1962), whereas the first addition theorem for spherical waves did not appear until the mid-1950s. It was with the work of Twersky (1952) that a generalized formalism for cooperative acoustic and electromagnetic (EM) scattering by cylinders began to emerge. Olaofe (1970) solved the two-cylinder problem in a manner analogous to the multiple-reflection model for the Fabry–Perot cavity. A more recent treatment of parallel cylinders is given by Yousif and Köhler (1988) wherein they solve the problem of scattering by a pair of parallel cylinders of different size and composition that are illuminated by obliquely incident plane wave radiation of arbitrary polarization. Those authors also calculate scattering matrices for selected angles of incidence. A brief review of the history of dependent scattering by cylinders is given by Fuller (1991).

B. Scattering by Interacting Spheres

To our knowledge, the first to attempt a thorough investigation of the problem of EM scattering by interacting spheres was Trinks (1935). Explicit solutions can be gleaned from that work for a pair of identical Rayleigh spheres, restricted in orientation with respect to the incident field. Germogenova (1963) extended Trinks' method to account for beam angles of arbitrary incidence on a pair of dissimilar Rayleigh spheres.

Friedman and Russek (1954) derived an addition theorem for the spherical scalar waves. Shortly thereafter, Stein (1961) and Cruzan (1962) presented addition theorems for spherical vector wave functions. Stein's work was first submitted in 1959 and includes an addition theorem for coordinate rotation as well as translation. The works of Stein and Cruzan provided, for the first time, the mathematical underpinnings for the transformation of vector spherical waves based in one coordinate frame to another, arbitrarily located frame. Adding to the observations of Logan (1965) regarding the early contributions of Clebsch (1863) to elastic scattering by single spheres, we note that the addition theorem for spherical wave functions depends on the so-called Clebsch–Gordan coefficients derived nearly 100 years earlier. (It is also worth noting that Gordan was a contemporary of Clebsch's and is not of the Klein–Gordon fame as some believe today.) Recently, Xu has compared the numerical efficiency of different implementations of these theories (Xu, 1996).

The landmark investigations into dependent scattering by spheres clearly were made by Liang and Lo (1967) and, especially, Bruning and Lo (1971). This constituted the first comprehensive, computationally viable solution to the problem of EM scattering by two spheres and included experimental verification via microwave scattering measurements. These authors assimilated the relevant theorems into a workable theory, which was then applied to the calculation of radar and bistatic cross sections of linear chains of two or three spheres. Calculations by Ludwig (1991) indicate that there was probably an error in the computer program Bruning and Lo used for their calculations for three spheres.

Borghese *et al.* (1987) have worked extensively in this field, having derived their theoretical framework independently from Lo and his colleagues. Gérardy and Ausloos (1982) also derive an independent, rigorous solution to the problem of light scattering by clusters of spheres and provide what appears to be the first correct expression for extinction cross sections based on an integral of the Poynting flux. This work has been recently extended to planar arrays of spheres by Quinten and Kreibig (1993). Mackowski (1991, 1994) and Fuller (1994a) derive expressions for extinction and scattering cross sections of multiple sphere configurations, as well, and Mackowski has also provided a direct solution for absorption cross sections that does not directly involve scattering and extinction cross sections.

Crucial experimental work continued in the late 1970s when Schuerman and Wang established a microwave analog facility at which dielectric bodies of a wide variety of shapes could be constructed and their scattering properties measured (Schuerman, 1980b; Wang, 1980; Wang *et al.*, 1981). This valuable work has been continued by Gustafson and others (Zerull *et al.*, 1993; Xu and Gustafson, 1997). Kattawar initiated a major extension of the work of Lo and his colleagues. This led to the theoretical confirmation (Kattawar and Dean, 1983; Fuller *et al.*, 1986) of the so-called specular resonances observed by Wang *et al.* (1981), the generalization of Bruning and Lo's treatment to arbitrarily configured spheres (Fuller, 1987), and the successful order-of-scattering (OS) technique (Fuller and Kattawar, 1988a, b). In addition to the studies discussed previously, work in this field has also been carried out by

Peterson and Ström (1973), Tsang *et al.* (1991), Ohtaka *et al.* (e.g., Inoue and Ohtaka, 1985), Varadan *et al.* (1983), Hamid *et al.* (1991), and Ioannidou *et al.* (1995). Additional historical information is given by Fuller (1991). This chapter deals with particles that can be divided into unique spherical domains. Recently, Videen *et al.* (1996) have reported results for light scattering by overlapping domains.

Theories of scattering by eccentrically stratified spheres have been derived by Fikioris and Uzunoglu (1979), Borghese *et al.* (1992), Fuller (1993a, b), and Mackowski and Jones (1995). Treatments for multiple inclusions have been rendered by Fuller (1993a, b) and Borghese *et al.* (1994). An approximate treatment of the problem of scattering by structured spheres has recently been developed that allows the study of nonspherical inhomogeneities and that can also accommodate large numbers of inclusions (Hill *et al.*, 1995). An outline of the exact theory is provided in the next section, including an application of the OS method to the concentric and single eccentric inclusion problems. Early measurements on eccentrically stratified spheres were made by Swarner and Peters (1963), and current comparisons are underway between the authors of this chapter and the author of Chapter 13.

Approaches to the compound sphere problem have also been made based on the Purcell and Pennypacker (1973) method, that is, the discrete dipole approximation (DDA) covered in Chapter 5. Druger *et al.* (1979) applied this technique to eccentrically stratified spheres, and before being made aware of Bruning and Lo's work (in a private communication with M. Kerker) Kattawar and Humphreys (1980) calculated the scattering properties of two pseudospheres, each comprising 32 point dipoles. West and Smith (1991) have used the DDA to model the optical properties of aggregates of tholin particles.

III. SCATTERING AND ABSORPTION OF LIGHT BY HOMOGENEOUS AND CONCENTRICALLY STRATIFIED SPHERES

A. Vector Spherical Harmonics

Time-harmonic fields scattered by a sphere must satisfy the vector Helmholtz equation

$$\nabla^2 \mathbf{E}^{\text{sca}} + k^2 \mathbf{E}^{\text{sca}} = \mathbf{0}, \tag{1}$$

which is itself a direct consequence of Maxwell's equations for an infinite, linear, isotropic medium containing no free charges or currents. The basis functions that are to be used to mathematically represent the fields must therefore satisfy

$\nabla^2 \mathbf{M} + k^2 \mathbf{M} = \mathbf{0}$. Because $\mathbf{M}$ must also satisfy $\nabla \cdot \mathbf{M} = \mathbf{0}$, a simple choice for this function is $\mathbf{M} = \nabla \times \mathbf{v}\psi$. Substitution of this expression for $\mathbf{M}$ into Eq. (1) and manipulation of the resulting expressions with the use of identities from vector calculus lead one to the conclusion that it is sufficient that $\mathbf{M}$ satisfies Eq. (1) if $\mathbf{v} = \mathbf{R}$ and ψ satisfies the scalar wave equation $\nabla^2 \psi + k^2 \psi = 0$. The preceding discussion is somewhat specific to spherical coordinates and follows Stratton (1941, Chapter 7) and Bohren and Huffman (1983, Chapter 4). Morse and Feshbach (1953, Chapter 13) provide an elegant treatment for identifying $\mathbf{v}\psi$ in generalized curvilinear coordinates.

A second set of functions that is orthogonal to $\mathbf{M}$ and likewise satisfies Eq. (1) is $\mathbf{N} = \nabla \times \mathbf{M}/k$. The scalar wave equation is separable in spherical coordinates and the separation constants are integers m and n, and give rise to a complete set of mutually orthogonal functions ψ_{mn}. A construction of these functions is given by Bohren and Huffman (1983, Chapter 4). The functions $\mathbf{M}$ and $\mathbf{N}$ comprise a complete set of mutually orthogonal functions and any other function that satisfies Eq. (1) may therefore be projected onto the $\mathbf{M}, \mathbf{N}$ basis. To streamline the notation, which becomes rather ornate as one proceeds to more complicated particle morphologies, we introduce the index $p = 1, 2$ for reference to the TM ($\mathbf{N}$) and TE ($\mathbf{M}$) modes of the electric field.

Introducing the angle-dependent functions τ_{mnp},

$$\tau_{mn1}(\cos\vartheta) = \frac{d}{d\vartheta} P_n^m(\cos\vartheta), \qquad \tau_{mn2}(\cos\vartheta) = \frac{m}{\sin\vartheta} P_n^m(\cos\vartheta), \quad (2)$$

where P_n^m are associated Legendre functions, along with the position vector $\mathbf{R} = R\mathbf{e}_R$ ($\mathbf{e}_R = \mathbf{R}/|\mathbf{R}|$), the vector spherical harmonics (VSH), otherwise known as vector spherical wave functions, may be written as

$$\mathbf{N}_{mnp}(\mathbf{R}) = e^{im\varphi}\left[(2-p)\frac{z_n(kR)}{(kR)^2}n(n+1)P_n^m(\cos\vartheta)\mathbf{e}_R \right.$$
$$+ \frac{z_n^{[2-p]}(kR)}{kR}\tau_{mnp}(\cos\vartheta)\boldsymbol{\vartheta}$$
$$\left. + \frac{i^p z_n^{[2-p]}(kR)}{kR}\tau_{mn(3-p)}(\cos\vartheta)\boldsymbol{\varphi}\right], \quad (3)$$

where $\boldsymbol{\vartheta}$ and $\boldsymbol{\varphi}$ are unit vectors in the directions of the polar and azimuthal angles ϑ and φ, respectively. The $z_n^{[0]} \equiv z_n$ are Ricatti–Bessel functions, with prime derivatives denoted $z_n^{[1]}$, which will be expressed in terms of either spherical Bessel functions as $z_n(kR) = kRj_n(kR)$ or spherical Hankel functions of the first kind as $z_n(kR) = kRh_n^{(1)}(kR)$. The reader is cautioned when referring to the literature that τ_{mn1} and τ_{mn2} are often written as τ_{mn} and π_{mn} (or $-\pi_{mn}$ and τ_{mn}; cf. Fuller, 1987; Kerker, 1969; Bruning and Lo, 1971) and that z_n often represents spherical Bessel functions. The VSH are given in a more conventional form by Fuller (1994a).

B. RESPONSE OF A SPHERE TO HARMONIC PLANE WAVES

The ability of a particle to scatter light depends on its characteristic size s, its refractive index m(λ), and the wavelength λ of the incident light in the surrounding medium. Parameters of the form ks and mks, $k = 2\pi/\lambda$, are typically used to describe the "optical" dimensions of the scattering system. For a sphere, s is simply its radius, r.

To quantify the response of a sphere to a train of electromagnetic waves with planar surfaces of constant phase and unit amplitude, it is best to interpret those waves into the "native language" of the sphere, that is, to expand the plane wave in vector spherical harmonics:

$$\mathbf{E}^{\text{inc}} = \mathbf{E}_0 \exp[ik\mathbf{R} \cdot \mathbf{n}^{\text{inc}}] = \sum_{n=1}^{\infty} \sum_{m=-n}^{n} \sum_{p=1}^{2} p_{mnp} \mathbf{N}_{mnp}^{(1)}, \qquad (4)$$

where $\mathbf{n}^{\text{inc}}$ is the unit vector in the direction of incidence. Hansen (1935, 1937) was the first to formulate such an expansion, and the $\mathbf{N}_{mnp}$ are sometimes referred to as Hansen's functions.

The expansion for the scattered field may be written as

$$\mathbf{E}^{\text{sca}} = \sum_{m,n,p} p_{mnp} \bar{a}_{np} \mathbf{N}_{mnp}^{(3)} = \sum_{m,n,p} a_{mnp} \mathbf{N}_{mnp}^{(3)}. \qquad (5)$$

The superscripts on the VSH are a standard notation (cf. Stratton, 1941, Chapters 7 and 9; Morse and Feshbach, 1953, Chapter 13), where (1) and (3) indicate the use of $z_n(kR) = kRj_n(kR)$ and $z_n(kR) = kRh_n^{(1)}(kR)$, respectively. The expansion in Eq. (4) allows us to decompose the plane wave into spherical partial waves with amplitudes p_{mnp}. This process is illustrated in Fig. 1. For practical purposes the expansion of $\mathbf{E}^{\text{inc}}$ can be truncated after enough partial waves are included to reproduce $\mathbf{E}_0 \exp[ik\mathbf{R} \cdot \mathbf{n}^{\text{inc}}]$ at the surface of the sphere with sufficient accuracy, thus allowing one to match boundary conditions and extract the coefficients of the scattered field. A frequently used estimate of the maximum order $n_{\max}$ required in the expansion is obtained from the criterion developed by Wiscombe (1980), that is,

$$n_{\max} = \varrho + 4.05\varrho^{1/3} + 2, \qquad \varrho = kr. \qquad (6)$$

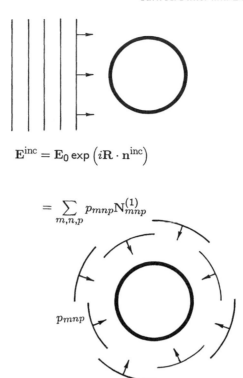

$$\mathbf{E}^{\mathrm{inc}} = \mathbf{E}_0 \exp\left(i\mathbf{R}\cdot\mathbf{n}^{\mathrm{inc}}\right)$$

$$= \sum_{m,n,p} p_{mnp}\mathbf{N}^{(1)}_{mnp}$$

Figure 1 Decomposition of an infinite plane wave into vector spherical waves with amplitudes p_{mnp}.

Adopting the notation $\psi_n(\varrho) = \varrho j_n(\varrho)$ and $\xi_n(\varrho) = \varrho h_n^{(1)}(\varrho)$, the coefficients $\bar{a}_{np}$ are found from the boundary conditions at the surface of the sphere to be

$$\bar{a}_{n1} = -\frac{m\psi_n'(\varrho)\psi_n(\eta) - \psi_n(\varrho)\psi_n'(\eta)}{m\xi_n'(\varrho)\psi_n(\eta) - \xi_n(\varrho)\psi_n'(\eta)},$$

$$\bar{a}_{n2} = -\frac{m\psi_n(\varrho)\psi_n'(\eta) - \psi_n'(\varrho)\psi_n(\eta)}{m\xi_n(\varrho)\psi_n'(\eta) - \xi_n'(\varrho)\psi_n(\eta)},$$
(7)

with $\eta = m\varrho$. These coefficients are variously named after Lorenz (1898), Mie (1908), and Debye (1909), and we refer to them here as LM coefficients (most authors have written $\bar{a}_{n1}$ and $\bar{a}_{n2}$ as $-a_n$ and $-b_n$, respectively). The $\bar{a}_{np}$ coefficients are complex amplitudes that quantify the response of the sphere to the nth partial wave in the plane wave decomposition. They are completely analogous to the Fresnel coefficients for a plane wave reflecting from a flat surface and follow from the same simple algebra once the boundary conditions are imposed. The more difficult part of this process is finding the coefficients p_{mnp}, which is discussed in Stratton (1941, Chapter 7) and Morse and Feshbach (1953, Chapter 13). (One commonly sees p_{mnp} written as p_{mn} and q_{mn} for $p = 1$ and 2, respectively.) For particles with spherical symmetry, the coordinate system can be chosen so that the incident field propagates parallel to the z axis of the body-centered coordinate frame, that is, $\mathbf{n}^{\mathrm{inc}} = \mathbf{e}_z$, and this will lead to important simplifications that allow us to write the scattered electric field in the standard notation:

$$\mathbf{E}^{\mathrm{sca}} = \sum_{n=1}^{\infty} i^n \frac{2n+1}{n(n+1)} \left[i a_n \mathbf{N}_{e1n}^{(3)} - b_n \mathbf{M}_{o1n}^{(3)} \right] \tag{8}$$

(Bohren and Huffman, 1983, Chapter 4)

$$= \sum_{n=1}^{\infty} \sum_{p=1}^{2} \left[a_{1np} \mathbf{N}_{1np}^{(3)} \right]. \tag{9}$$

C. Representations with the Rotation Functions $\mathcal{D}_{mn}^s$

Recursion relations for the quantities p_{mnp} and τ_{mnp} are known (Tsang *et al.*, 1985; Fuller, 1987). However, it is useful to express them here in terms of the rotation functions $\mathcal{D}_{mn}^s$. The $\mathcal{D}_{mn}^s$ are equivalent, to a normalizing factor, to the matrix elements of finite rotations defined by Edmonds (1974) and Mishchenko *et al.* (1996b), and methods for calculating them are given in Section VIII.C.

The functions τ_{mnp} may be expressed as

$$\tau_{mnp} = -\frac{1}{2} \left[n(n+1)\mathcal{D}_{1n}^m + (-1)^p \mathcal{D}_{-1n}^m \right]. \tag{10}$$

In the most general sense, the incident field $\mathbf{E}^{\mathrm{inc}}(\mathbf{R})$ represented by the expansion of Eq. (4) is taken to propagate in the z' direction and is polarized in the x' direction, where z' and x' are obtained by the rotation of the sphere coordinate frame through the Euler angles α, β, and γ (Fig. 2 of Chapter 1). This gives

$$p_{mnp} = -(-1)^m i^{n+1} \frac{2n+1}{2} e^{-im\alpha}$$

$$\times \left(\frac{1}{n(n+1)} \mathcal{D}_{mn}^1(\beta) e^{-i\gamma} + (-1)^p \mathcal{D}_{mn}^{-1}(\beta) e^{i\gamma} \right). \tag{11}$$

Note that this representation corresponds to an incident propagation direction of $\vartheta = \beta$, $\varphi = \alpha$, and a polarization direction of γ relative to the sphere x axis in the

x–z plane. The expansion coefficients for the incident field at an origin $\mathcal{O}_\ell$, denoted p^ℓ_{mnp}, are obtained from a simple phase shift between initial and translated origins, that is,

$$p^\ell_{mnp} = \exp(i\mathbf{k} \cdot \mathbf{R}_\ell) p_{mnp}$$
$$= \exp\left[i R_\ell(\cos\Theta_\ell \cos\beta + \sin\Theta_\ell \sin\beta \cos(\Phi_\ell - \alpha))\right] p_{mnp}, \quad (12)$$

in which R_ℓ, Θ_ℓ, and Φ_ℓ denote the spherical coordinates of $\mathcal{O}_\ell$ relative to the initial origin.

D. Radiometric Quantities

For an unpolarized beam of unit intensity incident along the z axis, the scattered power into the solid angle $d\Omega$ is proportional to the Z_{11} element of the sphere's phase matrix, otherwise known as the differential scattering cross section $dC_{\text{sca}}/d\Omega$. The scattering cross section may be written as

$$C_{\text{sca}} = \int_\Omega \frac{\text{Energy scattered/unit time/unit solid angle}}{\text{Incident energy flux (energy/unit area/unit time)}}\, d\Omega$$
$$= \int_{4\pi} \frac{dC_{\text{sca}}}{d\Omega}\, d\Omega. \quad (13)$$

The phase function is defined to be

$$a_1(\vartheta^{\text{sca}}) = \frac{4\pi}{C_{\text{sca}}} \frac{dC_{\text{sca}}}{d\Omega} \quad (14)$$

and the scattered intensity at a distance R from the particle is then simply

$$I^{\text{sca}} = \frac{C_{\text{sca}} a_1(\vartheta^{\text{sca}}) I^{\text{inc}}}{4\pi R^2}. \quad (15)$$

The free-space Maxwell's equations for time-harmonic fields allow us to write the scattered magnetic field as $\mathbf{H}^{\text{sca}} = -i\nabla \times \mathbf{E}^{\text{sca}}/\omega\mu_0$, where μ_0 is the permeability of the free space and ω is the free-space angular frequency, and the total scattering cross section, C_{sca}, can be found from an integral of the time-averaged Poynting flux of scattered radiant energy $\mathbf{E}^{\text{sca}} \times (\mathbf{H}^{\text{sca}})^*$ over all directions (ϑ^{sca}, φ^{sca}):

$$C_{\text{sca}} = \frac{1}{|\mathbf{E}^{\text{inc}} \times (\mathbf{H}^{\text{inc}})^*|} \frac{1}{2} \text{Re} \int_0^\pi \sin\vartheta^{\text{sca}}\, d\vartheta^{\text{sca}} \int_0^{2\pi} d\varphi^{\text{sca}}\, \mathbf{E}^{\text{sca}} \times (\mathbf{H}^{\text{sca}})^* \cdot \mathbf{n}^{\text{sca}} R^2, \quad (16)$$

where $\mathbf{E}^{\text{inc}} \times (\mathbf{H}^{\text{inc}})^*$ is the Poynting flux of the incident field, and time-averaging over an interval that is much longer than the period of the stimulating radiation is accomplished with the complex conjugation ($*$) of $\mathbf{H}$ (Stratton, 1941, p. 135). By integrating Eq. (16) over an imaginary spherical surface that is concentric with the scatterer, it can be shown that the scattering cross section of a sphere is

$$C_{\text{sca}} = \frac{2\pi}{k^2} \sum_{n=1}^{\infty} (2n+1)\left(|a_n|^2 + |b_n|^2\right) = \frac{2\pi}{k^2} \sum_{n=1}^{\infty} \sum_{p=1}^{2} (2n+1)|\bar{a}_{np}|^2. \quad (17)$$

Essential to polarimetry is the amplitude matrix **S**. As defined in Chapter 1, it is seen that when the incident beam is directed along the z axis and the xz plane serves as the scattering plane

$$\mathbf{S}(\vartheta^{\text{sca}}) = \frac{1}{ik} \sum_{n} \sum_{p=1}^{2} \frac{2n+1}{n(n+1)} \begin{pmatrix} \bar{a}_{np}\tau_{1np}(\vartheta^{\text{sca}}) & 0 \\ 0 & \bar{a}_{np}\tau_{1n3-p}(\vartheta^{\text{sca}}) \end{pmatrix}. \quad (18)$$

The extinction cross section may be found from the optical theorem, namely,

$$C_{\text{ext}} = \frac{2\pi}{k} \,\text{Im}[S_{11}(0) + S_{22}(0)] = -\frac{2\pi}{k^2} \sum_{n=1}^{\infty} \sum_{p=1}^{2} (2n+1)\,\text{Re}(\bar{a}_{np}). \quad (19)$$

Conservation of energy then provides the absorption cross section $C_{\text{abs}} = C_{\text{ext}} - C_{\text{sca}}$ and the single scattering albedo $\varpi = C_{\text{sca}}/C_{\text{ext}}$. An expansion for the absorption cross section can be obtained by integrating the radial component of the Poynting vector for the internal field, evaluated at the surface of the sphere. The same expansion also follows from substitution of Eqs. (19) and (17) into $C_{\text{ext}} - C_{\text{sca}}$ (Kattawar and Eisner, 1970). The expansion for C_{abs} is

$$C_{\text{abs}} = \frac{2\pi}{|m|^2 k^2} \sum_{n=1}^{\infty} (2n+1)\,\text{Re}\, i\psi_n'(\eta)\psi_n^*(\eta)\left(m|\bar{c}_{n1}|^2 + m^*|\bar{c}_{n2}|^2\right), \quad (20)$$

where $\bar{c}_{np}$ are the expansion coefficients for the internal fields, which can be written in terms of the LM coefficients as

$$\bar{c}_{n1} = \frac{im\bar{a}_{n1}}{m\psi_n'(\varrho)\psi_n(\eta) - \psi_n(\varrho)\psi_n'(\eta)},$$

$$\bar{c}_{n2} = \frac{im\bar{a}_{n2}}{m\psi_n(\varrho)\psi_n'(\eta) - \psi_n'(\varrho)\psi_n(\eta)}. \quad (21)$$

Note that $\bar{c}_{n1}$ and $\bar{c}_{n2}$ are interchanged in the notation of others (cf. Bohren and Huffman, 1983, p. 100; Mackowski, 1991).

E. CONCENTRICALLY STRATIFIED SPHERES

1. Reflection and Transmission of an Outgoing Spherical Wave at a Concentric, Concave Spherical Boundary

Consider now the case of an *outgoing* electric partial wave, represented by $\mathbf{E}_{mn} = \mathbf{N}_{mn}^{(3)} + \mathbf{M}_{mn}^{(3)}$, that crosses a spherical dielectric discontinuity of radius r_1 centered about the coordinate system to which the VSH are referenced. Let the refractive index of the interior region be m_1 and that of the exterior be unity. The outgoing field will be partially reflected and partially transmitted at the interface. The tangential components of the exterior (transmitted) partial field must match the sum of the interior (incident + reflected) partial fields at the boundary and we may therefore write

$$\left[\check{c}_n^1\mathbf{N}_{mn}^{(3)} + \check{d}_n^1\mathbf{M}_{mn}^{(3)}\right]_{\vartheta,\varphi} = \left[\check{a}_n^1\mathbf{N}_{mn}^{(1)} + \check{b}_n^1\mathbf{M}_{mn}^{(1)} + \mathbf{N}_{mn}^{(3)} + \mathbf{M}_{mn}^{(3)}\right]_{\vartheta,\varphi}. \tag{22}$$

Because of the orthogonality properties of the VSH, there is no coupling of the mn to the $m'n'$ normal modes (nor is there coupling between the TE and TM modes) at the boundary. Thus the angular dependence of the ϑ and φ components of the partial fields cancel algebraically in Eq. (22). Because of the different refractive indices on either side of the boundary, the radial functions do not cancel and we have

$$m_1\check{c}_n^1\xi_n'(\varrho_1) = \xi_n'(\eta_1) + \check{a}_n^1\psi_n'(\eta_1), \tag{23}$$

where $\varrho_1 = kr_1$ and $\eta_1 = m_1\varrho_1$. From the magnetic counterpart of Eq. (23) we obtain

$$\check{c}_n^1\xi_n(\varrho_1) = \xi_n(\eta_1) + \check{a}_n^1\psi_n(\eta_1). \tag{24}$$

Thus

$$\check{c}_n^1 = \frac{\psi_n(\varrho_1)\xi_n'(\varrho_1) - \xi_n(\varrho_1)\psi_n'(\varrho_1)}{m_1\xi_n'(\varrho_1)\psi_n(\eta_1) - \xi_n(\varrho_1)\psi_n'(\eta_1)}. \tag{25}$$

As in Eq. (21), by virtue of the Wronskian,

$$W\{j_n(z), y_n(z)\} = z^{-2}i \qquad (y_n \text{ represents the spherical Neuman functions}), \tag{26}$$

we can write

$$\check{c}_n^1 = \frac{-i}{m_1\xi_n'(\varrho_1)\psi_n(\eta_1) - \xi_n(\varrho_1)\psi_n'(\eta_1)}. \tag{27}$$

Precisely the same arguments produce the expressions

$$\breve{a}_n^1 = -\frac{m_1 \xi_n'(\varrho_1)\xi_n(\eta_1) - \xi_n(\varrho_1)\xi_n'(\eta_1)}{m_1 \xi_n'(\varrho_1)\psi_n(\eta_1) - \xi_n(\varrho_1)\psi_n'(\eta_1)}, \tag{28}$$

$$\breve{d}_n^1 = -\frac{\xi_n'(\eta_1)\psi_n(\eta_1) - \xi_n(\eta_1)\psi_n'(\eta_1)}{m^1 \xi_n(\varrho_1)\psi_n'(\eta_1) - \xi_n'(\varrho_1)\psi_n(\eta_1)}, \tag{29}$$

$$\breve{b}_n^1 = \frac{m_1 \xi_n(\varrho_1)\xi_n'(\eta_1) - \xi_n'(\varrho_1)\xi_n(\eta_1)}{m_1 \xi_n(\varrho_1)\psi_n'(\eta_1) - \xi_n'(\varrho_1)\psi_n(\eta_1)}, \tag{30}$$

and for incoming spherical waves, the standard LM coefficients are given as $\bar{a}_{n1}^1 \equiv -a_n$, $\bar{a}_{n2}^1 \equiv -b_n$, $\bar{c}_{n1}^1 (= -m_1 \breve{c}_n^1) \equiv -c_n$, and $\bar{c}_{n2}^1 (= -m_1 \breve{d}_n^1) \equiv -d_n$.

For incoming spherical waves, the LM coefficients of order n for the scattered and internal partial fields are the spherical wave analogs to the Fresnel coefficients for reflection and refraction of a plane wave at a planar boundary. With the derivation of the "˘" (háček) coefficients (Nussenzveig, 1992, Chapter 9; Fuller, 1993a, b; Mackowski and Jones, 1995), the analogy can now be extended to the case of *outgoing* spherical waves reflected and transmitted by *concave* spherical surfaces. Application of these concepts leads to an elegant solution of the standard concentrically stratified sphere problem. This solution is summarized in the next subsection and will also serve as an introduction to the OS technique applied in Sections IV and V.

2. Scattering Coefficients of a Concentrically Coated Sphere

The following treatment is an alternative to the original derivation by Aden and Kerker (1951). Perhaps the simplest coefficients for reflection and transmission by multilayered structures are those based on coefficients for a thin film. Their derivation is illustrated in Fig. 2, where $\bar{r}_{\ell p}$, $\breve{r}_{\ell p}$, $\bar{t}_{\ell p}$, and $\breve{t}_{\ell p}$ represent the Fresnel coefficients for reflection and transmission at the surfaces indicated for orthogonal polarizations p. In this case we may express the amplitude of the electric field reflected by the film as

$$E_p^{\text{ref}} = \bar{r} E_p^{\text{inc}}, \tag{31}$$

$$E_p^{\text{ref}} = E_p^{\text{inc}} \left(\bar{r}_{1p} + \bar{t}_{1p} \bar{r}_{2p} \breve{t}_{1p} \sum_{k=0}^{\infty} [\bar{r}_{2p} \breve{r}_{1p}]^k \right) = E_p^{\text{inc}} \left(\bar{r}_{1p} + \frac{\bar{t}_{1p} \bar{r}_{2p} \breve{t}_{1p}}{1 - [\bar{r}_{2p} \breve{r}_{1p}]} \right). \tag{32}$$

Posing the problem of electromagnetic scattering by concentrically coated spheres as one of multiple reflection of spherical partial waves, with amplitudes given by the coefficients for coupling of plane waves to spherical partial waves, as illustrated in Fig. 3, one is led to a surprisingly simple solution (Fuller, 1993a, b): The

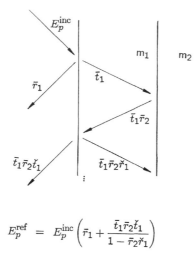

$$E_p^{\text{ref}} = E_p^{\text{inc}}\left(\bar{r}_1 + \frac{\bar{t}_1\bar{r}_2\check{t}_1}{1 - \bar{r}_2\check{r}_1}\right)$$

Figure 2 Development of the reflected electric field amplitude from a dielectric thin film in terms of the Fresnel coefficients.

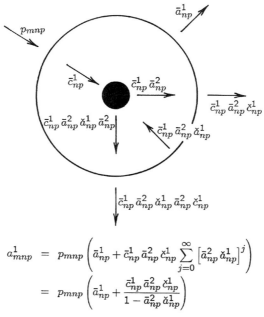

$$a_{mnp}^1 = p_{mnp}\left(\bar{a}_{np}^1 + \bar{c}_{np}^1\,\bar{a}_{np}^2\,\check{c}_{np}^1 \sum_{j=0}^{\infty}\left[\bar{a}_{np}^2\,\check{a}_{np}^1\right]^j\right)$$

$$= p_{mnp}\left(\bar{a}_{np}^1 + \frac{\bar{c}_{np}^1\,\bar{a}_{np}^2\,\check{c}_{np}^1}{1 - \bar{a}_{np}^2\,\check{a}_{np}^1}\right)$$

Figure 3 Development of the scattered electric field partial wave amplitude from concentric spheres in terms of the LM coefficients.

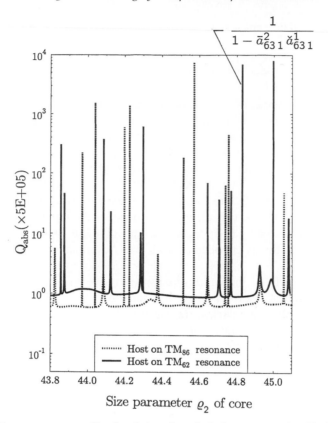

Figure 4   Resonance spectra, offset for clarity, of a spherical core concentric with shells of two different thicknesses. The dotted curve depicts the absorption spectrum of a weakly dye-doped core inside a much larger host, and all features of that spectrum can be explained in terms of morphology-dependent resonances in the core. The solid curve is for a shell that is thin enough to allow its resonant modes to couple with those of the core. Several features in the latter spectrum cannot be explained in terms of the independent whispering gallery modes of the core or shell, but are found to arise in the "coefficient of finesse" associated with this coupling. For example, the peak at a core size parameter of just over 44.8 is due to the $[1 - \bar{a}_{63\,1}^2 \check{a}_{63\,1}^1]^{-1}$ term in Eq. (33).

scattering coefficients can be expressed simply as

$$
a_{1np}^1 = p_{1np}\left(\bar{a}_{np}^1 + \bar{c}_{np}^1 \bar{a}_{np}^2 \,{}^1\check{c}_{np} \sum_{k=0}^{\infty} [\bar{a}_{np}^2 \check{a}_{np}^1]^k \right)
$$

$$
= p_{1np}\left(\bar{a}_{np}^1 + \frac{\bar{c}_{np}^1 \bar{a}_{np}^2 \check{c}_{np}^1}{1 - [\bar{a}_{np}^2 \check{a}_{np}^1]} \right), \tag{33}
$$

and the scattered field may be written as

$$\mathbf{E}^{\text{sca}} = \sum_{n,p} a^1_{1np} \mathbf{N}^{(3)}_{1np} \longrightarrow -\sum_{n=1}^{\infty} \left(a^T_n \mathbf{N}^{(3)}_{e1n} + b^T_n \mathbf{M}^{(3)}_{o1n} \right). \tag{34}$$

Because the structure of Eq. (34) is the same as that of Eq. (8), all of the radiometric properties discussed in the previous section will be of the same form, but with a^T_n and b^T_n replacing a_n and b_n. That structure also suggests features, perhaps observed by Essien *et al.* (1993), that cast the stratified sphere as a concave–convex spherical resonator. Calculations illustrating such behavior are presented in Fig. 4. In addition to better physical insight, Eq. (33) also allows implementation of a coated sphere algorithm with the addition of only a few lines of code to an existing LM program. This algorithm also avoids the problems of numerical instability encountered by Bohren and Huffman (1983, p. 484). Another numerically stable algorithm is provided by Toon and Ackerman (1981) and Bhandari (1985).

Recursive algorithms for multilayered spheres have been developed by Bhandari (1985) and Mackowski *et al.* (1990), and the treatment in this section also readily lends itself to a recursive treatment that parallels that for multilayered films. An example (out of many) of the use of multilayered spheres to model the optical properties of droplets with radially varying refractive index is provided by Massoli (1998).

IV. ECCENTRIC TWO-SPHERE SYSTEMS

In the previous section, we proceeded from the case of a homogeneous sphere to that of a system composed of two concentrically stratified spheres. The next step to be taken in laying the foundations for the theory of systems of compound spheres is to address the case of eccentric two-sphere systems. This includes pairs of distinct spheres as well as spheres that are eccentrically stratified (i.e., an arbitrarily placed spherical inclusion in an otherwise homogeneous spherical host).

To match boundary conditions at the surfaces of each sphere, one must provide, in addition to the transformation of the incident plane wave, a transformation that will represent the outgoing spherical waves of one sphere in terms of the regular basis functions of the other. This is accomplished with the use of the addition theorem for vector spherical harmonics, with which we may write

$$\mathbf{N}^{(3)}_{mnp}(\mathbf{R}_{\ell'}) = \sum_{l=1}^{\infty} \sum_{k=-l}^{l} \sum_{q=1}^{2} \left\{ \begin{matrix} A^{\ell\ell'}_{klq\,mnp} \\ \tilde{A}^{\ell\ell'}_{klq\,mnp} \end{matrix} \right\} \mathbf{N}^{\{(1)\}}_{klq}(\mathbf{R}_{\ell}), \qquad R_{\ell\ell'} \left\{ \begin{matrix} \geq \\ \leq \end{matrix} \right\} R_{\ell}. \tag{35}$$

$\left\{\begin{smallmatrix} A \\ \tilde{A} \end{smallmatrix}\right\}$ are translation coefficients that represent a projection of one VSH basis, centered about the ℓth origin, onto the VSH basis associated with the ℓ'th origin a distance $R_{\ell\ell'}$ away, and are analogous to the projections p_{mnp} of a plane wave onto the normal modes of a sphere in standard LM theory. The strengths of the couplings of fields scattered from sphere ℓ to normal modes associated with sphere ℓ' are gauged by the translation coefficients. Continued progress in this field has hinged, in part, on advances in the treatment of the translation coefficients $A^{\ell\ell'}_{klq\,mnp}$ and $\tilde{A}^{\ell\ell'}_{klq\,mnp}$, and a novel derivation of the addition theorem, along with methods for computing the translation matrix elements, are given in Section VIII.

A. ORDER-OF-SCATTERING SOLUTION

1. Two-Sphere Clusters

The processes that arise when pairs of spheres scatter EM radiation may be described as follows: An infinite train of plane waves stimulates the normal modes of each sphere in the manner delineated by LM theory. Unlike the case for concentrically stratified spheres, however, all of the multipole fields emanating from one sphere stimulate each of the normal modes in the other because the modes of a particular sphere are not orthogonal to those of one centered about a different origin. Each sphere responds to the field incident on it from the other, scattering radiation to the field point and back to the other sphere. The total scattered field may be written as

$$\mathbf{E}^{\mathrm{sca}} = \mathbf{E}^{\mathrm{sca}}_1 + \mathbf{E}^{\mathrm{sca}}_2 = \sum_{j=0}^{\infty} \left[{}^{(j)}\mathbf{E}^{\mathrm{sca}}_1 + {}^{(j)}\mathbf{E}^{\mathrm{sca}}_2 \right], \tag{36}$$

where the jth-order partial fields ${}^{(j)}\mathbf{E}^{\mathrm{sca}}_\ell$ are in turn expressed as

$$ {}^{(j)}\mathbf{E}^{\mathrm{sca}}_\ell = \sum_{n,m,p} {}^{(j)}a^\ell_{mnp} \mathbf{N}^{(3)}_{mnp}(\mathbf{R}_\ell), \tag{37}$$

with

$$ {}^{(j)}a^\ell_{mnp} = \bar{a}^\ell_{np} \sum_{\nu,\mu,q} {}^{(j-1)}a^{\ell'}_{\mu\nu q} A^{\ell,\ell'}_{\mu\nu q\,mnp}, \qquad j \neq 0,$$

$$ {}^{(0)}a^\ell_{mnp} = \bar{a}^\ell_{np} P^\ell_{mnp}. \tag{38}$$

The expansion of the total scattered field from either sphere is then

$$\mathbf{E}_\ell^{\text{sca}} = \sum_{n,m,p} \sum_{j=0}^{\infty} {}^{(j)}a_{mnp}^\ell \mathbf{N}_{mnp}^{(3)}(\mathbf{R}_\ell) = \sum_{n,m,p} a_{mnp}^\ell \mathbf{N}_{mnp}^{(3)}(\mathbf{R}_\ell). \tag{39}$$

2. Nonconcentric Spheres

Now let a wavefront be incident on a composite particle made from a sphere of refractive index m_2 and size parameter ϱ_2 located arbitrarily within a spherical and otherwise homogeneous host of refractive index m_1 and size parameter ϱ_1. The incident wavefront will still couple to the normal modes of the host with a strength determined by p_{mnp}. The shell will produce a scattered and a transmitted field with associated partial wave amplitudes ${}^{(0)}a_{mn1}^1 (= -p_{mn}a_n^1)$, ${}^{(0)}a_{mn2}^1 (= -q_{mn}b_n^1)$, and ${}^{(0)}c_{mn1}^1 (= -p_{mn}c_n^1)$, ${}^{(0)}c_{mn2}^1 (= -q_{mn}d_n^1)$, respectively. The coefficients for the first-order partial fields of the inclusion are

$$^{(1)}a_{mnp}^2 = \bar{a}_{np}^2 \sum_{\nu,\mu,q}^{(0)} c_{\mu\nu q}^1 \tilde{A}_{\mu\nu q\,mnp}^{12}, \tag{40}$$

where the $\bar{a}_{np}^2$ are the LM coefficients of a core particle immersed in an infinite medium of refractive index m_1. The fields scattered by the inclusion are then partially transmitted beyond the mantle or internally reflected by its surface with the respective amplitudes ${}^{(2)}\check{c}_{mnq}^1$ and ${}^{(2)}\check{a}_{mnq}^1$. The internally reflected fields will in turn be scattered by the inclusion and so on. The coefficients for the total scattered field may thus be constructed as

$$a_{mnp}^T = \sum_{j=1}^{\infty} {}^{(j)}a_{mnp}^1, \tag{41}$$

where

$$^{(j)}a_{mnp}^1 = \check{c}_{np}^1 \sum_{\nu,\mu,q} {}^{(j)}a_{\mu\nu q}^2 \tilde{A}_{\mu\nu q\,mnp}^{12},$$

$$^{(j+1)}a_{mnp}^2 = \bar{a}_{np}^2 \sum_{\nu,\mu,q} {}^{(j)}c_{\mu\nu q}^1 \tilde{A}_{\mu\nu q\,mnp}^{12}, \tag{42}$$

$$^{(j)}c_{mnp}^1 = \check{a}_{np}^1 \sum_{\nu,\mu,q} {}^{(j)}a_{\mu\nu q}^2 \tilde{A}_{\mu\nu q\,mnp}^{12}$$

for $j > 0$. The coefficients $\check{a}_{np}^1$ and $\check{c}_{np}^1$ were introduced in Section III.E as, respectively, the reflection and transmission coefficients for outgoing spherical waves (concentric with the host) at its concave outer surface. As with the case of external aggregation, the fields from the other sphere must first be represented in the basis functions that are concentric with the surface being considered, but then that surface will respond with the same set of amplitudes as it would for a plane wave. This process is illustrated in Fuller (1995b). Calculations based on this order-of-scattering treatment are in agreement with those obtained by Borghese *et al.* (1992), who solved a system of linear equations to obtain the scattering coefficients. For a concentric inclusion, Eq. (41) reduces to Eq. (33).

V. AGGREGATES OF N_S ARBITRARILY CONFIGURED SPHERES

A. EXTERNAL AGGREGATION

The scattering model is now expanded to represent an ensemble of N_S separate (i.e., external to each other) spheres. The spheres are characterized by size parameters $x_\ell = kr_\ell$ and refractive indices $m_\ell = m_{r\ell} + im_{i\ell}$, and are located at positions X_ℓ, Y_ℓ, and Z_ℓ on a coordinate frame fixed to the cluster. The incident field consists of a plane, linearly polarized wave. As opposed to the single-sphere case, we adopt here a more general representation of the incident field, in that it is taken to propagate in the z' direction and is polarized in the x' direction, where z' and x' represent the rotation of the cluster coordinates z and x through the Euler angles α, β, and γ. Note that β and α represent the polar and azimuth angles of the incident field direction relative to the cluster frame, and γ represents the rotation of the electric field vector from the x–z plane. The incident field $\mathbf{E}_0^{\text{inc}}(\mathbf{R}_0)$ is represented by Eq. (4) referenced to the origin of the cluster. Because of the linearity of Maxwell's equations, scattering from an ensemble of N_S spheres can be formulated as the superposition of fields that are scattered from the individual spheres,

$$\mathbf{E}^{\text{sca}} = \sum_{\ell=1}^{N_S} \mathbf{E}_\ell^{\text{sca}}. \tag{43}$$

The field of each sphere can, in turn, be represented as a VSH expansion centered about the origin of that sphere,

$$\mathbf{E}_\ell^{\text{sca}} = \sum_{n,m,p}^{\infty} a_{mnp}^\ell \mathbf{N}_{mnp}^{(3)}(\mathbf{R}_\ell). \tag{44}$$

As before, the scattering expansion coefficients for sphere ℓ, a_{mnp}^ℓ, will be given by

$$a_{mnp}^\ell = \bar{a}_{np}^\ell f_{mnp}^\ell, \tag{45}$$

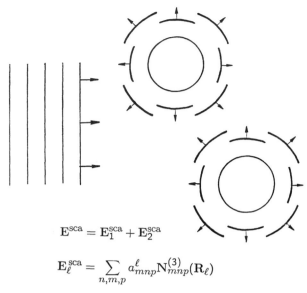

$$\mathbf{E}^{\text{sca}} = \mathbf{E}_1^{\text{sca}} + \mathbf{E}_2^{\text{sca}}$$

$$\mathbf{E}_\ell^{\text{sca}} = \sum_{n,m,p} a_{mnp}^\ell \mathbf{N}_{mnp}^{(3)}(\mathbf{R}_\ell)$$

Figure 5 Any sphere in a cluster is stimulated by the field incident on the cluster plus the scattered fields of all other spheres in the cluster.

where f_{mnp}^ℓ now denote the coefficients for the VSH expansion, centered about the origin of sphere ℓ, of the exciting field at ℓ. As illustrated in Fig. 5 for two spheres, the exciting field will consist of the incident plane wave at ℓ as well as the scattered fields that originate from all other spheres in the ensemble and arrive at ℓ, represented by the application of Eq. (35), where the $A_{mnp\,klq}^{\ell\ell'}$ are chosen. Note that this leads to the representation of the scattered field from ℓ' in terms of regular (as opposed to outgoing) harmonics centered about ℓ. By inserting Eq. (35) into Eq. (44) (now written for ℓ') and summing over all spheres except ℓ, the exciting field at ℓ is described by

$$\mathbf{E}_\ell^{\text{ex}} = \sum_{n,m,p} \left(p_{mnp}^\ell + \sum_{\substack{\ell'=1 \\ \ell' \neq \ell}}^{N_S} \sum_{l,k,q} A_{mnp\,klq}^{\ell\ell'} a_{klq}^{\ell'} \right) \mathbf{N}_{mnp}^{(1)}. \tag{46}$$

By replacing Eq. (46) into Eq. (45), a linear relationship for the scattered field expansion coefficients is obtained

$$a_{mnp}^\ell - \bar{a}_{np}^\ell \sum_{\substack{\ell'-1 \\ \ell' \neq \ell}}^{N_S} \sum_{l,k,q} A_{mnp\,klq}^{\ell\ell'} a_{klq}^{\ell'} = \bar{a}_{np}^\ell p_{mnp}^\ell. \tag{47}$$

In practice, the expansion for the scattered field from ℓ is truncated after $n = n_{\max,\ell}$ orders. The total number of complex-valued equations represented by Eq. (47) will then be given by

$$M = 2 \sum_{\ell=1}^{N_S} n_{\max,\ell} (n_{\max,\ell} + 2).$$

(48)

Several previous investigations have found that $n_{\max,\ell}$ depends almost entirely on the size parameter of sphere ℓ and is not affected by the number, sizes, and locations of surrounding spheres. Two situations in which Eq. (6) appears to underestimate the required number of orders are known. One (discussed in the next section) involves whispering gallery modes of bispheres (Fuller, 1991) and the other is for highly conducting, Rayleigh-limit spheres in contact ($|m_\ell|^2 \gg 1$ and $x_\ell \ll 1$). In the latter case, it was shown (Mackowski, 1995) that $n_{\max} = 10$ orders were required to describe the scattered fields of a pair of identical spheres in contact, with $x_\ell = 0.01$ and m $= 3 + 2i$. [Similar problems have been noted by Fuller (1987), p. 93, in relation to metal spheres.] Clusters of dielectric Rayleigh-limit spheres, on the other hand, can usually be modeled with an electric dipole approximation, for which the TE mode is neglected and the expansions are truncated after one order. For this case, it can be shown that the formulation given here becomes equivalent to the dipole iteration models developed by Purcell and Pennypacker (1973) and Jones (1979) and results in $3N_S$ complex-valued equations for a^ℓ_{m11}, with $m = -1, 0$, and 1.

In a compact matrix form, Eq. (47) can be represented by

$$\mathbf{a}^\ell - \bar{\mathbf{a}}^\ell \sum_{\substack{\ell'=1 \\ \ell' \neq \ell}}^{N_S} \mathbf{A}^{\ell\ell'} \mathbf{a}^{\ell'} = \bar{\mathbf{a}}^\ell \mathbf{p}^\ell,$$

(49)

in which $\mathbf{a}^\ell$ and $\mathbf{p}^\ell$ are length-$2n_{\max,\ell}(n_{\max,\ell} + 2)$ vectors for the scattering and incident field coefficients for sphere ℓ, $\mathbf{A}^{\ell\ell'}$ is the translation matrix between origins ℓ' and ℓ, and $\bar{\mathbf{a}}^\ell$ is a diagonal matrix representing the LM coefficients for sphere ℓ. Equation (49) can also be generalized to clusters of nonspherical particles, in which $\bar{\mathbf{a}}^\ell$ would be replaced by the T matrix for particle ℓ (see Chapter 6). It is important to note, however, that this general formulation relies on an accurate T-matrix representation of the scattered field (as opposed to the far field) in the near zone. In general, this would require that the smallest circumscribing spheres of the particles do not overlap—which implies that the superposition method could not be used to calculate, for example, the scattering properties of densely packed clusters of ellipsoidal particles.

B. SPHERES WITH MULTIPLE SPHERICAL INCLUSIONS

The multiple-inclusion problem closely parallels the problem of scattering by sphere clusters. In addition to the standard expansions for the total fields exterior to the host and interior to the inclusions, it is assumed that the total electric field at points inside the host but outside the inclusions is (Fuller, 1993a, b, 1995b; Mackowski and Jones, 1995)

$$\mathbf{E}^{\text{host}} = \sum_{n,m,p} \left[c_{mnp}^1 \mathbf{N}_{mnp}^{(1)}(\mathbf{R}^1) + \sum_{\ell>1} a_{\mu\nu p}^\ell \mathbf{N}_{mnp}^{(3)}(\mathbf{R}^1) \right]. \tag{50}$$

The boundary conditions of the multiple-inclusion system can also be satisfied with the aid of the addition theorem for VSH, and one is led to a self-consistent set of linear equations for the scattering coefficients

$$a_{mnp}^1 = \bar{a}_{np}^1 p_{mnp} + \check{c}_{np}^1 \sum_{\ell\neq1} \sum_{\nu,\mu,q} a_{\mu\nu q}^\ell \tilde{A}_{mnp\,\mu\nu q}^{1\ell}, \tag{51}$$

where

$$a_{mnp}^\ell = \bar{a}_{np}^\ell \sum_{\nu,\mu,q} \left(c_{\mu\nu q}^1 \tilde{A}_{\mu\nu q\,mnp}^{\ell1} + \sum_{\ell'\neq\ell} a_{\mu\nu q}^{\ell'} A_{mnp\,\mu\nu q}^{\ell\ell'} \right), \tag{52}$$

$$c_{mnp}^1 = \bar{c}_{np}^1 + \check{a}_{np}^1 \sum_{\ell\neq1} \sum_{\nu\mu} \sum_q a_{\mu\nu q}^\ell \tilde{A}_{mnp\,\mu\nu q}^{1\ell}. \tag{53}$$

This latter system of equations is then solved and the coefficients a_{mnp}^1 of the external field can be calculated. The summation in Eq. (52), delineating the total exciting field at the ℓth inclusion, is physically the same as Eq. (47), but with the plane wave coefficients p_{mnp} replaced by the coefficients of the host field, expanded in the ℓth basis. There is now a circumscribing spherical surface that interacts with the inclusions, but this interaction can be described as before and poses no significant complication to the basic solution, which will be described in the next sections. All radiometric quantities can then be derived from the expansion of the external field

$$\mathbf{E}^{\text{sca}} = \sum_{n=1}^\infty \sum_{m=-n}^n \sum_p a_{mnp}^1 \mathbf{N}_{mnp}^{(3)}(\mathbf{R}^1). \tag{54}$$

C. Fixed-Orientation Properties

1. Solution Methods

Given a cluster configuration and incident beam direction and polarization, Eq. (49) is solved most efficiently using iteration methods. Perhaps the most widely used method—and the most conceptually straightforward—is the OS technique, introduced in the earlier discussion of two-sphere systems. This corresponds to the Born expansion of Eq. (49), in which

$$\mathbf{a}^\ell = \overline{\mathbf{a}}^\ell \left(\delta_{\ell\ell'}\mathbf{I} + \sum_{\substack{\ell'=1 \\ \ell' \neq \ell}}^{N_S} \mathbf{A}^{\ell\ell'}\overline{\mathbf{a}}^{\ell'} + \sum_{\substack{k=1 \\ k \neq \ell}}^{N_S} \mathbf{A}^{\ell k}\overline{\mathbf{a}}^k \sum_{\substack{\ell'=1 \\ \ell' \neq k}}^{N_S} \mathbf{A}^{k\ell'}\overline{\mathbf{a}}^{\ell'} + \cdots \right)\mathbf{p}^{\ell'}, \quad (55)$$

where $\mathbf{I}$ is the unit matrix. The first term in the series, which is equivalent to the Rayleigh–Gans–Debye (RGD) approximation, represents the interaction of the sphere ℓ solely with the incident wave (i.e., no multiple scattering). The second term represents interaction with waves that have undergone one "reflection" from the neighboring spheres; the third represents interaction with doubly reflected waves, and so on. The rate of convergence of Eq. (55) depends on the spectral radius of Eq. (49)—which is affected most strongly by the magnitude of the elements in $\overline{\mathbf{a}}^\ell$. In particular, when the spheres sustain cavity resonances, the Born series diverges at the resonance frequency of the ℓ'th monomer. Interestingly, it has been seen that two nearly identical spheres will exhibit a resonance spectrum characteristic of the bisphere that evolves through splitting or shifting of the monomer resonance and the OS series can converge at the frequencies of the bisphere resonance, but the number of terms in the series can be at least two orders of magnitude larger than when no resonance is involved. Under conditions of such system resonances, the truncation order will be about 50% larger when calculating the scattering coefficients because that many more modes are needed to match boundary conditions in the near field. On the other hand, once the scattering coefficients have been determined, the truncation index for the expansion of the scattered fields is once again given by Eq. (6) (cf. Fuller, 1989, 1991).

The OS solution technique is implemented by successive substitution into Eq. (55), which yields

$$\mathbf{a}^\ell = \sum_{j=0}^\infty {}^{(j)}\mathbf{a}^\ell, \qquad {}^{(0)}\mathbf{a}^\ell = \overline{\mathbf{a}}^\ell\mathbf{p}^\ell, \qquad {}^{(j)}\mathbf{a}^\ell = \overline{\mathbf{a}}^\ell \sum_{\substack{\ell'=1 \\ \ell' \neq \ell}}^{N_S} \mathbf{A}^{\ell\ell'}\big[{}^{(j-1)}\mathbf{a}^{\ell'}\big]. \quad (56)$$

The biconjugate gradient method has also been used to solve iteratively Eq. (49). Although this method typically involves over twice the number of operations as

OS to perform a single iteration, it has been reported that the method can converge significantly faster than OS when the spheres are near resonance.

Scattering properties of the cluster for different orientations relative to the incident wave can be obtained either by fixing the incident direction and polarization and realigning the sphere positions in the cluster or by fixing the sphere positions and realigning the incident wave. The latter method is computationally more efficient than the former, because it involves changing only the right-hand-side vector in Eq. (49). To calculate the scattering matrix (described later), this approach does require a rotation transformation on the calculated scattering coefficients, so that the scattering angles ϑ and φ become defined relative to the incident beam. The transformation is not required if only the total cross sections of the spheres are needed.

2. Cross Sections

To determine the total power, W, radiated from an imaginary surface Σ enclosing an ensemble of N_S scatterers, the integrals

$$
W = \sum_{\ell=1}^{N_S} \int_\Sigma \left[\mathbf{E}^{\mathrm{inc}} \times \left(\mathbf{H}_\ell^{\mathrm{sca}} \right)^* + \mathbf{E}_\ell^{\mathrm{sca}} \times \left(\mathbf{H}^{\mathrm{inc}} \right)^* \right] \cdot \mathbf{e}_{R_\ell} R_\ell^2 \sin \vartheta \, d\vartheta \, d\varphi
$$

$$
+ \int_\Sigma \left(\sum_\ell \left[\mathbf{E}_\ell^{\mathrm{sca}} \times \left(\mathbf{H}_\ell^{\mathrm{sca}} \right)^* \right] + \sum_\ell \sum_{\ell' \neq \ell} \left[\mathbf{E}_\ell^{\mathrm{sca}} \times \left(\mathbf{H}_{\ell'}^{\mathrm{sca}} \right)^* \right] \right)
$$

$$
\times \mathbf{e}_{R_\ell} R_\ell^2 \sin \vartheta \, d\vartheta \, d\varphi \tag{57}
$$

must be evaluated.

The first integral in Eq. (57) represents the interference between the incident electromagnetic field and the fields scattered by each of the monomers in the system. This interference term corresponds to the total power removed from the incident beam. The extinction cross section of sphere ℓ in the cluster is obtained from that integral or by application of the optical theorem to the partial field scattered from the sphere, giving

$$
C_{\mathrm{ext}}^\ell = -\frac{2\pi}{k^2} \sum_{n,m,p} \mathcal{E}_{mn} \operatorname{Re} a_{mnp}^\ell \left[p_{mnp}^\ell \right]^*,
$$

$$
\text{with } \mathcal{E}_{mn} = \frac{n(n+1)}{2n+1} \frac{(n+m)!}{(n-m)!}. \tag{58}
$$

The second integral is equal to the total power scattered by the ℓth sphere, whereas the third integral accounts for interference between the fields of the ℓth and ℓ'th scatterers, which contributes further to the scattering cross section of the ensemble. The surface of integration is henceforth assumed to be a sphere centered about the principal origin.

The equation for the total scattering cross section can be written as

$$C_{sca}^{\ell} = \frac{2\pi}{k^2} \sum_{n,m,p} \mathcal{E}_{mn} \left[|a_{mnp}^{\ell}|^2 + \mathrm{Re}\left([a_{mnp}^{\ell}]^* \sum_{\substack{\ell'=1 \\ \ell' \neq \ell}}^{N_S} \sum_{l,k,q} A_{mnp\, klq}^{\ell\ell'} a_{klq}^{\ell'} \right) \right]. \quad (59)$$

The first and second terms in the preceding equation can be loosely viewed as "self" and "interference" contributions to scattering. The interference contributions to scattering were not properly accounted for in the early work of Borghese *et al.* (1979).

It can be shown that Eq. (59) is equivalent to

$$C_{sca} = \frac{4\pi}{k^2} \sum_{m,n,p} \mathcal{E}_{mn} |a_{mnp}|^2, \quad (60)$$

where a_{mnp} are the cluster-centered expansion coefficients for the scattered field, to be discussed presently. This illustrates that, as in the case of a single sphere, C_{sca} is an inherently real quantity—the real part of Eq. (57) need not be taken because the integral itself is real valued. The structure of this equation is identical to that for single spheres, where the amplitude coefficients of the natural modes of the sphere have been replaced with amplitude coefficients that may be associated with the natural modes of the cluster. These latter modes depend, in turn, on the amplitudes of the normal modes of the monomers in the cluster and on the precise information on the geometry of the cluster that is contained in the translation coefficients.

The absorption cross section of sphere ℓ can be obtained from integration of the internal field Poynting vector over the surface of the sphere, which yields

$$C_{abs}^{\ell} = -\frac{2\pi}{k^2} \sum_{n,m,p} \mathcal{E}_{mn} \left(\mathrm{Re}\, \frac{1}{[\bar{a}_{np}^{\ell}]^*} + 1 \right) |a_{mnp}^{\ell}|^2. \quad (61)$$

Alternatively, the preceding form could be deduced by multiplying the complex conjugate of Eq. (47) by $\mathcal{E}_{mn} a_{mnp}^{\ell} / [\bar{a}_{np}^{\ell}]^*$; summing over n, m, and p; and using the definitions in Eqs. (58) and (59). The total extinction and absorption cross sections of the cluster are obtained simply from the sum of the sphere cross sections.

The absorption cross section has a clear physical interpretation; that is, $I^{inc} C_{abs}^{\ell}$ is the rate of radiant energy absorption by sphere ℓ. The extinction and scattering cross sections for the sphere, on the other hand, are less physically meaningful. It is entirely possible for $C_{ext}^{\ell} < C_{abs}^{\ell}$, or even $C_{ext}^{\ell} < 0$, for a sphere in a cluster. The latter case would correspond to the forward-scattered wave from sphere ℓ constructively interfering with the incident wave. By the same token, $C_{\text{sca}}^{\ell} \leq 0$ is also possible, which would imply that the net scattered intensity above the surface of ℓ is directed inwards, rather than outwards. Such "anomalous" results on the sphere level—which disappear on the cluster level—are merely mathematical artifacts of the superposition method used to solve the wave equations, that is, the separation of the scattered field into partial fields from each sphere.

3. Amplitude and Scattering Matrices

Two basic approaches can be taken to calculate the scattering matrix elements, as a function of the scattering angle, from the cluster as a whole. The first method retains the superposition of partial fields that are scattered from each sphere in the cluster. Because these fields emanate from different origins, it is necessary in this approach to explicitly account for the phase differences among the partial fields as they interfere in the far-field region—much in the same way as RGD theory. The second approach transforms the separate partial field expansions into a single expansion based on a single origin of the cluster. This approach obviates the need to account for phase differences in the scattered field, yet it does increase the order of the expansions.

The sphere-centered (or partial field superposition) method will be discussed first. Here, the scattered field is given by Eq. (43), with each partial field represented by Eq. (44). The vector harmonic expansions for the scattered field in Eq. (44), however, are defined relative to the cluster coordinate system. In calculating the amplitude and scattering matrix of the cluster, it is more appropriate to transform the solution so that the harmonics are defined relative to the incident field coordinate system. Because the incident field system is defined by the Euler rotation of the cluster system through α, β, and γ, an inverse rotation will align the harmonics relative to the incident field system. The appropriate transformation on the scattering coefficients is given by

$$a_{mnp}^{\prime\ell} = \exp(im\gamma) \sum_{k=-n}^{n} \mathcal{D}_{mn}^{k}(\beta) \exp(ik\alpha) a_{knp}^{\ell}. \tag{62}$$

When the transformed coefficients $a_{mnp}^{\prime\ell}$ are used in Eq. (44), the angles ϑ and φ are defined relative to the incident field frame; that is, when $\vartheta = 0$ is the direction of the incident wave, $\varphi = 0$ is the direction of the incident electric field vector.

Using the asymptotic form of the spherical Hankel functions for large argument, the far-field components of the scattered wave from sphere ℓ will appear as

$$E_{\vartheta,\ell}^{sca} = \frac{i}{kR} \exp(ikR_\ell) \sum_{n,m,p} (-i)^{n+1} a_{mnp}^{\prime\ell} \tau_{mnp}(\vartheta) \exp(im\varphi),$$

(63)

$$E_{\varphi,\ell}^{sca} = -\frac{1}{kR} \exp(ikR_\ell) \sum_{n,m,p} (-i)^{n+1} a_{mnp}^{\prime\ell} \tau_{mn\,3-p}(\vartheta) \exp(im\varphi).$$

To combine the waves from the different spheres in the far-field region, it is necessary to account for phase differences that appear in the $\exp(ikR_\ell)$ term. This can be done by referencing the fields to a common origin of the cluster and employing the relation

$$\exp[ik(R_\ell - R_0)] = \exp\left[ikR_{\ell0}(\cos\Theta_{\ell0}\cos\vartheta + \sin\Theta_{\ell0}\sin\vartheta\cos(\Phi_{\ell0} - \varphi))\right]$$

$$\equiv \Delta_\ell(\vartheta, \varphi).$$

(64)

The total scattered field in the far-field region can now be described by

$$E_\vartheta^{sca} = \sum_{\ell=1}^{N_S} E_{\vartheta,\ell}^{sca} \Delta_\ell = \frac{i}{kR_0} \exp(ikR_0) \sum_{n,m,p} (-i)^{n+1} F_{mnp}(\vartheta, \varphi) \tau_{mnp}(\vartheta),$$

(65)

$$E_\varphi^{sca} = \sum_{\ell=1}^{N_S} E_{\varphi,\ell}^{sca} \Delta_\ell = -\frac{1}{kR_0} \exp(ikR_0) \sum_{n,m,p} (-i)^{n+1} F_{mnp}(\vartheta, \varphi) \tau_{mn\,3-p}(\vartheta),$$

in which

$$F_{mnp} = \exp(im\varphi) \sum_{\ell=1}^{N_S} a_{mnp}^{\prime\ell} \Delta_\ell.$$

(66)

This is similar to the formulation given by Bruning and Lo (1971).

To identify the elements of the amplitude scattering matrix, a scattering plane is first defined by $\varphi = 0$ (i.e., the x'–z' plane). Components of the scattered field that are polarized parallel and perpendicular to this plane are given by E_ϑ^{sca} and E_φ^{sca}, respectively. The solution to Eq. (49), for a given incident state defined by α, β, and γ, will correspond to incident radiation polarized parallel to the scattering plane. A second solution, in which γ is incremented by $\pi/2$ in Eq. (11), will correspond to incident radiation polarized perpendicular to the scattering plane. Using both solutions, the four elements of the amplitude matrix can be expressed as

$$S_{11} = \frac{i}{k} \sum_{n,m,p} \tau_{mnp}(\vartheta)(-i)^{n+1} F_{mnp}^1(\vartheta, \varphi),$$

$$S_{22} = \frac{i}{k} \sum_{n,m,p} \tau_{mn\,3-p}(\vartheta)(-i)^n F_{mnp}^2(\vartheta, \varphi),$$

$$\tag{67}$$

$$S_{12} = \frac{i}{k} \sum_{n,m,p} \tau_{mnp}(\vartheta)(-i)^{n+1} F_{mnp}^2(\vartheta, \varphi),$$

$$S_{21} = \frac{i}{k} \sum_{n,m,p} \tau_{mn\,3-p}(\vartheta)(-i)^n F_{mnp}^1(\vartheta, \varphi),$$

in which the superscripts 1 and 2 on F correspond to scattering coefficients calculated for parallel or perpendicular incident field polarization, respectively. The elements of the phase matrix can be directly calculated from the amplitude matrix elements (Chapter 1).

The cluster-centered approach to calculating the amplitude and phase matrices begins by combining the partial scattered fields into a single expansion about a common cluster origin. The "total" scattering coefficients in this expansion, denoted a_{mnp}, are obtained from the addition theorem translation given by

$$a_{mnp} = \sum_{\ell=1}^{N_S} a_{mnp}^{0\ell},$$

$$\tag{68}$$

$$a_{mnp}^{0\ell} = \sum_{l,k,q} \tilde{A}_{mnp\,klq}^{0\ell} a_{klq}^\ell, \qquad n = 1, 2, \ldots, n_{\max,\ell 0}.$$

The total coefficients can then be transformed according to Eq. (62) to align them with respect to the incident field. The far-field components of the electric field will be given by Eq. (65), in which the partial coefficients are replaced by the total coefficients. The amplitude matrix elements are also given by Eqs. (67), with F^1 and F^2 replaced by a^1 and a^2.

The maximum order n on the translated scattering coefficient $a_{mnp}^{\ell 0}$, which is denoted $n_{\max,\ell 0}$, will, in general, be greater than the partial field truncation limit of $n_{\max,\ell}$. This is because the expansion of the partial field of sphere ℓ about the cluster origin 0 must now include the phase shift effects as described in Eq. (64). The number of orders required to describe this effect will increase as the distance between the sphere and the cluster origins, $R_{\ell 0}$, increases. It has been shown that $n_{\max,\ell 0}$ can be conservatively estimated by applying Eq. (6) to the size parameter based on the smallest sphere, centered about the cluster origin, that encloses sphere ℓ, that is, $x_{\ell 0} = k(R_{\ell 0} + a_\ell)$. It should also be emphasized that the re-expansion of the partial fields about a common origin occurs after a solution to Eq. (49) has been obtained—and thus does not affect the truncation limits $n_{\max,\ell}$ that are used on the partial field expansions.

VI. CLUSTER T MATRIX AND RANDOM-ORIENTATION PROPERTIES

A. SPHERE-CENTERED AND CLUSTER-CENTERED T MATRICES

Calculation of the random-orientation cross sections and scattering matrix elements of a cluster can proceed obviously by numerical quadrature of the fixed-orientation properties over the incident direction and polarization states as defined by α, β, and γ. Although such quadrature-based methods can be relatively efficient when used to calculate the total cross sections of sphere clusters—especially clusters that possess a random structure (i.e., fractal-like aggregates)—they are less efficient when accurate values for the scattering matrix elements are needed. Recent investigations of the polarimetric scattering properties of sphere clusters have shown that the depolarized scattering components can be highly sensitive to the orientation of the cluster with respect to the incident field (e.g., Mishchenko *et al.*, 1995). Because of this, accurate numerical calculation of the polarimetric scattering properties of a cluster can require an extremely fine quadrature scheme. For such situations, it can be computationally advantageous to calculate the T matrix of the sphere cluster and employ established relations to extract the random-orientation cross sections and scattering matrix elements from the T matrix (Mackowski, 1994; Mackowski and Mishchenko, 1996).

The inversion of Eq. (49) identifies the sphere-centered $\mathbf{T}^{\ell\ell'}$ matrix for sphere ℓ, that is,

$$\mathbf{a}^\ell = \sum_{\ell'=1}^{N_S} \mathbf{T}^{\ell\ell'} \mathbf{p}^{\ell'}. \tag{69}$$

The sphere-centered $\mathbf{T}^{\ell\ell'}$ matrix retains the partial field representation of the scattered field—and is therefore subject to the same phase difference constraints in the calculation of the scattered field that were encountered in the previous section. On the other hand, the random-orientation cross sections of the individual spheres, and of the cluster as a whole, can be obtained directly from the $\mathbf{T}^{\ell\ell'}$ matrices. In particular, the random-orientation extinction and absorption cross sections of sphere ℓ are given by

$$
\begin{aligned}
\langle C_{\mathrm{ext}}^{\ell} \rangle &= -\operatorname{Re} \frac{2\pi}{k^2} T_{mnp\,klq}^{\ell\ell'} \tilde{A}_{klq\,mnp}^{\ell'\ell}, \\
\langle C_{\mathrm{abs}}^{\ell} \rangle &= -\frac{2\pi}{k^2} \frac{\mathcal{E}_{mn}}{\mathcal{E}_{uw}} \operatorname{Re} \left(\frac{1}{[\bar{a}_{np}^{\ell}]^*} + 1 \right) T_{mnp\,klq}^{\ell\ell'} \tilde{A}_{klq\,uvw}^{\ell'\ell''} \left[T_{mnp\,uvw}^{\ell\ell''} \right]^*.
\end{aligned}
\tag{70}
$$

In the preceding and in what follows we adopt a tensorial approach, in that summation over all subscripts/superscripts not appearing on the left-hand side of the equation is implied.

The cluster T matrix represents the scattered wave from the cluster by a single expansion, centered about the origin of the cluster. It is defined so that

$$
a_{mnp} = T_{mnp\,klq} p_{klq}, \tag{71}
$$

in which the incident field coefficients p_{klq} are referenced with respect to the cluster origin. The transformation of the sphere-centered T matrices to the cluster T matrix is accomplished by two successive addition theorem transformations, given by

$$
\begin{aligned}
T_{mnp\,klq} &= \tilde{A}_{mnp\,uvw}^{0\ell} T_{uvw\,klq}^{\ell}, \\
T_{mnp\,klq}^{\ell} &= T_{mnp\,uvw}^{\ell\ell'} \tilde{A}_{uvw\,klq}^{\ell'0}.
\end{aligned}
\tag{72}
$$

The cluster T matrix, defined by the preceding equations, is completely equivalent to that which would be calculated via the extended boundary condition method (Chapter 6).

B. ITERATIVE SOLUTION OF THE CLUSTER T MATRIX

An efficient scheme for calculating the cluster T matrix, which obviates the need to directly invert Eq. (49) for the $T^{\ell\ell'}$ matrices, is obtained by transforming Eq. (49) according to Eq. (72). This results in a system of equations for the n, m, and p components of $T_{mnp\,klq}^{\ell}$, given by

$$
T_{mnp\,klq}^{\ell} = \bar{a}_{np}^{\ell} (1 - \delta_{\ell\ell'}) A_{mnp\,uvw}^{\ell\ell'} T_{uvw\,klq}^{\ell'} + \bar{a}_{np}^{\ell} \tilde{A}_{mnp\,klq}^{\ell 0}. \tag{73}
$$

The basic procedure for calculating the cluster T is to solve the preceding system for the row vectors of T^{ℓ} using successive values of l, k, and q. Note that, for each l, k, and q, Eq. (73) contains the same number of unknowns as Eq. (49). Following each solution, the obtained vector is translated and added into T following Eq. (69). The maximum value of l, for a given origin ℓ, can be identified in the course of the calculations by examining the convergence of the independent scattering cross section of sphere ℓ, that is,

$$\langle C_{\text{sca,ind}}^{\ell} \rangle = \frac{2\pi}{k^2} \frac{\mathcal{E}_{mn}}{\mathcal{E}_{kl}} |T_{mnp\,klq}^{\ell}|^2. \tag{74}$$

Once the relative change in $\langle C_{\text{sca,ind}}^{\ell} \rangle$ for successive values of l decreases below an acceptable limit, the T^{ℓ} matrix for sphere ℓ can be removed from Eq. (73). Spheres that are closer to the cluster origin will have their T^{ℓ} matrices converge faster than those that are farther removed. In this sense, the number of equations to be solved decreases as the entire T matrix approaches convergence.

C. RECURSIVE T-MATRIX ALGORITHM

Wang and Chew (1993) proposed a recursive T-matrix algorithm (RTMA), which makes use of a successive transformation of sphere clusters into single-particle T-matrix representations. To describe the method, consider the example when a cluster T matrix is calculated for a two-sphere system. Denote the matrix $T^{(2)}$, and now add a third sphere to the system. If the third sphere lies outside of a circumscribing sphere about the two-sphere system, the resulting scattering problem can be viewed as a two-body problem, in which the third sphere interacts solely with the original bisphere. In the form of Eq. (73), the problem would be represented as

$$T_{mnp\,klq}^{3} - \bar{a}_{np}^{3} A_{mnp\,uvw}^{30} T_{uvw\,klq}^{2} = \bar{a}_{np}^{3} \tilde{A}_{mnp\,klq}^{30},$$

$$T_{mnp\,klq}^{2} - T_{mnp\,rst}^{(2)} A_{rst\,uvw}^{03} T_{uvw\,klq}^{3} = T_{mnp\,klq}^{(2)}, \tag{75}$$

in which the coordinate origin for the three-sphere system is taken to coincide with the origin of the two-sphere system. Following solution of Eq. (75), the three-body T matrix would be obtained from

$$T_{mnp\,klq}^{(3)} = \tilde{A}_{mnp\,uvw}^{30} T_{uvw\,klq}^{3} + T_{mnp\,klq}^{2}. \tag{76}$$

A fourth sphere could now be added and the process repeated, until the T matrix for the entire cluster has been calculated.

The apparent advantage of this process is that it reduces calculation of the cluster T matrix into a sequence of two-body calculations—the latter can be performed efficiently by realigning (via rotational transformations, discussed later) the two bodies to share a common z axis. However, for reasons alluded to previously, the process would not be expected to be accurate for arbitrary clusters of spheres. The T matrix for a cluster will provide a valid description of the scattered field only for regions that are outside of a circumscribing sphere of the cluster. This constraint alone would appear to limit the RTMA to relatively simple clusters, for example, linear chains of spheres. In addition, the number of harmonic orders needed to accurately describe the near-field scattered wave will, in general, be larger than that needed to describe the far-field wave. Because of this, application of the RTMA procedure would involve increasingly larger truncation limits on the "partial" cluster T matrices.

VII. MEASUREMENTS AND APPLICATIONS

It is to be understood that the subheadings in this section serve only as general groupings and that there can be considerable overlap between them. It has been difficult to divide this section into subsections because light scattering is a quintessentially multidisciplinary study, and it is unnatural to compartmentalize applications of scattering and absorption by compounded spheres into exclusive categories.

Several general classes of compounded structures are intermingled in the specific applications discussed in the following. At present these classes can be represented as (1) agglomerates of similar particles, (2) composites assembled from dissimilar particles, (3) media containing high volume fractions of grains, and (4) fabricated arrays.

A. Particle Characterization

Advances in light-scattering instruments for real-time characterization or discrimination of micrometer-sized particles are important in environmental and process monitoring. Knowledge of the chemical and physical nature of such particles is often a fundamental step in the analysis of their production and implications, but many current interpretations of light-scattering data are based on simplifying assumptions with regard to particle morphology and internal structure. Compounded spheres can be inherently nonspherical, nonhomogeneous particles, and reliable calculations for their optical properties provide a valuable means of enhancing the information content of optical particle measurements.

1. Optics of Soot Agglomerates

Much of the development of multiple-sphere radiative models has been motivated by the need to better predict the absorption and scattering properties of flame-generated carbonaceous soot particles. Soot typically consists of aggregates of small (radius $\sim$ 10–30 nm) primary spheres. The average number of spheres in an aggregate can range from a few 10s for soot formed in rich premixed flames to 10^4 for large-scale turbulent diffusion flame soot. Because the primary spheres coagulate by Brownian diffusion, the structure of soot aggregates can be well described by fractal relationships. Numerical simulations of diffusion-limited aggregation typically result in fractal dimensions D_f between 1.7 and 1.9 and prefactors k_f between 1.3 and 1.6 (Mountain and Mulholland, 1988). In addition to its relevance to atmospheric radiation (to be addressed in Section B), it is well recognized that heat transfer from flames can be dominated by thermal emission from soot particles, which makes an accurate understanding of their radiative properties important in combustion diagnostics and heat transfer modeling. Furthermore, soot can be associated with mutagenic organic carbon compounds, and exact optical models will be useful in any upgrades to emissions monitoring resulting from revised U.S. Environmental Protection Agency standards.

The relative simplicity of the RGD model has enabled several researchers to use the method to extract, via laser light-scattering measurements, the structural properties of soot aggregates (Dobbins and Megaridis, 1991; Sorensen *et al.*, 1992, 1997; Köylü and Faeth, 1992; Cai *et al.*, 1993). Typically, these methods involve measurement of the vertical–vertical polarized component of the scattering distribution for which the primary particle contribution will be a constant.

As the first term in the Born expansion of Eq. (49), the RGD approximation neglects intracluster scattering. Most of the attempts to explicitly model multiple scattering (i.e., near-field interactions) among the primary spheres in an aggregate have been performed using various dipole interaction models (Purcell and Pennypacker, 1973; Jones, 1979; Chen *et al.*, 1991; Charalampopoulos and Chang, 1991; Sorensen *et al.*, 1992; Markel *et al.*, 1997). Because of their treatments of dipole self-interactions and field translations, some of these models are not completely equivalent to that obtained when the exact, multipole formulation is reduced to the dipole limit. In particular, the models of Jones (1979) and Chen *et al.* (1991) have been shown to violate energy conservation, whereas the multipole formulation obeys energy conservation at all levels of approximation—including the dipole limit.

A key difference between interacting dipole and RGD models is in the prediction of depolarized scattering effects (e.g., vertical–horizontal polarization scattering distributions)—which is a manifestation solely of the sphere interactions. Recent calculations by Singham and Bohren (1993) and Charalampopoulos and Chang (1991) have indicated that measurement of such depolarization effects could be used, in combination with interacting sphere models, to extract higher-order structure information on soot aggregates.

Aggregates that have overall sizes significantly smaller than the radiation wavelength can be modeled using an electrostatics approach. Here, the electric field is represented by the gradient of a potential, with the potential, in turn, satisfying Laplace's equation. Analytical solutions for sphere clusters in the electrostatics limit have been developed by Gérardy and Ausloos (1980) and Mackowski (1995) using multipole techniques analogous to those used for electromagnetic interactions among spheres. Application of the electrostatics solution to fractal-like aggregates has been performed by Mackowski (1995), in which a series approximation for the aggregate polarizability was developed. It was shown that the dipole (i.e., single electric order) approximation is insufficient to accurately describe the electric fields in contacting, Rayleigh-limit spheres that have relatively large |m|—which will typically be the case for carbonaceous soot in near- to mid-infrared wavelengths. As a result, the absorption cross sections of soot aggregates in the infrared can be a factor of 2 greater than that predicted by the RGD model.

In addition to validating or correcting assumptions regarding the RGD approximation, it is not uncommon that aggregates are composed of primaries with sizes lying well out of the Rayleigh range, in which case the rigorous theory becomes even more important, or that aggregates and composite particles of other materials, such as fumed silica or TiO_2, are of interest. Journals on colloid and surface science, as well as analytical chemistry, are replete with such examples.

2. Immunoassay

Flow cytometry is often fluorescence based, but of recent interest in medical diagnostics is the determination, by elastic scattering, of the aggregation of micro- and nanospheres by bioconjugate-mediated agglutination. Models of the radiant power scattered into a given detector angle based on exact theories of scattering by aggregates have proven useful in this regard (e.g., Fuller *et al.*, 1998). (A bit more on immunoassay will be covered in Section C.)

3. Two-Dimensional Scattering

Analysis of two-dimensional (2D) scattering is of increasing utility in optical characterization. Kaye *et al.* (1992) have demonstrated that it can aid in the characterization of nonspherical particles. Hirst *et al.* (private communication) have also found that 2D scattering images may discriminate between concentric and eccentrically stratified spheres in accelerating flows, and Holler *et al.* (1998) are exploring its ability to characterize aggregates and other particle shapes. Calculations made in concert with these recent works are displayed in Fig. 6.

B. REMOTE SENSING AND CLIMATOLOGY

1. Atmospheric Aerosols

It has become clear that a sufficiently accurate understanding of climate change depends on a more complete understanding of aerosol composition and geologic distribution (National Research Council, 1996). Limits in the sensitivity and accuracy of satellite-based aerosol measurements will be pushed back as sensor technology evolves; instrumentation on the NASA Earth Observing System fleet being deployed presently and in the near future (e.g., Mishchenko *et al.*, 1997c; Gordon, 1997) is an example. To an ever-increasing degree, scalar radiative transfer and its implicit reliance on single scattering will no longer provide an adequate means of interpreting such measurements (Lacis *et al.*, 1998). These inadequacies will, in turn, require that treatments of multiple scattering, particle absorption, and vector radiative transfer be employed more extensively than in the past. Spectral multiangle radiometry (Kahn *et al.*, 1997) and polarimetry (e.g., Mishchenko and Travis, 1997; Herman *et al.*, 1997; Bréon *et al.*, 1997) will provide unprecedented sensitivity to combinations of particle shapes, size distribution, and composition. On the other hand, existing standard aerosol optics models (Shettle and Fenn, 1979; d'Almeida *et al.*, 1991) that are used for atmospheric radiation calculations (e.g., Gordon, 1997) likely do not properly account for chemical speciation, aerosol absorption, and particle morphology.

Through the mid-1990s, considerable attention was paid to sulfate aerosol scattering with relatively little focus on the effects (light absorption, in particular) of carbonaceous (soot and organic) particles on Earth's radiation balance; yet in some regions their mass concentrations are often comparable to or greater than those of sulfates (Malm *et al.*, 1994). Atmospheric soot occurs as aggregates of black carbon spherules coated with unburned or partially oxidized fuel. In anthropogenic haze, individual haze elements frequently occur in the form of sulfate/soot *composite* particles (e.g., Podzimek, 1990; Parungo *et al.*, 1992). The theory for compounded spheres is thus a valuable tool for understanding the optical properties of a major class of atmospheric particulates and in assessing the impact of such particles on radiation balance, as well as on visual air quality. Calculations indicate that externally and internally mixed aerosols can have significantly different radiative properties (Fuller *et al.*, 1994a; Fuller, 1994b, 1995a, b; Chýlek *et al.*, 1996). Fuller *et al.* (1999) have modeled soot absorption for more realistic scenarios than those considered in the preceding references and the impacts of these different properties on climate forcing and boundary layer heating rates are being investigated, but earlier work indicates that instantaneous column heating in an aerosol layer is quite sensitive to changes in absorption (Fuller *et al.*, 1994b).

Airborne dust can be a significant aerosol component and the complex morphology and composition of dust particles renders their optical properties difficult to model (Sokolik *et al.*, 1997). The theory can also be applied to the study of radiometric properties of wind-blown dust and soil and can augment existing scattering models by significantly extending the range of morphologies, sizes, and compositions that can be studied.

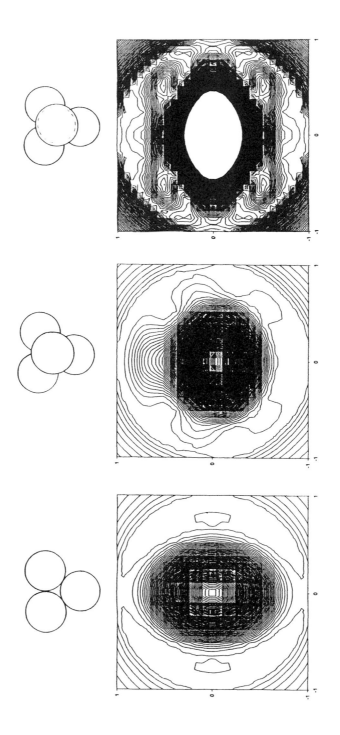

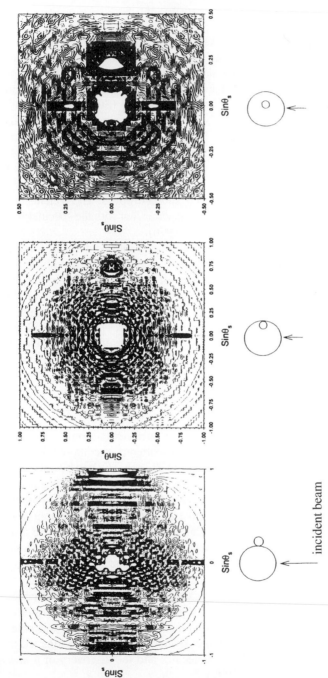

incident beam

Figure 6 Two-dimensional scattering patterns. Upper panel: The 2D intensity pattern is calculated for close-packed clusters of identical spheres for (a) a triangular array illuminated by radiation incident in the plane of the triangle, at one of the vertices; (b) with the addition of a fourth sphere to form a tetrahedron; and (c) with the addition of a fifth sphere to form a hexahedron. Lower panel: A pair of very different spheres, illuminated at broadside incidence where (a) a satellite droplet resides on the surface of a larger host, (b) the smaller droplet resides just inside the host, and (c) an inclusion is displaced by a distance equal to its radius from the center of the host. The size parameters of the spheres are about 3 for the calculations displayed in the upper panel. For the lower panel, the size parameter of the host is about 50 and that of the smaller sphere is about 20.

261

2. Cloud Absorption

Carbon/sulfate composite aerosols can serve as cloud condensation nuclei, and through either nucleation or impact scavenging, carbon can be incorporated into cloud droplets. The occurrence of carbon on or in cloud droplets is, in turn, of interest as a possible contributor to light absorption by clouds. The so-called cloud absorption anomaly has been given considerable attention (Stephens and Tsay, 1990; Ramanathan *et al.*, 1995; Stephens, 1996), and even though carbonaceous material often appears to be of secondary importance, in some cases it may increase absorption by a few percent (Heintzenberg and Wendisch, 1996) and so, play a role in regional cloud forcing. Figure 7 demonstrates the effects of trace amounts of carbon, typical of remote marine loadings, on the absorption efficiency (the ratio of C_{abs} to geometric cross section) of cloud droplets. The conclusions that can be drawn from these calculations contrast somewhat with the speculations by Chýlek and Hallett (1992) and Chýlek *et al.* (1996).

Cloud absorption may also play a role in monitoring indirect climate forcing by aerosols: It is anticipated that changes in cloud albedo, which would be difficult to detect, at best, could be linked to aerosol-induced changes in droplet size distribution. These size distributions would then be used to estimate modifications of cloud lifetime and of cloud radiative forcing by pollution. Because a reduction in albedo by absorbing carbon could offset increases in albedo resulting from changes in droplet size, a rigorous model of droplet absorption is necessary if one is to infer actual changes in droplet size distributions from changes in cloud albedo induced by anthropogenic aerosols and biomass burning.

3. Astronomy

In planetary astronomy, the radiative properties of surfaces and some cloud layers can be strongly influenced by the occurrence of scatterers as compounded particles. It has been shown that surface albedos can be strongly affected by how scatterers are mixed with absorbers in planetary regoliths (Hapke, 1993; Hapke *et al.*, 1975; Hapke *et al.*, 1998). There is also evidence, based on radiometric properties of aggregates, that such particles may exist in certain layers of extraterrestrial atmospheres (West and Smith, 1991). Calculations of aggregates have also been used to help interpret data from radio occultations by planetary rings (Gresh, 1990). As discussed elsewhere in this book (Chapters 13 and 19), interplanetary and interstellar grains are often nonspherical and may frequently occur as aggregated particles. An ability to relate their optical and structural properties enables scientists to better understand the astrophysical processes involved in early stellar and planetary evolution (Poynting–Robertson drag being an example). It is noteworthy that attempts to understand ultraviolet scattering by carbon grains led to major progress in the production of C_{60} molecules and the discovery of a new crystalline form of carbon (Huffman, 1991).

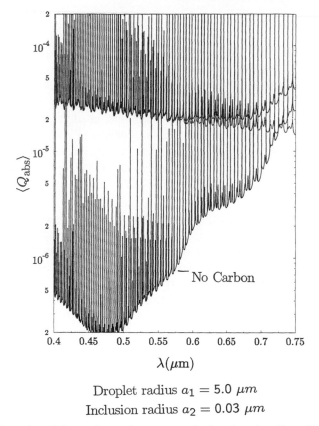

Droplet radius $a_1 = 5.0$ μm

Inclusion radius $a_2 = 0.03$ μm

Figure 7   Absorption efficiency of monodisperse water droplets, 5 μm in radius, with randomly located 0.03-μm carbon inclusions. The mass ratios are equivalent to about 400 ng of light-absorbing carbon per gram of water, representative of polluted cloud water. In the visible, the lower curve is for pure water and the upper curve includes absorption by encapsulated carbon. The band-averaged absorption efficiency of the droplets rises from 1.6×10^{-5} for pure water to 1.4×10^{-4} for droplets inoculated with carbon grains. Approximately 20% of the absorption efficiency of the inoculated droplets is due to carbon grains exposed to high fields associated with the droplet whispering gallery modes.

C. Photonics

1. Morphology-Dependent Resonances in Microparticles

One of the most interesting features of light scattering by microspheres is their ability to act as resonance cavities, sustaining whispering gallery modes with quality (Q) factors reaching nearly 10^{10} (Gorodetsky *et al.*, 1996; Vernooy *et al.*, 1998). This produces low-threshold nonlinear optical effects such as stimulated

Raman scattering, lasing, spectral hole burning, and cavity quantum electrodynamic effects.

There is, at this time, an accelerating interest in the role of systems of compound spheres in photonics. Work continues with concentric core–shell morphologies, especially in relation to lasing emission from layered microspheres
and cladded fibers (e.g., Lock, 1990; Essien *et al.*, 1993; Knight *et al.*, 1993). Referring back to Fig. 3, it has also been shown that, in addition to the whispering
gallery modes of the component spheres in systems of concentrically stratified
spheres, resonances produced by poles in the "coefficient of finesse" in Eq. (33)
can arise that are analogous to the resonant modes of a Fabry–Perot cavity. (Obviously, this same analysis can be readily applied to coaxial fibers and multilayered
structures.)

Because many microdevices based on microcavity resonances will likely involve spheres residing in contact with a substrate or other optical elements, the
theory of dependent scattering is important in understanding how coupling to
these other objects will affect their resonance properties (e.g., Arnold *et al.*, 1991,
1992). One interesting feature of interparticle coupling is an alteration of resonance spectra that depends on the incident direction of the beam (Fuller, 1991;
Arnold *et al.*, 1994).

A theoretical understanding of inclusions is of interest because of observed
reduced threshold lasing in microdroplets inoculated with nanometer-sized scatterers (e.g., Taniguchi *et al.*, 1996), as well as other effects of seeding on microdroplet optical properties. Because the mode volumes of resonances in microspheres are concentrated in equatorial or sagittal planes of the spheres, studies
of mode coupling in structures such as multisphere or stratified sphere systems
should be useful in the study of microdisk lasers. This may be especially true
with regard to improvements to input and output coupling to the resonances (e.g.,
Nöckel and Stone, 1997; Djaloshinski and Orenstein, 1998).

It has been found that microparticles can be organized by intense optical fields
(Burns *et al.*, 1990). To a first approximation, the organization may be described
in terms of potential wells related to the local intensity in and near the particles. These potentials can be modified, however, by electromagnetic interactions
between the particles. The organization of microparticles is also relevant to optical band gap materials—dielectric "crystals" composed of periodic arrays of
spheres that have photonic properties analogous to those of semiconductors (Romanov *et al.*, 1997). The study of cooperative scattering by arrays of dielectric
particles may provide a deeper understanding of these phenomena. Photon localization and coherent backscatter (Mishchenko, 1996; Wiersma *et al.*, 1997) are
other active areas of research where mutual polarization effects on interparticle
scattering are of interest. Interestingly, and somewhat counterintuitively, strongly
scattering media have been found that act as high-gain media for laser emission

(Oliveira *et al.*, 1997; Beckering *et al.*, 1997), and the role of particle interactions under such circumstances is poorly understood at this time.

2. Biosensors and Biomedical Optics

Closely related to such fields as photon localization and diffuse reflectometry of planetary regoliths is the subject of diffuse reflectance and transmittance through biological tissue. These topics are addressed through the study of wave propagation through discrete random media—a subject of considerable experimental and theoretical research. For example, Zurk *et al.* (1995) and Tsang *et al.* (1998) have modeled extinction rates as functions of particulate volume fraction for collections of several thousand Rayleigh spheres and ellipsoids, respectively. Work is also being carried out in the area of 2D imaging of Mueller matrix elements for light backscattered from turbid media (Cameron *et al.*, 1998; Raković and Kattawar, 1998). Our understanding of radiative transfer in such media would be advanced if one could determine effective values for C_{ext}, ϖ, and $\langle \cos \Theta \rangle$ that account for coupling between neighboring particles.

Surface-enhanced Raman spectroscopy (SERS) utilizes the strong local fields produced by dipolar plasmon resonances in nanometer-sized metal particles (Moskovits, 1985). Because of the biological compatibility of gold and silver, nanoparticles made of these metals are used in the detection and spectroscopy of biomaterials. The ability to control the quality of such Raman-active surfaces depends on how the metal particles couple to substrates and to each other (Grabar *et al.*, 1995; Freeman *et al.*, 1995; Westcott *et al.*, 1998; Quinten *et al.*, 1998). Oldenburg *et al.* (1998) have also been active in the study of hybrid particles similar to those used in the immunoassay scheme mentioned previously. This area can clearly benefit from applications of the theory described in this chapter, though one is cautioned that the interactions between small metal spheres can, once again, be problematic (Fuller, 1987).

3. Photon Correlation Spectroscopy

In addition to the photon correlation spectroscopy of bulk colloidal suspensions (e.g., Ovod *et al.*, 1998), there is considerable interest in the optical properties of micrometer-sized particles that entrain micrometer- or submicrometer-sized inhomogeneities. One focus of this interest is on the effects that the inhomogeneities might have on the phenomena associated with the morphology-dependent resonances of their hosts. Investigations have been carried out in this context on fluctuations in the intensity of elastically scattered light (Bronk *et al.*, 1993) and in the resonance spectra (Ngo and Pinnick, 1994; Fuller, 1994c, 1995b).

D. Numerical Test Bed for Other Techniques

Compounded spheres can form inherently nonspherical, nonhomogeneous scatterers. As techniques for exact or approximate solutions for scattering by nonspherical particles are introduced, the exact theories described in this chapter will continue to enjoy use as a means of testing those techniques, as they have in the past (Mulholland *et al.*, 1994; West, 1991; Flatau *et al.*, 1993; Hill *et al.*, 1995). It is anticipated that the analysis discussed in this chapter will be of use in determining the accuracies of calculations based on the finite difference time domain (Chapter 7; Cole, 1998) and geometric optics approximations [see Chapters 10 and 15; Roll *et al.* (1998) for examples of recent advances in geometric optics applied to nonspherical and spherical particles, respectively].

There is also a need to validate effective medium theories (Bohren, 1986; Chapter 9) and recent calculations have been made that compare exact calculations for absorption by internally mixed soot to selected effective medium approximations (Fuller *et al.*, 1999).

Scattering by clusters has enjoyed a synergism with microwave measurements (Kattawar and Dean, 1983; Fuller *et al.*, 1986; Fuller and Kattawar, 1988a, b; Xu, 1997; Chapter 13). The availability of microwave data was instrumental in the evolution of the theoretical work of Kattawar *et al.*, and that theory has, in turn, been used to calibrate measurements at other microwave facilities.

E. Extensions of the Theory

1. Focused Beams

As long as one is able to express the incident field as an expansion in vector spherical harmonics, it is possible to solve the boundary value problem for incident waveforms other than plane waves. Applications of this approach to single spheres can be found in the works of Grehan *et al.* and Barton *et al.* (e.g., Barton *et al.*, 1989; Barton, 1998; Wu *et al.*, 1997) on generalized LM theory. Recently the beam shape coefficients of Gouesbet and Grehan have been cast in terms of the VSH addition coefficients (Doicu and Wriedt, 1997c).

2. Pulses

In this case, one works with the Fourier transform of the pulse to express the incident beam in the frequency domain. Provided that the range and resolution of the scattering spectra are not too great, it should be a relatively straightforward matter to investigate interactions between multisphere systems, analogous to the studies carried out for cases of single spheres (Rheinstein, 1968; Inada, 1974; Eiden, 1975; Shifrin and Zolotov, 1995).

3. Higher Degrees of Compound Structure

There are structures that are somewhat more complicated than those discussed in this chapter, which, with fairly straightforward extensions of the theory, can be addressed. Among these are multiple concentric layers (Bhandari, 1985; Mackowski et al., 1990), aggregates of coated particles (Borghese et al., 1987; Hamid et al., 1992; Fuller et al., 1999), higher levels of eccentric layering (Videen and Ngo, 1998), and aggregates of compounded spheres.

VIII. VECTOR ADDITION THEOREM

A. DERIVATION

Stein (1961) and Cruzan (1962) developed the vector harmonic addition theorem by a painstaking application of the scalar harmonic addition theorem, developed by Friedman and Russek (1954), to the vector components of an outgoing spherical wave. A more direct approach, demonstrated here, is to consider the vector addition theorem as applied to a plane, linearly polarized wave.

As before, the propagation and polarization directions of the incident wave are defined by rotation of a z-propagating wave, with electric field pointed in the x direction, through the Euler angles (α, β, γ). The expansion coefficients for the plane wave, in terms of the vector harmonics about origin ℓ, are related to the coefficients about origin 0 by the phase shift given in Eq. (12). The expansion coefficients about origin ℓ can also be obtained by application of the addition theorem, which appears as

$$p_{mnp}^{\ell} = \sum_{l,k,q} \tilde{A}_{np\,klq}^{\ell 0} p_{klq}. \tag{77}$$

Equating Eq. (77) with Eq. (12) and using the orthogonality relation for the plane wave expansion coefficients,

$$\frac{1}{8\pi^2} \int_0^{2\pi} d\alpha \int_{-1}^{1} d(\cos\beta) \int_0^{2\pi} d\gamma \, p_{mnp} p_{klq}^* = \frac{2n+1}{n(n+1)} \frac{(n-m)!}{(n+m)!} \delta_{nl}\delta_{mk}\delta_{pq} \tag{78}$$

gives

$$\tilde{A}_{mnp\,klq}^{\ell 0} = \frac{1}{8\pi^2} \frac{n(n+1)}{2n+1} \frac{(n+m)!}{(n-m)!} \int_0^{2\pi} d\alpha \int_{-1}^{1} d(\cos\beta) \int_0^{2\pi} d\gamma \, p_{mnp} p_{klq}^*$$
$$\times \exp\left[ikR_{\ell 0}(\cos\Theta_{\ell 0}\cos\beta + \sin\Theta_{\ell 0}\sin\beta\cos(\Phi_{\ell 0} - \alpha)) \right]. \tag{79}$$

The integrations can be performed by employing the rotation function representation of $p_{mnp}(\alpha, \beta, \gamma)$ given in Eq. (11). The product of two rotation functions, which occurs when $p_{mnp} p^*_{klq}$ is expanded, can be linearized into a series of rotation functions by the relation

$$
\begin{aligned}
\mathcal{D}^s_{mn} \mathcal{D}^{s'}_{kl} &= (-1)^{k+s'} \frac{(l-k)!(l+s')!}{(l+k)!(l-s')!} \mathcal{D}^s_{mn} \mathcal{D}^{-s'}_{-kl} \\
&= (-1)^{k+s'} \frac{(l-k)!(l+s')!}{(l+k)!(l-s')!} \\
&\quad \times \sum_w (-1)^{n+l+w} \hat{C}^w_{-mn\,kl} \hat{C}^w_{sn\,-s'l} \mathcal{D}^{s-s'}_{m-k\,w},
\end{aligned} \tag{80}
$$

in which the $\hat{C}$ are equivalent (to a normalizing factor) to the vector-coupling coefficients. The sum over w includes values from $w = \max(|n-l|, |m-k|, |s-s'|)$ to $w = n+l$. By recognizing that

$$
l(l+1)\hat{C}^w_{1n\,-1l} = (-1)^{n+l+w} n(n+1)\hat{C}^w_{-1n\,1l} \tag{81}
$$

and

$$
\mathcal{D}^0_{m-kw}(\beta) = (-1)^{k-m} P^{k-m}_w(\cos \beta), \tag{82}
$$

it can readily be shown that

$$
\begin{aligned}
\frac{1}{2\pi} \int_0^{2\pi} p_{mnp} p^*_{klq}\, d\gamma &= -\frac{1}{2} i^{l-n}(-1)^k(2n+1) \frac{2l+1}{l(l+1)} \frac{(l-k)!}{(l+k)!} \\
&\quad \times \hat{C}^w_{-mn\,kl} \hat{C}^w_{-1n\,1l} P^{k-m}_w(\cos \beta) \exp[i(k-m)\alpha]. \tag{83}
\end{aligned}
$$

The sum over w spans the same range as before, yet now includes only even $n+l+w$ if $p = q$ and odd $n+l+w$ if $p \neq q$. The phase factor in Eq. (79) is now expanded using addition theorems for spherical Bessel functions and associated Legendre functions:

$$
\begin{aligned}
&\exp\left[ikR_{\ell 0}(\cos\Theta_{\ell 0}\cos\beta + \sin\Theta_{\ell 0}\sin\beta\cos(\Phi_{\ell 0}-\alpha))\right] \\
&= \sum_{v=0}^\infty i^v(2v+1)j_v(kR_{\ell 0}) \\
&\quad \times \sum_{u=-v}^v P^u_v(\cos\Theta_{\ell 0})\exp(iu\Phi_{\ell 0})P^{-u}_v(\cos\beta)\exp(-iu\alpha). \tag{84}
\end{aligned}
$$

The orthogonality properties of the Legendre functions can now be used to perform the remaining integrations over α and β in Eq. (79). The final result is

$$
\tilde{A}_{mnp\,klq}^{\ell 0} = -i^{l-n}(-1)^m(2n+1)\sum_w i^w \hat{C}_{-mn\,kl}^w \hat{C}_{-1n\,1l}^w j_w(kR_{\ell 0})
$$
$$
\times P_w^{k-m}(\cos\Theta_{\ell 0})\exp[i(k-m)\Phi_{\ell 0}],
$$
$$
w = \max(|n-l|,|m-k|),\dots,n+l. \tag{85}
$$

Again, if $p = q$ then $w + n + l$ takes on even values; otherwise $w + n + l$ is odd. The addition coefficients for translation of spherical waves, $A_{mnp\,klq}^{\ell 0}$ with $R_{\ell 0} > R_\ell$, will be identical in form to Eq. (85), except with the spherical Bessel function $j_w(kR_{\ell 0})$ replaced by the spherical Hankel function $h_n(kR_{\ell 0})$.

It has been shown (using the symbolic package Mathematica) that the result in Eq. (85) is completely equivalent to the previously derived results of Stein and Cruzan. Unlike the previous formulas, however, the result here shows that the same-mode and cross-mode coupling coefficients (i.e., $p = q$ and $p \neq q$) have the same basic formula.

B. SYMMETRY AND ROTATIONAL DECOMPOSITION

The following relations are useful in calculation of the addition coefficients:

$$
A_{mnp\,klq}^{\ell\ell'} = (-1)^{m+n+p+k+l+q}\frac{(2n+1)l(l+1)}{(2l+1)n(n+1)}A_{-klq\,-mnp}^{\ell\ell'}
$$
$$
= (-1)^{m+k}\frac{(2n+1)l(l+1)}{(2l+1)n(n+1)}A_{-klq\,-mnp}^{\ell'\ell}
$$
$$
= \frac{(2n+1)l(l+1)}{(2l+1)n(n+1)}\frac{(n-m)!(l+k)!}{(n+m)!(l-k)!}[A_{klq\,mnp}^{\ell'\ell}]^*. \tag{86}
$$

The same relations apply to the plane wave addition coefficient $\tilde{A}$. In the last relation, the conjugate applies to all complex quantities in the addition coefficient with the exception of the basis function $h_w(kR_{\ell\ell'})$. This exception need not be made for $\tilde{A}$ when the medium has a real refractive index.

In the course of the iterative solution of Eq. (47), it is necessary to perform the matrix–vector multiplication

$$
a_{mnp}^{\ell\ell'} = A_{mnp\,klq}^{\ell\ell'}a_{klq}^{\ell'}. \tag{87}
$$

This operation requires on the order of $(n_{\max})^4$ multiplications to transform $a^{\ell'}$ into $a^{\ell\ell'}$. The process is accelerated considerably if the translation from ℓ' to ℓ corresponds to the z axis of the cluster; that is, the translation is along the symmetry axis. For this case the azimuthal degrees become decoupled, and the addition coefficient is nonzero only for $k = m$. The multiplication process is now reduced to the order of $(n_{max})^3$ steps. This savings in computational time can be exploited for general translations by decomposing the addition coefficient into three steps: the first involving a coordinate rotation of ℓ' toward ℓ, the second involving an axial translation of ℓ' to ℓ, and the last involving an inverse rotation back to the original frame. The steps involved in this process are

$$
\begin{aligned}
a_{mnp}^{\ell\ell',1} &= \exp(ik\Phi_{\ell\ell'})\mathcal{D}_{mn}^k(\Theta_{\ell\ell'})a_{knp}^{\ell'}, \\
a_{mnp}^{\ell\ell',2} &= A'^{\ell\ell'}_{mnp\,mlq}a_{mlq}^{\ell\ell',1}, \\
a_{mnp}^{\ell\ell'} &= (-1)^{m+k}\exp(-im\Phi_{\ell\ell'})\mathcal{D}_{mn}^k(\Theta_{\ell\ell'})a_{knp}^{\ell\ell',2},
\end{aligned}
\tag{88}
$$

in which the addition coefficient in the second step is calculated for axial translation between ℓ' and ℓ, that is, $\Theta'_{\ell\ell'} = \Phi'_{\ell\ell'} = 0$. Because each step in the process involves on the order of $(n_{max})^3$ steps, it can be significantly faster when n_{max} is large than the direct process of Eq. (87). In addition, the rotational and axial translation matrices can be stored in less memory space (by a factor of approximately n_{max}) than the undecomposed $A^{\ell\ell'}$.

C. ROTATION FUNCTIONS AND VECTOR COUPLING COEFFICIENTS

The rotation coefficients $\mathcal{D}_{kn}^m$ appearing in this work are related to the generalized spherical functions $d_{km}^{(n)}$ by

$$
\mathcal{D}_{kn}^m(\beta) = (-1)^{m+k}\left[\frac{(n-k)!(n+m)!}{(n+k)!(n-m)!}\right]^{1/2}d_{km}^{(n)}(\beta).
\tag{89}
$$

In addition, they are also related to the Jacobi polynomials by

$$
\begin{aligned}
\mathcal{D}_{kn}^m(\beta) &= (-1)^{m+k}\frac{(n-k)!}{(n-m)!}\left(\cos\left(\frac{\beta}{2}\right)\right)^{k+m}\left(\sin\left(\frac{\beta}{2}\right)\right)^{k-m} \\
&\quad \times \mathcal{P}_{n-k}^{(k-m),(k+m)}(\cos\beta).
\end{aligned}
\tag{90}
$$

A recurrence relation for calculating the functions upwards in m is

$$
\begin{aligned}
\mathcal{D}_{kn}^0(\beta) &= (-1)^k P_n^{-k}(\cos\beta), \\
\mathcal{D}_{kn}^{m+1} &= \cos^2\left(\frac{\beta}{2}\right)\mathcal{D}_{k-1\,n}^m - (n-k)(n+k+1)\sin^2\left(\frac{\beta}{2}\right)\mathcal{D}_{k+1\,n}^m \\
&\quad - k\sin(\beta)\mathcal{D}_{kn}^m.
\end{aligned}
\tag{91}
$$

The symmetry relations are

$$\mathcal{D}_{kn}^{-m}(\beta) = (-1)^{m+k}\frac{(n-m)!(n-k)!}{(n+m)!(n+k)!}\mathcal{D}_{-kn}^{m}(\beta)$$

$$= \mathcal{D}_{mn}^{-k}(\beta) = (-1)^{k+m}\mathcal{D}_{mn}^{-k}(-\beta). \tag{92}$$

The $\hat{C}$ coefficients appearing in this work are related to the vector coupling coefficients by

$$\hat{C}_{mn,kl}^{w} = \left(\frac{(n+m)!(l+k)!(w-m-k)!}{(n-m)!(l-k)!(w+m+k)!}\right)^{1/2}C_{mn,kl}^{m+kw}. \tag{93}$$

The $\hat{C}$ coefficients can be calculated by first defining

$$\hat{C}_{mn,kl}^{w} = g_{nlw}(n+m)!(l+k)!(w-m-k)!S_{mn,kl}^{w}, \tag{94}$$

where

$$g_{nlw} = \left(\frac{(2w+1)(n+l-w)!(w+n-l)!(w+l-n)!}{(n+l+w+1)!}\right)^{1/2}. \tag{95}$$

The S coefficients, in turn, obey the three-term downwards recurrence relation

$$S_{mn,kl}^{w-1} = b_w S_{mn,kl}^{w} + c_w S_{mn,kl}^{w+1}, \tag{96}$$

$$b_w = \frac{(2w+1)\{(m-k)w(w+1)-(m+k)[n(n+1)-l(l+1)]\}}{(w+1)(n+l-w+1)(n+l+w+1)},$$

$$\tag{97}$$

$$c_w = -\frac{w(w+n-l+1)(w+l-n+1)(w+m+k+1)(w-m-k+1)}{(w+1)(n+l-w+1)(n+l+w+1)},$$

with starting values of

$$S_{mn,kl}^{n+l+1} = 0,$$

$$\tag{98}$$

$$S_{mn,kl}^{n+l} = \frac{1}{(n-m)!(l+k)!(n+m)!(l-k)!}.$$

The minimum value of w will be the larger of $|n-l|$ or $|m+k|$. Recurrence formulas for S in the n and l indices can be obtained from the formulas

$$S_{mn,kl}^{w} = (-1)^{w+n+k}S_{m+kw,-kl}^{n} = (-1)^{w+n+k}S_{kl,-m-kw}^{n}$$

$$= (-1)^{n+m}S_{mn,-m-kw}^{l} = (-1)^{n+m}S_{m+kw,-mn}^{l}. \tag{99}$$

ACKNOWLEDGMENTS

This work was supported by the Geosciences Program of the Cooperative Institute for Research in the Atmosphere (CIRA), by NASA's Global Aerosol Climatology Project, and by the Dupont Educational Assistance Program.

Chapter 9

Effective Medium Approximations for Heterogeneous Particles

Petr Chýlek
Atmospheric Science Program
Departments of Physics and Oceanography
Dalhousie University
Halifax, Nova Scotia
Canada B3H 3J5

Gorden Videen
U.S. Army Research Laboratory
Adelphi, Maryland 20783

D. J. Wally Geldart
Atmospheric Science
Program
Department of Physics
Dalhousie University
Halifax, Nova Scotia
Canada B3H 3J5

J. Steven Dobbie
Atmospheric Science
Program
Department of Physics
Dalhousie University
Halifax, Nova Scotia
Canada B3H 3J5

H. C. William Tso
Atmospheric Science
Program
Department of Physics
Dalhousie University
Halifax, Nova Scotia
Canada B3H 3J5

Light Scattering by Nonspherical Particles: Theory, Measurements, and Applications

273

I. INTRODUCTION

The optical properties of heterogeneous particles have been of considerable interest and importance in several different branches of science. They are needed in various applications in atmospheric science, oceanography, astronomy, and other fields of geophysical science. Individual aerosol particles may be a composed mixture (black carbon in ammonium sulfate, black carbon in quartz, clay mineral in quartz, black carbon in water, sulfuric acid with a crustal core, ammonium sulfate in sulfuric acid, silica and black carbon agglomerates, metals in sulfuric acid, etc.). Such particles can have a very complicated morphology (Gillette and Walker, 1977; Chýlek *et al.*, 1981; Pinnick *et al.*, 1985; Sheridan and Musselman, 1985; Sheridan, 1989a, b; Sheridan *et al.*, 1993; Reitmeijer and Janeczek, 1997). The problem of scattering and absorption of electromagnetic radiation by these composite particles is so complicated that the exact solution of Maxwell's equations with appropriate electromagnetic boundary conditions is not practical. Moreover, the exact shapes, sizes, positions, orientations, and numbers of inhomogeneities are usually unknown. Even if we could spend an enormous amount of computer time to obtain the numerical solution, the lack of information concerning the detailed structure of atmospheric particles prevents us from obtaining an appropriate solution.

For these reasons it seems attractive to have a way to determine some average optical properties of heterogeneous materials that would enable us to treat that heterogeneous material in much the same way as we treat homogeneous substances. This problem is not limited to problems of atmospheric science or the geosciences. The first attempts to find a way to simplify heterogeneous materials by describing them by average or effective material constants is almost as old as electromagnetic theory itself (Maxwell-Garnett, 1904). The Maxwell-Garnett mixing rule is still frequently used today in the geophysical sciences and other fields.

Other prescriptions for the average optical properties of heterogeneous materials followed soon thereafter and, as of today, we have a rather long list of so-called mixing rules or effective medium approximations (EMAs). They all were derived under various sets of restrictive conditions. However, they have one basic assumption in common, namely, that the typical dimension d of an inhomogeneity must be much smaller than the wavelength λ of the considered radiation, $d \ll \lambda$.

Practical physical and geophysical situations frequently present problems that do not conform to the essential requirement of smallness of the regions of inhomogeneity compared to the wavelength. Consequently, there is an ongoing search for new ways to extend the region of applicability of EMAs to slightly larger sizes of inclusions and inhomogeneties. EMAs that are constructed to allow larger sizes of inclusions (but still smaller than the wavelength) are generally called extended effective medium approximations (EEMAs). The targeted region of their applicability is $d < \lambda$ instead of $d \ll \lambda$, as required for ordinary effective medium approximations.

Effective and extended effective medium approximations (for which we will at times simply use the term effective medium approximations to cover both) are not approximations in a strict mathematical sense. It is not generally possible to estimate the accuracy of a given approximation by considering the magnitude of neglected terms with respect to those that are kept. EMAs are often based on an *ad hoc* assumption that leads to a simplified, solvable model of a real, complicated, and usually unsolvable situation. As a result, one is able to derive a simple or only moderately complicated prescription (e.g., the mixing rule) of how to calculate the average optical properties of a heterogeneous composite material from the known properties and amounts of its individual components. Because there are no specific algebraic terms neglected and because the exact solution of the problem is usually unknown, the accuracy of such derived effective material constants (effective dielectric constants or effective refractive indices of material) and the precise conditions for their permissible use are not easy to assess.

II. EFFECTIVE MEDIUM APPROXIMATIONS

The EMAs for composite materials are traditionally developed for electrostatic fields. The starting point is the basic electrostatic linear relation between the electric displacement $\mathbf{D}$ and electric field $\mathbf{E}$. For an ideal homogeneous material, the relation is $\mathbf{D} = \varepsilon\mathbf{E}$, provided that the material is also isotropic in space.

In the case of heterogeneous particles with macroscopic inclusions, the appropriate linear relation is

$$\mathbf{D}(x, y, z) = \varepsilon(x, y, z)\mathbf{E}(x, y, z), \tag{1}$$

where $\varepsilon(x, y, z)$ is a dielectric constant at the point (x, y, z) within the particle. The dependence of ε on position accounts for the inhomogeneity. More general linear relations could be used to account for possible anisotropic and nonlocal correlation in space, but those are not required for our purposes.

For the case of an inhomogeneous medium, the effective dielectric constant ε_{eff} of a composite material is defined by (Landau and Lifshitz, 1960)

$$\int_V \mathbf{D}(x, y, z) \, dx \, dy \, dz = \int_V \varepsilon(x, y, z) \mathbf{E}(x, y, z) \, dx \, dy \, dz$$

$$= \varepsilon_{eff} \int_V \mathbf{E}(x, y, z) \, dx \, dy \, dz, \tag{2}$$

where $\mathbf{E}(x, y, z)$, $\mathbf{D}(x, y, z)$, and $\varepsilon(x, y, z)$ are the local electric field, local electric displacement and local dielectric constant, respectively.

Using a definite set of assumptions concerning the shapes of inclusions and the topology of a mixture, one can obtain an appropriate analytical expression for an effective dielectric constant as a function of volume fractions, f_1 and f_2, and dielectric constants, ε_1 and ε_2, of individual components. The famous formulas for an effective dielectric constant derived in this way include, among others, the Lorentz–Lorenz (Lorentz, 1880; Lorenz, 1880),

$$f_1 \frac{\varepsilon_1 - 1}{\varepsilon_1 + 2} = \frac{\varepsilon_{eff} - 1}{\varepsilon_{eff} + 2}, \tag{3}$$

the Maxwell-Garnett (1904),

$$f_1 \frac{\varepsilon_1 - \varepsilon_2}{\varepsilon_1 + 2\varepsilon_2} = \frac{\varepsilon_{eff} - \varepsilon_2}{\varepsilon_{eff} + 2\varepsilon_2}, \tag{4}$$

$$f_2 \frac{\varepsilon_2 - \varepsilon_1}{\varepsilon_2 + 2\varepsilon_1} = \frac{\varepsilon_{eff} - \varepsilon_1}{\varepsilon_{eff} + 2\varepsilon_1}, \tag{5}$$

and the Bruggeman (1935)

$$f_1 \frac{\varepsilon_1 - \varepsilon_{eff}}{\varepsilon_1 + 2\varepsilon_{eff}} + f_2 \frac{\varepsilon_2 - \varepsilon_{eff}}{\varepsilon_2 + 2\varepsilon_{eff}} = 0 \tag{6}$$

mixing rules.

In the case of the Lorentz–Lorenz mixing rule, the second component is vacuum with $\varepsilon_2 = 1$. The Maxwell-Garnett expression [Eq. (4)] assumes that inclusions with a dielectric constant ε_1 are embedded in a host material with a dielectric constant ε_2. When the roles of the inclusion and the host material are reversed the inverse Maxwell-Garnett [Eq. (5)] is obtained. Bruggeman treats both materials on an equal basis and his mixing rule, Eq. (6), is symmetric with respect to an interchange of materials.

These classical EMAs can be generalized for the case of an n-component mixture and written in a simple generic form (Aspens, 1982),

$$\frac{\varepsilon_{\text{eff}} - \varepsilon_0}{\varepsilon_{\text{eff}} + 2\varepsilon_0} = f_1 \frac{\varepsilon_1 - \varepsilon_0}{\varepsilon_1 + 2\varepsilon_0} + f_2 \frac{\varepsilon_2 - \varepsilon_0}{\varepsilon_2 + 2\varepsilon_0} + f_3 \frac{\varepsilon_3 - \varepsilon_0}{\varepsilon_3 + 2\varepsilon_0} + \cdots, \tag{7}$$

where f_i and ε_i are the volume fraction and dielectric constant of the ith component and ε_0 is the dielectric constant of the host material. For the case of the Lorentz–Lorenz EMA, the host material is taken to be vacuum, $\varepsilon_0 = 1$. The Maxwell-Garnett EMAs are obtained by choosing one of the dielectric constants of the components to be the host medium. The Bruggeman EMA is obtained by assuming that the inclusions are embedded in an effective medium with $\varepsilon_0 = \varepsilon_{\text{eff}}$.

Other simple mixing rules used occasionally in various applications include volume averages of the dielectric constants,

$$\varepsilon_{\text{eff}} = f_1\varepsilon_1 + f_2\varepsilon_2 + \cdots, \tag{8}$$

the refractive indices, $m = \sqrt{\varepsilon}$,

$$\sqrt{\varepsilon_{\text{eff}}} = f_1\sqrt{\varepsilon_1} + f_2\sqrt{\varepsilon_2} + \cdots, \tag{9}$$

or the cube roots of the dielectric constants (Landau and Lifshitz, 1960),

$$\varepsilon_{\text{eff}}^{1/3} = f_1\varepsilon_1^{1/3} + f_2\varepsilon_2^{1/3} + \cdots. \tag{10}$$

In the case of time-dependent fields the response of the medium depends on how fast the changes of the field are. If the changes of the strength of the electric field are sufficiently slow compared to the time required for the response of the medium, the medium will be nearly in equilibrium with the electric field at any moment and the dielectric constant of the medium can be taken to be equal to the electrostatic dielectric constant. There is no difference between the medium response to the static or to a slowly changing time-dependent electric field. In this quasistatic region the relation represented by Eq. (1) remains valid. The extension of the electrostatic mixing rules into the quasistatic regime is accomplished by assuming that the mixing rules obtained from electrostatic considerations are valid also for slowly time varying fields.

III. FREQUENCY-DEPENDENT DIELECTRIC FUNCTION

When the changes of the electric field are fast compared to the time required for atoms and molecules of the medium to reach an equilibrium position and orientation with respect to the changing electric field, the electric displacement **D** will not be in phase with the electric field **E**. A simple linear relation between **D** and $\mathbf{E}$ as given by Eq. (1) will no longer be valid. Different physical processes will play a role for the case of a conductor as compared to an insulator.

A. IDEAL INSULATOR

An ideal insulator is characterized by vanishing conductivity, $\sigma = 0$. Let the time dependence of a harmonic electric field be given by

$$\mathbf{E}(t) = \mathbf{E}_o \cos \omega t. \tag{11}$$

When the induced polarization is not able to follow the changes in the electric field fast enough to stay in phase with it, the electric displacement vector will have a time lag with respect to the electric field vector. The displacement vector at any given time t will depend on the past values of the electric field and on the ability of a medium to respond to the time-varying electric field. The response of the medium is conveniently characterized by a time-delayed dielectric response function $\varepsilon(t - t')$ and the relation between $\mathbf{D}$ and $\mathbf{E}$ is written in the form (Bottcher and Bordewijk, 1978)

$$\mathbf{D}(t) = \int_{-\infty}^{t} \varepsilon(t - t')\mathbf{E}(t') \, dt' = \mathbf{E}_o \int_{-\infty}^{t} \varepsilon(t - t') \cos \omega t' \, dt'. \tag{12}$$

Substituting $\tau = t - t'$, the electric displacement vector of Eq. (12) can be expressed as

$$\mathbf{D}(t) = \mathbf{E}_o(\varepsilon_1 \cos \omega t + \varepsilon_2 \sin \omega t), \tag{13}$$

where the notation

$$\varepsilon_1(\omega) = \int_0^\infty \varepsilon(\tau) \cos \omega \tau \, d\tau \tag{14}$$

and

$$\varepsilon_2(\omega) = \int_0^\infty \varepsilon(\tau) \sin \omega \tau \, d\tau \tag{15}$$

is introduced.

We note that both the electric field $\mathbf{E}$ and the electric displacement $\mathbf{D}$ are real. However, the $\mathbf{D}$ field is out of phase with the $\mathbf{E}$ field. This manifests itself in a $\sin \omega t$ term in Eq. (13).

Alternatively this out-of-phase behavior can be characterized by a phase difference δ in an equivalent form

$$\mathbf{D}(t) = \mathbf{D}_o \cos(\omega t - \delta), \tag{16}$$

with

$$\cos \delta = \frac{E_0}{D_0} \varepsilon_1(\omega) \tag{17}$$

and

$$\sin \delta = \frac{E_0}{D_0} \varepsilon_2(\omega). \tag{18}$$

B. Absorption of Electromagnetic Radiation by a Perfect Insulator

The work δW done by one cycle of an electromagnetic radiation field per unit volume of a medium while increasing its electric displacement from $\mathbf{D}$ to $\mathbf{D} + \delta \mathbf{D}$, for the case of an isotropic and homogeneous medium, is given by (Jackson, 1975; Bottcher and Bordewijk, 1978)

$$\delta W = \frac{1}{4\pi} \int_{D(0)}^{D(T)} E \, dD, \tag{19}$$

where $T = 1/\nu$ is the time period of a cycle. Using Eq. (13), dD is obtained in the form

$$dD = \omega E_0(-\varepsilon_1 \sin \omega t + \varepsilon_2 \cos \omega t) \, dt. \tag{20}$$

Substituting Eqs. (13) and (20) into Eq. (19) and performing the integration, one finds that the work done by electromagnetic radiation per one cycle is

$$\delta W = \frac{1}{4} \varepsilon_2 E_0^2. \tag{21}$$

Finally, the work W done by an electromagnetic field on the medium per second is obtained by multiplying the last expression by a number of cycles per second, $\omega/2\pi$:

$$W = \frac{\omega}{8\pi} \varepsilon_2 E_0^2. \tag{22}$$

This work done by radiation on the medium is equal to the amount of radiation energy being absorbed by a unit volume of a medium per second.

We note that the absorption occurs even in perfect insulators because of a lag in the electric displacement with respect to an applied time-variable electric field. Both fields, $\mathbf{E}$ and $\mathbf{D}$, are real fields and the value of the material constant $\varepsilon_2(\omega)$ is fully determined by the time lag of the $\mathbf{D}$ field [Eq. (18)].

C. Complex Notation

Although the electric field and electric displacement are real field vectors, it is customary for mathematical convenience to introduce complex fields. This is achieved by adding a suitable imaginary part to the real fields so that the time dependence becomes exponential. If we choose $\exp(-i\omega t)$ for the time dependence, the electric field will have a form

$$\mathbf{E}(t) = \mathbf{E}_o e^{-i\omega t}, \tag{23}$$

with the understanding that only the real parts of $\mathbf{E}$ and $\mathbf{D}$ have physical significance. We refer to the real coefficient $\mathbf{E}_o$ in Eq. (23) as the electric field amplitude.

In analogy with the electrostatic relation given by Eq. (1), the complex form of the electric displacement and electric field can be related by

$$\mathbf{D}(t) = \varepsilon(\omega)\mathbf{E}_o e^{-i\omega t}, \tag{24}$$

with complex frequency-dependent dielectric function

$$\varepsilon(\omega) = \varepsilon_R(\omega) + i\varepsilon_I(\omega), \tag{25}$$

where $\varepsilon_R(\omega)$ and $\varepsilon_I(\omega)$ are the real and imaginary parts of the complex dielectric function $\varepsilon(\omega)$.

Finally, the real and imaginary parts of the electric displacement $\mathbf{D}$ can be written as

$$\operatorname{Re}\mathbf{D}(t) = \mathbf{E}_o(\varepsilon_R \cos \omega t + \varepsilon_I \sin \omega t) \tag{26}$$

and

$$\operatorname{Im}\mathbf{D}(t) = \mathbf{E}_o(\varepsilon_I \cos \omega t - \varepsilon_R \sin \omega t). \tag{27}$$

Comparing the real part of the complex dielectric displacement from Eq. (26) with the original real electric displacement in Eq. (13), we find

$$\varepsilon_R(\omega) = \varepsilon_1(\omega) = \int_0^\infty \varepsilon(\tau) \cos \omega \tau \, d\tau, \tag{28}$$

$$\varepsilon_I(\omega) = \varepsilon_2(\omega) = \int_0^\infty \varepsilon(\tau) \sin \omega \tau \, d\tau. \tag{29}$$

We emphasize that the physical fields themselves, both $\mathbf{E}(t)$ and $\mathbf{D}(t)$, are real. The complex notation is introduced only for mathematical convenience. In the low-frequency limit $\omega \to 0$ we have $\varepsilon_R = \varepsilon$ and $\varepsilon_I = 0$. Thus in the zero-frequency limit the complex dielectric function $\varepsilon(\omega)$ reduces to the electrostatic dielectric constant ε.

D. Absorption of Electromagnetic Radiation by a Conductor

If, due to free charges, the material has a nonzero conductivity σ an additional dissipation of energy will occur as a result of an electric current density

$$\mathbf{I}(t) = \sigma \mathbf{E}(t). \tag{30}$$

The amount of energy dissipated per unit volume of material per one cycle of an electromagnetic field with circular frequency ω is (Jackson, 1975)

$$\begin{aligned} \delta W &= \int_0^T I(t) E(t)\, dt \\ &= \sigma E_0^2 \int_0^T \cos^2 \omega t\, dt = \frac{\pi}{\omega} \sigma E_0^2. \end{aligned} \tag{31}$$

The electromagnetic energy dissipated as a result of a nonzero conductivity of a medium per unit volume and unit time is then

$$W = \frac{1}{2} \sigma E_0^2. \tag{32}$$

By summing the rates of electromagnetic energy absorbed because of the out-of-phase part of the electric displacement in the dielectrics [Eq. (22)] and because of a nonzero value of conductivity [Eq. (32)], we obtain the total rate of energy absorption as

$$W_T = \frac{\omega E_0^2}{8\pi} \left(\varepsilon_2 + \frac{4\pi\sigma}{\omega} \right). \tag{33}$$

Thus, we can formally account for the amount of energy dissipation resulting from the electric current density by adding the $4\pi\sigma/\omega$ term to ε_2. The material constants $\varepsilon_1(\omega)$, $\varepsilon_2(\omega)$, and σ are all real. When the conductivity-dependent term $4\pi\sigma/\omega$ is added to the imaginary part of the complex dielectric function $\varepsilon(\omega)$, Eq. (29) is modified to

$$\varepsilon_I(\omega) = \varepsilon_2(\omega) + \frac{4\pi\sigma}{\omega}. \tag{34}$$

IV. DYNAMIC EFFECTIVE MEDIUM APPROXIMATION

The effective dielectric function of a mixture of materials can be defined using a relation of a form similar to Eq. (2). For a harmonic field of circular frequency ω, we can write

$$\int_V \mathbf{D}(x, y, z, \omega)\, dx\, dy\, dz = \int_V \varepsilon(x, y, z, \omega)\mathbf{E}(x, y, z, \omega)\, dx\, dy\, dz$$

$$= \varepsilon_{\text{eff}}(\omega) \int_V \mathbf{E}(x, y, z, \omega)\, dx\, dy\, dz, \qquad (35)$$

where $\mathbf{E}(x, y, z, \omega)$ and $\mathbf{D}(x, y, z, \omega)$ denote the local electric field and displacement amplitudes at circular frequency ω and $\varepsilon(x, y, z, \omega)$ is the local complex dielectric function at the point (x, y, z) and frequency ω. Equation (35) can be supplemented by additional simplifying assumptions and used as a starting point for the analytical derivation of classical effective medium approximations. It can also be used in combination with a suitable numerical technique to find the solution of Maxwell's equations to determine numerically the effective dielectric constant of a given composite medium.

A. NUMERICAL APPROACH

For numerical evaluation we subdivide the considered volume V of the composite medium into N small volume elements ΔV_i in such a way that each volume element is filled by only one of the materials of a composite medium. We replace the electric field amplitude $\mathbf{E}(x, y, z, \omega)$ within each volume element by its average value $\mathbf{E}_i$ and replace the integral by the summation over all volume elements. Thus, at a given frequency ω we have

$$\varepsilon_{\text{eff}} \sum_{i=1}^N \mathbf{E}_i\, \Delta V_i - \sum_{i=1}^N \varepsilon_i \mathbf{E}_i\, \Delta V_i = 0. \qquad (36)$$

This vector equation comprises three separate scalar equations: one for each of the vector components of $\mathbf{E}$. Because we have assumed the composite medium to be isotropic, the three scalar equations are identical. Therefore, for an isotropic composite medium, we can apply Eq. (36) to just one of the vector components of $\mathbf{E}$ to obtain

$$\varepsilon_{\text{eff}} \sum_{i=1}^N E_i\, \Delta V_i - \sum_{i=1}^N \varepsilon_i E_i\, \Delta V_i = 0, \qquad (37)$$

where E_i stands for the considered component of an average electric field within a volume element ΔV_i. The summation is over all volume elements ΔV_i with $\sum \Delta V_i = V$.

For a composite medium consisting of n materials with dielectric constants $\varepsilon_1, \varepsilon_2, \ldots, \varepsilon_n$, we can write

$$\varepsilon_{\text{eff}} \left(\sum_{i=1}^{N_1} E_{1i} \Delta V_{1i} + \sum_{j=1}^{N_2} E_{2j} \Delta V_{2j} + \sum_{k=1}^{N_3} E_{3k} \Delta V_{3k} + \cdots \right)$$

$$= \varepsilon_1 \sum_{i=1}^{N_1} E_{1i} \Delta V_{1i} + \varepsilon_2 \sum_{j=1}^{N_2} E_{2j} \Delta V_{2j} + \varepsilon_3 \sum_{k=1}^{N_3} E_{3k} \Delta V_{3k} + \cdots, \quad (38)$$

where the summation index i runs over N_1 volume elements filled with material of dielectric constant ε_1, j runs over N_2 volume elements filled with material of dielectric constant ε_2, and so on. $E_{1i}, E_{2j}, \ldots$ are the corresponding average components of electric fields within volume elements $\Delta V_{1i}, \Delta V_{2j}, \ldots$.

Now, we use a suitable numerical method for solution of Maxwell's equations to calculate the average electric field in each of the volume elements and evaluate numerically the effective dielectric constant of a given n-component mixture:

$$\varepsilon_{\text{eff}} = \frac{\varepsilon_1 \sum_{i=1}^{N_1} E_{1i} \Delta V_{1i} + \varepsilon_2 \sum_{j=1}^{N_2} E_{2j} \Delta V_{2j} + \varepsilon_3 \sum_{k=1}^{N_3} E_{3k} \Delta V_{3k} + \cdots}{\sum_{i=1}^{N_1} E_{1i} \Delta V_{1i} + \sum_{j=1}^{N_2} E_{2j} \Delta V_{2j} + \sum_{k=1}^{N_3} E_{3k} \Delta V_{3k} + \cdots}. \quad (39)$$

The effective dielectric constant of a composite medium depends in general on all the details of all the components. It depends on the sizes and shapes of the individual inclusions and on the way the individual grains are distributed throughout the composite material. In solving numerically Maxwell's equations to obtain the electric fields **E** within each volume element, the electromagnetic boundary conditions must be satisfied on the interfaces between all grains as well as on the surface of the composite particle. Therefore, the effective dielectric constant of a composite particle generally depends also on the size and the shape of the composite particle itself.

The advantage of the numerical approach is that all known details of a composite medium can be taken into account in the evaluation of the effective dielectric constant. The obvious disadvantage is that each specific case must be considered separately. This requires a large amount of computer time and is impractical for many applications, particularly because some details of the particle's composition may be unknown. For the extension of such numerical results to heterogeneous particles in practice, additional assumptions concerning the boundary conditions and other details of the composition usually have to be made and averaging over a large number of possible distributions of grains within a composite material must be performed. A parameterization of the numerical results obtained in this way may be useful for a limited range of applications. However, because the details of the distribution of the materials in the mixture are often unknown, there is no guarantee that the numerical solution of the effective dielectric constant, using Eq. (39), will provide more accurate and useful data than some of the analytical mixing rules.

B. Mixture of Nonmagnetic Materials Remains Nonmagnetic

An effective magnetic permeability of a mixture μ_{eff} may be defined as a function of the distribution of the local magnetic permeability $\mu(x, y, z, \omega)$ by an equation analogous to Eq. (35). Replacing ε by μ and $\mathbf{E}$ by $\mathbf{H}$, we have

$$\int_V \mu(x, y, z, \omega)\mathbf{H}(x, y, z, \omega)\, dx\, dy\, dz$$

$$= \mu_{\text{eff}}(\omega) \int_V \mathbf{H}(x, y, z, \omega)\, dx\, dy\, dz. \tag{40}$$

In the special case of nonmagnetic individual components in a mixture, that is, if $\mu(x, y, z) = 1$ for all components, we always obtain $\mu_{\text{eff}} = 1$. That means that a mixture of nonmagnetic material remains nonmagnetic. This is to be expected when materials are treated on a macroscopic level, that is, when their magnetic and electric properties are characterized by bulk material constants and interface effects are negligible. Such a treatment seems to be appropriate for naturally occurring materials in atmospheric and geophysical applications.

It has been suggested (Bohren, 1986; Perrin and Lamy, 1990; Ossenkopf, 1991) that a definite effective magnetic permeability different from unity should be assigned to a mixture of nonmagnetic materials in order to obtain better agreement with the transmission and reflection of radiation by a plane-parallel layer consisting of composite particles. Because nonmagnetic components produce only a nonmagnetic composite particle according to Eq. (40), such an assignment is equivalent to using different definitions and different values of an effective dielectric constant for calculation of the transmission and the reflection by a layer of composite particles. As a practical tool for obtaining more accurate parameterizations this may be an acceptable procedure; however, an expectation that a mixture of nonmagnetic materials is really magnetic is not supported by theoretical or experimental evidence.

C. VANISHING FORWARD SCATTERING AMPLITUDE

The usual derivation of EMAs is based on electrostatic considerations of the electric field and electric displacement within the heterogeneous material (Maxwell-Garnett, 1904; Bruggeman, 1935; Aspens, 1982). After appropriate expressions for effective dielectric constants are derived, it is assumed that they are also valid at finite (nonzero) frequencies as long as the size of the inclusions is much smaller than the wavelength of radiation within the host material. This basic limitation of EMAs is usually expressed in the form of the size parameter x that relates the inclusion's characteristic dimension d to the wavelength of radiation in the host medium:

$$x = \frac{\pi d\, \text{Re}(m)}{\lambda} \ll 1, \tag{41}$$

where $\text{Re}(m)$ is the real part of the refractive index of the matrix (host material) and λ is the wavelength of the electromagnetic radiation in vacuum.

An alternative derivation of the classical EMAs can be obtained by imposing the requirement that the scattering amplitude S for scattering of radiation by spherical inclusions placed in an effective medium vanishes in the forward direction (Stroud and Pan, 1978; Niklasson *et al.*, 1981; Ossenkopf, 1991)

$$S_{11}(\mathbf{n}, \mathbf{n}) = S_{22}(\mathbf{n}, \mathbf{n}) = \frac{i\lambda}{2\pi} S(\Theta = 0) = 0, \tag{42}$$

where Θ is the scattering angle (Chapter 1). It has been shown (Stroud and Pan, 1978) that the definition of the effective dielectric constant given by Eq. (35) implies the vanishing of the forward scattering amplitude [Eq. (42)]. The advantage of starting with the $S(0) = 0$ condition over that of an electrostatic approach is appreciated when one considers generalizing one of the EMAs beyond the inclusion size limitation given by Eq. (41). This point will be expanded upon in the section dealing with extended effective medium approximations.

For simplicity, let us consider a heterogeneous material consisting of a binary mixture of grains of materials with complex dielectric constants ε_1 and ε_2 and volume fractions f_1 and f_2. For the case of spherical grains we use the Lorenz–Mie scattering formalism and explicitly write down the forward scattering amplitude for an individual grain embedded in an effective medium with an effective refractive index m_{eff}. We obtain (van de Hulst, 1957; Bohren and Huffman, 1983)

$$S(0) = \frac{1}{2} \sum_{n=1}^{\infty} (2n + 1)(a_n + b_n), \tag{43}$$

where a_n and b_n are the usual Lorenz–Mie scattering coefficients, which depend on the size of the inclusions and their relative refractive indices with respect to the medium in which they are embedded.

For particles in the Rayleigh scattering regime, all but the first Lorenz–Mie scattering amplitude a_1 can be neglected. The electric dipole contribution a_1 of a homogeneous sphere can be expanded in a power series of the size parameter. Keeping only the leading term, proportional to the third power of the expansion parameter, the scattering amplitude of grains of refractive indices m_1 and m_2, with the corresponding volume fractions f_1 and f_2 embedded in an effective medium with an effective dielectric constant ε_{eff}, can be written as a sum of contributions of individual grains.

Under these conditions, the vanishing forward scattering amplitude,

$$S(0) = -i\left(\frac{2\pi r}{\lambda}\right)^3 \varepsilon_{\text{eff}}^{3/2}\left[f_1 \frac{\varepsilon_1 - \varepsilon_{\text{eff}}}{\varepsilon_1 + 2\varepsilon_{\text{eff}}} + f_2 \frac{\varepsilon_2 - \varepsilon_{\text{eff}}}{\varepsilon_2 + 2\varepsilon_{\text{eff}}}\right] = 0, \qquad (44)$$

leads directly to the Bruggeman mixing rule of Eq. (6).

If the inclusions of the material of dielectric constant ε_1 and volume fraction f_1 are surrounded by a host (matrix) material of dielectric constant ε_2 and volume fraction f_2, we use a layered sphere model instead of a homogeneous one. We assume that the inhomogeneity embedded in an effective medium has the structure of a layered sphere with the host material forming the outer layer and the inclusion material forming the core. The volume fraction of the inclusions is simply the cube of the ratio of the core and shell radii. Neglecting all partial waves (a_n and b_n) except a_1 and keeping only the leading term in the size parameter expansion of a_1, we obtain the Maxwell-Garnett expression of Eq. (4) for the effective dielectric constant for inclusions of material ε_1 within a host of material ε_2. Similarly, the inverted Maxwell-Garnett [Eq. (5)] and the Lorentz–Lorenz mixing rules are also obtained from the condition of the vanishing forward scattering amplitude with the appropriate host and inclusion specifications.

D. Numerical Results

We limit the comparison of various mixing rules to the case of two-component mixtures, containing either water and ice or black carbon and water. These cases are of interest for remote sensing of melting snow or hail and for the evaluation of the effect of black carbon on the absorption of solar radiation by atmospheric aerosols and cloud droplets.

The Maxwell-Garnett rules provide extreme cases of the topological distribution (Fig. 1) of materials, where the grains of one of the components are completely isolated from each other. For the case of real dielectric constants, the Maxwell-Garnett mixing rules of Eqs. (4) and (5) provide the upper and the lower bounds on possible values of the effective dielectric constants of a two-component mixture. The random distribution of grains, as considered by the Bruggeman mixing rule given by Eq. (6) or volume averaging of dielectric constants and their

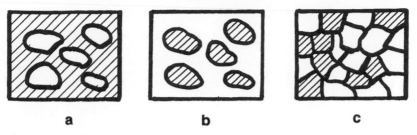

Figure 1   Schematic representation of a two-component mixture. The Maxwell-Garnett topology of white grains embedded in a shaded matrix (a) and of shaded grains embedded in a white matrix (b) are modeled in a dynamic EMA approach by layered spheres with an inclusion material forming a core and the matrix forming the outer layer. The Bruggeman topology (c) is modeled by an ensemble of homogeneous spheres.

square or cubic roots [Eqs. (8)–(10)], lead to effective dielectric constants that are always between the two extreme values given by the Maxwell-Garnett results.

The dependence of the effective dielectric constant on the topological distribution of the components is demonstrated in Figs. 2 and 3 for the case of a water ($\varepsilon_w = 42.6 + 41.3i$) and ice ($\varepsilon_i = 3.2 + 0.01i$) mixture at $\lambda = 3.17$ cm. The numerical results suggest that the effective dielectric constant of the mixture, and especially its imaginary part, is highly sensitive to the topological distribution of its components. For a wide range of volume fractions the dependence of the effective dielectric constant on the topological structure, demonstrated by the differences between the two Maxwell-Garnett curves, is stronger than the dependence on the volume fractions of the components.

A study of the liquid water distribution in hailstones (Chýlek *et al.*, 1984b) suggests that the water and ice distribution does not follow any of the basic topological patterns of Fig. 1 exactly. In certain regions liquid water is confined into isolated droplets separated by an ice matrix, whereas water in other regions forms a maze of connected continuous veins and sheets. Apparently, some fraction of liquid water could be described by the Maxwell-Garnett topology with isolated water droplets within an ice matrix, whereas the remaining part of the liquid water distribution resembles isolated ice grains within a water matrix or a random distribution of water and ice. This situation may be typical for the general structure of a water and ice mixture. It suggests that a volume average of the Bruggeman and the Maxwell-Garnett results (Chýlek *et al.*, 1991) may be a more realistic approximation than the assumption of the validity of one or the other mixing rule.

The dielectric constants of water and black carbon at visible wavelengths are closer to each other than water and ice dielectric constants at centimeter wavelengths. Consequently, the differences between the two versions of the Maxwell-Garnett, Bruggeman, and other mixing rules are not so large for this case.

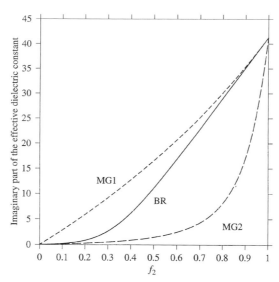

Figure 2   Imaginary part of the effective dielectric constant as a function of the liquid water volume fraction for the case of an ice–water mixture at a wavelength of $\lambda = 3.17$ cm. The Maxwell-Garnett topology with ice inclusions in a water matrix (MG1) leads to considerably higher values of the imaginary part of ε_{eff} than the Maxwell-Garnett topology with the water inclusions in an ice matrix (MG2). The Bruggeman results (BR) are close to the Maxwell-Garnett values for water inclusions in an ice matrix (MG2) at low liquid water volume fractions and approach the Maxwell-Garnett values for ice inclusions in a water matrix (MG1) at high liquid water volume fractions.

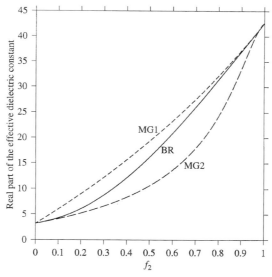

Figure 3   Same as Fig. 2 but for the real part of the effective dielectric constant.

E. GRAIN SIZE LIMITATION OF EFFECTIVE MEDIUM APPROXIMATIONS

All derivations of the classical EMAs beginning with electrostatic considerations contain explicitly or implicitly the requirement that the electric field is uniform throughout the inclusion. In the framework of the dynamic approximation and application to time-varying fields, this leads to the basic condition that the inclusion's size parameter (in the material of the host medium) must satisfy $x \ll 1$. Similar grain size parameter restrictions are obtained from the $S(\Theta = 0) = 0$ approach. Considering a host material of refractive index 1.33 at a visible wavelength of 0.55 μm and taking $x = 0.1$ as an upper limit on x that satisfies $x \ll 1$, we get a maximum allowable inclusion radius of 0.0066 μm. Obviously, the grain radius in many geophysical applications is well beyond the limit of 0.0066 μm. Restricting oneself to the formal requirement used in the derivations of the EMAs would make these approximations of little use for atmospheric and other geophysical applications.

However, it has been noticed that in many cases both the Bruggeman and the Maxwell-Garnett EMAs provide accurate results (within 5% error) up to an inclusion size parameter around 0.5 (Niklasson and Granqvist, 1984). According to the conditions under which the approximations were derived, there is no reason to expect them to provide accurate results at this size of inclusions. It seems that the EMAs can provide reasonable results beyond the specified limit set by their derivations. It should be remembered that the conditions and limitations used in the derivations of these approximations are sufficient for them to be valid, but they are not necessary. Thus, it may eventually be possible to derive these approximations using less restrictive conditions, although nobody has yet succeeded in doing so.

Defining the limits of applicability of EMAs is a nontrivial task even for spherical inclusions, and there is always a danger of two extremes. One is the potential use of these approximations without any restrictions, with their application being so far beyond appropriate limits that the results are meaningless. The other is the advocacy of a very restrictive use of EMAs limited to the cases of slab transmission through heterogeneous media with inclusion sizes satisfying the criterion $x \ll 1$. The latter stance would essentially eliminate the use of EMAs from most applications in the field of atmospheric science.

Our approach to the range of applicability of EMAs is based on practical empiricism. Comparison of effective medium approximation results with those of more accurate numerical methods when available, such as the discrete dipole approximation (DDA, Chapter 5), the finite difference time domain method (Chapter 7), or the exact solution of scattering and absorption by spherical particles containing one or more arbitrarily located spherical inclusions (Chapter 8), can provide guidance on the range of applicability. In specific cases, depending on the desired accuracy of the optical characteristics, one may determine how far the $x = \pi d \, \text{Re}(m)/\lambda \ll 1$ limit may be extended.

In addition to the issue of accuracy for a given application, there is also a question with respect to which optical properties the EMAs can be applied. One school of thought claims that EMAs and their corresponding effective refractive indices can be applied only to problems of transmission through a heterogeneous slab of suspended particles, but not to other optical properties of interest such as absorption or scattering cross sections. In atmospheric science applications we are interested in applying EMAs to problems of composite aerosol particles or contaminated cloud droplets. We need to calculate a minimum set of single-scattering parameters required in solving the radiative transfer equation. For the simplest approximations used in radiative transfer problems, this set consists of at least the extinction cross section, single-scattering albedo (ratio of the scattering to extinction cross section), and asymmetry parameter $\langle \cos \Theta \rangle$ characterizing the differential scattering cross section [i.e., the $(1, 1)$ element of the phase matrix; see Chapter 1].

The classical EMAs can be derived in such a way that their effective refractive index can be treated on an equal footing with the refractive index of a homogeneous particle only for the case when both the host particle and the inclusions are in the Rayleigh scattering limit. However, as mentioned before, this does not imply that the Rayleigh scattering limit is a necessary condition for the region of applicability of these approximations; it is only a sufficient condition. We demonstrate in the following sections that the EMAs provide accurate results, within a few percent, for absorption and differential scattering cross sections for cases of geophysical interest, even when both the host particle and the inclusions are well outside the Rayleigh scattering limit.

V. EXTENDED EFFECTIVE MEDIUM APPROXIMATIONS

The basic goal of extended effective medium approximations (EEMAs) is to obtain generalizations of the classical approach with less restrictive requirements on the inclusion size (Stroud and Pan, 1978; Wachniewski and McClung, 1986; Grimes, 1991). The path to such approximations is the use of the vanishing forward scattering amplitude $S(\Theta = 0) = 0$ without the restriction of the Rayleigh scattering limit.

One of the first approximations of this type, proposed by Stroud and Pan (1978), includes an electric dipole term, the usual Rayleigh approximation, and adds a magnetic dipole term to account for the induced magnetic dipole moment in metallic particles. The authors emphasize that their EEMA, developed for highly absorbing inclusions, is not restricted to inclusions much smaller than the wavelength of the radiation in the host material. Chýlek and Srivastava (1983) further modified the EEMA approach for larger spherical highly absorbing grains by keeping all terms contributing to the Mie-type series expansion of the forward scattering amplitude. Wachniewski and McClung (1986) modified the approach and included larger nonspherical inclusions.

All versions of the EEMA have been constructed under an explicit or implied assumption that at least one of the components is highly absorbing. The extension of inclusion sizes up to the order of the wavelength is heuristically justified by the fact that electromagnetic waves or photons usually cannot distinguish details of a medium that are smaller than about one half of the wavelength (this would correspond to the size parameter of inclusion up to $x \leqslant \pi/2$, instead of $x \ll 1$). Bohren (1986) criticized the use of EEMAs and pointed out that when used outside their intended region of application, for example, in the case when none of the components is absorbing, the EEMAs may lead to incorrect results. However, this is not the intended domain of applications, as previously stated (Chýlek and Srivastava, 1983).

Stroud and Pan (1978) have suggested that the EEMA should be applicable to the case of absorbing inclusions as long as $(\mathrm{Im}\, k_{\mathrm{eff}})^{-1}$ is large compared to the characteristic size d of inclusions. The effective wavenumber is $k_{\mathrm{eff}} = 2\pi m_{\mathrm{eff}}/\lambda$, where $m_{\mathrm{eff}} = \mathrm{Re}(m_{\mathrm{eff}}) + i\,\mathrm{Im}(m_{\mathrm{eff}})$ is a complex effective refractive index. Thus, we have a new, less restrictive, condition on the allowable size of inclusions in the form

$$\frac{2\pi d\,\mathrm{Im}(m_{\mathrm{eff}})}{\lambda} \ll 1. \tag{45}$$

For practical applications in atmospheric and geophysical sciences involving strongly absorbing materials such as black carbon at visible and infrared (IR) wavelengths or liquid water at millimeter and centimeter wavelengths, it is useful to rewrite this condition using the size parameter, $x = \pi d\,\mathrm{Re}(m_{\mathrm{eff}})/\lambda$, of inclusions within an effective medium:

$$\frac{2x\,\mathrm{Im}(m_{\mathrm{eff}})}{\mathrm{Re}(m_{\mathrm{eff}})} \ll 1. \tag{46}$$

If we allow the size parameter to be of order 1, the condition of applicability of the EEMA can still be satisfied, as long as the imaginary part of the effective refractive index, $\mathrm{Im}(m_{\mathrm{eff}})$, remains small, satisfying the relation

$$\frac{2\,\mathrm{Im}(m_{\mathrm{eff}})}{\mathrm{Re}(m_{\mathrm{eff}})} \ll 1. \tag{47}$$

If the real part of the refractive index of the host medium is not much larger than 2, the last condition can be written simply as

$$\mathrm{Im}(m_{\mathrm{eff}}) \ll 1. \tag{48}$$

Thus, for modest concentrations of highly absorbing inclusions such that $\text{Im}(m_{\text{eff}}) \ll 1$, the EEMAs allow us to extend the range of applicability of the approximations to inclusion size parameters of order 1.

We recall that the starting point in the development of the formalism of the EEMAs is the requirement of a vanishing forward scattering amplitude $S(\Theta = 0) = 0$. Because we wish to develop an approximation that is applicable to highly absorbing inhomogeneities with size parameters on the order of unity, we can no longer include only the electric and magnetic dipole contributions to the scattering amplitudes. Once we consider such larger inclusions, we must include in the forward scattering amplitude all contributing electric and magnetic multipoles.

If we assume the spherical inclusions described by a distribution $n(r)$ of their radii, the forward scattering amplitude can be readily written using the Lorenz–Mie scattering formalism. The effective refractive index of a heterogeneous medium is obtained from the $S(\Theta = 0) = 0$ condition, which leads (Chýlek and Srivastava, 1983) to an iterative scheme to solve for the effective refractive index,

$$\left(m_{\text{eff}}^2\right)_{k+1} = m_1^2 \frac{A_k(1 - f) + f B_k}{A_k(1 - f) - 2 f B_k}, \tag{49}$$

where

$$A_k = i \frac{12\pi^2}{\lambda^3} \left(m_{\text{eff}}^3\right)_k \tag{50}$$

and

$$B_k = \int \sum_{n=1}^{\infty} (2n + 1) \left\{ a_n[r, m_2/(m_{\text{eff}})_k] + b_n[r, m_2/(m_{\text{eff}})_k] \right\} n(r) \, dr, \tag{51}$$

with f being the volume fraction occupied by the highly absorbing inclusions. The topological structure corresponding to the Bruggeman EMA has been assumed. However, the symmetry typical of the Bruggeman EMA with respect to the interchange of materials has been broken by the assumption that the material of refractive index m_1 is divisible into grains that are much smaller than the wavelength (the Rayleigh scattering region), whereas the size distribution of grains of material with refractive index m_2 is prescribed by the size distribution $n(r)$ that may contain grains with size parameters of order 1. Consequently, the mixing rule defined by the set of Eqs. (49)–(51) is not symmetric under the interchange of m_1, f_1 and m_2, f_2.

When the size parameter x of the grains of material m_2 is much smaller than 1, the EEMA reduces to the classical Bruggeman EMA. The Bruggeman effective medium dielectric constant may be conveniently used as an initial guess for A_k and B_k in the iterative process.

It is not easy to determine how accurately the effective refractive index obtained using the EEMA represents the optical properties of the composite medium. We also do not know to what optical properties such an index can be applied (Bohren, 1986; Perrin and Lamy, 1990; Ossenkopf, 1991; Chýlek and Videen, 1998). Is it only transmission through a slab of particles that can be described by an effective refractive index? Or can we also use it to calculate the absorption by a composite medium? And what about the differential scattering cross section, or phase function? One may make a strong case for the pragmatic point of view that the region of applicability of the effective and the extended effective medium approximations depends on the required accuracy, provided that the basic conditions of the derivation are satisfied.

If both the size of the inclusions and the size of a host particle are much smaller than the wavelength, then the effective refractive index of a composite particle can be used in the same way as the ordinary refractive index of a homogeneous material. That is, it can be used to calculate the absorption, scattering, extinction, and differential scattering cross sections. However, as the size of the inclusions increases, the expected accuracy of the individual optical characteristics is less obvious.

Scattering by larger particles is dominated by a forward scattering peak. The requirement that the scattering amplitude vanishes in the forward direction for inclusions placed in an effective medium significantly reduces, but does not eliminate, the scattering. The scattering is further reduced by strongly absorbing inclusions (the EEMA constructed for the case of a highly absorbing medium is not intended and cannot be expected to work for nonabsorbing grains in a nonabsorbing matrix). The accuracy of the extended as well as the ordinary effective medium approximations can be tested only for a few cases when the results can be compared with more accurate solutions or experimental measurements. In the following section the differential scattering cross section and the absorption, scattering, and extinction cross sections are obtained using different effective medium approximations and are compared with those obtained using the DDA, the exact solution of the scattering problem of a spherical host containing an arbitrarily located spherical inclusion, and laboratory measurements.

VI. COMPARISON WITH OTHER APPROXIMATIONS, MODELS, AND MEASUREMENTS

The question of the applicability of EMAs for predicting scattering and absorption cross sections and angular scattering characterization is a question of the degree of the required accuracy. If the required accuracy is on the order of tenths of a percent, then EMAs cannot be used to obtain the absorption and angular scattering patterns. However, in most geophysical situations such a high degree of accuracy is not even meaningful. Information about the mass, shapes, composition, and microstructure of composite particles is often quite meager. Consequently, a prediction of their optical characteristics that is accurate to within a few percent is acceptable.

To establish the accuracy of the optical properties of individual heterogeneous particles obtained using EMAs, we compare the EMA predictions with (1) laboratory measurements of the refractive index of a composite material, (2) laboratory measurements of the differential scattering cross section of a spherical composite particle, (3) numerical results obtained using a model of an arbitrarily located spherical inclusion within a host sphere, and (4) numerical results obtained using the DDA technique.

As the first application, we consider scattering from a heterogeneous sphere at centimeter wavelengths, where we compare the EMA results with experimental measurements and with calculations for a simplified model. The model consists of a host sphere containing only one arbitrarily located spherical inclusion. Model results are averaged over several hundred thousand inclusion locations within the host sphere to obtain the scattering cross section.

As a second example, multiple black carbon particles are considered within a water host sphere. The DDA is used to establish the accurate results with which the EMAs are compared.

A. SCATTERING AND ABSORPTION AT MICROWAVE WAVELENGTHS: LABORATORY MEASUREMENTS

Test materials consisting of water droplets embedded in an acrylic matrix were prepared with two liquid water fractions of $f_W = 1.6\%$ and 2.7%. The refractive indices of acrylic and water at the wavelength $\lambda = 3.17$ cm are $m_A = 1.686$ and $m_W = 7.7 + 2.48i$. The large difference in optical properties of the materials provides a suitable test for EMAs, because under these circumstances we can expect large differences between the results of individual EMAs. The refractive index of the mixtures was measured using the waveguide attenuation method (Chýlek *et al.*, 1988). The obtained results are $m_1 = 1.72 + 0.011i$ and $m_2 = 1.75 + 0.013i$ for the 1.6% and 2.7% water volume fractions, respectively.

The mean water inclusion radius was estimated using an optical microscope to be $r_W = 0.02$ cm. This corresponds to a mean size parameter of $x_W = 0.07$ in the acrylic medium. In this region one may expect accurate EMA results if an appropriate EMA describing the topological distribution of water droplets in an acrylic matrix is used.

Table I

Values of the Real and Imaginary Parts of the Effective Refractive Index of a Water–Acrylic Mixture with Water Volume Fraction f_W Determined Using the Experimental Waveguide Method (Exp.), EEMA, Bruggeman EMA (BR), Maxwell-Garnett EMA for Water Inclusions in an Acrylic Matrix (MG-1), Maxwell-Garnett EMA for Acrylic Inclusions in a Water Matrix (MG-2), and Volume Averages of Refractive Indices ($\langle m \rangle$), Dielectric Constants ($\langle \varepsilon \rangle$), and the Third Root of Dielectric Constants ($\langle \varepsilon^{1/3} \rangle$). The $\delta \operatorname{Re}(m)$ and $\delta \operatorname{Im}(m)$ Are Relative Errors of Individual EMA Results with Respect to Experimental Measurements of the Refractive Index m

f_W	m	Exp.	EEMA	BR	MG-1	MG-2	$\langle m \rangle$	$\langle \varepsilon \rangle$	$\langle \varepsilon^{1/3} \rangle$
1.6%	$\operatorname{Re}(m)$	1.72	1.72	1.72	1.72	1.85	1.78	1.92	1.76
	$\operatorname{Im}(m)$	0.011	0.010	0.009	0.009	0.116	0.045	0.165	0.031
	$\delta \operatorname{Re}(m)$		< 1%	< 1%	< 1%	7.5%	3.5%	11%	2.3%
	$\delta \operatorname{Im}(m)$		< 9%	18%	18%	950%	310%	1400%	180%
2.7%	$\operatorname{Re}(m)$	1.75	1.75	1.75	1.75	1.95	1.85	2.06	1.81
	$\operatorname{Im}(m)$	0.013	0.011	0.011	0.011	0.183	0.073	0.254	0.048
	$\delta \operatorname{Re}(m)$		< 1%	< 1%	< 1%	12%	5.6%	18%	3.4%
	$\delta \operatorname{Im}(m)$		< 15%	15%	15%	1300%	460%	1800%	270%

The effective refractive index of the mixture was calculated using several different EMAs (Table I). At low water volume fractions the individual water droplets are isolated from each other by the acrylic matrix. This suggests the use of the Maxwell-Garnett mixing rule for water inclusions within an acrylic matrix. Also the Bruggeman mixing rule and the EEMA should provide the most accurate results, because these approximations are based on the topological distribution of inclusions that is, at low water volume fractions, in agreement with the actual distribution of water droplets within the acrylic matrix.

As expected, the Bruggeman, the Maxwell-Garnett (for water inclusions in an acrylic matrix), and the EEMA provided results much closer to the measured values than other EMAs. The real part of the refractive index given by these three approximations is within 1% of the measured value. Although the relative error in the imaginary part of the refractive index is up to 18%, the absolute error remained small, within 0.002 of the measured value (Table I).

The results of the other EMAs (including the Maxwell-Garnett approximation for acrylic inclusions in a water matrix) are much worse. The errors in the real part of the refractive index are up to 18% and the imaginary part of the refractive index has errors up to a factor of 18 (1800%). This clearly indicates that the most important criterion for the use of a specific EMA is the topological distribution of individual grains.

Table II

Values of the Extinction, Scattering, and Absorption Efficiencies (Q) of a Water–Acrylic Mixture with Water Volume Fraction f_W Calculated Using the Effective Refractive Index Determined by the Experimental Waveguide Method, EEMA, Bruggeman EMA, Maxwell-Garnett EMA for Water Inclusions in an Acrylic Matrix (MG-1), Maxwell-Garnett EMA for Acrylic Inclusions in a Water Matrix (MG-2), and Volume Averages of Refractive Indices ($\langle m \rangle$), Dielectric Constants ($\langle \varepsilon \rangle$), and the Third Root of Dielectric Constants ($\langle \varepsilon^{1/3} \rangle$). The δQ Are Relative Errors of Individual EMA Results with Respect to the Calculation Using Experimental Measurements of the Refractive Index m

f_W	Q	Exp.	EEMA	BR	MG-1	MG-2	$\langle m \rangle$	$\langle \varepsilon \rangle$	$\langle \varepsilon^{1/3} \rangle$
1.6%	Q_{ext}	2.46	2.47	2.47	2.46	2.75	2.74	2.62	2.59
	Q_{sca}	2.09	2.14	2.14	2.14	1.52	1.81	1.37	1.81
	Q_{abs}	0.37	0.33	0.33	0.32	1.20	0.93	1.25	0.78
	δQ_{ext}		< 1%	< 1%	< 1%	12%	11%	6%	5%
	δQ_{sca}		< 3%	< 3%	< 3%	24%	13%	34%	13%
	δQ_{abs}		11%	11%	13%	220%	150%	240%	110%
2.7%	Q_{ext}	2.50	2.50	2.49	2.49	2.57	2.82	2.51	2.80
	Q_{sca}	2.09	2.12	2.12	2.13	1.32	1.78	1.26	1.83
	Q_{abs}	0.41	0.37	0.37	0.36	1.25	1.03	1.25	0.97
	δQ_{ext}		< 1%	< 1%	< 1%	3%	13%	< 1%	12%
	δQ_{sca}		< 2%	< 2%	< 2%	37%	15%	40%	12%
	δQ_{abs}		10%	10%	12%	200%	150%	200%	140%

A key question is whether EMAs can be used only for extinction, or whether they can also describe the other scattering characteristics. To address this question, we consider a spherical particle of a radius $r = 3.17$ cm made from a water–acrylic mixture. Table II shows the calculated extinction, scattering, and absorption efficiencies (i.e., respective cross sections normalized by geometrical cross sections, denoted by Q_{ext}, Q_{sca}, and Q_{abs}) using the measured refractive index of the mixture and using the effective refractive indices obtained from the EMAs considered. The EMAs with the correct topology describing isolated water droplets within an acrylic matrix (Bruggeman, EEMA, and Maxwell-Garnett for water inclusions in acrylic matrix) provide extinction values with an error smaller than 1%. The scattering cross sections have only slightly poorer accuracy. The absorption cross section, which is calculated as the difference between the extinction and scattering cross sections, has larger errors, up to 13%. The other EMAs show errors in the scattering and extinction cross sections up to 40% and in the absorption cross section up to 240%.

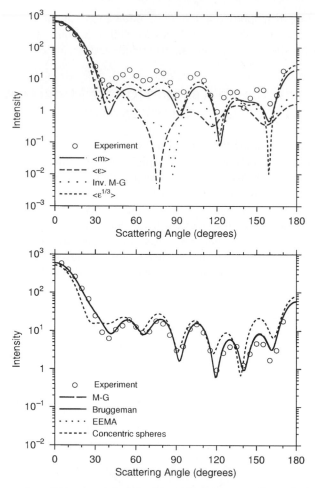

Figure 4   Comparison of the measured scattered intensity for the case of water inclusions (volume fraction 1.6×10^{-2}) in an acrylic matrix at $\lambda = 3.17$ cm with various EMA calculations. Only the Bruggeman EMA, the EEMA, and the Maxwell-Garnett EMA for water inclusions in an acrylic matrix provide acceptable agreement with measurements.

The EMAs with the correct topology can also provide a reasonably accurate description of the angular scattering. The measured differential scattering cross section (Chýlek *et al.*, 1988) is compared with the Lorenz–Mie calculations using the effective refractive index provided by individual EMAs (Fig. 4). A reasonable agreement is again obtained using the Bruggeman, EEMA, and the proper form of the Maxwell-Garnett approximation; these three approximations give essentially identical results. The other mixing rules as well as a concentric coated sphere model (with acrylic core and water shell) provide less accurate results.

B. SCATTERING AND ABSORPTION AT MICROWAVE WAVELENGTHS: MODEL CALCULATIONS

Next we consider a model of a composite particle consisting of one arbitrarily located spherical inclusion within a spherical host. This is one of the cases described in Chapter 8. We use the computer code developed by Gorden Videen (Videen *et al.*, 1995a, b; Ngo *et al.*, 1996).

The case of a spherical host particle containing only one or very few inclusions is of practical interest in atmospheric science. Such a composite particle may be formed by condensation of matrix material onto a partially nonsoluble condensation nucleus. The position of the inclusion within a host sphere is specified by the distance d and the illumination incidence angle α as shown in Fig. 5. The

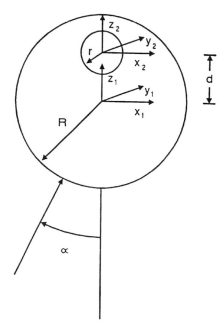

Figure 5 Geometry of a simple model consisting of an arbitrarily positioned spherical inclusion of radius r within a host sphere of radius R. The location of the inclusion within the host is specified by the distance d and illumination incidence angle α.

absorption and the angular distribution of the scattered radiation depend on the exact position of the inclusion within the host sphere as well as the inclusion size.

The wavelength of the incoming radiation is taken to be $\lambda = 3.17$ cm, corresponding to that of the laboratory measurements. The radius of the host sphere is $R = 3.17$ cm and the matrix refractive index is $m_A = 1.686$. The refractive index of the inclusion is $m_W = 7.70 + 2.48i$ and the inclusion radius varies depending on the specified water volume fraction. The specified refractive indices are those of acrylic and water at the wavelength $\lambda = 3.17$ cm (Chýlek *et al.*, 1988; Chýlek and Videen, 1998).

To compare the results of the EMAs with the exact model solution, we consider inclusion material volume fractions ranging from 10^{-5} to 10^{-1}, which correspond to inclusion radii varying from 0.068 to 1.471 cm, or size parameters of 0.23 to 4.9 in the host medium. The numerical results are averaged over 810,000 random inclusion locations within the host. Such a large number of individual locations is needed to keep the standard deviation of the calculated averages of the absorption, scattering, and extinction cross sections below 0.5%.

The absorption, scattering, and extinction efficiencies are plotted as a function of the inclusion volume fraction in Fig. 6. The results obtained using the six selected EMAs are compared with the exact model calculations. Because we have only one inclusion in the host sphere, the volume fraction is directly related to the inclusion radius and size parameter.

At an inclusion volume fraction of 10^{-5}, the inclusion radius is 0.068 cm and its size parameter with respect to the wavelength in the host medium $x = 2\pi r m_A/\lambda = 0.23$. According to the small particle limit used in the derivation of the Bruggeman and Maxwell-Garnett formulas, the value $x = 0.23$ can be considered a little over the upper limit of allowable inclusion sizes (the condition $x \ll 1$ is used in the derivations). Nevertheless, in the region $x < 0.23$ the Bruggeman, the Maxwell-Garnett, and the EEMA formulas do provide accurate values within 1% for the scattering and extinction cross sections.

At an inclusion volume fraction of 10^{-4}, the inclusion size parameter with respect to the wavelength of radiation in the host sphere is $x = 0.49$. At this value, the condition $x \ll 1$ is definitely not satisfied and therefore there is no reason to expect the Bruggeman and the Maxwell-Garnett EMAs to provide accurate values for the extinction. The application of EMAs to the scattering cross section at this size parameter could also be questioned. In spite of that the scattering and extinction cross sections calculated using the Bruggeman and Maxwell-Garnett approximations and the EEMA are still within 1% of the exact value obtained by the solution of Maxwell's equations for the boundary conditions of a sphere with an arbitrarily located spherical inclusion. It is only the averaging of dielectric constants and refractive indices and the inverted Maxwell-Garnett formula (with the materials of matrix and inclusions interchanged) that deviate significantly from the correct values. The differential scattering cross section is also reproduced

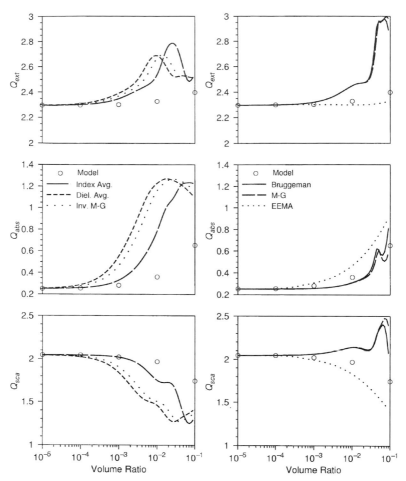

Figure 6 Extinction, absorption, and scattering efficiencies at a wavelength of $\lambda = 3.17$ cm as a function of the water volume fraction in the acrylic matrix are calculated using different EMAs and compared to the model results for an arbitrarily located spherical water inclusion within a host acrylic sphere.

very accurately up to size parameter $x = 0.5$ by the classical Bruggeman and Maxwell-Garnett approximations, as well as by the EEMA.

For larger size parameters the errors of the Bruggeman and Maxwell-Garnett approximations start to increase, indicating the breakdown of the approximations. With inclusion size parameters in the range, $1 < x < 2$, typical errors in the scattering and extinction cross sections are between 5 and 15%. The extended effective medium approximation predicts scattering and extinction cross sections that are within 2% of the exact value up to the size parameter $x = 1$ and within 8% up to $x = 2$.

The absorption cross section is calculated as a difference between the extinction and scattering cross sections and has considerably larger errors (up to 30% around $x = 1$) and the use of EMAs to calculate absorption by water and acrylic mixtures at centimeter wavelength is questionable.

The differential scattering cross section is reasonably well reproduced by the EMAs (Fig. 7). The larger differences between the model and the EMAs occur

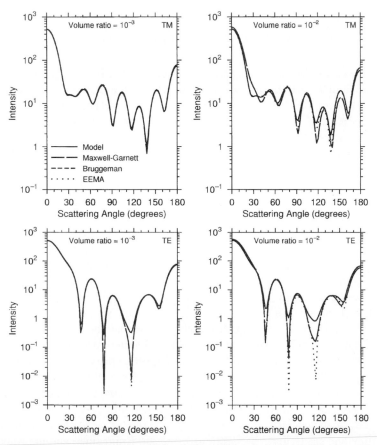

Figure 7 Angular scattered intensity (TM and TE modes) calculated using the Bruggeman and Maxwell-Garnett EMAs and the EEMA for the case of water inclusions within an acrylic host sphere compared to model results for an arbitrarily located spherical water inclusion within a host acrylic sphere. The model results are averaged over 810,000 different inclusion positions within the host sphere. The wavelength is $\lambda = 3.17$ cm.

only close to the deep minima of the cross section. These differences are really an artifact of the model calculations, where the averages are taken over 810,000 individual inclusion locations. Such averaging will necessarily smooth out any deep dips occurring in the differential cross section for a fixed location of an inclusion. On the other hand, the EMA results for a sphere with a single effective dielectric constant will retain these deep minima always seen for Mie scattering by monodisperse spheres. Although the relative errors are large at the minima (Fig. 8), the average cosine of the scattering angle (asymmetry parameter) and the single-scattering albedo are generally within 5 to 10% of the model values (Fig. 9).

C. BLACK CARBON IN WATER DROPLETS

For the case of black carbon particles inside cloud or haze droplets at a visible wavelength of $\lambda = 0.55$ μm, we take the refractive indices of water and black carbon to be $m_1 = 1.33$ and $m_2 = 1.75 + 0.44i$, respectively. The real and imaginary parts of the effective refractive indices were obtained using the selected EMAs for several black carbon volume fractions.

The extinction, scattering, and absorption cross sections have been calculated using the DDA (Draine, 1988; Draine and Flatau, 1994) that we have adapted for treatment of heterogeneous particles (a configuration of 48^3 dipoles was used). The values obtained using the DDA for various sizes of black carbon inclusions are considered to be the correct values and the results of individual EMAs are compared to those values to determine their accuracy. The size parameter of black carbon inclusions within an $R = 0.35$-μm-radius host water droplet is varied between $x = 0.33$ and 1.44. The size of black carbon grains is related to the number of dipoles n across the diameter of a spherical inclusion (Table III).

The results obtained using the volume averages of the refractive indices and dielectric constants or $\varepsilon^{1/3}$ and using the "inverted" Maxwell-Garnett approximations are poor (not shown), considerably worse than those obtained using the Bruggeman, Maxwell-Garnett, and extended effective medium approximations. The results of the Bruggeman and Maxwell-Garnett approximations are generally close to each other. To illustrate graphically the differences between the DDA and the EMAs, we present plots only for the Maxwell-Garnett (black carbon embedded in water matrix) and the extended effective medium approximations (Fig. 10).

At the size parameter $x = 2\pi r \, \mathrm{Re}(m_{\mathrm{eff}})/\lambda = 0.33$ ($n = 3$), the accuracy of the extinction and scattering cross sections using the Bruggeman and the Maxwell-Garnett EMAs is between 1 and 2% (Fig. 10). At the same size parameter the accuracy of Q_{ext} and Q_{sca} of the EEMA is better than 0.5%.

At the size parameter $x = 0.55$ ($n = 5$), the errors of the Q_{ext} and Q_{sca} using the Bruggeman and the Maxwell-Garnett EMAs are around 4%, whereas

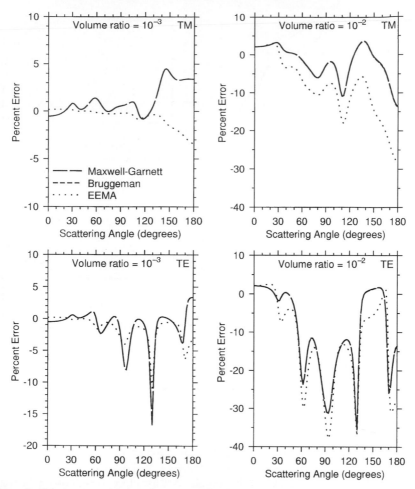

Figure 8   Relative error of the angular scattered intensity (TM and TE modes) calculated using the Bruggeman and Maxwell-Garnett EMAs and the EEMA for the case of water inclusions within an acrylic host sphere. The model results for an arbitrarily located spherical water inclusion within the host acrylic sphere are considered to be the correct values.

the errors of the EEMA remain below 1%. At the largest size parameter $x =$ 1.44 ($n = 13$), the errors of the Bruggeman and Maxwell-Garnett EMA–derived extinction and scattering cross sections are between 10 and 12%, whereas the EEMA errors remain between 3 and 4%. There is a significant improvement in the accuracy of the calculated extinction and scattering cross sections when the

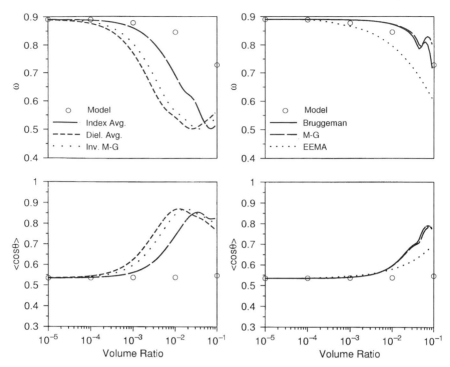

Figure 9 The single-scattering albedo ϖ and the asymmetry parameter $\langle \cos \Theta \rangle$ at a wavelength of $\lambda = 3.17$ cm as a function of the water volume fraction in an acrylic matrix are calculated using different EMAs and compared to model results for an arbitrarily located spherical water inclusion within the host acrylic sphere.

classical effective medium approximations are replaced by the extended effective medium approximation at size parameter values $x \geqslant 0.5$.

When the extinction and scattering efficiencies are close to each other, the relative error in absorption will be large. Consequently, the absorption efficiency cal-

Table III

Size Parameter $x = 2\pi r m_1/\lambda$ and Number of Inclusions N for Black Carbon Volume Fraction of 0.1 and Given Number of Dipoles n across the Diameter of an Inclusion. The Refractive Index of the Water Matrix at $\lambda = 0.55$ μm Is Taken to Be $m_1 = 1.33$

n	3	4	5	6	7	8	9	10	11	12	13
x	0.33	0.44	0.55	0.66	0.77	0.88	0.99	1.11	1.21	1.33	1.44
N	410	172	88	51	33	22	15	11	8	6	5

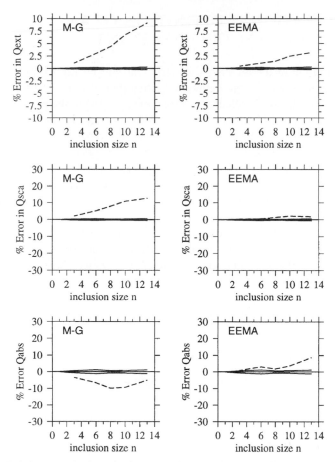

Figure 10   Relative errors in the extinction, scattering, and absorption efficiencies calculated using the Maxwell-Garnett EMA and the EEMA. The results obtained using the discrete dipole approximation are considered to provide the correct values. The two lines symmetric with respect to the horizontal axis represent the uncertainty due to statistics of the black carbon inclusion distribution within a water matrix. The inclusion size is specified by the number of dipoles n across the diameter of an inclusion. The diameter of the host sphere is spanned by 48 dipoles. The inclusion size parameter within the water matrix varies between $x = 0.33$ ($n = 3$) and $x = 1.44$ ($n = 13$).

culated using the EMAs will have errors considerably larger that those involved in Q_{ext} and Q_{sca}. In the case of about 10% black carbon mixed with water (Fig. 10), the errors in the absorption efficiency are between 8 and 10%, with the EEMA being slightly more accurate than the Maxwell-Garnett and Bruggeman EMAs.

VII. OPERATIONAL DEFINITION OF AN EFFECTIVE DIELECTRIC CONSTANT

The definition of the effective dielectric constant by Eq. (35) or (39) shows that at a given frequency the defined effective dielectric constant ε_{eff} is a volume average of the local dielectric constant $\varepsilon(x, y, z)$ with the local electric field $\mathbf{E}(x, y, z)$ used as a weighting function. Although this E-field weighted average is well justified for the case of extinction of a wave propagating through a composite medium, it may not be the best kind of average for the calculation of the absorption and scattering properties of small composite particles. Comparison with the measurements, models, and other approximations presented in the previous section suggests that absorption is not well approximated by various EMAs.

For the absorption of electromagnetic radiation by small composite particles, a more appropriate approximation of an effective dielectric constant may be a volume average of a local dielectric function $\varepsilon(x, y, z)$ weighted by $|E(x, y, z)|^2$. The defining equation

$$\int_V \varepsilon(x, y, z)|E(x, y, z)|^2 \, dx \, dy \, dz = \varepsilon_{\text{eff}} \int_V |E(x, y, z)|^2 \, dx \, dy \, dz \quad (52)$$

has a simple physical interpretation as requiring that the total electromagnetic energy stored inside a homogeneous particle of dielectric constant ε_{eff} is the same as the amount of energy stored inside an original inhomogeneous composite particle.

In a form discretized for numerical calculation, Eq. (52) can be rewritten analogously to Eq. (39) as

$$\varepsilon_{\text{eff}} = \frac{\varepsilon_1 \sum_{i=1}^{N_1} |E_{1i}|^2 \Delta V_{1i} + \varepsilon_2 \sum_{j=1}^{N_2} |E_{2j}|^2 \Delta V_{2j} + \varepsilon_3 \sum_{k=1}^{N_3} |E_{3k}|^2 \Delta V_{3k} + \cdots}{\sum_{i=1}^{N_1} |E_{1i}|^2 \Delta V_{1i} + \sum_{j=1}^{N_2} |E_{2j}|^2 \Delta V_{2j} + \sum_{k=1}^{N_3} |E_{3k}|^2 \Delta V_{3k} + \cdots}.$$

$$(53)$$

Another plausible alternative for the definition of an effective dielectric function from the point of view of absorption and scattering of radiation by small particles is a $|E(x, y, z)|^2$ weighted average of $[\varepsilon(x, y, z)]^{1/2}$, which is a $|E|^2$ weighted average of refractive indices

$$\int_V \sqrt{\varepsilon(x, y, z)}|E(x, y, z)|^2 \, dx \, dy \, dz$$

$$= \sqrt{\varepsilon_{\text{eff}}} \int_V |E(x, y, z)|^2 \, dx \, dy \, dz \quad (54)$$

or

m_{eff}

$$
= \frac{m_1 \sum_{i=1}^{N_1} |E_{1i}|^2 \Delta V_{1i} + m_2 \sum_{j=1}^{N_2} |E_{2j}|^2 \Delta V_{2j} + m_3 \sum_{k=1}^{N_3} |E_{3k}|^2 \Delta V_{3k} + \cdots}{\sum_{i=1}^{N_1} |E_{1i}|^2 \Delta V_{1i} + \sum_{j=1}^{N_2} |E_{2j}|^2 \Delta V_{2j} + \sum_{k=1}^{N_3} |E_{3k}|^2 \Delta V_{3k} + \cdots}.
$$

$$(55)$$

The judgment on the suitability of the proposed effective dielectric functions defined by Eq. (53) or (55) for the calculation of absorption by heterogeneous particles will have to be postponed until more numerical results become available.

VIII. CONCLUSIONS

The classical effective medium approximations as represented by the Bruggeman and Maxwell-Garnett mixing rules are usually derived under the assumption that the grain size parameter $x = 2\pi r/\lambda \ll 1$, where λ is the wavelength of radiation in an effective medium and $2r$ is a typical grain dimension. The extended effective medium approximations are supposed to extend the region of validity up to x on the order of 1. For the considered cases of black carbon grains in water droplets at visible wavelengths and water inclusions in an acrylic matrix at centimeter wavelengths, we conclude that:

1. The choice of which effective medium approximation (Bruggeman, Maxwell-Garnett, or inverted Maxwell-Garnett) should be used in any specific application is determined by the geometrical arrangement of the grains. Large errors in all radiative characteristics of composite particles can occur if an effective medium approximation is used that is not consistent with the specific grain distribution.

2. In the inclusion size parameter region of $x < 0.1$ the Bruggeman, the Maxwell-Garnett, and the EEMA provide accurate values for extinction and scattering cross sections that are within 1% of the "correct" values obtained using a more accurate method (DDA), a model that can be solved exactly (a spherical inclusion arbitrarily located within a spherical host), and laboratory measurements. They also provide accurate numerical values for the differential scattering cross section (or the phase function) and asymmetry parameter $\langle \cos \Theta \rangle$. The errors in the absorption cross section are considerably larger, between 10 and 15% for the considered cases of spherical water inclusions in an acrylic matrix and of black carbon in water droplets.

3. For an inclusion size parameter ranging up to $x = 2$, the Bruggeman and Maxwell-Garnett EMAs provide extinction and scattering cross sections with errors between 10 and 15%. In the same size parameter range, the errors of the EEMA are between 3 and 5%.

4. The EMAs and the EEMA provide a suitable tool for treatment of heterogeneous and composite particles in geophysical problems. The considered cases suggest that the EMAs and the EEMA can be used to obtain the single-scattering parameters (extinction, single-scattering albedo, and asymmetry parameter) needed for radiative transfer calculations, provided that the previously mentioned restrictions on the inclusion size parameter are satisfied.

5. The EMA results for the extinction and scattering cross sections and the asymmetry parameter are usually more accurate than the results for the absorption cross section. There is a need to develop new approaches designed specifically for calculations of the absorption cross section or other characteristics of the interaction of electromagnetic radiation with composite materials.

Chapter 10

Monte Carlo Calculations of Light Scattering by Large Particles with Multiple Internal Inclusions

Andreas Macke

Institut für Meereskunde
24105 Kiel, Germany

I. INTRODUCTION

The difficulty of calculating the scattering of electromagnetic radiation by many naturally occurring particles is due to a large extent to their nonsphericity, but is also attributable to inhomogeneities within the particles. Examples in remote-sensing and climatology studies are water droplets in the terrestrial atmosphere that contain various insoluble inclusions, ice particles with internally trapped air bubbles, inhomogeneous composites of mineral aerosols, and planetary regolith particles. Internal scattering also complicates the optical techniques for material testing. Examples are the detection of contamination on silicon wafer surfaces (Ivakhnenko *et al.*, 1998) or the optical particle sizing in industrial spray drying of food (Göbel *et al.*, 1997).

The calculation of the scattering properties of particles with arbitrary shapes and internal structures is rather difficult. Exact solutions exist for a few symmetric particles with layered structures (e.g., Toon and Ackerman, 1981) as well as for spherical particles with arbitrarily located spherical inclusions (Fuller, 1995b; Chapters 8 and 9). Furthermore, a number of algorithms based on the discrete dipole approximation (Purcell and Pennypacker, 1973; Chapter 5) have been developed to treat scattering by nonspherical particles with arbitrary internal optical structures (Goedecke and O'Brien, 1988). However, the relatively high demands for computer time and memory limit applications of these techniques to moderate size parameters.

In this chapter, a relatively simple hybrid technique combining ray optics and Monte Carlo radiative transfer is presented, which permits the treatment of light scattering by arbitrarily shaped host particles containing spherical and nonspherical inclusions and is valid for host particles that are large compared to the wavelength of the incoming radiation. It, therefore, extends the treatment of nonspherical, inhomogeneous particles to the large size parameter range.

Section II describes the ray-tracing/Monte Carlo model. Applications to light scattering by atmospheric ice crystals and lunar soil grains are discussed in Section III. A simplified treatment of the scattering problem using an independent superposition of the scattering properties of the host and internal particles is presented in Section IV. Finally, some concluding remarks and a brief outline of further potential applications are given in Section V.

II. RAY-TRACING/MONTE CARLO TECHNIQUE

The scattering of a light ray entering a particle containing discrete inclusions is simulated by a combination of ray-tracing and Monte Carlo techniques (hereafter referred to as the RT/MC technique). The ray-tracing program takes care of the individual reflection and refraction events at the outer boundary of the particle (e.g., Muinonen *et al.*, 1989; Chapters 11 and 15) and the MC routine simulates internal scattering processes.

After an incident photon is refracted into the host particle, it is allowed to travel a free path length l given by

$$l = -\bar{l} \log R(0, 1), \tag{1}$$

where $\bar{l}$ is the mean free path length between two subsequent scattering events and $R(0, 1)$ is an equally distributed random number within the interval $(0, 1)$.

If the photon has not reached one of the boundaries of the medium, its previous direction is changed along the local zenith Θ and azimuth φ scattering angles

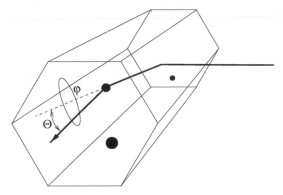

Figure 1   Illustration of the internal scattering geometry in hexagonal ice crystal with inclusions.

(Fig. 1) according to

$$\int_0^\Theta a_1^{\mathrm{incl}}(\Theta') \sin \Theta' \, d\Theta' = R(0, 1) \int_0^\pi a_1^{\mathrm{incl}}(\Theta') \sin \Theta' \, d\Theta', \qquad (2)$$

$$\varphi = R(0, 2\pi), \qquad (3)$$

where a_1^{incl} denotes the scattering phase function (Section XI of Chapter 1) of the internal scatterer. Absorption is taken into account by multiplying the photon energy with the single-scattering albedo ϖ^{incl} of the internal scatterer. Together with the processes described by Eqs. (1)–(3), this represents a direct Monte Carlo solution of the radiative transfer equation for a scattering and absorbing medium. These processes are repeated until the photon enters the host boundary surface, where it is again subject to reflection and refraction events. The entire procedure is repeated again for the internally reflected component each time until the photon energy falls below a specified threshold.

Note that calculations of the internal scattering properties require a nonabsorbing surrounding host medium. However, absorption within the host medium can still be taken into account by adding purely absorbing inclusions.

The scheme outlined previously produces the ray-tracing single-scattering albedo ϖ^{RT} and the ray-tracing scattering phase function a_1^{RT} for the host particle. Additionally, diffraction on the host particle's projected area S must be taken into account. For polyhedral particles, the projected area can be described by a closed polygon, from which the diffraction phase function a_1^{D} can be calculated analytically (Cai and Liou, 1982; Macke *et al.*, 1996b). For large spherical host particles with size parameters $x = 2\pi r/\lambda \gg 1$, diffraction is given by

(van de Hulst, 1957)

$$
a_1^{\mathrm{D}}(\Theta) = \begin{cases} 4x^2 \left(\dfrac{J_1(x \sin \Theta)}{x \sin \Theta} \right)^2, & \Theta \in \left[0, \dfrac{\pi}{2}\right], \\ 0, & \text{otherwise,} \end{cases} \tag{4}
$$

where $J_1(y)$ is the Bessel function of the first kind, r is the particle radius, and λ is the wavelength.

Ray-tracing and diffraction properties are added, weighted by their individual scattering cross sections. By definition, the ray-tracing extinction cross section is equal to the geometric cross section of the host particle, $C_{\mathrm{ext}}^{\mathrm{RT}} = S$, and the ray-tracing scattering cross section is equal to $C_{\mathrm{sca}}^{\mathrm{RT}} = \varpi^{\mathrm{RT}} S$. The diffraction extinction cross section $C_{\mathrm{ext}}^{\mathrm{D}} = S$ is always equal to the diffraction scattering cross section $C_{\mathrm{sca}}^{\mathrm{D}}$. Therefore, the total cross sections, single-scattering albedo, and phase functions are given by

$$
C_{\mathrm{ext}} = C_{\mathrm{ext}}^{\mathrm{RT}} + C_{\mathrm{ext}}^{\mathrm{D}} = 2S, \tag{5}
$$

$$
C_{\mathrm{sca}} = C_{\mathrm{sca}}^{\mathrm{RT}} + C_{\mathrm{sca}}^{\mathrm{D}} = \left(1 + \varpi^{\mathrm{RT}}\right) S, \tag{6}
$$

$$
\varpi = \frac{C_{\mathrm{sca}}}{C_{\mathrm{ext}}} = \frac{\varpi^{\mathrm{RT}} + 1}{2}, \tag{7}
$$

$$
a_1(\Theta) = \frac{C_{\mathrm{sca}}^{\mathrm{RT}} a_1^{\mathrm{RT}}(\Theta) + C_{\mathrm{sca}}^{\mathrm{D}} a_1^{\mathrm{D}}(\Theta)}{C_{\mathrm{sca}}^{\mathrm{RT}} + C_{\mathrm{sca}}^{\mathrm{D}}} = \frac{\varpi^{\mathrm{RT}} a_1^{\mathrm{RT}}(\Theta) + a_1^{\mathrm{D}}(\Theta)}{\varpi^{\mathrm{RT}} + 1}. \tag{8}
$$

The anisotropy of the scattered radiation is described by the asymmetry parameter

$$
\langle \cos \Theta \rangle = \int_0^\pi a_1(\Theta) \cos \Theta \sin \Theta \, d\Theta = \frac{\varpi^{\mathrm{RT}} \langle \cos \Theta \rangle^{\mathrm{RT}} + \langle \cos \Theta \rangle^{\mathrm{D}}}{\varpi^{\mathrm{RT}} + 1}. \tag{9}
$$

For large size parameters, the diffraction asymmetry parameter is very close to 1 so that

$$
\langle \cos \Theta \rangle \approx \frac{\varpi^{\mathrm{RT}} \langle \cos \Theta \rangle^{\mathrm{RT}} + 1}{\varpi^{\mathrm{RT}} + 1}. \tag{10}
$$

Note that the use of RT and MC techniques entails the following two assumptions. First, the distance between nearest neighbor internal scatterers must be larger than a few times their radii in order to treat them as independent scatterers. Second, the distance between the host particle boundary and internal scatterers must also exceed a certain value in order to assure the validity of Snell's law and Fresnel's formulas. According to Mishchenko *et al.* (1995), both conditions require a mean free path length $\bar{l}$ larger than about four times the radius of the internal scatterers. Therefore, even for a small number density of internal scatterers, the random nature of the Monte Carlo process will lead to occasional violations of this requirement. However, setting $\bar{l}$ to 20 times the particles' radii ensures that possible near-field effects play only a minor role in the total scattering simulation.

It should be noted that this computational scheme in its present state does not include interference mechanisms such as coherent backscattering (Tsang *et al.*, 1985, Chapter 5; Barabanenkov *et al.*, 1991). This mechanism results from constructive interference of so-called self-avoiding reciprocal multiple-scattering paths and causes a narrow intensity peak centered at exactly the backscattering direction. Although coherent backscattering does not change the total optical cross section of the composite particle and is unlikely to modify noticeably the total asymmetry parameter, it may increase the total backscattering phase function by as much as 50–60% (Mishchenko, 1992b) and, therefore, can significantly affect the results of active remote-sensing retrievals such as lidar measurements.

III. RESULTS

The following two sections describe example applications of the RT/MC technique. Section A describes light scattering by inhomogeneous atmospheric ice crystals and is partly based on the paper by Macke *et al.* (1996a). The second example (Section B) examines the effect of internal structures on the asymmetry parameter of composite particles and summarizes the results derived by Mishchenko and Macke (1997).

A. ATMOSPHERIC ICE PARTICLES

The study of light scattering by atmospheric ice crystals is largely motivated by the strong variability of ice particle shapes (e.g., Takano and Liou, 1995; Macke *et al.*, 1996b; Chapter 15). In most studies, the ice material itself is assumed to be homogeneous. In fact, although the shape problem is obvious from numerous *in situ* aircraft measurements (e.g., Heymsfield *et al.*, 1990), little is known about the internal structure of the ice particles. Considering the possible increase of the concentration of aerosol particles in the upper troposphere, caused either by natural phenomena such as volcanic eruptions (Sassen *et al.*, 1995) or by anthropogenic causes such as high-altitude aircraft exhausts (Schumann, 1994) or high-reaching convective transport of industrial combustions (Raes *et al.*, 1995), we might expect to find that increased scavenging and aggregation processes may lead to a large number of trapped particles inside ice crystals. Air bubbles may also be trapped inside rapidly growing ice particles or inside suddenly frozen supercooled water droplets. Furthermore, particle growth by riming also leads to highly inhomogeneous internal structures.

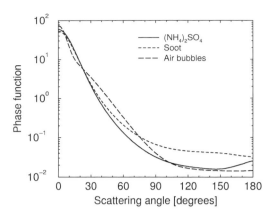

Figure 2 Scattering phase functions of spherical inclusions made of ammonium sulfate, soot, and air bubbles embedded in a hexagonal ice crystal. Reprinted from A. Macke, M. I. Mishchenko, and B. Cairns (1996a), The influence of inclusions on light scattering by large ice particles. J. Geophys. Res. **101**, 23,311–23,316; © 1996 by the American Geophysical Union.

Three types of internal scatterers are considered here: ammonium sulfate aerosols [(NH$_4$)$_2$SO$_4$], soot particles, and air bubbles. Figure 2 shows the scattering phase functions for these types of internal scatterers. Both ammonium sulfate particles and air bubbles are transparent in the visible. Therefore, these particles only affect the scattering properties of the ice crystal scattering phase function. Soot, on the other hand, is strongly absorbing and, therefore, has the potential of increasing the ice particles' absorption. The sizes of all types of impurities are assumed to obey a standard gamma distribution defined by an effective radius r_{eff} and an effective variance v_{eff} (Hansen and Travis, 1974). The values chosen for the three scatterers are given in Table I. The refractive indices $m = m_r + im_i$ are taken from Toon *et al.* (1976) for (NH$_4$)$_2$SO$_4$ and from Nilsson (1979) for soot. For air bubbles m is set to 1. These values are divided by the refractive index of ice (Warren, 1984), for which the small imaginary part ($\sim 10^{-9}$) is neglected, to obtain the respective relative refractive indices (Table I). Because the impurities are assumed to be spherical, the Lorenz–Mie theory was used to obtain the size-distributionally averaged optical properties of the internal scatterers relative to the surrounding ice material. Calculations were performed at a wavelength of $\lambda = 0.55$ μm, corresponding to maximum solar irradiation. The resulting extinction efficiencies $Q_{ext}(\text{incl}) = C_{ext}(\text{incl})/(\text{average projected area})$, single-scattering albedo ϖ_{incl}, and asymmetry parameters $\langle \cos \Theta \rangle_{incl}$ are shown in Table I. Figure 2 compares the phase functions of the internal scatterers.

The number density of the inclusions determines the mean free path length $\bar{l}$ [Eq. (1)] or, equivalently, the volume extinction coefficient $k_{ext} = 1/\bar{l}$. For an

Table I

Effective Radius r_{eff} and Effective Variance v_{eff} for Three Types of Inclusions, Inclusion Relative Refractive Indices (m_r, m_i), and Respective Size-Averaged Extinction Efficiencies $Q_{ext}(incl)$, Single-Scattering Albedos ϖ_{incl}, and Asymmetry Parameters $\langle \cos \Theta \rangle_{incl}$

Type	r_{eff} (μm)	v_{eff}	(m_r, m_i)	$Q_{ext}(incl)$	ϖ_{incl}	$\langle \cos \Theta \rangle_{incl}$
$(NH_4)_2SO_4$	0.5	0.2	(1.15, 0.0)	1.354	1.0000	0.9213
Soot	0.5	0.2	(1.18, 0.38)	2.122	0.4257	0.9033
Air bubbles	1.0	0.1	(0.75, 0.0)	1.977	1.0000	0.8817

Source: A. Macke, M. I. Mishchenko, and B. Cairns (1997), The influence of inclusions on light scattering by large hexagonal and spherical ice crystals, *in* "IRS'96 Current Problems in Atmospheric Radiation" (W. L. Smith and K. Stamnes, Eds.), pp. 226–229, Deepak, Hampton, VA.

ensemble of n_0 particles per unit volume element with sizes obeying the standard gamma distribution, k_{ext} is given by (Lacis and Mishchenko, 1995)

$$k_{ext} = \pi r_{eff}^2 (1 - v_{eff})(1 - 2v_{eff}) Q_{ext} n_0. \tag{11}$$

It is often more convenient to describe the optical density of the inclusions in terms of the optical thickness τ. To this end, the volume extinction coefficient must be multiplied by a length H, which represents a characteristic dimension of the host particle. For the purpose of this study, H is defined by the maximum dimension of the host particle. However, it should be noted that this choice does not necessarily reflect the typical length scale for spherical and nonspherical particles. See Macke *et al.* (1996a) for a brief discussion of this problem.

The effects of different types of internal impurities are studied for a hexagonal ice column with shape defined by a length to (hexagonal) diameter ratio of 200 μm/100 μm. This choice roughly corresponds to the mean particle dimensions observed in natural cirrus clouds.

Figure 3 shows the total scattering phase function (ice column with inclusions) for optical thicknesses $\tau = 0.5, 2.5$, and 5. With increasing τ, all three inclusions cause a noticeable broadening of the forward scattering features from 0° to 22°, as well as a decrease in the magnitude of the 22° halo maximum. The broadening results from the predominantly forward scattering phase functions of the individual impurities, which spreads the light rays that are directly transmitted through plane-parallel crystal facets. The same mechanism reduces the magnitude of the halos and the magnitude of the backscattering peak. The nonabsorbing inclusions cause an increase in side scattering, whereas side scattering decreases in magnitude for ice crystals containing soot particles. For both absorbing and nonabsorb-

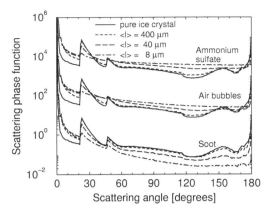

Figure 3 Scattering phase functions of a hexagonal column with internal ammonium sulfate (multiplied by 10^4), air bubble (multiplied by 10^2), and soot inclusions. The optical thickness of the internal scatterers is $\tau = 0.5, 2.5,$ and 5. Reprinted with changes from A. Macke, M. I. Mishchenko, and B. Cairns (1996a), The influence of inclusions on light scattering by large ice particles. J. Geophys. Res. **101**, 23,311–23,316; © 1996 by the American Geophysical Union.

ing inclusions, the departures from the scattering phase function for the pure ice crystal increases as the inclusion number density increases. However, additional absorption caused by soot particles reduces the contribution of refracted rays to the total phase function [Eq. (8)], which explains the opposite side scattering effects for ice particles with absorbing versus nonabsorbing contaminations.

Ice particles in the atmosphere are often quite irregular in shape rather than a simple crystal and thus the assumption of symmetric hexagonal particles is not valid in this case. In this regard, an interesting question is whether internal scatterers significantly change the overall scattering properties of an irregularly shaped host particle, which has no characteristic ray paths because of its randomized geometry. Figure 4 shows the scattering phase functions of a randomized, fractal-type host particle (see Macke *et al.*, 1996b, for details) with increasing optical thickness of air bubble inclusions. A comparison with Fig. 3 indeed reveals that the inclusions have a weaker effect on the overall scattering properties for more irregularly shaped particles. However, the changes in the total scattering phase function are still significant, even for optical thicknesses as small as 0.5. Unlike the situation for a hexagonal host particle, internal scattering in the random polycrystal leads to a sharper forward scattering pattern. This result may be explained by the fact that the forward scattering for a homogeneous polycrystal is determined by a superposition of halos resulting from a minimum deviation for various component ice prisms, whereas internal scattering processes complicate the conditions for this minimum deviation and apparently increase the fraction of

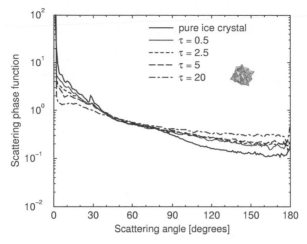

Figure 4   Phase function versus scattering angle for a random polycrystal with increasing optical thickness of air bubble inclusions.

internal reflections at the expense of direct transmissions. We note that this random polycrystal with air bubble inclusions may serve as a model for graupel and hailstone particles, which possess irregular external geometries as well as internal inhomogeneities in the form of air bubble inclusions (e.g., Pruppacher and Klett, 1997).

B. PLANETARY REGOLITH PARTICLES

Because large, homogeneous particles always have positive asymmetry parameters, it has been hypothesized that the negative asymmetry parameters of planetary regolith particles retrieved with an approximate bidirectional reflectance model (Hapke, 1993) result from their presumably complicated internal structure. The diameter of lunar regolith particles is typically about 50 μm, so that scattering of visible light by such particles can be described by means of the geometric optics approximation.

In the following, spherical host particles with a diameter of 50 μm and refractive indices at $\lambda = 0.55$ μm equal to 1.55 and 1.31 are assumed, corresponding to silicate and pure ice, respectively. Both materials commonly cover surfaces of bodies in the solar system. For the scattering inclusions, refractive indices are chosen to represent either voids inside the host particle ($m = 1$) or refractive grains with large and small contrast compared to the host particle ($m = 2$, $m = 1.65$). Table II shows the four types of host/inclusion pairs used in this study.

Table II

**Refractive Indices of the Host Medium
and Inclusions**

Type	m_{host}	m_{incl}
1	1.55	1
2	1.55	2
3	1.55	1.65
4	1.31	1

Source: M. I. Mishchenko and A. Macke (1997), Assymetry parameters of the phase function for isolated
and densely packed spherical particles with multiple internal inclusions in the geometric optics limit,
J. Quant. Spectrosc. Radiat. Transfer **57**, 767–794.

The size distribution of the inclusions is assumed to follow the standard gamma distribution (Hansen and Travis, 1974) with an effective radius of 0.5 μm and an effective variance of 0.1. The single-scattering properties of the inclusions have been calculated using the Lorenz–Mie theory. Figure 5 shows the corresponding scattering phase functions.

Equation (10), which assumes $\langle \cos \Theta \rangle^D \approx 1$, already suggests that the total asymmetry parameter for a large, isolated, composite particle cannot be negative. The only way to make the total asymmetry parameter equal to 0 is to have only nonabsorbing inclusions and a backward delta-function-like ray-tracing phase function, so that $\varpi^{RT} = 1$ and $\langle \cos \Theta \rangle^{RT} = -1$. In fact, Fig. 6 shows that the ray-tracing asymmetry parameter $\langle \cos \Theta \rangle^{RT}$ for the nonabsorbing composite particle does decrease systematically with increasing optical thickness. However, in none of the cases does it approach -1 or even become noticeably smaller than 0. The smallest $\langle \cos \Theta \rangle^{RT}$ value is equal to -0.0122 and corresponds to the type 1 composite particle and an extremely large optical thickness of $\tau = 25$. As a con-sequence, the total asymmetry parameter values are always positive and never drop below 0.4922.

It has been suggested by Hapke (1993) that the ray-tracing asymmetry param-eter can be reduced significantly by increasing absorption inside the composite particle because in this case it is only the light scattered to the sides and rear by the inclusions near the back-facing surface of the composite particle that readily es-capes. However, because the inclusions scatter light predominantly in the forward direction (Fig. 5), backscattering by the composite particle requires (multiple) in-ternal scattering events and thus longer ray paths than the direct transmittance that causes the forward scattering. Therefore, increasing absorption suppresses the backscattering component of the ray-tracing phase function. Furthermore, it

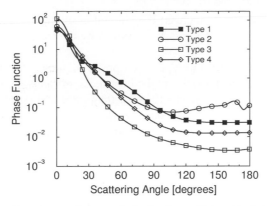

Figure 5   Phase function versus scattering angle for four types of internal inclusions. Reprinted from M. I. Mishchenko and A. Macke (1997), Assymetry parameters of the phase function for isolated and densely packed spherical particles with multiple internal inclusions in the geometric optics limit, *J. Quant. Spectrosc. Radiat. Transfer* **57**, 767–794, © 1997, with permission from Elsevier Science.

follows from Eq. (9) that increasing absorption reduces the ray-tracing contribution to the total asymmetry parameter so that the strongly forward scattering diffraction component dominates, and the asymmetry parameter of the composite particle increases. Therefore, the only way to considerably reduce the ray-tracing

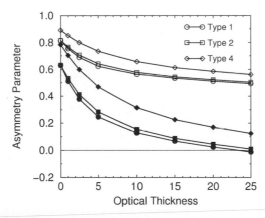

Figure 6   Ray-tracing (filled symbols) and total (open symbols) asymmetry parameters versus scattering optical thickness τ for type 1, type 2, and type 4 composite particles. Reprinted from M. I. Mishchenko and A. Macke (1997), Assymetry parameters of the phase function for isolated and densely packed spherical particles with multiple internal inclusions in the geometric optics limit, *J. Quant. Spectrosc. Radiat. Transfer* **57**, 767–794, © 1997, with permission from Elsevier Science.

asymmetry parameter is to increase the amount of multiple scattering inside the composite particle. This can only be achieved by decreasing rather than by increasing the absorption.

Similar calculations for the type 3 composite particle show that the reduced contrast between the inclusions and the host medium causes significantly larger ray-tracing asymmetry parameters than for the type 1 and type 2 particles. This readily can be explained by the fact that the type 3 inclusion has a stronger forward scattering and a weaker backscattering component than the phase functions for the type 1 and type 2 inclusions (Fig. 5).

Figure 6 shows that the ray-tracing asymmetry parameter for the composite ice particle (type 4) is systematically larger than that for silicate particles. This can be explained by the fact that the phase function for vacuum bubbles in ice is less backscattering than that for both vacuum bubbles and highly refractive grains in the silicate host medium (Fig. 5). Furthermore, homogeneous particles with a lower refractive index (ice) tend to be more forward scattering than those with a larger refractive index (silicate). It thus appears that it is more difficult to make an ice particle backscattering by filling it with multiple voids than it is for a silicate particle.

The qualitative results of this section also hold for densely packed large composite particles, even for isotropic scattering internal inclusions. A detailed discussion can be found in Mishchenko and Macke (1997) and Hillier (1997).

IV. ANALYTIC APPROXIMATION

From the practical point of view, it is interesting to compare RT/MC results with those obtained by an independent scattering approximation (ISA), where the scattering properties of the host particle and the inclusions are treated separately, that is, where radiative interactions between the ice particle surface and the embedded inclusions are neglected. Note that the ISA still requires that the refractive indices of the inclusions have to be given with respect to the refractive index of the host particle.

Because the internal scatterers do not affect the diffraction pattern of the host particle, only the ray-tracing phase function needs to be modified. The composite ray-tracing phase function of the homogeneous host particle and the inclusions is given by

$$a_1^{RT}(\text{host} + \text{incl}) = \frac{C_{sca}^{RT}(\text{host})a_1^{RT}(\text{host}) + NC_{sca}(\text{incl})a_1^{incl}}{C_{sca}^{RT}(\text{host}) + NC_{sca}(\text{incl})}. \tag{12}$$

Here, $C_{sca}^{RT}(\text{host})$ and $a_1^{RT}(\text{host})$ denote the scattering cross section and the ray-tracing phase function of the homogeneous host particle. The number, scattering

Table III

Asymmetry Parameter $\langle \cos \Theta \rangle$ for Ice Crystals Containing Ammonium Sulfate, Air Bubble, and Soot Inclusions as a Function of τ (See Text). The Changes in the Single-Scattering Albedo ϖ for Crystals with (Absorbing) Soot Inclusions are Also Shown. Results Based on the Independent Scattering Approximation are Given in Parentheses

τ	$\langle \cos \Theta \rangle$, $(NH_4)_2SO_4$	$\langle \cos \Theta \rangle$, air bubbles	$\langle \cos \Theta \rangle$, soot	ϖ, soot
0	0.8153	0.8153	0.8153	1.0000
0.5	0.8113 (0.8370)	0.8060 (0.8340)	0.8372 (0.8484)	0.9249 (0.9307)
2.5	0.7841 (0.8832)	0.7722 (0.8740)	0.8905 (0.9253)	0.7128 (0.7372)
5	0.7588 (0.9079)	0.7385 (0.8953)	0.9233 (0.9686)	0.5906 (0.6125)
10	0.7215 (0.9284)	0.6935 (0.9131)	0.9512 (0.9945)	0.5164 (0.5253)

Source: A. Macke, M. I. Mishchenko, and B. Cairns (1997), The influence of inclusions on light scattering by large hexagonal and spherical ice crystals, *in* "IRS'96 Current Problems in Atmospheric Radiation" (W. L. Smith and K. Stamnes, Eds.), pp. 226–229, Deepak, Hampton, VA.

cross section, and scattering phase function of the inclusions are represented by N, $C_{sca}(incl)$, and a_1^{incl}.

Similarly, adding the single-scattering albedo of the host and the inclusions yields

$$\varpi^{RT}(host + incl) = \frac{C_{sca}^{RT}(host) + NC_{sca}(incl)}{C_{ext}^{RT}(host) + NC_{ext}(incl)}$$

$$= \frac{C_{ext}^{RT}(host)\varpi_{host}^{RT} + NC_{ext}(incl)\varpi_{incl}}{C_{ext}^{RT}(host) + NC_{ext}(incl)}. \qquad (13)$$

Substituting Eqs. (12) and (13) into Eq. (8) provides the total phase function of the host/inclusions combination in the approximation of independent scattering.

Table III shows the approximate asymmetry parameter and single-scattering albedo (the latter, for soot only) together with the RT/MC results for a hexagonal host particle as discussed in Section III.A. The ISA results are given in parentheses. Obviously, the ISA does not give a satisfactory estimate of the true scattering behavior, even for internal scatterers with small optical thickness. For the nonabsorbing inclusions, $\langle \cos \Theta \rangle$ increases with increasing τ, contrary to the RT/MC results. An independent superposition of the phase functions of the ice crystal and inclusions mostly affects the forward scattering region, while side and backscattering are little influenced because both constituents are already predominantly forward scattering. As a result, the combined forward scattering peak becomes broader, the side scattering and backscattering remain almost the same, and the asymmetry parameter increases. On the other hand, a change in the internal ray paths as simulated in the Monte Carlo procedure not only influences the forward scattering behavior, but also all subsequent internal reflections, thereby leading to a broadening of the forward scattering peak and a systematic increase in side scattering and backscattering. In other words, scattering by the host particle and the internal inclusions is strongly coupled because of the ability of the inclusions to systematically change the internal ray paths. In case of absorption either by the ice crystal or by the inclusions, the length of the effective ray paths and, thus, the coupling decreases. This is demonstrated by the case of soot inclusions, for which the RT/MC and ISA results agree better than for nonabsorbing inclusions.

V. CONCLUSIONS

The demonstrated influence of different types of inclusions on the single-scattering behavior of a composite particle provides an additional motivation for studying the internal structures of naturally occurring particles. A more accurate characterization of the impurity effects requires a more precise knowledge of realistic number densities and of inclusion sizes.

Although the present study assumes uniformly distributed inclusions, it should be noted that the Monte Carlo technique applied can be easily extended to nonuniform distributions of internal scatterers.

Finally, natural inclusions are not always spherical in shape, thereby potentially preventing the use of the Lorenz–Mie theory to solve for their single-scattering properties. Therefore, future applications of the RT/MC technique should take advantage of results from scattering theories such as the T-matrix method (Mishchenko, 1993; Chapter 6), the finite difference time domain method (Yang and Liou, 1995; Chapter 7), or the discrete dipole approximation (Draine and Flatau, 1994; Chapter 5), which are suitable for nonspherical particle shapes.

A standard Fortran 90 implementation of the RT/MC technique is freely available from the author upon request.

ACKNOWLEDGMENT

The author is grateful to M. Mishchenko for valuable discussions and substantial motivations to the work presented in Section III.B.

Chapter 11

Light Scattering by Stochastically Shaped Particles

Karri Muinonen

Observatory
University of Helsinki
FIN-00014 Helsinki, Finland

I. INTRODUCTION

Theoretical light-scattering studies for stochastically shaped particles can be divided into a handful of partly overlapping groups. Extensive studies have been made on scattering by spheres with surface deformations much smaller than the radius. Rather recently studies have begun on spheres with surface deformations of arbitrary size. Scattering by random crystal particles have been analyzed in the ray optics approximation and recent advances further include results on scattering by fractal-shaped particles. Random ellipsoids and Chebyshev particles have been utilized for the derivation of ensemble-averaged scattering characteristics. Finally,

Light Scattering by Nonspherical Particles: Theory, Measurements, and Applications

random cylinders have been introduced to further understanding of scattering by elongated particles.

As for scattering by spheres with small surface deformations, among the early treatments of light scattering by stochastically shaped particles was the work of Hiatt *et al.* (1960) on surface roughness and its effects on the backscattering characteristics of spheres. Barrick (1970) offered a perturbation series approach to scattering by rough spheres.

Schiffer and Thielheim (1982a, b) provided a ray optics treatment for convex particles with a Gaussian distribution of surface normals. They accounted for single reflection, multiple surface scattering, transmission, forward diffraction, and shadowing by surface roughness. Forward diffraction was computed in the scalar Kirchhoff approximation by ensemble averaging using Gaussian statistics. Surface roughness was seen to produce an increasing slope in the phase function toward the backscattering direction. Mukai *et al.* (1982) and Perrin and Lamy (1983) studied scattering by large rough particles.

Schiffer (1985) focused on the effect of surface roughness on the spectral reflectance of dielectric particles. Convex particles with roughness scales smaller than the wavelength of incident light were assumed and both Gaussian and exponential correlation functions for the surface heights were made use of. Large-scale roughness was suggested as being responsible for the gross properties of angular scattering but was not included in the analysis. It was found that certain color effects could be caused by the small-scale surface roughness.

Schiffer (1989) studied light scattering by perfectly conducting, statistically irregular particles by superposing stochasticity on a sphere. The statistical parameters required were the mean and the covariance function. Deformation from the spherical shape was assumed to be small and the perturbation series approach of Yeh (1964) and Erma (1968a, b, 1969) was developed up to the second order. Schiffer noticed a divergence of results for the exponential correlation function. As an example, he validated the perturbation series approach for randomly oriented spheroidal particles. Schiffer (1990) then extended the earlier perturbation series work to irregular particles of arbitrary material. Again, up to the second order, he provided largely analytical results for the scattering characteristics. Similar work by César *et al.* (1994) also required surface deformations much smaller than the radius of the sphere.

Bahar and Chakrabarti (1985) applied the so-called full-wave theory to large rough conducting spheres assuming that the correlation length of the surface deformations was small as compared to the circumference of the sphere. For roughness scales of order a wavelength, roughness was shown to have a significant effect on the scattering characteristics. The results were compared to those from the perturbation series approach of Barrick (1970) and the reformulated current method of Abdelazeez (1983). Schertler and George (1994) provided the backscattering cross section of a perfectly conducting, roughened sphere.

Peltoniemi *et al.* (1989) studied scattering of light by stochastically rough particles in the ray optics approximation using a Markovian ray-tracing algorithm. No longer were the surface deformations required to be small as compared to the radius. They generated example two-dimensional silhouettes of rough particles using multivariate lognormal statistics. Continuing that work, Muinonen (1996a) and Muinonen *et al.* (1996) completed the mathematical model for what they called the Gaussian (or lognormal) random particle (or sphere) and published extensive light-scattering computations using the ray optics approximation. Muinonen *et al.* (1997) then studied analytically the regime of validity of the ray optics approximation for Gaussian random spheres. Muinonen (1996b) put forward the Rayleigh-volume and Rayleigh–Gans approximations for solving the integral equation of electromagnetic scattering. Lumme and Rahola (1998) and Peltoniemi *et al.* (1998) provided tentative results from close-to rigorous solutions of the integral equation. Nousiainen and Muinonen (1999) applied the Gaussian random sphere in ray optics studies of light scattering by oscillating raindrops.

As to random crystals, Muinonen *et al.* (1989) studied light scattering by randomly oriented and randomly distorted crystal particles, obtaining scattering characteristics resembling those of Peltoniemi *et al.* (1989). Macke *et al.* (1996b) and Hess *et al.* (1998) derived ray optics results for distorted hexagonal columns and plates and imperfect hexagonal crystals, respectively, and Yang and Liou (1998a) assumed surface roughness for the crystal faces in their ray optics computations.

Chiappetta (1980b) published a high-energy, phenomenological formulation of scattering by irregular absorbing particles and concluded that the eikonal and shadow functions described reasonably well the effects of surface roughness on light scattering. Light scattering by fractal particles was studied by Bourrely *et al.* (1986) for particles large compared to the wavelength. The particle shapes were assumed to be axially symmetric, with the incident wave propagating in the direction of the symmetry axis. Uozumi *et al.* (1991) and Uno *et al.* (1995) carried out diffraction studies for Koch fractals. Macke *et al.* (1996b) performed ray-tracing computations for complicated deterministic and random fractals.

Distributions of ellipsoids, spheroids, and Chebyshev particles have been incorporated in analyses of ensemble scattering. Huffman and Bohren (1980) studied absorption spectra of nonspherical particles in the Rayleigh-ellipsoid approximation, whereas Hill *et al.* (1984) studied scattering by shape distributions of soil particles and spheroids. Mugnai and Wiscombe (1986) and Wiscombe and Mugnai (1988) carried out ensemble averaging for Chebyshev particles, and Mishchenko (1993; Chapter 2) studied shape distributions of randomly oriented axially symmetric particles.

As to light scattering by stochastically deformed cylinders, Bahar and Fitzwater (1986) carried out studies for rough conducting cylinders in a fashion similar to that of Bahar and Chakrabarti (1985) for rough conducting spheres. They published straightforward formulas based on the full-wave theory and compared scattering by smooth and rough cylinders. Ogura *et al.* (1991) studied scattering by a slightly random cylindrical surface for horizontal polarization. Videen and Bickel (1992) measured the scattering matrix for a rough quartz fiber and presented theoretical models based on a multiconcentric fiber bundle and a Rayleigh system. Muinonen and Saarinen (1998) formulated the stochastic geometry of the Gaussian random cylinder and established the ray optics approximation for infinite Gaussian cylinders.

As to experiments for understanding scattering by stochastically shaped particles (see Chapters 12 and 13), for example, Giese *et al.* (1978) studied scattering by compact and fluffy particles large compared to the wavelength in order to understand the optical properties of interplanetary dust. They concluded that the Lorenz–Mie theory for spherical particles was unable to explain the observations and showed by model calculations and microwave analog measurements that large, fluffy particles were possibly an important component of the interplanetary dust cloud.

Holland and Gagne (1970) measured scattering matrices for polydisperse systems of irregular particles and compared the scattering characteristics to those from spherical particles with the same optical properties and size distribution. They found considerable differences in the backscattering characteristics of spheres and irregular particles.

Perry *et al.* (1978) measured scattering matrices for small, rounded, irregular ammonium sulfate particles and deformed cubic sodium chloride particles. They were able to explain the scattering characteristics of the rounded particles using Lorenz–Mie calculations for an equivalent size distribution, but noted large differences between the Lorenz–Mie calculations and the scattering characteristics of the cubic particles. Kuik *et al.* (1991) measured scattering matrices for small water droplets and larger irregular quartz particles and considerable differences were evident in the scattering patterns.

Sasse and Peltoniemi (1995) measured the angular scattering characteristics of rough carbon particles and successfully applied the computational methods of Peltoniemi *et al.* (1989) to explain the measurements. Sasse *et al.* (1996) and Piironen *et al.* (1998) measured albedos for rough carbon particles and meteoritic dust particles and applied the ray optics methods of Muinonen *et al.* (1996) to analyze their experimental results.

As for applying the theoretical methods to light scattering by solar system dust particles, Muinonen (1994) summarizes how coherent backscattering can explain the backscattering enhancement and polarization reversal observed for many solar system bodies. In particular, coherent backscattering can manifest itself in scattering by rough, solid surfaces, single particles, and particulate media.

This chapter summarizes the theoretical studies of electromagnetic scattering by randomly shaped nonspherical particles. Section II describes the stochastic geometry of Gaussian spheres and cylinders and the maximum likelihood estimator for inverting the statistical shape parameters from sample shapes. In the order of increasing significance of particle shape, Section III addresses scattering by Gaussian spheres in the Rayleigh-volume and Rayleigh–Gans approximations and using the second-order perturbation approximation, the variational volume integral equation method, and the discrete dipole approximation in the resonance region. Ray optics results follow for both Gaussian spheres and cylinders. Section IV concludes the review by outlining some future prospects in studies of scattering by Gaussian particles.

II. STOCHASTIC GEOMETRY

A. GAUSSIAN SPHERE

The Gaussian random sphere is described in detail by Muinonen (1998). The size and shape of the Gaussian sphere are specified by the mean radius and the covariance function of the logarithmic radius. The covariance function is given as a series of Legendre polynomials with nonnegative coefficients. For each degree, these coefficients provide the spectral weights of the corresponding spherical harmonics components in the Gaussian sphere. Weighting the spectrum toward higher-degree harmonics will result in sample spheres with larger numbers of hills and valleys per solid angle. Increasing the variance of the logarithmic radius will enhance the hills and valleys radially. For detailed information on Gaussian and lognormal statistics, the reader is referred to Vanmarcke (1983) and Aitchison and Brown (1963). Stoyan and Stoyan (1994) provide a modern treatise on random shapes.

The shape of a Gaussian random sphere $r = r(\vartheta, \varphi)$ is described in spherical coordinates (r, ϑ, φ) by the spherical harmonics series for the so-called logradius $s = s(\vartheta, \varphi)$ (Muinonen, 1996a):

$$r(\vartheta, \varphi) = a \exp\left[s(\vartheta, \varphi) - \frac{1}{2}\beta^2\right],$$

$$s(\vartheta, \varphi) = \sum_{l=0}^{\infty} \sum_{m=-l}^{l} s_{lm} Y_{lm}(\vartheta, \varphi),$$

(1)

where a and β are the mean radius and the standard deviation of the logradius and Y_{lm}'s are the orthonormal spherical harmonics. The logradius is real valued so that

$$s_{l,-m} = (-1)^m s_{lm}^*, \qquad l = 0, 1, \ldots, \infty,$$
$$m = -l, \ldots, -1, 0, 1, \ldots, l, \qquad (2)$$

where the asterisk denotes a complex conjugate value, thereby implying $\text{Im}(s_{l0}) \equiv 0$.

The covariance functions of the radius and logradius $a^2\Sigma_r$ and Σ_s, respectively, and the corresponding variances $a^2\sigma^2$ and β^2 are interrelated through

$$\Sigma_r = \exp(\Sigma_s) - 1, \qquad \sigma^2 = \exp(\beta^2) - 1. \qquad (3)$$

The relative standard deviation of radius $\sigma = \sqrt{\exp(\beta^2) - 1}$ depends solely on β.

The real and imaginary parts of the spherical harmonics coefficients s_{lm}, $m \geq 0$, are independent Gaussian random variables with zero means and variances

$$\text{Var}[\text{Re}(s_{lm})] = (1 + \delta_{m0})\frac{2\pi}{2l+1}C_l,$$
$$\text{Var}[\text{Im}(s_{lm})] = (1 - \delta_{m0})\frac{2\pi}{2l+1}C_l, \qquad (4)$$
$$l = 0, 1, \ldots, \infty, \qquad m = 0, 1, \ldots, l.$$

The coefficients $C_l \geq 0$, $l = 0, \ldots, \infty$, are the coefficients in the Legendre polynomial expansion of the logradius covariance function Σ_s:

$$\Sigma_s(\gamma) = \beta^2 C_s(\gamma) = \sum_{l=0}^{\infty} C_l P_l(\cos\gamma), \qquad \sum_{l=0}^{\infty} C_l = \beta^2, \qquad (5)$$

where γ is the angular distance between two directions (ϑ_1, φ_1) and (ϑ_2, φ_2) and C_s is the logradius correlation function:

$$C_s(\gamma) = \sum_{l=0}^{\infty} c_l P_l(\cos\gamma), \qquad \sum_{l=0}^{\infty} c_l = 1. \qquad (6)$$

In practice, the series representations in Eqs. (1) and (5)–(6) must be truncated at a certain degree l_{max} sufficiently high to maintain good precision in the generation of sample spheres.

The two perpendicular slopes

$$s_\vartheta = \frac{r_\vartheta}{r}, \qquad \frac{1}{\sin\vartheta}s_\varphi = \frac{r_\varphi}{r\sin\vartheta} \qquad (7)$$

are independent Gaussian random variables with zero means and standard deviations

$$\rho = \sqrt{-\Sigma_s^{(2)}(0)}\,,\tag{8}$$

where $\Sigma_s^{(2)}$ is the second derivative of the covariance function with respect to γ. The correlation length ℓ and correlation angle Γ are defined by

$$\ell = \frac{1}{\sqrt{-C_s^{(2)}(0)}}\,, \qquad \Gamma = 2\arcsin\left(\frac{1}{2}\ell\right).\tag{9}$$

For example, the modified Gaussian correlation function

$$C_s(\gamma) = \exp\left(-\frac{2}{\ell^2}\sin^2\frac{1}{2}\gamma\right)\tag{10}$$

can be utilized in the generation of sample shapes. It is a proper correlation function, because its Legendre coefficients are all nonnegative:

$$c_l = (2l+1)\exp\left(-\frac{1}{\ell^2}\right)i_l\left(\frac{1}{\ell^2}\right), \qquad l = 0, \ldots, \infty,\tag{11}$$

where i_l is a modified spherical Bessel function. When using the modified Gaussian correlation function, the random shape is parameterized by two parameters only, σ and ℓ. Gaussian sample spheres are shown in Fig. 1.

B. GAUSSIAN CYLINDER

Gaussian random cylinders $\varrho = \varrho(\varphi, z)$ can be conveniently generated in cylindrical coordinates (ϱ, φ, z) using a two-dimensional Fourier series for the logarithmic radius $s = s(\varphi, z)$ (Muinonen and Saarinen, 1999):

$$\varrho(\varphi, z) = a\exp\left[s(\varphi, z) - \frac{1}{2}\beta^2\right],$$

$$s(\varphi, z) = \sum_{m=-\infty}^{\infty}\sum_{k=-\infty}^{\infty} s_{mk}\exp\left[i(m\varphi + kKz)\right]\tag{12}$$

and, as for Gaussian random spheres, a and β are the mean radius and the standard deviation of the logradius. The logradius is real valued so it is required that

$$s_{-m,-k}^* = s_{mk}, \qquad m, k = 0, 1, \ldots, \infty.\tag{13}$$

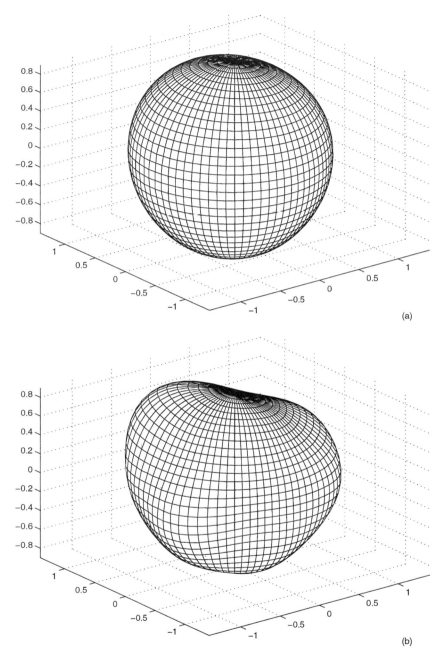

Figure 1   Sample Gaussian spheres with relative standard deviation of radius $\sigma = 0.1$ and correlation angle (a) $\Gamma = 90°$, (b) $\Gamma = 30°$, and (c) $\Gamma = 10°$ assuming a modified Gaussian correlation function.

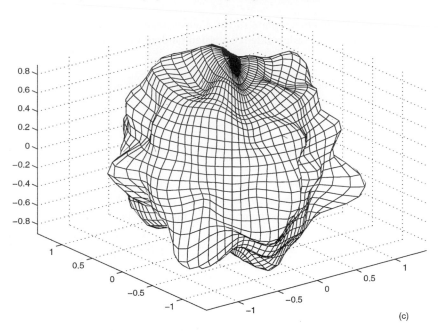

(c)

Figure 1   (Continued.)

For example, a necessary and sufficient set of independent coefficients consists of those s_{mk} for which $m \geq 0$, $-\infty < k < \infty$, excluding s_{0k} with $k < 0$. In addition, $\mathrm{Im}(s_{00}) \equiv 0$.

The covariance function Σ_s can be expressed as a two-dimensional Fourier series with nonnegative coefficients C_{mk}, $m, k = 0, \ldots, \infty$,

$$\Sigma_s(\gamma, \zeta) = \beta^2 C_s(\gamma, \zeta) = \sum_{m=0}^{\infty} \sum_{k=0}^{\infty} C_{mk} \cos(m\gamma) \cos(kK\zeta),$$

(14)

$$K = \frac{\pi}{L_z}, \qquad \sum_{m=0}^{\infty} \sum_{k=0}^{\infty} C_{mk} = \beta^2,$$

where $2L_z$, the period in the axial direction, must approach infinity or, in practice, must be chosen large enough. In Eq. (14), γ and ζ are the azimuthal and axial distances of two points (φ_1, z_1) and (φ_2, z_2).

The covariance functions Σ_ϱ and Σ_s and the variances σ^2 and β^2 are interrelated through

$$\Sigma_\varrho = \exp(\Sigma_s) - 1, \qquad \sigma^2 = \exp(\beta^2) - 1.$$

(15)

If the real and imaginary parts of the Fourier coefficients s_{mk} (recalling the previous conditions) are independent Gaussian random variables with zero means and variances

$$\text{Var}[\text{Re}(s_{mk})] = \frac{1}{8}(1 + \delta_{m0} + \delta_{k0} + 5\delta_{m0}\delta_{k0})C_{mk},$$

$$\text{Var}[\text{Im}(s_{mk})] = \frac{1}{8}(1 + \delta_{m0} + \delta_{k0} - 3\delta_{m0}\delta_{k0})C_{mk},$$

(16)

the logradius will be normally distributed with zero mean and covariance function Σ_s.

The Fourier coefficients C_{mk}, $m, k = 0, 1, \ldots, \infty$, can be taken as free parameters, subject to the conditions in and preceding Eq. (14). However, a convenient example correlation function is the cylindrical modified Gaussian function

$$C_s(\gamma, \zeta) = \exp\left(-\frac{2}{\ell_\varphi^2}\sin^2\frac{1}{2}\gamma - \frac{1}{2\ell_z^2}\zeta^2\right),$$

(17)

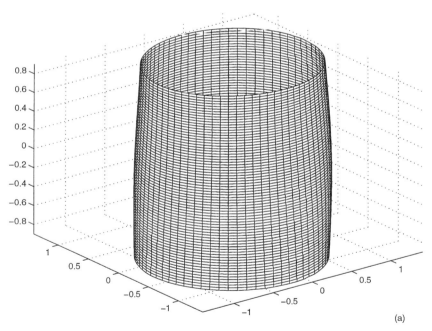

(a)

Figure 2 Sample Gaussian cylinders with relative standard deviation of radius $\sigma = 0.1$ and azimuthal correlation angle and axial correlation length (a) $\Gamma_\varphi = 90°$, $\ell_z = 1.414$; (b) $\Gamma = 30°$, $\ell_z = 0.518$; and (c) $\Gamma = 10°$, $\ell_z = 0.174$ for a modified Gaussian correlation function in the cylindrical geometry.

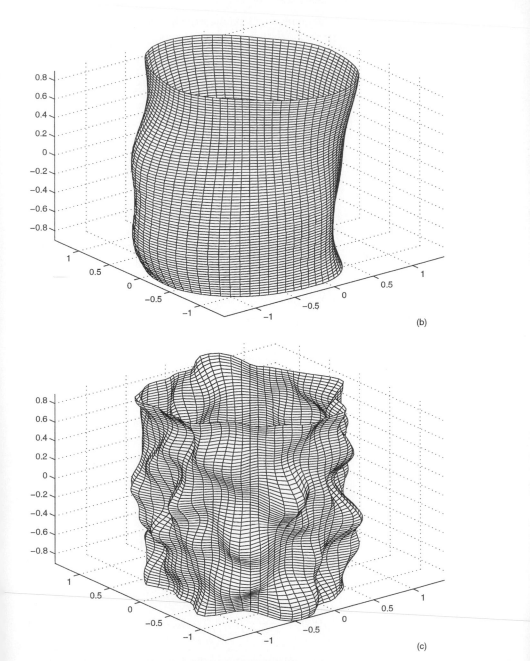

Figure 2   (Continued.)

where ℓ_φ and ℓ_z are the azimuthal and axial correlation lengths. Furthermore, we define the correlation angle Γ_φ so that

$$\Gamma_\varphi = 2\arcsin\left(\frac{1}{2}\ell_\varphi\right). \tag{18}$$

We obtain

$$C_{mk} = \beta^2 \left[(2 - \delta_{m0}) \exp\left(-\frac{1}{\ell_\varphi^2}\right) I_m\left(\frac{1}{\ell_\varphi^2}\right) \right]$$
$$\times \left[(2 - \delta_{k0}) \frac{\ell_z}{L_z} \exp\left(-\frac{1}{2}k^2\pi^2\frac{\ell_z^2}{L_z^2}\right) \right], \qquad m, k = 0, \ldots, \infty, \tag{19}$$

I_m being a modified Bessel function. The stochastic shape is thus parameterized by σ, Γ_φ, and ℓ_z, that is, the relative standard deviation, the azimuthal correlation angle, and the axial correlation length, respectively. Figure 2 shows sample Gaussian cylinders generated with the help of the modified Gaussian correlation function.

C. INVERSE PROBLEM

The inverse problem of determining the statistical parameters from a finite number of sample shapes (spheres or cylinders) has been solved using the maximum likelihood estimator by Lamberg *et al.* (1999) and is briefly described for the Gaussian sphere by Muinonen and Lagerros (1998). Among the parameters to be determined are the origins of the sample shapes, closely resembling the one-dimensional thresholds described by Aitchison and Brown (1963).

Following Lamberg *et al.* (1999), but omitting the mean radius estimation as in Muinonen and Lagerros (1998), the probability density (or likelihood) function for the $l_{\max} + 1$ Legendre coefficients $\{C_l\}$ and the origins $\{\mathbf{r}_n\}$ of the N sample shapes can be written as

$$p(\{C_l\}, \{\mathbf{r}_n\}) \propto \prod_{n=1}^{N} \exp\left[-\frac{1}{2}\sum_{l=0}^{L}(2l + 1)\left(\frac{\widetilde{C}_{ln}(\mathbf{r}_n)}{C_l} + \log_e C_l\right) \right], \tag{20}$$

where the Legendre coefficients $\widetilde{C}_{ln}(\mathbf{r}_n)$ pertain to individual sample shapes,

$$\widetilde{C}_{ln}(\mathbf{r}_n) = \frac{2l + 1}{4\pi}\frac{1}{l + 1}\sum_{m=0}^{l} |s_{lm}(\mathbf{r}_n)|^2. \tag{21}$$

A constraint on the location of the origin is the requirement that the shape be mathematically starlike with respect to the origin.

The inverse problem of estimating the model parameters consists of finding those $\{C_l\}$ and $\{\mathbf{r}_n\}$ that maximize the likelihood function in Eq. (20). The total number of parameters thus amounts to $l_{\max} + 1 + 3N$. For given origins $\{\mathbf{r}_n\}$, it is straightforward to show that the Legendre coefficients maximizing the likelihood function coincide with the means of the coefficients for the individual sample shapes,

$$C_l = \frac{1}{N} \sum_{n=1}^{N} \tilde{C}_{ln}(\mathbf{r}_n). \tag{22}$$

In practice, the final maximum likelihood solution for the parameters can be obtained iteratively by first computing estimates for the Legendre coefficients from Eq. (22), then improving the origins by using the new Legendre coefficients in Eq. (20), and repeating the procedure until changes in the origins and the Legendre coefficients become negligible. Whenever the origin of a sample shape changes, new spherical harmonics coefficients are needed for the computation of the Legendre coefficients in Eq. (22), which constitutes a considerable computational challenge in the application of the maximum likelihood estimator.

III. SCATTERING BY GAUSSIAN PARTICLES

A. RAYLEIGH-VOLUME APPROXIMATION

For particles small compared to the wavelength, the extinction, scattering, and absorption parameters are approximately determined by the volume of the particle. In what we call the Rayleigh-volume approximation, the polarizability α is assumed to be scalar:

$$\alpha = \frac{3}{4\pi} \frac{m^2 - 1}{m^2 + 2}, \tag{23}$$

where m is the complex relative refractive index. In the Rayleigh-ellipsoid approximation, now also established for Gaussian particles, the polarizability becomes a tensor (Battaglia *et al.*, 1999).

The Rayleigh-volume approximation is valid for small size parameters and small phase shifts across the particle (van de Hulst, 1957). Expressing the validity criteria with the help of the mean-volume size parameter ($\langle \; \rangle$ denotes ensemble averaging)

$$x_V = k \sqrt[3]{\frac{3\langle V \rangle}{4\pi}} = x \exp(\beta^2), \tag{24}$$

where k is the wavenumber, we obtain (Muinonen, 1996b)

$$x \ll \exp(-\beta^2) = \frac{1}{1 + \sigma^2},$$

$$2x|m - 1| \ll \exp(-\beta^2) = \frac{1}{1 + \sigma^2}.$$

(25)

Larger variances yield larger particles, as is explicitly manifested in the validity criterion.

The ensemble-averaged absorption and scattering cross sections and the scattering matrix assume the forms (Muinonen, 1996b)

$$C_{abs} = 4\pi k \langle V \rangle \operatorname{Im} \alpha, \qquad C_{sca} = \frac{8\pi}{3} k^4 \langle V^2 \rangle |\alpha|^2,$$

(26)

$$\mathbf{F}^R = \frac{3}{2} \begin{bmatrix} \frac{1}{2}(1 + \cos^2 \Theta) & -\frac{1}{2}\sin^2 \Theta & 0 & 0 \\ -\frac{1}{2}\sin^2 \Theta & \frac{1}{2}(1 + \cos^2 \Theta) & 0 & 0 \\ 0 & 0 & \cos \Theta & 0 \\ 0 & 0 & 0 & \cos \Theta \end{bmatrix},$$

where Θ is the scattering angle,

$$\langle V \rangle = \frac{4\pi}{3} a^3 \exp(3\beta^2),$$

$$\langle V^2 \rangle = \langle V \rangle^2 \frac{1}{2} \int_{-1}^{1} d\xi \, \exp[9\Sigma_s(\xi)], \qquad \xi = \cos \gamma.$$

(27)

The second volume moment and thus the scattering cross section can be computed analytically for covariance functions with nonzero C_0 or C_1 only:

$$\langle V^2 \rangle = \langle V \rangle^2 \exp(9C_0) \frac{\sinh 9C_1}{9C_1}.$$

(28)

Note that the ensemble-averaged volume is independent of the correlation function so that Gaussian particles with equal-radius standard deviations but different correlation angles exhibit similar absorption properties in the Rayleigh-volume approximation. It is evident that the scattering matrix and the vanishing asymmetry parameter coincide with those for spheres in the Rayleigh-volume approximation.

The scattering cross section can differ substantially from that for monodisperse equal-volume spheres because it depends on the second volume moment. The parameter

$$\frac{\langle V^2 \rangle}{\langle V \rangle^2} = \frac{1}{2} \int_{-1}^{1} d\xi \, \exp[9\Sigma_s(\xi)] \geq 1 \qquad (29)$$

represents the ratio of the scattering cross section of Gaussian particles to that for a monodisperse distribution of spheres with volumes equal to the ensemble-averaged volume of the Gaussian particles and that ratio is always larger than unity. However, for given β, the distribution of perfect spheres with the same β,

$$C_l = \beta^2 \delta_{l0}, \qquad l = 0, 1, 2, \ldots, \qquad (30)$$

always maximizes the scattering cross section.

The Rayleigh-volume approximation is equivalent to approximating scattering by Gaussian spheres with Rayleigh scattering by a lognormal size distribution of spheres with renormalized mean radius $\tilde{a}$ and standard deviation $\tilde{\beta}$ given by

$$\tilde{a} = a \exp(\beta^2 - \tilde{\beta}^2), \qquad \tilde{\beta}^2 = \frac{1}{9} \log_e \left\{ \frac{1}{2} \int_{-1}^{1} d\xi \, \exp[9\Sigma_s(\xi)] \right\}. \qquad (31)$$

Such a choice of parameters guarantees that the first and second moments of volume are equal for the two distributions of particles.

B. RAYLEIGH–GANS APPROXIMATION

The Rayleigh–Gans approximation is valid for small polarizability contrasts and small phase shifts (van de Hulst, 1957; Muinonen, 1996b):

$$|\alpha| \ll 1, \qquad 2x_V |m - 1| \ll 1, \qquad (32)$$

or in terms of the size parameter x,

$$|\alpha| \ll 1, \qquad 2x |m - 1| \ll \exp(-\beta^2). \qquad (33)$$

In particular, the approximation can be applied to particles of arbitrary sizes when the latter of the two conditions holds.

In the Rayleigh–Gans approximation, the ensemble-averaged absorption and scattering cross sections and the scattering matrix take the forms (Muinonen, 1996b)

$$C_{\text{abs}} = 4\pi k \langle V \rangle \, \text{Im} \, \alpha,$$

$$C_{\text{sca}} = \frac{2}{3} |\alpha|^2 k^4 \int_{4\pi} d\Omega F(\mathbf{q}) F_{11}^{\text{R}}, \qquad (34)$$

$$\mathbf{F}^{\text{RG}} = 4\pi \, F(\mathbf{q}) \mathbf{F}^{\text{R}} \Big/ \left[\int_{4\pi} d\Omega F(\mathbf{q}) F_{11}^{\text{R}} \right],$$

where $\mathbf{F}^R$ is the Rayleigh scattering matrix in Eq. (26) and where

$$\mathbf{q} = k(\mathbf{n}^{\text{inc}} - \mathbf{n}^{\text{sca}}), \qquad q = |\mathbf{q}| = 2k \sin \tfrac{1}{2}\Theta, \tag{35}$$

represents the functional dependence on the incident and scattered waves propagating in the $\mathbf{n}^{\text{inc}}$ and $\mathbf{n}^{\text{sca}}$ directions, respectively. The function $F(\mathbf{q})$ derives from the form factor $f(\mathbf{q})$,

$$F(\mathbf{q}) = \langle |Vf(\mathbf{q})|^2 \rangle, \qquad f(\mathbf{q}) = \frac{1}{V} \int_V d^3\mathbf{r} \exp[-i\mathbf{q} \cdot \mathbf{r}]. \tag{36}$$

Obviously, the absorption cross section coincides with that of the Rayleigh-volume approximation and can be computed from Eq. (26). The other scattering parameters require ensemble averaging for the square form factor $F(\mathbf{q})$ that contains the information on particle size and shape. Note that the scattering matrix does not depend on the refractive index in the Rayleigh–Gans approximation.

We can develop Eq. (36) as

$$F(\mathbf{q}) = 8\pi^2 \int_{-1}^{1} d\xi \int_0^\infty dr\, r^2 \int_0^\infty dr'\, r'^2$$
$$\times Q_2[s, s', \Sigma_s(\xi)] \frac{\sin q\sqrt{r^2 + r'^2 - 2rr'\xi}}{q\sqrt{r^2 + r'^2 - 2rr'\xi}}, \tag{37}$$

where Q_2 is the bivariate complementary Gaussian cumulative distribution function with zero means and covariance matrix Σ_s,

$$Q_2[s, s', \Sigma_s(\xi)] = \int_s^\infty ds \int_{s'}^\infty ds' n_2[s, s', \Sigma_s(\xi)], \tag{38}$$

and s and s' are logradii as in Eq. (1). Equation (37) needs to be evaluated numerically, a task that becomes demanding for particles large compared to the wavelength.

As for the scattering cross section and the asymmetry parameter, the integration over the scattering angle can be carried out analytically and we obtain

$$C_{\text{sca}} = 8\pi^3 k^4 |\alpha|^2 \int_{-1}^{1} d\xi \int_0^\infty dr\, r^2 \int_0^\infty dr'\, r'^2$$
$$\times Q_2[s, s', \Sigma_s(\xi)] A_1\left(2k\sqrt{r^2 + r'^2 - 2rr'\xi}\right),$$

$$\langle \cos\Theta \rangle = \frac{8\pi^3 k^4 |\alpha|^2}{C_{\text{sca}}} \int_{-1}^{1} d\xi \int_0^\infty dr\, r^2 \int_0^\infty dr'\, r'^2$$
$$\times Q_2[s, s', \Sigma_s(\xi)] A_2\left(2k\sqrt{r^2 + r'^2 - 2rr'\xi}\right), \tag{39}$$

where we have introduced the functions

$$A_1(z) = \frac{8}{z^6}\big[z^4(1 - \cos z) + 4z^3 \sin z + z^2(4 + 20\cos z)$$
$$- 48z \sin z + 48(1 - \cos z)\big],$$

$$A_2(z) = \frac{8}{z^8}\big[z^6(1 + \cos z) - 8z^5 \sin z + z^4(8 - 56\cos z) + 336z^3 \sin z$$
$$+ z^2(144 + 1296\cos z) - 2880z \sin z + 2880(1 - \cos z)\big].$$

(40)

Again, Eq. (39) requires numerical integration.

Figure 3 shows the Rayleigh–Gans scattering phase functions for a size parameter $x = 10$ and the statistical parameters $\sigma = 0.05, 0.10$, and 0.20 and $\Gamma = 10°$, $30°$, and $90°$ [correlation function as in Eq. (10)]. The resonance structure is evident for the smallest σ and Γ values and is gradually averaged out for increasing σ and Γ, when the size distribution becomes broader (in terms of particle volume).

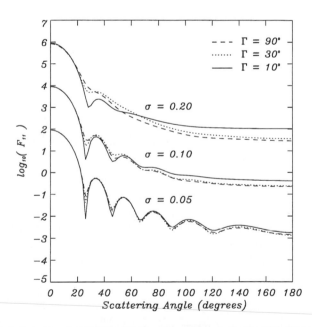

Figure 3 Scattering phase function F_{11} for Gaussian spheres in the Rayleigh–Gans approximation for a modified Gaussian correlation function and varying relative standard deviation of radius σ and correlation angle Γ. The size parameter is $x = ka = 10$, where k is the wavenumber and a is the mean radius.

C. METHODS FOR THE RESONANCE REGION

Of utmost importance in the resonance region is the analytical, second-order perturbation approximation (Schiffer, 1989, 1990) for light scattering by statistically irregular particles with surface deformations much smaller than the mean radius of the particle and much smaller than the wavelength.

To make the perturbation approximation applicable to Gaussian random particles, we must rephrase the shape as

$$r(\vartheta, \varphi) \equiv a\big[1 + f(\vartheta, \varphi)\big],$$

$$f(\vartheta, \varphi) = \exp\Big[s(\vartheta, \varphi) - \frac{1}{2}\beta^2\Big] - 1,$$

(41)

so that

$$\langle f(\vartheta, \varphi)\rangle = 0,$$

$$\langle f(\vartheta_1, \varphi_1)f(\vartheta_2, \varphi_2)\rangle = \Sigma_r(\gamma).$$

(42)

Note that the random variable f is lognormally distributed with zero mean and threshold equal to -1. Furthermore, the Legendre expansion of the radius covariance function is required:

$$\Sigma_r(\gamma) = \sum_{l=0}^{\infty} D_l P_l(\cos\gamma),$$

$$D_l = \Big(l + \frac{1}{2}\Big)\int_{-1}^{1} d\xi\, P_l(\xi)\Sigma_r(\xi), \qquad \xi = \cos\gamma.$$

(43)

The Legendre coefficients of the radius covariance function can be calculated analytically using the 3j symbols (Edmonds, 1974) from the coefficients of the logradius covariance function. For small surface deformations, the D_l coefficients come close to the C_l coefficients of the logradius covariance function.

The absorption and scattering cross sections can be computed from the extinction cross section and the incoherent and coherent scattering cross sections:

$$C_{\text{abs}} = C_{\text{ext}} - C_{\text{sca}},$$

$$C_{\text{sca}} = C_{\text{sca}}^{\text{incoh}} + C_{\text{sca}}^{\text{coh}}.$$

(44)

With the help of the 3j symbols, Schiffer's $J^{(m)}$ and $H^{(m)}$ functions (superscript denoting refractive index), and the zeroth-order Lorenz–Mie scattering coefficients $a_{l,1}^{(0)}$ and $b_{l,1}^{(0)}$, the cross sections are

$$
\begin{aligned}
C_{\text{ext}} &= \frac{2\pi}{k^2} \operatorname{Re} \sum_{nlL} (-1)^l \rho_L (2n+1)(2l+1) \begin{pmatrix} n & l & L \\ -1 & 1 & 0 \end{pmatrix}^2 \\
&\quad \times \left[J_1^{(m)}(n, l, L) + J_2^{(m)}(n, l, L) \right], \\
C_{\text{sca}}^{\text{incoh}} &= \frac{\pi}{k^2} \sum_{nlL} \rho_L (2n+1)(2l+1) \begin{pmatrix} n & l & L \\ -1 & 1 & 0 \end{pmatrix}^2 \\
&\quad \times \left[\left| H_1^{(m)}(n, l, L) \right|^2 + \left| H_2^{(m)}(n, l, L) \right|^2 \right], \\
C_{\text{sca}}^{\text{coh}} &= \frac{2\pi}{k^2} \operatorname{Re} \sum_{nlL} (-1)^l \rho_L (2n+1)(2l+1) \begin{pmatrix} n & l & L \\ -1 & 1 & 0 \end{pmatrix}^2 \\
&\quad \times \left[a_{l,1}^{(0)*} J_1^{(m)}(n, l, L) + b_{l,1}^{(0)*} J_2^{(m)}(n, l, L) \right].
\end{aligned}
\tag{45}
$$

The scattering matrix elements divide into zeroth-order, incoherent, and coherent contributions:

$$
\begin{aligned}
a_\nu(\Theta) &= a_\nu^{(0)}(\Theta) + a_\nu^{\text{incoh}}(\Theta) + a_\nu^{\text{coh}}(\Theta), \qquad \nu = 1, 2, 3, 4, \\
b_\nu(\Theta) &= b_\nu^{(0)}(\Theta) + b_\nu^{\text{incoh}}(\Theta) + b_\nu^{\text{coh}}(\Theta), \qquad \nu = 1, 2,
\end{aligned}
\tag{46}
$$

where the detailed formulas can be found in Schiffer (1990).

In Fig. 4, the extinction, scattering, and absorption cross sections are shown for the size parameter $x \le 10$ and refractive index $m = 1.5 + i0.05$, standard deviation $\sigma = 0.07$, and the polynomial (P_2, P_3, P_4) and modified Gaussian correlation functions ($\Gamma = 30°$). Noticeable deviations are realized for size parameters $x > 6$. It is interesting to note the proximity of the results for the P_2 and modified Gaussian correlation function, evidently resulting from the proximity of the correlation angles (for P_2, we have $\Gamma = 33.6°$). The 3j symbols were computed numerically using the program by Clark (1978).

Peltoniemi *et al.* (1998) performed preliminary computations for light scattering by Gaussian particles using the variational volume integral equation method (which we will call Jscat; see Peltoniemi, 1996), the discrete dipole approximation (DDA; Draine and Flatau, 1994; Chapter 5), and the Rayleigh–Gans approximation (RGA; Muinonen, 1996b). The computations were in fair mutual agreement for varying refractive indices, but turned out to require vast amounts of computer time for ensemble averaging as compared to the second-order perturbation approximation (PS2), for instance. Note that, in the integral equation method of Peltoniemi (1996), special attention is directed toward correctly accounting for the singularity in the Green's function.

Figures 5 and 6 compare preliminary results for Gaussian particles obtained using the different techniques. The mean-radius size parameter was fixed at $x = 1$,

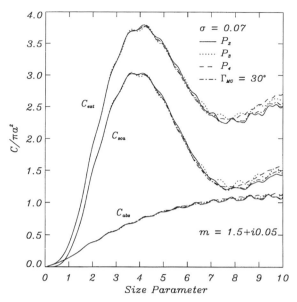

Figure 4 Extinction, scattering, and absorption cross sections for Gaussian spheres using the second-order perturbation approximation. Here, $\sigma = 0.07$, and results are shown for polynomial (P_2, P_3, P_4) and modified Gaussian (MG) correlation functions as a function of mean-radius size parameter $x = ka$.

the refractive index was either $m = 1.5$ or $m = 3 + i4$, and a modified Gaussian correlation function was assumed with $\sigma = 0.1$ and $\Gamma = 30°$. For $m = 1.5$ (Fig. 5), the DDA, PS2, and Lorenz–Mie results (using the computer code in Bohren and Huffman, 1983) are in good agreement, whereas the Jscat is offset probably as a result of the small number of sample shapes. Even though the RGA is strictly not valid for these parameters, it produces a reasonable approximation for the scattering phase function and degree of linear polarization. As for $m = 3 + i4$ (Fig. 6), the Jscat and PS2 scattering phase functions overlap closely, being clearly different from the Lorenz–Mie result. However, the Jscat and Lorenz–Mie polarizations are close to each other, whereas the PS2 polarization is markedly more neutral. It is evident that more computations are needed using the Jscat and DDA techniques in order to make more definitive conclusions.

Michel (1995) applied the strong-permittivity fluctuations theory to scattering by irregular particles. By solving the Dyson equation for the first statistical moment of the electric field, he obtained the ensemble-averaged extinction cross section. In the future, it appears promising to apply the methods developed by Michel to scattering by Gaussian random spheres in the resonance region.

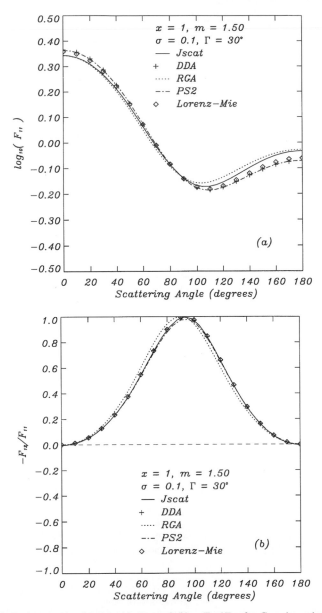

Figure 5 Scattering matrix elements (a) F_{11} and (b) $-F_{12}/F_{11}$ for Gaussian spheres (modified Gaussian correlation function) computed using the variational volume integral equation method (Jscat), discrete dipole approximation (DDA), Rayleigh–Gans approximation (RGA), and second-order perturbation approximation (PS2). The Lorenz–Mie results are included for an equal-volume sphere.

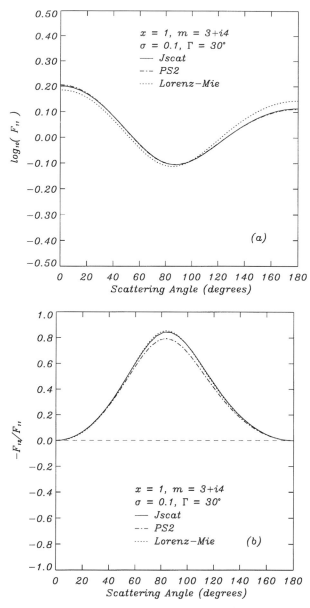

Figure 6 Scattering matrix elements (a) F_{11} and (b) $-F_{12}/F_{11}$ for Gaussian spheres (modified Gaussian correlation function) computed using the variational volume integral equation method (Jscat) and second-order perturbation approximation (PS2). The Lorenz–Mie results are included for an equal-volume sphere.

D. RAY OPTICS APPROXIMATION

The ray optics treatment for Gaussian spheres and cylinders here derives from the work of Muinonen *et al.* (1989, 1996), Muinonen (1989, 1996a, b), and Peltoniemi *et al.* (1989). As for the geometric optics part, a Mueller matrix is related to every ray and at a boundary surface, reflection and refraction take place according to Snell's law and Fresnel's reflection and refraction matrices. As for the forward diffraction part, the two-dimensional silhouette is numerically computed for each sample shape and diffraction is then ensemble averaged in the Kirchhoff approximation.

Before applying the ray optics approximation to Gaussian particles, a general summary of the approximation is given. As for the more advanced geometrical theory of diffraction, Kirchhoff approximation, and ray-by-ray integration method that account for certain wave-optical effects, the reader is referred to the work of Keller (1962), Muinonen (1989), and Yang and Liou (1996b), respectively.

For ray optics to be valid, the curvature radii R_c on the particle surface must be much larger than the wavelength of incident light λ and the central phase shifts across the surface irregularities must be much larger than one (van de Hulst, 1957):

$$kR_c \gg 1, \qquad 2kR_c|m - 1| \gg 1, \tag{47}$$

where $k = 2\pi/\lambda$. The curvature radius can be defined with the help of the Gaussian total curvature K (Struik, 1961; Muinonen *et al.*, 1997):

$$R_c \equiv \frac{1}{\sqrt[4]{\langle K^2 \rangle}}. \tag{48}$$

By making use of the size parameter $x = ka$, we can now recast Eq. (47) as

$$x \gg \sqrt[4]{a^4 \langle K^2 \rangle}, \qquad 2x|m - 1| \gg \sqrt[4]{a^4 \langle K^2 \rangle}. \tag{49}$$

The ray optics regime depends on the mean radius and $\sqrt[4]{a^4 \langle K^2 \rangle}$. The validity criteria in Eq. (49) could be further refined by utilizing the entire probability density of K.

The scattering matrix **F** relates the Stokes vectors of the incident and scattered light $\mathbf{I}^{\text{inc}}$ and $\mathbf{I}^{\text{sca}}$; denoting the directions of incidence and scattering by $\Omega^{\text{inc}} = (\theta, \phi)$ and $\Omega^{\text{sca}} = (\Theta, \Phi)$, respectively, and the distance between the particle center and the observer by R,

$$\mathbf{I}^{\text{sca}}(\Omega^{\text{sca}}, \Omega^{\text{inc}}) = \frac{C_{\text{sca}}(\Omega^{\text{inc}})}{4\pi R^2} \mathbf{F}(\Omega^{\text{sca}}, \Omega^{\text{inc}}) \cdot \mathbf{I}^{\text{inc}}(\Omega^{\text{inc}}),$$

$$\int_{4\pi} \frac{d\Omega^{\text{sca}}}{4\pi} F_{11}(\Omega^{\text{sca}}, \Omega^{\text{inc}}) = 1, \tag{50}$$

where C_{sca} is the scattering cross section and F_{11} is the scattering phase function, both depending in general on the incident direction (Ω^{inc}). Note that the incident direction $\Omega^{inc} = (0, 0)$ in the incident ray coordinate system.

For particles large compared to the wavelength, the scattering cross section and scattering matrix can be divided into the forward diffraction and geometric optics parts (denoted by the superscripts D and G):

$$C_{sca}(\Omega^{inc}) = C_{sca}^D(\Omega^{inc}) + C_{sca}^G(\Omega^{inc}),$$

$$\mathbf{F}(\Omega^{sca}, \Omega^{inc}) = \frac{1}{C_{sca}(\Omega^{inc})}[C_{sca}^D(\Omega^{inc})\mathbf{F}^D(\Omega^{sca}, \Omega^{inc})$$

$$+ C_{sca}^G(\Omega^{inc})\mathbf{F}^G(\Omega^{sca}, \Omega^{inc})], \tag{51}$$

$$\int_{4\pi} \frac{d\Omega^{sca}}{4\pi} F_{11}^D(\Omega^{sca}, \Omega^{inc}) = \int_{4\pi} \frac{d\Omega^{sca}}{4\pi} F_{11}^G(\Omega^{sca}, \Omega^{inc}) = 1.$$

In the ray optics approximation, we strictly require

$$C_{sca}^D(\Omega^{inc}) = A_\perp(\Omega^{inc}),$$

$$C_{ext}(\Omega^{inc}) = C_{abs}(\Omega^{inc}) + C_{sca}(\Omega^{inc}) = 2A_\perp(\Omega^{inc}), \tag{52}$$

where C_{ext} and C_{abs} are the extinction and absorption cross sections and $A_\perp$ is the cross-sectional area. The absorption cross section is due solely to geometric optics: $C_{abs} = C_{abs}^G$. The asymmetry parameter can be divided into the forward diffraction and geometric optics parts as in Eq. (51) for the scattering matrix.

For the geometric optics contribution, for external incidence, Snell's law is applied in the form

$$\sin \iota = \text{Re}(m) \sin \tau, \tag{53}$$

where ι and τ are the angles of incidence and refraction, respectively. When determining τ, we thus make use of $\text{Re}(m)$ only and assume that $\text{Im}(m)$ either has negligible influence on τ or is large enough to entirely eliminate internal ray propagation. To be more accurate, Snell's law can be rigorously generalized to complex refractive indices (e.g., Modest, 1993).

For external incidence, the Mueller matrices of the reflected and refracted rays (denoted by the superscripts r and t) can be obtained from

$$\mathbf{M}^r = \mathbf{R} \cdot \mathbf{L} \cdot \mathbf{M}^{inc},$$

$$\mathbf{M}^t = \mathbf{T} \cdot \mathbf{L} \cdot \mathbf{M}^{inc}, \tag{54}$$

where **L** is the rotation to the plane of incidence and **R** and **T** are the Fresnel reflection and transmission matrices:

$$\mathbf{L} = \begin{bmatrix} 1 & 0 & 0 & 0 \\ 0 & \cos 2\eta & \sin 2\eta & 0 \\ 0 & -\sin 2\eta & \cos 2\eta & 0 \\ 0 & 0 & 0 & 1 \end{bmatrix},$$

$$\mathbf{R} = \frac{1}{2} \begin{bmatrix} r_\parallel r_\parallel^* + r_\perp r_\perp^* & r_\parallel r_\parallel^* - r_\perp r_\perp^* & 0 & 0 \\ r_\parallel r_\parallel^* - r_\perp r_\perp^* & r_\parallel r_\parallel^* + r_\perp r_\perp^* & 0 & 0 \\ 0 & 0 & 2\,\mathrm{Re}(r_\parallel r_\perp^*) & 2\,\mathrm{Im}(r_\parallel r_\perp^*) \\ 0 & 0 & -2\,\mathrm{Im}(r_\parallel r_\perp^*) & 2\,\mathrm{Re}(r_\parallel r_\perp^*) \end{bmatrix}, \quad (55)$$

$$\mathbf{T} = \frac{1}{2} \begin{bmatrix} t_\parallel t_\parallel^* + t_\perp t_\perp^* & t_\parallel t_\parallel^* - t_\perp t_\perp^* & 0 & 0 \\ t_\parallel t_\parallel^* - t_\perp t_\perp^* & t_\parallel t_\parallel^* + t_\perp t_\perp^* & 0 & 0 \\ 0 & 0 & 2\,\mathrm{Re}(t_\parallel t_\perp^*) & 2\,\mathrm{Im}(t_\parallel t_\perp^*) \\ 0 & 0 & -2\,\mathrm{Im}(t_\parallel t_\perp^*) & 2\,\mathrm{Re}(t_\parallel t_\perp^*) \end{bmatrix},$$

where η is the rotation angle and $r_\parallel$, $r_\perp$, $t_\parallel$, and $t_\perp$ are Fresnel's coefficients:

$$r_\parallel = \frac{m\cos\iota - \cos\tau}{m\cos\iota + \cos\tau}, \qquad r_\perp = \frac{\cos\iota - m\cos\tau}{\cos\iota + m\cos\tau},$$

$$t_\parallel = \frac{2\cos\iota}{m\cos\iota + \cos\tau}, \qquad t_\perp = \frac{2\cos\iota}{\cos\iota + m\cos\tau}. \qquad (56)$$

Because the Mueller matrices in Eq. (55) interrelate flux densities that are not conserved, in practical ray tracing, the energy conservation is established by renormalizing the refraction coefficients in **T** so that

$$|r_\parallel|^2 + \frac{\mathrm{Re}(m^*\cos\tau)}{\cos\iota}|t_\parallel|^2 = 1, \qquad |r_\perp|^2 + \frac{\mathrm{Re}(m\cos\tau)}{\cos\iota}|t_\perp|^2 = 1. \qquad (57)$$

The internal incidence is treated analogously except that inside the particle, rays can be totally reflected and attenuated because of absorption. The condition for total internal reflection is $\sin\iota > 1/\mathrm{Re}(m)$, again depending on $\mathrm{Re}(m)$ only. Exponential absorption $\exp[-2\,\mathrm{Im}(m)k\Delta r]$ is assumed along the ray path (Δr is the path length) and the exponential attenuation factor is applied to the entire Mueller matrix.

To summarize, rays are traced until the flux decreases below a specific cutoff value or the ray has undergone a specific number of internal or external reflections. Scattered rays carry Mueller matrices that contribute to the geometric optics scattering matrix.

For particles much larger than the wavelength, the forward diffraction contribution can be approximated by the Dirac delta function. More detailed computations can be made using the Kirchhoff approximation (e.g., Muinonen, 1989). In the spherical geometry, the forward diffraction scattering matrix can be written as

$$\mathbf{F}^{D}(\Omega^{sca}, \Omega^{inc}) \propto \frac{k^2}{4\pi A_{\perp}(\Omega^{inc})} |u(\Omega^{sca}, \Omega^{inc})|^2 (1 + \cos\Theta)^2 \mathbf{1},$$

$$u(\Omega^{sca}, \Omega^{inc}) = \int_0^{2\pi} d\phi' \int_0^{r(\phi', \Omega^{inc})} dr' \, r' \qquad (58)$$

$$\times \exp[-ikr' \sin\Theta \cos(\Phi - \phi')],$$

where $\mathbf{1}$ is the 4×4 unit matrix and $r(\phi', \Omega^{inc})$ describes the particle silhouette for a given incident direction Ω^{inc}. In particular, for a spherical particle with equal projected area $A_{\perp}$, the diffraction pattern is

$$u(\Omega^{sca}, \Omega^{inc}) = A_{\perp} \frac{2J_1(k \sin\Theta)}{k \sin\Theta}, \qquad (59)$$

where J_1 is the Bessel function of the first kind.

In the cylindrical geometry, for particles of finite projected length $2Z$, the forward diffraction scattering matrix can be approximated by (e.g., Muinonen and Saarinen, 1999)

$$\mathbf{F}^{D}(\Omega^{sca}, \Omega^{inc}) \propto \frac{k^2}{4\pi \cdot 2ZA_{\perp}(\Omega^{inc})} |u(\Omega^{sca}, \Omega^{inc})|^2 (1 + \cos\Theta)^2 \mathbf{1},$$

$$u(\Omega^{sca}, \Omega^{inc}) = \int_{-Z}^{Z} dx \int_{y_1(x, \Omega^{inc})}^{y_2(x, \Omega^{inc})} dy \qquad (60)$$

$$\times \exp[-ik(x \sin\Theta \cos\Phi + y \sin\Theta \sin\Phi)],$$

where y_1 and y_2 describe the silhouette for given incident direction Ω^{inc}. The forward diffraction can be approximated by the diffraction pattern from a rectangular obstacle of length $2Z$ and width $2c$ defined so that the cross-sectional area coincides with that of the Gaussian sample cylinder:

$$u(\Omega^{sca}, \Omega^{inc}) = \frac{2 \sin(kc \sin\Theta \sin\Phi)}{k \sin\Theta \sin\Phi} \frac{2 \sin(kZ \sin\Theta \cos\Phi)}{k \sin\Theta \cos\Phi}. \qquad (61)$$

It is important to note that singularities can now show up in the scattering matrix elements. For example, for an infinite circular cylinder and normal incidence, all the scattered radiation is strictly confined to the plane perpendicular to the cylinder axis.

For Gaussian spheres and cylinders, the primary goal is the computation of ensemble-averaged extinction, scattering, and absorption cross sections and scattering matrices. For example, for Gaussian spheres, the scattering cross section and scattering matrix are

$$C_{\text{sca}}^{\text{G}} = \langle C_{\text{sca}}^{\text{G}}(\{s_{lm}\}) \rangle,$$

$$\mathbf{F}^{\text{G}} = \frac{1}{C_{\text{sca}}^{\text{G}}} \langle C_{\text{sca}}^{\text{G}}(\{s_{lm}\}) \mathbf{F}^{\text{G}}(\{s_{lm}\}) \rangle,$$

(62)

where the random variables $\{s_{lm}\}$ are the spherical harmonics coefficients of the logradius [Eq. (1)].

Figure 7 shows the scattering phase functions F_{11}^{G}, degrees of linear polarization $-F_{12}^{\text{G}}/F_{11}^{\text{G}}$, and ratios $F_{22}^{\text{G}}/F_{11}^{\text{G}}$ for Gaussian spheres with refractive index $m = 1.5$ and a modified Gaussian correlation function for varying standard deviation of the radius and correlation angle. For small σ and/or large Γ, the scattering phase functions resemble those for perfect spheres, but become markedly smoother with increasing σ and/or decreasing Γ (Fig. 7a). Similar trends are clearly visible in the degree of linear polarization (Fig. 7b). Strong depolarization is evident from Fig. 7c; for perfect spheres, the matrix element ratio would be equal to one. The current results were computed using 10^6 rays and sample shapes (i.e., one ray per shape) and using an angular resolution of $4°$.

Figure 8 illustrates ray-tracing results for Gaussian cylinders with varying cylinder obliquity with respect to the scattering plane defined by the incident direction and the direction normal to the incident and cylinder axis directions (Muinonen and Saarinen, 1998). The angle $\theta = 90°$ corresponds to normal incidence. Increasing tilt results in fewer and fewer rays being scattered toward the backscattering domain and into the scattering plane. The results were computed using 10^6 rays and shape realizations and using an angular resolution of $5°$ over the full solid angle. The refractive index is $m = 1.55 + i10^{-4}$.

IV. CONCLUSION

The future prospects of light scattering by Gaussian particles are promising. For example, for Gaussian random spheres small compared to the wavelength, the Rayleigh-volume approximation has already been extended to a Rayleigh-ellipsoid approximation, that is, to finding best-fit ellipsoids for Gaussian sample particles and approximating their scattering by the electrostatics approximation for ellipsoids (Battaglia *et al.*, 1999). In the resonance region, the second-order perturbation approximation can be extended to higher orders and, upon convergence, such analytical techniques can be significantly faster than volume integral equation methods. However, the latter provide important means for verifying the validity of the perturbation approximations and, in addition, are applicable to opti-

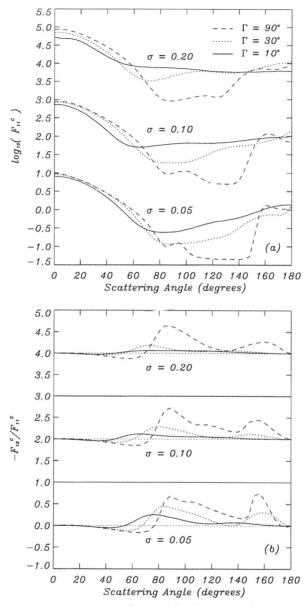

Figure 7 Scattering matrix elements (a) F_{11}^{G}, (b) $-F_{12}^{G}/F_{11}^{G}$, and (c) F_{22}^{G}/F_{11}^{G} for Gaussian spheres in the geometric optics approximation for a modified Gaussian correlation function and varying relative standard deviation of radius σ and correlation angle Γ. The refractive index is $m = 1.5$. Curves for $\sigma = 0.10$ and $\sigma = 0.20$ are shifted up by 2.0 and 4.0 vertical units, respectively.

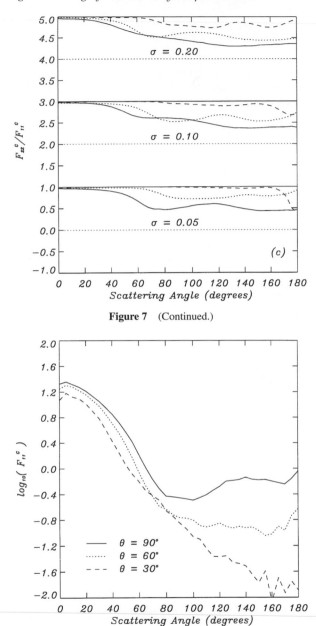

Figure 7   (Continued.)

Figure 8   Scattering matrix elements F_{11}^{G} for Gaussian cylinders in the geometric optics approximation for a modified Gaussian covariance function with $\sigma = 0.1$, $\Gamma_{\varphi} = 30°$, and $\ell_z = 0.5$, with varying incidence angle θ. For $\theta = 90°$, the cylinder axis is normal to the scattering plane. The refractive index is $m = 1.55 + i10^{-4}$.

cally inhomogeneous Gaussian particles. It would be tempting to devise a second-order perturbation approximation for light scattering by Gaussian cylinders.

This chapter has summarized the theoretical studies of light scattering by stochastically shaped particles with particular emphasis on Gaussian random particles. Such studies are essential for making overall conclusions about scattering by many kinds of natural and artificial nonspherical particles. It is important to examine what extinction, absorption, and scattering characteristics can and cannot be explained by the Gaussian shape model. The various methods help us to understand light scattering by small solar system particles and to derive the physical properties of asteroids, comets, and other solar system bodies.

ACKNOWLEDGMENTS

I am grateful to J. I. Peltoniemi, T. Nousiainen, and A. Battaglia for numerous enlightening discussions on light scattering by stochastically shaped particles, and to M. I. Mishchenko and L. D. Travis for their constructive criticism on an early version of the chapter.

Part IV

Laboratory Measurements

Chapter 12

Measuring Scattering Matrices of Small Particles at Optical Wavelengths

Joop W. Hovenier

Department of Physics and Astronomy
Free University
1081 HV Amsterdam
and
Astronomical Institute "Anton Pannekoek"
University of Amsterdam
1098 SJ Amsterdam, The Netherlands

I. INTRODUCTION

A survey of experimental techniques for measuring scattering characteristics is presented in Sections I and IV of Chapter 2. The main purpose of this chapter is to present a concise description of an advanced experimental setup for measuring all elements of the scattering matrix of an ensemble of particles, as functions of the scattering angle, in the visible part of the spectrum.

Measuring the full scattering matrix, $\mathbf{F}$, instead of only some of its elements (e.g., F_{11} and F_{21}) has several advantages. For example, as experience has shown us, some experimental errors are barely or not at all discernible when only a few

Light Scattering by Nonspherical Particles: Theory, Measurements, and Applications
Copyright © 2000 by Academic Press. All rights of reproduction in any form reserved.

elements are measured. Furthermore, for several important applications it is the full scattering matrix that is needed. A typical example is provided by multiple scattering in planetary atmospheres when polarization is taken into account (see, e.g., van de Hulst, 1980).

The experimental setup described in this chapter was built in the 1980s and subsequently improved and extended, mainly by P. Stammes, F. Kuik, and H. Volten, at the Department of Physics and Astronomy, Free University, Amsterdam, where the setup is still operating. The treatment in this chapter is necessarily brief; further details can be found in the literature, in particular, in Stammes (1989), Kuik *et al.* (1991), Kuik (1992), and Volten *et al.* (1998).

II. MUELLER MATRICES AND POLARIZATION MODULATION

In our experimental setup a beam with wavelength λ produced by a light source (a laser) passes through a linear polarizer and a polarization modulator before it illuminates randomly oriented particles contained in a jet stream (aerosols) or a sample holder (hydrosols). Light scattered by the particles at a scattering angle Θ may optionally pass through a quarter-wave plate and a polarization analyzer before its flux is measured by a detector. Using Stokes parameters (van de Hulst, 1957; Hovenier and van der Mee, 1983), the Stokes vector of the light reaching the detector can be written as

$$\mathbf{I}'(\Theta) = c_1 \mathbf{A}(\gamma_A)\mathbf{Q}(\gamma_Q)\mathbf{F}(\Theta)\mathbf{M}(\gamma_M)\mathbf{P}(\gamma_P)\mathbf{I}, \tag{1}$$

where $\mathbf{I}$ is the Stokes vector of the beam leaving the light source; c_1 is a real constant; and $\mathbf{A}$, $\mathbf{Q}$, $\mathbf{M}$, and $\mathbf{P}$ are 4×4 Mueller matrices of the analyzer, quarter-wave plate, modulator, and polarizer, respectively. The orientation angles γ_A, γ_Q, γ_M, and γ_P of the corresponding optical components are the angles between their optical axes and the reference plane, measured anticlockwise from the reference plane when looking in the direction of beam propagation. It is assumed that the scattering plane acts as the plane of reference for the Stokes parameters. This plane coincides with the horizontal plane in the experimental setup. Because the particles are in random orientation in three-dimensional space, the scattering matrix, $\mathbf{F}(\Theta)$, depends only on the scattering angle as far as directions are concerned. Note that because our treatment of polarization follows van de Hulst (1957) and Hovenier and van der Mee (1983), we use the $\exp(j\omega t)$ time convention, where $j^2 = -1$, rather than the $\exp(-i\omega t)$ convention adopted in Chapter 1. As explained in Section VI of Chapter 1, this causes a sign change in the numerical values of the scattering matrix elements in the fourth row and fourth column except the F_{44} element.

The Mueller matrices $\mathbf{A}$, $\mathbf{Q}$, and $\mathbf{P}$ occurring in Eq. (1) are, for an arbitrary orientation angle γ, as follows (see, e.g., Shurcliff, 1962):

$$\mathbf{A}(\gamma) = \mathbf{P}(\gamma) = \frac{1}{2} \begin{bmatrix} 1 & C & S & 0 \\ C & C^2 & SC & 0 \\ S & SC & S^2 & 0 \\ 0 & 0 & 0 & 0 \end{bmatrix}, \tag{2}$$

where

$$C = \cos 2\gamma, \tag{3}$$

$$S = \sin 2\gamma, \tag{4}$$

and

$$\mathbf{Q}(\gamma) = \begin{bmatrix} 1 & 0 & 0 & 0 \\ 0 & C^2 & SC & -S \\ 0 & SC & S^2 & C \\ 0 & S & -C & 0 \end{bmatrix}. \tag{5}$$

The modulator introduces a sinusoidal modulation in time of the polarization of the light before scattering. Combined with lock-in detection, this increases the accuracy of the measurements and yields the capability to deduce several elements of the scattering matrix from only one detected signal. For this purpose, we use the linear electrooptic effect (also called the Pockels effect), that is, the phenomenon that certain crystals become birefringent when an electric field is applied. The voltage over the crystal is varied sinusoidally in time and this creates a varying phase shift between the parallel and perpendicular components of the electric field, which can be written as

$$\phi = \phi_0 \sin \omega t, \tag{6}$$

where ω is the angular frequency of the modulator voltage. The modulator introduces a phase shift even when no voltage is applied, but the voltage is adjusted in each experiment to compensate for this by adding a constant direct-current (dc) component. We now have, for an arbitrary orientation angle γ,

$$\mathbf{M}\gamma = \begin{bmatrix} 1 & 0 & 0 & 0 \\ 0 & C^2 + S^2 \cos \phi & SC(1 - \cos \phi) & -S \sin \phi \\ 0 & SC(1 - \cos \phi) & S^2 + C^2 \cos \phi & C \sin \phi \\ 0 & S \sin \phi & -C \sin \phi & \cos \phi \end{bmatrix}, \tag{7}$$

where ϕ varies in time as given by Eq. (6).

Using Bessel functions of the first kind, $J_k(x)$, we can write

$$\sin\phi = \sin(\phi_0 \sin \omega t) = 2 \sum_{k=1}^{\infty} J_{2k-1}(\phi_0) \sin(2k-1)\omega t, \tag{8}$$

$$\cos\phi = \cos(\phi_0 \sin \omega t) = J_0(\phi_0) + 2 \sum_{l=1}^{\infty} J_{2l}(\phi_0) \cos 2l\omega t. \tag{9}$$

It is convenient to adjust the amplitude of the modulation voltage so that $J_0(\phi_0) = 0$, that is, $\phi_0 = 2.40483$ rad. As a result, we obtain

$$\sin\phi = 2J_1(\phi_0) \sin \omega t + \text{terms with } k \geq 2, \tag{10}$$

$$\cos\phi = 2J_2(\phi_0) \cos 2\omega t + \text{terms with } l \geq 2, \tag{11}$$

where $2J_1(\phi_0) = 1.03830$ and $2J_2(\phi_0) = 0.86350$.

Combining Eqs. (1)–(11), we find for the flux reaching the detector

$$I'(\Theta) = c_2[DC(\Theta) + 2J_1(\phi_0)S(\Theta) \sin \omega t + 2J_2(\phi_0)C(\Theta) \cos 2\omega t], \tag{12}$$

where c_2 is a constant for a specific optical arrangement and higher order terms with $\sin(2k-1)\omega t$ and $\cos 2l\omega t$ [cf. Eqs. (10) and (11)] have been omitted. The coefficients $DC(\Theta)$, $S(\Theta)$, and $C(\Theta)$ contain elements of the scattering matrix. Table I shows the results for eight different optical arrangements. By using lock-in detection the $\sin \omega t$ and $\cos 2\omega t$ components can be separated from the total detected signal. In our setup these two components plus the constant (dc) part of the detected signal are sufficient to determine the angular distributions of all elements of the scattering matrix, apart from one normalization factor for the matrix. To fa-

Table I

Eight Combinations of the Orientation Angles γ_P, γ_M, γ_Q, and γ_A of, Respectively, the Polarizer, the Modulator, the Quarter-Wave Plate, and the Analyzer, Used during the Measurements. A Bar for γ_Q or γ_A Means That the Optical Component Was Not Used. The Coefficients $DC(\Theta)$, $S(\Theta)$, and $C(\Theta)$ Correspond to the dc, the $\sin \omega t$, and the $\cos 2\omega t$ Component of the Detected Signal, Respectively

Combination	γ_P	γ_M	γ_Q	γ_A	$DC(\Theta)$	$S(\Theta)$	$C(\Theta)$
1	$0°$	$-45°$	—	—	F_{11}	$-F_{14}$	F_{12}
2	$0°$	$-45°$	—	$0°$	$F_{11} + F_{21}$	$-(F_{14} + F_{24})$	$F_{12} + F_{22}$
3	$0°$	$-45°$	—	$45°$	$F_{11} + F_{31}$	$-(F_{14} + F_{34})$	$F_{12} + F_{32}$
4	$0°$	$-45°$	$0°$	$45°$	$F_{11} + F_{41}$	$-(F_{14} + F_{44})$	$F_{12} + F_{42}$
5	$45°$	$0°$	—	—	F_{11}	$-F_{14}$	F_{13}
6	$45°$	$0°$	—	$0°$	$F_{11} + F_{21}$	$-(F_{14} + F_{24})$	$F_{13} + F_{23}$
7	$45°$	$0°$	—	$45°$	$F_{11} + F_{31}$	$-(F_{14} + F_{34})$	$F_{13} + F_{33}$
8	$45°$	$0°$	$0°$	$45°$	$F_{11} + F_{41}$	$-(F_{14} + F_{44})$	$F_{13} + F_{43}$

cilitate the discussion of Table I, we consider the ratios $F_{ij}(\Theta)/F_{11}(\Theta)$, except for the phase function $F_{11}(\Theta)$. This function is only obtained on a relative scale in our experiments, but the ratios $F_{ij}(\Theta)/F_{11}(\Theta)$ are not hampered by constants of proportionality if we proceed as follows for each scattering angle.

Combination 1 of Table I yields F_{11} and by dividing $S(\Theta)$ and $C(\Theta)$ by $DC(\Theta)$, also F_{12}/F_{11} and F_{14}/F_{11}. Similarly, combination 5 gives F_{13}/F_{11} and another value for F_{14}/F_{11}. Because of reciprocity, we have for an ensemble of randomly oriented particles (see, e.g., van de Hulst, 1957, Section 5.22)

$$\frac{F_{21}}{F_{11}} = \frac{F_{12}}{F_{11}}, \tag{13}$$

$$\frac{F_{31}}{F_{11}} = -\frac{F_{13}}{F_{11}}, \tag{14}$$

$$\frac{F_{41}}{F_{11}} = \frac{F_{14}}{F_{11}}. \tag{15}$$

We can now find F_{24}/F_{11}, because combination 2 yields

$$\frac{F_{14} + F_{24}}{F_{11} + F_{21}} = \frac{F_{14}/F_{11} + F_{24}/F_{11}}{1 + F_{12}/F_{11}} \tag{16}$$

and F_{24}/F_{11} is the only unknown ratio on the right-hand side of this equation. Similarly, we find F_{34}/F_{11} and F_{44}/F_{11} from combinations 3 and 4. The same three ratios may also be determined from combinations 6–8. Using the $\cos 2\omega t$ components of the detected signals, we find, in a completely analogous manner, F_{22}/F_{11}, F_{32}/F_{11}, and F_{42}/F_{11} from combinations 2–4, as well as F_{23}/F_{11}, F_{33}/F_{11}, and F_{43}/F_{11} from combinations 6–8. Thus, all element ratios are found.

Checks are provided by elements or ratios of elements that are determined in more than one way in the previous procedure and in addition by the relations

$$\frac{F_{32}}{F_{11}} = -\frac{F_{23}}{F_{11}}, \tag{17}$$

$$\frac{F_{42}}{F_{11}} = \frac{F_{24}}{F_{11}}, \tag{18}$$

$$\frac{F_{43}}{F_{11}} = -\frac{F_{34}}{F_{11}}, \tag{19}$$

which are also a result of reciprocity. It should be noted that sensitivity to errors may cause one way to obtain a certain element or ratio of elements to be more useful than another. The measurements with a particular combination of optical components are conducted at different scattering angles in order to determine the angular distribution of the scattering matrix over a large scattering angle range. The relative phase function is often normalized to the value at a fixed scattering angle or to the value of an equivalent sphere phase function at a particular angle.

When particles and their mirror particles are present in equal numbers or when all particles have a plane of symmetry, the scattering matrix has a simple block-diagonal form [see Eq. (61) of Chapter 1]. This, of course, provides more checks, which may help to detect experimental errors, such as misalignments of optical components. The block-diagonal structure of the scattering matrix is often assumed *a priori* in experimental work. In that case fewer than eight optical arrangements are sufficient. For example, we can then use combinations 1, 2, 7, and 8 to obtain the relative phase function and all ratios of elements, whereas combination 1 alone gives the relative phase function and the angular distribution of the degree of linear polarization for unpolarized incident light, that is, $-F_{21}(\Theta)/F_{11}(\Theta)$.

III. EXPERIMENTAL SETUP

In this section we will describe some of the main features of the experimental setup. A schematic overview is shown in Fig. 1. The light source is a laser, for example, a He–Ne laser ($\lambda = 633$ nm) or a He–Cd laser ($\lambda = 442$ nm). The laser beam first passes through a linear polarizer and then travels through an electrooptic modulator (Gsänger, model LM 0202). The modulator, together with the

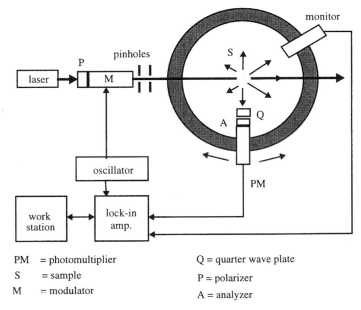

Figure 1   Schematic overview of the experimental setup as seen from above.

polarizer, can be rotated about its longitudinal axis. The modulator operates at an alternating-current (ac) voltage with an angular frequency of $\omega = 2\pi \times 10^3$ rad/s and an amplitude of about 100–200 V. To avoid stray light, the modulated laser beam passes through two pinholes. The beam of light leaving the pinholes has Stokes parameters proportional to $(1, \cos\phi, 0, -\sin\phi)$ in combinations 1–4 and $(1, 0, \cos\phi, -\sin\phi)$ in combinations 5–8 of Table I, where $\sin\phi$ and $\cos\phi$ obey Eqs. (10) and (11), respectively. This light is scattered by a sample of particles at the center of a goniometer ring with a diameter of 100 cm. For aerosols, a vertically directed jet is used, which keeps the particles in the sample in random orientation. Hydrosols are contained in a sample holder, for example, a cylindrically shaped cuvette of Pyrex glass with a diameter of 30 mm. To reduce the influence of undesirable reflections in the latter case, the cuvette is placed in the center of a cylindrical Pyrex glass basin with a diameter of 22 cm and flat entrance and exit windows. This basin is filled with glycerin having the same refractive index as glass, so that strong reflections occur farther away from the scattering sample. A magnetic stirrer in the cuvette homogenizes the hydrosols continuously. The sampler holder and basin are not shown in Fig. 1.

The unscattered part of the incident beam can be absorbed by a beam stop. The light scattered in a particular direction may pass a quarter-wave plate and analyzer (cf. Table I) before it reaches the detector. The latter is a photomultiplier tube mounted on a trolley that runs on the goniometer ring. The field of view of the detector can be restricted by means of pinholes. Another photomultiplier is used to monitor the flux of the scattered light at a fixed scattering angle (see Fig. 1). If, for whatever reason, the flux of scattered radiation fluctuates during a measurement, the signal of the monitor can be used to correct for such fluctuations. The detector can measure the scattered light at all scattering angles between about 5° and 175° for aerosols and between 20° and 160° for hydrosols. Several black screens (not shown in Fig. 1) are employed in order to avoid unwanted contributions to the signals measured. The detected signals are led to lock-in amplifiers, where they are separated into a dc signal and two signals varying in time as $\sin\omega t$ and $\cos 2\omega t$, respectively. Finally, the signals are processed on a workstation. The entire setup is fully computerized.

While measuring at any given scattering angle, 720 measurements are conducted in about 2 s. Consequently, the data points are in fact the average of at least 720 separate measurements.

IV. TESTS

The experimental setup was tested by employing homogeneous isotropic spherical particles and then comparing the experimental results for the scattering matrix with the results of Lorenz–Mie computations. For this purpose water droplets in air, produced by a nebulizer, as well as commercially available latex spheres suspended in water, have been used. The size distribution of the water droplets was not known *a priori*, but the experimentally determined values of the scattering matrix elements were fitted to the values computed with the Lorenz–Mie theory for several size distributions of spheres, until good agreement was obtained.

Further tests were provided by

1. Using general relationships for the elements of a scattering matrix of an ensemble of randomly oriented particles, including the nonnegativity of the eigenvalues of the Cloude coherency matrix (see Chapter 3)
2. Checking the block-diagonal structure of the scattering matrix, which should exist when randomly oriented particles and their mirror particles are present in equal numbers or when each particle has a plane of symmetry
3. Determining certain scattering matrix elements in more than one way (see Section II)

V. RESULTS

In addition to test measurements with spherical particles, the experimental setup has been used for determining the scattering matrix of randomly oriented ice crystals (Stammes, 1989; Kuik, 1992), irregularly shaped quartz particles (Kuik *et al.*, 1991; Kuik, 1992), different types of coastal and inland water phytoplankton species, as well as two types of estuarine sediments (Volten *et al.*, 1998) and rutile particles in water (Volten *et al.*, 1999). Furthermore, work is in progress on a variety of mineral aerosols such as feldspar, red clay, quartz, loess, Sahara sand, volcanic ashes, and olivine.

To illustrate the capabilities of our experimental setup, we show some results for hydrosols in Fig. 2 and for aerosols in Fig. 3, both for $\lambda = 633$ nm. Figure 2 shows the angular dependence of the phase function and the degree of linear polarization for unpolarized incident light for five different types of phytoplankton species. The error bars are based on several series of measurements, each series containing 720 measurements, as explained in Section III. The results of the measurements for the phase function are compared with "the standard scattering function" of San Diego Harbor water (Petzold, 1972). Lorenz–Mie calculations, using *a priori* known or estimated input parameters, did not produce good approximations, as can be seen in Fig. 2. A more detailed discussion, involving 15 different types of coastal and inland water phytoplankton species, as well as two types of estuarine sediments, is given by Volten *et al.* (1998).

Figure 3 shows the angular dependence of six scattering matrix elements for three types of mineral aerosols, namely, feldspar, red clay, and quartz. The elements F_{13}, F_{14}, F_{23}, and F_{24}, were found to be zero over the entire range of

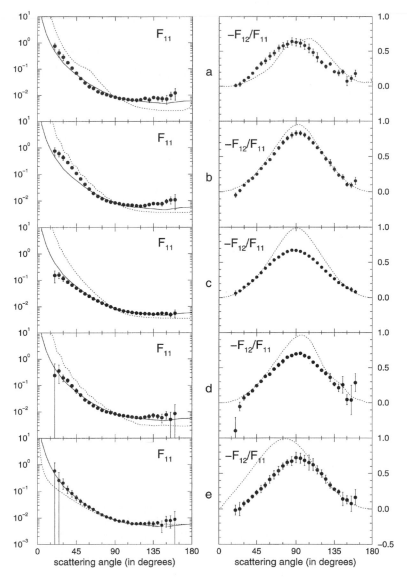

Figure 2   Measured phase functions F_{11} and ratios $-F_{12}/F_{11}$ versus scattering angle Θ are shown in the left and right panels, respectively (filled circles) for (a) *Microcystis aeruginosa* without gas vacuoles, (b) *Microcystis aeruginosa* with gas vacuoles, (c) *Microcystis* sp., (d) *Phaeocystis*, and (e) *Volvox aureus*. Also plotted are the standard phase function for San Diego Harbor water (solid curves, left panels) and the results of Lorenz–Mie calculations (dashed curves, left and right panels). The phase functions $F_{11}(\Theta)$ are scaled at 90° to the phase function of San Diego Harbor water. Experimental errors are smaller than the size of the circles if no error bar is shown. Reprinted from H. Volten, J. F. de Haan, J. W. Hovenier, *et al.* (1998), Laboratory measurements of angular distributions of light scattered by phytoplankton and silt, *Limnol. Oceanogr.* **43**, 1180–1197.

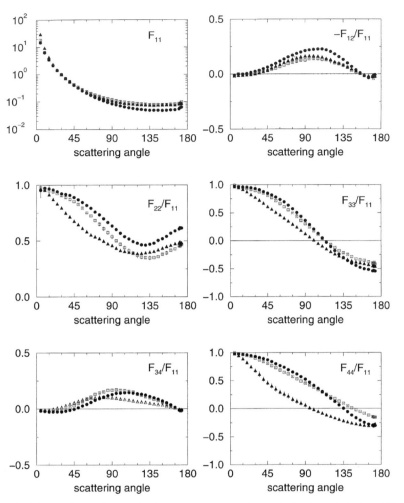

Figure 3 Measured scattering matrix elements versus scattering angle (in degrees) for $\lambda = 633$ nm. Circles correspond to the results for feldspar, triangles to red clay, and squares to quartz. The measurements are presented together with their error bars. If no error bars are shown, the errors are smaller than the size of the symbols.

scattering angles within the accuracy of the measurements. Scanning electron microscope photographs showed the particles to be clearly nonspherical, with mean diameters of 3.0, 5.1, and 9.7 μm for the feldspar, red clay, and quartz particles, respectively. The experimental results will be compared with computed scattering matrices, using methods described in other chapters of this book.

ACKNOWLEDGMENTS

It is a pleasure to express my gratitude to Hester Volten and Johan de Haan for many fruitful discussions and also for their comments on a first draft of this chapter. I am grateful to Kari Lumme of the University of Helsinki for providing some of the mineral aerosol samples.

Chapter 13

Microwave Analog to Light-Scattering Measurements

Bo Å. S. Gustafson

Department of Astronomy
University of Florida
Gainesville, Florida 32611

I. INTRODUCTION

New and old theoretical solutions to the light-scattering problem need to be submitted to state-of-the-art rigorous experimental tests. There are also many light-scattering problems that still cannot be solved theoretically and therefore must be explored using experimental techniques. This chapter addresses the problem of systematic exploration and testing of theoretical solutions. Such experi-

Light Scattering by Nonspherical Particles: Theory, Measurements, and Applications
367

ments are best done for a single particle in a single orientation at a time. The optical constants of the particle's material, its shape, internal structure, and orientation must be accurately known.

It is hard to manufacture and characterize micrometer-sized particles accurately, which limits optical single-particle scattering measurements, and the particle's location and orientation in the beam are usually completely unknown. However, no absolute dimensions are encountered in classical electrodynamics. For example, the dimension of a spherical particle is usually given by the circumference to wavelength ratio or size parameter $x = 2\pi a/\lambda$, where a is the particle radius and λ is the wavelength of the incident light. The light-scattering problem can therefore be scaled to any convenient dimension. In addition, scaling up or down permits a single facility to be used to study electromagnetic radiation of any wavelength as long as the wavelength is sufficiently long so that quantum effects may be neglected and a classical description of matter applies both at the modeled and at the analog wavelength. This is usually true at visual and longer wavelengths.

A factor of 10^3 to 10^5 translates optical wavelengths to the millimeter or centimeter microwave range, a convenient scale that usually enables much improved control over the scatterer. The microwave range typically also allows for higher measurement accuracy than in the optical range because of improved beam control and mature signal-handling technology. Greenberg *et al.* (1961) built the first single-frequency unit in the 1950s using the X band (3.18-cm wavelength) and Giese and co-workers built a Ka-band (8 mm) facility in the 1970s (Zerull, 1985). Although these laboratories are now defunct, the University of Florida built a modern w-band facility (Gustafson, 1996) spanning the 2.7–4-mm range that became operational in 1995. The microwave analog method is well suited to test theoretical solutions and for systematic explorations of scattering properties of particles for which theoretical solutions are either unreliable, computationally taxing, or nonexistent. The method is less well suited to characterization of the scattering by natural particles in broad surveys. This is because of the need to manufacture analog models and the broad parameter space involved in the characterization of the scattering problem.

II. ANALOG MATERIALS

Materials in optical applications are usually described using their complex refractive index $m = m_r + im_i$ or equivalently, in electrodynamics, through the dielectric constant $\varepsilon = \varepsilon' + i\varepsilon''$, where $i = \sqrt{-1}$. (The relations $\varepsilon' = m_r^2 - m_i^2$ and $\varepsilon'' = 2m_r m_i$ hold for nonmagnetic materials.) Although m and ε are often referred to as optical constants, they depend on the angular frequency, ω, so that an analog material must usually be substituted as part of the scaling process. To see this, it is useful to note that the material properties could also be represented using a combination of harmonic oscillators of strength proportional to $(\omega_0^2 - \omega^2)^{-1}$, where the resonance frequency ω_0 is higher than microwave frequencies and often closer to the visual range (Bohren and Huffman, 1983). Based on this representation, optical properties vary only slowly across the microwave range because it is far removed in frequency from ω_0 whereas the visual range usually is much closer to the resonance so that the oscillator strength may change across the range. The oscillator strength and loss are related through the Kramers–Kronig relations (Bohren and Huffman, 1983) so that the loss as well as the oscillator strength depend on the frequency. For example, silicates typically have refractive indices in the $1.55 + i \cdot 0.0001$ to $1.7 + i \cdot 0.03$ range in the visible. They reach one or several extreme values of order $1.1 + i \cdot 1$ in the 10-μm region and then stabilize around $3.5 + i \cdot 0.03$ in the millimeter microwave range and at longer wavelengths. Similarly, water ice has $m \approx 1.31 + i \cdot 3 \cdot 10^{-9}$ in the visible; however, the real part dips below unity near $\lambda = 3$ μm, where the imaginary part peaks around $m_i = 0.6$. The real refractive index approaches an asymptotic value close to 1.78 with the imaginary part in the 10^{-2}–10^{-4} range at wavelengths exceeding a few hundred micrometers.

We tested several methods to measure the refractive index at microwave frequencies and found that a comparison between Lorenz–Mie theory and measurements of the angular distribution of the scattering from a sphere made from the material in question is reliable. It appears to be similar in accuracy to the slotted waveguide method of Roberts and von Hippel (1946) that Schuerman *et al.* (1981) used on the acrylic resin polymethyl methacrylate, also known as Lucite, to obtain $m \approx 1.61 + i \cdot 0.004$ at 9.417 GHz ($\lambda = 3.18$ cm). We obtained $1.605 + i \cdot 0.003$ for the same material at 85 equally spaced frequencies across the 75–110-GHz interval (2.7 mm $< \lambda <$ 4 mm). Zerull *et al.* (1993) also used a fit to the angular scattering by a sphere to obtain $m \approx 1.735 + i \cdot 0.007$ for nylon at 35 GHz ($\lambda = 8$ mm). We obtained $m \approx 1.74 + i \cdot 0.005$ across the 75–110-GHz interval. We conclude that these plastics are remarkably consistent in their refractive index even though they are produced by different manufacturers and that there is indeed no measurable frequency dependence.

These easy-to-machine common plastic compounds, including Delrin with $m \approx 1.655 + i \cdot 0.00$, are excellent analogs for silicates. A variety of other plastics can be mixed with pigments to represent a broad range of rocks, ices, and organics. We have developed plastic compounds to produce any refractive index in the range $m_r = 1.195$–1.7 and $m_i = 0.02$–0.08 with relative ease. We have also made individual samples with imaginary parts as high as 0.2 and real parts near 2. It is usually practical to use commercially available compounds to approach but not precisely duplicate a desired refractive index. In this way discrete and highly reproducible values can be obtained including $m \approx 2.517 + i \cdot 0.017$ for BK7 glass or $m \approx 3.08 + i \cdot 0.001$ for alumina glass. We have also used the metals aluminum, steel, and copper to reach very high real and imaginary refractive indexes. We note that use of analog materials sometimes has great advantages; for example, water ice and many other volatiles can be represented with ease by plastic compounds that can be machined and that do not melt at room temperatures.

Although material optical "constants" are often wavelength dependent in the visual range, that is, materials may have intrinsic color, we have seen that they seldom change appreciably with the wavelength in the microwave region so that most materials are "gray" in microwaves. The wavelength independence in the microwave range allows us to separate color effects induced by the particle size, shape, porosity, and so on from the material-induced colors, whereas materials whose refractive index changes with wavelength in the visual range must be represented using a set of microwave analog models. Again, we see that the microwave analog method has important advantages that are well suited for systematic studies. Color measurements are discussed in Section III.C.

III. MEASUREMENT PRINCIPLES

The ability to separate the wanted scattered radiation from the incident radiation and scattered stray radiation is a key feature that sets the microwave analog method apart from its optical counterparts. The direct illumination and scattering by the support as well as most of the background radiation can be measured separately, that is, without the scattering body in place. This unwanted component can then be subtracted from a measurement that includes the scattering body. The process is complicated by the fact that there are usually definite phase relations between the components so that they interfere. We therefore need to know both the amplitude and the phase to subtract the signals. However, these cannot be measured directly, because only intensities are measured using any type of detector. The amplitude and phase must be inferred from a set of intensity measurements. Several arrangements have been developed for this purpose (e.g., Greenberg *et al.*, 1961; Lind *et al.*, 1965; Zerull *et al.*, 1993). Modern network analyzers achieve comparable accuracy and stability in automated and commercially available packages. The network analyzer-based University of Florida equipment has been described in detail elsewhere (Gustafson, 1996).

Unlike the two preceding facilities, the Florida facility routinely yields amplitude and phase at all scattering angles. This is made possible by dimensional stability that reduces fluctuations in the background signal. Although it is possible to derive the scattering components based on the ratio of the scattered intensity to the intensity measured in the unobstructed direct beam, we found that higher accuracy can be obtained using a sphere and Lorenz–Mie theory for calibration. The scattering quantities that we refer to as "measured" are thus not obtained directly but are processed, first through the subtraction of the background signal and then through calibration.

Although the scattered signal can be separated out, captured, and calibrated, the idealized conditions of light-scattering that are commonly assumed in theoretical works cannot be fully reproduced in any laboratory. For example, the incident wavefront is necessarily finite and suffers from imperfections; the distance to other objects is also finite and the scattering sample must be supported against gravity so some supporting mechanism is necessarily nearby. Gustafson (1996) describes how the Florida facility is designed for stability with its thermal and mechanical layout, how its geometric configuration minimizes the studied model's interaction with the support, and how the use of lenses compensates for the finite beam width to produce a flat wavefront in the central part of the beam.

A. MEASURED AND INFERRED SCATTERING QUANTITIES: THE COMPLETE SET

When, as in the Florida facility, the change in both intensity and phase suffered by a scattered wave can be deduced in the four combinations of polarization along the scattering plane and perpendicular to the plane, we characterize the scattering process completely. This may be easier to visualize using a geometric description of the polarization ellipse for arbitrarily polarized light than it is using the Stokes parameters directly. We recall that the Stokes parameters are used to describe the intensity and polarization state in nearly all modern works because they lead to a convenient mathematical representation, but a geometrical representation in terms of the shape and orientation of the polarization ellipse is equally valid as long as we consider a single frequency at a time. To describe an arbitrary electromagnetic plane wave propagating through empty space along the z axis of a Cartesian coordinate system, it is sufficient to describe the amplitude components of the real electric field vector

$$
\begin{aligned}
E_x &= A\cos(d_1 - d), \\
E_y &= B\cos(d_2 - d), \\
E_z &= 0,
\end{aligned}
\tag{1}
$$

where the orientation of the ellipse is determined by the constant parts of the phases d_1 and d_2 and the shape by the amplitudes A and B and the constant phase difference $d_1 - d_2$. Besides the constant parts, the phase factors consist of a variable part, $d = \omega t - kz$. Here ω is the angular frequency, t time, $k = 2\pi\omega/c$ the propagation constant or wavenumber, and c the speed of light. The magnetic field $\mathbf{H}$ follows from $\mathbf{S} = \mathbf{E} \times \mathbf{H}$, where $\mathbf{S}$ is the Poynting vector. The corresponding Stokes vector is [see Eqs. (A6)–(A9) of Chapter 1]

$$
\begin{aligned}
I &= A^2 + B^2, \\
Q &= A^2 - B^2, \\
U &= -2AB\cos(d_1 - d_2), \\
V &= 2AB\sin(d_1 - d_2).
\end{aligned}
\tag{2}
$$

The squared amplitudes are intensities that can be measured directly, in principle, but are processed in practice to remove background interference and then calibrated. The phases can be obtained using a network analyzer that compares the received signal to a stabilized reference signal of known phase. In the Florida facility, the signal originates in the network analyzer so that the transmitted signal can also provide the phase reference. We note that there are only three independent parameters describing the wave: the two amplitudes A and B and the phase difference $d_1 - d_2$. There exists a relation $I^2 = Q^2 + U^2 + V^2$ for a strictly monochromatic wave so that the Stokes vector also has only three independent parameters. When both the incident and the scattered waves are described using their respective Stokes parameters, the scattering can be described by a linear transformation, mathematically a 4×4 matrix $\mathbf{Z}$ [see Eq. (13) of Chapter 1]:

$$
(I, Q, U, V)^{\mathrm{T}} = \frac{1}{R^2}\mathbf{Z}(I_0, Q_0, U_0, V_0)^{\mathrm{T}},
\tag{3}
$$

where T denotes the matrix transpose, R is the distance from the particle to the detector, and the subscript 0 denotes the incident beam parameters. The 16-element $\mathbf{Z}$ matrix contains only seven independent parameters. Eight parameters are found in the intensity and phase of the mutually perpendicular components measured in the laboratory or, equivalently, in the corresponding complex amplitudes. Of these eight, only the phase differences, not the absolute phases, are relevant (Bohren and Huffman, 1983), which reduces the number to seven.

The principle of optical equivalence states that the Stokes parameters contain the complete set of quantities needed to characterize the intensity and state of polarization of a beam of light, in practical analyses. It follows that the transformation matrix $\mathbf{Z}$ between the incident and the scattered Stokes vectors fully describes the scattering process to the same level of detail. This principle is based on the fact that optical measurements involve linear transformations only. Section IV shows an example of phase as well as intensity measurements in four polarizations, so that all scattering parameters can be determined. This (Gustafson, 1999) may be the first accurate measurement of an elusive light-scattering parameter and the first *complete* experimental characterization of a scattering particle.

B. SCATTERED INTENSITIES AND POLARIZATION: THE SIMPLIFIED SET FOR NATURAL LIGHT

In most practical situations, particles are illuminated by "natural light" with $Q_0 = U_0 = V_0 = 0$ so that there is no definite phase relation between the four polarization components. We see from the form of Eq. (3) that in this case we may concentrate on a subset of the Z matrix. We may rewrite this part of the matrix in terms of quantities measured "directly" in the laboratory. These are the intensities I_1 polarized perpendicular to the scattering plane and I_2 polarized along the plane. Their units are those of the corresponding incident intensities $I_{1,0}$ and $I_{2,0}$. We may then define the transformation matrix as follows:

$$\begin{bmatrix} I_2 \\ I_1 \end{bmatrix} = \frac{1}{k^2 R^2} \begin{bmatrix} i_{22} & i_{21} \\ i_{12} & i_{11} \end{bmatrix} \begin{bmatrix} I_{2,0} \\ I_{1,0} \end{bmatrix}. \tag{4}$$

From these matrix elements, we compute the total scattered intensity as

$$I(\Theta) = \frac{i_{11} + i_{12} + i_{21} + i_{22}}{2k^2 R^2} I_0, \tag{5}$$

and the dimensionless degree of linear polarization

$$P(\Theta) = \frac{(i_{11} + i_{12}) - (i_{21} + i_{22})}{i_{11} + i_{12} + i_{21} + i_{22}}, \tag{6}$$

both as a function of the scattering angle Θ.

C. COLOR

We have seen in Section II that because the resonant frequencies of most materials are far removed from microwave frequencies, their refractive index changes only slowly across the microwave range and is usually constant across a single microwave band. The broadband facility in Florida thus allows us to study the dependency on frequency while the refractive index remains constant. This allows us to separate the color induced by the particle geometry (e.g., size, shape, and internal structure such as porosity) from the intrinsic color of the material because it is neutral, or gray, by definition when the refractive index is constant. The color range covered in the Florida w-band facility is nearly as broad as the human visual range.

Practically any number and combination of frequencies can be used at the Florida facility ranging from 75 to 110 GHz. We integrate the total intensity obtained from Eq. (5) across the waveband, which is represented by the sum over intensities normalized to the width of the band, $\Delta\omega$, so that the plotted values

$$\frac{\sum I(\Theta)}{\Delta\omega} \tag{7}$$

are proportional to the incident intensity and inversely proportional to the distance squared. To give a quantitative measure of the color in intensity and in polarization, we split the laboratory spectrum into "blue" and "red" portions. We represent the integral of matrix elements in Eq. (4) illuminated by the solar spectrum across the $[\omega_a, \omega_b]$ interval using

$$\langle i \rangle_{[\omega_a,\omega_b]} = \frac{\sum_{freq=\omega_a}^{\omega_b} I_{0,freq} i_{freq}}{\sum_{freq=\omega_a}^{\omega_b} I_{0,freq}}, \tag{8}$$

where the indices indicating polarization have been dropped and $I_{0,freq}$ is the solar illumination integrated across the small frequency interval corresponding to one frequency step.

This mimics the usual format of reported observations where the numerator is the observed flux across the waveband and the denominator is a normalization to the solar spectrum. The matrix elements are $\langle i \rangle_{red}$ and $\langle i \rangle_{blue}$, respectively. The averages $\langle I \rangle_{color}$ and $\langle P \rangle_{color}$ can now be computed by substituting the corresponding averages in Eqs. (5) and (6). In addition, we quantify color through the ratio

$$Color(\Theta) = \frac{I_{red} - I_{blue}}{I_{red} + I_{blue}}. \tag{9}$$

We can choose the intervals $[\omega_a, \omega_b]$ for either red or blue colors to simulate a variety of optical filters used in astronomy. In particular, for the comparison of laboratory colors with colors obtained from comet observations, we simulate narrow-band continuum filters 443 and 642 nm using the frequency bands 75–85 GHz and 100–110 GHz.

D. SIZE DISTRIBUTION

Stepping the wavelength during measurements can also be translated to steps in units of the size parameter, x. The measured scattering signals, when weighted by any series of numbers representing a size distribution, could then yield the scattering at a single wavelength from the specified size distribution of otherwise identical particles. In principle, a combination of color and size distribution could also be obtained. However, the design of a microwave facility sufficiently broad to represent most natural size distributions is not possible using standard microwave techniques. The range covered in the Florida facility corresponds to a narrow size interval ranging from $a\lambda/4$ to $a\lambda/2.7$, where a is a characteristic particle dimension in millimeters. For example, a 5-mm particle, which is among the smallest dielectric particles that can be accurately measured in the laboratory, corresponds to the 0.69–1.02-μm size range at $\lambda = 0.55$ μm. The upper size limit is given by the beam width and confines particle models to less than 200–250 mm across, corresponding to the 27.4–40.8-μm range at the same frequency. We can therefore simulate size distributions from 0.7 to 41 μm at $\lambda = 0.55$ μm by using multiple-particle models.

E. ANGULAR SCATTERING PROCEDURES AND ACCURACIES

The mechanical and thermal, as well as electronic, stability of the Florida facility depicted in Fig. 1 allows precise measurements of the angular scattering under automated operation lasting for several weeks. A single-particle

Figure 1 The microwave analog light-scattering facility at the Laboratory for Astrophysics. The transmitting antenna is to the right with the mobile receiving antenna in the foreground. The particle model at the center is supported by a platform made of lightweight plastics consisting of a Styrofoam base and a commercially available thin-walled straw. The platform is usually custom made to suit a specific particle model. Once the model is in place, the fully automated facility operates without human intervention.

orientation requires approximately 16 h of measurement to obtain 0.1° angular resolution from 5°–165° at 85 frequencies and four combinations of polarization, with equal time needed for a background measurement. Because each quantity is sampled 500 times to reduce electric and sampling noise, there is a total of 6.4×10^6 measurements per orientation including the background measurement and an average rate of 56 measurements per second. The particle can be precisely and automatically rotated about an axis perpendicular to the scattering plane through the rotation of the support. The background includes the scattering by the support, which does not have perfect symmetry so that a background measurement is made for each orientation of the support. The background usually does not need to be remeasured except when the target support is replaced. The repeatability and accuracy of angular measurements usually are comparable to errors introduced through uncertainties in the target parameters including the refractive index, shape, and orientation. The discussion of these errors by Gustafson (1996) is still valid, although some improvements have been achieved through mechanical refinement of the scattering arm positioning mechanism and through improved stability of the high-frequency microwave mixers.

Scattering angles of 168°–180° cannot be measured because the antennas would overlap in this angular interval (Fig. 2). However, it is possible to add a backscattering measurement capability through the addition of a receiving circuitry in the transmitting antenna. The 0°–5° interval is plagued by direct illumination of the receiver antenna by way of the first side lobes and therefore would require a tedious measurement procedure similar to that at 0°, which is discussed in the next section.

F. FORWARD SCATTERING PROCEDURES AND ACCURACIES

The forward scattering amplitude and phase are, in principle, measured in a manner similar to those at other scattering angles. However, for small particles, extra care must be taken so that the illumination remains stable between the times of the measurements of the scattered signal plus background and the background alone. This is because the intensity of the scattered signal may constitute a small fraction of the direct beam background signal: approximately 10^{-3} for an $x \approx 10$ sphere. The measurement error is, to a first approximation, proportional to the measured signal strength, which in forward scattering is dominated by the background. This means that the error is expected to be practically independent of the particle's cross section so that the forward scattering by large scatterers can be measured with much higher relative accuracy than that by small particles. The large background signal is a direct consequence of the large beam size, which,

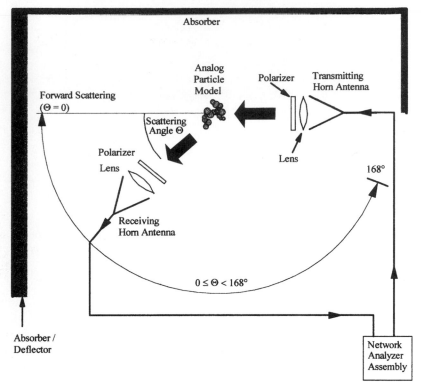

Figure 2   Layout of the University of Florida microwave facility in which the particle model and the wavelength are scaled by the same factor. The scaling may be used to simulate any part of the electromagnetic spectrum; for example, a factor of 6000 corresponds to visual light and scales micrometer-sized particles to the centimeter range. The lens and polarization filter equipped transmitting antenna creates an approximately 100λ diameter ($\sim$30 cm) linearly polarized beam featuring a flat wavefront near the target. An identical assembly makes up the receiver end and can be positioned at any scattering angle from $0°$ to $168°$. The plane of polarization of the transmitted and received beams can be rotated independently. Adapted from B. Å. S. Gustafson (1996).

at about 100 wavelengths across, is approximately a factor of 10 wider (approximately 100 times larger in cross section) than those of the two preceding laboratories. This compromise was made to accommodate particle size parameters up to $x = 200$ or 250 in the beam.

Because the background is so large, our current procedure calls for the measurement of the background signal before and after each measurement with the scattering body in the beam. Direct measurement errors are typically less than one part in 5×10^4. This corresponds to approximately $0.2\lambda^2$ or less when translated

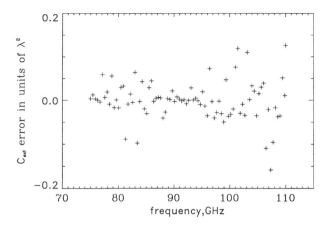

Figure 3 Direct measurement errors in forward scattering given as an equivalent extinction cross section. The errors are to a large extent due to calibration errors and are expected to be improved upon.

into a perturbation in the extinction cross section (see Fig. 3) and is comparable to the geometric cross section of an $x = 1.6$ sphere. Actual errors in the scattering amplitudes are found to be on the order of one part in 10^3 of the measured signal although the particle is suspended on a string rather than supported on a platform as in our normal procedure for angular measurements. This unexpectedly large error appears to result from imperfect calibration; the measurements reported here are therefore expected to be improved upon.

IV. MEASUREMENTS

We use measurements from the microwave laboratory at the University of Florida to help guide theory development (e.g., Xu, 1997) and to test exact (Xu and Gustafson, 1996, 1997) and approximate (Xu and Gustafson, 1999) solutions to the scattering problem and to establish their limit of validity. We systematically explore the optical properties of aggregated dust particles (Gustafson *et al.*, 1999) and cubes (in preparation). An emerging database of color and polarization measurements is used in an interpretation of comet observations (Kolokolova *et al.*, 1998; Gustafson and Kolokolova, 1999) and for an interpretation of the atmosphere surrounding Saturn's satellite Titan (Thomas-Osip, 2000). Representative examples are discussed next.

A. THE SPHERE

The scattering by a homogeneous spherical particle has been a standard test in light scattering since the boundary condition solution, now known as the Lorenz–Mie theory (Lorenz, 1898; Mie, 1908; Debye, 1909), was found about a century ago. We use a commercially available BK7 glass sphere ($m \approx 2.517 + i \cdot 0.017$) intended for use as spherical lenses to illustrate in practice the performance of the microwave laboratory. The specific sphere used in the experiment was 9.998 mm in diameter. We show both amplitude and phase measurements at 0.1° resolution from 5° to 165° and at forward scattering, $\Theta = 0$.

In parts a and b of Fig. 4, we compare the experimentally determined angular distribution of the matrix elements i_{11} and i_{22} at 80 GHz to the Lorenz–Mie solution. The agreement is very good, especially in view of remaining uncertainties in the refractive index and the geometry of the target. The Lorenz–Mie theory predicts no cross-polarization components, i_{12} and i_{21}, because of the symmetry of the sphere. The experiment returned cross-polarized values below 1 at all scattering angles and below 0.01 at most angles, in good agreement with the theory.

For polarizations 11 and 22, respectively, parts c and d of Fig. 4, show the phase of the scattered radiation relative to the geometric center of the sphere. The agreement is excellent considering that a 100-μm error in the placement of the sphere corresponds to nearly 20° in phase shift at backscattering angles. This measurement (reported by Gustafson, 1999) may be the first accurate phase measurement of the angular distribution of scattered radiation and the first to estimate measurement errors at 0° in detail. The agreement constitutes new evidence of the accuracy of both the experimental data and the Lorenz–Mie theory.

Integration over the scattering angle yields the scattering cross section and the asymmetry factor

$$C_{\text{sca}} = \frac{\pi}{k^2} \int_0^\pi \left[i_{11}(\Theta) + i_{21}(\Theta) + i_{12}(\Theta) + i_{22}(\Theta) \right] \sin \Theta \, d\Theta,$$

$$\langle \cos \Theta \rangle = \frac{\pi}{k^2} \int_0^\pi \frac{\left[i_{11}(\Theta) + i_{21}(\Theta) + i_{12}(\Theta) + i_{22}(\Theta) \right] \cos \Theta \sin \Theta \, d\Theta}{C_{\text{sca}}}.$$

(10)

We obtain the scattering efficiency factor $Q_{\text{sca}} = C_{\text{sca}}/\pi a^2 = 1.74$, which is 4% below the Lorenz–Mie value 1.82. Similarly, $\langle \cos \Theta \rangle = 0.658$, which is 2% below the Lorenz–Mie value 0.671. The measured forward scattering intensity and phase were 2100 and $-13.37°$ compared to the Lorenz–Mie values 1970 and $-12.68°$, respectively. We use the fundamental extinction formula (sometimes called the optical theorem; Bohren and Huffman, 1983; Section VIII of Chapter 1)

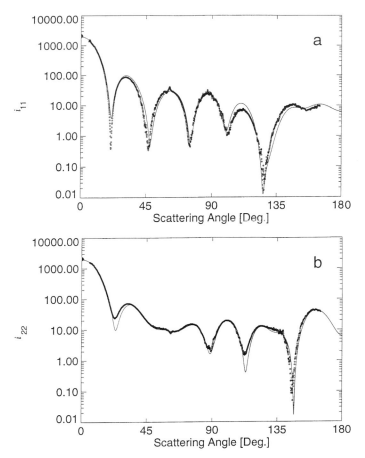

Figure 4a, b Matrix elements i_{11} and i_{22} for a sphere. Solid curves represent the Lorenz–Mie solution and dots show microwave laboratory data. Reprinted from B. Å. S. Gustafson (1999), Scattering by complex systems. I. Methods, *in* "Formation and Evolution of Solids in Space" (J. M. Greenberg and A. Li, Eds.), pp. 535–548. Kluwer Academic, Dordrecht, with kind permission of Kluwer Academic Publishers.

to calculate the extinction cross section C_{ext} and the extinction efficiency Q_{ext} as

$$C_{ext} = \frac{4\pi}{k} \, \mathrm{Im} \left[S(0) \right], \qquad Q_{ext} = \frac{C_{ext}}{\pi a^2}, \tag{11}$$

where $S(0)$ is the complex scattering amplitude at scattering angle $0°$ calculated from the intensity and phase. The measured efficiency is 2.54, which is 3% higher than the Lorenz–Mie value 2.47. The error in the absorption cross section and

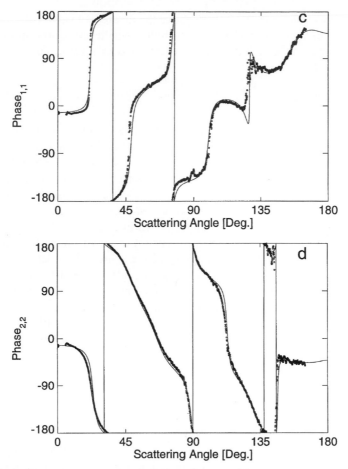

Figure 4c, d Phase of the scattered components polarized perpendicular and parallel to the plane of scattering, respectively. Solid curves represent the Lorenz–Mie solution and dots show microwave experimental data. Reprinted from B. Å. S. Gustafson (1999), Scattering by complex systems. I. Methods, *in* "Formation and Evolution of Solids in Space" (J. M. Greenberg and A. Li, Eds.), pp. 535–548. Kluwer Academic, Dordrecht, with kind permission of Kluwer Academic Publishers.

efficiency jumps to 23% because it is obtained from the difference between the extinction and scattering, that is, $Q_{abs} = Q_{ext} - Q_{sca}$. The radiation pressure is also a compound quantity given by

$$C_{pr} = C_{ext} - \langle \cos \Theta \rangle C_{sca}, \qquad Q_{pr} = \frac{C_{pr}}{\pi a^2}. \qquad (12)$$

The laboratory value for Q_{pr} is 12% above the theoretical value 1.25. The measurements were repeated at 75 GHz with slightly smaller deviations from the theory: $\Delta_{sca} = 2\%$, $\Delta_{\langle\cos\Theta\rangle} = 0\%$, $\Delta_{ext} = -2\%$, $\Delta_{abs} = -12\%$, and $\Delta_{pr} = -5\%$.

B. AGGREGATES

Low-density aggregates are representative of a broad class of particles that initially grow by condensation to form individual grains. Then, as much of the condensable gas is depleted and particles collide and stick together, they grow primarily through aggregation. The size parameter of the constituent grains depends on the ratio of condensables to the number of condensation centers and on the wavelength of observation, whereas the packing depends largely on the conditions and mode of aggregation (e.g., Wurm and Blum, 1998). We distinguish small aggregates that necessarily consist of small grains from the more complex scattering involving large aggregates or aggregates of larger grains. All aggregates discussed here have nominal packing factors near 10% with 90% void unless stated otherwise. This packing is in the range expected from aggregation under a broad range of microgravity conditions and in cometary materials (e.g., Greenberg and Gustafson, 1981; Wurm and Blum, 1998). Examples of scattering by more densely packed aggregates arc discussed by Xu and Gustafson (1997), Kolokolova *et al.* (1998) and Gustafson and Kolokolova (1999).

1. Coherent Scattering by Small Aggregates

The original paper on the scattering by "bird's nest" type of structures at visual wavelengths (Greenberg and Gustafson, 1981) is an investigation into the scattering by 250–500 randomly oriented silicate cylinders representing classical interstellar grains that are loosely assembled into an aggregate of approximately 10% packing. The conceptual models are based on the evolutionary sequence that interstellar grains may have undergone as they became incorporated into primitive solar system bodies and eventually were released as cometary and interplanetary dust. The fact that the scattering by structures of this type and size shows very little evidence of interaction between the individual grains in the structure was known from a comparison between the coherent scattering approximation and microwave experimental data (Gustafson, 1980; Greenberg and Gustafson, 1981; Gustafson, 1983) that were primarily obtained at the X-band facility, then at the State University of New York at Albany.

When the scattering from N identical grains is added coherently, the angular distribution of the scattered intensities peaks at N^2 times the scattering by the individual grains in the forward dircction ($\Theta = 0$) where the scattering from all grains is in phase. In the backscattering hemisphere, where the scattering adds at nearly random phase, the intensity is on the order of N times the scattering by individual grains. The requirement is that the aggregates be at least a few wavelengths across, as indeed was the case. The intensity goes through a series of interference minima in the transition from forward to backscattering. Because coherent addition is found to be a good approximation for all polarization components, their relative intensities are nearly unaffected by the fact that the grains are in an aggregate. The degree of linear polarization [Eq. (6)] therefore closely mimics that of individual grains at all scattering angles. That is, when normalized to the scattering by individual grains, the forward scattering peaks at the value N and then drops to near unity. This sharp drop takes place at smaller scattering angles for larger aggregates compared to the wavelength. For 250–500 aggregated grains at 10% packing factor, the forward interference pattern extends to 30°–60° scattering angle. When the degree of linear polarization is similarly normalized to the polarization produced by the constituent particles, the ratio is near unity in all scattering directions. This ratio is unity by definition in the coherent scattering approximation.

Zerull *et al.* (1993) confirmed this basic result a decade later using similar targets at the Ka-band facility at Bochum University. Loosely packed aggregates of approximately 500 or fewer grains of the "bird's nest" type thus have polarization properties similar to those of the constituent grains scattering independently of each other, whereas the intensity distribution is reminiscent of much larger structures. The targets were then coated by a simulated organic refractory mantle and the measurements repeated as successive mantles were grown. A transition from coherent scattering to a mode where a more complex interaction takes place occurs as the mantles grow sufficiently thick.

A particle may exhibit color although the refractive index of its material is nearly independent of the wavelength. Examples are silicates at visual wavelengths or nearly all materials in the microwave range. Neutral to red color is expected at small scattering angles, whereas the color should be close to that of the constituent grains, which usually is blue, in the backscattering region as long as coherent scattering applies. The color at small scattering angles should be shifted to the red because the interference peak extends to larger scattering angles at longer wavelengths.

However, because measurements were made at a single frequency at the X- and Ka-band facilities, the only way to simulate color required the manufacturing of sets of scale models and all measurements had to be repeated using a different scale for each color. The manufacturing of sets of precisely scaled models of intricate aggregates proved to be a prohibitive task. Color therefore remained a largely unexplored issue until 1995. The study of color and scattering by large aggregates was in part the motivation for the design of the microwave laboratory in Florida as a broad-band facility. The desire to investigate the scattering by larger structures in comparison to the wavelength than before led to the choice of a relatively short wavelength despite the technical difficulties associated with higher frequencies and the outfitting with relatively large antennas. Using this facility, Gustafson *et al.* (1999) confirmed that the color of small aggregates of the "bird's nest" type is blue, as was expected, whereas the color from large aggregates is much weaker.

2. Intensity, Polarization, and Colors from Large Aggregates

Only the smaller types of aggregates could be studied in the X-band and Ka-band facilities and, like color, the scattering by larger aggregates was essentially unexplored until 1995. Outstanding questions were therefore, at what conditions (e.g., size of constituent particles, size of aggregate, refractive index, packing) the simple coherent scattering approximation breaks down and what the consequences of the interactions are. Although the coupled dipole approximation by Purcell and Pennypacker (1973) had been used for decades, it was not until the emergence of contemporary computers that a modern implementation of this method (Draine and Flatau, 1994; Chapter 5) became capable of addressing the scattering by smaller structures. However, even for these, the required computing time remains prohibitive for many applications and the results are not necessarily reliable (Xu and Gustafson, 1999). Once again, we therefore turn our attention to results from microwave analog experiments.

Figure 5a shows an aggregate of 1450 ± 20 polystyrene spheres 0.63 ± 0.35 mm in diameter. Sets of two or three linear aggregates of these size spheres with refractive index near $m = 1.615+i\cdot0.03$ are a possible representation of classic size interstellar grains when observed in the visible. The refractive index was selected to approximate the silicate core and also to be sufficiently close to the value for organic refractory materials so that it may also approximate a mantle, given the uncertainties. Overall, the aggregate is related to those investigated by Greenberg and Gustafson (1981). The primary difference is in the larger number of aggre-

Figure 5a   Aggregate of 1450 ± 20 polystyrene ($m = 1.615 + i \cdot 0.03$) spheres of 0.63 ± 0.35 mm diameter resting on a thin sheet of expanded plastic of low refractive index used to support the delicate assembly during the laboratory experiments. Sets of two to three spheres represent silicate interstellar dust cores in the visual range.

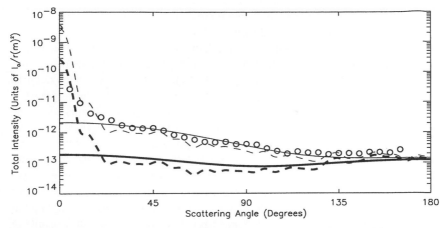

Figure 5b   Angular distribution of scattered intensity averaged over orientations and frequency (circles). Thick curves are for the coherent scattering approximation (dashed curve) and the incoherent scattering (solid curve) where any phase relations are neglected. The thin curves are for the coherent (incoherent) approximation where spheres are grouped in sets of 12 and each group is approximated by an equal-volume sphere. Reprinted from B. Å. S. Gustafson *et al.* (1999), Scattering by complex systems. II. Results from microwave measurements, *in* "Formation and Evolution of Solids in Space" (J. M. Greenberg and A. Li, Eds.), pp. 549–564. Kluwer Academic, Dordrecht, with kind permission of Kluwer Academic Publishers.

gated particles so that the dimensions are on the order of 10 wavelengths across or 5 μm in the visible as opposed to three wavelengths or approximately 1.5 μm across.

In Fig. 5b, the angular distribution of intensity is averaged over 36 orientations and over all 85 frequencies. Although a larger number of orientations is desired for the simulation of a cloud of randomly oriented particles, the number of orientations needed decreases rapidly with the particle size and we do not expect a true random average to be significantly different. Two sets of theoretical curves that were generated assuming randomness are shown for comparison. The set of heavy curves represents the coherent scattering approximation (thick dashed curve) and the incoherent scattering (thick solid curve) where any phase relations are neglected. In incoherent scattering, the intensities can be added directly. This is the way the spheres would scatter if dispersed in a tenuous cloud. Phase relations resulting from differences in the geometric path are only accounted for in our coherent scattering comparison. Both solutions are comparable except at scattering angles smaller than approximately 20° where definite phase relations are established. Evidently, the aggregate is sufficiently dispersed so that differences in the geometric path lead to nearly random phase differences at the larger scattering angles. The experimental data indicate that constructive interference is a good approximation close to the forward direction and incoherent scattering is approached in the backscattering direction. However, unlike results for the aggregates previously reported, the angular distribution of scattered intensities is considerably brighter at all other scattering angles. This could be due to the detailed structure of the aggregate and is not necessarily intrinsic to large-sized aggregates. For practical reasons, the spheres were assembled in groups or clumps where the spheres are in contact with their neighbors on all sides. The aggregate tends to behave as an assembly of these groups rather than as an aggregate of the individual spheres. This is illustrated using the set of thin curves where the spheres are grouped in sets of 12 and each group is approximated by an equal-volume sphere. This approximation probably exaggerates the coupling between closest neighbors and neglects coupling to distant spheres. We expect less pronounced coupling between nearby particles in a more uniform aggregate but the qualitative results of interactions should be the same because coupling between particles is strongly dependent on the interparticle distance. Equivalently, we expect these same effects to show up as a uniform aggregate grows larger. As before, the coherent (thin dashed curve) and incoherent (thin solid curve) solutions differ only at the smallest scattering angles.

We note that when the particles are cooperating in small groups, they scatter significantly more per particle than when they are dispersed as in a cloud. Similarly, the computation of scattering and radiation pressure cross sections showed that the average values for the aggregated grains is much higher than for the same grains dispersed.

The degree of linear polarization produced by the constituent particles when dispersed so that they do not interact (thick solid curve in Fig. 5c) is close to the well-known polarization by isotropic Rayleigh particles. Similar to smaller particles, the polarization is always positive and the maximum is near 90° and near unity. There is only a weak asymmetry shifting the maximum to higher scattering angles. The magnitude of this small shift is practically the only color dependency of the polarization across the frequency range and it is too small to be seen in the figure. Because the coherent scattering approximation ignores possible phase velocity differences in the orthogonal polarization directions, polarization is identical in our coherent and incoherent scattering approximations.

Figure 5c illustrates the color dependency of polarization. The thin dashed and solid curves corresponding to the coherent and incoherent scattering by the aggregated groups of particles are for the red and blue polarization, respectively. They were obtained using the same parameters that generated the thin curves in Fig. 5b. Because Lorenz–Mie spheres represent the groups, any cross-polarization term vanishes in this approximation, which helps explain the high polarization. The shift in the maximum polarization with wavelength range is now obvious. We notice that the polarization shifts from red in the forward hemisphere to blue

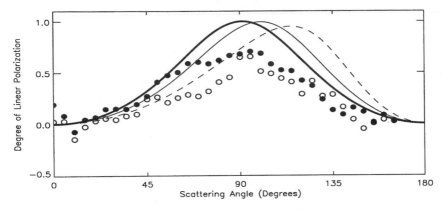

Figure 5c   Angular distribution of the degree of linear polarization. Dots represent experimental results for the red half of the frequency range and circles for the blue half. The thick solid curve represents both color ranges and both coherent and incoherent scattering by the constituent particles. The thin solid curve is for groups of 12 spheres in the red whereas the dashed curve is for the blue. Reprinted from B. Å. S. Gustafson *et al.* (1999), Scattering by complex systems. II. Results from microwave measurements, *in* "Formation and Evolution of Solids in Space" (J. M. Greenberg and A. Li, Eds.), pp. 549–564. Kluwer Academic, Dordrecht, with kind permission of Kluwer Academic Publishers.

at high angles as a result of the well-understood shift in the maximum originating in the Lorenz–Mie calculations. The experiment shows that the actual polarization is nearly neutral in the backscattering hemisphere, whereas the same overall trend in polarization color as in the approximation is still discernible. In Fig. 5c, the experimentally determined polarization values are plotted separately for the red and blue halves of the spectral range. Although both curves have approximately the same shape as the Rayleigh polarization, the maximum polarization is significantly lower. This is primarily due to the cross-polarization terms i_{12} and i_{21} in the scattering matrix. The polarization is for the most part slightly more positive in the red than in the blue. Although this color effect is not strong, the red polarization is seen in all experimental data we have obtained for aggregated small particles. Reddening is seen even if the cross-polarization terms were ignored. Our interpretation is that coupling between particles in the blue is stronger than in the red, because the particles are larger compared to the wavelength. Stronger interparticle dependence for larger particles is often seen both in the laboratory and in theoretical calculations.

   An opinion sometimes voiced about the "bird's nest" model dust is that if the coherent scattering approximation applies to intensities, the color of the scattered light might be expected to be close to the color of the constituent particles. Because the constituent particles derive from classical interstellar grains, the color

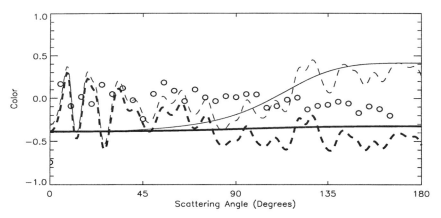

Figure 5d   Color as a function of scattering angle. Experimental data are less blue than the isotropic blue incoherent scattering by constituent spheres (thick solid curve). The thin solid curve for incoherent scattering by groups of 12 spheres shows that the reduction in color in the backscattering hemisphere may be due to scattering on groups. The thick dashed curve is for coherent scattering by the constituent spheres and the thin dashed curve is for groups of 12 spheres. The reduction in the forward hemisphere may be caused by interference resulting from coherent scattering. Coherent scattering by the groups (thin dashed curve) shows reduced color at all angles resulting from a combination of these effects. Reprinted from B. Å. S. Gustafson *et al.* (1999), Scattering by complex systems. II. Results from microwave measurements, *in* "Formation and Evolution of Solids in Space" (J. M. Greenberg and A. Li, Eds.), pp. 549–564. Kluwer Academic, Dordrecht, with kind permission of Kluwer Academic Publishers.

should be blue. We show that the coherent scattering approximation predicts a color shift at small scattering angles, and while the shift can be seen, the coherent scattering approximation is insufficient to accurately predict the color. The consistently blue color of the Rayleigh-like constituent particles is illustrated by the thick solid curve in Fig. 5d and the color using the coherent scattering approximation with the constituent spheres is the thick dashed curve. Although the first few oscillations are from a broadening in the forward diffraction pattern with increasing wavelength leading to a shift in the interference pattern, the oscillations in the backscattering hemisphere average out from a cloud of similar particles. The overall shift toward the red at small scattering angles is therefore due to the broader forward peak created by coherent scattering in the red and would persist independent of the number of orientations. At high scattering angles the shift is toward the blue for this size aggregate. However, the direction and magnitude of the shift is expected to depend on the dimensions of the aggregate in a periodic way. The forward scattering always shifts toward red colors, whereas the backscattering can shift in either direction. If we were to observe a cloud of aggregates with a broad size distribution, the forward scattering would be shifted toward the red, whereas the shifts in the backscattering region would average out and we would expect to see the same color as that from the constituent particles in this approximation.

However, the dots in Fig. 5d show that the experimental results exhibit an almost neutral color at all scattering angles. A hint as to the origin of the color is given by the coherent scattering from the groups of particles shown by the thin dashed curve. The arrangement into groups leads to a redshift at high scattering angles. Together with the shift from forward coherent scattering at small angles, this approaches the color obtained in the experiment. No matter what the explanation for the redshift is, we see that at least this particular "bird's nest" type of aggregate shows nearly neutral color and not the blue color of its constituent particles. Gustafson *et al.* (1999) show that while the scattering differs from one model to the other, the general trend is that color and polarization are much lower than when they are produced by independently scattering constituent particles. Both color and polarization are also lower than those due to independent groups of 12 particles that we used to represent clusters of constituent particles. The effect of aggregation is to lower the polarization and cause a reddening of the forward scattering. Backscattering is also reddened when compared to the color of the constituent particles but the amount varies, probably depending on the importance of coupling to the nearest particles. When the aggregates are 10 wavelengths across or larger, the polarization is no longer dominated by the constituent particles.

We have separated the effects of small constituent spheres, which have blue color because of their size, from the effect of aggregation, which is to redden the scattering or to reduce the size-induced blue color effect. We note that color can be further modified by the color of the bulk material out of which the grains are made. The intrinsic color of the bulk material of actual grains may range from gray (neutral) to brownish yellow or red.

V. DISCUSSION

Microwave analog measurements fill a special need because of their versatility, accuracy, and capability of systematic explorations of electromagnetic scattering at any wavelength. The high degree of control over the scattering experiment allows its use as a guide in theory development; for example, we have tested incomplete solutions to the scattering by large dielectric cubes by covering faces of the cube with absorbing or reflecting materials at will. It is the only resort in cases where no theoretical solution exists or when theoretical solutions are unreliable. It is also useful in the many cases when numerical computing is even more demanding.

Analog measurements require the construction of scale models. Because of this and because of the large parameter space to explore, microwave measurements are tedious. Given the many tasks for which an accurate microwave laboratory is uniquely suited, it may often be hard to justify its use in the exploration of scattering by broad varieties of natural particles that can be studied using direct optical techniques and clouds of particles. There are, however, a broad range of systematic surveys that do not necessarily require the high accuracy of our w-band facility and that could best be done using a microwave analog facility that trades some of the accuracy for higher data log rates.

ACKNOWLEDGMENTS

The data obtained in the microwave laboratory are the result of a team effort and the author is grateful to L. Kolokolova, J. Loesel, J. Thomas-Osip, T. Waldemarsson, and Y.-l. Xu for their collaboration. I have also profited greatly from many years of collaboration at various times with J. M. Greenberg, R. T. Wang, R. H. Giese, and R. H. Zerull, who introduced me to the microwave analog method of light-scattering measurements. I am also indebted to L. Kolokolova and the editors for making several suggestions that led to significant improvements in the manuscript and to S. F. Dermott and the rest of the faculty at the Department of Astronomy for continual encouragement. This work was supported by NASA's Planetary Atmospheres Program through Grant NAGW-2482.

Part V

Applications

Chapter 14

Lidar Backscatter Depolarization Technique for Cloud and Aerosol Research

Kenneth Sassen

Department of Meteorology
University of Utah
Salt Lake City, Utah 84112

I. INTRODUCTION

The interaction of light with aerosols, or cloud and precipitation particles (i.e., hydrometeors), as often vividly revealed in a variety of optical displays, has long intrigued scientific observers and led to fundamental advances in the field of optical physics. The appearance of halo arcs, rainbows, and coronas, to name but a few impressive celestial occurrences, actually represents concentrations of reflected, refracted, or diffracted natural light in the scattering phase functions of atmospheric particles of various shapes, sizes, and orientations. Often, because of the scattering geometry peculiar to each display, the reflections from randomly polarized sunlight viewed by an observer can become partially or even completely polarized (Sassen, 1987a; Können and Tinbergen, 1998). This basic principle is a powerful passive remote-sensing tool that has even been useful in identifying the composition of the outer clouds of Venus from scattered sunlight (Hansen and Hovenier, 1974).

In the backscattering direction, we have the specter of the Brocken (i.e., glory), the heiligenschein, and the rare anthelion, but these retroreflection phenomena have not adequately predicted the potential of polarized laser light-scattering measurements at the 180° scattering angle. Because most types of lasers inherently generate linearly polarized radiation, some of the earliest atmospheric tests of the new light detection and ranging (lidar) technologies in the 1960s were naturally aimed at assessing the information content of the lidar backscatter depolarization technique (LBDT) when probing clouds. This application was even more predictable in view of the history of the development of microwave radar remote sensing, which was certainly prominent in the minds of the developers of lidar. (Including this author, and so frequent references to analogous radar research are given here; see also Chapter 16.) With radar, it was known that nonspherical particles, such as aerodynamically distorted raindrops, melting snowflakes, and hailstones, produced easily measurable amounts of linear or circular depolarization. (Because of the weak Rayleigh scattering of small ice crystals, however, it generally was not possible to detect the cross-polarized components from ice clouds with then-available technologies.) Moreover, differential reflectivity (i.e., horizontal versus vertical copolar returns), phase change during propagation, and depolarization data collected during radar elevation angle scans revealed information indicative of hydrometeor shape and orientation, including, impressively, the realignment of ice crystals in thunderstorm tops caused by the transitory electric fields associated with lightning discharges (Hendry *et al.*, 1976).

On the other hand, it was soon apparent that at optical frequencies the amounts of laser depolarization sensed from many atmospheric targets was considerably greater than using the analogous radar method (Schotland *et al.*, 1971). Obviously, the differences in the refractive indices and sizes of hydrometeors, relative to the probing wavelength, resulted in very different backscattering behaviors. Basic scattering theories dictate that spherical, homogeneous particles do not produce any depolarization during single backscattering, whereas nonspherical particles with the hexagonal shape and refractive index (in the visible) of ice crystals, for example, could generate strong depolarization during the internal refraction and reflection events contributing mainly to backscattering. Thus, it was indicated that polarization lidars operating in the visible and near-infrared portions of the electromagnetic spectrum should be capable of unambiguously distinguishing water droplet from ice crystal clouds. As discussed later, although some experimental complications had to be understood and overcome, this capability inherent in lidar is still unique among remote-sensing techniques. And it has since been shown that it is possible to learn much more of aerosol and cloud microphysical properties using polarization lidar.

In this chapter, we review the working theories, such as they are, that have supported this application, the laboratory and field experiments that established its utility, often with the help of model simulations, and discuss the future applications and role that this method will play in helping to settle major current uncertainties in climate research. Only recently are we coming to grips with comprehending all the distinctions between radar-Rayleigh and lidar-Mie hydrometeor scattering, as we will illustrate using the rigorous example of the bright-band phenomenon where snowflakes melt to form raindrops. Here the full complement of scattering laws come into play. In addition, we will use the case study of a spectacular halo/arc-producing cirrus cloud to illustrate the capabilities of modern polarization diversity lidar systems for sensing detailed ice cloud content.

II. THEORETICAL BACKGROUND

Before laying the theoretical foundation for this lidar application, it is useful to establish some definitions of importance to understanding laser light scattering in the atmosphere. There are essentially three scattering regimes that apply: (1) Rayleigh scattering, principally for air molecules; (2) the Rayleigh–Mie transition zone for the majority of aerosols and just-formed cloud particles; and (3) the Mie and geometrical optics domains for cloud and precipitation particles. (Note that because "Mie" or "Lorenz–Mie" scattering strictly refers to the electromagnetic interaction with homogeneous isotropic spheres, in dealing with arbitrarily shaped particles we will refer to the Mie scattering "zone" or "domain" only in the context of having a particle size that is on the order of the wavelength of the illuminating light.) Implicit in our discussion is that we confine ourselves to visible and near-visible laser wavelengths (ostensibly from ~0.3–3.0 μm), for at longer wavelengths Rayleigh scattering clearly applies to small cloud particles, and the overwhelming absorption of midinfrared radiation by the water substance inhibits internal scattering and the production of significant depolarization (Eberhard, 1992).

It is useful to mention here that hydrometeors may be distinguished from aerosols by their generally larger size and water/ice-dominated composition, or more effectively, by their growth activity, that is, whether they are actively growing/evaporating or, for dry or deliquesced particles, in equilibrium with their environment. The rapid transition from aerosol to hydrometeor commonly occurs in the cloud base region as hygroscopic particles swell in updrafts until further growth above vapor saturation is assured. In other words, they have gotten over the hump in their individual Köhler curve. Thus, young hydrometeors scatter light in the Rayleigh–Mie transition zone until continued growth elevates them to the quiescent geometrical optics realm, where scattering is related to the cross-sectional area. As we shall see, this transition has recently been the subject of considerable theoretical work, along with pioneering lidar research.

Basic Stokes $\{I, Q, U, V\}$ parameterization has been discussed in Chapter 1. For our purposes, we begin with the representation of radiation scattered into the exact backscattering direction from polarized laser light, where, after simplifying assumptions, the scattering matrix is given by (Mishchenko and Hovenier, 1995)

$$\mathbf{F}(180°) = \text{diag}[F_{11}(180°), F_{22}(180°), F_{33}(180°), F_{44}(180°)]. \tag{1}$$

The assumptions are that the nonspherical particles are randomly arrayed in three-dimensional (3D) space and that they display a reasonable symmetry and/or shape diversity. (Obviously, this is violated when ice crystals assume preferred orientations, as they often do, but the alternative leads potentially to the consideration of all 16 matrix elements.) Next we define a scattering plane with reference to the laser beam, which is 100% linearly polarized parallel to this plane. Then, the Stokes vector is proportional to $\{1, 1, 0, 0\}$ and the linear depolarization ratio δ (also expressed as LDR), or the ratio of the perpendicular-to-parallel polarization components of backscattered light, is given by

$$\delta = \frac{F_{11}(180°) - F_{22}(180°)}{F_{11}(180°) + F_{22}(180°)}. \tag{2}$$

Because for spheres $F_{11}(180°) = F_{22}(180°)$, we get the well-known finding of $\delta = 0$ for single scattering by cloud droplets.

Finally, although not yet well exploited (Woodard *et al.*, 1998), lasers transmitting circularly polarized light can measure the circular depolarization ratio δ_c (or CDR), given in terms of δ as (Mishchenko and Hovenier, 1995)

$$\delta_c = \frac{2\delta}{1 - \delta}. \tag{3}$$

A range of other depolarization combinations, such as linear analyses from circular polarization (as also pioneered in microwave radar studies), are possible and can be similarly defined. It should also be mentioned that direct measurements of Stokes parameters have been attempted in both the field (Houston and Carswell, 1978) and the laboratory (Griffin, 1983).

We show in Fig. 1 a schematic view of the distinct backscattering mechanisms for a sphere and a hexagonal ice crystal model according to ray optics theory. Here the physical optics explanation of the Stokes theory becomes apparent. In the geometric optics domain, spheres backscatter through a combination of surface waves (trapped at the dielectric interface) and axial reflections off the front and far drop faces, none of which produce depolarization. An ice crystal prism, on

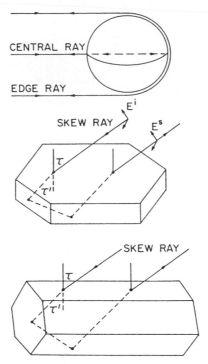

Figure 1   Geometric optics view of the light ray paths responsible for backscattering from a spherical water drop, contrasted to the internal skew rays for simple plate and column ice crystal models. Reprinted from K.-N. Liou and H. Lahore (1974), Laser sensing of cloud composition: A backscattered depolarization technique, *J. Appl. Meteorol.* **13**, 257–263. Copyright © 1974 American Meteorological Society.

the other hand, may produce a nondepolarizing specular reflection when a crystal face is fortuitously aligned perpendicular to the laser beam direction, but it is considerably more likely that (except for scattering geometries involving particles with fixed orientations) internally refracted and reflected ray paths will be chiefly responsible for backscattering. These processes result in the reorientation of the incident polarization vector at every interface, leading to depolarization when the backscattered ray is transposed into the initial plane of polarization.

Finally, as an overview of the range of hydrometeor depolarizing behaviors, we present in Fig. 2 data collected by a helium–neon (HeNe) continuous-wave (CW) laser–lidar analog device in the laboratory and field during the early 1970s to help evaluate the potential of lidar for cloud physics research (see the review by Sassen, 1991). Although we will often revisit the issue of hydrometeor identification in

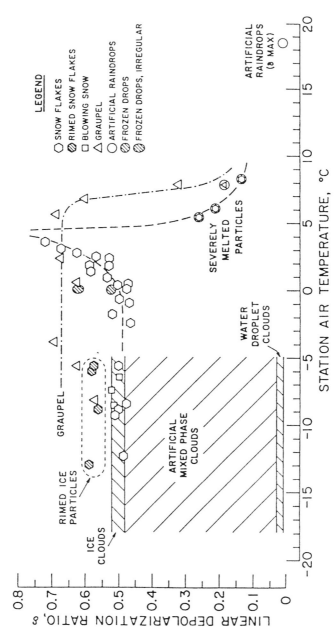

Figure 2 Results of early laboratory and field studies using a CW laser–lidar analog device, showing the great range of linear depolarization ratios encountered from various types of hydrometeors. Reprinted from K. Sassen (1991), The polarization lidar technique for cloud research: A review and current assessment, *Bull. Am. Meteorol. Soc.* **72**, 1848–1866. Copyright © 1991 American Meteorological Society.

later sections, at this point it is useful to consider backscattering from nonspherical and irregular particles from the geometric optics viewpoint. The hatched regions in Fig. 2 are from early laboratory experiments and confirm the basic utility of the LBDT, that is, cloud phase discrimination. Whereas laboratory supercooled water clouds produce near-zero δ, the simple ice clouds artificially nucleated from them generate $\delta \approx 0.5$; not surprisingly clouds of mixed water–ice composition generated intermediate values.

Importantly, when precipitating ice particles were probed from out of a laboratory window, the LBDT was shown to enable the separation of various ice particle types. Snowflakes, composed essentially of randomly oriented dendritic ice crystals, tend to produce the same $\delta \approx 0.5$ as randomly oriented laboratory crystals of generally mixed habits. However, as frozen cloud droplets begin accumulating on the ice crystal faces, the increase in surface complexity leads to a depolarization increase. Ultimately, the droplet accretion process results in the formation of low-density graupel particles, or even hailstones under the proper conditions. Graupel data are shown by the triangle symbols in Fig. 2, where it can be seen that these opaque aspherical particles generate $\delta \approx 0.65$. The action of the final microphysical process, hydrometeor melting, is revealed to produce a strong increase in depolarization for snowflakes, but graupel are unaffected. This again appears to be due to changes in surface complexity; whereas graupel internally absorb melt water and do not appear to change shape much, snowflake surfaces become rounded, compacted, and coated unevenly by water. The additional changes that occur during the final stages of melting show the expected transition from inhomogeneous to pure water drops, but hidden scattering effects explainable by geometric optics theory will be revealed later.

It is not inappropriate to suggest here that the LBDT is akin to human perception in its ability to clearly identify differences in particle shape—certainly the wavelengths used for probing are the same.

III. POLARIZATION LIDAR DESIGN CONSIDERATIONS

Although a detailed discussion of polarization lidar technology is clearly beyond the scope of this chapter, lidar design considerations have consequences for both data collection and interpretation. Unlike some lidar techniques that rely on advanced spectroscopic hardware, polarization diversity requires only the assembly of off-the-shelf components, namely, commercial laser sources (chiefly Nd:YAG and ruby), telescopes, and detector packages that are tailored to the laser wavelength. Only the simple addition of a polarization beam-splitting prism just behind the focal plane aperture along with an extra detector is needed for two-channel polarization measurements. In Fig. 3 is an example of the design

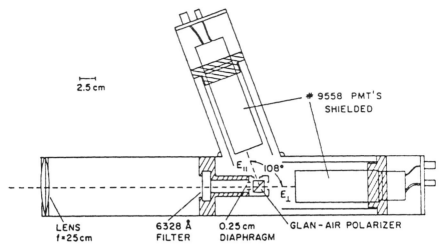

Figure 3 Typical polarization lidar receiver design using a polarizing prism and dual photomultiplier tube (PMT) detectors to permit simultaneous depolarization measurements. This example is for a laboratory lidar analog receiver, and so has a forward collecting lens. Reprinted from K. Sassen (1974), Depolarization of laser light backscattered by artificial clouds, *J. Appl. Meteorol.* **13**, 923–233. Copyright © 1974 American Meteorological Society.

of a dual-polarization receiver, in this case from a CW laser–lidar analog device (Sassen, 1974), which substitutes a light-collecting lens for the reflector used in pulsed lidar systems. The standard receiver components are a laser line interference filter (to block out background solar scattering), a field-of-view (FOV) limiting aperture and polarizer prism (in this case a Glan-air calcite cube) placed at the receiver focal point, and the dual detectors. In this example, two extended S-20 photomultiplier tubes were used to detect the backscattered 0.633-µm HeNe laser light. Although this design dates from the early 1970s, it remains in vogue today.

For polarization applications, the use of steering mirrors for directing the laser beam is unwise in order to avoid unnecessarily corrupting the transmitted and received polarization properties. Thus, polarization lidars usually involve placing the laser and telescope side by side on a stable table that is pointed into the vertical direction, after assuring the parallel alignment of the two beams. (Alternatively, a 45° mirror can be used to direct the beam into the receiver FOV.) Shown in Fig. 4 is a photograph of our turnkey ruby (0.694-µm) lidar system at the University of Utah Facility for Atmospheric Remote Sensing (FARS), which uses the receiver design of Fig. 3 and the side-by-side configuration on a (manually) steerable table supported by a yoke. The ability to steer the lidar is important to aid in beam alignment (with the use of a distant target board and steerable collimator), but also for allowing the collection of off-zenith measurements from ice clouds to

Figure 4 Photograph of a basic polarization lidar installation, at the University of Utah Facility for Atmospheric Remote Sensing. The PC-controller station is visible at right; at left are the laser transmitter, telescope receiver, and infrared radiometer units mounted on a table supported by a yoke to permit scanning.

identify anisotropic scattering conditions, as explained later. Note from Fig. 4 the placement of the collimator directly in front of the laser, and adjacent to it, the coaligned PRT-5 infrared radiometer. Data acquisition and storage for this system is accomplished simply, using a digital storage oscilloscope and a personal computer (PC) with 8-mm tape backup. This facility was built specifically for remote-sensing studies and has an opening skylight for the lidar and a roof parapet for mounting a variety of supporting radiometers and cameras.

The divergence of the laser beam and the FOV of the receiver defined by the aperture have important consequences for polarization data interpretation. Owing to the finiteness of these angles, the effects of laser pulse attenuation differ from that predicted by the Beer–Lambert law for an infinitely narrow divergence angle. This not only complicates quantitative data analysis for standard lidar systems, but also impacts on depolarization analysis through the viewing of multiply scattered radiation by the receiver. On the other hand, this aspect could be used to help interpret lidar signals when the FOV is rapidly changed. Other improvements could be the addition of multiwavelength depolarization measurements for

studying particles in the Rayleigh–Mie transition region and for differential reflectivity observations. In general, the widely used Nd:YAG laser transmitters are
well suited for aerosol and cloud research. With frequency doubling, simultaneous
10-Hz outputs at 1.06 and 0.532 μm are on the order of 1 J, with an approximately
10-ns pulse length and 1.0-mrad divergence angle. Fast (up to 100 MHz) data dig-
itization boards to take advantage of such high spatial and temporal resolutions
are commercially available, although they are currently limited to 8-bit resolution. Details of the design of a modern lidar system combining these attributes,
the University of Utah Polarization Diversity Lidar (PDL), are given in Sassen
(1994).

Still more advanced lidar systems incorporating spectroscopic technologies
have added polarization diversity, such as the turnkey Raman water vapor lidar system routinely operated at the Southern Great Plains Cloud and Radiation
Testbed (SGP CART) site in north-central Oklahoma (Goldsmith *et al.*, 1998).
Because it is relatively straightforward to add depolarization receiver channels,
the amount of extra information gained is economical. Methods that can increase
cloud/aerosol information, and improve spectroscopic data quality in some cases,
include Raman scattering, high spectral resolution lidar (HSRL), differential absorption lidar (DIAL), and even the new near-infrared Doppler lidars (for reviews,
see Carswell, 1983; Ansmann *et al.*, 1993; Sassen, 1995). Moreover, considerable
expertise is required to fabricate eye-safe, low-power, unattended laser ceiliome-
ters to routinely monitor cloud heights, and dual-polarization capabilities also
have considerable advantages in this regard as well (Spinhirne, 1993).

The basic lidar equation for polarization applications was provided early on in
Schotland *et al.* (1971) for the case of isotropic scattering media:

$$P_{\perp,\parallel}(R) = P_0 \left(\frac{ctA_r}{2R^2} \right) \beta_{\perp,\parallel}(R) \exp\left[-2\eta \int \sigma(R)\,dR \right], \tag{4}$$

where the subscripts $\perp$ and $\parallel$ refer to the planes of polarization perpendicular
and parallel to the incident (vertical by convention) polarization plane, R is the
range, P_0 the linearly polarized output power, c the speed of light, t the laser pulse
length, A_r the receiver collecting area, β the backscattering coefficients per unit
volume (in units of per length per steradian), η the multiple-scattering correction
factor that accounts for "captive diffraction" (usually taken out of the integral for
simplicity), and σ is the extinction coefficient per unit volume (per length). The
integral is taken over the range of R_0 ($R = 0$) to R.

The linear depolarization ratio δ can now be expressed from the ratio of the
polarized lidar equations as

$$\delta(R) = \frac{P_\perp(R)}{P_\parallel(R)} = \frac{\beta_\perp(R)}{\beta_\parallel(R)}, \tag{5}$$

which is reduced simply to the ratio of the backscattering coefficients.

Although not recently given much attention, Eqs. (4) and (5) can be violated under conditions associated with the anisotropic scattering medium of uniformly oriented ice crystal populations, and δ values even exceeding unity have been reported in the field (Derr *et al.*, 1976; Sassen, 1976). Could the assumption that σ is independent of the polarization state be incorrect owing to the bire- fringence property of ice and pulse propagation effects? Although Takano and Jayaweera (1985) have indicated that $\delta \gtrsim 1.0$ is theoretically possible for some combinations of lidar observation angle and crystal axis ratio and orientation, the great δ increases noted to occur with penetration depth into some ice clouds sug- gest that more than single backscattering is involved. Perhaps, the transmission of light directly through horizontally oriented ice plates (see Sassen, 1987a) at most lidar angles, coupled with birefringence-induced depolarization, creates an increasingly altered pulse polarization state owing to this form of forward scat- tering. At microwave radar frequencies, propagation effects through regions of oriented particles can have major impacts, and represent a mainstay of attempts to infer target composition. To account for such effects, Schotland *et al.* (1971) gave the following broader definition for the linear depolarization ratio in terms of the atmospheric transmission terms τ in the two polarization planes:

$$\delta(R) = \frac{\beta_\perp(R)}{\beta_\parallel(R)} \exp(\tau_\parallel - \tau_\perp). \tag{6}$$

Because lidar depolarization may be influenced by the polarization corruption of the forward and backscattered laser pulses during transmission, this is a research area that will no doubt receive more attention using scanning lidars.

Finally, it is important to note that the lidar backscattering and extinction coef- ficients actually represent the range-dependent sums of the contributions from all the atmospheric constituents present, including molecular, aerosol, and hydrom- eteor scatterers, although when present cloud scattering would typically domi- nate. Moreover, and importantly, the preceding equations apply only to the single- scattering case, and so additional backscattering terms (Sassen and Zhao, 1995) must be considered in the probing of most clouds in our atmosphere.

IV. AEROSOL RESEARCH

Surprisingly, relatively little lidar research has been directed toward the study of nonspherical or inhomogeneous aerosols in the troposphere using the depolar- ization method. This is partly the result of the great difficulties inherent in sepa- rating the weakly depolarizing ($\delta \sim 0.02$) molecular returns (Hohn, 1969) from the aerosol signal using simple two-channel lidars (see the following for a dis- cussion of the promise of more advanced techniques), but may also reflect a lack of recognition of the potential benefits. Polarization studies of the background and polluted lower troposphere, particularly in conjunction with HSRL or Raman measurements to separate out the molecular constituents, hold particular promise for characterizing the content of such climatically important aerosols.

Boundary layer aerosol research is further complicated by the effects of relative humidity, which causes hygroscopic aerosols to deliquesce and swell into spherical haze particles. Such wetted particles may have solid inclusions and odd refractive indices, but will generate near-zero δ because of the relatively minute particle sizes. Thus, the δ values from hygroscopic particles will decrease as the relative humidity increases (Murayama *et al.*, 1997). As for the dry particulates, their depolarizing properties will depend on their size, composition (i.e., refractive index and uniformity), and shape (McNeil and Carswell, 1975; Kobayashi *et al.*, 1985, 1987). Because boundary layer aerosols may comprise a mixture of non- and hygroscopic particles of various materials over a large size range, particularly in urban areas, this scattering "soup" presents significant problems in quantitative lidar signal analysis, but polarization diversity in combination with spectroscopic techniques could be of great benefit.

Occasionally, dense (i.e., relative to molecular scattering) aerosol layers in the free troposphere are created by catastrophic local or regional events, such as fierce dust storms or fires. Plate 14.1 (see color Plates 14.1–14.5) shows an image of an Asian (Kosa) dust storm, as the particulates were swept up into a developing weather system and advected over a polarization lidar in Tokyo, Japan (Murayama *et al.*, 1998). Present in the returned energy display are sporadic strongly scattering cirrus clouds between 9 and 12 km, streaks of Kosa dust descending from approximately 11 km, and the aerosol of the boundary layer in the lowest kilometer. In terms of depolarization, the cirrus produced $\delta \approx 0.45$, the dust between 0.1 and 0.2, and the low-level aerosol less than 0.1. Note that the strongest returns in the boundary layer correspond to $\delta > 0.05$, which are regions of activated haze particles. In contrast, the significant depolarization in the Kosa dust, despite their relatively weak backscattering, indicates that they are crystalline particles. One may ask whether these potential cloud nuclei may have interacted with the cirrus clouds clearly embedded in the dust plume.

The example of a smoke plume created by a brush/forest fire approximately 25 km upwind of the FARS site in Plate 14.2 depicts a more dynamic microstructure in the dense aerosol cloud, which had a milky, bluish appearance. The layer is quite irregular, indicative of variations in the fire conditions that created the plume, and also perhaps preserving gravity waves. Toward the end of the display, a tongue of aerosols descends toward the surface (at 1.52 km above mean sea level, MSL) during a period of plume fumigation. The corresponding δ values decrease with height in the smoke layer: $\delta < 0.025$ are found at the layer top, whereas values up to 0.10 occur at the bottom, particularly near the end of the period. We conclude that the backscattering was dominated by spherical haze particles from condensed water and organic vapors in the upper regions of the layer, but that relatively large and irregular ash particles were in the process of sedimenting out of the layer, perhaps aided by coagulation. Following these observations, a light coating of gray ash was noted on the lidar telescope window.

It is particularly with regard to researching the aerosols of the stratosphere that polarization lidar has played a leading role (Iwasaka and Hayashida, 1981). Because the distinction between stratospheric clouds and aerosols is sometimes vague, as the particles remain small in response to limited growth opportunities and rapid fallout in the thin air, we will consider both here. Principally, stratospheric scatterers can be divided into (i) the background aerosol dominated by (spherical) aqueous sulfuric acid droplets generated photochemically *in situ*; (ii) the apparently uncommon polar stratospheric clouds (PSCs) of various types often connected to ozone depletion; and (iii) the occasional volcanically injected aerosols composed of sulfuric acid droplets, ash, and possibly frozen particles. The major aerosol/cloud forming materials include compounds derived from sulfuric and nitric acids admixed with water, providing a rich tableau for theoretical chemists.

Spherical, homogeneous acid droplets must generate $\delta \approx 0$, and so are easily identified (Sassen and Horel, 1990). Frozen multisubstance particles, however, will depolarize visible laser light not only to the degree dictated by their amount of nonsphericity but also according to their size: Stratospheric particles tend to reside in the Rayleigh–Mie transition region with regard to light scattering. A number of theoretical studies have shown that variations in δ measured at a given wavelength are highly sensitive to the particle size parameter, which represents a significant complication to single-wavelength lidar analyses (Mishchenko and Sassen, 1998). Based largely on lidar studies, three distinct types of PSCs have been identified, although the question of the precise corresponding particle compositions has not been settled by theoretical chemists (Poole *et al.*, 1990). Recently, it has also been demonstrated that ice crystal clouds generating strong depolarization can also inhabit the Antarctic middle stratosphere (Gobbi *et al.*, 1998).

The fate of volcanically produced aerosols in the lower stratosphere (LS) and upper troposphere (UT) also involve chemical processes. Based on laboratory studies (Sassen *et al.*, 1989a), the neutralization and partial crystallization of sulfuric acid drops from the absorption of ammonia gas of lower-tropospheric origin was shown to generate $\delta \sim 0.1$ for micrometer-sized particles, which were similar to lidar measurements in an unusual volcanic aerosol layer injected into the UT over Central America and later studied over Salt Lake City, Utah (Sassen and Horel, 1990). Also unexpected were the δ values of approximately 0.05 measured in the LS following the 1991 Mt. Pinatubo volcanic eruptions at times of unusually cold midlatitude tropopause temperatures in conjunction with Bishop's ring observations (Sassen *et al.*, 1994a). It was conjectured that nonspherical submicrometer-sized frozen sulfuric acid tetrahydrate (SAT) particles could have been responsible for the optical display and nonzero δ values.

Finally, the case of aircraft condensation trail (contrail) formation in the UT bridges the Rayleigh–Mie transition and geometric optics regimes, and so provides a strict test of polarization lidar analysis methods, as well as a great deal of promise for remote sensing of particle characteristics. A theoretical study specifically addressing this problem using a uniquely wide selection of nonspherical particle shapes and aspect ratios (Mishchenko and Sassen, 1998) has revealed that polarization lidar, particularly dual-wavelength systems, has significant potential for sizing the rapidly growing contrail particles. For nearly all particles modeled, a backscatter depolarization resonance region was found for effective size parameters of approximately 10–15 (corresponding to an $\sim$1-μm particle radius at the 0.532-μm laser wavelength), which produced the highest δ values. Because $\delta < 0.01$ were predicted for effective size parameters of less than approximately 5, the rapid δ changes as the contrail particles grow through the scattering regime transition zone should allow for valuable research using sufficiently high resolution polarization lidars (Sassen and Hsueh, 1998).

V. WATER AND MIXED-PHASE CLOUD RESEARCH

The simplest application of the LBDT is the identification of clouds composed of spherical cloud droplets. This follows from the null detection of the cloud base region in the depolarized channel, before the buildup of multiply scattered depolarization gradually is manifested with penetration depth. The cause of the depolarization is related mainly to the polarization properties of azimuthally scattered light from spheres at near-backscattering angles, in combination with double scattering that redirects the light into the receiver FOV (Carswell and Pal, 1980). This process is controlled essentially by the size of the instantaneous scattering volume (i.e., FOV and range to cloud) and the cloud droplet number size distribution. Although most theoretical simulations based on the lidar geometry show a rather monotonic δ increase, this appears to be a consequence of the unrealistic treatment of water cloud content as vertically homogeneous targets. Consideration of the evolution of cloud droplet sizes with height yields important insights into multiple scattering in water clouds (Sassen and Zhao, 1995). By significantly restricting the lidar FOV on the order of 160 μrad these depolarizing effects can be largely negated (Eloranta and Piironen, 1994).

We provide in Plate 14.3 an example of the results of varying the receiver FOV while the laser illuminates a pure water phase layer studied by our PDL system at the Oklahoma CART site. Depicted is a height–time display of parallel-polarized relative 0.532-μm backscattering of the thin ($\sim$140 m) stratus cloud, along with the corresponding linear depolarization ratio display (note inserted color δ key). Over the indicated 5-min period the receiver FOV was gradually decreased in

10 steps from 3.8 to 0.28 mrad (the transmitter divergence was 0.45 mrad). Despite some variability that reflects the microphysical changes associated with the cellularity in cloud structure, it is clear that the peak depolarization steadily decreases with decreasing FOV from approximately 0.35 to 0.05. Also note that δ decreases rapidly near the cloud top in response to evaporation, as droplet sizes decrease during the mixing process with the dry air above.

To the polarization lidar experimentalist, spheres may be boring, but the situation becomes more interesting when both water and ice exist in a supercooled cloud layer. As illustrated in mixed-phase cloud model studies (Sassen *et al.*, 1992), supercooled water clouds cannot contain a relatively large amount of ice for long, for the competing and more rapid ice phase growth occurs at the expense of the cloud droplets and the water vapor needed for their growth. As a result, as the ice content in a supercooled mixed-phase cloud increases, the LDR within the cloud actually decreases because of a lessening of the effects of multiple scattering using the usual (1–3-mrad) lidar receiver FOVs in the model. Thus, simple polarization techniques do not show much promise in quantitatively separating the ice and water contents in mixed-phase clouds.

Although clouds that produce precipitation reaching the ground are discussed later, polarization lidar has unique capabilities for studying thin supercooled cloud layers that produce *virga*, or precipitating particles that do not reach the ground. Under such conditions, lidar detects the weak signals from sedimenting ice particles below the much stronger returns upon entering the water cloud. In terms of δ values, the ice virga may produce the typical 0.4–0.5 values for randomly oriented ice crystals, but often displays the near-zero ratios (when probed in the zenith direction) indicative of horizontally oriented planar crystals (see the following discussion). Of course, water clouds display the characteristic multiple-scattering δ-value increasing trend with penetration depth as a function of the lidar FOV, the signature of this process. Thus, scanning polarization lidar can unambiguously identify these common midlevel cloud conditions, which pose a special problem in attempts to use radar units to characterize clouds. Because even research-grade millimeter wave radars would have great difficulty in detecting the supercooled cloud droplets, the radar would only detect the diffuse ice virga composed of relatively large ice particles, and hence misidentify the cloud system by ignoring the radiatively more important water cloud.

Perhaps the most trying meteorological conditions to interpret involve determining the structure and composition of winter mountain storm cloud systems, which have received much attention because of their potential for artificial cloud seeding to increase snowfall amounts (Sassen *et al.*, 1990a). Snowfall and/or fog are common at mountain side field sites during the cloud seeding experiment periods, and so simple cloud retrieval algorithms relying on subcloud clear air conditions are useless. Lidar returns from supercooled liquid water (SLW) cloud layers embedded in such diverse media are made difficult to identify because of the potentially strong attenuation in the snowfall below, and of course may often go undetected with lidar if the attenuation is strong enough. Nonetheless, the versatility of the LBDT has been amply illustrated by lidar studies of orographic storm clouds. In addition to identifying liquid-dominated SLW and mixed-phase clouds, applying the principles of hydrometeor shape, surface complexity governing backscatter depolarization, and orientation allows the discrimination of graupel, rimed and pristine snowflakes, and uniformly oriented ice crystals. Such capabilities are important for cloud seeding research because the locations (and temperatures) of SLW clouds and the growth mechanism of precipitation particles are closely tied to the likelihood of the successful introduction of artificial ice nuclei.

VI. CIRRUS CLOUD RESEARCH

As demonstrated early on (Schotland *et al.*, 1971), the differences in δ between water and ice clouds were so dramatic that there was little doubt that cloud thermodynamic phase discrimination was inherent in the LBDT, which is quite important because such clouds at high altitudes have distinct radiative effects (Sassen *et al.*, 1985). This recognition came at a critical time, as it was becoming apparent that the extensive cirrus cloud layers, which covered a significant portion of the globe, would likely have a major impact on our climate and on how our atmosphere would respond to the hypothesized changes brought about by the buildup of greenhouse gases through cloud feedback mechanisms (Liou, 1986). The next step was to evaluate the information content of depolarization variations in ice clouds, and this effort gained impetus in the 1980s as scientific attention became focused on cirrus clouds in order to better understand their impact on the planet's radiation balance (Sassen *et al.*, 1990b). Ray-tracing theory has clearly demonstrated that δ values depend critically on the hexagonal ice crystal axis ratio (Takano, 1987; Takano and Liou, 1995) and that preferentially oriented crystals generate lidar elevation angle-dependent δ values (Takano and Jayaweera, 1985). For example, it can be seen from Table I that δ can vary from about 0.3 for randomly oriented thin plates to 0.6 for solid columns. In other words, δ increases with increasing axis ratio, and when the effects of the birefringence property of ice are also considered, the δ_b values in Table I tend to be somewhat stronger. Thus, according to theory, polarization lidar should have the capability of discriminating between basic ice crystal types. Experimental differences depending on crystal habit were also apparent in the field and laboratory (see the review in Sassen, 1991).

However, the situation in nature seems to have limited these laser remote-sensing opportunities, because most cirrus (off-zenith) measurements reveal δ in the 0.4–0.5 range. It appears that, as indicated by direct *in situ* particle sampling

Table I

Backscatter Linear Depolarization Ratio Values at Visible Wavelengths Computed Through Ray Tracing for Randomly Oriented Solid Ice Crystals with the Indicated Length L to Radius a Ratios (in μm). The Two Columns Show the Results Computed Ignoring (δ) and Including (δ_b) Ice Birefringence Effects

$L/2a$	δ	δ_b
8/80 (thin plate)	0.339	0.399
16/80 (plate)	0.355	0.396
32/80 (thick plate)	0.394	0.508
64/80 (short column)	0.382	0.500
200/80 (column)	0.550	0.616
400/80 (long column)	0.563	0.611

Source: Adapted from Takano (1987).

(Sassen *et al.*, 1994b), cirrus ice crystal collections often show a great diversity in particle habits and axis ratios reflecting dynamic cloud processes involving a combination of vertical redistribution (from vertical air motions and sedimentation), turbulent mixing, and new particle generation, as shown by two-dimensional (2D) model simulations (e.g., Khvorostyanov and Sassen, 1998). Thus, opportunities to probe homogeneous ice cloud compositions may be uncommon, although speculations as to the content of particular regions of cirrus clouds have been widespread. For example, in Sassen *et al.* (1989b), regions associated with Doppler radar-detected updrafts yielded rather low δ ($\sim$0.2–0.3), which were attributed either to rapidly growing haze particles, as precursors to new ice particles, or to the peculiar shapes of newly formed ice crystals. Atypically high δ ($\sim$0.5–0.8) were measured at the tops of unusually cold, corona-producing mid-latitude cirrus layers extending slightly into the LS and also in contact with tropopause folds (Sassen *et al.*, 1995). These high δ values were attributed to the effects on ice particle nucleation and growth of homogeneously frozen sulfuric acid droplets of stratospheric/volcanic origin. Recently, *in situ* sampling of a cold corona-producing cirrus layer (Sassen *et al.*, 1998) revealed the presence of small ($\sim$20 μm diameter) simple ice crystals, but the question of the high-δ particle shape has not yet been fully resolved.

Microphysically, a ubiquitous ice cloud polarization lidar application deals with the anisotropic scattering behavior of horizontally oriented planar ice crystals, which initially led to the misidentification of an ice altostratus layer as mixed phase using zenith lidar (Platt, 1977), but soon offered promise in understanding the fall attitudes of atmospheric ice plates (Platt, 1978; Platt *et al.*, 1978). Because

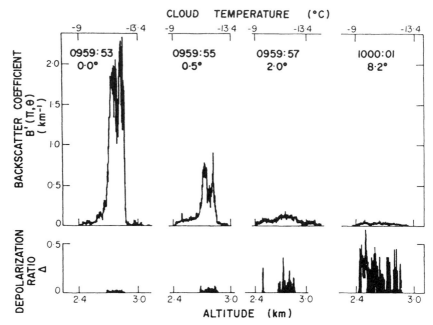

Figure 5 Changes in lidar backscattering and depolarization in a medium of horizontally oriented ice plate crystals produced as the lidar is scanned away from the zenith direction by the indicated amounts. Reprinted from K. Sassen (1991), The polarization lidar technique for cloud research: A review and current assessment, *Bull. Am. Meteorol. Soc.* **72**, 1848–1866. Copyright © 1991 American Meteorological Society.

of this orientation effect and other temperature- and humidity-dependent factors controlling ice crystal shape that could yield increased depolarization (as in Table I), δ-value climatologies of cirrus clouds typically show a trend of increasing δ values with decreasing temperature/increasing height (Platt *et al.*, 1987, 1998). Provided in Fig. 5 is an example of the lidar pointing angle-dependent backscattering and depolarization properties caused by oriented plate crystals. Because of the typically small crystal wobble angles from the horizontal plane (typically $\sim 2.5°$), it is only necessary to tip the lidar a few degrees off the zenith to observe often greatly decreased backscatter and increased depolarization. This behavior is undoubtedly due to the coexistence of randomly oriented crystals of different size or habit (Takano and Jayaweera, 1985). Because the specular reflection is approximately 360 times stronger than the backscattering from a sphere of equivalent cross section according to Sassen (1977a), there need not be many of the oriented plates to generate noticeable effects. For example, assuming a $\delta = 0.5$ value for an ice cloud, a simple model predicts that the ratio of unoriented-to-oriented crystals need be only $1/2000$ to produce $\delta \approx 0.45$, $1/200$ for $\delta \approx 0.25$, and $1/75$ for $\delta \approx 0.15$.

Finally, in Plate 14.4 is a unique view of the disorientation of aerodynamically oriented plate crystals apparently caused by turbulent air motions associated with breaking Kelvin–Helmholtz waves in a thin layer embedded within a cirrus cloud system derived from the blowoff of hurricane Nora in September 1997, which basked in the unusually warm El Niño waters off the southern California coast before sending inland a huge cirrus cloud shield that rapidly crossed the continent. This cirrus cloud layer generated vivid and relatively uncommon optical displays, including the 22° halo, perihelia of 22° and 120°, a parhelic circle, and an upper tangent arc/Parry arc combination (see inserted photograph). And what of the ice crystal shapes that were responsible for the near-zero δ values and optical displays? A group of ice crystals collected *in situ* under these conditions is included in Plate 14.4: The hexagonal and rare triagonal particles are solid, have sharp edges, and are often large enough ($\gtrsim 100$ µm) to maintain their preferred horizontal orientations, all necessary halo arc-generating virtues. In addition, some plates show an asymmetrical division of prism faces, and if this also occurs for the large columns (imperfectly preserved in the replicator fluid), then it may be possible to explain how the crystals could maintain the so-called Parry arc orientation (Können and Tinbergen, 1998).

VII. PRECIPITATION AND THE PHASE CHANGE

It is no coincidence that this appraisal of polarization lidar applications for cloud research ends with the consideration of the physics of precipitation, for in the study of various precipitation mechanisms we are confronted with the most stringent tests for understanding the backscatter depolarization behaviors of the full range of hydrometeors. Take the case of raindrops at the surface that began their descent as snowflakes aloft. In passing through the freezing level the low-density ice particles gradually melt to produce irregular, mixed-phase particles, followed by the collapse of wet snowflakes into inhomogeneous, ice-containing raindrops, which may be spherical to aspherical depending on the diameter. This is the environment of the microwave radar "bright band." Rainfall and drizzle can also be produced entirely through the liquid phase droplet coalescence process, which should produce depolarization only through multiple scattering. Drizzle precipitation should produce especially low amounts of depolarization, because the concentrations of the drops are typically low and the particles are small enough ($\lesssim 100$ µm) to scatter as perfect spheres (unlike aerodynamically distorted raindrops). Finally, snowfall at the surface can be composed of individual ice crystals, which may display uniform orientations, ice crystal aggregates (snowflakes), and near-spherical graupel particles that grow by collecting frozen cloud droplets in convective updrafts like their larger hailstone relatives. Clearly, these principal precipitation-generating mechanisms should be easily separable on the basis of the LBDT.

The bright-band phenomenon warrants further examination. Shown in Fig. 6 and Plate 14.5 are vertical profiles of returned lidar and W-band radar power and depolarization (averaged over 2 min) and height–time displays of ruby lidar backscattering and depolarization obtained from light rainshowers. Note in Fig. 6 that with increasing height the nearly nondepolarizing raindrops (which have not yet reached the ground) give way to the much more strongly scattering and depolarizing snowflakes aloft. The rapid backscattering increase in the snow starts where severely melted snowflakes collapse into raindrops, but is soon followed by a rapid decrease in signal owing to the strong attenuation from the relatively large low-density snowflakes. Although model results have shown that the width and strength of the peak in snow backscattering above the melting layer depends on the precipitation intensity (Sassen, 1977b), the appearance of the signal on an oscilloscope display often resembles that of the microwave radar "bright band." Because the lidar "bright band" analog is related to the strong optical attenuation in snow, however, its cause is completely foreign to that at microwave frequencies, which results from dielectric effects in wet snowflakes. Finally, also note the conspicuous lidar "dark band" that separates the raindrops and snowflakes, which went unexplained until recently despite its dramatic appearance (Uthe, 1978).

As for the cause of this lidar dark band, Sassen and Chen (1995) offered an explanation of this phenomenon using ray optics theory combined with laboratory experiments of melting ice drops suspended in a laser beam (Sassen, 1977c). Lorenz–Mie theory dictates that water spheres backscatter light exclusively through surface waves and axial retroreflections that include contributions from both the front and the rear drop faces (Bryant and Cox, 1966). Laser experiments have also illustrated that the rear face reflection is a significant contributor to the total backscattering, because the front surface acts as a lens to focus light on the rear face (Ro *et al.*, 1968). However, an inhomogeneous raindrop behaves differently because the irregular ice core of a collapsed snowflake in effect blocks the internal paraxial return and so diminishes backscattering. It is clear then that a sudden increase in backscattering would occur as the melting process nears completion. Thus, the lidar dark band appears to owe its existence to two microphysical events in the melting zone: the structural collapse of the severely melted snowflake, which suddenly decreases laser backscattering owing to the combination of decreased particle cross sections and concentrations (from increased fall speeds), and the near completion of the melting process, which suddenly increases the returned power by allowing the full complement of spherical particle backscattering mechanisms to come into play.

The approximately 1-h height versus time display in Plate 14.5, however, indicates that the lidar-observed behavior of the melting region can be quite variable.

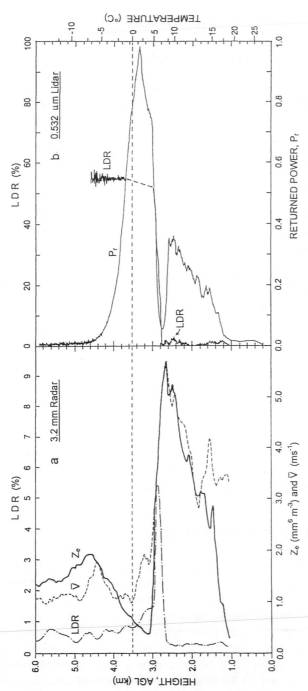

Figure 6 Comparison of vertical profiles of 3.2-mm microwave radar and 0.532-μm PDL data obtained from the melting layer, where LDR is the linear depolarization ratio, V the Doppler velocity, Z_e the radar reflectivity factor, and P_r the returned laser power in relative units. The lidar LDR data could not be calculated near the freezing level (dashed line) because of off-scale signals in the depolarized channel. Reprinted from K. Sassen and T. Chen (1995), The lidar dark band: An oddity of the radar bright band analogy, *Geophys. Res. Lett.* **22**, 3505–3508.

These zenith ruby lidar returned energy and depolarization displays depict the passage of a modest thunderstorm that briefly produced light rain at the FARS site around 2212 UTC. This storm was generated over the nearby Oquirrh Mountains during the summer monsoon season, and only brushed our location to yield a trace of rain, thus allowing the lidar to capture a cross section of the propagating convective shower. Initially, anvil layers up to 9 km and pendant mammatta structures were interrogated, but as the main thunderstorm anvil and precipitation advected over the site, the range of lidar probing was limited by the strong optical attenuation (seen from ~2200–2225). Note the sheared ice particle fall streak that descends from approximately 7 km at 2200 to 4 km at 2210, to be transformed below into a near-vertical rain shaft as a result of the higher raindrop fall speeds.

Although much of the cloud can be seen to produce typical 0.45–0.5 ice δ values, two notable exceptions are apparent. The first occurs between 6.5 and 7 km where low ratios occur in the anvil owing to the presence of horizontally oriented ice plates, as was confirmed by tilting the lidar a few degrees off the zenith at 2158:20. More notable is the depolarizing behavior of the particles in the melting zone, which show some surprising features in addition to the expected ice–water δ-value transition. The lidar dark band at 3.5 km separating the snow/rain regimes is particularly obvious from 2205 to 2215. However, note the absence of the dark band in the rain shafts after 2215, the wide range of depolarization in the rain varying from approximately 0 to 0.2, and the gap in the ice returns just above the rain around 2205. In this last respect this bright-band resembles more its microwave radar analog (a result of water/ice dielectric constant differences), in contrast to the melting layer study of Sassen and Chen (1995). It is obvious that understanding the bright band phenomena requires an improved characterization of the microphysics in the melting zone, including ice particle type (single crystals versus aggregates, graupel, or hail), and raindrop size and susceptibility to oscillate in response to turbulence and collisions. Such shape deformations could generate the relatively significant depolarization noted at about 2210 (Sassen, 1977c), or alternatively melting hailstones could have been responsible.

VIII. CONCLUSIONS AND OUTLOOK

We have attempted to illustrate here how lidars employing the backscatter depolarization technique have successfully exploited the sensitivity of light scattering to remotely determine the characteristics of nonspherical atmospheric particles. Starting in the early 1970s, the capabilities of polarization lidar for differentiating between the great range of atmospheric targets was comprehensively examined in the field and laboratory. Lidars operating at visible and near-infrared wavelengths sense depolarization arising from various sources: Rayleigh-scattering molecules, aerosols in the Rayleigh and Rayleigh–Mie transition zone, and cloud and precipitation particles mostly in the geometric optics domain, which when present typically dominate the backscattered signal.

Polarization lidar studies of aerosols have made significant strides in analyzing the exotic clouds of the stratosphere, but relatively little has been accomplished with respect to the aerosols in the lower tropospheric boundary layer. The problems associated with the quantitative analysis of aerosol properties are numerous, but with the use of polarization diversity coupled with intrinsically calibrated lidar techniques such as Raman and HSRL, basic uncertainties can be overcome and attention focused on better understanding particle shape, size, and composition, taking into account humidity effects. We look forward to such lidar applications for the improved study of tropospheric aerosols.

As for large and nonabsorbing (relative to the laser wavelength) hydrometeors, the results gained early on (see, e.g., Fig. 2), along with ray-tracing theory, clearly imply that laser depolarization is sensitive to the exact shape of the hydrometeors. This is very unlike Rayleigh scattering, which essentially senses nonspherical particles as equivalent ellipsoids regardless of the presence of surface or internal structures. Rather, photons interact with every nook and cranny of a complex particle to a scale on the order of the incident wavelength. As reviewed in Sassen (1991), backscatter depolarization depends on the balance between the sums of nondepolarizing specular reflections and the internally refracted/reflected ray paths and intracrystal element scatterings responsible for depolarization. Take the example of a spatial dendritic-branched ice crystal: The 3D aspect promotes interbranched and internal scatterings characteristic of randomly oriented hexagonal particles, but at the same time, as such particles acquire frozen cloud droplets through the accretion process, each near hemisphere acts like a miniature diffusing element until in the limit, a graupel particle, very high δ values approaching the diffusing medium of ground glass are encountered.

As useful as polarization lidar measurements of clouds and aerosols have been, however, their utility for atmospheric research can be enhanced when the data are collected as part of a coordinated active and passive remote-sensing ensemble, often resulting in a striking synergy of observations. In addition to lidars, the ensemble currently in vogue includes short-wavelength Doppler radars, dual-channel microwave radiometers, and various visible and infrared radiometers. This multiple remote sensor approach has increased steadily in importance in field research programs, and as illustrated here, polarization lidar plays a crucial role in a number of applications. Most notably, polarization diversity allows for the unambiguous identification of liquid water clouds, the accurate determination of cloud base and (cirrus) cloud top heights, and a measure of the microphysical content of ice clouds and precipitation. This applies to lidars either on the ground or in high-flying aircraft (Spinhirne *et al.*, 1983), as well as potentially from Earth's orbit (Winker and Trepte, 1998). Modern lidar/radar "bright band" research is an excellent example of how fundamental scattering principles can be applied to comprehending the intricacies of the hydrometeor phase change process. As a matter of fact, the induced scattering variations are so intriguing that a new feature, the lidar "dark band," was only recently recognized.

We look forward to the continued integration of polarization diversity into other lidar probing techniques for probing clouds, particularly those methods that effectively separate the molecular and cloud/aerosol backscattering constituents (Ansmann *et al.*, 1992). Not only will depolarization data enhance and improve the interpretation of cloud and aerosol quantities, but they can also be used as a quality check on the precision of some spectrally separated channels (e.g., "pure" molecular for which Rayleigh-predicted values of δ are known). The approach of using predictions from detailed cloud microphysical models to help evaluate lidar returns is clearly promising. We also believe that a more concerted focus given to evaluating additional laser backscatter methods such as circular depolarization and complete Stokes parameterization may contain unique, and as yet unknown attributes, particularly with regard to utilizing propagation effects for increasing our knowledge of hydrometeor content. The addition of polarization diversity is economical in terms of hardware and indispensable in terms of dealing with nonspherical atmospheric scatterers.

ACKNOWLEDGMENTS

Our recent polarization lidar research program has been funded by NSF Grant ATM-9528287, NASA Grants NAG-1-1314 and NAG-2-1106, and DOE Grant DEFG0394ER61747 from the Atmospheric Radiation Measurement program.

Chapter 15

Light Scattering and Radiative Transfer in Ice Crystal Clouds: Applications to Climate Research

K. N. Liou, Yoshihide Takano, and Ping Yang
Department of Atmospheric Sciences
University of California, Los Angeles
Los Angeles, California 90095

Light Scattering by Nonspherical Particles: Theory, Measurements, and Applications
Copyright © 2000 by Academic Press. All rights of reproduction in any form reserved.

I. INTRODUCTION

Understanding the radiation budget of Earth and the atmosphere system, and hence its climate, must begin with an understanding of the scattering and absorption properties of cloud particles. A large number of cloud particles are nonspherical ice crystals. Basic scattering, absorption, and polarization data for the type of nonspherical ice crystals that occur in cirrus clouds are required for reliable modeling of their radiative properties for incorporation in climate models; for interpretation of the observed bidirectional reflectances, fluxes, and heating rates from the air, the ground, and space; and for development of remote-sensing techniques to infer cloud optical depth, temperature, and ice crystal size. Moreover, because of the limitation of our present knowledge and understanding, fundamental investigation of the light-scattering and polarization characteristics of nonspherical ice crystals is also an important scientific subject in its own right.

Laboratory experiments reveal that the shape and size of an ice crystal are governed by temperature and supersaturation, but it generally has a basic hexagonal structure. In the atmosphere, if the ice crystal growth involves collision and coalescence, its shape can be extremely complex. Recent observations based on aircraft optical probes and replicator techniques for midlatitude, tropical, and contrail cirrus show that these clouds are largely composed of bullet rosettes, solid and hollow columns, plates, aggregates, and ice crystals with irregular surfaces with sizes ranging from a few micrometers to 1000 μm. In addition to the nonspherical shape problem, a large variation of size parameters at the solar and thermal infrared wavelengths also presents a basic difficulty in light-scattering calculations.

We wish to address the issue of the variability of size parameter for nonspherical ice crystals in fundamental electromagnetic scattering and present a unified theory for light scattering by ice crystals covering all sizes and shapes that can be defined mathematically or numerically. Further, we shall illustrate the importance of the basic scattering, absorption, and polarization data for ice crystals in climate and remote-sensing research.

II. UNIFIED THEORY FOR LIGHT SCATTERING BY ICE CRYSTALS

The scattering of light by spheres can be solved by the exact Lorenz–Mie theory and computations can be performed for the size parameters that are practical for atmospheric applications. However, an exact solution for the scattering of light by nonspherical ice crystals covering all sizes and shapes that occur in Earth's atmosphere does not exist in practical terms. It is unlikely that one specific method can be employed to resolve all the scattering problems associated with nonspherical ice crystals. In the following, we present a unified theory for light scattering by ice crystals by means of a combination of geometric optics and finite difference time domain methods.

A. GEOMETRIC RAY TRACING

The principles of geometric optics are the asymptotic approximations of the fundamental electromagnetic theory, valid for light-scattering computations involving a target whose dimension is much larger than the incident wavelength. The geometric optics method has been employed to identify the optical phenomena occurring in the atmosphere, such as halos, arcs, and rainbows. In addition, it is the only practical approach for the solutions of light scattering by large nonspherical particles at this point. In this section we shall review the conventional and improved approaches, the methodology dealing with absorption in the context of geometric ray tracing, and the numerical implementation by the Monte Carlo method. As we have published a series of papers on this subject (Liou and Coleman, 1980; Liou, 1980, 1992; Cai and Liou, 1982; Takano and Liou, 1989a, b, 1995; Liou and Takano, 1994; Yang and Liou, 1995, 1996b, 1997, 1998a), only the fundamentals and the associated equations will be presented here. References of the relevant works can be found in these papers.

1. Conventional Approach

When the size of a scatterer is much larger than the incident wavelength, a light beam can be thought of as consisting of a bundle of separate parallel rays that hit the particle. Each ray will then undergo reflection and refraction and will pursue its own path along a straight line outside and inside the scatterer with propagation directions determined by the Snell law only at the surface. In the context of geometric optics, the total field is assumed to consist of the diffracted rays and the reflected and refracted rays, as shown in Fig. 1a. The diffracted rays pass around the scatterer. The rays impinging on the scatterer undergo local reflection and refraction, referred to as Fresnelian interaction. The energy that is carried by the diffracted and the Fresnelian rays is assumed to be the same as the energy that is intercepted by the particle cross section projected along the incident direction. The intensity of the far-field scattered light within the small scattering-angle interval $\Delta\Theta$ in the scattering direction Θ can be computed from the summation of the intensity contributed by each individual ray emerging in the direction between $\Theta + \Delta\Theta/2$ and $\Theta - \Delta\Theta/2$. Except in the method presented by Cai and Liou (1982), all the conventional geometric ray-tracing techniques have not accounted for phase interferences between relevant rays. It is usually assumed that

(a)

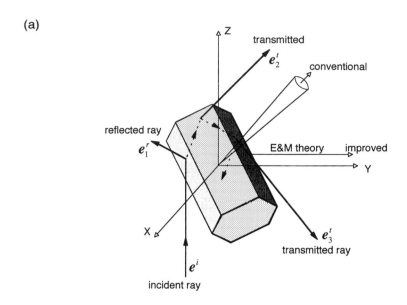

(b)

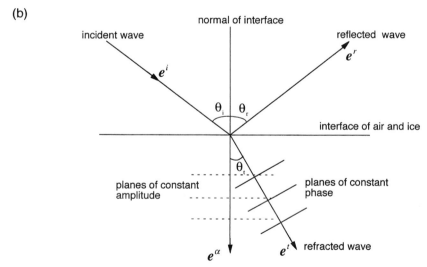

Figure 1 (a) Geometry of ray tracing involving a hexagon in three-dimensional space. Conventional and improved methods are also indicated in the diagram. (b) Geometric ray tracing in a medium with absorption. The planes of constant amplitude of the refracted wave are parallel to the interface, whereas the direction of the phase propagation for the inhomogeneous wave inside the medium is determined via Snell's law.

the interference is smoothed out when the particles are randomly oriented. In this case the extinction efficiency (the ratio of the extinction cross section to the average projected area of the particle) of the scatterer is 2. On the basis of Babinet's principle, diffraction by a scatterer may be regarded as that by an opening on an opaque screen perpendicular to the incident light, which has the same geometric shape as the projected cross section of the scatterer. The well-known Fraunhofer diffraction formula can be employed to compute the diffraction component for hexagonal ice particles.

In the geometric ray-tracing method, the directions of the rays are first determined. In reference to Fig. 1a, they can be defined by the following unit vectors:

$$\mathbf{e}_p^r = \mathbf{x}_p - 2(\mathbf{x}_p \cdot \mathbf{n}_p)\mathbf{n}_p, \qquad p = 1, 2, 3, \ldots, \tag{1a}$$

$$\mathbf{e}_p^t = \frac{1}{m_p}\{\mathbf{x}_p - (\mathbf{x}_p \cdot \mathbf{n}_p)\mathbf{n}_p - [m_p^2 - 1 + (\mathbf{x}_p \cdot \mathbf{n}_p)^2]^{1/2}\mathbf{n}_p\},$$

$$p = 1, 2, 3, \ldots, \tag{1b}$$

$$\mathbf{x}_p = \begin{cases} \mathbf{e}^i, & p = 1, \\ \mathbf{e}_1^t, & p = 2, \\ \mathbf{e}_{p-1}^r, & p \geq 3, \end{cases} \tag{1c}$$

where $m_p = m$ for $p = 1$ and $m_p = 1/m$ for $p > 1$, with m being the refractive index, and $\mathbf{n}_p$ denote the unit vectors normal to the surface. When $m_p^2 < 1 - (\mathbf{x}_p \cdot \mathbf{n}_p)^2$, total reflection occurs and there will be no refracted ray. The electric fields for two polarization components associated with the rays can be computed from the Fresnel formulas [see Eqs. (18a) and (18b)]. Summing the energies of the rays that emerge within a preset small scattering-angle interval in a given direction, the phase function can be obtained for this part. Let the normalized phase functions [i.e., (1, 1) element of the scattering matrix] for the parts of reflection and refraction and diffraction be F_{11}^r and F_{11}^d, respectively. Then the normalized phase function is $F_{11} = (1 - f_d)F_{11}^r + f_d F_{11}^d$, where $f_d = 1/2\varpi(1 - f_\delta)$ with f_δ being the delta transmission associated with 0° refraction produced by two parallel prismatic faces and ϖ being the single-scattering albedo, which can be determined from the absorption of individual rays and the constant extinction efficiency.

2. Improved Geometric Optics Approach

The laws of geometric optics are applicable to the scattering of light by a particle if its size is much larger than the incident wavelength so that geometric rays can be localized. In addition to the requirement of the localization principle, the conventional geometric ray-tracing technique assumes that the energy attenuated by the scatterer may be decomposed into equal extinction from diffraction and

Fresnel rays. Moreover, the Fraunhofer diffraction formulation used in geometric ray tracing does not account for the vector property of the electromagnetic field and requires a Kirchhoff boundary condition, which cannot take into consideration the effects of the changes along the edge contour of the opening. Finally, calculations of the far field directly by ray tracing will produce a discontinuous distribution of the scattered energy, such as the delta transmission noted by Takano and Liou (1989a).

To circumvent a number of shortcomings in the conventional geometric optics approach, an improved method has been developed (Yang and Liou, 1995, 1996b). It is simple in concept in that the energies determined from geometric ray tracing at the particle surface are collected and mapped to the far field based on the exact electromagnetic wave theory. In this manner, the only approximation is on the internal geometric ray tracing. This differs from the conventional approach, which collects energies produced by geometric reflections and refractions directly at the far field through a prescribed solid angle.

The tangential components of the electric and magnetic fields on surface S that encloses the scatterer can be used to determine the equivalent electric and magnetic currents for the computation of the scattered far field on the basis of the electromagnetic equivalence theorem (Schelkunoff, 1943). In this theorem, the electromagnetic field detected by an observer outside the surface would be the same as if the scatterer were removed and replaced by the equivalent electric and magnetic currents given by

$$\mathbf{J} = \mathbf{n}_S \times \mathbf{H}, \tag{2a}$$

$$\mathbf{M} = \mathbf{E} \times \mathbf{n}_S, \tag{2b}$$

where $\mathbf{n}_S$ is the outward unit vector normal to the surface. For the far-field region, we have

$$\mathbf{E}^s(\mathbf{r}) = \frac{\exp(ikr)}{ikr} \frac{k^2}{4\pi} \left(\frac{\mathbf{r}}{r}\right)$$
$$\times \iint_S \left[\mathbf{M}(\mathbf{r}') + \left(\frac{\mathbf{r}}{r}\right) \times \mathbf{J}(\mathbf{r}') \right] \exp\left(-ik\mathbf{r} \cdot \frac{\mathbf{r}'}{r}\right) d^2\mathbf{r}', \tag{3}$$

where $\mathbf{r}/r$ denotes the scattering direction, $\mathbf{r}$ is the reference position vector, $\mathbf{r}'$ is the position vector of the source point, k is the wavenumber, and $i = \sqrt{-1}$. The far-field solution can also be determined by a volume integral involving the internal field.

By means of geometric ray tracing, the electric field on the surface of a particle can be evaluated after the successive application of Fresnel reflection and refraction coefficients parallel and perpendicular to a defined reference plane at the point of interaction taking into account the path length in the three-dimensional geometry. If an ice crystal shape is of great complexity such as an aggregate, the surface can be defined as a cubic box so that the computation of the electric field can be conducted on a regularly shaped surface. The electric field can be defined on the illuminated and shadowed sides as follows:

$$\mathbf{E}(\mathbf{r}) = \begin{cases} \mathbf{E}_a(\mathbf{r}) + \mathbf{E}_b(\mathbf{r}), & \mathbf{r} \in \text{illuminated side,} \\ \mathbf{E}_b(\mathbf{r}), & \mathbf{r} \in \text{shadowed side,} \end{cases} \tag{4a}$$

where

$$\mathbf{E}_a(\mathbf{r}) = \mathbf{E}_i(\mathbf{r}) + \mathbf{E}_1^r(\mathbf{r}), \tag{4b}$$

$$\mathbf{E}_b(\mathbf{r}) = \sum_{p=2}^{\infty} \mathbf{E}_p^t(\mathbf{r}). \tag{4c}$$

In these equations, $\mathbf{E}_i$ is the incident electric field, $\mathbf{E}_1^r$ is the electric field for external reflection, and $\mathbf{E}_p^t$ are the electric fields produced by two refractions and internal reflections ($p \geq 2$). Because the transverse electromagnetic wave condition is implied in ray tracing, the magnetic field for each reflection and refraction can be obtained from

$$\mathbf{H}_p^{r,t}(\mathbf{r}) = \mathbf{e}_p^{r,t} \times \mathbf{E}_p^{r,t}(\mathbf{r}) \qquad \text{for } \mathbf{r} \in \text{ outside the particle.} \tag{5}$$

In practice, the mapping of the near-field solution to the far field can be done in its entirety for $\mathbf{E}_a$ in Eq. (4b). But for $\mathbf{E}_b$ in Eq. (4c), the mapping is done ray by ray and the results will include the diffraction pattern. Full account of phase interferences is taken in this mapping process in the determination of the phase function.

In accord with the conservation principle for electromagnetic energy concerning the Poynting vector (Jackson, 1975), the extinction and absorption cross sections of the particle can be derived as follows:

$$C_{\text{ext}} = \text{Im}\left\{ \frac{k}{|\mathbf{E}_i|^2}(\varepsilon - 1) \iiint_V \mathbf{E}(\mathbf{r}') \cdot \mathbf{E}_i^*(\mathbf{r}') \, d^3\mathbf{r}' \right\}, \tag{6a}$$

$$C_{\text{abs}} = \frac{k}{|\mathbf{E}_i|^2}\varepsilon_i \iiint_V \mathbf{E}(\mathbf{r}') \cdot \mathbf{E}^*(\mathbf{r}') \, d^3\mathbf{r}', \tag{6b}$$

where the asterisk denotes the complex conjugate, ε_i is the imaginary part of the permittivity, and V is the particle volume.

Finally, when the ray-tracing technique is applied to obtain the surface field, one must properly account for the area elements from which the externally reflected and transmitted localized waves make a contribution to the surface field. If the cross section of the incident localized wave is $\Delta\sigma_i$, the area on the particle surface for external reflection is

$$\Delta\sigma_1^r = -\Delta\sigma_i\,(\mathbf{n}_i \cdot \mathbf{e}^i)^{-1}. \tag{7a}$$

For the transmitted rays, the area is given by

$$\Delta\sigma_p^t = -\Delta\sigma_i\,(\mathbf{n}_1 \cdot \mathbf{e}_1^t)\big[(\mathbf{n}_1 \cdot \mathbf{e}^i)(\mathbf{n}_p \cdot \mathbf{e}_p^t)\big]^{-1}, \qquad p = 2, 3, 4, \dots, \tag{7b}$$

where all unit vectors have been defined in Eqs. (1a)–(1c). The radius of the cross section of a ray should be on the order of k^{-1} so that the phase change over the ray cross section is not significant and permits proper account of the phase interference of the localized waves by using the phase information at the centers of the rays. Because the phase variation over the ray cross section can be neglected, the numerical results are not sensitive to the shape of the ray cross sections. We may use a circular shape in the calculations.

3. Absorption Effects in Geometric Optics

The geometric optics approach that has been used in the past generally assumes that the effect of absorption within the particle on the propagating direction of a ray can be neglected so that the refracted angle and the ray path length can be computed from Snell's law and the geometry of the particle. This is a correct approach if absorption is weak, such as that of ice and water at most solar wavelengths. For strong absorption cases, rays refracted inside the particle are almost totally absorbed so that the geometric optics method can also be used to compute diffraction and external reflection as long as the particle size is much larger than the incident wavelength. Although the preceding argument is physically correct in the limits of weak and strong absorption, we shall consider the general absorption effect in the context of geometric optics based on the fundamental electromagnetic wave theory. Note that the effect of the complex refractive index on geometric optics has been formulated only for the Fresnel coefficients (Stratton, 1941; Born and Wolf, 1970).

Consider the propagation of the incident wave from air into ice (Fig. 1b). The wave vectors associated with the incident and reflected waves are real because these waves, which are outside the ice medium, must have the same properties. However, the wave vector of the refracted wave is complex; this is referred to as the inhomogeneity effect. These wave vectors can be represented by

$$\mathbf{k}_i = k\mathbf{e}^i, \qquad \mathbf{k}_r = k\mathbf{e}^r, \qquad \mathbf{k}_t = k_t\mathbf{e}^t + ik_\alpha\mathbf{e}^\alpha, \tag{8}$$

where $\mathbf{e}^i$, $\mathbf{e}^r$, $\mathbf{e}^t$, and $\mathbf{e}^\alpha$ are unit vectors; the subscripts i, r, and t denote the incident, reflected, and refracted waves, respectively; $k = 2\pi/\lambda$ in which λ is the wavelength in air; and k_t and k_α are two real parameters that determine the complex wave vector of the refracted wave. For nonabsorptive cases, k_α is zero. The corresponding electric vectors can be expressed by

$$\mathbf{E}_i(\mathbf{r}, t) = \mathbf{A}_i \exp\left[i(k\mathbf{r} \cdot \mathbf{e}^i - \omega t)\right], \tag{9a}$$

$$\mathbf{E}_r(\mathbf{r}, t) = \mathbf{A}_r \exp\left[i(k\mathbf{r} \cdot \mathbf{e}^r - \omega t)\right], \tag{9b}$$

$$\mathbf{E}_t(\mathbf{r}, t) = \mathbf{A}_t \exp\left[i(k_t\mathbf{r} \cdot \mathbf{e}^t + ik_\alpha \mathbf{e}^\alpha - \omega t)\right], \tag{9c}$$

where $\mathbf{A}_i$, $\mathbf{A}_r$, and $\mathbf{A}_t$ are the amplitudes and ω is the circular frequency. Further, we define the following parameters:

$$N_r = \frac{k_t}{k}, \qquad \tilde{N}_i = \frac{k_\alpha}{k}. \tag{10}$$

At the interface of the two media, at which the position vector is denoted as $\mathbf{r}_S$, the phases of the wave vibration must be the same for the incident, reflected, and refracted waves. Thus from Eqs. (8) and (10) we obtain

$$\mathbf{e}^i \cdot \mathbf{r}_S = \mathbf{e}^r \cdot \mathbf{r}_S = N_r(\mathbf{e}^t \cdot \mathbf{r}_S) + i\tilde{N}_i(\mathbf{e}^\alpha \cdot \mathbf{r}_S). \tag{11}$$

Because the wave vectors for the incident and reflected waves are real, we must have

$$\mathbf{e}^i \cdot \mathbf{r}_S = \mathbf{e}^r \cdot \mathbf{r}_S = N_r(\mathbf{e}^t \cdot \mathbf{r}_S), \qquad \mathbf{e}^\alpha \cdot \mathbf{r}_S = 0. \tag{12}$$

Based on the geometry defined by Eq. (12), a generalized form of the Snell law can be derived and is given by

$$\sin\theta_i = \sin\theta_r, \qquad \sin\theta_t = \frac{\sin\theta_i}{N_r}, \tag{13}$$

where θ_i, θ_r, and θ_t denote the incident, reflected, and refracted angles, respectively (Fig. 1b). The vector $\mathbf{e}_\alpha$ in Eq. (12) is normal to the interface of the two media. It follows that the planes of constant amplitude of the refracted wave are parallel to the interface. To determine N_r and N_i, we use the electric field of the refracted wave, which must satisfy the wave equation in the form

$$\nabla^2 \mathbf{E}_t(\mathbf{r}, t) - \frac{(m_r + im_i)^2}{c^2} \frac{\partial^2 \mathbf{E}_t(\mathbf{r}, t)}{\partial t^2} = 0, \tag{14}$$

where c is the speed of light in vacuum and m_r and m_i are the real and imaginary parts of the refractive index, respectively. Substituting Eq. (9c) into Eq. (14) and using Eq. (10) lead to

$$N_r^2 - \tilde{N}_i^2 = m_r^2 - m_i^2, \qquad N_r\tilde{N}_i \cos\theta_t = m_r m_i. \tag{15}$$

Let $N_i = \tilde{N}_i \cos \theta_t$. Then from Eqs. (12) and (15), we obtain

$$N_r = \frac{\sqrt{2}}{2} \Big\{ m_r^2 - m_i^2 + \sin^2 \theta_i$$
$$+ \big[(m_r^2 - m_i^2 - \sin^2 \theta_i)^2 + 4m_r^2 m_i^2 \big]^{1/2} \Big\}^{1/2}, \qquad (16a)$$

$$N_i = \frac{m_r m_i}{N_r}. \qquad (16b)$$

These two parameters are referred to as the adjusted real and imaginary refractive indices.

After determining N_r and N_i, the refracted wave given in Eq. (9c) can be rewritten in the form

$$\mathbf{E}_t(\mathbf{r}, t) = \mathbf{A}_t \exp(-k N_i l_a) \exp\big[i (k N_r \mathbf{e}^t \cdot \mathbf{r} - \omega t) \big], \qquad (17)$$

where $l_a = (\mathbf{e}^\alpha \cdot \mathbf{r}) / \cos \theta_t$ is the distance of the propagation of the refracted wave along the direction $\mathbf{e}^t$. It is clear that the direction of the phase propagation for the inhomogeneous wave inside the medium is determined by N_r via Snell's law, whereas the attenuation of the wave amplitude during the wave propagation is determined by N_i. Consequently, the refracted wave can be traced precisely. Following Yang and Liou (1995), the Fresnel reflection and refraction coefficients in terms of the adjusted real and imaginary refractive indices are given by

$$R_l = \frac{N_r \cos \theta_i - \cos \theta_t}{N_r \cos \theta_i + \cos \theta_t}, \qquad T_l = \frac{2 \cos \theta_i}{N_r \cos \theta_i + \cos \theta_t}, \qquad (18a)$$

$$R_r = \frac{\cos \theta_i - N_r \cos \theta_t}{\cos \theta_i + N_r \cos \theta_t}, \qquad T_r = \frac{2 \cos \theta_i}{\cos \theta_i + N_r \cos \theta_t}, \qquad (18b)$$

where the subscripts l and r denote the horizontally and vertically polarized components, respectively.

4. Monte Carlo Method for Ray Tracing

Use of the Monte Carlo method in connection with geometric ray tracing was first developed by Wendling *et al.* (1979) for hexagonal ice columns and plates. Takano and Liou (1995) further innovated a hit-and-miss Monte Carlo method to trace photons in complex ice crystals, including absorption and polarization.

Let a bundle of parallel rays, representing a flow of photons, be incident on a crystal from a direction denoted by a set of two angles with respect to the crystal principal axis. Consider a plane normal to this bundle of incident rays and the geometric shadow of a crystal projected onto this plane. Further, let a rectangle (defined by X and Y) enclose this geometric shadow such that the center of this

rectangle coincides with the center of the crystal. One of the sides, X, is parallel
to the geometric shadow of the crystal principal axis. A point (x_i, y_i) is selected
inside this rectangle using random numbers, RN, whose range is from 0 to 1 such
that

$$x_i = X\left(\text{RN} - \frac{1}{2}\right),$$ (19a)

$$y_i = Y\left(\text{RN} - \frac{1}{2}\right).$$ (19b)

In this manner, x_i is from $-X/2$ to $X/2$, whereas y_i is from $-Y/2$ to $Y/2$. If the
point is inside the geometric shadow, it is regarded as an incident point on the
crystal. Otherwise it is disregarded. If there are more than two crystal planes for
a photon, the point closer to the light source is regarded as the incident point. The
coordinates of an incident point (x_i, y_i) can be transformed to the coordinates
(x, y, z) with respect to the body-framed coordinate system using the method
described by Takano and Asano (1983) for efficient geometric ray-tracing procedures. Once the incident coordinates are determined, the photons are traced with a
hit-and-miss Monte Carlo method. The Fresnel reflection coefficients, R_l and R_r,
are first calculated and compared with a random number, RN. If $(|R_l|^2 + |R_r|^2)/2$
is greater than RN, the photon is reflected. Otherwise, it is transmitted. When a
photon traverses a particle, it can be absorbed. One can account for absorption
by means of stochastic procedures. When a photon enters a crystal, an absorption
path length l_a is generated with a random number such that

$$\text{RN} = \exp(-2kN_i l_a), \qquad \text{i.e., } l_a = -\ln\left(\frac{\text{RN}}{2kN_i}\right).$$ (20)

The random number represents the probability of the transmission of a photon.
The absorption path length l_a denotes a distance traversed by a photon in the
crystal before the photon is absorbed. An actual path length, l, between an incident point and the next internal incident point can then be calculated on the basis
of Snell's law and the specific ice crystal geometry. The transmission is then given
by $T = \exp(-k_i l)$. If $T \le \text{RN} \le 1$, then the photons associated with these RNs
are absorbed. Equivalently, if l is greater than l_a, then the photon is absorbed. Otherwise, it is transmitted without absorption. This procedure is repeated whenever
photons travel inside the crystal.

After a photon is transmitted out of the crystal or reflected externally, it can
reenter the crystal depending on the crystal shape. In this case, a new incident
direction can be calculated using the direction cosine of the scattered beam. The
new incident coordinates can also be determined from the new incident direction
and the coordinates of an emergent point of the photon on the crystal surface.
The foregoing procedure is repeated until the photon escapes from the crystal.

When a photon reenters the crystal, the scattering angle and the scattering matrix are computed with respect to the original incident direction. In the conventional method, the number of scattered photons per unit solid angle, $2\pi \sin \Theta \, \Delta\Theta$, is counted as the phase function. The single-scattering albedo is obtained from the ratio of the number of scattered photons to the number of incident photons. The Monte Carlo method allows us to treat complicated ice crystals effectively and can be employed in connection with the improved geometric ray tracing.

The surface of ice crystals may not be exactly smooth particularly if they undergo collision processes. Also, a careful examination of some polycrystalline ice crystals reveals rough structures on the surfaces (Cross, 1968). Halo and arc patterns that are absent from some cirrus clouds could be caused by deviations of the ice crystal surfaces from defined hexagonal structures. Incorporation of some aspects of the ice crystal surface roughness in geometric ray tracing has been recently undertaken by Takano and Liou (1995), Muinonen *et al.* (1996), Macke *et al.* (1996b), and Yang and Liou (1998a). Our approach follows the idea developed by Cox and Munk (1954) for wavy sea surfaces. A rough surface may be thought of as consisting of a number of small facets that are locally planar and randomly tilted from the flat surface. We may use a two-dimensional Gaussian probability function to define the surface tilt as follows:

$$p(z_x, z_y) = \frac{1}{\pi \sigma^2} \exp\left[-\frac{z_x^2 + z_y^2}{\sigma^2}\right], \tag{21a}$$

with

$$z_x = \frac{\partial z}{\partial x} = \left[(\cos\theta)^{-2} - 1\right]^2 \cos\varphi, \tag{21b}$$

$$z_y = \frac{\partial z}{\partial y} = \left[(\cos\theta)^{-2} - 1\right]^2 \sin\varphi, \tag{21c}$$

where z_x and z_y are the slopes defined for a facet of rough surface along two orthogonal directions, θ and φ are the local polar angles defining the position of the tilt of the surface facet, and σ is a parameter controlling the degree of roughness. In general, effects of the surface roughness on ice particles are to smooth out the scattering maxima that occur in the phase function (see Fig. 4).

B. Finite Difference Time Domain Method

The geometric ray-tracing method with a modification in the mapping of the near field to the far field can be applied to size parameters on the order of about 15–20. We have developed the finite difference time domain (FDTD) method for light scattering by small ice crystals with specific applications to size parameters smaller than about 20 (Yang and Liou, 1995, 1996b; Chapter 7). Details of this

method have been elaborated on in Chapter 7. For the continuity of this presentation, however, we shall address the physical fundamentals of the methodology.

The FDTD technique is a direct implementation of the Maxwell curl equations
to solve the temporal variation of electromagnetic waves within a finite space
containing the scatterer given by

$$\nabla \times \mathbf{E}(\mathbf{r}, t) = -\frac{\mu}{c} \frac{\partial \mathbf{H}(\mathbf{r}, t)}{\partial t}, \tag{22a}$$

$$\nabla \times \mathbf{H}(\mathbf{r}, t) = \frac{\varepsilon}{c} \frac{\partial \mathbf{E}(\mathbf{r}, t)}{\partial t} + \frac{4\pi}{c} \sigma \mathbf{E}(\mathbf{r}, t), \tag{22b}$$

where μ, ε, and σ are the permeability, permittivity, and conductivity of the
medium, respectively.

First, the three-dimensional scatterer must be discretized by a number of suit-
ably selected rectangular cells, referred to as grid meshes, at which the optical properties are defined. Discretizations are subsequently carried out for the
Maxwell curl equations by using the finite difference approximation in both time
and space. The propagation and scattering of the excited wave in the time domain
can be simulated from the discretized equations in a manner of time-marching
iterations.

Second, in numerical computations, scattering of the electromagnetic wave by
a particle must be confined to finite space. It is therefore required in the application of the FDTD technique to impose artificial boundaries so that the simulated
field within the truncated region would be the same as that in the unbounded case.
Implementation of an efficient absorbing boundary condition to suppress spurious
reflections is an important aspect of the FDTD method associated with numerical
stability and computer time and memory requirements.

Third, the solution of the finite difference analog of the Maxwell curl equations
is in the time domain. To obtain the frequency response of the scattering particle,
we require an appropriate transformation. The discrete Fourier transform technique can be employed to obtain the frequency spectrum of the time-dependent
signals if a Gaussian pulse is used as an initial excitation. Correct selection of the
pulse is required to avoid numerical aliasing and dispersion.

Finally, mapping of the near-field results to the far field must be performed
to derive the scattering and polarization properties of the particle. A surface integration or a volume integration technique, mentioned in Section II.A.2, can be
employed to obtain the far-field solution. Fundamental problems of the FDTD
method in numerical calculations include the staircasing effect in approximating
the particle shape and the absorbing boundary condition used to truncate the computational domain. We have shown in Chapter 7 that the FDTD approach can be
applied to size parameters smaller than about 20 with adequate accuracies.

C. ESSENCE OF THE UNIFIED THEORY AND COMPARISON WITH MEASUREMENTS

It is unlikely that one specific method can be satisfactorily used to tackle the scattering of light by nonspherical ice crystals covering all size parameters. However, by unifying the improved geometric ray-tracing and FDTD methods discussed previously, we are now in a position to resolve the intricate problems involving light scattering and absorption by nonspherical ice crystals. This approach is referred to as the unified theory for light scattering by ice crystals covering all sizes and shapes that commonly occur in the atmosphere. Demonstration of this unified theory is shown in Fig. 2 in terms of the extinction efficiency as a function of size parameter kL for randomly oriented columns, where L is the column length. The improved geometric optics method breaks down at size parameters smaller than about 15, whereas the FDTD method is computationally reliable for size parameters smaller than about 20 because of numerical limitations. Also illustrated is a verification of the improved geometric ray tracing for size parameters from 15 to about 20.

Figure 3 displays the commonly occurring ice crystal shapes in cirrus clouds generated from computer programs, along with the phase function patterns at a wavelength of 0.63 μm computed from the geometric ray-tracing method. The size parameters for these ice crystals are on the order of 100. Irregular shapes,

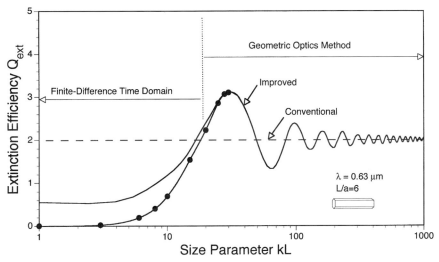

Figure 2 Presentation of a unified theory for light scattering by ice crystals using the extinction efficiency as a function of size parameter as an example (see text for further explanations).

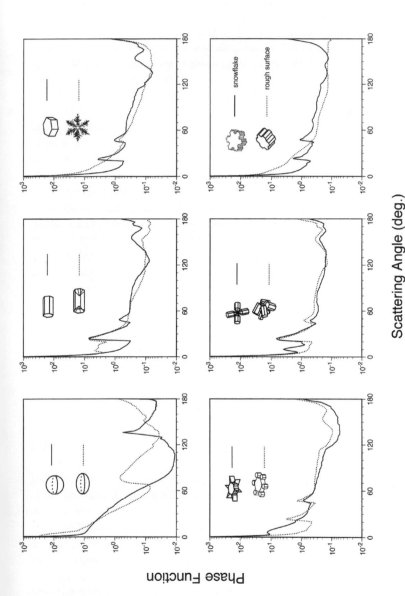

Scattering Angle (deg.)

Figure 3 Commonly occurring ice crystal shapes in cirrus clouds generated from the computer program, along with the phase function patterns for the 0.63-μm wavelength computed from a geometric ray-tracing method. The patterns for snowflakes, dendrites, and plates with attachments are produced from a fractal shape generation program. Results for a sphere and spheroid are also shown for comparison purposes.

such as hollow column, dendrite, and fernlike plate, and rough surface ice crystals do not produce well-defined halo patterns that are common to hexagonal-based crystals such as bullet rosettes and aggregates. Results for small size parameters less about 20 can be computed from the FDTD method.

Measurements of the scattering and polarization patterns for ice crystals have been performed in cold chambers (e.g., Sassen and Liou, 1979a; Volkovit-skiy *et al.*, 1980). Desirable ice crystal sizes and shapes, however, are difficult to generate and sustain for a period of time to perform light-scattering experiments. A light-scattering experimental program has been recently conducted using hexagonal icelike crystals as measured in the analog manner so that optical experiments can be performed over a relatively long period of time for complex-shaped particles (Barkey *et al.*, 1999). The experiment consisted of a polarized laser beam at $\lambda = 0.63$ μm and an array of 36 highly sensitive photodiode detectors arranged between the scattering angles 2.8° and 177.2° mounted in a linear array on a half dome, which can rotate to vary the azimuthal angle. After careful calibration and signal acquisition, this system was used to measure the phase functions for a glass sphere and a glass fiber configured to scatter light like an infinite cylinder. The experimental results match closely those computed from the Lorenz–Mie theory. The crystals used were made out of sodium fluoride (NaF), which has an index of refraction (1.33) close to ice in the visible. The crystal was mounted on top of a small pedestal and its orientation position was controlled by a rotator. Angular integrations in the experiment can follow the computational procedures in theory.

Figure 4a shows a comparison between measurements and theory for an aggregate that was assembled from NaF columns with small glass fiber attachments glued onto small holes. To simulate random orientation, a 1° increment was used for all possible orientation angles. General agreement between measurements and theory is shown but with several discrepancies. Most notable is that the experimental results are lower than the theory in backscattering directions, which are dominated by internal reflections. This difference could be caused by absorption of small glass fibers and glues that connect the columns. Comparison results for a rough-surface plate are shown in Fig. 4b. All eight sides were sanded with small scratches evenly distributed across the crystal surface. Between 25° and 180° scattering angles, the measurements closely follow the theoretical results. For scattering angles less than 20°, the experimental results are higher, however. The scanning electron photomicrographs reveal features on the roughened crystal surface on the order of 0.5–1 μm. More light could have been scattered through them as compared with the defined cross-sectional area used in diffraction calculations.

The electrodynamic levitation technique has also been used recently to suspend and grow an individual ice crystal for light-scattering experiments (Bacon *et al.*, 1998). The apparatus consists of an electrodynamic balance with an internally mounted thermal diffusion chamber, a laser beam, a 1024-element linear photo-

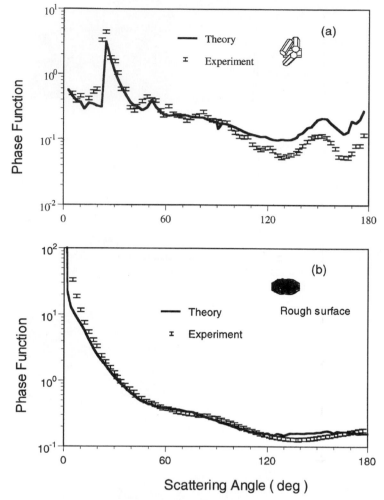

Figure 4 Phase functions for randomly oriented aggregate and rough surface plate crystals made from NaF with an index of refraction of 1.33 in the visible. The experiment used a polarized laser beam at $\lambda = 0.63$ μm as a light source and the positions of the detector and the crystal were controlled by automatic mechanical devices (Barkey *et al.*, 1999). The theoretical results are derived from the geometric ray-tracing/Monte Carlo method.

diode array, and two cameras for top and side views of the ice crystal. Shown in Fig. 5 are experimental results for two ice crystal sizes and shapes defined by the depicted photos (courtesy of N. J. Bacon). Theoretical results computed from conventional geometric ray tracing, which does not account for phase interferences,

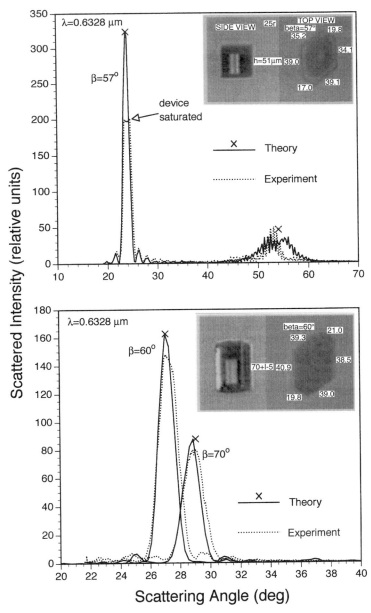

Figure 5 Phase function measurements for a single ice crystal suspended by the electrodynamic levitation technique (Bacon *et al.*, 1998). The sizes and shapes are determined from the top and side views of two cameras. The angle β denotes the ice crystal orientation with respect to the incident laser beam. Theoretical results are computed from the conventional ($\times$) and modified geometric ray-tracing methods.

show discrete maxima associated with the halo pattern. The modified geometric optics method generates closely matched patterns, except some deviations in the 55° scattering angle region in the top diagram. Differences between theoretical and experimental results can be attributed to the uncertainty in the measurement of the ice crystal size ($\sim$4 μm) and in the computation of the near field based on the geometric ray-tracing approximation.

III. APPLICATION TO REMOTE SENSING AND CLIMATE RESEARCH

Determination of the composition and structure of clouds and aerosols from the ground, the air, and space based on remote sensing is an important task in climate studies. In the following, we wish to demonstrate the applicability of the basic scattering, absorption, and polarization data for nonspherical ice crystals to various types of remote sensing of cirrus clouds and to climate studies.

A. BIDIRECTIONAL REFLECTANCE

Solar radiances reflected from clouds can be used to determine their composition and structure. The nondimensional bidirectional reflectance, the ratio of reflected and incident radiances for given positions of the Sun and observer and an underlying surface, is primarily a function of the cloud optical depth and particle size and shape. Development of reliable remote-sensing techniques from satellites for the detection of cirrus clouds and retrieval of their optical and microphysical properties using bidirectional reflectances must begin with an understanding of the fundamental scattering and absorption properties of ice particles.

Figure 6 shows measurements of the bidirectional reflectances of cirrus that were obtained with the scanning radiometer on board *ER-2* over Oklahoma on November 24 and 25, 1991, presented by Spinhirne *et al.* (1996), who also derived the best-fit cloud optical depths and surface albedos from concurrent lidar and spectral radiometric observations. For interpretation, we used a typical cirrostratus size distribution having a mean effective size of 42 μm and three ice crystal models: spheres, defined hexagons, and irregular ice particles (aggregates with rough surfaces). For the same optical depth, ice spheres reflect much less radiation than nonspherical ice crystals. The best matches for the three cases presented appear to be irregular particles. Because the measured data were about 20° apart, it is possible that some scattering maxima could be missed in the observations. It appears that the ice crystals in these developed cirrus must contain a combination of hexagonal and irregular ice crystals. In the visible, the bidirectional reflectance

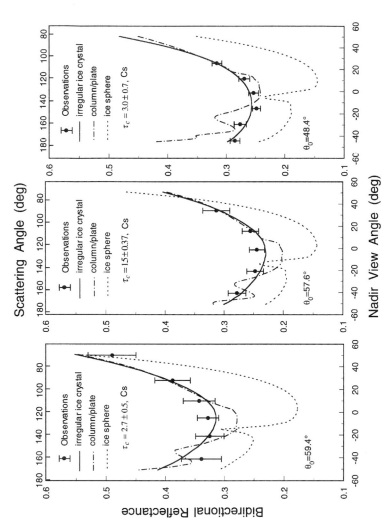

Figure 6 Measurements of visible bidirectional reflectances of cirrus obtained from the scanning radiometer on board *ER-2* over the area of Oklahoma on November 24 and 25, 1991 (Spinhirne *et al.*, 1996), and interpretations using light-scattering results for spheres, defined hexagons, and aggregates with rough surfaces based on the adding–doubling method for radiative transfer. A typical ice crystal size distribution with a mean effective size of 42 μm and predetermined optical depths are employed in the interpretation.

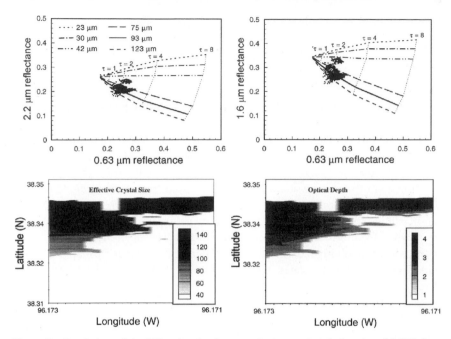

Figure 7   Correlations of the bidirectional reflectances in the wavelength domains of 0.63/1.6 μm and 0.63/2.2 μm. Six representative ice crystal size distributions with mean effective sizes ranging from 23 to 123 μm and optical depths from 0.1 to 8 are used in the construction of these curves. The viewing geometry includes $\theta_0 = 31.5°$, $\theta_0 = 12.5°$, and $\Delta\phi = 100°$. Also shown are the bidirectional reflectances obtained from MAS during the SUCCESS experiment on April 26, 1996. The bottom panels illustrate the retrieved optical depth and ice crystal mean size based on a statistical searching method (Rolland and Liou, 1998).

is largely dependent on the optical depth and ice crystal shape. The size information has been found from measurements at near infrared wavelengths where substantial absorption by ice occurs (King *et al.*, 1997).

In the following, we show the potential of determining the optical depth and ice crystal size based on correlation of bidirectional reflectance data in the domain of $\lambda = 0.63/1.6$ μm and 0.63/2.2 μm (Fig. 7). In the construction of the correlation diagram, six representative midlatitude ice crystal size distributions were used along with optical depths ranging from 0.1 to 8. The mean effective size ranges from 23 to 123 μm. The adding–doubling method for radiative transfer was employed to compute the bidirectional reflectances for cirrus cloud layers. Also shown are bidirectional reflectances obtained from the MODIS Airborne Simulator (MAS) for a sample viewing geometry occurring on April 26, 1996, during the SUCCESS experiment (Rolland and Liou, 1998). The retrieved optical depth ranges from about 2 to 4 and the retrieved mean effective ice crystal sizes are about 40–120 μm, as shown in the following maps. Validation of these retrievals has not been made at this point, however.

B. LINEAR POLARIZATION OF REFLECTED SUNLIGHT

Next, we present the applicability of the scattering data for nonspherical ice crystals to the interpretation of polarization of the reflected sunlight from cirrus clouds. Figure 8 shows the linear polarization pattern in the solar principal plane as a function of scattering angle that was measured from a cirrus cloud using a wavelength of 2.22 μm (Coffeen, 1979) at which the Rayleigh scattering contribution is minimum. The measured polarization values are less than about 6% and are positive from 50° and 150° scattering angles. The theoretical results based on the adding–doubling radiative transfer program (Takano and Liou, 1989b) include spheres, columns, plates, and a mixture of dendrites, bullet rosettes, and plates.

Results from the spherical model deviate significantly from observations in which the rainbow feature does not exist. For plates and columns, negative polarization results in the scattering angle region from 20° to 40° produced by halo patterns show general agreement with the observed data. The results for columns appear to match the observations, except in the backscattering direction from about 150° to 180°. With the inclusion of dendrites the backscattering polarization decreases and there is a general agreement between theoretical results and observed data in the entire scattering-angle range. It appears that the polarization patterns of the reflected sunlight can be used to infer the shape of cloud particles, which otherwise cannot be accomplished by other remote-sensing techniques.

C. LIDAR BACKSCATTERING DEPOLARIZATION

The depolarization technique using lidar backscattering returns has been developed to differentiate between ice and water clouds. It is based on the fundamental scattering properties of nonspherical ice crystals and spherical water droplets. The incident polarized light beam from spheres will retain its polarization state in the backscattering direction, if multiple scattering can be neglected. However, a cross-polarized component, referred to as depolarization, will be produced by nonspherical particles because of their deviation from the spherical geometry. In the geometric optics region, Liou and Lahore (1974) showed that depolarization is the result of internal reflections and refractions by hexagonal ice particles. To

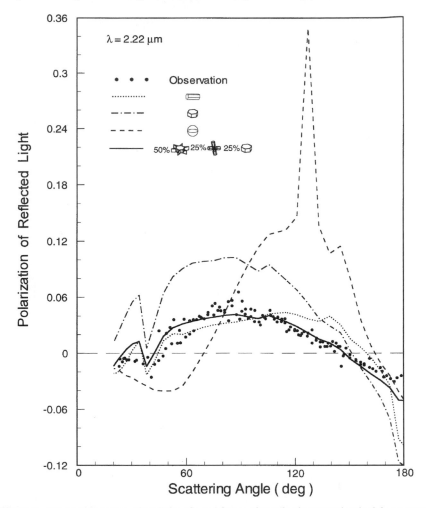

Figure 8   Linear polarization of sunlight reflected from a cirrus cloud measured at the 2.2-μm wavelength (Coffeen, 1979). The solar zenith angle is 70°. The theoretical polarization results are computed for ice spheres, columns, plates, and a mixture of dendrites, bullet rosettes, and plates as a function of the scattering angle.

quantify the amount of depolarization, a parameter called the depolarization ratio, defined as the ratio of the cross-polarized return power to the return power of the original polarization state, is introduced. It has been used to differentiate between ice and water clouds, as well as to determine some aspects of the physical characteristics of ice clouds (Sassen, 1991; Chapter 14).

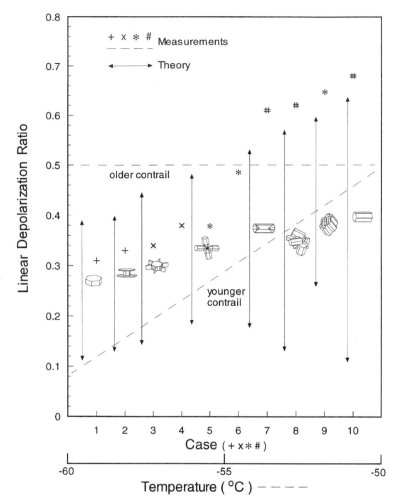

Figure 9 Depolarization ratios determined from high-resolution polarization lidar for contrail cirrus (Sassen and Hsueh, 1998, case; Freudenthaler *et al.*, 1996, temperature) and computed from the unified theory for light scattering by ice crystals with shapes ranging from single and double plates, solid and hollow columns, dendrites, bullet rosettes, aggregates, and irregular surface particles, the sizes of which span from a few micrometers to the geometric optics limit.

Figure 9 shows the depolarization ratios determined from high-resolution 0.532/1.06-μm polarization lidar for contrail cirrus presented by Sassen and Hsueh (1998) and Freudenthaler *et al.* (1996). The former authors showed that the lidar depolarization ratio in persisting contrails ranged from about 0.3 to 0.7, whereas the latter authors observed this ratio from 0.1 to 0.5 for contrails with temperatures ranging from $-60°$ to $-50°C$ depending on the stage of their growth. For interpretation, we have carried out backscattering depolarization calculations for various sizes and shapes displayed in this figure employing the unified theory for light scattering by ice crystals described in Section II. The vertical bars indicate the results for ice crystals of a few micrometers to the geometric optics region. Depolarization generally becomes larger for larger ice particles and reaches a maximum of about 0.6 for size parameters in the geometric optics limit. One exception is for columns, which produce a depolarization of about 0.65 for size parameters of about 10 because of resonance effects.

D. INFORMATION CONTENT OF 1.38-μM AND THERMAL INFRARED SPECTRA

Water vapor exhibits a number of absorption bands in the solar spectrum. Bidirectional reflectance at the top of the atmosphere in these bands will contain information of high-level clouds. Specifically, the 1.38-μm band has been found to be useful for the detection of cirrus clouds (Gao and Kaufman, 1995). The line spectra in this band have also been shown to contain rich information on the composition and structure of clouds and were a subject for a small-satellite proposal (Liou *et al.*, 1996).

To investigate the line formation in cirrus in the 1.38-μm band, we use a radiation model with a 1-cm^{-1} resolution containing 10 equivalent absorption coefficients based on the correlated k-distribution method for water vapor and other greenhouse gases derived from the updated 1996 HITRAN data (Liou *et al.*, 1998). The adding–doubling radiative transfer program including all Stokes parameters is used to perform the transfer of monochromatic radiation in vertically inhomogeneous atmospheres decomposed into a number of appropriate homogeneous layers. This program incorporates line absorption, scattering and absorption by nonspherical ice crystals, Rayleigh and background aerosol scattering, and surface reflection, and accounts for both direct solar flux and thermal emission contributions. In the calculations, we employ cirrostratus and cirrus uncinus models having mean effective ice crystal sizes of 42 and 123 μm, respectively, with a shape composition of 50% aggregates/bullet rosettes, 25% hollow columns, and 25% plates.

Figure 10a illustrates the bidirectional reflectances in the 1.38-μm water vapor line spectrum from 6600–7500 cm^{-1} for clear and cirrus cloudy conditions. The line structure of the water vapor absorption exhibits significant fluctuations. At about 7100–7400 cm^{-1}, the reflectances from the clear atmosphere are extremely small as a result of strong water vapor absorption. Multiple scattering produced by ice particles contributes to the strength of reflectances in the line

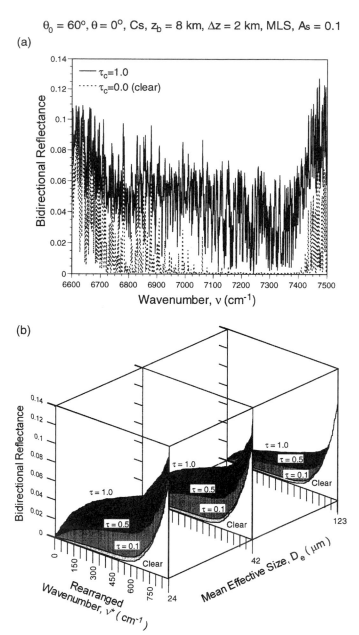

Figure 10 (a) Bidirectional reflectances for clear and cirrus cloudy atmospheres as a function of wavenumber in the 1.38-μm water vapor band. (b) Monotonically increasing bidirectional reflectances as a function of rearranged wavenumber in the domain of optical depth and mean effective ice crystal size. The solar and emergent zenith angles are denoted in the figure and the calculations were carried out in the solar principal plane.

wing regions. Figure 10b shows the reflectance spectra whose wavenumbers are ordered according to their magnitudes so that monotonically increasing functions are displayed in the domain of optical depth and mean effective ice crystal size. Low values indicate that the reflectances are associated with line centers, whereas high values are related to line wings. Reflectances are dependent on the optical depth and ice crystal size, as clearly demonstrated in this example. Consequently, a retrieval procedure can be constructed for the determination of these two parameters. Moreover, we find that the cloud position can also be inferred from the spectra because the reflectances are determined by the amount of water vapor above the cloud. The preceding example clearly demonstrates that the 1.38-μm line spectra contain rich information about the cirrus composition and structure. Of course, the question of uniqueness of the solution of cloud parameters within the broad range of spectral lines is one that requires further investigations and numerical experimentations.

Information on thin cirrus in the tropics has been noted from the analysis of satellite Infrared Radiation Interferometer Spectrometer (IRIS) data (Prabhakara *et al.*, 1993), particularly in the 8–12-μm window region. Recent technological advancements have led to the development of the High-Resolution Interferometer Sounder (HIS), a Michelson interferometer covering a broad spectral region in the infrared (3.5–19 μm) with high spectral and spatial resolutions (Smith *et al.*, 1998). Information on thin cirrus containing small ice crystals appears between $\lambda = 10$ and 12 μm. Interpretation of the line structure in the thermal infrared for cirrus cloudy atmospheres and exploration of the information content with respect to the cloud optical depth, ice crystal size, and position would be an exciting project.

In summary, because of the spatial and temporal variabilities of ice crystal sizes and shapes in cirrus clouds, remote sensing of their optical and microphysical properties from space presents an unusual challenge in atmospheric sciences.

E. SOLAR ALBEDO

Reflection of solar radiation by clouds determines the amount of solar energy absorbed within the atmosphere and by the surface. Thus, understanding the broad-band solar albedo (reflection) is fundamental in the analysis of the cloud radiative forcing associated with climate studies. We wish to illustrate the importance of the nonsphericity of ice particles on the interpretation of observed solar albedo determined from radiometric measurements. Figure 11 shows the broad-band solar albedo as a function of the ice water path (IWP) derived from broad-band flux aircraft observations for cirrus clouds during the FIRE experiment in Wisconsin, October–November, 1986 (Stackhouse and Stephens, 1991). The extensions of the vertical and horizontal lines through the data points represent the uncertainty of measurements.

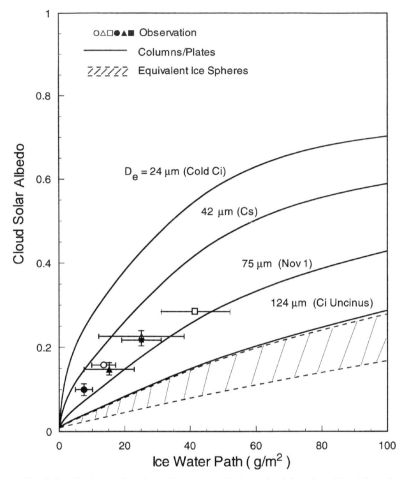

Figure 11 Solar albedo as a function of ice water path determined from broad-band flux observations from aircraft for cirrus clouds that occurred during the FIRE experiment, Wisconsin, November–December, 1986 (Stackhouse and Stephens, 1991). The solid lines represent theoretical results computed from a line-by-line equivalent solar model using observed ice crystal sizes and shapes for a range of mean effective ice crystal diameters. The dashed lines are corresponding results for equivalent spheres.

The solid lines are theoretical results computed from the line-by-line equivalent solar radiative transfer model mentioned previously using a set of observed ice crystal size distributions for columns and plates. The dashed lines denote the results based on these size distributions converted into equivalent spheres. Regardless of the input parameters for spheres, the theoretical results significantly underestimate the observed values primarily because of the nature of stronger forward scattering for spherical particles and stronger absorption for spheres at near infrared wavelengths. Using the mean effective ice crystal size defined in Liou *et al.* (1998), we show that the size that best fits the observed data lies between 50 and 75 μm, typical ice crystal sizes at the top portion of midlatitude cirrus cloud systems.

Further, we have also investigated from a theoretical perspective the effects of ice crystal shape on solar albedo by using a mean effective size of 16 μm, representing a typical ice crystal size for contrail cirrus. In this study, four shapes are used in which bullet rosettes have both smooth and rough surfaces. Cloud albedo, not shown here, becomes progressively smaller for hollow columns, plates, and equal-area spheres relative to that of bullet rosettes, primarily because their asymmetry parameters become increasingly larger to allow stronger forward scattering to take place. The effect of ice crystal surface roughness does not appear to alter the solar albedo values for nonspherical particles. It does, however, affect the phase function pattern, a critical parameter in remote-sensing applications.

F. TEMPERATURE SENSITIVITY TO ICE CRYSTAL NONSPHERICITY

Many dynamic and thermodynamic factors and feedbacks affect temperature perturbations. Nevertheless, we wish to demonstrate that the scattering and absorption properties of nonspherical ice crystals are relevant and important in the modeling of the role of clouds in climate. The potential effect of ice crystal nonsphericity in light scattering on climatic temperature perturbations is studied by using a one-dimensional cloud–climate model developed by Liou and Ou (1989). Perturbation calculations were performed by varying the cloud cover and IWP for a typical cirrostratus cloud model with a thickness of 1.7 km and a base height of 9 km. Cloud positions and covers for middle and low clouds were prescribed and other parameters in the model remained unchanged in the perturbation runs. The radiative properties of columns/plates and area-equivalent ice spheres were incorporated into the climate model to investigate sensitivity to the surface temperature change. The present climate condition is defined at 288 K, which corresponds to a cirrus cloud cover of 20% and an IWP of 20 g/m^2 based on the column/plate model.

The left panel of Fig. 12 shows the variation of surface temperature as a function of cloud cover when the IWP is fixed. Because the greenhouse effect produced by the trapping of thermal infrared radiation outweighs the solar albedo effect, increasing the cloud cover increases the surface temperature. If a spherical model is used, a significant increase of the surface temperature occurs because

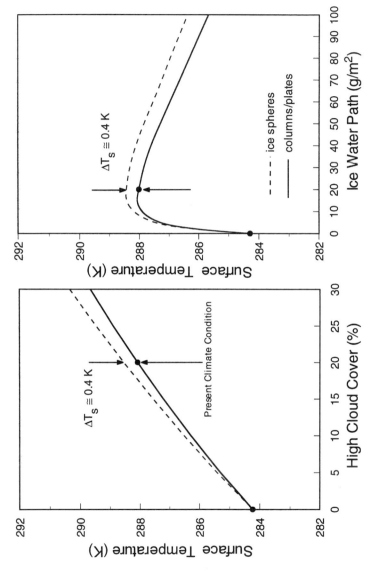

Figure 12 Surface temperatures determined from a one-dimensional cloud and climate model using a radiative transfer parameterization based on the scattering and absorption properties of hexagonal columns/plates and equivalent ice spheres. The model has a present climate condition corresponding to the surface temperature of 288 K involving a typical cirrostratus located at 9 km with a thickness of 1.7 km and an ice water content of 10^{-2} g/m^3 and the cirrus has 20% fractional coverage. Perturbations are performed for both cloud cover and IWP.

equivalent spheres reflect less solar radiation as shown in Fig. 11. At the present climate condition, the increase amounts to about 0.4 K, which appears to be substantial. Variation of the IWP when the cloud cover is fixed is shown in the right panel of Fig. 12. An increase of the surface temperature occurs at IWPs up to 15 g/m^2 after which a decrease occurs. This is because an increase in the infrared (IR) emissivity is relatively smaller as compared with an increase in the solar albedo, thereby leading to cooling effect. Using the spherical model, the cloud radiative forcing increases by about a factor of 2 because of a reduction in the solar albedo. For this reason, larger surface temperatures are produced relative to the case involving the column/plate model.

In view of the preceding discussion, a sufficient sensitivity of climatic temperature perturbations can be observed when the shape of ice particles (spheres vs hexagons) is accounted for in radiative transfer calculations. Thus a physically based cloud microphysical model is required in the parameterization of the radiative properties of cirrus clouds for climate models.

IV. SUMMARY

We have presented a unified theory for light scattering by ice crystals of all sizes and shapes that can be defined mathematically or numerically. This theory is a combination of a geometric optics approximation for size parameters larger than about 20 and a finite difference time domain method for size parameters smaller than about 20. Conventional geometric ray tracing was first reviewed, followed by a discussion of the physical fundamentals of the improved method involving the mapping of the tangential components of the electric and magnetic fields on the ice crystal surface to the far field on the basis of the electromagnetic equivalence theorem. By virtue of this mapping, the only approximation is in the calculation of the surface electric fields by means of the Fresnel coefficients and the applicable Snell's law based on the geometry. Phase interference and wave diffraction are both accounted for in the method.

The issue of absorption in the medium in the context of geometric ray tracing, referred to as the inhomogeneous effect, was subsequently discussed. We showed that the phase propagation of a wave inside the medium is determined by an adjusted real part of the refractive index through the Fresnel and Snell laws, whereas attenuation of the wave amplitude is determined by an adjusted imaginary part of the refractive index. The adjusted refractive indices are derived on the basis of the fundamental electromagnetic wave theory. We further described an efficient way of performing geometric ray tracing in complex-shaped ice particles via the Monte Carlo method and presented a methodology to treat the possibility of irregularity of the ice crystal surface using the stochastic approach. Comparisons of the phase function results derived from the theory and laboratory-controlled experiments were also made.

For size parameters smaller than about 20, we adopted a finite difference time domain technique for light scattering by small nonspherical ice crystals, which solves the Maxwell equations by finite difference numerical means in the time domain by discretizing the scatterer with given optical properties. The solution requires the imposition of a numerically stable absorbing boundary condition. The frequency spectrum of the time-dependent results can be obtained by using a suitable Gaussian pulse via the discrete Fourier transform technique. The far-field solution can be derived by employing a surface or a volume integration approach. The method that we have developed was verified through comparisons with the exact Lorenz–Mie results for spheres and infinite circular cylinders and was shown to be efficient and accurate for size parameters on the order of 20. It can also be effectively applied to small inhomogeneous particles such as aerosols.

We then presented a number of examples demonstrating the application of the unified theory for light scattering by ice crystals to remote sensing of ice crystal clouds and to investigation of the climatic effect of cirrus. We showed that interpretations of the bidirectional reflectance and polarization patterns measured from aircraft and satellites require the correct scattering, absorption, and polarization data for nonspherical ice crystals. Lidar backscattering observations of cirrus and contrail, particularly those utilizing the depolarization technique, also require the correct scattering information on nonspherical ice particles. Based on observations and appropriate ice cloud models, the ice crystal size and shape and optical depth information for cirrus can be inferred from the reflected solar intensity and polarization. Moreover, we illustrated that rich information on cirrus cloud composition and structure is contained in the 1.38-μm solar line spectra. Indeed, determination of the optical and microphysical properties of cirrus, subvisual cirrus, and contrails based on remote sensing presents a great challenge in view of the substantial variability of ice crystal sizes and shapes in space and time.

Finally, we discussed the importance of the scattering and absorption properties of nonspherical ice particles in conjunction with studies of cloud radiative forcing and climatic temperature perturbations resulting from uncertainties in the cirrus cloud parameters. We used the models of ice columns/plates and equivalent spheres to illustrate the effect of nonspherical shapes on the broad-band solar albedo and showed that cloud albedo is much smaller for equivalent spheres because of stronger forward scattering. Although the temperature responses to climate change are complex and involve numerous dynamic and thermodynamic factors and feedbacks, we illustrate that the physically based single-scattering properties of nonspherical ice crystals are relevant and significant in the modeling of cirrus cloud radiative transfer for climate studies.

ACKNOWLEDGMENTS

Support of this research work includes National Science Foundation Grant ATM-97-96277, NASA Grants NAG-5-6160 and NAG-1-1966, and DOE Grant DE-FG03-95ER61991.

Chapter 16

Centimeter and Millimeter Wave Scattering from Nonspherical Hydrometeors

Kültegin Aydın
Department of Electrical Engineering
Pennsylvania State University
University Park, Pennsylvania 16802

I. INTRODUCTION

This chapter deals with electromagnetic wave scattering from hydrometeors at centimeter and millimeter wavelengths, also known as the microwave (1–40 GHz) and millimeter wave (40–300 GHz) bands. The radar meteorology community refers to millimeter wave radars as those operating above 30 GHz. Radar engineers have designated different frequency bands with the following letters (Skolnik, 1990): L band (1–2 GHz), S band (2–4 GHz), C band (4–8 GHz), X band (8–12 GHz), K_u band (12–18 GHz), K band (18–27 GHz), K_a band (27–40 GHz), V band (40–75 GHz), and W band (75–110 GHz). Presently, operational and research radars used for the remote sensing of storms and clouds operate at frequencies in the S, C, X, K_u, K_a, and W bands (Hobbs *et al.*, 1985; Bringi and Hendry, 1990; Lhermitte, 1987, 1990; Meneghini and Kozu, 1990; Pazmany *et al.*, 1994; Clothiaux *et al.*, 1995; Kropfli *et al.*, 1995; Sekelsky and McIntosh, 1996; Testud *et al.*, 1996).

Oguchi (1983) provides an excellent review of electromagnetic wave propagation and scattering in rain and other hydrometeors. This chapter reviews the developments over the past 15 years, focusing on selected studies dealing with polarimetric scattering properties of nonspherical hydrometeors. Applications are directed toward the identification and quantification of hydrometeors using polarimetric radars (Atlas, 1990; Doviak and Zrnic, 1993). Estimation of rainfall rate; detection of hail; discrimination of snow, rain, and graupel; and estimation of cloud ice water content are among these applications, which should have significant impact on fields such as meteorology, hydrology, climatology, and radiowave communications.

Many terrestrial and satellite communication systems operate at microwave frequencies and are affected by hydrometeors in their paths. A medium filled with hydrometeors will attenuate the signal, cause interference between different links because of bistatic scattering, and in frequency reuse systems using transmission on orthogonal polarizations cause crosstalk as a result of depolarization (Awaka and Oguchi, 1982; de Wolf and Ligthart, 1993; Roddy, 1996). Because of space limitations these topics will not be covered. The focus of this chapter will be on polarimetric radar remote-sensing applications over the frequency range 3–300 GHz.

II. POLARIMETRIC RADAR PARAMETERS

In this chapter we will use the linear polarization basis. A simple linear transformation can be used to convert from the linear polarization basis to another basis such as circular or elliptical (Azzam and Bashara, 1987; Chandrasekar *et al.*, 1994). Over the past 15 years most polarization studies in radar meteorology have focused on linear polarization techniques. Results from earlier work on circular polarization can be found in the papers by McCormick and Hendry (1975) and Hendry and Antar (1984). A linearly polarized wave propagating in rain experiences depolarization, which is the conversion of energy from the initial to the orthogonal polarization as a result of raindrop canting, differential phase shift, and differential attenuation between the horizontally (h) and vertically (v) polarized components [see Eq. (1)]. At S-band frequencies differential phase shift is the dominant factor causing depolarization (Humphries, 1974; Seliga *et al.*, 1984). Methods have been developed to recover circular polarization parameters affected by differential phase shift (Bebbington *et al.*, 1987; Jameson and Davie, 1988; Torlaschi and Holt, 1993).

The notation in Chapter 1 is based on the forward scattering alignment (FSA) convention (Ulaby and Elachi, 1990). The notation used in this chapter is based on the backscattering alignment (BSA) convention (Ulaby and Elachi, 1990), with the exception of forward scattering parameters involving attenuation and propagation phase shift. The amplitude matrix **S** for the BSA convention relates the incident and scattered wave fields as follows:

$$\begin{bmatrix} E_v^{sca} \\ E_h^{sca} \end{bmatrix} = \frac{e^{ikR}}{R} \begin{bmatrix} S_{vv} & S_{vh} \\ S_{hv} & S_{hh} \end{bmatrix} \begin{bmatrix} E_v^{inc} \\ E_h^{inc} \end{bmatrix} \quad [\text{V m}^{-1}]. \tag{1}$$

The amplitude matrix in the FSA convention [Eq. (4) of Chapter 1] is related to that in Eq. (1) as follows:

$$\begin{bmatrix} S_{vv} & S_{vh} \\ S_{hv} & S_{hh} \end{bmatrix} = \begin{bmatrix} S_{11} & S_{12} \\ -S_{21} & -S_{22} \end{bmatrix} \quad [\text{m}]. \tag{2}$$

The phase matrix elements in the BSA (denoted by an overbar) and FSA [Eqs. (14)–(29) in Chapter 1] conventions are related as

$$\overline{Z}_{ij} = Z_{ij}, \qquad i = 1, 2 \text{ and } j = 1, 2, 3, 4 \quad [\text{m}^2], \tag{3a}$$

$$\overline{Z}_{ij} = -Z_{ij}, \qquad i = 3, 4 \text{ and } j = 1, 2, 3, 4 \quad [\text{m}^2]. \tag{3b}$$

The backscattering and bistatic scattering cross sections can be expressed in terms of the amplitude matrix elements as

$$\sigma_{ij} = 4\pi |S_{ij}|^2, \qquad i, j = \text{v, h} \quad [\text{m}^2]. \tag{4}$$

For backscattering we have $\sigma_{hv} = \sigma_{vh}$ because of reciprocity (see Chapter 1). The effective reflectivity factors (Battan, 1973) at v and h polarizations and the difference reflectivity [note that Golestani *et al.*, 1989, define $Z_{DP} = 10 \log(Z_h - Z_v)$ for $Z_h > Z_v$] are defined as

$$Z_v = C n_0 \langle \sigma_{vv} \rangle = 2\pi C n_0 \langle \overline{Z}_{11} + \overline{Z}_{12} + \overline{Z}_{21} + \overline{Z}_{22} \rangle \quad [\text{mm}^6 \text{ m}^{-3}], \tag{5a}$$

$$Z_h = C n_0 \langle \sigma_{hh} \rangle = 2\pi C n_0 \langle \overline{Z}_{11} - \overline{Z}_{12} - \overline{Z}_{21} + \overline{Z}_{22} \rangle \quad [\text{mm}^6 \text{ m}^{-3}], \tag{5b}$$

$$Z_{DP} = |Z_h - Z_v| = 4\pi C n_0 |\langle \overline{Z}_{12} + \overline{Z}_{21} \rangle| \quad [\text{mm}^6 \text{ m}^{-3}]. \tag{5c}$$

The brackets $\langle \cdot \rangle$ indicate ensemble averaging and n_0 is the total number of particles per cubic meter (see Chapter 1). The coefficient $C = 10^{18} \lambda^4 / (\pi^5 |K|^2)$, where λ is the free-space wavelength in meter units, $K = (\varepsilon - 1)/(\varepsilon + 2)$, and ε is the complex dielectric constant of water or ice (Battan, 1973). Unless specified otherwise, ε will be that of water even if the medium contains ice particles. The effective reflectivity factors are also expressed as $10 \log(Z_v$ or Z_h or $Z_{DP})$ in dBZ units. Other radar parameters are defined later.

The differential reflectivity (Seliga and Bringi, 1976)

$$Z_{DR} = \frac{Z_h}{Z_v} = \frac{\langle \overline{Z}_{11} - \overline{Z}_{12} - \overline{Z}_{21} + \overline{Z}_{22} \rangle}{\langle \overline{Z}_{11} + \overline{Z}_{12} + \overline{Z}_{21} + \overline{Z}_{22} \rangle}. \tag{6}$$

The linear depolarization ratios

$$LDR_v = \frac{\langle \sigma_{hv} \rangle}{\langle \sigma_{vv} \rangle} = \frac{\langle \overline{Z}_{11} + \overline{Z}_{12} - \overline{Z}_{21} - \overline{Z}_{22} \rangle}{\langle \overline{Z}_{11} + \overline{Z}_{12} + \overline{Z}_{21} + \overline{Z}_{22} \rangle}, \tag{7a}$$

$$LDR_h = \frac{\langle \sigma_{vh} \rangle}{\langle \sigma_{hh} \rangle} = \frac{\langle \overline{Z}_{11} - \overline{Z}_{12} + \overline{Z}_{21} - \overline{Z}_{22} \rangle}{\langle \overline{Z}_{11} - \overline{Z}_{12} - \overline{Z}_{21} + \overline{Z}_{22} \rangle}. \tag{7b}$$

For backscattering we have $Z_{DR} = (LDR_v)/(LDR_h)$. The ratio parameters in Eqs. (6) and (7) are generally expressed as $10 \log(\cdot)$ in dB units. Note that the depolarization ratios for circular and elliptical polarizations are denoted as CDR and EDR, respectively.

The copolarized correlation coefficient (note that the $e^{-i\omega t}$ time dependence is used throughout this chapter)

$$\rho_{hv} \exp(i\delta_{hv}) = \frac{\langle S_{hh} S_{vv}^* \rangle}{[\langle |S_{hh}|^2 \rangle \langle |S_{vv}|^2 \rangle]^{1/2}}, \tag{8a}$$

$$\rho_{hv} = \frac{[\langle \overline{Z}_{33} + \overline{Z}_{44} \rangle^2 + \langle \overline{Z}_{43} - \overline{Z}_{34} \rangle^2]^{1/2}}{[\langle \overline{Z}_{11} - \overline{Z}_{12} - \overline{Z}_{21} + \overline{Z}_{22} \rangle \langle \overline{Z}_{11} + \overline{Z}_{12} + \overline{Z}_{21} + \overline{Z}_{22} \rangle]^{1/2}}, \tag{8b}$$

$$\delta_{hv} = \arg(\langle S_{hh} S_{vv}^* \rangle) = \arctan\left[\frac{\langle \overline{Z}_{43} - \overline{Z}_{34} \rangle}{\langle \overline{Z}_{33} + \overline{Z}_{44} \rangle}\right]. \tag{8c}$$

The cross-polarized correlation coefficients

$$\rho_v \exp(i\delta_v) = \frac{\langle S_{vv} S_{hv}^* \rangle}{[\langle |S_{vv}|^2 \rangle \langle |S_{hv}|^2 \rangle]^{1/2}}, \tag{9a}$$

$$\rho_v = \frac{[\langle \overline{Z}_{31} + \overline{Z}_{32} \rangle^2 + \langle \overline{Z}_{41} + \overline{Z}_{42} \rangle^2]^{1/2}}{[\langle \overline{Z}_{11} + \overline{Z}_{12} + \overline{Z}_{21} + \overline{Z}_{22} \rangle \langle \overline{Z}_{11} + \overline{Z}_{12} - \overline{Z}_{21} - \overline{Z}_{22} \rangle]^{1/2}}, \tag{9b}$$

$$\delta_v = \arg(\langle S_{vv} S_{hv}^* \rangle) = \arctan\left[\frac{\langle \overline{Z}_{41} + \overline{Z}_{42} \rangle}{\langle -\overline{Z}_{31} - \overline{Z}_{32} \rangle}\right] \tag{9c}$$

and

$$\rho_h \exp(i\delta_h) = \frac{\langle S_{hh} S_{vh}^* \rangle}{[\langle |S_{hh}|^2 \rangle \langle |S_{vh}|^2 \rangle]^{1/2}}, \tag{10a}$$

$$\rho_h = \frac{[\langle \overline{Z}_{32} - \overline{Z}_{31} \rangle^2 + \langle \overline{Z}_{42} - \overline{Z}_{41} \rangle^2]^{1/2}}{[\langle \overline{Z}_{11} - \overline{Z}_{12} - \overline{Z}_{21} + \overline{Z}_{22} \rangle \langle \overline{Z}_{11} - \overline{Z}_{12} + \overline{Z}_{21} - \overline{Z}_{22} \rangle]^{1/2}}, \tag{10b}$$

$$\delta_h = \arg(\langle S_{hh} S_{vh}^* \rangle) = \arctan\left[\frac{\langle \overline{Z}_{42} - \overline{Z}_{41} \rangle}{\langle \overline{Z}_{32} - \overline{Z}_{31} \rangle}\right]. \tag{10c}$$

Note that a similar parameter (the correlation between the co- and cross-polarized elements) for the circular polarization basis is denoted as ρ_c. In addition to the parameters defined previously, a bistatic radar system can also measure the ratios of the bistatic-to-backscattering reflectivities BBR_v and BBR_h (Aydin *et al.*, 1998):

$$BBR_v = \frac{Z_v(\text{bistatic})}{Z_v(\text{back})}, \tag{11a}$$

$$BBR_h = \frac{Z_h(\text{bistatic})}{Z_h(\text{back})}. \tag{11b}$$

Finally, there are several parameters of interest that relate to the forward scattering characteristics of the particles; these are defined next based on the FSA convention of Chapter 1.

The specific differential phase

$$K_{DP} = 10^3 \left(\frac{180}{\pi}\right)\left(\frac{2\pi}{k}\right) n_0 \, \text{Re}\left[\langle S_{22}(\mathbf{n}, \mathbf{n}) - S_{11}(\mathbf{n}, \mathbf{n}) \rangle\right]$$

$$= 10^3 \left(\frac{180}{\pi}\right) n_0 \langle K_{34}(\mathbf{n}) \rangle \qquad [\text{deg km}^{-1}], \tag{12}$$

where K_{34} is an element of the extinction matrix and k is the wavenumber in m^{-1} units (see Chapter 1).

The specific attenuation at v and h polarizations and the specific differential attenuation

$$A_v = 4.343 \times 10^3 n_0 \langle \sigma_{ev} \rangle = 8.686 \times 10^3 \left(\frac{2\pi}{k}\right) n_0 \, \text{Im}\left[\langle S_{11}(\mathbf{n}, \mathbf{n}) \rangle\right]$$

$$= 8.686 \times 10^3 n_0 \frac{\langle K_{11}(\mathbf{n}) + K_{12}(\mathbf{n}) \rangle}{2} \qquad [\text{dB km}^{-1}], \tag{13a}$$

$$A_h = 4.343 \times 10^3 n_0 \langle \sigma_{eh} \rangle = 8.686 \times 10^3 \left(\frac{2\pi}{k}\right) n_0 \, \text{Im}\left[\langle S_{22}(\mathbf{n}, \mathbf{n}) \rangle\right]$$

$$= 8.686 \times 10^3 n_0 \frac{\langle K_{11}(\mathbf{n}) - K_{12}(\mathbf{n}) \rangle}{2} \qquad [\text{dB km}^{-1}], \tag{13b}$$

$$\Delta A = A_h - A_v = 8.686 \times 10^3 n_0 \langle -K_{12}(\mathbf{n}) \rangle \qquad [\text{dB km}^{-1}], \tag{13c}$$

where σ_{ev} and σ_{eh} are the extinction cross sections at v and h polarizations. Attenuation resulting from hydrometeors, especially raindrops and melting snow and ice particles, can be significant at C-band frequencies and above (as well as at the S band for special circumstances; e.g., Ryzhkov and Zrnic, 1994).

Note that optimal polarization parameters will not be discussed here and can be found in the papers by McCormick (1985), Kwiatkowski *et al.* (1995), and Hubbert and Bringi (1996). Also, issues related to measurement techniques and the statistical properties of these parameters will not be considered. The reader is referred to the following publications: Bringi *et al.* (1983), Sachidananda and Zrnic (1986), Chandrasekar and Bringi (1988a, b), Chandrasekar *et al.* (1990), Doviak and Zrnic (1993), Kostinski (1994), and Jameson and Kostinski (1999).

III. HYDROMETEOR MODELS

As methodologies are developed to extract more information about hydrometeors utilizing polarimetric radars, it becomes increasingly important to model the hydrometeors more accurately. Their shape, size, fall behavior, and composition must be represented realistically with enough detail to capture the dominant features influencing the measurable radar parameters at the frequencies of interest. For example, raindrops are generally modeled as oblate spheroids with size-dependent axis ratios (defined as the ratio of the vertical axis to the horizontal axis with the symmetry axis along the vertical direction; this ratio is less than 1 for oblate spheroids and greater than 1 for prolate spheroids). However, their actual equilibrium shapes are somewhat more flat on the bottom as shown in Fig. 1 (Beard and Chuang, 1987). This is different from the elliptical cross section of

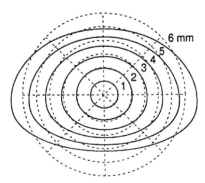

Figure 1 Computed shapes of raindrops in air for $D = 1, 2, 3, 4, 5$, and 6 mm with origin at center of mass. Shown for comparison are dashed circles of diameter D divided into 45° sectors. Reprinted from K. V. Beard and C. Chuang (1987), A new model for the equilibrium shape of raindrops, *J. Atmos. Sci.* **44**, 1509–1524. Copyright © 1987 American Meteorological Society.

an oblate spheroid. Comparison of the actual equilibrium shape with an oblate spheroidal model having the same volume and axis ratio shows that the backscattering and extinction cross sections are not significantly different (within 3%) up to 15 GHz. The difference begins to become noticeable at 35 GHz for backscattering (at side incidence) from drops with $D > 3$ mm. For example, the oblate spheroidal model with $D = 5$ mm has about 10% lower σ_{vv} than the equilibrium shape model. A brief overview of hydrometeor characteristics is presented next.

A. SHAPE

Raindrops can oscillate and deviate from their equilibrium shape. Oscillating drops have been modeled, on average, as oblate spheroids (axis ratio < 1) with larger axis ratios compared to the equilibrium shape model (Beard *et al.*, 1983, 1991; Chandrasekar *et al.*, 1988; Beard and Kubesh, 1991; Kubesh and Beard, 1993; Tokay and Beard, 1993; Keenan *et al.*, 1997; Pruppacher and Klett, 1997; Bringi *et al.*, 1998). Ice phase hydrometeors such as hailstones, graupel particles, and snowflakes may have irregular shapes as well as shapes that may be modeled as spheroids (Pruppacher and Klett, 1997). Hailstones collected at ground level in Colorado have oblate spheroidal (predominant shape) and conical shapes with axis ratios varying between 0.5 and 1, with 0.77 being the approximate mean value (Matson and Huggins, 1980; Knight, 1986).

There are a great variety of snow crystal shapes (Magono and Lee, 1966; Pruppacher and Klett, 1997). In terms of their scattering characteristics for radar remote-sensing applications, they can be grouped under three basic forms: columnar, planar, and spatial crystals. The spatial crystals have the most complex and difficult shapes to deal with in terms of scattering calculations, especially at the higher millimeter wave frequencies such as 94 and 220 GHz, which show sensitivity to the details of their structure. Bullet rosettes, capped columns, and side planes are among such crystal types.

B. COMPOSITION

The composition of hydrometeors affects their dielectric constants and, as a result, their scattering characteristics. Raindrops are composed of water, which has a temperature-dependent dielectric constant (Ray, 1972). This temperature dependence has a negligible effect on the backscattering cross sections at the S band. At higher frequencies the effect becomes noticeable for drop sizes at or near resonance scattering. On the other hand, ice phase hydrometeors are generally inhomogeneous particles composed of ice and air mixtures (except for certain pristine crystals such as hexagonal plates, which are essentially solid ice). These can be treated as homogeneous particles for scattering calculations by using effective dielectric constants determined by the Maxwell-Garnett or Bruggeman theories (Bohren and Battan, 1980; Chapter 9). The same theories can also be applied to particles with water in their mixture, such as spongy graupel particles and hailstones (Bohren and Battan, 1982; Chýlek *et al.*, 1991; Fujiki *et al.*, 1994; Meneghini and Liao, 1996). Melting hailstones and other ice particles may also acquire a coating of water or water–ice mixture on top of a solid ice or ice–air mixture core (Rasmussen and Heymsfield, 1987). Such particles would have to be treated as inhomogeneous scatterers.

C. FALL BEHAVIOR (ORIENTATION)

A raindrop falls in the atmosphere with its symmetry axis (which is the minor axis of the oblate spheroid model) aligned in the vertical direction with very little deviation (Pruppacher and Klett, 1997). The tilt of the symmetry axis from the vertical direction is called canting. The mean canting angle for raindrops has been estimated to be near 0° with a standard deviation less than 3°. The fall behavior of hailstones and graupel is not as well understood. Conical shapes fall with their apex pointed upward most frequently, but the apex can also be pointed downward (Pruppacher and Klett, 1997). For radar observations at near side incidence, it does not make a difference either way, as long as the canting angle distributions are the same. On the other hand, it is important to know if the symmetry axis of oblate and prolate spheroidal models is along the vertical or horizontal direction. It has been suggested that oblate spheroidal hailstones may fall with their minor axes, on average, vertically aligned (Knight and Knight, 1970; Roos and Carte, 1973; List *et al.*, 1973; Matson and Huggins, 1980) as well as horizontally aligned (Knight and Knight, 1970; Kry and List, 1974). They can also tumble, gyrate, and wobble. Such uncertainties in the falling pattern of hailstones make it difficult to obtain quantitative information about these particles using polarimetric techniques. Columnar and planar (larger than 100 μm) snow crystals fall with their large dimensions horizontal and deviations from the horizontal plane exhibit a Gaussian distribution (Pruppacher and Klett, 1997; Sassen, 1980). There are no established relationships between size and canting angle distributions for water and ice phase hydrometeors.

D. SIZE DISTRIBUTION

The gamma model size distribution is widely used for all types of hydrometeors and can be expressed as

$$N(D) = N_0 D^\mu e^{-\Lambda D} \qquad [\text{mm}^{-1}\,\text{m}^{-3}], \tag{14}$$

where D is a representative dimension of the particle such as equivalent-volume spherical diameter or maximum dimension and $N(D)\,dD$ is the number of particles per unit volume with sizes from D to $D + dD$. Raindrops have equivalent-volume spherical diameters ranging from about 0.1 to about 8 mm. The well-known exponential drop size distribution is a special case of Eq. (14) with $\mu = 0$, which has been used for raindrops (Marshall and Palmer, 1948) as well as for hailstones (Federer and Waldvogel, 1975; Cheng and English, 1983; Musil *et al.*, 1976). The gamma model with varying values of μ is also used for raindrops and snow crystals (Ulbrich, 1983; Kosarev and Mazin, 1991; Mitchell *et al.*, 1996; Tokay and Short, 1996). A bimodal size distribution composed of two gamma model distributions has also been suggested for ice crystals (Mitchell *et al.*, 1996).

IV. SCATTERING CHARACTERISTICS OF HYDROMETEORS

Before dealing with the polarimetric radar signatures of hydrometeors it is worthwhile to understand their single-scattering characteristics. The following sections take a look at some of the important results reported in the literature.

A. RAINDROPS

Seliga *et al.* (1982a, 1984) and Jameson (1983) first investigated the effects of oscillations on the backscattering cross sections of raindrops at the S band (10-cm wavelength). The oscillating drops were modeled as spheroids with varying axis ratios making them oblate at one extreme and prolate at the other (axisymmetric oscillation). The differential reflectivity Z_{DR} decreased by about a few tenths of a decibel up to 1 dB, depending on the oscillation and the drop size distribution model parameters (Seliga *et al.*, 1984). Jameson (1983) showed that for a given axis ratio (r) distribution, Z_{DR}^{-1} (expressed as a ratio) was proportional to the power $[|S_{hh}|^2]$ weighted average of $r^{7/3}$. His results also indicated that Z_{DR} decreases because of drop oscillations and that the effects of oscillations could be represented by axis ratios larger than the equilibrium values. Zrnic and Doviak (1989) showed that oscillation magnitudes that produce less than 10% change in the equilibrium axis ratios are insufficient to produce measurable changes in S-band Z_{DR} or K_{DP}, yet they can lead to a detectable increase in the side bands of Doppler spectra. K_a-band extinction cross sections of oscillating raindrops (axisymmetric oscillations) were shown to be well represented by average axis ratio models (Aydin and Daisley, 1998). Several such models have been developed and are used in polarimetric radar measurements of rainfall (Chandrasekar *et al.*, 1988; Beard *et al.*, 1991; Beard and Kubesh, 1991; Kubesh and Beard, 1993;

Tokay and Beard, 1993; Keenan *et al.*, 1997). It should be noted that representing drop oscillations with average axis ratios may be feasible for radar applications at the S and C bands, but this approach may not yield acceptable results at higher frequencies. Furthermore, the effects of transverse mode (different from the axisymmetric mode) drop oscillations on polarimetric radar observables have yet to be determined. These effects could be significant at higher frequencies such as the K_a band (8-mm wavelength), where it has been shown that the backscattering parameters such as Z_{DR}, CDR, and LDR are very sensitive to variations in the shapes of ice and liquid water particles with size parameters ranging from about 1 to 16 and axis ratios from about 0.66 to 1.3 (Sturniolo *et al.*, 1995).

B. Hailstones and Graupel Particles

As ice phase hydrometeors descend through the 0° C level they begin to melt and acquire a coating of water. Herman and Battan (1961) first studied the effects of water coating on the backscattering cross sections of spherical ice particles. They showed that the backscattering cross sections (at 3.21-, 4.67-, and 10-cm wavelengths) for particles small compared to the wavelength increased with water coating thickness. However, for larger sizes the trend became more complicated, decreasing in some cases and increasing in others followed by oscillations. These results showed good qualitative agreement with experimental measurements of backscattering from melting hail in the laboratory (Atlas *et al.*, 1960). Aydin *et al.* (1984) considered oblate spheroidal and conical models of hailstones and the effects of water coating on σ_{hh} (which they denote σ_H), Z_{DR}, and CDR at a wavelength of 10 cm (Fig. 2). Figure 3 shows their results for the oblate spheroid model with an axis ratio of 0.6 as a function of size and water coating thickness. It is clear that σ_H, Z_{DR}, and CDR increase with increasing water thickness. This trend changes at the larger sizes (beyond those shown in the figure). For smaller sizes the largest change in all these parameters takes place in the transition from ice to ice with a 0.05-mm water coating. Increasing the coating thickness by a factor of 10 beyond this does not create as much of a change. This trend does not hold for the larger sizes. Although not shown here, the conical model scattering parameters show similar behavior except that Z_{DR} is negative. This is because the axis ratio is greater than unity and its size (relative to the wavelength inside the particle) is below the resonance scattering regime. Increasing the water thickness makes Z_{DR} even more negative for these sizes.

Hailstones and graupel particles have also been modeled with conical shapes similar to the one shown in Fig. 2 but with the cone attached to an oblate spheroid rather than a sphere (Aydin *et al.*, 1984; Aydin and Seliga, 1984; Bringi *et al.*, 1984, 1986a, b). With this shape the axis ratio can be made less than unity, similar to some observations with particle imaging probes mounted on aircraft. Computa-

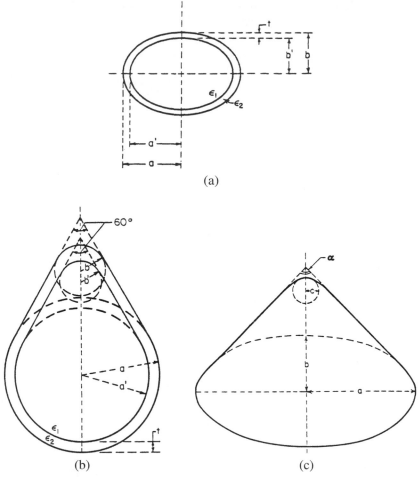

(a)

(b)   (c)

Figure 2   (a) Oblate spheroidal model of hail. A nonuniform water coating is also shown. Note that $b/a = b'/a'$ and $D = 2a(b/a)^{1/3}$, where D is the equivolume spherical diameter. (b) Sphere–cone–sphere model of hail. A nonuniform water coating is also shown. (c) Sphere–cone–oblate spheroid model of soft hailstones. Reprinted from K. Aydin, T. A. Seliga, and V. N. Bringi (1984), Differential radar scattering properties of model hail and mixed phase hydrometeors, *Radio Sci.* **19**, 58–66.

tional scattering results for this model compared favorably with the positive Z_{DR} values measured simultaneously by an S-band (3 GHz) radar (Bringi *et al.*, 1984, 1986a, b). Bringi *et al.* (1986a) utilized a melting model for graupel (Rasmussen and Heymsfield, 1987) and derived height profiles of Z_h and Z_{DR} at the S band and LDR_h (LDR in the graph) at the X band (Fig. 4). It is clear that all three pa-

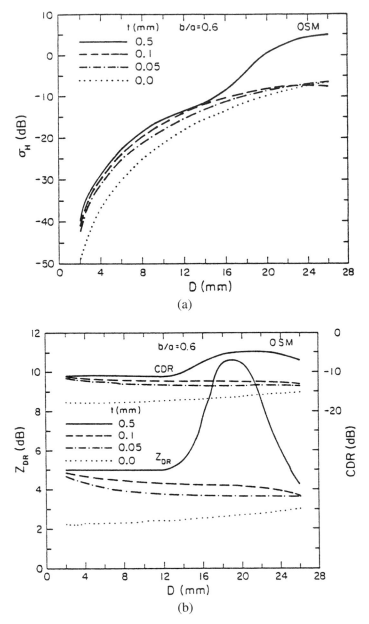

(a)

(b)

Figure 3 Profiles of (a) σ_H and (b) Z_{DR} and CDR of oblate spheroidal model of melting hail for axial ratio $b/a = 0.6$ and various water coating thickness t. Reprinted from K. Aydin, T. A. Seliga, and V. N. Bringi (1984), Differential radar scattering properties of model hail and mixed phase hydrometeors, *Radio Sci.* **19**, 58–66.

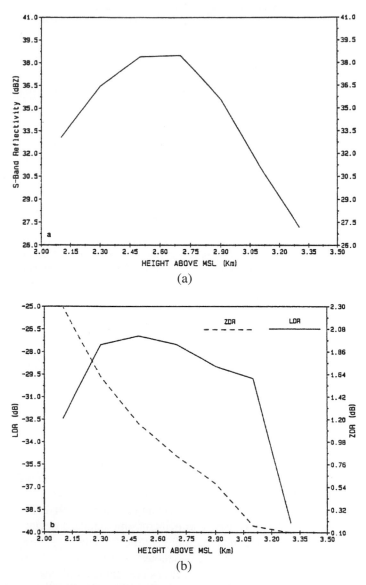

Figure 4 Vertical profile of (a) S-band reflectivity and (b) X-band LDR and S-band Z_{DR} using the graupel melting model for June 8, 1983. Note that the steady increase in Z_{DR} with decreasing altitude corresponds to the onset and progression of melting. The LDR "bright band" is due to soaking, water-coated conical graupel. Reprinted from V. N. Bringi, R. M. Rasmussen, and J. Vivekanandan (1986a), Multiparameter radar measurements in Colorado convective storms. I. Graupel melting studies, *J. Atmos. Sci.* **43**, 2545–2563. Copyright © 1986 American Meteorological Society.

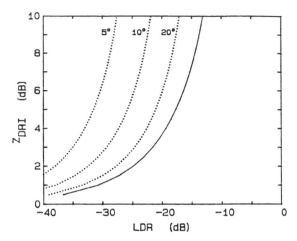

Figure 5   Values of LDR for particles having the same intrinsic Z_{DR} (Z_{DRI}). The solid line corresponds to randomly tumbling oblate spheroidal particles, and the dotted lines correspond to the Gaussian distribution of canting angles with the given standard deviation. Reprinted from I. R. Frost, J. W. F. Goddard, and A. J. Illingworth (1991), Hydrometeor identification using cross polar radar measurements and aircraft verification, *Prepr. Vol., Conf. Radar Meteorol., 25th, Paris, 1991,* 658–661.

rameters increase with decreasing height as a result of the melting of the graupel particles. Both Z_h and LDR_h exhibit a "bright band," where they reach a peak value at a certain altitude as a result of soaking and water-coated graupel particles. Z_{DR} reaches its maximum value when melting is complete and raindrops are formed, which have smaller axis ratios compared to graupel. LDR_h is reduced when melting is complete because raindrops have very little canting, unlike graupel particles.

Jameson (1987) derived relationships between linear and circular polarization scattering parameters and illustrated the sensitivity of these parameters to particle canting and radar elevation angle. Illingworth and Caylor (1989) and Frost *et al.* (1991) showed that LDR_h increases with increasing standard deviation of the canting angle for oblate spheroidal particles (in the Rayleigh scattering regime) having the same intrinsic Z_{DR}, which is denoted as Z_{DRI} in Fig. 5. Z_{DRI} is obtained when the v and h polarization directions are aligned with the oblate particle's minor and major axes, respectively. Notice that LDR_h (LDR in the graph) increases with Z_{DRI} and for a given Z_{DRI} reaches its maximum value when the particle is randomly oriented (in which case Z_{DR} would be 0 dB). Frost *et al.* (1991) showed, based on measurements at the S band and modeling results, that the side-looking LDR_h of wet aggregates is around −15 dB and for wet graupel it is −25 dB, which are quite different and can be used in differentiating them.

Several other studies on the scattering characteristics of melting hailstones have been published in the literature (Longtin *et al.*, 1987; Aydin and Zhao, 1990; Balakrishnan and Zrnic, 1990a, b; Vivekanandan *et al.*, 1990; Aydin and Girid-har, 1991; Fujiki *et al.*, 1994). All of these results illustrate the sensitivity of the scattering parameters on the composition of the hydrometeors (most importantly the amount of liquid water), their shape, orientation, and size. It is also empha-sized that realistic modeling (e.g., melting models) is important for the proper interpretation of radar measurements.

C. ICE CRYSTALS

Scattering from ice crystals is of interest for the interpretation of millimeter wave radar measurements of clouds. Higher frequencies are more sensitive to the smaller size ice crystals (typically less than 2 mm). The radar frequencies used for this purpose are around 35 and 94 GHz because they lie in propagation windows, that is, local minima in the absorption resulting from oxygen and water vapor (Wiltse, 1981). 220 GHz is another possible frequency for this type of application. However, it experiences significant attenuation in the atmosphere even though it is in a propagation window. The 220-GHz frequency may be more useful on airborne and spaceborne platforms. It should be noted that snowflakes formed from aggregates of dendrites have sizes in the range 2–12 mm (Pruppacher and Klett, 1997), in which case frequencies as low as 3 GHz would be suitable for radar remote-sensing applications.

O'Brien and Goedecke (1988a) computed the polarimetric scattering charac-teristics of a dendrite (with 4-mm maximum dimension) using the discrete dipole approximation (DDA; Chapter 5) at a wavelength of 10 mm. They found rea-sonable agreement for equivalent oblate spheroidal particles whose mass dis-tributions were most similar to that of the dendrite. Evans and Vivekanan-dan (1990) used the same method to study radar scattering from columns, needles, and plates for the S through K_a bands and radiative transfer mod-eling at 37, 85, and 157 GHz for microwave radiometry applications. They noted the significant difference in the elevation angle dependence of LDR for plates and columns. Vivekanandan *et al.* (1990) highlighted the elevation an-gle dependence of S- and C-band LDR and Z_{DR} for melting graupel parti-cles. Vivekanandan *et al.* (1991, 1994) used prolate and oblate spheroids to model ice crystals for scattering calculations at frequencies up to 35 GHz and to evaluate the effects of orientation and bulk density variations. Ma-trosov (1991) used the Rayleigh approximation to represent planar and colum-nar ice crystals (up to 1 mm in maximum dimension) with oblate and pro-late spheroidal models at a wavelength of 8 mm. He considered linear and circular polarization parameters and concluded that the elevation angle dependence of LDR or CDR can be used to differentiate oblate and prolate spheroidal particles.

Dungey and Bohren (1993) used the DDA to study scattering at 94 GHz from hexagonal columns and plates and noted a marked difference between the two. Aydin *et al.* (1994) compared radar measurements (12° elevation angle) at 95 GHz with scattering computations for stellar crystals based on shadow images obtained with an *in situ* two-dimensional cloud particle probe. They used the finite difference time domain (FDTD) method for the scattering computations (Kunz and Lubbers, 1993; Taflove, 1995; Chapter 7). The simulated Z_h compared very well with the radar measurements and was found less sensitive to particle canting than Z_v. Schneider and Stephens (1995) compared the Rayleigh approximation using spheroidal models with the DDA results for hexagonal columns and plates. They concluded that the difference is small when the maximum dimension of a particle is below 0.2, 0.6, and 1.4 mm for oblate spheroids (0.3, 0.8, and 2 mm for prolate spheroids) at wavelengths of 1, 3, and 8 mm, respectively. Liao and Sassen (1994) and Sassen and Liao (1996) used the conjugate gradient fast Fourier transform method to calculate the scattering characteristics of hexagonal plates, solid and hollow columns, solid and hollow bullets, and bullet rosettes at frequencies of 35 and 95 GHz. They derived relationships between the liquid water content and the reflectivity factor for a mixture of these ice crystal types.

Tang and Aydin (1995) presented results on scattering from hexagonal columns, hexagonal plates, and stellar crystals (Fig. 6). They considered ice crystals aligned such that the column axis and broad surfaces of plates and stellar crystals lie on the horizontal plane and are randomly oriented on this plane. They showed that only the hexagonal column creates any measurable LDR_h at vertical incidence (they also noted that LDR_h is higher for 220 GHz compared to 94 GHz). Measurement of LDR at vertical incidence was suggested as a means of differentiating columns from plates. They also showed that the phase matrix for this case is diagonal; this can also be deduced from Eqs. (61)–(64) of Chapter 1, and it can be obtained from LDR alone (for this case $LDR_h = LDR_v$):

$$\langle \overline{Z} \rangle = \frac{\langle \overline{Z}_{11} \rangle}{1 + \text{LDR}} \begin{bmatrix} 1 + \text{LDR} & 0 & 0 & 0 \\ 0 & 1 - \text{LDR} & 0 & 0 \\ 0 & 0 & 1 - \text{LDR} & 0 \\ 0 & 0 & 0 & 1 - 3\text{LDR} \end{bmatrix}. \quad (15)$$

This is identical to the form described for orientation-independent media such as trees, grass, and snow cover (Mead *et al.*, 1991, 1993). Resulting from this form of the phase matrix is the relation

$$\rho_{hv} = 1 - 2\text{LDR}, \quad (16)$$

which is also derivable from results based on a different formulation (Jameson, 1987).

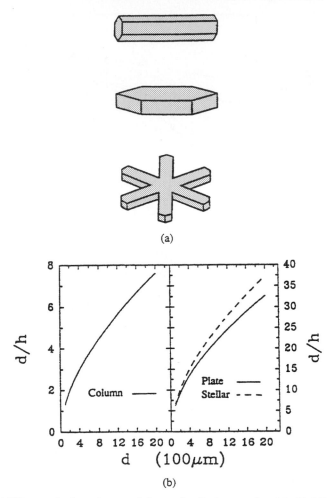

(a)

(b)

Figure 6   (a) Hexagonal column, hexagonal plate, and stellar ice crystal models. (b) d/h as a function of d, where d is the maximum dimension and h is the minimum dimension of the crystal. Reprinted from K. Aydin and C. Tang (1997a), Millimeter wave radar scattering from model ice crystal distributions, *IEEE Trans. Geosci. Remote Sens.* **35**, 140–146 (© 1997 IEEE).

The elevation angle dependence of LDR_h and ρ_{hv} for hexagonal columns and plates with a gamma model size distribution ($\mu = 1$ and $\Lambda = 5$ mm^{-1}) showed that the trends for columns and plates are the opposite of each other (Fig. 7; Aydin and Tang, 1997a). These two crystal types may be differentiated using the elevation angle dependence of LDR_h (or LDR_v) or ρ_{hv} or simply from their values at vertical incidence. The use of the elliptical depolarization ratio (EDR) was

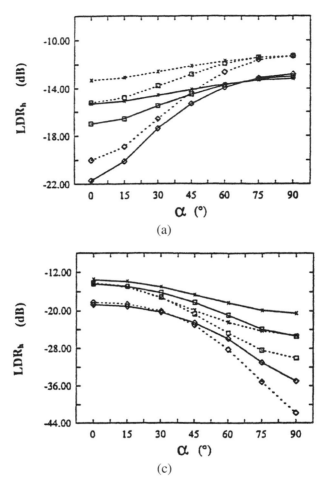

Figure 7 Radar observables for a distribution of hexagonal columns (a) LDR_h and (b) ρ_{hv} as a function of radar elevation angle α at 94 GHz (solid line) and 220 GHz (dashed line). The canting angle standard deviation is a parameter with values of 10° (diamond), 20° (square), and 30° (cross). Here a gamma model size distribution is used with $N_0 = 1$ $\mu m^{-2} m^{-3}$, $\mu = 1$, and $\Lambda = 5 \times 10^{-3}$ μm^{-1}. (c) and (d) are the same as (a) and (b), but for hexagonal plates. Reprinted from K. Aydin and C. Tang (1997a), Millimeter wave radar scattering from model ice crystal distributions, *IEEE Trans. Geosci. Remote Sens.* **35**, 140–146 (© 1997 IEEE).

suggested for particle identification with K_a-band radars (Matrosov and Kropfli, 1993; Matrosov *et al.*, 1996). EDR shows similar trends as CDR and LDR, but produces a stronger signal (with a smaller dynamic range), providing an advantage in low reflectivity regions where CDR and LDR may be difficult to measure.

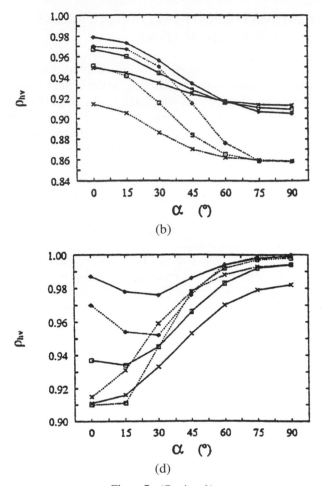

Figure 7   (Continued.)

The scattering characteristics of spatial crystal forms such as bullet rosettes, capped columns, and double plates have also been evaluated (Aydin and Walsh, 1996, 1998; Walsh and Aydin, 1997; Walsh, 1998). It is observed that aligned spatial crystals can exhibit different scattering characteristics compared to planar and columnar crystals such as hexagonal plates, stellar crystals, hexagonal columns, bullets, and needles. For example, at 220 GHz and for sizes larger than 0.4 mm the spatial crystals have LDR_h, Z_{DR}, and ρ_{hv} that do not change monotonically as a function of the elevation angle as they do for planar and columnar crystals.

V. DISCRIMINATION OF HYDROMETEORS WITH POLARIMETRIC RADAR

Polarimetric radars have been most useful for the discrimination of hydrometeor types. It was recognized early on that Z_{DR} (at the S band) might be used for distinguishing raindrops from ice particles (Hall *et al.*, 1980, 1984; Seliga, 1980; Seliga *et al.*, 1982b). At S-, C-, and X-band frequencies Z_{DR} resulting from rain is positive, ranging from a few tenths of a decibel up to 4 or 5 dB as a result of their oblate spheroidal shape oriented with the minor axis in the vertical direction. On the other hand, Z_{DR} for snowflakes and graupel (before they begin to melt) is close to 0 dB. This results from a much lower dielectric constant compared to water, together with a shape and orientation that project an average aspect ratio close to unity in the plane of polarization (the plane perpendicular to the incident wave direction). When a radar scans vertically through a storm the regions of ice and rain can be readily distinguished.

Radar observations of hail have shown both negative and positive Z_{DR} (dB) values. Regions of negative Z_{DR} are generally associated with high values of Z_h and are considered as indicators of hail (Seliga *et al.*, 1982a, b; Bringi *et al.*, 1984, 1986b; Aydin *et al.*, 1984, 1990b; Husson and Pointin, 1989). It was established through radar observations of pure rainfall (Leitao and Watson, 1984) and simulations based on raindrop size distribution measurements (Aydin *et al.*, 1986) that (Z_h, Z_{DR}) pairs caused by rainfall are clustered in a certain region of the Z_h–Z_{DR} plane. An upper bound can be obtained for the Z_h of rainfall as a function Z_{DR}. Aydin *et al.* (1986) devised a hail signal H_{DR} based on the difference between the measured Z_h and this upper bound (also called the rainfall boundary), which could be used to identify regions of hail below the melting level. Figure 8 shows the rainfall boundary and radar measurements in a hailstorm. The scatter plot contains data from hail (above the boundary), rain (below the boundary), and regions containing both rain and hail (a fuzzy region along the boundary). Using a similar approach, Balakrishnan and Zrnic (1990a) derived a rainfall boundary on the Z–K_{DP} plane (they define Z as the average of Z_h and Z_v) for differentiating rain and hail. A similar concept was developed earlier by Steinhorn and Zrnic (1988).

In regions of a storm where there is a mixture of rain and hail it is of interest to determine the fraction of the reflectivity caused by hail and rain. Golestani *et al.* (1989) proposed a method that can be applied for this purpose. It is based on establishing a relationship between Z_h and Z_{DP} (the difference reflectivity), which turns out to be linear on the dBZ scale. They show that radar measurements in rain lie close to this "rain line." Ice particles, because of their shapes and fall behavior, appear to be more spherical on average and, as a result, have a negligible contribution to Z_{DP} compared to raindrops. Therefore, the deviation from the "rain line" indicates the presence of ice phase hydrometeors and the amount of deviation can be used to estimate the fraction of Z_h resulting from ice and rain.

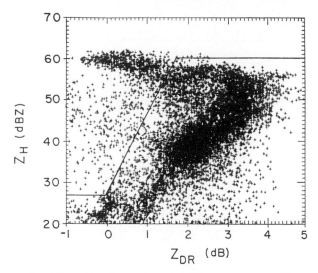

Figure 8   Scatter plot of Z_H vs Z_{DR} from radar measurements at a constant altitude of 400 m above ground level. The solid line is the rainfall boundary. Reprinted from K. Aydin, T. A. Seliga, and V. Balaji (1986), Remote sensing of hail with a dual linear polarization radar, *J. Clim. Appl. Meteorol.* **25**, 1475–1484. Copyright © 1986 American Meteorological Society.

The requirement of self-consistency among multiple parameters increases the level of confidence in their interpretation. For example, Balakrishnan and Zrnic (1990b) present vertical profiles of S-band Z (the average of Z_h and Z_v), Z_{DR}, K_{DP}, and ρ_{hv} for a hailstorm in Oklahoma on May 14, 1986, which produced hail smaller than 1 cm in size (Fig. 9). The profiles in Fig. 9a show that above the 0°C isotherm Z_{DR} is near 0 dB and K_{DP} is about 0° km^{-1}, indicating the presence of ice phase hydrometeors that are nearly spherical or are tumbling. The high value of ρ_{hv} is consistent with the other parameters for dry hailstones (Balakrishnan and Zrnic, 1990b; Aydin and Zhao, 1990). Below the 0°C isotherm Z_{DR} and K_{DP} begin to increase as a result of the onset of melting, while ρ_{hv} begins to decrease. The authors indicate that the decrease in ρ_{hv} is due to mixed-phase hydrometeors, which is consistent with melting hailstones. The presence of different hydrometeor types can lead to a reduction of ρ_{hv} (Jameson, 1989). The profiles in Fig. 9b correspond to a later time in the storm when the precipitation has turned to stratiform type. Below the 0°C isotherm a "bright band" is seen for both Z_h and Z_{DR}, and ρ_{hv} decreases slightly below the melting region and increases further below that, which are all consistent with the presence of rain at the ground.

An increasing number of studies are making use of multiparameter polarimetric radar measurements for obtaining the microphysical characteristics of storms

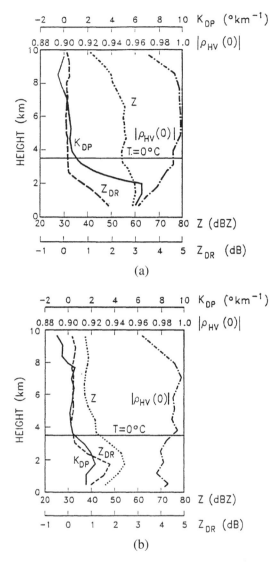

Figure 9 Vertical profiles of average reflectivity factor [$Z = (Z_H + Z_V)/2$], differential reflectivity (Z_{DR}), specific differential phase (K_{DP}), and the co-polarized correlation coefficient (ρ_{hv}) measured on May 14, 1986, at (a) 1559:32 CST at 40 km in range and 130° in azimuth. The 0°C isotherm is at 3.5 km, and the profiles are from averages over 2.1 km in range. (b) Vertical profiles as in (a) but for 1611:06 CST. Reprinted from N. Balakrishnan and D. S. Zrnic (1990b), Use of polarization to characterize precipitation and discriminate large hail, *J. Atmos. Sci.* **47**, 1525–1540. Copyright © 1990 American Meteorological Society.

(e.g., Caylor and Illingworth, 1987; Illingworth *et al.*, 1987; Bringi *et al.*, 1991; Meischner *et al.*, 1991; Hubbert *et al.*, 1993; Vivekanandan *et al.*, 1993a, b; Zrnic *et al.*, 1993a, b; Meneghini and Kumagai, 1994; Ryzhkov and Zrnic, 1994; Holler *et al.*, 1994; Russchenberg and Ligthart, 1996; Hubbert *et al.*, 1998). Some studies are focused on understanding storm electrification processes through spatial and temporal information on the hydrometeor characteristics obtained from polarimetric radars (Seliga *et al.*, 1983; Goodman *et al.*, 1988; Metcalf, 1995, 1997; Jameson *et al.*, 1996; Krehbiel *et al.*, 1996; Shao and Krehbiel, 1996; Caylor and Chandrasekar, 1996). Table I illustrates the range of values for polarimetric radar observables corresponding to different precipitation types. Straka *et al.*

Table I

Values of Polarimetric Radar Observables for Various Precipitation Types. D_0 Is the Median Volume Diameter of Raindrops. These Values Are for S-Band Radar Wavelengths. They Are Not Significantly Different at C Band Except for K_{DP}, Which Differs by about a Factor of 2.

	Z_h (dBZ)	Z_{DR} (dB)	K_{DP} (deg km^{-1})	LDR_h (dB)	ρ_{hv}	δ_{hv} (deg)
Rain ($D_0 < 1$ mm)	<28	0–0.7	0–0.03	<−32	>0.97	$|\delta_{hv}| < 1$
Rain ($D_0 > 1$ mm)	28–60	0.7–4	0.03–10	−32–−25	>0.95	$|\delta_{hv}| < 1$
Snow crystals, dry, high density	<35	0–6	0–0.6	<−28	>0.95	—
Snow aggregate, dry	<35	0–1	0–0.2	<−25	>0.95	—
Snow aggregate, wet	<45	0.5–3	0–0.5	−20–−10	0.5–0.9	$-5 < \delta_{hv} < 10$
Graupel, dry, low density	20–35	−0.5–1	−0.5–0.5	<−25	>0.95	$|\delta_{hv}| > 1$
Graupel, wet, high density	30–50	−0.5–2	−0.5–1.5	−30–−20	>0.95	$|\delta_{hv}| > 1$
Hail, dry (>5 mm)	45–60	−1–0.5	−0.5–0.5	−26–−18	>0.95	$|\delta_{hv}| > 1$
Hail, wet, small (<20 mm)	50–60	−0.5–0.5	−0.5–0.5	>−24	0.92–0.95	$|\delta_{hv}| < 15$
Hail, wet, large (>20 mm)	55–65	−1–0.5	−1–1	>−20	0.9–0.92	$5 < |\delta_{hv}| < 15$
Rain/small wet hail	45–60	−0.5–6	0–10	>−25	<0.95	$|\delta_{hv}| < 15$
Rain/large wet hail	55–65	−0.5–3	0–10	>−22	<0.92	$5 < |\delta_{hv}| < 15$

Source: Adapted from Doviak and Zrnic (1993) using results from Straka *et al.* (1999).

(1999) present a synthesis of information on multiparameter radar observables using results from observational and model studies reported in the literature. This information is utilized by Vivekanandan *et al.* (1999), who demonstrate a fuzzy logic algorithm for categorizing hydrometeor habits.

It was noted earlier that linear, circular, and elliptical depolarization ratios (LDR, CDR, and EDR) can be used for differentiating planar and columnar crystals by their trends as a function of elevation angle (e.g., Fig. 7) or by simply making observations at vertical incidence (Evans and Vivekanandan, 1990; Matrosov, 1991; Tang and Aydın, 1995; Matrosov *et al.*, 1996; Aydın and Tang, 1997a). Furthermore, ρ_{hv} can also be used in the same manner for this purpose (Aydın and Tang, 1997a). Zrnic *et al.* (1994) showed that ρ_{hv} can be used to identify the melting layer with a vertically pointing radar. These parameters should also be useful for distinguishing spherical cloud droplets from columnar and planar ice crystals. Reinking *et al.* (1997) report on K_a-band radar measurements of reflectivity, LDR, and EDR in winter storm clouds. They suggest that it is possible to separate graupel from drizzle and thick plates from graupel.

It should be noted that multiple-scattering effects could produce LDR and CDR even in a medium filled with spherical particles. Theoretical results at 10 and 34.5 GHz for a thin rain layer composed of spherical raindrops show trends similar to experimental observations, suggesting that part of the depolarization signal in the measurements is due to second-order multiple-scattering effects (Ito and Oguchi, 1994; Ito *et al.*, 1995).

Aydın and Walsh (1998) showed that using Z_h and Z_{DP} at side incidence ($0°$ radar elevation angle) it is possible to identify columnar and planar crystals, assuming that only one type is present or dominant. It should be noted that this technique could potentially be used at elevation angles up to about $75°$. Furthermore, based on these two parameters the presence of aggregates or spatial crystals, which produce similar reflectivities at both h and v polarizations, could be inferred. It should be possible to estimate the fraction of the reflectivity factor resulting from the aggregates. These are accomplished by establishing Z_h-Z_{DP} relations for columnar (hexagonal column) and planar (hexagonal plate, stellar crystal, planar rosette) crystals through size distribution simulations (Walsh, 1998). Figure 10 shows 95-GHz radar measurements obtained with the University of Wyoming airborne radar system. On board the aircraft was a cloud particle-imaging probe. The radar data shown here are for side incidence over the range 120–240 m away from the aircraft. The shadow images for the first case indicate columnar crystals and the radar data follow the column line. The shadow images for the second case indicate stellar crystals in part of the cloud and aggregates in another part. The data are closer to the stellar crystal line and follow its slope. Some of the data are distant from all of the lines and have smaller Z_{DP} values for a given Z_h, indicating the presence of particles other than planar or columnar crystals that do not contribute much to Z_{DP}.

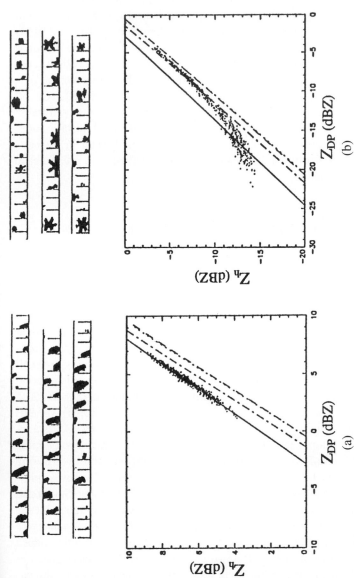

Figure 10 (a) Columnar crystals observed from an airborne cloud particle imaging probe (top) and airborne 95-GHz radar Z_h–Z_{DP} data for side incidence (below). (b) Same as in (a) but for stellar crystals and small aggregates. Note that the maximum vertical dimension of the particle-imaging data is 800 μm. The lines shown (in order from left to right) correspond to simulations for hexagonal column (solid), planar bullet rosette with four branches (dash-dotted), and stellar crystal (dotted). Note that the plate and stellar crystal lines appear to be identical. Z_h is the ice-based effective reflectivity factor. Reprinted from T. M. Walsh (1998), Polarimetric scattering characteristics of planar and spatial ice crystals at millimeter wave frequencies, Ph.D. Thesis, Pennsylvania State Univ, University Park, PA.

All of the applications discussed previously are for monostatic polarimetric radar systems in which the transmitters and receivers are colocated. Aydin *et al.* (1998) proposed the use of bistatic polarimetric radar measurements for detecting hail and estimating the median size of hail. A bistatic Doppler radar system has already been demonstrated for measuring dual-Doppler vector wind fields (Wurman, 1994). A polarimetric bistatic radar system should include a monostatic radar (single or dual polarized) and a dual-polarized bistatic receiver located at some distance from the radar (multistatic systems may have receivers at several different locations). Simulations indicate that pairs such as (LDR_v, ρ_v) and (BBR_v, Z_v), for v-polarized transmission or a similar set for h-polarized transmission, could be used to differentiate rain from hail. All of the bistatic radar parameters are significantly affected by the amount of liquid water in the hailstones. It is suggested that the melting layer may be identified with polarimetric bistatic radar measurements.

VI. QUANTITATIVE ESTIMATION WITH POLARIMETRIC RADAR

A. RAINFALL

Polarimetric radar methods for estimating the rainfall rate R were developed to overcome the problems associated with the traditional R–Z relationship method (Battan, 1973), which is still in use by the National Weather Service today. Problems associated with R–Z (where Z can be Z_h or Z_v) relationships are well known. Z is proportional to the integral of D^α over the drop size distribution (DSD), where D is the equivalent volume diameter of a raindrop. $\alpha \approx 6$ for Z at the S and C bands (and even at the X band for smaller raindrops), whereas $\alpha \approx 3.67$ for the rainfall rate and 3 for the water content (Battan, 1973). As a result, the R–Z relationship is strongly dependent on the DSD, which can be quite different from one rainfall event to another and can be highly variable even during the same event (Waldvogel, 1974). The differential reflectivity (Z_{DR}) was originally introduced for the purpose of improving radar rainfall estimates (Seliga and Bringi, 1976). At the S- and C-band frequencies Z_{DR} is a measure of the reflectivity weighted mean axis ratio (Jameson, 1983), which indicates that it is a measure of raindrop size because size and axis ratio are related for equilibrium-shaped drops (Pruppacher and Klett, 1997). If a two-parameter DSD is assumed, such as the gamma model with a fixed value for μ, then it is possible to estimate the parameters of the DSD using Z_{DR} and Z_h (Seliga and Bringi, 1976). In fact, it is not necessary to assume a specific DSD model to estimate the rainfall rate R or water content M. These parameters can be obtained through R–(Z_h, Z_{DR}) and M–(Z_h,

Z_{DR}) relationships derived from disdrometer measurements of DSD (Seliga *et al.*, 1986; Aydin and Giridhar, 1992) or simulations of a variety of DSDs.

The use of the specific differential phase has also been suggested for improving rainfall measurements with radar (Seliga and Bringi, 1978; Jameson, 1985, 1989; Sachidananda and Zrnic, 1986, 1987). K_{DP} at the S band is proportional to the integral of D^{α} over the DSD with $\alpha \approx 4.24$ (Sachidananda and Zrnic, 1987) for larger sizes and $\alpha \approx 5.6$ for smaller sizes (Ryzhkov and Zrnic, 1996). As a result, the R–K_{DP} relationship is not very sensitive to DSD variations for heavy rain involving larger drops and becomes more sensitive in light rain with smaller drops.

Comparisons of S-band radar rainfall rate estimates obtained using an R–(Z_h, Z_{DR}) relationship have shown good agreement with rain gauge and disdrometer measurements (Seliga *et al.*, 1979, 1981; Goddard *et al.*, 1982; Bringi *et al.*, 1982; Aydin *et al.*, 1987, 1990a). Using gamma model DSD simulations, Ryzhkov and Zrnic (1995) compared one and two-parameter relationships: R–Z_h, R–K_{DP}, R–(Z_h, Z_{DR}) or R–(Z_h, Z_v), R–(K_{DP}, Z_{DR}). They concluded that the R–(K_{DP}, Z_{DR}) relationship performs much better than the others; however, in a case study comparing radar and rain gauge measurements the R–K_{DP} relationship performed just as well (the rain rate was mostly between 10 and 45 mm h^{-1}). Simulations by Chandrasekar *et al.* (1990) showed that R–K_{DP} performs best for $R > 70$ mm h^{-1}; R–(Z_h, Z_{DR}) gives the least error among the three relationships for $20 < R < 70$ mm h^{-1} and is comparable to R–Z_h for $R < 20$ mm h^{-1}. Chandrasekar *et al.* (1993) developed a framework to optimally combine the estimates of rainfall rates from these three relationships.

Zrnic and Ryzhkov (1996) showed that the R–K_{DP} relationship is immune to beam blockage and ground clutter canceling. However, they also noted that K_{DP} estimates of R have lower resolution (about 2–4-km radial resolution) compared to methods using Z_h (several hundred meters radial resolution) and K_{DP} is affected more by nonuniform beam filling (Ryzhkov and Zrnic, 1998). On the other hand, the R–K_{DP} relationship provides much better estimates of the rain rate when mixed-phase precipitation is involved compared to those involving Z_{DR}. Balakrishnan and Zrnic (1990a) proposed using K_{DP} to estimate the rain rate in the presence of hail, based on the sensitivity of K_{DP} to anisotropic scatterers. Raindrops are highly aligned and considered anisotropic, whereas hailstones are considered isotropic because they produce similar radar returns at h and v polarizations as a result of their shape and fall behavior. Therefore, only the raindrops contribute to K_{DP}, because the averaged difference of the two polarization components in Eq. (12) will cancel for hail. This was experimentally demonstrated by comparing the rainfall rate obtained using K_{DP} with a ground-based rain gauge and *in situ* confirmation of the presence of hailstones with rainfall (Aydin *et al.*, 1995; Hubbert *et al.*, 1998).

The use of attenuation, differential attenuation, and differential phase shift along propagation links at microwave and millimeter wave frequencies have also been proposed for measuring the rainfall rate and water content (Atlas and Ulbrich, 1977; Giuli *et al.*, 1991; Jameson, 1993; Jameson and Caylor, 1994; Ruf *et al.*, 1996; Aydin and Daisley, 1998). Measurements of millimeter wave attenuation in rain and snow show promise for estimating rainwater and snow mass contents (Wallace, 1988; Nemarich *et al.*, 1988; Ruf *et al.*, 1996). However, more research is needed to determine the feasibility of techniques involving attenuation and phase shift for path integrated rainfall and snowfall measurements.

B. ICE AND MIXED-PHASE HYDROMETEORS

Quantitative estimation of ice and mixed-phase precipitation parameters is a difficult task owing to the significant variability in the shape, density (composition), and fall behavior of the hydrometeors involved. Polarimetric radars have not been able to provide significant improvement in this area because none of the polarimetric parameters is a measure of particle size. Some parameters have been used to categorize hailstone size to be greater than or less than 2 cm in diameter (Table I). However, this does not give the needed detail for estimating hailfall rate or hailfall kinetic energy flux (Waldvogel *et al.*, 1978a, b; Schmid and Waldvogel, 1986). Vertically pointing Doppler radars have been used to estimate exponential hailstone size distributions from which the hailfall parameters can be estimated (Ulbrich, 1978; Ulbrich and Atlas, 1982). Aydin *et al.* (1998) propose the use of bistatic radar measurements such as Z_v and BBR_v or Z_h and BBR_h for estimating the parameters of an exponential hailstone size distribution based on simulations using oblate spheroidal model hailstones. Earlier bistatic studies focused on rain and the estimation of exponential raindrop size distributions using the bistatic co- and cross-polarized reflectivities (Awaka and Oguchi, 1982) as well as Z_{DR}, LDR, and CDR (Dibbern, 1987).

R_s–Z_h relations (where R_s is the snowfall rate) for snowfall measurements have been available since the 1950s (Battan, 1973; Smith, 1984; Matrosov, 1992; Xiao *et al.*, 1998), but their successful use has been limited owing to the variability in the size distribution and the density of snowflakes. Relationships between ice water content (IWC) or ice mass content and reflectivity factor have been proposed for estimating ice cloud content (Sassen, 1987b; Vivekanandan *et al.*, 1994; Atlas *et al.*, 1995; Schneider and Stephens, 1995; Sassen and Liao, 1996; Aydin and Tang, 1997b; Ryzhkov *et al.*, 1998). Large scatter in IWC estimates from Z_h measurements is predicted (Atlas *et al.*, 1995), although some of this scatter may be reduced in certain cases if the crystal types can be identified and the corresponding IWC–Z_h relations are used (Aydin and Tang, 1997b). Polarimetric techniques have also been suggested for this purpose, but they require further investigation and experimental verification (Aydin and Tang, 1997b).

The ratio of reflectivity factors at two different frequencies (DFRs) has been suggested for gauging hailstone size (Atlas, 1964). The idea was to make use of the differences in the reflectivity factors corresponding to the Rayleigh and resonance scattering regimes such as 3 and 10 GHz for hailstones. This idea was extended to gauging ice crystal size using frequency pairs such as 35 and 94, 35 and 220, and 94 and 220 GHz (Matrosov, 1993; Tang and Aydin, 1995; Sekelsky and McIntosh, 1996; Lohmeier *et al.*, 1997). Meneghini and Kumagai (1994) used a DFR (from 10 and 34.5 GHz) together with the 10-GHz reflectivity factor to estimate the snow size distribution in clouds. More studies involving experiments with *in situ* confirmation are needed to better assess the potential of these techniques for ice crystal measurements.

Chapter 17

Microwave Scattering by Precipitation

Jeffrey L. Haferman
Fleet Numerical Meteorology and Oceanography Center
Monterey, California 93943

481

I. INTRODUCTION

Because microwaves readily penetrate clouds and do not depend on the sun as a source of illumination, microwave radiometry provides an excellent compliment to infrared and visible remote-sensing techniques. In contrast to *active* microwave remote sensing (Chapter 16) in which a pulse is *actively* emitted from a source and then backscattered from a target, *passive* microwave radiometers measure radiation that is naturally emitted by some source, such as the atmosphere and surface of Earth. This naturally emitted radiation may interact with other matter and be scattered, absorbed, and reemitted before being detected by a radiometer. In this chapter, microwave scattering by nonspherical hydrometeors will be discussed in the context of remote sensing of precipitation.

The potential of using passive microwave radiometry for observing precipitation from space was first explored mathematically by Buettner (1963) over three decades ago. In 1972, the first Electrically Scanning Microwave Radiometer (ESMR) was launched aboard *Nimbus-5*, and some of the first spaceborne passive microwave precipitation retrievals were performed. Soon thereafter, passive microwave precipitation observations were made using radiometers aboard *Skylab* in 1973, the *Nimbus-6* ESMR in 1975, and the Scanning Multichannel Microwave Radiometer (SMMR) aboard *Nimbus-7* and *Seasat-1* in 1978.

Since these early missions, several projects dedicated to the accurate measurement of global precipitation have been developed. For example, the first Special Sensor Microwave/Imager (SSM/I) was launched in June 1987 as part of the Defense Meteorological Satellite Program. Another project, the Global Precipitation Climatology Project established by the World Climate Research Program, was created to produce monthly global rainfall totals on a $2.5°$ latitude by $2.5°$ longitude scale for the period 1986–95 using data returned from satellite observations. In November 1997 the Tropical Rainfall Measuring Mission (TRMM) satellite was launched as part of a joint program between the United States and Japan to measure rainfall from space (Simpson *et al.*, 1996).

Some satellite methods of precipitation estimation use visible or infrared measurements to infer rainfall amounts. However, at these frequencies, the radiation observed by satellite radiometers originates from cloud tops and, therefore, is an indirect observation of the precipitation below. In contrast, measurements using microwaves that can penetrate clouds provide a more direct means of observing and estimating precipitation. To derive a mathematical relationship between the radiation measured by a satellite radiometer and the underlying precipitation structure, physically based microwave rainfall retrieval algorithms typically apply radiative transfer theory to microphysical cloud models to obtain brightness temperature (T_B) versus rain rate (R) relationships. In terms of the four-component Stokes intensity vector introduced in Section XII of Chapter 1, the brightness temperature is defined as the equivalent blackbody temperature of a given Stokes intensity vector component. Current and past sensors have measured brightness temperatures at two orthogonal polarizations; future sensors may (and some current airborne sensors do) include polarimetric channels to measure the third (plane of polarization) and possibly the fourth (ellipticity) Stokes parameters (see Gasster and Flaming, 1998; Yueh, 1997).

Mathematically, the brightness temperature for the first Stokes parameter is obtained by inverting Planck's law for a fixed wavelength λ to yield

$$T_{B,I}(\vartheta, \varphi) = \frac{C_2}{\lambda \ln[1 + C_1/\{\lambda^5 | I(\vartheta, \varphi)|\}]} \; \text{sgn}[I(\vartheta, \varphi)], \tag{1}$$

where $C_1 = 3.74126 \times 10^8$ W μm^4 m^{-2}, $C_2 = 1.4388 \times 10^4$ μm K, and I is the radiant intensity propagating in the direction described by the polar and azimuthal angles ϑ and φ.[1] The function $\text{sgn}(x)$ is equal to 1 when its argument is positive, equal to 0 when its argument is 0, and equal to -1 when its argument is negative. Brightness temperatures for the other Stokes parameters (Q, U, and V) are obtained by switching the symbol I with the symbol for the Stokes parameter under consideration. Brightness temperatures for vertical and horizontal polarizations (Section III of Chapter 1) are obtained by first defining the alternate Stokes parameters

$$I_v = \frac{I + Q}{2}, \tag{2}$$

$$I_h = \frac{I - Q}{2}, \tag{3}$$

and then inserting I_v or I_h into Eq. (1) to obtain $T_{B,v}$ or $T_{B,h}$, respectively, where lowercase subscripts have been used to avoid confusion between vertically polarized brightness temperatures ("v") and the fourth Stokes parameter brightness temperature ("V").

Panegrossi *et al.* (1998) have noted that there is often a lack of consistency between model-generated T_Bs and those observed by actual satellite radiometers. They suggest that some of this mismatch may be attributed to (1) using one-dimensional radiative transfer models that do not account for three-dimensional precipitation structures and (2) approximating all hydrometeor species as spherical scatterers. Currently, it is not entirely clear how important these two issues are in terms of retrieving accurate precipitation quantities from satellite-based microwave radiometers. The three-dimensional issue has received some attention (e.g., Roberti *et al.*, 1994) throughout the years, but is still an outstanding problem. The assumption of spherical scatterers has also received some attention in

---

[1]The constants C_1 and C_2 are defined as $C_1 = 2\pi h c_0^2$ and $C_2 = h c_0 / k$, where c_0 is the speed of light in a vacuum, h is Planck's constant, and k is the Boltzmann constant. In some literature, the constant C_1 is defined as $h c_0^2$.

the passive microwave remote-sensing literature, but, even though the theory is well known (Tsang *et al.*, 1985), it is not entirely clear how the assumption of sphericity impacts microwave precipitation retrievals.

The purpose of this chapter is to review the relevant literature that deals with the effect of nonspherical randomly and nonrandomly oriented hydrometeors on microwave brightness temperatures and to review the mathematical theory necessary to model scattering by nonspherical hydrometeors in passive microwave precipitation retrieval. It is known that small raindrops are nearly spherical, but as raindrop size increases, the shape can depart significantly from spherical (Chuang and Beard, 1990). Ice in the atmosphere can take on a variety of shapes and orientations (Pruppacher and Klett, 1997; Matson and Huggins, 1980). Thus, although no attempt is made here to quantify the effect of nonsphericity on passive microwave precipitation retrieval, it is hoped that the review presented will allow interested researchers to pursue this issue on their own.

Because some researchers have speculated that hydrometeor shape influences microwave polarization signatures, this issue is also touched upon. For example, Spencer *et al.* (1989) observed polarization differences[2] of about 5–10 K at the 85-GHz frequency of the SSM/I near convective cores over land and speculated that they were due to oriented ice within stratiform precipitation areas. Heymsfield and Fulton (1994a, b) noted SSM/I 85-GHz polarization differences between 8 and 13 K and 37-GHz polarization signatures up to 7 K in mesoscale convective systems and in tornadic storms over land. Petty and Turk (1996) observed similar 85-GHz signatures in moderate to heavy precipitation in tropical cyclones over the ocean. These authors attributed the polarization signatures to nonspherical hydrometeors with a preferred orientation and proposed that this only happens in stratiform-type precipitation (for convective precipitation, convective motion produces random orientation and larger, nearly spherical graupel particles, presumably leading to smaller polarization signatures). In addition, Kutuza *et al.* (1998) have observed large polarization signatures for downwelling microwave radiation propagating through precipitation.

The chapter is organized as follows. Section II reviews previous work in microwave radiative transfer modeling of nonspherical hydrometeors. Section III discusses the mathematical formulation for performing microwave radiative transfer calculations with nonspherical hydrometeors. Section IV describes a simple atmospheric model containing nonspherical hydrometeors and presents some microwave radiative transfer results for this model atmosphere. Finally, in Section V, conclusions and suggestions for future work are given.

---

[2]The terms "polarization difference" and "polarization signature" are taken throughout the text to mean the difference between vertically and horizontally polarized brightness temperatures, that is, $T_{B,v} - T_{B,h}$ $(= T_{B,Q})$.

II. REVIEW OF PREVIOUS WORK

A. OVERVIEW

The intent of this section is to present an overview of previous work done in the area of scattering by nonspherical hydrometeors in passive microwave precipitation retrieval studies. The following studies have been performed within the last couple of decades and are reviewed here to provide a framework for understanding the developments and the state of the art in this area. The following sections are organized according to a common author (not necessarily the lead author) of a set of papers. Some of the older studies (e.g., Oguchi, 1960) are not reviewed, but the interested reader may piece together a more complete historical account by referring to the citations in the studies reviewed here.

B. WEINMAN

1. Wu and Weinman (1984)

Wu and Weinman (1984) presented results showing upwelling[3] brightness temperatures at a zenith viewing angle of 50° for two orthogonal polarizations and using nonspherical particles for frequencies of 6.6, 10.7, 18, 21, and 37 GHz. They modeled polydispersions of oblate water drops and oriented oblate ice particles, over both water and land surfaces. The scalar form of the radiative transfer equation was solved by using the Eddington approximation for each polarization state. That is, horizontal and vertical extinction coefficients were defined and the radiative transfer equation was solved separately for each polarization. A linear anisotropic phase function defined by a polarization-independent asymmetry parameter was used, and the polarized extinction coefficients of the particles were computed for an angle of 50° with respect to the particle semi-minor axis of symmetry. To approximate an optically smooth ocean surface, a constant emittance value of 0.333 was used for horizontal polarizations, and a value of 0.605 was used for vertical polarizations. Over land, the surface was modeled as a Lambertian surface (see Section III.I) with an emittance of 0.9.

Despite the lack of rigor used in the formulation of the radiative transfer equation, several important observations were made by Wu and Weinman concerning the effect of hydrometeor shape on the computed upwelling brightness temperatures. First, spherical particles depolarize the strongly polarized signature emitted

[3]The term "upwelling brightness temperature" is taken throughout the text to mean the brightness temperature that represents the net radiation emitted in an upward direction by the entire atmospheric path between the ground and the observation point in the direction under consideration. "Downward brightness temperature" is defined as the net radiation emitted in a downward direction from a semi-infinite atmosphere along a given line of sight to an observation point on the ground.

by an optically smooth ocean surface. The amount of depolarization increases with frequency and with rain rate. For example, at 6.6 GHz and rain rates of 32 and 64 mm/h, the polarization difference is about 28.5 and 7.5 K, respectively, while at 37 GHz the polarization differences are 6.6 and 0.6 K at rain rates of 2 and 4 mm/h, respectively, and no polarization difference occurs for higher rain rates. Over land, the signal is unpolarized for all frequencies and rain rates.

Second, for spherical ice and nonspherical water drops, the signal over land becomes slightly polarized (less than 1 K for all frequencies and rain rates). Over water, the signal becomes depolarized as for the case of spherical hydrometeors, but never becomes completely depolarized for the range of frequencies (6.6–37 GHz) and rain rates (2–64 mm/h) considered. For example, at 37 GHz the polarization differences are about 1 K for rain rates between 4 and 16 mm/h. However, this can be considered to be an almost negligible effect on the polarization signal.

Third, and most significantly, deforming the ice hydrometeors (from spheres to spheroids) had the largest impact on upwelling polarization signatures over land and water. For example, at 37 GHz and rain rates in excess of 48 mm/h, the polarization signal was about 11 K. The polarization signal decreases monotonically as frequency decreases (about 6 K at 21 GHz and rain rates greater than 32 mm/h).

2. Kummerow and Weinman (1988)

Kummerow and Weinman (1988) used the T-matrix method to compute the extinction coefficients of polydispersions of horizontally oriented oblate ice and raindrops with an incidence angle of $50°$ with respect to the particle semi-minor axis. The frequencies considered were 10.7, 18, 37, 50.3, and 85.6 GHz. Unfortunately, no radiative transfer computations were performed. However, the extinction coefficients were parameterized as a function of rain rate via

$$k_e \left[\text{km}^{-1}\right] = \kappa_1 R^{\kappa_2 + \kappa_3 \ln R}, \tag{4}$$

where k_e is a scalar extinction coefficient, $k_{e,v} = n_0[K_{11} + K_{22}]/2$ or $k_{e,h} = n_0[K_{11} - K_{22}]/2$, n_0 is the particle number density (see Chapter 1, Section VIII), and R is the rain rate in millimeters per hour. Values of κ_1, κ_2, and κ_3 were presented for liquid water and ice for the aforementioned frequencies.

The authors found that the horizontally polarized extinction coefficients were always larger than those for vertical polarization, but explained that this is because the hydrometeors were oriented with their long axis in horizontal alignment. An important finding from this study was demonstrated in a figure depicting the differences between the extinction coefficients of oblate spheroids and spheres of equal volume for horizontally and vertically polarized radiation as a function of rainfall rate for 18, 37, and 85 GHz. The figure showed that the horizontal (vertical) extinction coefficients for oblate particles are larger (smaller) than the extinction coefficients for spherical particles and that the differences grow with frequency and rainfall rate. For intermediate rainfall rates (10–50 mm/h) the absolute differences are between about 0.02 and 0.07 km^{-1} for 85 GHz and between about 0.005 and 0.03 km^{-1} for 37 GHz.

3. Schols *et al.* (1997)

Schols *et al.* (1997) performed vector radiative transfer computations on a model atmosphere that contained spherical rain and oriented oblate snow and melting particles. Upwelling brightness temperatures and polarization differences were computed for a frequency of 85 GHz and a 53° viewing angle. The results demonstrated that the melting particles elevated 85-GHz brightness temperatures in comparison to results that excluded melting particles, and that the oriented ice particles created a polarization signature. The model results were consistent with observations over a stratiform rain event.

C. EVANS

1. Evans and Vivekanandan (1990)

This study presented a simulation of radar and radiometer remote sensing of nonspherical atmospheric ice crystals. The study was geared toward the remote sensing of cirrus cloud properties, and the radiometric simulations were performed at frequencies of 37, 85, and 157 GHz. The discrete dipole approximation (DDA; see Chapter 5) was used to compute the radiative properties of horizontally oriented ice modeled as polydispersions of hexagonal plates, columns, and needles. These radiative properties were used as input to a vector radiative transfer model based on the adding–doubling method (Evans and Stephens, 1991). It was shown that, in general, ice particles of all shapes depress brightness temperatures (compared to the cases without ice) for all viewing angles at these frequencies, though the depression increases with frequency.

Some results from this study are shown in Figs. 1 and 2. Figure 1 shows the upwelling brightness temperature $T_{B,I}$ and polarization difference $T_{B,Q}$ as functions of zenith angle for 85 and 157 GHz and contrasts the base case (plates with density 0.92 g/cm^3 and polydispersion maximum size of 2 mm) with a low particle density (0.23 g/cm^3) case and lower maximum size (1 mm) case. All computations were performed using a three-layer atmosphere over a Lambertian surface (thus, polarization signatures are due to particle scattering only and not to surface effects). The figure shows the strong dependence of the brightness temperature depression on particle density and maximum size in the polydispersion. The po-

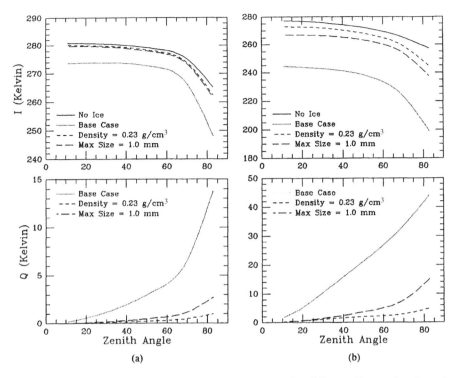

Figure 1   Upwelling brightness temperature $T_{B,I}$ and polarization difference $T_{B,Q}$ as functions of zenith angle for (a) 85 GHz and (b) 157 GHz. The base case (plates with density 0.92 g/cm³ and poly-dispersion maximum size of 2 mm) are contrasted with a low particle density (0.23 g/cm³) case and lower maximum size (1 mm) case. Reprinted from K. F. Evans and J. Vivekanandan (1990), Multi-parameter radar and microwave radiative transfer modeling of nonspherical atmospheric ice particles, *IEEE Trans. Geosci. Remote Sens.* **28**, 423–437 (© 1990 IEEE).

larization difference $T_{B,Q}$ increases with bulk ice density, maximum particle size, frequency, and viewing angle. The polarization difference increases with viewing angle because the particles are oriented in the horizontal plane and the path length is longer. Finally, $T_{B,Q}$ is always positive, indicating that the horizontally polarized radiation is being scattered out of the viewing direction, leaving more vertically polarized radiation.

Figure 2 shows the effect of ice particle shape on upwelling brightness temperature and polarization difference as a function of zenith angle for frequencies of 85 and 157 GHz. The total ice mass for each case is 0.1 g/m³, but the volume of the largest ice columns (with maximum dimension of 2 mm) is twice that of plates and 20 times that of needles. As shown in Fig. 2, the bright-

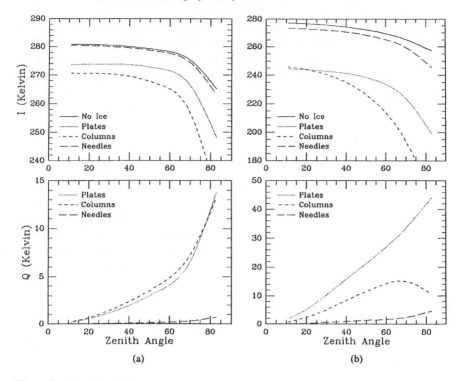

Figure 2   Upwelling brightness temperature $T_{B,I}$ and polarization difference $T_{B,Q}$ as functions of zenith angle for plates, columns, needles, and no ice layer: (a) 85 GHz and (b) 157 GHz. Reprinted from K. F. Evans and J. Vivekanandan (1990), Multiparameter radar and microwave radiative transfer modeling of nonspherical atmospheric ice particles, *IEEE Trans. Geosci. Remote Sens.* **28**, 423–437 (© 1990 IEEE).

ness temperature depression corresponds directly with the volume, with the more voluminous particles (columns, then plates, then needles) creating the largest depression.

2. Evans and Stephens (1995a, b)

Part I of this study (Evans and Stephens, 1995a) used the DDA to compute the single-scattering properties of five shapes of ice (solid and hollow columns, hexagonal plates, planar bullet rosettes, and spheres) at frequencies of 85.5, 157, 220, and 340 GHz. The large dimension of the particles ranged in size from 30 to 2000 μm, and the radiative properties of the polydispersion were computed using 18 different gamma size distributions (specified by characteristic particle size and distribution width) over these particles sizes. The long axes of the ice particles were assumed to be randomly oriented in the horizontal plane, and the particle aspect ratios varied according to empirical formulas.

Many important findings were presented in this paper. First, it was noted that, for the frequencies considered, there are three regimes for extinction as a function of particle volume: (i) Absorption dominates and varies as particle volume; (ii) Rayleigh scattering dominates varying as the square of the volume; and (iii) beyond the Rayleigh regime scattering increases more slowly than the square of the particle volume.

A second finding was that, for a given particle shape, the characteristic particle size of the distribution L_m (defined as the median of the distribution of the dimension L to the third power, where L is the particle maximum size) is important in determining the extinction properties of the distribution. Finally, it was found that particle shape also has a significant effect on polarization: Plates were found to be the most polarizing shape, followed by planar rosettes and then columns.

The scattering quantities were also fit as functions of particle size for several incident angles. Part II of the study (Evans and Stephens, 1995b) used these quantities as input to a radiative transfer model to examine the sensitivity of upwelling brightness temperature depressions at nadir and 49° as a function of ice water path. More recently, Evans *et al.* (1998) have used the DDA and polarized radiative transfer modeling to lay the theoretical foundations for cirrus property retrievals.

D. TURK

1. Turk and Vivekanandan (1995)

Turk and Vivekanandan (1995) performed microwave radiative transfer simulations by considering oriented ice plates, oblate raindrops, and conical graupel, over both ocean and land surfaces. Results were presented for both upwelling and downwelling viewing angles, for frequencies of 10, 20, 37, and 85 GHz.

Some of the results obtained in this study are shown in Fig. 3, which illustrates the behavior of the upwelling brightness temperature as a function of viewing angle for the case of a 4-km-thick layer of oblate raindrops with a constant rain rate of 5 mm/h, underlying a 4-km layer of conical graupel, all over the ocean, for frequencies of 20, 37, and 85 GHz. The effect of the graupel layer becomes more pronounced as frequency increases: At 20 GHz, the addition of the graupel layer is nearly negligible, but at 37 and 85 GHz, the graupel layer depresses the brightness temperatures considerably relative to the graupel-free cases. For all frequencies, the polarization difference can be quite large for off-nadir viewing angles. For example, at 85 GHz, the polarization difference $T_{B,Q}$ is about +20 K

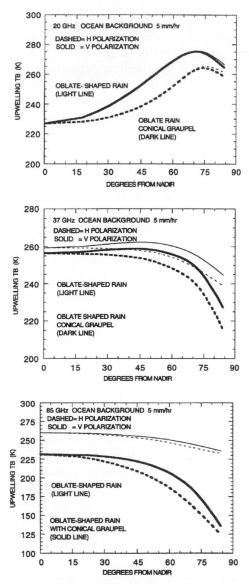

Figure 3   Upwelling brightness temperatures $T_{B,v}$ and $T_{B,h}$ at 20, 37, and 85 GHz as a function of view angle away from nadir. The thick curves denote a model consisting of a 4-km, 5-mm/h oblate spheroidally shaped rain layer beneath a 4-km layer of conical shaped graupel, and the thin curves denote a model consisting of a 4-km oblate rain layer only. Reprinted from J. Turk and J. Vivekanandan (1995), Effects of hydrometeor shape and orientation upon passive microwave brightness temperature measurements, *in* "Microwave Radiometry and Remote Sensing of the Environment" (D. Solomini, Ed.), pp. 187–196, VSP, Zeist, The Netherlands.

for viewing angles between 45° and 60°. For a fixed viewing angle of 48° (results not shown), the authors found polarization differences of about 20 K at 37 and 85 GHz for rain rates greater than 10 mm/h.

The authors also examined downwelling radiation at the surface and found that for a frequency of 85 GHz, the net downwelling brightness temperatures warmed with increasing ice plate mass content because of backscattering of the atmosphere and surface into the radiometer field of view. The polarization difference was found to be negative and had values of as much as −25 K for viewing angles between 45° and 60° from nadir. Computations at lower frequencies did not show these large polarization differences or brightness temperature warming because of the smaller albedo of ice at these frequencies. The results were also somewhat sensitive to the amount of cloud liquid water in the atmosphere.

2. Petty and Turk (1996)

This study used the model of Turk and Vivekanandan (1995) and compared the results to observed SSM/I brightness temperatures. The observed brightness temperatures were for situations corresponding to widespread, moderate to heavy rainfall in tropical cyclones. In general, the model results tended to overemphasize the polarization effects, though in many cases, the correct trends were produced. Furthermore, the observed SSM/I polarization signatures for rain rates between 5 and 50 mm/h (i) decreased linearly from about +10 K to 0 K at 19 GHz (presumably this is due to masking of the polarized surface as the atmosphere becomes opaque); (ii) were fairly constant with a mean value of +3.2 K for 37 GHz (this appears to be due to the scattering properties of the hydrometeors); and (iii) had values ranging from between +5 and +10 K for 85 GHz (again due to scattering by hydrometeors) in moderately scattering stratiform precipitation, and decreasing for more intensely scattering convective precipitation.

E. LIU

1. Liu *et al.* (1996b)

Apparently, the first fully three-dimensional polarized microwave code was developed by Liu *et al.* (1996b). The code combines forward and backward Monte Carlo techniques to solve the radiative transfer equation. Though the scheme does allow for the treatment of polarized radiation, it requires that the extinction matrix be diagonal, and, thus, it is not capable of modeling oriented particles, though it is capable of handling randomly oriented axisymmetric particles (Q. Liu, personal communication). This particular study considers only spherical particles and places emphasis on three-dimensional effects rather than concentrating on vector radiative transfer and polarization effects.

2. Liu and Simmer (1996)

This paper explored theoretically the effect of neglecting the vector radiative transfer approach when computing upwelling microwave brightness temperatures in the presence of spherical hydrometeors. The successive orders of scattering method was used for one-dimensional vector radiative transfer computations, and the Monte Carlo method (Liu *et al.*, 1996b) was used for three-dimensional vector radiative transfer computations. Though the formulation presented applies to randomly oriented axisymmetric particles (spheres, spheroids, ellipsoids, etc.), the findings of this study apply strictly to spherical particles. The formulation does not apply to preferentially oriented particles.

An important finding of this study applies to a homogeneous plane-parallel atmosphere with thermal sources and containing spherical particles. The finding is for polarized radiation emerging in an upward direction at the top of the atmosphere and is stated in terms of the phase matrix element Z_{21} (see Section VI of Chapter 1) as

$$Q \geq 0, \qquad \text{when } Z_{21} \leq 0 \text{ for all scattering angles}, \qquad (5)$$

$$Q \leq 0, \qquad \text{when } Z_{21} \geq 0 \text{ for all scattering angles}, \qquad (6)$$

$$Q = 0, \qquad \text{when } Z_{21} = 0 \text{ for all scattering angles}. \qquad (7)$$

If the preceding relations are only satisfied for some scattering angles, then the sign of Q depends on the single-scattering albedo, optical depth, and viewing angle. For most cases involving microwave radiation, relation (5) applies. Relation (6) sometimes applies to ice clouds and high microwave frequencies. For pure Rayleigh scattering, relation (5) applies.

The study also found that for finite (three-dimensional) clouds, the sign and magnitude of the polarization can change with cloud size. In addition, it was found that the upwelling polarization signature $T_{B,Q}$ due to spherical particles is typically only a couple of degrees Kelvin and never exceeds 10 K for microwave frequencies. Observed polarization signatures for precipitating clouds are sometimes larger, likely because of oriented nonspherical hydrometeors. Finally, it was found that the differences between using scalar and vector radiative transfer theory are negligibly small when considering spherical particles.

F. Czekala

1. Czekala and Simmer (1998)

Unlike the other studies reviewed in this section, Czekala and Simmer (1998) consider only nonspherical liquid water drops and do not examine the effects of nonspherical ice. This is important for understanding how nonspherical water drops can affect microwave radiation. The full Stokes vector formulation of the radiative transfer equation was used to model both upwelling and downwelling microwave radiation from rain modeled as rotationally symmetric ellipsoids with a size-dependent axis ratio. The particles were assumed to be oriented with their rotational axis aligned along the vertical, and azimuthal symmetry was assumed. The model computations were performed at frequencies of 10.7, 19.4, and 37 GHz, and a Marshall–Palmer drop size distribution (Section III.D) was used.

Figure 4 shows results of the difference (nonspherical minus spherical) of $T_{B,Q}$ computations for eight upward viewing angles between $0°$ and $90°$ as a function of rainfall rate. The surface emissivity is set to 1, so all polarization differences are due to scattering by the particles. The figure shows that for viewing angles from $8.3°$ to $51.8°$, the difference in the values of $T_{B,Q}$ between the nonspherical and the spherical particles increases with frequency and rain rate. At larger viewing angles, the difference between the two types of particles is frequency dependent: At high frequencies the brightness temperatures begin to saturate so the differences do not increase, but the differences at the lower frequencies continue to increase.

The authors noted that for the various cases examined, the total intensity (or, equivalently, the average of the horizontal and vertical brightness temperature) for both the upwelling and the downwelling cases did not differ significantly between the spherical and the nonspherical cases. In all upwelling cases, the nonspherical drops yield larger polarization signals than the spherical drops. For downwelling cases, the polarization difference between spherical and nonspherical drops is negative; that is, the spherical particles produce a larger downwelling polarization signature than the nonspherical particles.

2. Czekala (1998)

This study models oblate and prolate spheroidal ice particles with axial ratios of 5 and 0.2, respectively, and whose axis of symmetry is aligned along the vertical. The ice is positioned between 9 and 10 km height, using an exponential size distribution with a maximum equivalent sphere radius of 200 μm and a fixed ice water content of 0.05 kg/m². Results are computed for a frequency of 200 GHz and upwelling viewing angles, with the emphasis placed on near limb-viewing angles.

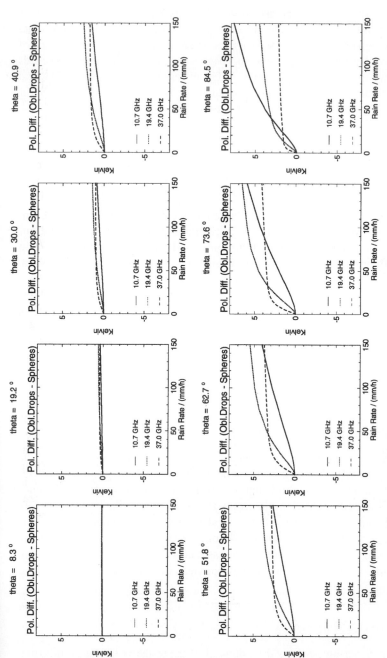

Figure 4 Results of the difference (nonspherical minus spherical) of polarization difference $T_{B,Q}$ computations for eight upward viewing angles between 0° and 90° as a function of rainfall rate. Oriented oblate and spherical liquid rain drops are considered. The surface emissivity is set to 1, so all polarization differences are due to scattering by the particles. Reprinted from H. Czekala and C. Simmer (1998), Microwave radiative transfer with nonspherical precipitating hydrometeors, *J. Quant. Spectrosc. Radiat. Transfer* **60**, 365–374.

The results show that the polarization difference $T_{\mathrm{B},Q}$ for oblate particles is positive for upward directions and negative for downward directions. For prolate particles, the sign of $T_{\mathrm{B},Q}$ is reversed for these situations. The largest magnitude of $T_{\mathrm{B},Q}$ is realized as the viewing angle approaches $90°$; for example, $T_{\mathrm{B},Q}$ exceeds 20 K for viewing angles greater than $85°$. The differences between spherical and nonspherical particle results for the total intensity $T_{\mathrm{B},I}$ are as large as 10 K for prolate and oblate particles, depending on the viewing direction.

G. HORNBOSTEL

1. Hornbostel *et al.* (1995)

Hornbostel *et al.* (1995) performed radiative transfer computations for down-welling microwave radiation through polydispersions of oriented oblate raindrops and compared the model results with ground-based polarimeter observations. In addition, ground-based measurements of rain rate, disdrometer (drop size) data, and other meteorological quantities were available for comparison with the po-larimeter and model data. The results show the potential of inferring drop size distribution parameters by inverting dual-polarized brightness temperatures. In addition, the authors pointed out that measurements of downwelling brightness temperatures from a hydrometeor layer often give larger polarization signatures than those obtained by physical modeling with spherical particles. However, they show that, by modeling raindrops as oblate spheroids, good agreement with mea-surements is obtained.

2. Hornbostel and Schroth (1995)

This paper uses the method of successive orders of scattering to model the downwelling 20-GHz brightness temperature received by a ground-based ra-diometer. The model results show that the polarization difference (vertical minus horizontal) is directly related to the drop size distribution (oblate raindrops with their axis of symmetry aligned to the vertical are used). The authors demonstrate that drop size distributions can be derived from measured radiometric polarization differences, and then show how these drop size distributions can be used as input parameters to compute the propagation properties of Earth–satellite communica-tion paths.

3. Hornbostel *et al.* (1997)

This short paper investigates the potential of using the polarization informa-tion from a ground-based radiometer to infer information about rainfall drop size distribution parameters. The vector radiative transfer equation is solved for frequencies of 13, 19, 37, and 90 GHz and rain rates ranging from 0 to 100 mm/h. The dependence of downwelling brightness temperatures (and brightness temperature differences) is shown as a function of rain rate for several drop size distributions.

4. Kutuza *et al.* (1998)

The paper by Kutuza *et al.* (1998) shows the possibility for a nonzero third Stokes parameter brightness temperature ($T_{B,U}$) due to scattering by hydrometeors. An important theoretical result of this paper is that for a homogeneous layer of nonspherical oriented raindrops, $T_{B,U}$ is dominated by the first azimuthal harmonic of the emission vector. For downwelling radiation, the model results give values of downwelling $T_{B,U}$ on the order of 2–3 K, depending on the rain rate, frequency, and mean drop canting angle. For a given frequency, downwelling $T_{B,U}$ increases with rain rate, reaches a saturation value, and then decreases. For a given rain rate, downwelling $T_{B,U}$ increases with frequency. Upwelling $T_{B,U}$ over an ice crystal layer reach values of about 4–5 K.

5. Schroth *et al.* (1998)

The paper by Schroth *et al.* (1998) is a follow-up study to the Kutuza *et al.* (1998) paper and shows the importance of including solar effects on modeling of downwelling $T_{B,U}$. For example, it is shown theoretically that, for microwave frequencies and realistic nonspherical raindrop canting angles, the resulting $T_{B,U}$ is about 2 or 3 K. However, including incident solar radiation increases these values by 5–7 times.

H. OTHER WORK

1. Smith and Mugnai (1989)

The paper by Smith and Mugnai (1989) is the third in a series of papers that investigates the effect of hydrometeors on upwelling microwave brightness temperatures. The study only considers spherical particles, but it is worthy of mention here because the series of papers was seminal in its use of a time-dependent three-dimensional cloud microphysics model as input to a plane-parallel radiative transfer model. From a physical perspective, the modeling effort in these studies was quite extensive and included several types of hydrometeors (rain, cloud, graupel, hail, ice crystals). In addition, the radiative effects of ice size, ice density, and mixed-phase layers were examined. The studies of scattering by nonspherical hydrometeors mentioned in the preceding sections have all used ad hoc atmospheric models; thus, it would be important to extend the microphysical cloud model approach of Smith and Mugnai (1989) to include nonspherical particles to gain a better understanding of the importance of nonsphericity on microwave scattering by precipitation. Roberti and Kummerow (1999) have recently made some progress on this challenge.

2. O'Brien and Goedecke (1988a, b)

These papers considered polarized microwave scattering by dendritic snow crystals at a 10-mm wavelength (30-GHz frequency). The studies examined the effect of complex refractive index mixing rules, equivalent particle shape and size, and mono- versus poly-disperse distributions on radiative cross sections and forward scattering matrix elements. Solutions of the radiative transfer equation were not performed, but the studies are fundamental in understanding the microwave scattering properties of snow.

3. Tsang

The textbook by Tsang *et al.* (1985) provides an extremely comprehensive and detailed coverage of the theory of microwave remote sensing, including treatments of the vector radiative transfer equation and scattering by nonspherical particles. The textbook includes several examples of scattering by precipitation. The paper by Tsang (1991) lays the theoretical foundation for exploring the potential of using passive microwave polarimetry (i.e., using all four components of the Stokes vector) on remote sensing of geophysical terrain. Though the paper is geared toward remote sensing of forests and vegetation, the formulation is completely general.

4. Three-Dimensional Studies

As mentioned previously in Section E, Liu *et al.* (1996b) and Liu and Simmer (1996) were the first to publish results of a three-dimensional polarized microwave radiative transfer model to compute upwelling brightness temperatures for precipitating atmospheres. Their results pertained to spherical particles only. Their Monte Carlo model is capable of treating randomly oriented axisymmetric particles, but cannot handle oriented particles. Haferman *et al.* (1997) presented a three-dimensional polarized radiative transfer model based on the discrete ordinates method and validated the results for plane-parallel atmospheres containing randomly oriented axisymmetric particles. In principle, the model can handle oriented particles, though it so far has not been applied to this type of situation.

Roberti and Kummerow (1999) have developed a three-dimensional "forward–backward" Monte Carlo code that treats spherical and/or randomly oriented nonspherical hydrometeors. They have also developed a forward Monte Carlo code that treats oriented nonspherical hydrometeors, but this code is for plane-parallel situations. However, the code should be extendable to three dimensions. These authors have intercompared some of their three-dimensional polarized computations with favorable results, but this continues to be an active area of research.

5. Other Related Studies

There are many other studies that have considered the propagation of microwave radiation through nonspherical particles. However, most of the studies considering passive remote sensing of precipitation have been reviewed in the preceding pages. Exceptions are the works by Huang and Liou (1983) and Bauer and Schluessel (1993). Both of these studies used the full vector form of the radiative transfer equation for plane-parallel atmospheres, but only considered spherical particles.

Other related studies that do not specifically deal with passive microwave remote sensing of precipitation include the following: several that considered the effect of precipitation on communication transmissions, for example, Oguchi (1960, 1981), Mishchenko (1992a), and Li *et al.* (1995); Macke and Großklaus (1998), who considered the effects of nonspherical particles on lidar remote sensing of precipitation; Jin (1998), who reviewed polarimetric scattering from nonspherical particles as applied to remote sensing of land surfaces; and, finally, Aydin (Chapter 16), who reviewed scattering by hydrometeors in the context of active remote sensing of precipitation.

III. MATHEMATICAL FORMULATION

A. RADIATIVE TRANSFER THEORY

The vector radiative transfer equation (VRTE) for polarized monochromatic radiation is given by Eq. (86) of Chapter 1. For media with thermal emission, for example, atmospheric hydrometeors in the microwave regime, the VRTE includes an additional emission term and is written for plane-parallel geometries as

$$\cos\vartheta \frac{d}{dz}\mathbf{I}(z;\vartheta,\varphi) = -n_0\langle\mathbf{K}(z;\vartheta,\varphi)\rangle\mathbf{I}(z;\vartheta,\varphi) + \mathbf{K}_a(z;\vartheta,\varphi)I_b(T)$$

$$+ n_0 \int_0^{2\pi}\int_{-1}^{1} \langle\mathbf{Z}(z;\vartheta,\varphi;\vartheta',\varphi')\rangle\mathbf{I}(z;\vartheta',\varphi')$$

$$\times d(\cos\vartheta')\,d\varphi', \tag{8}$$

where the vector $\mathbf{K}_a$ is a four-component absorption coefficient vector with components given by Eqs. (89)–(92) of Chapter 1 and $I_b(T)$ is the Planck blackbody radiance of particles at temperature T. The other terms are defined in Chapter 1, as is the laboratory reference frame in which the z axis and angles ϑ and φ are defined.

The VRTE given by Eq. (8) is completely general for the case of a plane-parallel medium and is the most rigorous and appropriate form of the radiative transfer equation to use for sparsely distributed, preferentially oriented, nonspherical particles. A simpler form of the VRTE is applicable to the situation of an azimuthally symmetric radiation field (note that solar radiation or rough surfaces may entail asymmetry). An example of this situation is a plane-parallel medium containing axisymmetric nonspherical particles aligned parallel to the horizontal $(x-y)$ plane of the laboratory frame (Section III of Chapter 1) and randomly aligned in azimuthal directions, with thermal emission as the only source of radiation. For such a situation, the third and fourth Stokes parameters are identically equal to zero, and the VRTE reduces to

$$\cos\vartheta \frac{d}{dz}\mathbf{I}(z;\vartheta) = -n_0\langle\mathbf{K}(z;\vartheta)\rangle\mathbf{I}(z;\vartheta) + \mathbf{K}_a(z;\vartheta)I_b(T)$$

$$+ n_0 \int_{-1}^{1} \langle\mathbf{Z}(z;\vartheta,\vartheta')\rangle\mathbf{I}(z;\vartheta')\,d(\cos\vartheta'), \qquad (9)$$

where the Stokes vector is now $\mathbf{I} = [I, Q]^T$ (T means transpose) and all vectors and matrices are of length 2 and dimension 2×2, respectively [however, the $(2, 2)$ element of the phase matrix is computed using the $(3, 3)$ element of the scattering matrix; see Eq. (68) of Chapter 1]. Finally, for the case of randomly oriented axisymmetric particles in a plane-parallel, azimuthally symmetric radiation field, the extinction matrix and absorption coefficient vector become scalars and the VRTE reduces to

$$\cos\vartheta \frac{d}{dz}\mathbf{I}(z;\vartheta) = -n_0 C_{ext}\mathbf{I}(z;\vartheta) + n_0 C_{abs}I_b(T)$$

$$+ n_0 \int_{-1}^{1} \langle\mathbf{Z}(z;\vartheta,\vartheta')\rangle\mathbf{I}(z;\vartheta')\,d(\cos\vartheta'), \qquad (10)$$

where the extinction and absorption cross sections C_{ext} and C_{abs} are defined in Section XI of Chapter 1, $\mathbf{I} = [I, Q]^T$, and $\mathbf{Z}$ has dimension 2×2 [see the note regarding the scattering matrix following Eq. (9)]. For unpolarized radiation, Eq. (10) reduces to the more familiar scalar radiative transfer equation.

Several methods for numerically solving the VRTE are described in the literature. Most of these methods use discrete angles and quadrature summation to simplify the integration term that appears in Eqs. (8)–(10). An exception is the Monte Carlo method, which uses probabilistic interaction laws to trace photons through a medium. Overviews of Monte Carlo methods for solving the VRTE are given by Adams and Kattawar (1993) and Marchuk *et al.* (1980). Descriptions of backward (or adjoint) Monte Carlo methods, which trace photons from a detector back to a source, are given by Liu *et al.* (1996b) and Roberti and Kummerow (1999).

Other numerical techniques for solving the plane-parallel VRTE include the method of successive orders of scattering (Hornbostel and Schroth, 1995; Liu and Simmer, 1996; Czekala and Simmer, 1998), the matrix operator method (Liu *et al.*, 1991), the method of invariant embedding (Adams and Kattawar, 1970), the Gauss–Seidel method (Herman *et al.*, 1995), the matrix-eigenvalue technique (Ishimaru, 1978), the iterative method (Tsang *et al.*, 1984; Jin, 1991; Kuga, 1991), the adding method (Hansen, 1971; Hovenier, 1971; Evans and Stephens, 1991), and the discrete ordinates method (Haferman *et al.*, 1997; Schulz *et al.*, 1999b). This list is not intended to be exhaustive, but is provided to give the reader an entry to the terminology and literature.

It should be noted that some VRTE solution methods with differing names are quite similar to one another, whereas some methods with the same names but different authors can be very different from one another. For example, the adding method of van de Hulst (1980) is a special case of the matrix operator method (Plass *et al.*, 1973). On the other hand, Evans and Stephens (1991) note that their adding method is very different from that described by Hansen (1971).

B. RADIATIVE (SINGLE-SCATTERING) PROPERTIES

The radiative properties for the VRTE in general depend on direction; position; wavelength; and particle size, composition, and orientation. For a small volume of independent scatterers, the radiative properties of the volume are computed by averaging the single-scattering properties of the volume of scatterers over size, shape, and orientation (Section X of Chapter 1). To define the scattering volume for single-scattering purposes, it suffices to specify:

a. The particle size distribution $N(r) \, dr$ that gives the number of particles per unit volume having an equivalent-sphere radius between r and $r + dr$ and satisfies the normalization $\int_{r_{min}}^{r_{max}} N(r) \, dr = n_0$ (where r_{min} and r_{max} are the minimum and maximum radii of the distribution)

b. A definition of an equivalent sphere (e.g., an equivalent sphere may be defined as a sphere having an equal volume or an equal average projected cross-sectional area to a given nonspherical particle)

c. A description of the particle shape, for example, the particle aspect ratio a/b, where $a/b < 1$ for prolate spheroids and $a/b > 1$ for oblate spheroids, with a and b representing the lengths of the nonrotational and symmetry axes, respectively

d. The refractive index of the particle

e. The wavelength under consideration

Given these specifications, the solution to Maxwell's equations for the scattered field of the polydispersion of particles in terms of the ensemble-averaged phase matrix (Chapter 1, Section VI) may be computed. The scattering properties needed for the VRTE may then be computed according to the formulas of Chapter 1.

C. SIZE DISTRIBUTION

In Chapter 1, the particle size distribution was written as a function of particle radius. For convenience, the particle size distribution is written in the remainder of this chapter as a function of particle diameter (for notes on converting size distributions between diameter and radius, see Appendix A). Using this convention, the particle size distribution $N(D)$ gives the number of particles per unit volume with diameters between D and $D + dD$. In the case of nonspherical particles, the distribution gives the number of particles per unit volume with *equivalent-sphere* (Appendix B) diameters between D and $D + dD$. The most general size distribution commonly employed in microwave precipitation studies is the modified gamma distribution for spherical particles, written as

$$N(D) = N_0(\Lambda D)^{P_1} \exp\left[-(\Lambda D)^{P_2}\right], \tag{11}$$

where N_0, Λ, P_1, and P_2 are parameters. Gasiewski (1993) provides expressions for computing quantities such as mean diameter, diameter variance, mode diameter, total particle number density, fractional volume, liquid and ice water density, and reflectivity, using this distribution. Setting $P_1 = 0$ and $P_2 = 1$ reduces Eq. (11) to an exponential distribution

$$N(D) = N_0 \exp[-(\Lambda D)]. \tag{12}$$

Of particular significance is the Marshall–Palmer distribution, which results by setting

$$\left. \begin{array}{l} N_0 = 8000 \text{m}^{-3}\,\text{mm}^{-1} \\[2mm] \Lambda = 4.1 R^{-0.21}\,\text{mm}^{-1} \end{array} \right\} \text{Marshall–Palmer}, \tag{13}$$

where R is the rain rate in millimeters per hour. The Marshall–Palmer distribution is very widely used in precipitation modeling to describe the size distribution of liquid raindrops. The drop size distribution can vary greatly over meteorological conditions, but the Marshall–Palmer distribution is a good representation of "average" rainfall. Bauer *et al.* (1998) have used the exponential distribution of

Table I

Sample Coefficients of Exponential Size Distribution for Four Hydrometeor Types [i.e., Eq. (12): $N(D) = N_0 \exp(-\Lambda D)$; $N_0 = c_1 R^{c_2}$; $\Lambda = c_3 R^{c_4}$; R Is the Rain Rate in mm/h]. D_{min} and D_{max} Are the Minimum and Maximum Diameters in the Size Distribution, and ρ Is the Particle Density

Hydrometeor type	c_1 $(m^{-3} \, mm^{-2} \, h)$	c_2	c_3 $(mm^{-3} \, h)$	c_4	D_{min} (mm)	D_{max} (mm)	ρ $(g \, m^{-3})$
Rain	7000	0.37	3.8	−0.14	0.2	2.5	1
Snow	1500	−0.38	2	−0.34	0.5	5	0.2
Graupel	3400	−0.01	2.2	−0.24	0.5	3	0.5
Hail	12.1	0	0.42	0	1	35	0.9

Source: Based on Bauer *et al.* (1998).

Eq. (12) to model four hydrometeor particles (rain, snow, graupel, and hail). Details are provided in Table I.

Another distribution that is commonly employed is the gamma distribution

$$N(D) = N_0 D^P \exp[-\Lambda D], \qquad (14)$$

which has the three parameters N_0, Λ, and P. Evans and Stephens (1995a) have used this distribution to model nonspherical ice crystals and suggest that the integration over a discrete size distribution [cf. Eq. (50) of Chapter 1] be performed using

$$\int_0^\infty N(D)k(D)\,dD \approx \sum_{i=1}^{N} k(D_i) \int_{D_i^l}^{D_i^u} N(D) \left(\frac{D}{D_i}\right)^\zeta dD = \sum_{i=1}^{N} n_i k(D_i), \quad (15)$$

where $k(D_i)$ represents a scattering property of a single particle in the ith bin, D_i^l and D_i^u are the lower and upper bounds of the ith bin, and the weight $(D/D_i)^\zeta$ is introduced to account for increasing contribution to the scattering properties with increasing size. For the gamma distribution, it follows that n_i may be written as

$$n_i = N_0 D_i^{-\zeta} \Lambda^{-(P+\zeta+1)} [\Gamma(P+\zeta+1, \Lambda D_i^u) - \Gamma(P+\zeta+1, \Lambda D_i^l)], \quad (16)$$

where $\Gamma(a, x)$ is the incomplete gamma function. A value of $P = 0$ yields an exponential distribution, though values of $P = 1$ and $P = 2$ are also typically used (as P increases, the width of the distribution decreases). Schneider and Stephens (1995) suggest a range of 0 to 6 for the parameter ζ.

D. Shape Distribution

1. Raindrops

Pruppacher and Pitter (1971) parameterized the shape of the surface of a raindrop using a series of Chebyshev polynomials that defines the radius of the raindrop with respect to the polar angle θ via

$$r(\theta) = r_0 \left(1 + \sum_{n=0}^{\infty} c_n \cos(n\theta) \right), \tag{17}$$

where r_0 is the radius of the undistorted spherical drop. This equation generates shapes that are consistent with observations showing that falling raindrops typically have a nearly oblate shape with a nearly round top and flattened bottom, where the flat bottom tends to become more extreme as the particle increases in size. For spheroids, shape can be characterized by defining the particle aspect ratio a/b, where a and b are the lengths of the nonrotational and symmetry axes (sometimes in the literature, a and b refer instead to the particle semi-major and semi-minor axes, or vice versa). The spheroid is "oblate" when $a/b > 1$, "prolate" when $a/b < 1$, and spherical for $a/b = 1$. In some of the equations that follow, the ratio b/a will be used to uphold the conventions defined in this paragraph.

A recently published set of coefficients for Eq. (17) is presented by Chuang and Beard (1990), who parameterize the aspect ratio b/a as a function of equivalent spherical raindrop diameter D (in centimeters) using (the coefficient of the second term has been corrected according to personal communication with K. Beard)

$$\frac{b}{a} = 1.01668 - 0.098055D - 2.52686D^2 + 3.75061D^3 - 1.68692D^4. \tag{18}$$

This equation is useful for approximating raindrops as oblate particles and is valid for the diameter range 0.1–0.9 cm. Other parameterizations of the particle aspect ratio for raindrops have been reviewed by Chandrasekar *et al.* (1988). The most recent work by Beard and co-workers (Andsager *et al.*, 1999) gives the following formula for nonoscillating drops (e.g., raindrops in the presence of an electric field) (D in centimeters):

$$\frac{b}{a} = 1.0048 + 0.0057D - 2.628D^2 + 3.682D^3 - 1.677D^4, \tag{19}$$

valid for the diameter range 0.1–0.9 cm, and, for oscillating raindrops,

$$\frac{b}{a} = 1.012 - 0.1445D - 1.028D^2, \tag{20}$$

valid for the diameter range 0.1–0.44 cm. The preceding formula is from a fit to measurements from several published results.

2. Ice

Ice crystals (e.g., plates, dendrites, and columns), which tend to assume a preferential orientation, take on a wide variety of shapes. Matrosov *et al.* (1996) have compiled a list of coefficients to describe ice crystal shapes for use in the formulas

$$d = a_2 L^f, \tag{21}$$

$$\rho = a_1 L^c, \tag{22}$$

where d (cm) is the crystal thickness, L (cm) is the maximum particle dimension, and ρ is the ice crystal density (g cm^{-3}). The coefficients a_1, a_2, f, and c are listed in Table II. Wang (1997) has used elementary mathematical functions to

Table II

Coefficients of the Bulk Density—Size and Thickness—Major Dimension Relationships [i.e., Eq. (21) Thickness: $d = a_2 L^f$; Eq. (22) Density: $\rho = a_1 L^c$] for Different Ice Crystal Classes. Except for Aggregates, Dimensions Are in Centimeters and Densities Are in Grams per Cubic Centimeter. For Aggregates, $\rho(\text{g cm}^{-3}) = 0.07 S^{-1.1}$, where S Is the Particle Major Dimension in mm, and the Particle Shape Can Be Modeled as Prolate and Oblate Spheroids

Ice crystal class	a_1	a_2	f	c
Dendrites	0.25	0.009	0.377	−0.377
Solid thick plates	0.916	0.138	0.778	0.0
Hexagonal plates	0.916	0.014	0.474	0.0
Solid columns ($L/d \leq 2$)	0.916	0.578	0.958	0.0
Solid columns ($L/d > 2$)	0.916	0.260	0.927	0.0
Hollow columns ($L/d \leq 2$)	0.53	0.422	0.892	−0.092 (cold)
	0.82			−0.014 (warm)
Hollow columns ($L/d > 2$)	0.53	0.263	0.930	−0.377 (cold)
	0.82			−0.014 (warm)
Long solid columns	0.916	0.035	0.437	0.0
Solid bullets ($L \leq 0.03$ cm)	0.916	0.153	0.786	0.0
Hollow bullets ($L > 0.03$ cm)	0.77	0.063	0.532	−0.0038
Elementary needles ($L < 0.05$ cm)	0.916	0.030	0.611	0.0
Aggregates (see table caption)				

Source: Based on Matrosov *et al.* (1996).

represent the size and shape of ice crystals. The relationship between a given ice crystal and the diameter of a sphere of equivalent volume is discussed by Evans and Stephens (1995a).

Hail and graupel are ice hydrometeors that are in general larger than the ice crystals mentioned previously. Hail and graupel are usually spheroidal or conical in shape and tend to tumble or wobble when falling and do not generally assume a preferential orientation. Wang (1982) has proposed a very general function to describe the shape of hail and graupel and has also provided expressions for cross-sectional area, volume, and surface area of the particle. Turk and Vivekanandan (1995) modeled the shape of graupel as a cone with a 45° half angle connected to a half-oblate spheroidal bottom, which is related to an equal volume sphere according to

$$D^3 = \frac{b^2}{2}h + a.$$ (23)

In Eq. (23), a and b are the semi-minor and semi-major axes of the spheroidal bottom, h is the cone height, and D is the diameter of the equal volume sphere. Matson and Huggins (1980) studied hailstones falling in their natural environment and found that 84% of the observed hailstones were spherical and the remainder were roughly conical. However, in one storm, they found roughly half of the hailstones to be conical. Overall, they found the mean ratio of minimum to maximum hailstone dimensions to be about 0.77. The results of Bringi *et al.* (1986a, b) imply that the relations for rain given by Eqs. (18)–(20) are reasonable approximations for hail and graupel.

For mixed-phase particles (e.g., melting particles), Schols *et al.* (1997) used the linear relationship

$$\frac{a}{b} = f\left(\frac{a}{b}\right)_{\text{liquid}} + (1 - f)\left(\frac{a}{b}\right)_{\text{ice}},$$ (24)

where f is the fraction of the particle that is melted and $(a/b)_{\text{liquid}}$ and $(a/b)_{\text{ice}}$ are the aspect ratios for liquid and ice phase particles, respectively, for a given equivalent spherical diameter.

E. ORIENTATION DISTRIBUTION

The orientation distribution (see Section X of Chapter 1) used in the majority of studies of nonspherical hydrometeor scattering for passive microwave precipitation retrieval has been such that the particles are either randomly oriented [Eq. (54) of Chapter 1] or perfectly aligned along an axis of symmetry [Eq. (55) of Chapter 1] (with the exception of the work by Kutuza *et al.*, 1998). As mentioned in Section D, observational evidence indicates that raindrops fall with a flattened bottom (Pruppacher and Pitter, 1971), ice crystals tend to fall with their long axis aligned perpendicular to the direction of fall (but because of electrification sometimes orient themselves with their long axis parallel to the fall direction; see Caylor and Chandrasekar, 1996), and graupel and hail tumble and wobble as they fall (Matson and Huggins, 1980). Examples of nonrandom orientation distributions are the truncated Gaussian distribution used by Aydin and Tang (1997a)

$$p(\beta) = \frac{\exp[-(\beta - \overline{\beta})^2/2\sigma^2]}{\int_0^{\pi/2} \exp[-(\beta - \overline{\beta})^2/2\sigma^2] \sin \beta \, d\beta}, \tag{25}$$

where $\overline{\beta}$ is the mean drop canting angle and σ is the standard deviation, and the harmonic oscillator distribution of Vivekanandan *et al.* (1991)

$$p(\beta) = \frac{1}{\pi \beta_m}\left[1 - \left(\frac{\beta - \overline{\beta}}{\beta_m}\right)^2\right]^{-1/2}, \tag{26}$$

where β_m is the amplitude. Equation (26) must be normalized to unity over the range $\overline{\beta} \pm \beta_m$. Equations (25) and (26) are one-dimensional probability distributions in β. Other examples of one-dimensional orientation distributions are given by Mishchenko (1991b) and Karam *et al.* (1995). Multidimensional orientation distributions are given by Metcalf (1988), Holt (1984), and Tsang (1991).

F. COMPLEX REFRACTIVE INDEX

The amplitude matrix (defined in Section IV of Chapter 1) requires specification of a particle complex refractive index

$$m \equiv m_r + im_i, \tag{27}$$

where the real part m_r is related to how fast light propagates through the species and the imaginary part m_i is related to how much radiation the species absorbs. For liquid water, Ray (1972) has used an extension of Debye theory to obtain refractive indices of water ranging from -20 to $50°C$ and ice from -20 to $0°C$. A more recent tabulation of the complex refractive index of ice has been given by Warren (1984). Both the Ray (1972) and the Warren (1984) studies cover a large portion of the electromagnetic spectrum, including the microwave. For the refractive index of ice in the microwave, m_r has an approximately constant value of 1.78 for all frequencies and temperatures, whereas m_i is weakly frequency and temperature dependent and ranges from about 10^{-4} to 10^{-2} (very weak absorption). The refractive index for liquid water in the microwave has a stronger frequency and temperature dependency, with m_r ranging from between about 3 and 10 and m_i ranging from approximately 0.5 to 3 (strong absorption) for microwave frequencies and common atmospheric temperatures.

The complex refractive index of mixtures (e.g., porous ice consisting of ice and air, or spongy ice consisting of ice and water) is treated in detail in Chapter 9. Bohren and Battan (1982) also cover this topic and compare the Maxwell-Garnett (treatment of a mixture as a nonsymmetric host and inclusion) versus the Bruggeman (treatment of the substances as symmetric) theories of the effective dielectric constant ($\varepsilon \equiv m^2$) for a two-component mixture. Smith and Mugnai (1989) proffer the following mixing formula for porous ice:

$$m_{\text{mix}} = \left[f_{\text{air}} m_{\text{air}}^2 + (1 - f_{\text{air}}) m_{\text{ice}}^2 \right]^{1/2}, \tag{28}$$

where the fraction of air is defined as

$$f_{\text{air}} = 1 - \frac{\rho_p}{\rho_i}, \tag{29}$$

with ρ_p set to the density of porous ice [0.45 gm cm^{-3} in the Smith and Mugnai (1989) study] and ρ_i set to the density of pure ice (0.91 gm cm^{-3}). The refractive index of moist air is given by Battan (1973) as

$$m_{\text{air}} = 1 + 10^{-6} \left[\frac{77.6}{T} \left(P + 4810 \frac{e_v}{T} \right) \right], \tag{30}$$

where P is the air pressure and e_v is the partial pressure of water vapor, both in millibars. An alternate relationship for the complex refractive index of an ice–air mixture is given by Matrosov *et al.* (1996) as

$$\frac{m_{\text{mix}}^2 - 1}{m_{\text{mix}}^2 + 2} = (1 - f_{\text{air}}) \frac{m_{\text{ice}}^2 - 1}{m_{\text{ice}}^2 + 2}. \tag{31}$$

Finally, a formulation for computing the refractive index of most snow based on Debye theory is given by Sadiku (1985).

G. Hydrometeor Terminal Velocity

Some of the size distributions described in Section C use the rain rate R as a parameter. This value of R should be considered a *nominal* rain rate; the true rain rate (ignoring vertical air velocities) is given by

$$R = \int v(D) N(D) V(D) \, dD, \tag{32}$$

where $v(D)$ is the terminal velocity of a particle with equivalent spherical diameter D and $V(D)$ is the particle volume. Because nonspherical particles could have significantly more or less drag force exerted on them as they fall, the rain rate computed using Eq. (32) could differ for an identical volume of spheres versus nonspheres. Expressions for $v(D)$ for various precipitation particles are given by Mitchell (1996).

H. Modeling the Atmosphere

In general, solution of the VRTE requires specification of a model atmosphere, including profiles of density, pressure, temperature, and atmospheric gases. Ulaby *et al.* (1981) provide formulas for these quantities for the U.S. standard atmosphere. Most researchers prescribe an ad hoc atmosphere for performing simulations, though several studies have used cloud models (e.g., Smith and Mugnai, 1989; Roberti *et al.*, 1994).

In the microwave region of the spectrum, gaseous absorption and emission are due to water vapor and oxygen; no other species contribute significantly in this spectral region. For water vapor, the model by Liebe *et al.* (1993) with the updates suggested by Rosenkranz (1998) is arguably the best microwave water vapor model available. A state-of-the-art model for microwave attenuation by gaseous oxygen is described by Liebe *et al.* (1992).

I. Modeling the Surface

1. Boundary Conditions

The boundary conditions for the VRTE may be written as

$$\mathbf{I}^+(z_0) = \psi I_b(T_0) + (1 - f_d)\mathbf{R}_s\mathbf{I}_m^- + \frac{f_d}{\pi}\left[\mathbf{R}_d\mathbf{I}_c^- + \int_0^{2\pi} v\mathbf{R}_d\mathbf{I}^- \, d\varphi'\right], \quad (33)$$

$$\mathbf{I}^o(z_1) = \mathbf{I}_{ba}^o + \mathbf{I}_c^o \, \delta(\vartheta - \vartheta') \, \delta(\varphi - \varphi'), \quad (34)$$

where the superscripts $+$, $-$, and o represent radiation moving from the lower boundary at $z = z_0$ toward the domain, from the domain toward the lower boundary, and from outside the upper boundary at $z = z_1$ toward the domain, respectively. The vectors $\mathbf{I}_{ba}$ and $\mathbf{I}_c$ denote background and collimated radiation, respectively, $\mathbf{I}_m^-$ is the Stokes vector in the direction mirroring that of $\mathbf{I}^+$, and $\delta(\ )$ is the Dirac delta function. The symbol f_d is set to 1 for the case of diffuse reflection and is set to 0 for the case of specular reflection. The symbol ψ denotes the emission vector, $I_b(T_0)$ is the Planck blackbody radiance for the surface at temperature T_0, and the symbols $\mathbf{R}_s$ and $\mathbf{R}_d$ are the bidirectional reflectance matrices for specular and diffuse reflection, respectively. Finally, v is the cosine of the angle between the propagation direction and the normal to the boundary under consideration. Equation (33) states that the Stokes vector for radiation leaving the surface at $z = z_0$ is due to emission plus contributions from specularly and diffusely reflected radiation. Equation (34) states that radiation transmitted from outside the boundary at $z = z_1$ is due to background and collimated radiation (collimated radiation is generally ignored for microwave applications).

2. Lambertian (Diffuse) Surfaces

A Lambertian, or diffuse, surface is an idealized surface that is "perfectly" rough and therefore emits and reflects equally in all directions. For such a surface, in which the reflected radiation is unpolarized, the bidirectional reflection matrix is given by

$$\mathbf{R}_d = \begin{pmatrix} \rho_0 & 0 & 0 & 0 \\ 0 & 0 & 0 & 0 \\ 0 & 0 & 0 & 0 \\ 0 & 0 & 0 & 0 \end{pmatrix}, \tag{35}$$

where ρ_0 is the hemispherical reflectance at the boundary. The emission vector for a Lambertian surface is given by

$$\psi = \begin{pmatrix} \psi_0 \\ 0 \\ 0 \\ 0 \end{pmatrix}, \tag{36}$$

where ψ_0 is the unpolarized hemispherical emittance of the boundary. For the case of a diffusely reflecting boundary, the emission vector given by Eq. (36) is used in Eq. (33) with $f_d = 1$.

3. Fresnel (Specular) Surfaces

A Fresnel, or specular, surface is an idealized surface that is "perfectly" smooth. For such a surface, the reflected radiation is polarized and angle dependent, and the bidirectional reflection matrix is given by (Tsang *et al.*, 1985)[4]

$$
\mathbf{R_s} = \begin{pmatrix}
\frac{1}{2}\left(|R_v|^2 + |R_h|^2\right) & \frac{1}{2}\left(|R_v|^2 - |R_h|^2\right) & 0 & 0 \\
\frac{1}{2}\left(|R_v|^2 - |R_h|^2\right) & \frac{1}{2}\left(|R_v|^2 + |R_h|^2\right) & 0 & 0 \\
0 & 0 & \mathrm{Re}\left(R_v R_h^*\right) & \mathrm{Im}\left(R_v R_h^*\right) \\
0 & 0 & -\mathrm{Im}\left(R_v R_h^*\right) & \mathrm{Re}\left(R_v R_h^*\right)
\end{pmatrix},
\tag{37}
$$

where the vertical and horizontal reflection coefficients are given by the Fresnel relations

$$
R_v(\mu) = \frac{m^2\mu - \sqrt{m^2 + \mu^2 - 1}}{m^2\mu + \sqrt{m^2 + \mu^2 - 1}},
\tag{38}
$$

$$
R_h(\mu) = \frac{\mu - \sqrt{m^2 + \mu^2 - 1}}{\mu + \sqrt{m^2 + \mu^2 - 1}},
\tag{39}
$$

where $\mu = \cos\vartheta$. The radiation emitted by a purely specularly reflecting surface is also polarized and angle dependent, and the emission vector is given by

$$
\psi = \begin{pmatrix}
1 - \frac{1}{2}\left(|R_v|^2 + |R_h|^2\right) \\
-\frac{1}{2}\left(|R_v|^2 - |R_h|^2\right) \\
0 \\
0
\end{pmatrix}.
\tag{40}
$$

For the case of a specularly reflecting boundary, the emission vector given by Eq. (40) is used in Eq. (33) with $f_d = 0$.

4. Rough Surfaces

The boundary conditions stated in the preceding sections assume that the surface is perfectly rough or smooth. However, realistic surfaces require a more complicated surface emission vector and reflection matrix. Tsang *et al.* (1985, pp. 70, 203) draw upon the classic work of Beckmann and Spizzichino (1963) to develop upper and lower limits for the reflection matrices for random rough surfaces for all four Stokes parameters. More recently, Gasiewski and Kunkee (1994) have developed a tilted-facet geometrical optics model to obtain expressions for the emission vector and reflection matrix for the first three Stokes parameters, and Yueh (1997)

[4]The sign of U differs between Chapter 1 and Tsang *et al.* (1985). Therefore, the sign of the [3, 4] and [4, 3] elements in the reflection matrix given here also differs from Tsang *et al.* (1985).

has presented a two-scale scattering model (Bragg scattering by small-scale waves plus coherent reflection by large-scale waves) to accurately describe the emission and reflection for all four Stokes parameters.

IV. EXAMPLES OF MODEL ATMOSPHERE SIMULATIONS AND RESULTS

In this section, a simple atmospheric model containing nonspherical hydrometeors is described and polarized microwave radiative transfer results are presented. These examples are simple enough so that interested researchers may use the results as benchmarks.

A. DESCRIPTION OF MODEL ATMOSPHERE

The atmospheric system used for the examples in this section is based on the microwave example of Evans and Stephens (1991), but is modified to incorporate spheroidal particles. The system is summarized in Table III. The system consists of two 4-km-thick layers, where the upper layer contains spheroidal ice particles, the lower layer contains spheroidal raindrops, and the temperature profile varies linearly within each layer such that $T = 300$ K at the surface, $T = 273$ K at the ice/rain interface, and $T = 245$ K at $z = 8$ km. The atmospheric gaseous absorption coefficient is set to $k_{gas} = 0.01$ and $k_{gas} = 0.15$ km^{-1} for the lower

Table III

Summary of Parameters Used for Model Atmosphere Radiative Transfer Simulations. The Temperature Indicated Is for the Layer Interface; within Each Layer, the Temperature Varies Linearly. The Gaseous Absorption Coefficient k_{gas} Is Constant within Each Layer. T_{back} Is the Equivalent Blackbody Temperature of the Cosmic Background, and m Is the Complex Refractive Index of the Fresnel Surface

$T_{back} = 2.7$ K			
z (km)	T (K)	$k_{gas,19}$ (km^{-1})	$k_{gas,85}$ (km^{-1})
8	245	0.005	0.01
4	273	0.03	0.15
0	300		
$m = 3.724 + i2.212$			

and upper layers, respectively, at 85 GHz and $k_{gas} = 0.005$ and $k_{gas} = 0.03$ km^{-1} for these layers at 19 GHz.

The two-layer atmosphere is placed above a specularly reflecting surface with an index of refraction of $m = 3.724 + i2.212$, representing a smooth ocean surface at a temperature of 300 K. Unpolarized, diffuse, background radiation at a blackbody equivalent temperature of 2.7 K, representative of cosmic background radiation, is incident from above the atmosphere. These boundary conditions are stated as

$$\mathbf{I}^+(z = 0 \text{ km}) = \psi I_b(300 \text{ K}) + \mathbf{R}_s \mathbf{I}_m^-, \tag{41}$$

$$\mathbf{I}^0(z = 8 \text{ km}) = I_b(2.7 \text{ K}) \times (1, 0, 0, 0)^T, \tag{42}$$

where the various terms are defined following Eqs. (33) and (34) in Section III.I.

B. SINGLE-SCATTERING PROPERTIES OF HYDROMETEORS

For the numerical examples described in the subsequent section, frozen hydrometeors are placed in the upper atmospheric layer and liquid hydrometeors in the lower layer. The radiative properties for a Marshall–Palmer distribution (Section III.C) of randomly oriented axisymmetric particles are computed using the T-matrix method (Chapter 6). The Marshall–Palmer distribution is transformed into a power law distribution so that the integrations required for averaging over size distribution may be done analytically (see Appendix C). The radiative properties for preferentially oriented axisymmetric particles are computed using the DDA (Chapter 5), and the integration over the Marshall–Palmer distribution is performed using Eq. (16). For all of the cases presented, equivalent-volume sphere diameters are used in the Marshall–Palmer distribution to describe the polydispersion. The complex refractive index of ice is computed according to Warren (1984) and of liquid water according to Ray (1972). All frozen hydrometeors are taken as pure ice with a density of 0.91 g cm^{-3}. For all species the lower and upper limits of integration are taken as $r_{min} = 0.054$ mm and $r_{max} = 6$ mm, respectively.

The rain rate R in the Marshall–Palmer distribution is hereafter referred to as the *nominal* rain rate, because it is a proxy parameter used to specify the drop size distribution and is not necessarily the *actual* rain rate that would result from Eq. (32). Some results for the radiative properties of randomly oriented liquid water drops at 19 GHz are shown as a function of the *nominal* rain rate R in Fig. 5. Because the particles are randomly oriented and axisymmetric, the radiative properties for the VRTE reduce to scalars. The results in Fig. 5 consider axis ratios of 0.5, 1, and 2 and show that, for a given volume of liquid water, all of the proper-

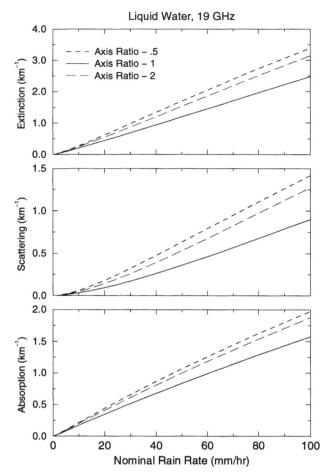

Figure 5 Radiative properties of a Marshall–Palmer distribution of randomly oriented axisymmetric water drops at a frequency of 19 GHz.

ties (extinction, scattering, and absorption coefficient) are largest for the prolate $(a/b = 0.5)$ particles, are slightly smaller for the oblate $(a/b = 2)$ particles, and are the smallest for the spherical $(a/b = 1)$ particles. The trends are the same for liquid water at 85 GHz, though the differences are most extreme for the 19-GHz case.

For ice, the scattering properties of ensembles of randomly oriented axisymmetric particles are not significantly different from those of an equivalent volume of spherical particles (results not shown). On the other hand, the scattering ma-

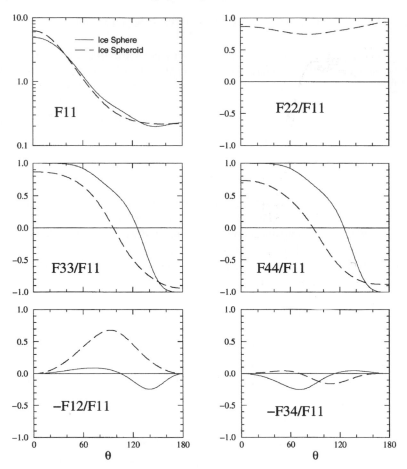

Figure 6   Scattering matrix elements for a Marshall–Palmer distribution of randomly oriented axisymmetric ice particles at a frequency of 85 GHz.

trix elements (Section XI of Chapter 1) can differ significantly. For example, the scattering matrix components shown in Fig. 6 for spheres versus oblate spheroids with $a/b = 5$ demonstrate that there are large differences at certain scattering angles for elements other than the F_{11} element.

For preferentially oriented nonspherical particles, very large differences in the single-scattering properties can occur for equivalent volumes of different particle shapes. For additional details, the reader should refer to Evans and Stephens (1995a).

C. Radiative Transfer Results and Discussion

For the results that follow, a Marshall–Palmer distribution of hydrometeors (using equivalent-volume sphere diameters) is placed into each layer as described, and the adding–doubling method (Evans and Stephens, 1991, 1995b) is used to solve the VRTE. The resulting Stokes parameters are transformed into equivalent blackbody temperatures, and results are presented for an upwelling viewing direction of 53° and frequencies of 19 and 85 GHz.

1. Randomly Oriented Axisymmetric Particles

The first example considers randomly oriented axisymmetric ice and liquid hydrometeors. A Marshall–Palmer distribution of ice described by the *nominal* rain rate R is placed into the upper layer, and the same distribution of liquid hydrometeors is placed into the lower layer. The shapes considered are prolate ($a/b = 0.5$), spherical ($a/b = 1$), and oblate ($a/b = 2$), with all particles within the distribution assuming the same shape (i.e., no distribution of shapes is considered). According to the formulas presented in Section III.D, raindrop axial ratios are typically between 0.5 and 1, so giving the entire distribution an axial ratio of 0.5 can be considered an extreme case. Ice particles can take on larger aspect ratios, so for this example, ice axial ratios of $a/b = 0.25$ and 4 are also considered.

In addition to the "base case" that considers rain and ice, cases with rain only and ice only are also considered. Several quantities are defined here for discussion purposes:

$$T_{v,\Delta} = T_{v,\text{sphere}} - T_{v,\text{spheroid}}, \qquad (43)$$

$$T_{h,\Delta} = T_{h,\text{sphere}} - T_{h,\text{spheroid}}, \qquad (44)$$

$$T_{Q,\Delta} = T_{Q,\text{sphere}} - T_{Q,\text{spheroid}}, \qquad (45)$$

where the subscripts v, h, and Q indicate brightness temperatures for vertical, horizontal, and vertical minus horizontal polarizations, and the subscripts "sphere" and "spheroid" indicate the particle shape used in the model atmosphere. In general, at 85 GHz, $|T_{v,\Delta}|$ and $|T_{h,\Delta}|$ are less than 2 K for all cases considered except for the extreme case that considers ice spheroids with axial ratios of 0.25 and 4. In this case, $|T_{v,\Delta}|$ and $|T_{h,\Delta}|$ range from about 5 to 10 K, but $|T_{Q,\Delta}|$ is always less than 2 K.

The most significant departure from spheres is for 19 GHz considering rain only or rain and ice, where $|T_{v,\Delta}|$ and $|T_{h,\Delta}|$ can reach values in excess of 5 K. Figure 7 shows plots of T_{19h} and $T_{19v,\Delta}$ as a function of R. The plot of T_{19h} shows that the brightness temperature increases more rapidly with rain rate for spheroids than for spheres. This is because the radiative characteristics for this case (Fig. 5) are larger for the spheroidal particles and, thus, the emission at a fixed low rain

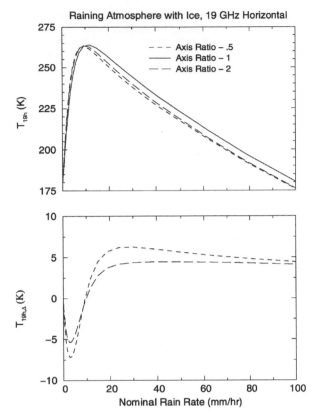

Figure 7   Upwelling 19-GHz brightness temperatures at a viewing angle of 53° for a raining atmosphere with randomly oriented axisymmetric ice and water particles. The atmosphere is described in the text and in Table III. T_{19h} is the upwelling horizontal brightness temperature, and $T_{19h, \Delta}$ is the difference between spheres and an equivalent volume of spheroids for the 19-GHz horizontally polarized brightness temperature.

rate is stronger for the spheroidal particles ($T_{19h, \Delta} < 0$). At larger rain rates, the opacity and scattering is greater for spheroidal particles, which leads to a larger depression than for spherical particles ($T_{19h, \Delta} > 0$). The results for vertical polarization are similar. Figure 8 shows T_Q for this case, and it also shows the difference $T_{Q, \Delta}$, which exceeds 2 K (to a maximum of nearly 6 K) for rain rates between about 1 and 10 mm/h. Because the raindrop axial ratios considered are extreme, the results should be interpreted as limiting cases.

A final set of cases were run using the previous set of atmospheres as input to the scalar RTE. For randomly oriented axisymmetric particles, the extinction, ab-

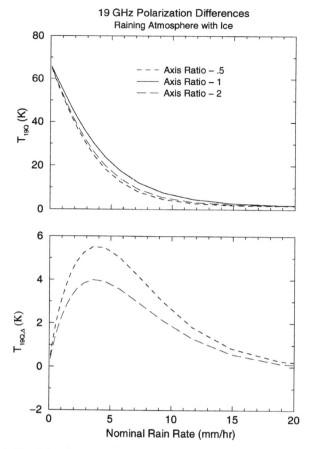

Figure 8 As in Fig. 7, but the polarization difference T_Q ($= T_v - T_h$) is shown. Thus, $T_{19Q,\Delta}$ is the difference of polarization signatures between spheres and an equivalent volume of spheroids for a frequency of 19 GHz.

sorption, and scattering coefficients are scalar quantities. For input to the RTE, the phase function (the [1, 1] element of the scattering matrix) is used, and the vertical or horizontal reflection coefficients (as appropriate) are used at the bottom boundary in order to compute T_v and T_h, respectively. For all of the cases considered, the results from the RTE were always within 2 K of the more rigorous VRTE results as long as the appropriate spheroidal particle properties were used.

It should be noted that this example only considered randomly oriented particles at two frequencies and over a Fresnel surface. It appears that randomly oriented spheroidal particles do not significantly impact radiative transfer results in comparison to results obtained using spherical particles. However, it would be worthwhile to perform similar numerical experiments using other surface types, additional microwave frequencies, and mixed-phase particles.

2. Preferentially Oriented Axisymmetric Particles

The second example considers ice particles oriented with their long dimension aligned parallel to the ocean surface in the layer from 4 to 8 km. The rain layer from 0 to 4 km is filled with spherical water drops with a constant *nominal* rain rate of $R = 2$ mm/h. Results are compared as a function of ice *nominal* rain rate for spherical ice versus solid ice columns. The aspect ratio for the columns is that from Eq. (21) and Table II, namely, $d = 0.260L^{0.927}$, though the density is fixed at 0.91 gm cm^{-3} for both particle shapes.

Results of the VRTE solution are depicted in Fig. 9. For a given ice *nominal* rain rate, a given percentage of spherical ice is converted to oriented columnar ice for particles with an equivalent spherical diameter of 1 mm or less [see also Evans and Vivekanandan (1990) and Roberti and Kummerow (1999) for similar modeling experiments]. The top panel of Fig. 9 shows how T_{85h} decreases for a fixed rain rate as the fraction of oriented ice columns is increased. The results for T_{85v} (not shown) are largely unaffected by converting spheres to columns. Presumably, this is due to the columns being oriented with their long dimension aligned with the horizontal plane, and thus, the relatively larger scattering and extinction properties along the line of sight (see Evans and Stephens, 1995a) lead to a more depressed horizontally polarized brightness temperature.

Because T_{85h} decreases as the fraction of columns is increased and T_{85v} remains approximately fixed, the polarization difference T_{85Q} (bottom panel of Fig. 9) increases with the fraction of columns. For the case of all spheres, the polarization difference is about 2 K or less for all values of R. However, this value increases by converting spheres to columns and reaches a value of nearly 25 K for a rain rate of about 1.7 mm/h with 100% of the spheres converted to columns.

V. CONCLUSIONS AND RECOMMENDATIONS

This chapter has considered scattering by nonspherical hydrometeors in the problem of passive microwave remote sensing, with particular emphasis placed on modeling microwave radiation propagating through layers of precipitation. Several studies dealing with this topic were reviewed in Section II. In general, most of the previous work has considered spheroidal ice and water particles oriented along an axis of symmetry. Some of the studies considered the effects of conical shaped hail and a few of the others considered dendritic ice crystals of various

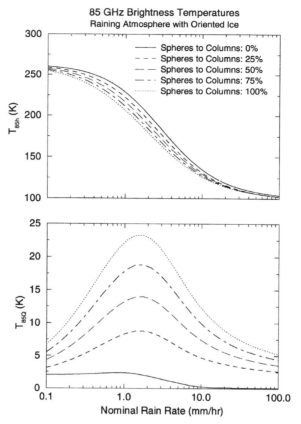

Figure 9 As in Fig. 7, but for oriented ice particles at a frequency of 85 GHz. The lower layer contains spherical water drops at a fixed rain rate of 2 mm/h. The percentage "Spheres to Columns" indicates the percentage of ice spheres at the given *nominal* ice rain rate that are converted to equivalent-volume horizontally oriented ice columns. The upper panel shows the upwelling horizontal brightness temperature T_{85h}, and the lower panel shows the upwelling polarization signature T_{85Q}.

shapes such as columns, plates, and needles. All of the studies found that, theoretically, the oriented nonspherical particles give rise to brightness temperatures that can differ considerably from models that use spherical hydrometeors only. However, the magnitude and sign of the difference between spherical and nonspherical model results depends heavily on frequency, viewing angle, and properties of the underlying surface. A seemingly important practical finding of these studies is that oriented ice and water particles can produce larger polarization signatures (the difference between the vertical and horizontal brightness temperature) than can be obtained from spherical particles. Observations of polarization signatures up to 13 K from the 85-GHz channel of the SSM/I within stratiform precipitation regions have been reported previously and there are accounts of 37-GHz polarization signatures up to 7 K from airborne radiometers observing tornadic storms over land. If these signatures are truly due to hydrometeors, the studies reviewed here indicate that modeling of such signatures requires the use of oriented nonspherical particles.

Section III provided the necessary mathematical framework for interested researchers to develop their own models of microwave radiative transfer through atmospheres containing nonspherical hydrometeors. The topics covered included the vector radiative transfer equation (incorporating thermal emission and multiple scattering); details regarding single-scattering properties; definition of size, shape, and orientation distributions; and information on the complex refractive indices of water and ice, modeling the surface, and modeling the atmosphere. Details on the numerical methods for solving the equation of transfer are not covered, but references are supplied. The computation of the particle radiative properties are covered in some of the other chapters in this book.

Finally, Section IV presented two examples of microwave radiative transfer through atmospheres containing nonspherical particles. The first example considered an atmosphere with randomly oriented axisymmetric ice and water. The results for this example indicated that, for 19 and 85 GHz over an ocean surface, nonsphericity has a small impact on the radiative transfer results and that there is not much difference between using the scalar and vector radiative transfer theory as long as the appropriate radiative properties are used. The second example considered an atmosphere containing horizontally oriented ice columns and spherical ice over a raining layer. The results for this example showed that the upwelling horizontally polarized 85-GHz brightness temperatures are up to nearly 20 K lower for a case with mixed spheres and columns versus a case with spheres alone. In addition, residual polarization signatures of 5–20 K were found. The results from this second example indicate that polarization signatures that are sometimes observed in precipitation events are very likely due to oriented hydrometeors. Furthermore, to correctly model this type of situation, one must use the rigorous vector radiative transfer theory.

Though several researchers have investigated the topic of nonsphericity in microwave radiative transfer in scattering precipitating atmospheres, there are still several important outstanding issues that require investigation. Foremost among these is the question "What impact does nonsphericity actually have on microwave precipitation algorithms?" The answer to this question, however, depends largely on the ability to correctly model nonspherical particles within microwave radiative transfer models.

One of the important modeling issues that still needs to be developed further is the incorporation of nonspherical particles into three-dimensional, time-dependent, cloud microphysical structures. In addition, it would be interesting to examine cases that include vertically aligned ice particles, which occur in lightning-producing thunderstorms. Another modeling issue worthy of consideration is the determination of whether detailed particle shape is important in microwave radiative transfer modeling [as an example, it might be possible to model dendritic ice columns with an "equivalent" spheroid—Schneider and Stephens (1995) have touched upon this issue for active remote-sensing applications]. Finally, it is extremely important to continue to gather microwave observations and coincident observations from an independent source to confirm model results. In this way, a better interpretation of microwave signals and an improved accuracy of geophysical retrievals will be attained.

VI. APPENDIX A. PARTICLE SIZE DISTRIBUTION: $N(r)$ VERSUS $N(D)$

To relate the independent variables r and D of a given distribution of particles, note that the total number of particles described by each distribution must be identical, that is,

$$\int N(r)\,dr = \int N(D)\,d(D), \tag{A1}$$

which implies that $N(r)\,dr = N(D)\,dD$. But, because $dD = 2dr$, we must have

$$N(r) = 2N(D). \tag{A2}$$

As an example, consider the Marshall–Palmer distribution given by $N(D) = N_0 \exp[-(\Lambda D)]$ with $N_0 = 8000 \text{ m}^{-3} \text{ mm}^{-1}$, $\Lambda = 4.1R^{-0.21} \text{ mm}^{-1}$, and R is the rain rate in millimeters per hour. Application of Eq. (A2) yields $N(r) = N_{0,r} \exp[-(\Lambda_r r)]$ with $N_{0,r} = 1.6 \times 10^4 \text{ m}^{-3} \text{ mm}^{-1}$ and $\Lambda_r = 8.2R^{-0.21} \text{ mm}^{-1}$.

VII. APPENDIX B. PARTICLE SIZE DISTRIBUTION: EQUIVALENT SPHERES

In this chapter, a particle is defined as being *equivalent* to a sphere if it has the same volume as a given sphere. However, other definitions of *equivalent* spheres appear in the literature, for example, definitions based on a sphere and spheroid having equal average projected areas. Let the quantity $N_c(r_c)$ be defined as the particle size distribution for an ensemble of spheroids, where r_c is the radius of a sphere that has a cross-sectional area equal to the average projected area of randomly oriented spheroids, and let $N(r_v)$ describe the same size distribution, but in terms of r_v, the radius of a sphere that has a volume equal to that of a given spheroid. The purpose of this appendix is to relate $N_c(r_c)$ to $N(r_v)$. First, the radius of a sphere that has a volume equal to that of a given spheroid is written as

$$r_v = \left(a^2 b\right)^{1/3}, \tag{B1}$$

where a and b are the lengths of the particle nonrotational and rotational (symmetry) semi-axes. For prolate spheroids, the radius of a sphere that has a cross-sectional area equal to the average projected area of randomly oriented spheroids is given by (Mishchenko and Travis, 1994c)

$$r_c = \frac{1}{2}\left[2a^2 + \frac{2ab}{e}\arcsin e\right], \tag{B2}$$

and for oblate spheroids by

$$r_c = \frac{1}{2}\left[2a^2 + \frac{b^2}{e}\ln\frac{1+e}{1-e}\right], \tag{B3}$$

where the particle eccentricity is defined as

$$e = \frac{(\xi^2 - 1)^{1/2}}{\xi}, \tag{B4}$$

and the particle axial ratio ξ is defined as $\xi = b/a$ for prolate spheroids and $\xi = a/b$ for oblate spheroids. The ratio of the equal-surface-area sphere radius to the equal-volume sphere radius for prolate spheroids is thus given by

$$\frac{r_c}{r_v} = \left[\frac{1}{2}\left(\xi^{-2/3} + \frac{\xi^{1/3}}{e}\arcsin e\right)\right]^{1/2}, \tag{B5}$$

and for oblate spheroids by

$$\frac{r_c}{r_v} = \left[\frac{1}{4}\left(2\xi^{2/3} + \frac{\xi^{-4/3}}{e}\ln\frac{1+e}{1-e}\right)\right]^{1/2}. \tag{B6}$$

To compute an expression involving $N(r_v)$ when given $N_c(r_c)$, r_c may be computed using Eq. (B5) or (B6) and then substituted into the expression involving $N(r_v)$. For example, the ice water content (IWC) is computed using

$$\text{IWC} = \rho_i \frac{4}{3}\int_{r_{v,\min}}^{r_{v,\max}} \pi r_v^3 N(r_v)\,dr_v, \tag{B7}$$

where ρ_i is the density of ice. Observing that

$$dr_v = \frac{r_v}{r_c}\,dr_c, \tag{B8}$$

the expression for ice water content may be rewritten in terms of $N_c(r_c)$ as

$$\text{IWC} = \rho_i \frac{r_v}{r_c} \frac{4}{3} \int_{r_{c,\min}}^{r_{c,\max}} \pi r_c^3 N_c(r_c)\, dr_c. \tag{B9}$$

VIII. APPENDIX C. USE OF POWER LAW DISTRIBUTION IN *T*-MATRIX METHOD

Mishchenko and Travis (1994c) suggested that the scattering properties of an ensemble of randomly oriented spheroids are nearly independent of the shape of the size distribution, but instead depend primarily on the effective radius and effective variance (r_{eff} and v_{eff}) of the distribution. Because the T-matrix method admits an efficient solution for integration over a power law size distribution, the resulting scattering computations for the power law distributions of spheroidal particles can be obtained much more rapidly than for other size distributions. For this paper, numerical experiments were performed to confirm whether this result also applies for microwave frequencies. This was done by specifying a gamma drop size distribution of spheroidal particles and then numerically integrating the scattering properties over the distribution to obtain the scattering properties of the ensemble. Simultaneously, r_{eff} and v_{eff} were computed and used as parameters of a power law distribution (as described by Mishchenko and Travis, 1994c), and the ensemble scattering properties were computed and compared with those obtained using the gamma distribution. It was found that, because the power law size distributions used in this paper had finite size limits, all radiative cross sections obtained using the two methods agreed well so long as those obtained using the power law distribution were multiplied by the ratio of average geometric cross-sectional areas computed using finite power law distribution versus infinite gamma distribution limits.

ACKNOWLEDGMENTS

Thanks to Frank Evans for providing his polarized radiative transfer and "dda" codes, to Mike Mishchenko for providing his T-matrix codes, and to Piotr Flatau and Bruce Draine for providing their "ddscat" code. Thorough reviews of the initial draft of this chapter were given by Joop Hovenier, Frank Evans, and Ted Smith. Witek Krajewski, Chris Kummerow, and Ted Smith have served as my mentors over the last several years. I also acknowledge valuable discussions with Harry Czekala, Joe Turk, Quanhua Liu, and K.-H. Ding. Kevin Berney helped with some of the computations and figures while he was a student in the 1996 NASA/GSFC Summer Institute on Atmospheric and Hydrospheric Sciences. Thanks to Jim Cornelius and others at FNMOC and the United States Navy for their support.

Chapter 18

Polarized Light Scattering in the Marine Environment

Mary S. Quinby-Hunt
Lawrence Berkeley National
Laboratory
University of California
Berkeley, California 94720

Patricia G. Hull
Department of Physics
Tennessee State University
Nashville, Tennessee 37209

Arlon J. Hunt
Lawrence Berkeley National
Laboratory
University of California
Berkeley, California 94720

I. INTRODUCTION

This chapter discusses polarized light scattering in the marine environment with emphasis on scattering from nonspherical particles. The marine environment is extensive; the oceans cover approximately 70% of Earth's surface. The marine environment encompasses not only the world's oceans, but the atmosphere above it as well as floating sea ice. Light propagating through the marine environment

Light Scattering by Nonspherical Particles: Theory, Measurements, and Applications

is scattered and absorbed by a variety of inhomogeneous components, including aerosols; suspended particles and bubbles in water; and occlusions, bubbles, and grain boundaries in sea ice. Many of these components are nonspherical in nature and their scattering deviates significantly from that predicted for spheres. An important key to understanding the presence and properties of nonspherical particles is to study the way they transform the polarization of light during scattering. This chapter contains a discussion of the methodology, measurement, and interpretation of polarized light scattering from nonspherical particles in the marine environment.

Aerosols in the marine atmosphere affect visibility, cloud formation, radiative transfer, and the heating and cooling of the earth. Aerosols alter the intensity and polarization of light reaching the sea and propagating over its surface. The effect aerosols have on light intensity and polarization depends on the shape, size, and refractive index of the aerosols and the angle between the incident light and the observer of the scattered light. Both the intensity and the polarization of light must be included for accurate radiative transfer calculations at the surface of and within the ocean (Plass *et al.*, 1981; Kattawar and Adams, 1989; Adams and Kattawar, 1993), calculations that are important in assessing the global heat budget. Kouzoubov *et al.* (1998) report that the polarization properties of light scattered from nonspherical particles significantly affect data interpretation for remote-sensing applications that rely on laser backscattering.

The angle and wavelength dependence of the radiant distribution of sunlight within the ocean is determined primarily by particulate scatterers. Ambient light in the ocean is significantly polarized with a strong dependence on the direction of observation (Waterman, 1954, 1988). The degree of polarization is generally the largest when viewed in the horizontal plane. It arises from the interaction of downwelling sunlight with biological, mineral, and anthropogenic scatterers. A major component of the scatterers are nonspherical, optically active, or exhibit some degree of nonrandom group orientation. This component includes marine organisms and their detritus as well as inorganic materials, such as sediments contributed by runoff and dust that has settled from the atmosphere. Polarization induced by scatterers not spherically symmetric also affects the propagation of synthetic coherent and incoherent light propagating in the ocean. For instance, populations of aligned asymmetric particles can reduce the intensity of linearly polarized light or induce phase delays in the propagation of coherent laser light. The difference between the polarization properties of suspended particles and viewed objects can be used to improve visibility and target discrimination by the use of linear or circular polarizers (Mertens, 1970). In addition to polarization effects, the scattering phase function for nonspherical particles can vary considerably from that predicted by spherical scatterers and lead to the misinterpretation of scattering and visibility data.

The propagation and scattering of sunlight in sea ice is critically important for a variety of life forms that live in the ice, on the ice, and under the ice. The variety and thickness of sea ice and the effects of snow, melt ponds, leads (open, linear cracks in sea ice), pressure ridges, and polynyas (open and freezing water areas surrounded by ice pack) have a considerable effect on the amount of sunlight reaching the water. Algae growing on the sea–ice interface contain a significant fraction of the biomass that forms the bottom of the food chain for all the organisms living in frozen seas. Sea ice itself is a complex material that varies tremendously depending on its formation, origin, and history. The scatterers in sea ice consist of brine pockets, air bubbles, and crystal boundaries, mostly nonspherical in nature. Much sea ice is in the form of large single crystals that are highly birefringent and can exhibit long-range ordering. Light-scattering studies of this complex medium are difficult and complicated by problems of sample preparation, sample preservation, and logistics.

In all three media, ensembles of nonspherical particles can exhibit nonrandom group orientation as a result of gravitational settling, fluid dynamic flows, phototaxis, turbulence, thermal or chemical gradients, freezing, biological interactions such as swarming behavior, and other orienting phenomena.

Nonspherical scatterers play important roles in the nature of light interaction and will be discussed in greater detail in Sections IV, V, and VI. Before discussing the measurements, Sections II and III discuss the formalisms most often used to describe polarized light scattering in the marine environment and the instruments used to measure it.

II. ANALYTICAL DESCRIPTION OF LIGHT SCATTERING

The effect of scattering on a beam of light propagating in a medium (in this case one of the marine environments) can be represented as a Mueller scattering matrix, hereafter "scattering matrix." This matrix is the primary tool used in this chapter to describe and analyze polarized light scattering in the marine environment.

A. SCATTERING MATRIX FORMALISM

Consider the scattering of polarized light from a suspension of particles. The intensity and polarization properties of any beam of light incident on or scattered by the suspension may be described by a four-element Stokes vector (Stokes, 1852) defined in terms of the complex electric fields $\mathbf{E}_l$ and $\mathbf{E}_r$ parallel and perpendicular to the scattering plane (van de Hulst, 1957; Kerker, 1969; Bohren and

Huffman, 1983):

$$I = \langle \mathbf{E}_l \mathbf{E}_l^* + \mathbf{E}_r \mathbf{E}_r^* \rangle \qquad \text{total intensity of light,}$$

$$Q = \langle \mathbf{E}_l \mathbf{E}_l^* - \mathbf{E}_r \mathbf{E}_r^* \rangle \qquad \pm 90° \text{ polarization,}$$

$$U = \langle \mathbf{E}_r \mathbf{E}_l^* + \mathbf{E}_r \mathbf{E}_l^* \rangle \qquad \pm 45° \text{polarization,}$$

$$V = \langle i \left(\mathbf{E}_l \mathbf{E}_r^* - \mathbf{E}_r \mathbf{E}_l^* \right) \rangle \qquad \text{circular polarization.}$$

The angular brackets indicate time averages and the asterisks denote complex conjugates. This definition is equivalent to that found in Chapter 1. However, it is not unique; other definitions are used (see, e.g., Jensen *et al.*, 1986). The effect of scattering by a suspension of particles on the Stokes vector may be represented by a 16-element scattering matrix that transforms the incoming light to scattered light. For example, the conventional scattering matrix for scattering by an ensemble of randomly oriented, nonoptically active particles is represented as [Eq. (61) of Chapter 1]

$$\begin{bmatrix} F_{11} & F_{12} & 0 & 0 \\ F_{21} & F_{22} & 0 & 0 \\ 0 & 0 & F_{33} & F_{34} \\ 0 & 0 & F_{43} & F_{44} \end{bmatrix}, \tag{1}$$

where $F_{21} = F_{12}$ and $F_{43} = -F_{34}$.

When there is no preferred orientation of the scatterers, each element of the scattering matrix is a function of the scattering angle, although not the orientation of the scattering plane. For the case of spherical particles the matrix element F_{22} equals F_{11} and F_{33} equals F_{44}. If the particles are very small or weakly scattering, F_{34} equals 0.

When the particles exhibit preferred orientation, the scattering matrix contains more independent terms. Each element of the scattering matrix may be a function of the orientation of the scattering plane as well as the scattering angle. With particle asphericity, F_{22} no longer equals F_{11}. Nonrandom orientation can result in nonzero off-diagonal block elements. The complex optical properties (birefringence, circular dichroism, optical rotary dispersion, etc.) of the scatterer also result in deviation of the off-diagonal block matrix elements from zero. In the most general case, all elements of the scattering matrix are nonzero. For example, the angle-dependent scattering from a single nonspherical object such as a dinoflagellate is extremely complex with 16 different matrix elements (Lofftus *et al.*, 1992).

The angle dependence of various elements of the scattering matrix either provide a direct measure of or are indicative of the physical properties of the scattering medium as is indicated by the following list:

F_{11} transformation of total intensity of incident light; element is proportional to the phase function; gives general size information;

F_{12} depolarization of linearly polarized light parallel and perpendicular to the scattering plane (F_{21} the converse); dependent on size, shape, and complex refractive index of the scatterers;

F_{13} depolarization of light linearly polarized $\pm 45°$ to the scattering plane (F_{31} the converse); orientation effects can cause nonzero elements;

F_{14} depolarization of circularly polarized light (F_{41} the converse); nonzero element indicates the presence of optical activity, helical structures; circular intensity differential scattering (CIDS) measures $(-F_{14}/F_{11})$; generally zero for nonoriented ensembles of particles;

F_{22} transformation of linearly polarized incident light ($\pm 90°$) to linearly polarized scattered light ($\pm 90°$); deviation from F_{11} is an important indication of the presence of nonspherical particles;

F_{33}, F_{44} transformation of linearly polarized incident light ($\pm 45°$) (circularly) to linearly polarized scattered light ($\pm 45°$) (circularly); deviation of F_{44} from F_{33} is indicative of nonspherical symmetry;

F_{34}, F_{43} transformation of circularly polarized incident light to linearly polarized scattered light ($\pm 45°$) (F_{43} is the converse); element is strongly dependent on size and complex refractive index of the scatterers.

The measurement and analysis of the form and details of the scattering matrix will comprise an important part of the remainder of this chapter.

B. MODELS

The most often used approach for predicting the intensity of light scattering and the scattering matrix elements is the Lorenz–Mie (hereafter Mie) calculation based on the solution to Maxwell's equations for a plane electromagnetic wave diffracted by a homogeneous sphere (Bohren and Huffman, 1983). Mie calculations are often used if the particles are sufficiently large and spherical to justify its use and sometimes even if they are not. Particles are considered to be non-spherical when F_{22}/F_{11} deviates significantly from unity at some angles and/or when the matrix elements that are identically zero for spheres deviate significantly from zero. Quinby-Hunt *et al.* (1994, 1997) found that if F_{22}/F_{11} is less than 0.9, agreement between the observation and the Mie calculation is poor.

The T-matrix method developed by Waterman (1965) and the coupled dipole method (otherwise known as the discrete dipole approximation; Chapter 5) developed by Purcell and Pennypacker (1973) have been used extensively for calculating the intensity and polarization of light scattering by nonspherical particles for many applications including the marine environment. The T-matrix method was used by Mishchenko and Travis (1994c) to calculate the scattering by a collection of spheroids. They concluded that even moderate asphericity leads to significant deviation from the results of Mie calculations. Because the deviations are more pronounced for larger scattering angles, Mie calculations are not appropriate in the analysis of remote-sensing data in which nonspherical particles play an important role. Kouzoubov *et al.* (1998) reviewed both the T-matrix and the coupled dipole methods, along with several others, for calculating the scattering matrix in laser remote sensing of ocean water. They, too, concluded that spheres are an inappropriate model for light scattering from ocean water because they are unable to mimic the behavior of the elements F_{22}, F_{33}, and F_{44}. They point out that the behavior of these elements for ocean water is consistent with asymmetric particle shapes, giving rise to depolarization of the scattered light. Variations on the coupled dipole approximation have been used by McClain and Ghoul (1986), Singham and Salzman (1986), Draine (1988), Singham (1988), Singham and Bohren (1988), Dungey (1990), Dungey and Bohren (1991), Hull *et al.* (1991, 1994), Shapiro *et al.* (1992, 1994b, c), and Draine and Flatau (1994). These and other mathematical models are discussed elsewhere in this volume (Chapters 4–7). Several approximations have been used for nonspherical scatterers including those used in the large-particle or geometric optics case (Chapter 2). However, although these approaches can predict diffraction intensities, they do not predict well the complex angle-dependent polarization transformation properties of large objects and are not considered here. It is worth mentioning that some limitations on the size of the particle to be modeled are relaxed if the particles are immersed in water because of the reduced relative index of refraction. For example, ocean microorganisms without carbonate or siliceous tests (skeletal debris of microorganisms) often have relative real indices of refraction that range from 1.02 to 1.15.

Methods for using the various models to analyze experimental data have been explored. Some earlier research on light scattering in the marine environment studied only the scattering and extinction efficiencies; other investigators fitted only the scattering phase function. In more rigorous approaches, the results of analytical models are compared with the data, by iterating model inputs (e.g., size, complex refractive index, shape) until the angular dependence of several elements of the scattering matrix are matched simultaneously. Techniques for simultaneously fitting the angle dependence of several curves are being developed using neural networks and other techniques based on the Levenburg–Marquant optimization (Press *et al.*, 1989; Hull and Quinby-Hunt, 1997; Hunt *et al.*, 1998; Ulanowski *et al.*, 1998).

C. OPTIMIZING DATA FOR MIE CALCULATIONS

If measurements indicate that the scatterers do not deviate severely from sphericity (i.e., if $F_{22} \approx F_{11}$) it is possible to quantify the effect of the non-spherical contributions on F_{11} and F_{12} and use this information to provide better fits to the measurements (Quinby-Hunt *et al.*, 1986). This can be seen by noting that the scattering matrix elements, F_{11}, F_{22}, and F_{12} are defined as real functions of the complex coefficients that transform the electric fields during scattering. (All the matrix elements are real functions, but not as simply defined.) The equations describing the transformation of light polarized parallel, $\mathbf{E}_l$, and perpendicular, $\mathbf{E}_r$, to the scattering plane are given as (van de Hulst, 1957; cf. Section 4 of Chapter 1)

$$\mathbf{E}'_l \propto A_2 \mathbf{E}_l + A_3 \mathbf{E}_r, \tag{2}$$

$$\mathbf{E}'_r \propto A_4 \mathbf{E}_l + A_1 \mathbf{E}_r. \tag{3}$$

The complex coefficients A_i are related to the Mueller matrix by the real quantities $m_i = |A_i|^2$. The three scattering matrix elements of interest for this calculation can be written as

$$F_{11} = \frac{1}{2}(m_1 + m_2 + m_3 + m_4), \tag{4}$$

$$F_{12} = \frac{1}{2}(m_2 - m_3 + m_4 - m_1), \tag{5}$$

$$F_{22} = \frac{1}{2}(m_2 - m_3 - m_4 + m_1). \tag{6}$$

For particles with spherical symmetry, $m_3 = m_4 = 0$; therefore $F_{11} = F_{22} = \frac{1}{2}(m_1 + m_2)$ and $F_{12} = \frac{1}{2}(m_2 - m_1)$. For an ensemble with no preferred orientation, $m_3 = m_4$. However, it is possible to adjust the measured matrix elements from slightly nonspherical particles to make them *consistent* with spherical scatterers. With this assumption, Eqs. (4)–(6) are solved for m_1, m_2, and m_3 in terms of the experimental values of F_{11}, F_{12}, and F_{22}. The new data sets are consistent with Mie scattering calculations; F_{22} is unity for all angles. This same procedure is difficult to apply for the other scattering matrix elements because they are derived from combinations of products of complex quantities that cannot be determined from the experimental data for ensembles of particles. This is illustrated in the development of inequality relationships between the matrix elements (Fry and Kattawar, 1981; Kattawar and Fry, 1982; Chapter 3). In addition to facilitating comparison of experimental data to the results of Mie calculations, this procedure can be used to assess the feasibility and propriety of attempting modeling with Mie calculations.

D. Physical Interpretation of Constraints on Scattering Matrices

A physically realizable Stokes vector must satisfy the condition $I^2 \geq (Q^2 + U^2 + V^2)$ (Born and Wolf, 1970). This inequality can be expressed as

$$P = \frac{\sqrt{Q^2 + U^2 + V^2}}{I} \leq 1, \qquad (7)$$

where P is the degree of polarization of the resulting light (Bohren and Huffman, 1983). When P equals 1, the light is entirely polarized. A scattering matrix is unphysical if it can produce an unphysical Stokes vector by operating on a physically realizable Stokes vector. Kostinski *et al.* (1993) developed a set of necessary conditions to test whether a scattering matrix is physically realizable. Hovenier and van der Mee (1996) provide an extensive investigation of scattering matrices. Fry and Kattawar (1981) and Kattawar and Fry (1982) derived a set of seven inequalities that can also be used for testing a given scattering matrix. Miller *et al.* (1997) also developed a set of three inequalities similar to those of Fry and Kattawar and used them to check whether the scattering matrices determined in their studies are physically realizable. However, although the inequalities of Miller *et al.* can be used for testing whether scattering matrix elements are physically real, they are chosen primarily to provide physical insight into the nature of the observed scattering.

The Stokes vector resulting from the interaction of linearly polarized light, $(1, \pm 1, 0, 0)^T$, with an arbitrary scattering matrix F_{ij} must satisfy $(F_{11} \pm F_{12})^2 \geq (F_{21} \pm F_{22})^2 + (F_{31} \pm F_{32})^2 + (F_{41} \pm F_{42})^2$ according to Eq. (7). (Hereafter ^{T} denotes the matrix transpose.) Fry and Kattawar explicitly derive these two inequalities by mathematically considering the consequences of ensemble averaging the scattering matrix of an individual scatterer. Four more inequalities can be generated by considering the interaction between light linearly polarized at $\pm 45°$ to the scattering plane, $(1, 0, \pm 1, 0)^T$, and circularly polarized light, $(1, 0, 0, \pm 1)^T$, with an arbitrary scattering matrix. We can express all these relationships as

$$P_{\text{lin}} = \sqrt{\frac{F_{21}^2 + F_{22}^2 + F_{31}^2 + F_{32}^2 + F_{41}^2 + F_{42}^2}{F_{11}^2 + F_{12}^2}} \leq 1, \qquad (8)$$

$$P_{\text{ellip}} = \sqrt{\frac{F_{21}^2 + F_{23}^2 + F_{31}^2 + F_{33}^2 + F_{41}^2 + F_{43}^2}{F_{11}^2 + F_{13}^2}} \leq 1, \qquad (9)$$

$$P_{\text{circ}} = \sqrt{\frac{F_{21}^2 + F_{24}^2 + F_{31}^2 + F_{34}^2 + F_{41}^2 + F_{44}^2}{F_{11}^2 + F_{14}^2}} \leq 1. \qquad (10)$$

These three inequalities provide necessary but not sufficient conditions as to whether a given scattering matrix is physically realizable. Violation of any or all of the inequalities implies a nonphysical scattering matrix.

These inequalities can *also* be interpreted physically because of the degree of polarization of the scattered light resulting from the interaction of fully polarized incident light with the scattering system. Therefore P_{lin}, P_{ellip}, and P_{circ}, which we refer to as the polarizance functions P_i, can be interpreted as the degree to which a given scattering system will depolarize linearly polarized light at $0°$, linearly polarized light at $45°$, and circularly polarized light, respectively. The function being equal to zero... (unity) indicates completely unpolarized (polarized) light. This technique is particularly useful when interpreting the scattering behavior of sea ice.

III. EXPERIMENTAL MEASUREMENT TECHNIQUES

There has been scientific interest in quantifying light scattering in the marine environment for centuries. The earliest reported measurements of the clarity of seawater relied on the determination of the maximum depth a standardized disk could be observed from the surface of the sea. The method was first used extensively by Secchi in the 1860s and is still occasionally used as a rough measure of sea clarity. The technique that now carries Secchi's name was purportedly originated by one Captain Bérnard. When sailing from Wallis Island to the Mulgraves, he observed a dish caught in a net at a depth of approximately 40 m (Cialdi, 1866; Tyler, 1968). More recently, a wide variety of extinction and angle-dependent scattering measurements have been performed by a number of investigators. Most measured only the phase function (total intensity or F_{11}) or fixed polarization as a function of angle. The reports are far too numerous to recount here. However, interest in the marine environment prompted some of the earliest and most sustained efforts to characterize the entire scattering matrix.

The angle dependence of the elements of the scattering matrix in seawater was measured first by subtraction methods (Beardsley, 1968; Carder *et al.*, 1971; Pak *et al.*, 1971; Kadyshevich *et al.*, 1971, 1976; Kadyshevich and Lyubovtseva, 1973). These early methods relied on pairs of intensities recorded with different polarizing elements; the results were subtracted to obtain the matrix elements. This technique lacks sensitivity because of the need to determine small differences between two large signals. It is also difficult to determine the absolute phase of the signal and hence the sign of the polarization. Hunt and Huffman first developed the technique of polarization modulation based on an acoustooptic modulator combined with intensity normalization in 1973. This technique is much more sensitive than subtraction methods and has been used by a number of researchers (Perry *et al.*, 1978; Thompson *et al.*, 1978, 1980; Bottiger *et al.*, 1980; Voss and Fry, 1984; Fry and Voss, 1985; Hunt and Huffman, 1973, 1975; Quinby-Hunt *et al.*, 1986, 1989, 1997; Quinby-Hunt and Hunt, 1988; Lofftus *et al.*, 1988, 1992; Hunt *et al.*, 1990; Miller *et al.*, 1994, 1996, 1997). Variations of this technique were also used for studying scattering from captured and cultured organisms in the sea (Maestre *et al.*, 1982) and two-dimensional representations of the scattering matrix being developed for cancer studies (Hielscher *et al.*, 1997).

The angle-scanning polarization-modulated nephelometer (Hunt and Huffman 1973, 1975; Hunt 1974) is described here because the description applies to many of the later instruments as they rely on similar principles. This instrument permits measurement of all 16 elements of the scattering matrix from near forward ($\sim$3°) to near backward ($\sim$172°) scattering. The nephelometer (Fig. 1a) uses a continuous-wave (CW) laser beam that passes through a linear polarizer and a polarization modulator and then to the sample region. The photoelastic modulator varies the polarization phase of the incoming light at a rate of 50 kHz. The scattered light is collected by a rotating arm containing selectable phase retarders and linear polarizers and passes into a photomultiplier tube. The alternating-current (ac) signals resulting from the modulator are detected by synchronous amplifiers tuned to the first and second harmonics of the modulator frequency. By varying the orientation of the modulator and its harmonic and the choice of filters and retarders, measurements of all 16 elements of the scattering matrix are made as a function of angle. Some elements are detected directly and others as linear combinations of measurements.

In this instrument, all the matrix elements except F_{11} are normalized by F_{11} by the electronics at the angle of measurement so as to range from $+1$ to -1, corresponding to the maximum value of the polarization transformation (the normalized elements are designated by italicized bold type as $\boldsymbol{F_{xy}}$). Unnormalized scattering signals typically range over many orders of magnitude because of the variation in the scattering phase function and scattering volume, making the variation in the signal due to polarization difficult to distinguish from the total intensity. After normalization, the value of $\boldsymbol{F_{22}}$ for spherical particles is $+1$ and drops well below this value for nonspherical particles.

Synchronous detection is important when the polarization signals are small. In cases where the off-diagonal block elements are nonzero because of optical activity, partial alignment of nonspherical particles, or particle birefringence, typical normalized signals for scattering from an ensemble are less than 5% and are very difficult to observe by subtraction methods because the signals often vary over many orders of magnitude. The ability of the instrument to measure the signal directly, instead of as a difference between two large signals, provides enhanced sensitivity. Synchronous detection also provides phase information (the sign of the signal) that is lost when using subtraction methods. This phase information

a. Monostatic nephelometer - cutaway

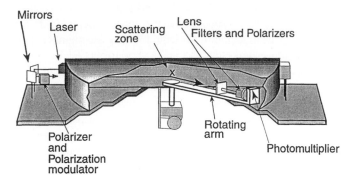

b. Bistatic Nephelometer - schematics, top, side views

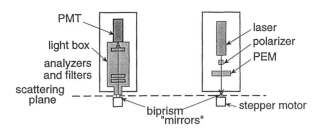

Laser beam travels into and detector views perpendicular to the plane of paper

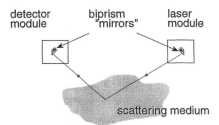

Figure 1 Experimental apparatus. (a) The scanning polarization-modulated nephelometer as designed by Hunt and Huffman (1973; Hunt, 1974; adapted from Hunt and Huffman, 1973). (b) Schematic and cross-sectional views of the bistatic nephelometer as configured for the study of sea ice (adapted from Miller *et al.*, 1996).

gives the sign of the change in polarization and is therefore very important in interpreting scattering data.

To probe at a distance, for example, into sea ice *in situ* or at a distance over the ocean, it is necessary to reconfigure the nephelometer. To do this, a "bistatic" nephelometer was designed and fabricated. This instrument (Fig. 1b), developed at the E. O. Lawrence Berkeley National Laboratory (LBNL), has been described in detail elsewhere (Miller *et al.*, 1994, 1996, 1997). Collimated laser light passes through a linear polarizer and photoelastic modulator that modulates the polarization state of the laser beam. Biprism optics moved by computer-controlled stepper motors direct the beam at varying angles into the medium where it is scattered. The light is collected by a second set of computer-controlled biprism optics and analyzing filters by a photomultiplier. The source and detector optics of the nephelometer may be separated to appropriate distances to permit a range of path lengths and scattering angles. Research at LBNL has shown that, with appropriate volume corrections, results using the bistatic nephelometer are in good agreement with those using the monostatic nephelometer although the changing nature of the scattering volume complicates interpretation of the phase function. The normalization procedure eliminates this difficulty with all other scattering matrix elements.

The bistatic nephelometer is designed for flexible sampling strategies because both the laser and the detector angles can be varied. The simplest case is to scan either the laser or the detector angle while leaving the other fixed. Another scanning strategy is to adjust the laser and detector mirror angles so that the optical path length traversed by the light is kept constant. The locus of scattering volumes then follows a pseudo-elliptical trajectory. An advantage of this scanning strategy is that the total optical path length is independent of scattering angle. Another scanning strategy permits the sampling point of the beam to traverse a line parallel to the instrument axis. Several of these strategies can be combined to scan over a horizontal plane of arbitrary height in the atmosphere to provide visibility/aerosol maps.

IV. POLARIZED LIGHT SCATTERING IN THE MARINE ATMOSPHERE

Holland and Gagne (1970) and Koepke and Hess (1988) have stated that nonspherical particles have an important contribution to polarized scattering behavior in the atmosphere, suggesting that this sort of behavior should be considered in a description of atmospheric scattering. Polarized light scattering in the marine atmosphere arises from aerosols that are largely spherical because of the prevalence of seawater-based particles; however, significant components of the aerosols can be nonspherical. Aerosols in the marine atmosphere result from the physical, chemical, and biological conditions of the underlying ocean; meteorological conditions (temperature, humidity, wind, and cloud cover); tropospheric settling; anthropogenic sources; and terrestrial (both natural and anthropogenic) inputs. Marine sources include condensed water vapor and condensed gases from biological ocean sources and aqueous salt aerosols from the wind–sea interface.

Several complex components make up the aerosols in the marine atmosphere. The distribution of particles is most often described mathematically as sums of log-normal distributions of the various components making up the marine aerosol (Shettle and Fenn, 1979; Hoppel *et al.*, 1990; Gathman and Davidson, 1993). The sea-salt component is believed to comprise the coarser fraction (r_{mode} generally >0.3 μm) of marine aerosols (Fitzgerald, 1991; Woolf *et al.*, 1988), whereas non-sea-salt sulfates, continental aerosols, and organic matter contribute to the smaller fractions ($r_{mode} \sim 0.03$ μm) (Mészáros and Vissy, 1974; Shettle and Fenn, 1979; Andreae, 1986; Charlson *et al.*, 1987). (r_{mode} is the radius of the particles whose number density is greatest in a log-normal distribution.) Under normal conditions, the main constituents of aerosols over the ocean include water, sea salt, non-sea-salt sulfate, mineral dust, a small fraction of nitrates, and organic matter (Hoppel *et al.*, 1990; Fitzgerald, 1991; Junge *et al.*, 1969; Junge, 1972; Toon and Pollack, 1976; Gras and Ayers, 1983; Koepke and Hess, 1988; Bates *et al.*, 1992; Novakov and Penner, 1992).

Of these, the most obviously nonspherical are mineral dust, anthropogenic particles and particles settling from the troposphere; however, under the right circumstances even the sea-salt component may become nonspherical. The sea-salt-containing (SSC) component of marine aerosols is produced primarily by bubble bursting and wave sheering (Duce and Woodcock, 1971; Wang and Street, 1978). Bubble bursting forms mostly spherical droplets only a few millimeters from the surface, but they are often advected up to 2 km above the surface (Junge, 1972; Woolf *et al.*, 1988). Humidity affects the size, shape, and homogeneity of these particles (Shettle and Fenn, 1979; Hänel, 1976; Kapustin and Covert, 1980). If the humidity becomes low enough, nonspherical particles may precipitate from the aqueous aerosols.

A. LIGHT SCATTERING FROM SEA-SALT-CONTAINING AEROSOLS

Few studies have investigated the polarization properties of light scattered by marine aerosols and even fewer used the fullest description of the polarization as embodied by the Mueller scattering matrix formalism (Deirmendjian, 1969; Perry *et al.*, 1978; Quinby-Hunt *et al.*, 1994, 1997). Because of the difficulty in calculating scattering from nonspherical particles, many researchers make the assumption that the particles making up the scattering medium are spherical. Deir-

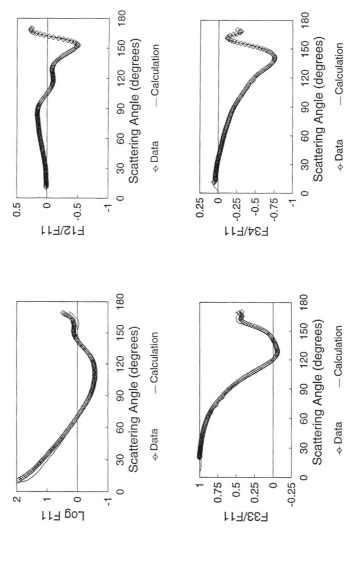

Figure 2 Angular distribution of the scattering matrix elements F_{11}, F_{12}, F_{33}, and F_{34} measured for seawater aerosols compared with calculations using an adaptation of the Mie model in Bohren and Huffman (1983) (adapted from Quinby-Hunt et al., 1997). The size distribution was modeled as a log normal with $r_{mode} = 0.7$ μm, $\sigma = 0.75$, and $m = 1.34$. The seawater was collected off northern California and had a salinity of 32.5 parts per thousand.

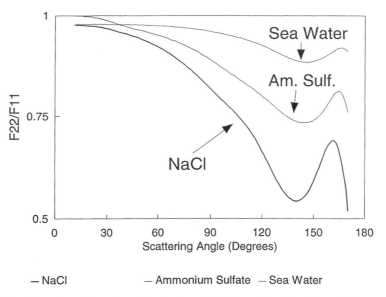

Figure 3   Comparison of experimental scattering matrix element F_{22} for water-containing aerosols of NaCl at 105°C, $(NH_4)_2SO_4$ at 180°C, and seawater at 140°C. The gas-aerosol temperature was elevated until scattering behavior indicated significant deviation from sphericity.

mendjian's (1969) compendium predicted scattering behavior from the assumed characteristics of hazes and fogs. In the marine environment under many conditions, this is reasonable. Figure 2 shows typical angular distributions of four scattering matrix elements for an SSC aerosol measured in the laboratory. The experimental results for F_{11}, F_{12}, F_{33}, and F_{34} (shown) are well described by an ensemble of log-normal distributions of spheres with $r_{mode} = 0.7$ μm (the standard deviation σ was 0.75), and the relative refractive index m of the aerosol particles was $1.34+0i$. F_{22} was greater than 0.95 for all scattering angles less than 150° and greater than 0.90 over the entire measurement range. The off-diagonal block elements were zero. Such scattering would be typical for this component in the marine atmosphere under most marine conditions.

Under the right conditions, the SSC component of aerosols can be nonspherical (Perry *et al.*, 1978). In laboratory studies and field observations with SSC component aerosols, Mészáros and Vissy (1974) found particles whose morphology indicated that they were mixtures of various seawater components that, as they dried, crystallized into different shapes. Laboratory studies of marine SSC aerosols demonstrated that decreasing humidity could strongly affect polarized light scattering. Quinby-Hunt *et al.* (1997) measured the polarized light scattering by aerosols of pure water, several salt solutions, and seawater. Figure 3 shows the extreme deviation of F_{22} from 1 observed when conditions simulating lower humidity are reached, causing formation of nonspherical particles.

B. NATURAL AND ANTHROPOGENIC TERRESTRIAL AEROSOLS

Natural and anthropogenic terrestrial aerosols contribute a large fraction of the nonspherical particles to the marine atmosphere. The limited research on the polarization properties of scattering by atmospheric aerosols has focused on terrestrial aerosols and most of the past measurements of the polarization properties of atmospheric aerosols involved lidar measurements that only consider backscatter and are not treated here (see Chapter 14). Hansen and Evans (1980) measured the angle dependence of four elements of the scattering matrix of urban aerosols. The aerosols were drawn into a laboratory nephelometer and presented without normalization or phase information. The authors did not discuss the data in detail, but reported that they generally agreed with that presented by Holland and Gagne (1970) and Perry *et al.* (1978). Veretennikov *et al.* (1979) reported the angle-dependent intensities and degree of linear polarization from wood smoke with varying humidity but did not address the issue of sphericity nor did they discuss the complete scattering matrix. Koepke and Hess (1988) and Mishchenko *et al.* (1997a) discussed and demonstrated the importance of shape when considering aerosol phase functions (proportional to the angular dependence of F_{11}), concluding that the shape of the scatterer is too often ignored.

Even less research has been conducted on the contributions of natural terrestrial or anthropogenic aerosols to polarized light scattering in the marine environment. That terrestrial aerosols must be considered as important contributors to light scattering in the marine atmosphere is demonstrated by studies that found such particles in the marine atmosphere, such as Sahara dust and volcanic ash, at great distances from the source (Koepke and Hess, 1988; Hoppel *et al.*, 1990; Fitzgerald, 1991; Novakov and Penner, 1992). Parungo *et al.* (1992) collected soot particles from Kuwaiti oil fires in 1991 at distances more than 500 km from the source.

In coastal areas and near shipping lanes, soot can be an important component of marine aerosols. Carbonaceous soot is one of the most prevalent contributors of particulates in the atmosphere and much of it is believed to derive from diesel engine combustion. Diesel contributions can be expected in the marine atmosphere as well because of the widespread use of diesel power by shipping, midsized pleasure and fishing craft, marine diesel generators, and compressors. This component may have an important effect on polarized light scattering in the atmosphere because the real and imaginary parts of the refractive index are large. To further complicate matters, the measured values for the refractive indices of soot vary considerably with the source of data, which probably indicates real variation among soots depending on the source. The real refractive indices of soots may vary from 1.33 to more than 2; the imaginary refractive indices can range from 0.12 to 0.9 (Hodkinson, 1964; Dalzell and Sarofim, 1969; Arakawa *et al.*, 1977; Pluchino *et al.*, 1980; Batten, 1985). This variation is probably due to determinations based on measurements of soots that are non-fully dense agglomerations of primary carbon particles or the presence of homogeneous low- to nonabsorptive products in the aerosol.

To explore the effect of this widely occurring and highly absorptive component in the marine atmosphere, we have calculated scattering from carbon soot using coupled dipole computer codes (Fig. 4). Each "soot" particle was represented by a cluster of smaller primary particles. The codes summed the contributions of 50 soot clusters, each cluster being composed of 200 randomly connected primary particles. Each of the 50 clusters differed from the others. The radius of gyration of the soot clusters was roughly 0.1–0.2 μm and the calculations were for a wavelength of 532 nm. At this wavelength, based on measurements of graphitic carbon, the refractive index of the primary particles that make up each particle of soot was estimated to be $1.8 + 0.7i$.

The calculations were compared to measurements made from the exhaust from a modern, one-cylinder, direct-injection diesel generator set (genset) under full load (3 kW maximum continuous power). Such a genset is typical of those used on small or midsized marine vessels. To determine r_{mode} and m, the data were simultaneously fit using the programs based on the Levenburg–Marquant optimization discussed previously (Hunt *et al.*, 1998) with computer codes that use ensembles of log-normal distributions of Mie spheres (such a computation is at present too computer intensive to use with the coupled dipole codes, although such a procedure is under development). These scattering data were fit with a log-normal distribution of spheres with $r_{mode} = 0.05$ μm and an m of $1.34 + 0.2i$ (Hunt *et al.*, 1998). The engine has a clean exhaust stream with no particles readily apparent. Probably because of the small size parameter, the particles generated by the modern direct-injection engine could be treated as if they originated from nearly spherical scatterers (electron micrographs show them to be extremely small agglomerations of even smaller primary particles forming a somewhat nonspherical cluster). In fact, Mie calculations predict F_{12} and F_{34} better than the coupled dipole calculations do. This apparent discrepancy can be attributed to the preliminary nature of the coupled dipole computations to date. Nonetheless, the data show that the soot clusters are not strictly spheres. F_{22} deviates from 1 (Fig. 4). This result is predicted by the coupled dipole calculation. Both the measurements and the coupled dipole calculations show that F_{33} differs from F_{44}, and neither is -1 at 180°, which are further indications of asphericity.

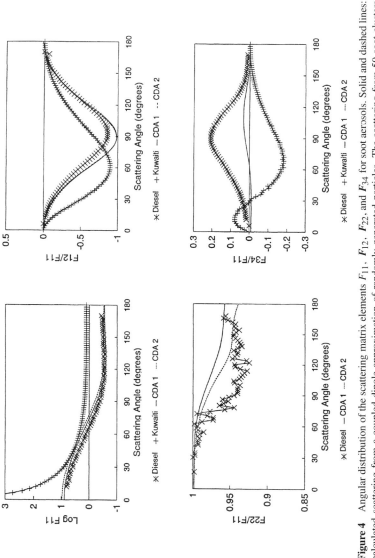

Figure 4 Angular distribution of the scattering matrix elements F_{11}, F_{12}, F_{22}, and F_{34} for soot aerosols. Solid and dashed lines: calculated scattering from a coupled dipole approximation of randomly generated particles. The scattering from 50 soot clusters composed of 200 primary particles was summed. The average radius of gyration for the particles was 0.1–0.2 μm. The wavelength was 532 nm and the complex refractive indices were $1.8 + 0.7i$ (solid line) and $1.8 + 0.9i$ (dashed line). Kuwaiti oil-fire soot was modeled as a log-normal distribution with a modal radius of 0.2 μm, standard deviation of 1.35, and $m = 1.75 + 0.45i$ (size distribution based on Parungo et al., 1992; refractive index based on Shettle and Fenn, 1979). Experimental data from diesel engine exhaust measured in the laboratory are also shown.

With older diesels under less favorable conditions, incomplete combustion might be expected, resulting in larger, more nonspherical soots, such as those reported by Lipkea *et al.* (1979) and Veretennikov *et al.* (1979) or the Kuwaiti soots. Therefore, polarized scattering was calculated using as input the distribution of particles measured downwind from the oil fires in Kuwait with the Mie codes (Quinby-Hunt *et al.*, 1997). Although a summation using the coupled dipole codes would be more appropriate, the computation was too computer intensive at the time. The results of the Mie calculations can provide useful insight into the expected effects of these large absorbing particles on polarized light scattering in the marine atmosphere with the caveat that perturbations resulting from their asphericity may be important. Parungo *et al.* (1992) reported a broad distribution of carbon particle sizes, which were approximated here as a broad log-normal distribution adjusted to roughly fit the number distribution reported, which was at 3.7 km altitude and 160 km from Kuwait. In this case, the modal radius for the log-normal distribution was 0.2 μm, $\sigma = 1.35$, and the refractive index $1.75 + 0.45i$. With the broad size distribution from the Kuwaiti fires and the significant contribution from large particles, the forward scattering component of the total intensity is significantly greater (Fig. 4). The peak linear polarization as seen in F_{12} is somewhat reduced and the maximum linear polarization occurs at 60° rather than at 90° as is seen in the diesel exhaust and the coupled dipole computation. Over most of the angular scan, F_{34} for the Kuwaiti soot is primarily negative, whereas both the coupled dipole calculations and the diesel exhaust are primarily positive.

V. POLARIZED LIGHT SCATTERING IN THE SUBMARINE ENVIRONMENT

In the underwater world, polarized light plays a complex and important role. A high density of particles (compared to the atmosphere) causes absorption, scattering, and other effects to produce an environment that contains an abundance of polarized light. The complex interaction between light and particles (living organisms, detritus, and inorganic matter) in the ocean has significant consequences for marine inhabitants and for underwater visibility and imagining.

An understanding of the effect of polarized light in the ocean on the behavior of marine organisms is an integral part of unraveling questions in marine biology. Some marine organisms are known to have vision that is sensitive to polarized light (Waterman, 1954). Crustaceans and cephaloids have built-in dichroic channels in their retinas that allow them to detect the linear polarization of light in the ocean and use it for orientation (Waterman, 1954, 1988; Stachnik, 1988). Waterman (1988) has shown that herring larvae and the shrimplike neomysis are also sensitive to light polarization. There is strong evidence that even marine productivity is affected by the presence of circularly polarized light. Right circularly polarized light increases net photosynthesis and chlorophyll *a* production, whereas left circularly polarized light has the opposite effect in *Dunaliella euchlora* (MacLeod, 1957). This effect may arise from inherent circular dichroism in the pigments of the algae studied. Circular dichroism in chlorophyll-containing organisms could increase circularly polarized light in the ocean.

Observation of various organisms illuminated by polarized light combined with knowledge of their polarization-sensitive vision suggests that nature has developed some interesting survival strategies to take advantage of the polarized light environment. Certain varieties of crustaceans including some shrimp and mysids have nearly transparent bodies that cause them to have lower visibility to their predators. These same organisms have a much greater visible contrast with the background when viewed between crossed polarizers (the authors, unpublished results, with Diane Carney, Moss Landing Marine Laboratory, San Jose State University, CA). Thus, such organisms illuminated with naturally polarized light and viewed with polarized eyesight may be much more visible to each other than they are to their predators. This could have important ramifications to the mating and swarming behavior of these organisms.

Polarized light scattering has been used to improve contrast in imaging, mapping, and underwater viewing systems (Briggs and Hatchett, 1965; Solomon, 1981; Hallock and Hallajian, 1983; Cariou *et al.*, 1990). Contrast is improved when the polarization states of light scattered from an object differ from those of the light scattered from the surrounding medium (Hallock and Hallajian, 1983). The use of linearly polarized light improved visibility by 20% in muddy waters (Briggs and Hatchett, 1965). Circularly polarized incident light sources improve underwater contrast further (Gilbert and Pernicka, 1966; Mertens, 1970). Polarization effects must be accounted for accurately in radiance calculations (Adams and Kattawar, 1993; Kattawar and Adams, 1989, 1990).

To understand the origin of polarized light in the ocean, we must first examine the scattering matrix for seawater. On-station measurements of all 16 elements of the scattering matrix were reported by Fry, Voss, and others (Voss and Fry, 1984; Fry and Voss, 1985; Thompson *et al.*, 1978) using an elaborate version of polarization-modulated nephelometry (some of their data are presented in Fig. 5). They suggest that the scattering properties of seawater resemble those of Rayleigh–Debye scattering (sometimes alternatively called Rayleigh–Gans or Rayleigh–Debye–Gans scattering); that is, they are characteristic of weakly interacting particles. In the scattering matrix this means that the *normalized* elements resemble those of Rayleigh scattering but with a much more forward scattering phase function than in the Rayleigh case. This leads to the significant polarization of light scattered from downwelling sunlight.

However, the *differences* between the observations and Rayleigh–Debye calculations reveal much about the nature of the particles in the ocean (refer to Fig. 5).

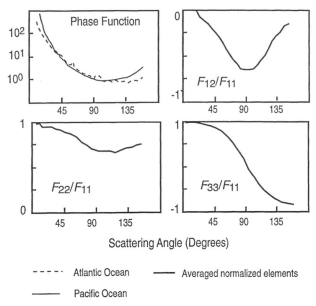

Figure 5   Experimental measurements of the scattering matrix of seawater averaged from the samples collected in the Atlantic and Pacific oceans (data from Voss and Fry, 1984). The off-diagonal block elements (not shown) were equal to zero within the sensitivity of the instrument.

The peak linear polarization is much less than 100%. F_{22} is not 1; in fact, it approaches 0.6, which is an indication that significant components of oceanic scatterers are not spherical. Deviation from sphericity is also indicated by F_{33} and F_{44} not reaching -1 in the backscatter direction. These measured differences can be used in models (Mie, coupled dipole, or other) to provide much more information about the particles responsible for the scattering.

A. BIOLOGICAL SCATTERERS: LIGHT SCATTERING FROM MARINE ORGANISMS

In the marine realm, large numbers of biological scatterers are nonspherical, but unlike scatterers often found in the atmosphere, large groups of microorganisms are self-similar. Marine organisms have many shapes; some are roughly spherical (e.g., *Chlorella*). Figure 6 shows the polarized light scattering of this ubiquitous (Malone, 1971a, b) organism (Quinby-Hunt *et al.*, 1989; Hunt *et al.*, 1990). Some organisms are ellipsoidal (many bacteria, picoplankton); others are helical (some bacteria, parts of dinoflagellates, octopus sperm heads); still others

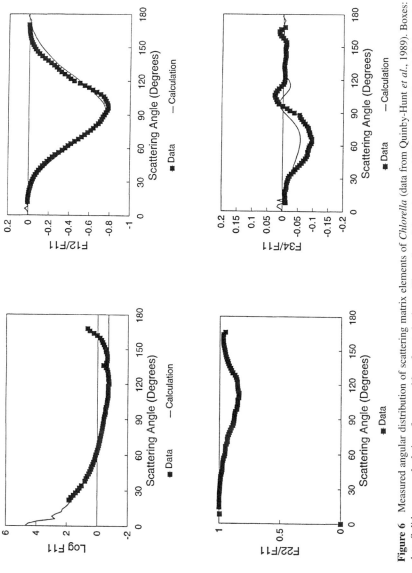

Figure 6 Measured angular distribution of scattering matrix elements of *Chlorella* (data from Quinby-Hunt *et al.*, 1989). Boxes: data. Solid curves: calculations for ensembles of coated spheres. The radius of the sphere was 1.75 μm with a 60-nm shell (hence the inner radius was 1.69 μm); the inner relative refractive index was 1.08 + 0.05*i*; the outer relative refractive index was 1.13 + 0.04*i*. Measurements were at a wavelength of 457 nm.

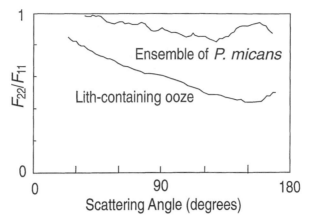

Figure 7   Angular distribution of F_{22} for *P. micans* and a foraminiferal ooze containing a large abundance of liths from coccolithophorids.

can have cylindrical character (*Ceratium extensum*, spines from *Globgerina*, and some species of cyanobacteria).

When a particle is nonspherical, the redundancy in the elements of the scattering matrix is reduced, and the F_{22} matrix element shows less spherical character. Such deviation is observed only to a small degree in *Chlorella* (Fig. 6; Quinby-Hunt *et al.*, 1989) and *Porphyridium cruentum* (red algae, size ~15 μm) (Fry and Voss, 1985), but to a much greater degree in the scattering from two clones of *Syncehococcus* sp., a cyanobacterium (Fry and Voss, 1985). Interestingly, the angular distribution of F_{22} for ensembles of the dinoflagellate *Prorocentrum micans* remains generally above 0.8 (Fig. 7), which is surprisingly high, given the asymmetry of the organism (the authors, unpublished data). On the other hand, in similar measurements of F_{22} for liths from coccolithophorids, a primary component of foraminiferal ooze, the value approaches 0.5 at angles greater than 90° (Fig. 7). The cup-shaped liths contain regions of single-crystal calcite, which is linearly birefringent. They dominate marine scattering during specific episodes (Balch *et al.*, 1991).

As noted previously, marine organisms are known to orient to various stimuli including gravity, direction of sunlight, and linear polarization of light; organisms and other particles orient to flow gradients from turbulence, electric and magnetic fields, and chemical gradients. Nonspherical, oriented particles can further polarize the underwater light field. Oriented ensembles may exhibit nonzero scattering matrix elements that for random orientations are identically zero.

Large deviations from the apparently simple scattering observed in seawater are observed when single organisms with complex symmetry are examined, as evidenced in the scattering from certain dinoflagellates such as *P. micans* and *Crypthecodinium cohnii* (Lofftus *et al.*, 1988, 1992; Shapiro *et al.*, 1990, 1991). Laboratory measurements on heads of the octopus sperm *Eledone cirrhosa*, a helical structure, showed significant nonzero F_{14} indicating circular-polarization effects (Wells *et al.*, 1986; Shapiro *et al.*, 1994a).

Dinoflagellates occur extensively in the ocean, are known to bloom periodically, and reside primarily in the photic zone. The unusual scattering observed from *P. micans* and *C. cohnii* (Lofftus *et al.*, 1992) has been postulated to be due to its helically organized chromosomes. The exact structure of the chromosomes of *P. micans* has yet to be agreed upon, but almost all proposed models involve some kind of helical structure (Bouligand *et al.*, 1968; Livolant and Bouligand, 1978; Gautier *et al.*, 1986; Dodge and Greuet, 1987). Because of the chiral organization of the DNA, the chromosome scatters right and left circularly polarized light differently resulting in a large circular dichroism (CD) and circular intensity differential scattering (CIDS, equivalent to $-F_{14}$) at 265–390 nm (Livolant and Bouligand, 1978, 1980; Livolant and Maestre, 1988; Rizzo, 1987; Soyer and Haapala, 1974).

The F_{14} signal is usually too small to be detected in randomly oriented ensembles; however, F_{14} signals measured at LBNL for small numbers of live *P. micans* were significant. The magnitude of the signal varied diurnally and reproducibly (Shapiro *et al.*, 1990, 1991); see Fig. 8. The signal increases strongly during a time that may correspond to a period in the cell cycle of the dinoflagellate when changes in its chromosome structure occur. Thus, a dinoflagellate bloom might induce unusual background circular polarization scattering properties in seawater depending on the time of day or organism cell cycle.

B. Mineralogical Scatterers

Although mineralogical scatterers make up a significant component of particles in the ocean, little research has been done to investigate their contribution to the polarization of scattered light in the submarine environment. To investigate what the contribution of minerals might be, scattering was measured from a variety of minerals and inorganic particles: muscovite (mica, monoclinic, plate-like morphology); kyanite (an orthosilicate, triclinic, elongated morphology); galena (lead sulfide, cubic, dimensions roughly equal); and synthetic spherical alumina shells for comparison. The scattering observed for all measured matrix elements differed significantly from sample to sample. Figure 9 demonstrates wide variation in scattering results for F_{12} (the degree of depolarization for linearly polarized light) and F_{22}. Even though the samples were roughly similar in size distribution ($\sim$1–6 μm), the linear polarization curves (F_{12}) differed dramatically in their magnitude and position of the peak. The F_{22} curves for the minerals deviate sig-

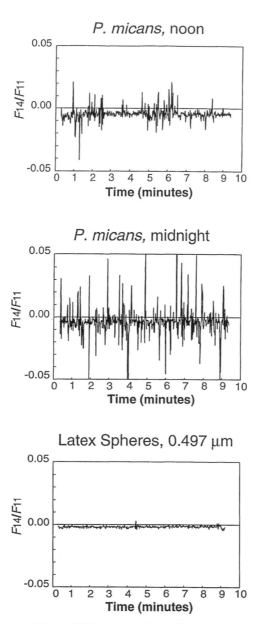

Figure 8 Measurements of F_{14} at 90° from suspensions of live *P. micans* taken at noon and midnight. Measurements of 0.497-μm-diameter latex spheres are shown to demonstrate the scattering that would be expected for spherical particles. Scattering for latex spheres did not vary with time. Adapted from Shapiro *et al.* (1991).

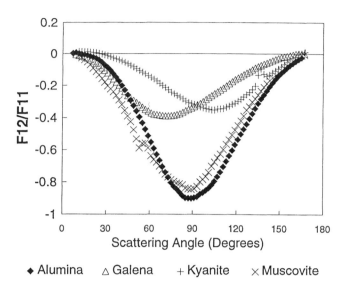

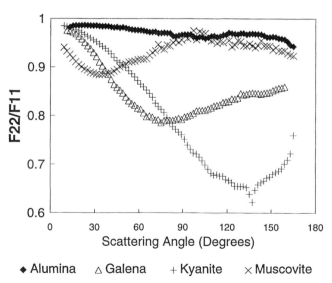

Figure 9   Angular distribution of F_{12} and F_{22} for three minerals. Results are compared to those for spherical alumina shells.

nificantly from 1, illustrating a definite nonspherical character, whereas the results for the alumina shell are close to 1. The data indicate the presence of certain non-spherical inorganic particles (e.g., particularly near sources of continental runoff) that could significantly modify the component of linearly polarized light in the sea and dramatically affect polarized backscatter measurements.

VI. POLARIZED LIGHT SCATTERING IN SEA ICE

Sea ice is a complex, heterogeneous material whose optical properties in the visible are dominated by scattering (Grenfell, 1983; Perovich and Grenfell, 1981). Scattering in sea ice is caused by geometrical inhomogeneities resulting from the presence of crystal boundaries, brine pockets, bubbles, and other inclusions and depends strongly on the initial growth conditions and age. Sea ice begins to freeze from liquid brine at $T = 1.8°C$, at which temperature, platelets of pure water ice form within the brine, increasing the salinity of the remaining brine (Weeks and Ackley, 1986). The platelets of pure water ice that form are birefringent (Onaka and Kawamura, 1983) and show varying degrees of orientation depending on the growth conditions. The shape and number density of the brine inclusions in sea ice are also strongly influenced by the initial growth conditions. Draining brine and trapped air bubbles originating in freezing seawater cause air-filled voids in the sea ice. Typically brine is trapped between vertical platelets of pure ice. The spacing between platelets and the initial ice salinity are primarily functions of the growth rate. As the temperature of the ice decreases, salts in the brine begin to precipitate (Nelson and Thompson, 1954).

Under certain growth conditions the platelets show considerable alignment. For example, as seawater begins to freeze, the ice platelets generally show no particular orientation, especially in the presence of moderate wave action (Onaka and Kawamura, 1983). As the ice sheet becomes thicker the increasing salinity of trapped brine pockets and anisotropic heat transfer favor platelet growth with the c axis parallel to the surface, causing trapped vertical brine pockets. (The c or optic axis corresponds to a direction about which the molecules are arranged symmetrically.) In this growth phase, which can occur at depths greater than approximately 15 cm, currents generated in the freezing seawater tend to align the platelets, with the c axis normal to the current flow (Weeks and Gow, 1978; Weeks and Ackley, 1986).

The angle- and polarization-dependent scattering properties of sea ice are believed to be an important component of realistic radiative transfer models. Most radiance calculations in an atmosphere–ocean system have been performed using a scalar theory approach where polarization effects are neglected. This approach is incomplete and, in the presence of significant polarization-dependent scattering, will be in error (Kattawar and Adams, 1990; Adams and Kattawar, 1993).

Radiance calculations are used to evaluate the effect of sunlight propagation in ice for issues such as climate change and biological processes under the ice pack. It is important to determine whether the polarization in sea ice is maintained or if multiple-scattering effects destroy the polarization so that a scalar theory is adequate for radiance calculations. Other reasons for studying marine ice scattering are to establish a connection between the light-scattering properties of sea ice and its morphology and to establish the usefulness of polarimetry in remote sensing.

To examine the scattering properties of ice with various morphologies, angle- and polarization-dependent light scattering was measured from samples of oriented first-year and multiyear sea ice from the Chukchi Sea near Pt. Barrow, Alaska, and from saline ice grown in an outdoor facility at the U.S. Army Cold Regions Research and Engineering Laboratory (CRREL) in Hanover, New Hampshire. Measurements of the full scattering matrix were made on samples prepared from ice cores from all three sources (Miller *et al.*, 1997). The properties of first-year sea ice were also measured *in situ* with the bistatic nephelometer at Pt. Barrow in 1994 (Miller *et al.*, 1994, 1996).

In situ measurements were made of only the first row of the scattering matrix because of the complex interaction of scattered light with the windows required to interface with the sea ice at Pt. Barrow. The *in situ* measurements revealed that linear polarization of light was partially preserved for path lengths up to 60 cm, whereas circular polarization effects were lost much more rapidly as a result of the birefringence of the ice crystals. Interpretation of the field measurements was complicated by multiple scattering.

More extensive measurements of light scattering in sea ice samples from Pt. Barrow as well as multiyear and saline ice could be performed in the laboratory (Miller *et al.*, 1997). Various geometries for the scattering samples were used. Significant sample-to-sample variation was encountered, but several general conclusions can be drawn from the data. The scattering intensity from thin slabs of all the ice samples showed large forward scattering and backscattering components that could be described qualitatively by scattering from pockets of brine with a diameter of about 30 μm in a medium of pure water ice. The phase functions for cylindrical samples are generally featureless implying the presence of multiple scattering. The orientation effects are manifested weakly in thin-slab c-axis-oriented samples and are primarily in the matrix elements F_{23} and F_{32}. This is consistent with a simple model that combines the effects of scattering from spherical inhomogeneities and the intrinsic birefringence of pure water ice.

The large observed values of the polarizance function P_{lin} (defined previously in Section II.D) for thin-slab samples illustrate the degree of preservation of linearly polarized light in the forward scattering and backscattering directions (Fig. 10). The figure also demonstrates that circular and elliptical polarization are not well preserved in any of the sea ice samples. Polarization is not well preserved in cylindrical samples because of multiple scattering (Miller *et al.*, 1997). Thus

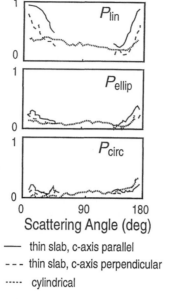

Figure 10   Polarizance functions for samples of first-year ice. Measurements were made in the laboratory. In all three cases the *c* axis of the ice was in the scattering plane. Solid line: thin-slab sample, *c* axis parallel to the laser, average of three samples. Dashed line: thin-slab sample, *c* axis perpendicular to the scattering plane, average of three samples. Dotted line: cylindrical sample, average of six samples. Data presented in Miller *et al.* (1997).

it may be concluded that only linear polarization will be preserved and only for relatively short distances (≤ 60 cm).

The sea ice phase functions measured in the laboratory were difficult to determine because of multiple scattering and large sample-to-sample variation. However, the results compared generally with those of Grenfell and Hedrick (1983) for NaCl and glacier ice except for showing larger forward scattering than they had reported.

VII. CONCLUSIONS

Light propagating through the marine environment is scattered and absorbed by a variety of inhomogeneous components, many nonspherical in nature. Analytical models for describing the scattering by nonspherical particles have shown that even moderate asphericity leads to significant deviation of the scattering matrix elements from the results of Mie calculations. In particular, spheres are an inappropriate model for light scattering from ocean water because they are unable to mimic the behavior of the elements F_{22}, F_{33}, and F_{44}. Experimental light-scattering measurements of the scattering matrix for seawater, marine organisms, minerals, and marine aerosols have confirmed such deviations (in some cases very large) from that predicted for spheres.

The angle dependence of various elements of the scattering matrix either provides a direct measure of or is indicative of the physical properties of the scattering medium. Particles are considered to be nonspherical when F_{22}/F_{11} deviates significantly from unity and/or when the matrix elements that are identically zero for spheres become measurable. For example, F_{14} is an indicator of optical activity (or helical structure), and F_{13} is an indicator of particle orientation. Because deviations of the scattering behavior from that predicted for spheres are often very small (e.g., nonzero off-diagonal block elements), measurement techniques must have sufficient sensitivity to detect them. One technique, angle-scanning polarization-modulated nephelometry (described in Section III), has the sensitivity and precision required to investigate the effects of polarized light scattering in the marine environment. It has had extensive use in the measurement of the scattering matrix for marine aerosols, ocean water and microorganisms, and sea ice. Light intensity and polarization scattering play an important role in the marine environment. In the marine atmosphere, it affects visibility, cloud formation, radiative transfer, and the heating and cooling of the earth. In the ocean, it has significant consequences for marine inhabitants and affects underwater visibility and imaging. In sea ice, it is critical for a variety of life forms that live in the ice, on the ice, and under the ice.

ACKNOWLEDGMENTS

The authors would like to thank the many researchers who have worked with us over the years especially Kevin Lofftus, Dany Kim-Shapiro, Lael Erskine, and David Miller. We would also like to acknowledge the Office of Naval Research, which has supported our marine research. Work for this chapter was partially supported by Office of Naval Research Contracts N00014-96-1-00307 (PH) and N00014-98-F-0021 (MSQ-H and AJH).

Chapter 19

Scattering Properties of Interplanetary Dust Particles

Kari Lumme

Observatory
University of Helsinki
FIN-00014 Helsinki, Finland

I. INTRODUCTION

   Although the number density of interplanetary dust particles (IDPs) at 1 astronomical unit (AU) is only a few particles per cubic kilometer (Leinert and Grün, 1990) the IDPs are still capable of producing a beautiful phenomenon at lower geographic latitudes. This so-called zodiacal light, known since the days of Cassini, is best seen either 1 hour before sunrise or 1 hour after sunset. The zodiacal cloud of IDPs permeates the solar system beyond the orbit of the planet Jupiter and

Light Scattering by Nonspherical Particles: Theory, Measurements, and Applications

comprises a mass of a major comet or about 10^{17} kg. Because this mass is distributed in a such huge volume, the resulting optical thickness at 1 AU is only about 10^{-7}–10^{-6}. In practical terms this means that all the light we see is only singly scattered by the IDPs.

There seems to be a consensus that the size range of the IDPs is from 1 μm to about 100 μm. The lower limit can basically be determined by the Poynting–Robertson effect, which effectively reduces the orbit of an IDP. Then in 10^4–10^5 years the particles reach distances where the solar heat evaporates them (Leinert and Grün, 1990). The upper limit of 100 μm is only poorly established but at least the particle detectors on board various spacecraft show a sharp decline in numbers beyond that limit.

The interplanetary dust cloud has an obvious three-dimensional (3D) distribution in the solar system. In the radial direction from the Sun a popular power law distribution of particle number density is normally found adequate; see, for example, Leinert and Grün (1990). The power law exponent γ deduced from observations is -1.3. It is not immediately clear which part of the exponent is due to the actual radial distribution of the number density and which part is due to the albedo of a typical IDP. Based on the *Helios 1* and *Helios 2* spacecraft data (Leinert *et al.*, 1981), Lumme and Bowell (1985) were able to decompose γ such that about -0.5 is due to the increase in albedo with decreasing heliocentric distance and about -0.8 is due to the number density variation. These numbers were based on the clear decrease in linear polarization of about 30% from 1 to 0.3 AU as observed by Helios spacecraft.

Observations of zodiacal light clearly show that the surface brightness steadily decreases out of the ecliptic plane. Various authors have tried to model the brightness decrease by adopting different empirical functions for the ecliptic latitude dependence. These were summarized in detail by Giese *et al.* (1986). Exactly the same problem as in the case of radial distribution is involved in latitude dependence: which part of that dependence is caused by the change in albedo and which part is due to the real change in number density. It is not unrealistic to assume that particles with higher orbital inclinations are systematically smaller than those concentrated in the ecliptic; see, for example, Grün *et al.* (1997).

In the ecliptic, two pronounced maxima are observed for the surface brightness of zodiacal light. The first one is close to the Sun and is referred to as the F corona. This can best be observed during a solar eclipse from the ground or from space because the steady increase starts at only a few degrees away from the Sun. The brightness of the F corona has often been interpreted as resulting from Fraunhofer diffraction. This would, in principle, seem to provide a method for estimating the particle size of a typical IDP. Unfortunately, however, there are two main obstacles for doing this reliably. First, what is observed along the line of sight (LOS) at a fixed elongation results from all scattering angles Θ, the main contribution being from those particles close to the Sun at $\Theta \approx 90°$. Second, if the particle sizes are a few micrometers, then diffraction is not very strong beyond a scattering angle of a few degrees.

The second pronounced maximum of zodiacal light is seen in the antisolar direction and is called the gegenschein. Inversion of the LOS observations in this region is much more reliable because there is only a narrow range of scattering angles close to 180° being involved. The gegenschein is a typical example of the so-called opposition effect of atmosphereless bodies in the solar system. Most of them show a pronounced nonlinear increase in brightness toward the planetary opposition geometry ($\Theta = 180°$). Since the days of Seeliger (1887) the opposition effect has been interpreted as resulting from mutual shadowing in an aggregate cloud of particles except very close to $\Theta = 180°$. However, the gegenschein shows unequivocally that single particles can also produce the effect without any shadowing mechanism.

Because the lifetime of IDPs is greatly limited by the Poynting–Robertson effect, there should be an efficient supply of new particles into the cloud. Prior to the IRAS mission it was widely believed that short-period comets provided the main supply. However, Lumme and Bowell (1985) strongly favored the C-asteroidal origin. This suggestion arose from the similarities in single-scattering albedo, color indices, polarization, and the distribution of dust as a function of ecliptic latitude between the C asteroids and the interplanetary dust cloud. Shortly after that suggestion the IRAS findings largely supported the idea. Nevertheless, the short-period comets certainly have their important, if not dominating role in the supply of new IDPs.

For a number of years stratospheric flights at altitudes greater than 20 km and balloons have been used to collect small, so-called "Brownlee particles." On the basis of chemical composition, a majority of particles less than 10 μm in diameter are deemed to be of terrestrial origin, whereas the bigger ones mostly come from space. There has been extensive analysis on the grain size, morphology, and chemistry of these particles (e.g., Brownlee, 1985; Gibson, 1992). Generally speaking, these can be subdivided into chondritic and nonchondritic particles. According to a series of cosmic dust catalogs (e.g., Warren *et al.*, 1994), the particle morphology can be crudely divided into porous aggregates and more solid, rough particles, both of which show quite random shapes.

As mentioned previously, IDPs almost certainly come from both asteroidal and cometary dust particles. It would therefore be quite natural that at least some observed properties of these objects (zodiacal light, comets, and asteroids) would be the same. Indeed, two ubiquitous phenomena have been observed. First, all of these objects sharply brighten toward the exact backscattering direction, that is, show the opposition effect. Second, an even more telltale feature is seen in their linear polarization. The reversal, a change from negative to positive polarization, takes place at about $\Theta = 160°$ and the whole polarization curve is very similar, only being modulated by albedo. There is certainly no *a priori* reason why all randomly chosen particles would behave this way. In Section II we review the observing geometry of zodiacal light and study the inversion in detail. We will study the complementary nature of the three objects in Section III. All these objects require different inversion techniques to yield physical information on the constituent particles and in this respect they complement each other. Obviously, all light-scattering methods need an explicit expression for the particle geometry and as can be seen later, this will cause quite different scattering patterns for the particles. Particle shapes will be studied in Section IV.

An extensive review of the existing light-scattering methods is given in Chapter 2. Therefore, we will not present any details of those in this chapter. We do, however, provide the essentials of those methods pertinent to light scattering by cosmic dust particles. We also explain in some detail our favorite method, the integral equation algorithm, which we are using in our work on light scattering. All this will be done in Section V.

II. OBSERVATIONS OF ZODIACAL LIGHT AND THEIR INTERPRETATION

The main optical data on the brightness and polarization of zodiacal light come from ground-based and spacecraft observations. The ground-based data have been summarized and compiled by Leinert (1975) and Leinert *et al.* (1998). The brightness is traditionally given in somewhat awkward units of S10, which corresponds to the brightness of one tenth magnitude star per square degree. The star is meant to be of solar type. The unit S10 translates into 6.61×10^{-12} (Leinert *et al.*, 1998) of the solar irradiance per steradian.

The ground-based data give the brightness and linear polarization as a function of the elongation ϵ (see Fig. 1) and the ecliptic latitude β. The reliable data range starts at an ϵ of a few degrees and extends all the way to the gegenschein region, $\epsilon > 170°$. There is almost complete coverage of the brightness in the ecliptic latitude β.

Linear polarization has been measured by at least 10 authors. The main results of these measurements are given in Leinert's (1975) compiled table. The measured maximum polarization of about 20% is seen at $\epsilon = 60°$. The well-known reversal of polarization takes place at about $\epsilon = 160°$. Unfortunately, there are not too many data points in the gegenschein region. Outside the ecliptic linear polarization is nearly constant at a constant ϵ. The wavelength dependence of linear polarization is not very well known. However, the few existing data sets suggest that this dependence is weak at most.

Valuable data on the brightness and linear polarization of zodiacal light were provided by two spacecraft, *Helios 1* and *Helios 2*, flown in the 1970s (Leinert *et al.*, 1981, 1982). Three independent photometer systems were mounted on

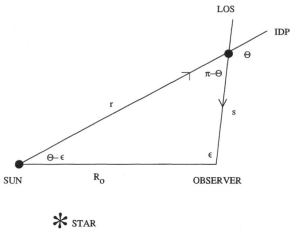

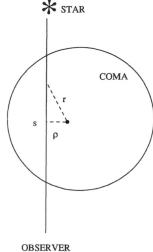

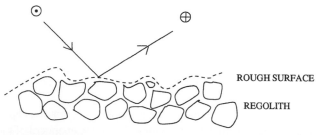

Figure 1   Basic observing geometries for the zodiacal light (top), a cometary coma (middle), and a planetary regolith (bottom).

these spacecraft to observe at a fixed β of 16°, 31°, and 90°. Each photometer recorded brightness and polarization in the U, B, and V spectral bands. Owing to the eccentric orbits of the Helios spacecraft, data were obtained at 0.3 and 1 AU. This spread in heliocentric distance provided a direct estimate for the dust number density, or more precisely, the product of that and the single-scattering albedo.

In the first approximation, the *Helios 1* and *Helios 2* observations were quite consistent with the ground-based data. The most prominent difference was found in the degree of linear polarization. At 0.3 AU, linear polarization was reduced to about two thirds of that at 1 AU (Leinert *et al.*, 1981, 1982). Also the brightness showed a small increase shortward of about $\epsilon = 60°$. In particular, the change in polarization strongly suggests that not only does the number density n_0 change with the radial distance from the Sun, but the particle properties, especially the single-scattering albedo, also change. We will model this later.

Next we will consider both the brightness and the polarization observations in the ecliptic. We will do this by following and modifying the work of Lumme and Bowell (1985). We first derive the brightness integral or the intensity along the LOS making an angle ϵ with the Sun–observer direction. A volume element dV along the LOS receives the amount of $\pi F n_0 \, dV/r^2$ of the incident solar flux πF where $n_0 = n_0(r)$ is the particle number density. The amount of $\pi F n_0 C_{\text{sca}} F_{11}(\Theta) \, dV/(4\pi s^2 r^2)$ is scattered in the direction of the observer with the scattering angle of Θ. Here F_{11} is the single-scattering phase function (Section XI of Chapter 1) and C_{sca} is the scattering cross section. The total amount of light (per steradian) received by the observer is therefore

$$I(\epsilon) = \frac{1}{4} F \int_0^\infty \frac{n_0(r)}{r^2} C_{\text{sca}}(r) F_{11}(\Theta) \, ds. \qquad (1)$$

From Fig. 1 we have

$$r = R_0 \frac{\sin \epsilon}{\sin \Theta}, \qquad s = R_0 \frac{\sin(\Theta - \epsilon)}{\sin \Theta}. \qquad (2)$$

Changing the variable s in Eq. (1) into the scattering angle Θ with the help of Eq. (2), we obtain

$$I(\epsilon, R_0) = \frac{F}{4R_0} \frac{1}{\sin \epsilon} \int_\epsilon^\pi n_0 \left(R_0 \frac{\sin \epsilon}{\sin \Theta} \right) C_{\text{sca}} \left(R_0 \frac{\sin \epsilon}{\sin \Theta} \right) F_{11}(\Theta) \, d\Theta, \qquad (3)$$

which is a Volterra integral equation of the first kind. The obvious approximation we have made previously is that F_{11} is assumed to be independent of the heliocentric distance r. Fortunately, however, even if the single-scattering albedo changes the general form of F_{11} need not change widely. It is evident from Eq. (3) that an independent solution for n_0 and C_{sca} is not possible. We can, therefore, set

$$n_0\left(R_0\frac{\sin\epsilon}{\sin\Theta}\right)C_{sca}\left(R_0\frac{\sin\epsilon}{\sin\Theta}\right) = K\left(R_0\frac{\sin\epsilon}{\sin\Theta}\right), \tag{4}$$

which is the kernel of the integral equation (3). There is no known analytical solution for Eq. (3) with this kernel.

In what follows, we try the direct method for the solution of Eq. (3). This requires that we assume some reasonable analytical forms for F_{11} and K. Our extensive computations of light scattering by nonspherical particles (Lumme *et al.*, 1997; Lumme and Rahola, 1998), have indicated that a good approximation for F_{11} would be

$$F_{11}(\Theta) = p_0\exp\left(a_1\Theta + a_2\Theta^2\right), \tag{5}$$

where the normalization coefficient p_0 assumes the form

$$p_0 = \frac{8i\sqrt{a_2/\pi}\exp(q_1^2)}{\mathrm{erfi}(q_1) - \mathrm{erfi}(q_1 - \sqrt{a_2}\pi) + \exp\left(\frac{ia_1}{a_2}\right)[\mathrm{erfi}(q_2) - \mathrm{erfi}(q_2 + \sqrt{a_2}\pi)]},$$

$$q_1 = -\frac{i + a_1}{2\sqrt{a_2}}, \tag{6}$$

$$q_2 = \frac{-i + a_1}{2\sqrt{a_2}},$$

where $i = \sqrt{-1}$ and erfi is the complex error function defined as $\mathrm{erfi}(z) = \mathrm{erf}(iz)/i$ in terms of the standard error function erf. Some fits of Eq. (5) to our light-scattering computations of stochastically rough spheres are shown in Fig. 2. As can be seen, the fits are very good.

We now concentrate on the functional form of the kernel function K. The classical approximation (see, e.g., Leinert and Grün, 1990) is a power law $r^{-\nu}$ distribution. The observations, particularly those by the Helios spacecraft (Leinert *et al.*, 1998), yield $\nu = 1.3 \pm 0.1$. The problem with the power law is that it predicts that the functional form of the intensity $I(\epsilon, R)$ remains independent of R; that is, $I(\epsilon, R_1)/I(\epsilon, R_2)$ is constant. This is in conflict with the Helios data, which show that the intensity increase when $\epsilon < 60°$ is more rapid at $r = 0.3$ AU than at $r = 1$ AU. We will now assume

$$K(R) = k_0\left[\frac{R_0}{R} + b_1\left(\frac{R_0}{R}\right)^2\right], \tag{7}$$

where k_0 is a constant and b_1 is a parameter to be found. This form has the extra bonus that the integration in Eq. (3) can be done analytically. Inserting now Eqs. (5), (6), and (7) into Eq. (3), we obtain after straightforward, though tedious

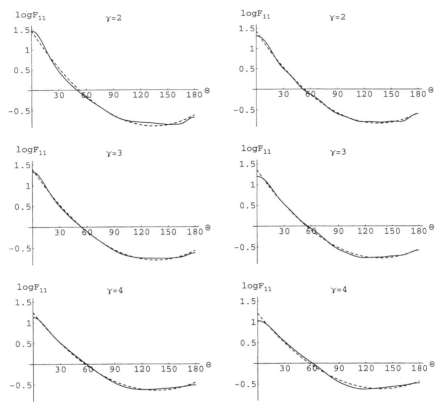

Figure 2 Some examples of discrete power law averaged single-scattering phase functions of stochastically rough spheres as a function of the power law exponent γ (solid curves) as compared to an analytical approximation, Eq. (5) (dashed curves).

algebra

$$I(\epsilon, R_0) = \sqrt{\frac{\pi}{a_2}} \frac{F k_0 p_0}{8i} \exp(-q_1^2) \frac{1}{\sin^2 \epsilon} \left\{ 2 \exp\left(\frac{ia_1}{a_2}\right) f_{-1}(\epsilon) - 2 f_1(\epsilon) \right.$$

$$+ \frac{ib_1 R_0}{\sin \epsilon} \left[\exp\left(\frac{3(1 + 2ia_1)}{4a_2}\right) f_{-2}(\epsilon) + \exp\left(\frac{3 - 2ia_1}{4a_2}\right) f_2(\epsilon) \right.$$

$$\left. \left. - 2 \exp\left(\frac{-1 + 2ia_1}{4a_2}\right) f_0(\epsilon) \right] \right\},$$

$$f_k(\epsilon) = \mathrm{erfi}\left(\frac{ki + a_1 + 2a_2 \epsilon}{2\sqrt{a_2}}\right) - \mathrm{erfi}\left(\frac{ki + a_1 + 2a_2 \pi}{2\sqrt{a_2}}\right),$$

(8)

where the intensity comes out real despite the appearance of the imaginary unit in the expression.

We are now left with three unknowns a_1, a_2, and b_1. To solve for these, we apply Eq. (8), in a nonlinear sense, to the data in the ecliptic (Leinert, 1975). The dependence of the color on the elongation ϵ is given by Leinert *et al.* (1982). Although the data extend from the exact backward direction, $\epsilon = 180°$, down to about $\epsilon = 1°$, the parameters a_1, a_2, and b_1 become very insensitive to the data at elongations below about $\epsilon = 30°$. This is a standard situation with the Volterra integral equation of the first kind. Accordingly, we cannot derive reliable information about the single-scattering phase function F_{11} in the forward (diffraction) direction. This is, of course, unfortunate because diffraction would be a direct size indicator for IDPs.

Our nonlinear least-squares algorithm provides the best-fit parameters a_1, a_2, and b_1. Those are given in Table I in the UBV bands with their 1 σ errors. The resulting F_{11} domains in these colors are shown in Fig. 3. The parameter b_1 = 0.2 translates into the parameter $v = 1.27$ between heliocentric distances 0.3 and 1 AU. As mentioned before, the region $\epsilon \leq 30°$ cannot be very reliable. The best-fit parameters produce now the predicted intensity in different spectral bands and heliocentric distances. These are also shown in Fig. 3. The fits are very good. As was mentioned previously, our kernel function, Eq. (7), also predicts a difference in the functional form of $I(\epsilon, R)$ with varying distance R. This is illustrated by the differences between the solid and the dashed curves for intensity in Fig. 3 and is verified with the Helios data.

We now turn to the observations of linear polarization of zodiacal light. Leinert (1975) has summarized the ground-based data of various authors. These data in the ecliptic are fairly consistent, showing a maximum polarization of about 20% at $\epsilon = 70°$. Unfortunately, there are only a few data points close to the backscattering (gegenschein) direction and the reversal of polarization is still fairly poorly observed. Most researchers, however, agree that the reversal is actually taking

Table I

Coefficients for the Exponential Phase Function F_{11} Approximation Together with Their Errors in the U, B, and V Spectral Bands

	a_1	a_2	b_1
V	-1.35 ± 0.20	0.37 ± 0.04	0.20 ± 0.05
B	-1.25 ± 0.20	0.37 ± 0.05	0.20 ± 0.05
U	-1.20 ± 0.30	0.37 ± 0.05	0.20 ± 0.05

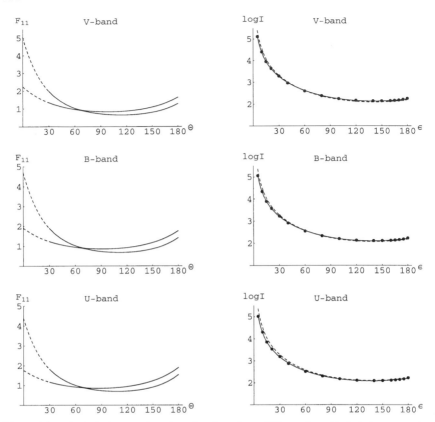

Figure 3 Derived single-scattering phase functions for an IDP together with their 1 σ uncertainty domains (left column) as compared to the resulting theoretical distribution of zodiacal light brightness and data (right column). Notice the predicted small brightness increase with the heliocentric distance, where solid curves correspond to 1 AU and dashed curves correspond to 0.3 AU.

place at about $\epsilon = 160°$. Measurements below $\epsilon = 25°$ are rather uncertain because the effect of interplanetary plasma must be removed.

The Helios spacecraft provided a large amount of polarization data of zodiacal light. Unfortunately, however, they did not make any measurements in the ecliptic but instead at the latitudes 16° and 31°. These data have been summarized by Leinert *et al.* (1982). Perhaps the most interesting finding was the clear reduction in the degree of polarization with decreasing heliocentric distance from 1 AU down to 0.3 AU. This reduction was about one third. Lumme and Bowell (1985) attributed this effect to the change of average properties of IDPs with heliocentric distance. More specifically, they thought that the single-scattering albedo of the particles increases with decreasing heliocentric distance. This idea is further supported by Levasseur-Regourd (1991), who found the same distance dependence from observations at thermal infrared wavelengths. The degree of linear polarization P is given by

$$P = \frac{-Q}{I},\tag{9}$$

where Q is the second Stokes parameter and when the coordinate system is chosen so that U vanishes (Section V of Chapter 1). Almost all atmosphereless bodies in the solar system show amazingly similar polarization behavior, that is, P being negative at scattering angles greater than 160° and reaching a maximum at about $\Theta = 90°$. Therefore, it is not surprising that all the observed data, for about 100 objects, can be described by a simple function, as shown by Lumme and Muinonen (1993),

$$P(\Theta) = (\sin\Theta)^{c_1}\left(\sin\frac{1}{2}\Theta\right)^{c_2}\sin(\Theta_1 - \Theta)$$

$$= p_2(\sin\Theta)^{c_1}\cos\Theta\left(\sin\frac{1}{2}\Theta\right)^{c_2} + p_1(\sin\Theta)^{1+c_1}\left(\sin\frac{1}{2}\Theta\right)^{c_2},\tag{10}$$

$$p_1 = -p_0\cos\Theta_1,$$

$$p_2 = p_0\sin\Theta_1.$$

For the approximately 40 asteroids analyzed, the best values for the parameters were $c_1 = 0.7$ and $c_2 = 0.35$, whereas for the other solar system objects $c_1 = 0.5$ and $c_2 = 0.35$. Once the parameters c_i are known, all the fits can be done linearly to get the inversion angle Θ_1 and the scale factor p_0.

Combining Eqs. (9) and (10) and in accordance with Eqs. (3) and (7), we can write for the second Stokes parameter

$$Q(\epsilon, R_0) = \frac{F}{R_0\sin\epsilon}\int_\epsilon^\pi K_2\left(R_0\frac{\sin\epsilon}{\sin\Theta}\right)P(\Theta)F_{11}(\Theta)\,d\Theta,\tag{11}$$

$$K_2(R) = k_0\left[\frac{R_0}{R} + b_2\left(\frac{R_0}{R}\right)^2\right],$$

where we have changed the kernel function K, Eq. (7), into a new one, K_2. We have done this because there is no reason to anticipate that the albedo, or C_{sca}, dependence would be the same for the intensity and the second Stokes parameter. Now the integral in Eq. (11) cannot be calculated analytically, as was the case for the intensity I. The observed degree of polarization now reads as $P_{obs}(\epsilon) = Q(\epsilon, R_0)/I(\epsilon, R_0)$. In the V band we use the compiled table of Leinert (1975) to solve for the parameters p_1, p_2, and b_2. There are no ground-based data on the

Table II

**Best-Fit Parameters of the Empirical Polarization Function
Together with the Inversion Angle in the U, B, and V Spectral
Bands**

	p_1	p_2	b_2	$\Theta_1(°)$
V	0.350	0.087	0.020	166.1
B	0.427	0.105	0.020	166.1
U	0.464	0.114	0.020	166.2

color dependence of polarization. We, therefore, assume that the color information of the Helios data, in spite of the noncliptic observations, can be used. From Leinert *et al.* (1982) we obtain

$$\left\langle \frac{P_{\text{obs}}(B)}{P_{\text{obs}}(V)} \right\rangle = 1.21, \qquad \left\langle \frac{P_{\text{obs}}(U)}{P_{\text{obs}}(V)} \right\rangle = 1.30. \tag{12}$$

The color, at least to the first approximation, does not seem to depend on the elongation. The best-fit parameters p_1, p_2, and b_2 together with the inversion angle Θ_1 are given in Table II. The fits, which turn out very well, are shown in Fig. 4. It can be seen that the change of polarization with heliocentric distance comes out automatically, shown by the dashed curves, and approximately what the Helios spacecraft observed. In deriving the three parameters we did not use the Helios data at 0.3 AU.

Next we consider polarization by a single IDP as a function of the heliocentric distance. This can easily be done by replacing the upper limit π of the integration in Eqs. (3) and (11) by $\epsilon + d\epsilon$, which has the effect that there is no LOS integration. We then obtain

$$P_{\text{IDP}}(\Theta) = \frac{K_2(R_0)}{K_1(R_0)} P(\Theta) = \frac{1 + b_2 \frac{R_0}{R}}{1 + b_1 \frac{R_0}{R}} P(\Theta). \tag{13}$$

We show P_{IDP} at distances $R = R_0$ and $R = 0.3R_0$ in Fig. 5. Quite understandably $P(\Theta)$ and $P_{\text{IDP}}(\Theta)$ are clearly different. It is interesting to compare $P_{\text{IDP}}(\Theta)$ for some other objects for which polarization has been observed in a sufficiently broad range of scattering angles. We now compare P_{IDP} with the cometary data. Most of the observed comets seem to fall into two categories in terms of linear polarization as shown by Levasseur-Regourd *et al.* (1996). Among those that have a higher degree of polarization we have chosen three well-observed comets, namely, P/Halley (Kikuchi *et al.*, 1987), P/Bradfield (Kikuchi *et al.*, 1989), and P/West (Michalsky, 1981). These data together with P_{IDP} are shown in Fig. 5.

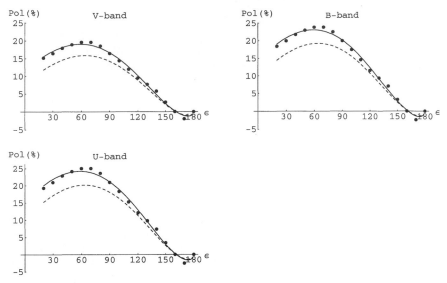

Figure 4 Polarization data of zodiacal light (dots) as compared to the theoretical predictions at 1 AU (solid curves). The dashed curves show computed linear polarization at 0.3 AU and are close to those observed by the Helios spacecraft at 0.3 AU.

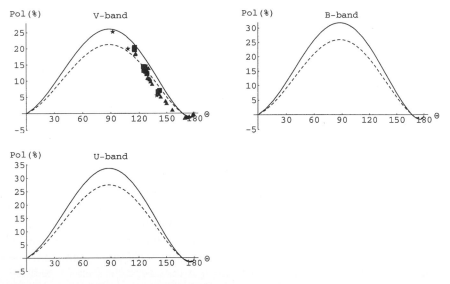

Figure 5 Local polarization of a typical IDP at 1 AU (solid curves) as compared to local polarization at 0.3 AU (dashed curves). For comparison we also show polarization in the V band for three comets.

The agreement is fairly good. The question of the color dependence of the polarization of comets is somewhat conflicting. There seems to be a slight tendency of the degree of polarization to increase with the wavelength. If true, this would be in direct conflict with the data on polarization of zodiacal light as found by Leinert *et al.* (1982).

Unfortunately, we cannot yet reliably calibrate the correlation between the degree of linear polarization and the single-scattering albedo. Most of the cometary data suggest that the albedo of the particles is low. Obviously, the data on zodiacal light seem to suggest the same.

III. DUST IN THE SOLAR SYSTEM: COMPLEMENTARY VIEW

The data obtained with the two basic methods in solar system research, photometry and polarimetry, clearly suggest that some properties of the small cosmic dust particles, generally speaking, must be similar. The ubiquitous opposition effect and the reversal of linear polarization are too similar to be completely accidental. There seems to be, however, a clear difference between the high-albedo icy satellites on one side and the regoliths on darker satellites and asteroids, comets, and zodiacal light on the other side. For the icy satellites the classical opposition increase in brightness starts only below phase angle $\alpha = 180° - \Theta = 2°$ contrary to $\alpha = 7°$ for the other objects. Also, this opposition surge is even more pronounced than the opposition effect. Polarization is quantitatively different in the sense that for the icy satellites the angle of polarization reversal often takes place at about $\alpha = 10°$, whereas for the other objects this angle is about $20°$.

Because of the similarities between the darker solar system objects, we will study in this section how complementary information can, at least in principle, be obtained. The common factor in the different environments is the question of the data inversion technique.

A. COMETS

When approaching the Sun, a comet normally becomes active and develops a coma around its nucleus. There are numerous measurements of both the intensity and the polarization of the coma as a function of the distance to the Sun. These measurements can, in principle, yield information about both the number density of coma particles and their light-scattering properties. Judging from the surface brightness of the coma, the number density n_0 in the inner coma seems to be fairly spherically symmetric, whereas at larger distances from the nucleus departures from this start to develop. These departures are caused by the radiation pressure on the small micrometer-sized particles. The slant optical thickness of the coma is normally small so that the observed surface brightness I at a given scattering angle $\Theta = 180° - \alpha$ and at a distance r from the nucleus can be written as

$$I(\rho, \Theta) = \pi F \int_{-\infty}^{\infty} n_0\left(\sqrt{\rho^2 + s^2}\right) Z_{11}\left(\sqrt{\rho^2 + s^2}, \Theta\right) ds,$$

$$(14)$$

$$F_{11} = \frac{4\pi}{C_{\text{sca}}} Z_{11},$$

where the definition for Z_{11} is given by Eq. (57) of Chapter 1 and where $\pi F = \pi F_0(R_0/R)^2$ is the scaled incident solar flux at a distance R and $R_0 = 1$ AU, Z_{11} is the first element of the phase matrix $\mathbf{Z}$ (Chapter 1). The basic geometry is shown in Fig. 1. We cannot solve for both the one-dimensional number density n_0 and the two-dimensional scattering function Z_{11} from the two-dimensional data I. We now assume that Z_{11} can be separated into the scattering cross section C_{sca} and the normalized angular dependence $F_{11}(\Theta)$ as

$$Z_{11}\left(\sqrt{\rho^2 + s^2}, \Theta\right) = \frac{C_{\text{sca}}(\sqrt{\rho^2 + s^2})}{4\pi} F_{11}(\Theta) \qquad (15)$$

and write Eq. (14) in the form

$$I(\rho, \Theta) = I_1(\Theta) K(\rho),$$

$$I_1(\Theta) = \frac{1}{4} F F_{11}(\Theta), \qquad (16)$$

$$K(\rho) = \int_{-\infty}^{\infty} k_s\left(\sqrt{\rho^2 + s^2}\right) ds,$$

where the scattering coefficient $k_s = n_0 C_{\text{sca}}$. We can still note that the approximation (15) would be exact if the coma particles would not evolve while coasting away from the nucleus. In a detailed analysis Baum *et al.* (1992) have shown that this is in fact not the case and that the albedo and/or scattering cross section must decrease while the cometary grains are flowing outward.

We next show how the integral equation (16) can be solved in the real inversion sense for k_s at any fixed configuration, that is, phase angle α. Taking the one-dimensional Fourier transform $\mathcal{F}$ of both sides of Eq. (16), we obtain

$$\mathcal{F}(K)(\omega) = \frac{1}{2\pi} \int_{-\infty}^{\infty} ds \int_{-\infty}^{\infty} k_s\left(\sqrt{t^2 + s^2}\right) e^{i\omega t} \, dt. \qquad (17)$$

We then change the variables s and t into u and ϕ as

$$s = u \sin \phi, \qquad t = u \cos \phi. \qquad (18)$$

Then Eq. (17) assumes the form

$$\mathcal{F}(K)(\omega) = \frac{1}{2\pi} \int_{-\infty}^{\infty} du \, k_s(u) u \int_0^{2\pi} \exp(i\omega u \cos\phi) \, d\phi \qquad (19)$$

and further

$$\mathcal{F}(K)(\omega) = \int_{-\infty}^{\infty} J_0(\omega u) k_s(u) u \, du, \qquad (20)$$

where J_0 is the zeroth-order Bessel function. By definition Eq. (20) is the zero-order Hankel transform $\mathcal{H}_0$ of the function $k_s(r)$. This can now be inverted to yield

$$k_s(r) = \mathcal{H}_0^{-1}[\mathcal{F}(K)](r), \qquad (21)$$

where the inverse transform $\mathcal{H}_0^{-1}$ has exactly the form of Eq. (20) except for the different variables in the integrand. The integral equations of the type of Eq. (16) are most often solved by transforming them into an Abel integral equation. Our method is, however, numerically more stable than the solution of the Abel integral equation because we do not have to take derivatives of the data.

Unfortunately, we cannot separate the number density and the scattering cross section from each other with the surface brightness data alone. However, a photometric measurement of any diminution of the starlight in traversing the coma would provide a remedy for this. Such measurements can be difficult but certainly not impossible. Assume now, as shown in Fig. 1, that a star with brightness I_0 is occulted by the coma and the brightness of the star varies as $I_*(\rho)$. Then

$$I_*(\rho) = I_0 \exp[-K_*(\rho)],$$

$$K_*(\rho) = \int_{-\infty}^{\infty} k_e\left(\sqrt{\rho^2 + s^2}\right) ds, \qquad (22)$$

$$k_e = n_0 C_{ext},$$

where C_{ext} is the extinction cross section and k_e is the extinction coefficient. Because K_* is a small quantity we obtain

$$K_*(\rho) \simeq 1 - \frac{I_*(r)}{I_0} = \int_{-\infty}^{\infty} k_e \sqrt{\rho^2 + s^2} \, ds. \qquad (23)$$

The solution of k_e in Eq. (23) can proceed in exactly the same way as that for k_s given previously. Taking now the ratio $K(\rho)/K_*(\rho)$, we obtain

$$\frac{I(\rho, \Theta)}{1 - \frac{I_*(\rho)}{I_0}} = \frac{n_0(\rho)C_{\text{sca}}(\rho)}{n_0(\rho)C_{\text{ext}}(\rho)} \frac{1}{4} F F_{11}(\Theta)$$

$$= \frac{1}{4}\varpi(\rho)F_{11}(\Theta)F, \qquad (24)$$

which is a directly observable quantity provided that we have a star occultation available. In Eq. (24), ϖ is the single-scattering albedo. A word of caution is in order: Both the numerator and the denominator of Eq. (24) are small quantities and, therefore, the error can easily be fairly large. We can still note that if the data extend close to the backscattering direction, $\Theta = 180°$, the derivable quantity $\varpi(\rho)F_{11}(180°)/4$ is exactly the geometric albedo, in the ray optics sense, for a single particle.

B. Planetary Regoliths

It is commonly believed that at least most atmosphereless bodies in the solar system are covered with a loose, optically thick dust or ice layer, called the regolith. As mentioned previously, the atmosphereless bodies exhibit just the same two ubiquitous phenomena as do the observations of zodiacal light and comets: the opposition effect and the reversal of linear polarization at about $\alpha = 20°$. We now briefly study how information about regolith dust particles can be retrieved from observations.

There are only a few solid bodies in the solar system that can be observed disk resolved; that is, the distribution of the surface brightness $I(\alpha, i, e)$ can be mapped as a function of the solar phase angle α and the local angles of incidence i and emergence e. Of course, by far the best object for detailed surface photometry is our Moon. Unfortunately, however, the best extensive photometric observations of the Moon have not been properly analyzed. Only quite recently Buratti *et al.* (1996) analyzed observations made during the Clementine mission and deduced a detailed picture of the well-known opposition effect.

There have been two main streams in interpretations of the available photometric data for atmosphereless bodies. Bowell *et al.* (1989) have summarized the approaches by Hapke and Lumme and Bowell. The basic difference between these two treatments is that it is the shadow hiding (or mutual shadowing) that, according to Hapke, causes the opposition effect and that the regolith particles are strongly backscattering and have low albedos. Lumme and Bowell maintain that, in principle, the shadow hiding and single-particle phase function, both functions of the phase angle, cannot be separated from the disk-integrated data alone. Buratti *et al.* (1996) assume that multiple scattering in the lunar regolith plays no role because the Moon does not show any clear limb darkening close to opposition. This conclusion, however, is based on an incorrect interpretation of the role of multiple scattering. Although these authors use the Henyey–Greenstein phase function for single scattering, they assume, according to Hapke's model, that multiple scattering can be assumed to be isotropic. This can easily be checked with strongly forward scattering phase functions with moderate single-scattering albedos in the range $0.5 \leq \varpi \leq 0.7$. Indeed, limb darkening with these parameters would be small. This effect would be further reduced if we compute limb darkening with the inclusion of surface roughness in terms of the generalized radiative transfer method (Lumme *et al.*, 1990).

Three main factors determine the distribution of surface brightness of a regolith: the average single-particle properties, ϖ and $F_{11}(\Theta)$; the mutual shadowing Φ_S at small ($\alpha \leq 10°$) phase angles; and the surface roughness of the regolith Φ_R. Without going any deeper into the subject, we write the basic equation in the form

$$\frac{I(\alpha, i, e)}{F \cos i} = R_1 + R_M,$$

$$R_1 = \frac{1}{4} \frac{\varpi}{\cos i + \cos e} F_{11}(\alpha) \Phi_S(\alpha; D) \Phi_R(\alpha, i, e; \rho), \qquad (25)$$

$$R_M = R_M(\alpha, i, e; \varpi, F_{11}),$$

where the reflection coefficients R_1 and R_M represent the contributions by single and multiple scattering, respectively, and the shadowing and roughness contributions Φ_S and Φ_R are normalized to unity at $\alpha = 0°$. The dominating parameters are explicitly shown in Eq. (25), where D denotes the packing density of the particles and ρ is the root mean square (rms) slope of a multivariate Gaussian surface.

To learn something about the cosmic dust particles in a regolith, we should somehow be able to separate ϖ and F_{11} from the effect of the other parameters. For a complete 3D data set $I(\alpha, i, e)$ this might be possible, provided that local texture variations are not too pronounced, but for disk-integrated data the task is just impossible.

We finally summarize, in Table III, the pros and cons of using the data on the three different sources (zodiacal light, comets, and regoliths) in order to learn some common properties of the cosmic dust particles. From this table it is obvious that all different sources have their merits and drawbacks, but if a synthesis could be made, a much more reliable picture of IDPs would arise.

IV. SHAPE MODELS FOR DUST PARTICLES

Before light scattering by small particles can seriously be studied, a method for conceptualizing their shapes must be formulated. In nature and in industrial applications, irregularly shaped small particles play such an important role that research on their properties is widely spread. One of the first questions is how

Table III

**Pros and Cons in Deriving Information on Cosmic Dust under Various
Circumstances**

Source	Pros	Cons
Zodiacal light	No mutual scattering effects between IDPs Any time variations very slow Plenty of 3D data available	Integration along LOS causes numerical instabilities in inversion Separation of number density and IDPs not straightforward All parameters functions of heliocentric distance
Comets	Isolated scattering angles If star occultation available separation of number density from properties is possible for IDPs No mutual effects between cometary particles	Temporal and spatial variations make interpretation unstable Star occultations rare, separation of number density from properties only rarely Scattering angle and time variations coupled
Regoliths	Isolated scattering angles No time variations Easy-to-gain data	Separation of particle properties from other variables very difficult and unreliable Limited scattering (phase) angle coverage Mutual effects (multiple scattering between particles) considerable

to classify particle shapes into a few basic categories. In a fairly broad sense, small particles could be divided into solid and aggregated or clustered particles. Lumme *et al.* (1997) classified particles into the following categories: polyhedral solids, stochastically rough (smooth) particles, and stochastic aggregates. The particles could also be divided into different sets in a mathematical sense. Then obvious classification categories would be solid and aggregated particles. Further, the solids could be subdivided into convex, star-shaped, and non-star-shaped forms, where the last two classes belong to concave shapes. A schematic picture of this classification is shown in Fig. 6.

A. CONVEX PARTICLES

By far the most common approach for describing convex shapes is to assume that they can be approximated by triaxial ellipsoids. It naturally depends on the particular application whether this approximation is sufficient. As far as light scat-

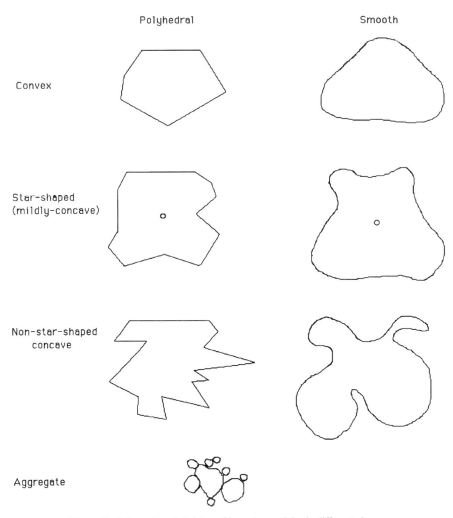

Figure 6   Schematic subdivision of irregular particles in different classes.

tering by small particles is concerned, general convex shapes can yield quite different results from equal-volume, best-fit ellipsoids. This can be judged from our recent exercise with spheres, equal-volume spheroids, and equal-volume stochastically rough spheres (Lumme and Rahola, 1998). In particular, linear polarization seems to be a good discriminator of particle shape.

   The standard procedure for describing convex shapes is to use the support function formalism, (see, e.g., Stoyan and Stoyan, 1994). Geometrically, the support function is the distance from the origin of a tangent plane to an arbitrary point on a convex surface. Thus, the support function $\rho(\vartheta, \varphi)$ is a two-dimensional (2D) scalar function uniquely associated with the radius vector $\mathbf{R}(\vartheta, \varphi)$ to that arbitrary point on a convex surface, where ϑ and φ are polar coordinates of the normal to the tangent plane. Why the support function formalism is superior to the standard radius vector approach becomes obvious in studies of particle shapes from their projections. There are many laboratory measurements with transmission electron microscope and scanning electron microscope instruments of profiles of both natural and industrial particles. With the support function formalism, the projection of any convex particle is given by $\rho(90°, \varphi)$, whereas in the radius vector concept there is no such simple way to describe the projection.

The basic properties of convex particles can be summarized in a few equations. First, the relation between the radius vector $\mathbf{R}$ and the support vector ρ (the support function with the notion of direction) is given by

$$\mathbf{R}(\vartheta, \varphi) = \mathbf{M} \cdot \rho(\vartheta, \varphi),$$

$$\mathbf{M} = \begin{pmatrix} \cos\vartheta \cos\varphi & -\sin\varphi & \sin\vartheta \cos\varphi \\ \cos\vartheta \sin\varphi & \cos\varphi & \sin\vartheta \sin\varphi \\ -\sin\vartheta & 0 & \cos\vartheta \end{pmatrix}.$$

(26)

An important quantity related to ρ is the Gaussian surface density $G(\vartheta, \varphi)$, defined as

$$G(\vartheta, \varphi) = AC - B^2,$$
$$A = \rho_{\vartheta\vartheta} + \rho,$$
$$B = \frac{1}{\sin\vartheta}\rho_{\vartheta\varphi} - \frac{\cos\vartheta}{\sin^2\vartheta}\rho_{\varphi},$$
$$C = \frac{1}{\sin^2\vartheta}\rho_{\vartheta\vartheta} + \frac{\cos\vartheta}{\sin\vartheta}\rho_{\vartheta} + \rho,$$

(27)

where the subscripts denote standard partial derivatives. A necessary and sufficient condition for an object to be convex is that $G(\vartheta, \varphi) \geq 0$ for all ϑ, φ.

The support function representation has an extra important property related to the choice of the origin. Assume, for instance, that ρ is expanded in spherical harmonics. If we change the origin O_1 to O_2 the expansion remains the same except the coefficients of the three terms of order 1. The radius vector representation does not have this property. In inversions of the shapes from projections, this special property in very important.

B. CONCAVE PARTICLES

Concave particles can be subdivided into two categories: star-shaped and non-star-shaped forms. Examples of these are depicted in Fig. 6. In mathematical set theory, a star-shaped form is defined as a geometry in which at least one point inside a particle can be found from which any boundary point is uniquely reached so that the radius vector intersects the boundary only once. In some sense star-shaped particles are mildly concave. To our knowledge, no universally accepted method exists of how to describe a non-star-shaped form. We will, accordingly, concentrate only on star-shaped surfaces.

A natural choice for the representation of star-shaped forms is the use of the radius vector **R** description. The convex hull (CH) of any concave shape represents, in a sense, the best approximation for concave figures and can be easily found numerically from

$$\mathrm{CH}(\Theta, \psi) = \rho(\Theta, \psi) = \max\big(\mathbf{R}(\vartheta, \varphi) \cdot \mathbf{n}(\Theta, \psi)\big), \tag{28}$$

where max denotes the maximum in the whole (ϑ, φ) set when (Θ, ψ) are kept fixed and where **n** is a unit vector and where we explicitly show that CH coincides with the support function ρ. In Fig. 7 we show some CHs for star-shaped figures.

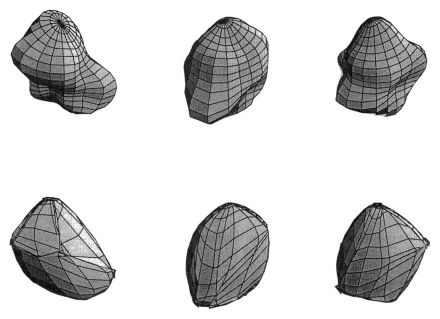

Figure 7   Three star-shaped figures (first row) and their convex hulls (second row).

The most versatile method for describing the shapes of naturally occurring random particles is to use the concept of stochastically rough spheres (SRSs). The method is explained in detail in Chapter 11. The basic concepts of this method have been known for a long time and have been applied to various problems. By far the most serious problem with the SRS technique is that the actual projections are almost impossible to obtain so that one must work with the planar intersections. In this respect the support function formalism is superior.

Our recent contribution to the SRS technique is the introduction of a new and flexible autocorrelation function $C(\gamma)$ between two radii (Lumme and Rahola, 1998). We suggested a function

$$C(\gamma) = q \frac{\sin(p \sin \frac{1}{2}\gamma)}{p \sin \frac{1}{2}\gamma} + (1 - q)P_2(\cos \gamma), \tag{29}$$

where γ is the angular distance between two radii, p is an arbitrary positive constant, $0 \leq q \leq 1$, and P_2 is the second-degree Legendre polynomial. C has a Legendre expansion with positive coefficients, as required (Lumme and Rahola, 1998). In the next section we will show some examples of light scattering by SRSs having the autocorrelation function of Eq. (29).

C. AGGREGATES

It is fairly straightforward to model the shape of a homogeneous, stochastically irregular particle. To model a cluster of closely packed particles is much more demanding and requires many more free parameters. First the properties, size distribution, and shapes of the constituent particles must be specified. Next the growing process of the cluster must be modeled.

Random close packing of particles is a classically unsolved problem. Even for equal-sized spheres only results based on computer simulations are available. The particle aggregates collected during stratospheric flights and shown in cosmic dust catalogs seem to indicate that the clusters are dense. A dense cluster is defined as a collection of touching particles where the packing density (filling factor) exceeds 0.05.

Lumme *et al.* (1997) analyzed light scattering by two nearly extreme geometries of clusters. In their first case the model consisted of several hundred cubic elements arranged in a regular pseudospherical geometry. This collection of constituent volume elements was then thinned by randonly removing elements but requiring that every remaining element continue to have at least one touching neighbor. In this model the filling factor remains constant as a function of the distance from the center of the cluster. The other extreme case studied by Lumme *et al.* (1997) was a fractal geometry, more specifically the so-called diffusion-limited aggregation (DLA) process. In this process the filling factor drops off very quickly with the distance from the center. Computer simulations have shown that the fractal dimension of the DLA is about 2.5.

V. LIGHT SCATTERING BY COSMIC DUST PARTICLES

Cosmic dust particles permeating the solar system are certainly not spherical in shape. Therefore, as explicitly shown by Lumme and Rahola (1998), the Lorenz–Mie theory is completely inadequate in explaining most of the light-scattering properties of these particles. A few exact theories and a large number of various approximate models exist, as summarized in Chapter 2, to account for light scattering by nonspherical particles. Some of the exact theories were applied to three different shapes by Hovenier *et al.* (1996) in order to test the compatibility of various methods. All the methods examined agreed within a reasonable accuracy.

Laboratory measurements of the light-scattering properties of irregular particles and particle clusters are very important. These measurements can be done at least in two ways. First, single noninteracting particles can be measured using some of the existing techniques such as electrostatic levitation, particles in flows, or microwave analog. Although there are plenty of published measurements of small particles, a detailed accompanying description of their shapes, sizes, and refractive indices is missing in most cases. The measurements should always provide maximal information to minimize the number of free adjustable parameters. Second, after the single-scattering properties are well known, the same particles could be measured in thick layers to mimic planetary regoliths. Then the effects of multiple scattering, close packing, and surface roughness could be separated from those of individual particles. In this regard perhaps the most promising set of laboratory measurements were acquired by Zerull *et al.* (1993), who used the microwave analog technique for clusters of up to 500 interacting spheres. Unfortunately, some crucial information was missing, although some of that could conceivably be recovered. Quite too often laboratory measurements concentrate only on the scattered intensity of light. This yields just one element of the whole 4×4 scattering matrix, whereas almost certainly all 16 elements are sensitive to different physical parameters of the sample.

A. INTEGRAL EQUATION TECHNIQUE FOR COSMIC DUST PARTICLES

We cannot know for sure the typical geometry of IDPs. Therefore, we need a flexible method that could be used to study light scattering by some basic classes of shapes as described in the previous section. For this we have chosen the integral equation (IE) formalism (Lumme and Rahola, 1998). We now summarize, very briefly, the key points of this technique.

The volume integral equation of electromagnetic scattering is discretized by assuming that the electric field is constant inside small cubic cells. By requiring that the integral equation be satisfied at the centers of the N cubes (point-matching or collation technique) and using a simple one-point integration, we obtain the basic set of coupled linear equations, which can be found in Lumme and Rahola (1998). In the present form our IE is mathematically and computationally very similar to the discrete dipole approximation (DDA; see Chapter 5). One of the great advantages of IE over DDA is that no effective medium theories (Chapter 9), such as Maxwell-Garnett or Bruggeman, are needed and thus no unnecessary free parameters are introduced.

IE requires the solution of a large system of linear equations where the coefficient matrix is not Hermitian but is complex symmetric. State-of-the-art methods for solving the equations most effectively have been studied by Rahola (1996). The quasi-minimal residual method (QMR) was shown to be the best one. QMR turned out to be three to four times faster than the classical conjugate gradient algorithm basically because the computationally expensive extra matrix–vector product is not needed. In our computations we systematically use the 3D convolution scheme to speed up the calculation of the matrix–vector product.

B. STOCHASTICALLY ROUGH SPHERES

We next apply our IE formalism to two basic classes of particles: SRSs described in detail in Chapter 11 and stochastic aggregates.

Although our code for the IE method is believed to be one of the fastest in the field, we are still, for computational reasons, restricted to volume-equivalent size parameters less than about 10. To a large extent this is a consequence of the large number of random orientations of a particle needed to obtain convergent results.

We assume two classes of SRSs defined in terms of the free parameters in the autocorrelation function of Eq. (29) and the standard deviation of the mean radius of an SRS. These classes are:

$$\text{Class 1:} \quad p = 5, \quad q = 0.5, \quad \sigma = 0.4,$$
$$\text{Class 2:} \quad p = 10, \quad q = 0.9, \quad \sigma = 0.2.$$

For the volume-equivalent size parameters we adopt values 1, 2, 4, and 6. Because we do not aim at any particular substance, we assume the refractive index $m = 1.5 + 0.005i$, which should not be too far from that of a typical IDP. Fortunately, the scattering matrix elements are not very sensitive to m in the range $1.5 \pm 0.2 + (0.005 \pm 0.005)i$.

Because IDPs must be distributed over a size distribution, we assume a discrete power law distribution with the exponent γ. We combine the light-scattering and linear polarization computations for average size parameters of 1, 2, 4, and 6 by this distribution. We show results for two values of $\gamma = 2$ and 3 in Fig. 8 for the intensity and in Fig. 9 for the degree of linear polarization. From these figures we can judge that the two ubiquitous phenomena, the opposition effect and the reversal of linear polarization, appear quite naturally. We emphasize that we have not tried to fit any observational data.

To see the effect of particle roughness on the intensity and linear polarization, we fit biaxial ellipsoids to our model SRSs using Mishchenko's T-matrix code (Chapter 6). The results are also given in Figs. 8 and 9, using dashed curves.

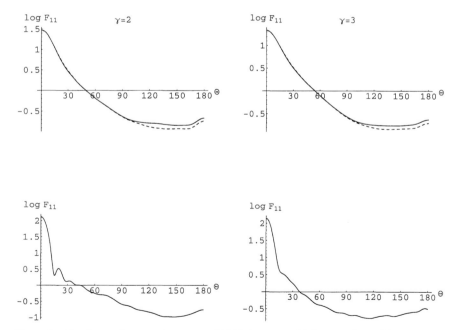

Figure 8 F_{11} element for power law averaged SRSs (upper panels) for two values of the power law exponent. The dashed curves show scattering by best-fit spheroids. In the lower panels we show F_{11} for two kinds of fluffy aggregates (defined in text).

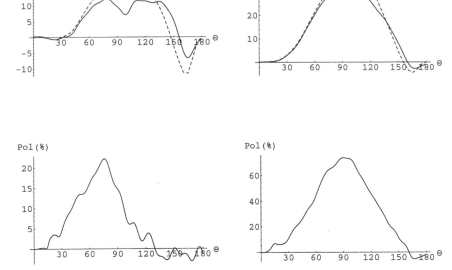

Figure 9   Linear polarization for power law averaged SRSs (upper panels). Note that the best-fit spheroids (dashed curves) are a much poorer approximation for polarization than for F_{11}. In the lower panels we show polarization for two kinds of fluffy aggregates (defined in text).

C. FLUFFY AGGREGATES

Lumme *et al.* (1997) analyzed light scattering by various aggregates. They computed all 16 elements of the scattering matrix by the DDA technique. There were two geometries for the aggregates. In the first geometry a sphere with a size parameter of 15 was subdivided into 100 or 200 cubic cells for the first model and 20 or 60 cells for the second model. In the process it was required that every remaining cell had at least one touching neighbor. In the second geometry we used the diffusion-limited aggregation process for fluffy aggregates. In this we created clusters of 100 and 200 constituent elements with a size parameter of 1.2. We used two refractive indices: $1.29 + 0.008i$ (ice) and $1.57 + 0.012i$ (silicate). In Fig. 8 (lower panels) we show results for F_{11} in the case of the cluster of 200 constituent elements in our first geometry (random removal) and in the DLA geometry. Linear polarization for the same geometries is shown in Fig. 9. It is clear that these models are also capable of producing both backward enhancement and negative polarization. However, the SRS model seems to be closer to the real data.

This might indicate that IDPs are predominantly rough particles rather than fluffy aggregates.

VI. DISCUSSION

In this chapter we have emphasized the importance of complementary observations and models of the different aspects of how cosmic dust is seen in the solar system. Because the two ubiquitous phenomena (opposition effect and reversal of linear polarization) are so similar for different objects even quantitatively, it is hard to believe that there would be more than just one physical reason for their existence. What this reason is remains unknown.

There are two mechanisms suggested to account for the opposition effect. Already in 1887 Seeliger introduced the famous mutual shadowing concept to explain the strong brightening of Saturn's rings toward the opposition geometry. This model and the several variants of it do require, however, a thick layer of interacting particles to work and cannot, therefore, explain the observed effect for cometary particles and IDPs. All these models implicitly assume that the single-particle phase function is basically flat in the backward direction. Several laboratory experiments and our recent studies in this field quite indisputably show that single particles can easily have strong opposition spikes. Another wave optical explanation for the opposition effect is the coherent backscatter mechanism (CBM). In this phenomenon waves propagating in opposite directions interfere constructively in the exact backscattering direction, which causes the strong increase in intensity. The CBM concept has the extra bonus that it could also explain the reversal of linear polarization as shown by Muinonen (1990). Unfortunately, a quantitative treatment of the CBM is still lacking.

To our knowledge, there are no systematic observations of cosmic dust extending beyond the first two elements of the whole scattering matrix. Such data would have the potential to considerably improve our knowledge of the dust. We realize that several elements of this matrix would be close to zero, but even then they could yield important upper and lower bounds for the physical properties of the particles. We continue to emphasize the importance of controlled laboratory work. Most important, all the relevant physical parameters of the measured samples should be reported and not be left for modeling only. Controlled measurements would mean, among other things, that the parameters should be changed one at a time to allow for the possible effects of the change.

Existing computer codes for light scattering by nonspherical particles seem to be in good order. In particular, the T-matrix algorithm (Chapter 6) is capable of handling size parameters exceeding 120 for rotationally symmetric particles and the integral equation formalism by Lumme and Rahola (1998) can compute light scattering by very complicated shapes up to equivalent size parameters of about 20. Because the computer capabilities constantly improve, the maximum size parameter increases accordingly.

Chapter 20

Biophysical and Biomedical Applications of Nonspherical Scattering

Alfons G. Hoekstra and Peter M. A. Sloot

Faculty of Mathematics, Computer Science, Physics, and Astronomy
University of Amsterdam
1098 SJ Amsterdam, The Netherlands

I. INTRODUCTION

The use of light scattering from single, nonspherical particles in biomedical and biophysical applications is most often concerned with scattering from single biological cells and thus deals with single scattering but not multiple scattering. Because the objective of such light-scattering measurements is the retrieval of specific characteristics of biological cells, we first address the nature of cells and how this may be expected to affect the scattering of light by them.

Biological cells have in general a very complicated structure, both in shape and in composition (Dyson, 1974). Shapes can range from nearly spherical to highly irregular dendritic structures. Cells can be procaryotic (without a nucleus; these are the bacteria and blue–green algae) or eucaryotic (with a nucleus). Nuclei may be nearly spherical or exhibit a complex shape. Cells can contain all kinds of organelles, which also may differ in size and shape. Furthermore, cells of a specific type often have a range in size, shape, and composition, resulting in a large biological variability. To complicate things even more, the relevant parameters may depend strongly on the age of the cells.

Of course, cells are living entities; they divide and actively respond to external stimuli. In many cases such cellular activity results in changes in cellular morphology and average or effective refractive index. Cells may enter pathological stages and, for example, become malignant cancer cells. Such changes usually show up as a change in cellular morphology.

In previous chapters of this book, it has been demonstrated that light scattering from single particles is very sensitive to their morphology and optical properties. We are therefore led to expect that light scattering can be a sensitive tool for discriminating between cell types or between healthy and malignant cells or for probing changes in cells resulting from stimuli. Indeed, physiological and morphological information on living cells can be retrieved using light scattering. Differences between cell populations, which are difficult to observe using microscopic methods, can be determined with light scattering. Furthermore, in contrast to microscopy, light-scattering techniques support rapid and noninvasive measurements of bulk quantities of cells. It is in such applications that single-scattering techniques have found widespread use in biomedicine and biophysics.

Let us consider a number of typical examples. The effect of penicillin on bacteria is seen within minutes in the scattered light (Berkman et al., 1970). Bacteria can be identified (Wyatt, 1968, 1969), changes in heat-treated bacteria can be observed (Berkman and Wyatt, 1970), and the size of rod-shaped bacteria can be measured (Bronk et al., 1995; van de Merwe et al., 1997). Human white blood cells can be identified (Salzman et al., 1975) and changes in the shape of platelets can be assessed (Michal and Born, 1971). Forward and sideward scattered light can be used to determine nuclear and cellular morphology (Benson et al., 1984) or cell activation may be probed (Kraus and Niederman, 1990). If cells undergo programmed cell death by a mechanism called apoptosis, they show morphological changes in cell size and granularity owing to loss of water. By measuring forward and sideward scattered light in a flowcytometer (see Section III.B), cells undergoing apoptosis can be discriminated from normal cells (Gorman et al., 1997; Ormerod et al., 1995). Many more fascinating examples of the use of light scattering to probe cellular properties can be found throughout the literature (e.g., Latimer, 1982), and we examine, in some detail, several more examples later.

Such an enormous amount of empirical data of light scattering from biological cells calls for theoretical tools to interpret measurements in terms of cell characteristics and changes (Latimer *et al.*, 1968). Furthermore, such theoretical tools are needed to help guide the experimenter in issues such as the choice of the optimal solid angle to be subtended by the light-scattering detector so that particular cell parameters may be best retrieved (Brunsting, 1974). A good interpretation of the light-scattering experiments or the development of new light-scattering experiments to probe relevant cellular properties by, for example, exploiting all the information that is present in the scattering matrix requires a suitable theoretical framework.

Today, with the availability of many light-scattering theories, we should in principle be able to find appropriate models for light scattering from biological cells. Furthermore, the availability of increasingly powerful computers also enables us to simulate models that require large-scale calculations. Without going into excessive detail, we will sketch the contours of a well-established theoretical framework for achieving all this and will present some recent case studies.

A number of experimental techniques that apply light scattering are in routine use in biomedical and clinical environments. Two of them, ektacytometry and flowcytometry, rely on measurements of single scattering and will be discussed here.

Cells aggregate to form tissues. Propagation of light through tissues is highly relevant for biomedical applications. However, these processes now involve multiple light scattering and radiative transfer, which is beyond the scope of this chapter. For more information we refer the reader to Welch and van Gemert (1992), Motamedi (1993), Das *et al.* (1997), and Tuchin (1998).

II. THEORETICAL FRAMEWORK

A. BIOLOGICAL CELLS AS NEAR INDEX MATCHING PARTICLES

In assessing which theories or approximations may be appropriate for calculating single scattering from biological cells, we must consider the key cell characteristics, namely, the range of sizes, refractive indices, and cell morphology. The size of biological cells may vary from only a few tenths of a micrometer for the smallest bacteria, up to dimensions of centimeters for certain marine algae (Dyson, 1974). Bacteria usually have a size of a few micrometers. Human red blood cells have a diameter of 6–8 μm, human white blood cells are in the range of 5–10 μm in diameter (see, e.g., Latimer *et al.*, 1968), and most other human cells are in the range of 5–20 μm in diameter. Taking $\lambda = 500$ nm as a typical wavelength, we find that for human cells the size parameter $x = 2\pi r/\lambda$ (where r is the cell radius) is in the range $30 < x < 130$.

In scattering experiments, the cells are usually suspended in a medium with a refractive index very close to that of water. The refractive index of the cells, or cell compartments, has a very limited range. For those wavelengths (usually in the visible region), where absorption by cellular molecules is absent or very small the refractive index is very close to the refractive index of the suspending medium. This means that the relative refractive index m is very close to 1. This is an important general rule and in the rest of this chapter we will assume that it is valid.

Many examples that support this rule can be found in the literature. For instance, Brunsting and Mullaney (1974) measured the refractive index of the cytoplasm and the nucleus of Chinese hamster ovary (CHO) cells and found $m_{cytoplasm} = 1.37$ and $m_{nucleus} = 1.39$. The relative refractive index m (relative to water, $m_{water} = 1.333$) is 1.03 and 1.04, respectively. The relative refractive index of human white blood cells is in the range $1.01 < m < 1.08$ (Keohane and Metcalf, 1959). More recently, Tuminello et al. (1997) measured the refractive index of *Bacillus subtilis* spores and over a large wavelength range the real part of the refractive index was between 1.5 and 1.6 (i.e., $1.13 < m < 1.2$), whereas the imaginary part was smaller than 0.01. Schmitt and Kumar (1996), who measured variations of refractive index in biological tissue, report a refractive index of 1.4–1.45 for structural fibers and nuclei ($m \sim 1.09$) and approximately 1.36 ($m \sim 1.02$) for the cytoplasm of cells of several types of tissue. Using phase contrast microscopy, Beuthan et al. (1996) measured the spatial variation of the refractive index in a single biological cell (L929 fibroblast). They found that only the nucleus and the cell membrane have a clearly different refractive index as compared to the mean value of the cell. This mean value is approximately 1.42 ($m \sim 1.07$), whereas the cell membrane has a refractive index of about 1.48 ($m \sim 1.11$) and the cytoplasm, about 1.38 ($m \sim 1.04$). More information on the refractive index of biological cells can be found in Bolin et al. (1989) and Duck (1990).

In general, we will assume that most biological cells are near index matching particles; that is, the relative refractive index m is very close to 1. We are therefore excluding any cases with large absorption at relevant visible wavelengths as well as situations in which the cells are not suspended in a medium with a refractive index close to that of water. Exact values of m are not well known and, in particular, information on the refractive index of the cell constituents is meager. However, remember that the biological variability of cellular morphology of a certain cell type introduces a significant spread in the differential scattering cross sections and as a consequence removes all detailed interference structure from the scattering matrix. Therefore, it is currently not clear to what accuracy the refractive index should be known in order to perform relevant calculations of single scattering from biological cells.

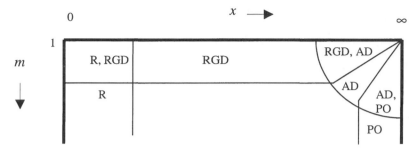

Figure 1   van de Hulst (x, m) plane in the limit of near index matching particles. R is Rayleigh scattering, RGD is Rayleigh–Gans–Debye scattering, AD is anomalous diffraction, and PO is physical optics.

In his renowned treatise on light scattering, van de Hulst (1957) shows a diagram of the (x, m) plane in which all light-scattering theories that apply to various regions of the plane are drawn. In Fig. 1 we show a part of van de Hulst's (x, m) plane diagram, adapted to our region of interest. For index matching particles, the possible scattering theories are limited to (excluding variational and numerical techniques to solve Maxwell's equations, which are considered in Section II.D) the following:

1. Approximate theories for which m must be close to 1 (see van de Hulst, 1957, Section 10.1; Chapter 2), that is, Rayleigh–Gans–Debye (RGD) scattering and anomalous diffraction (AD) scattering, and possible improvements of these theories, such as the Wentzel–Kramers–Brillouin (WKB) approach (see, e.g., Klett and Sutherland, 1992), modified RGD (Shimizu, 1983; Sloot and Figdor, 1986), or higher order RGD (Acquista, 1976)
2. Approximate theories for which m is not restricted, such as the physical optics (PO, ray tracing plus diffraction) method (see, e.g., Bohren and Huffman, 1983, Chapter 7; Muinonen, 1989; Peltoniemi *et al.*, 1989) or Rayleigh scattering (see van de Hulst, 1957, Section 6.4)
3. Exact analytical solutions of Maxwell's equations, such as Lorenz–Mie scattering by single homogeneous or concentric spheres (Bohren and Huffman, 1983), by stratified spheres (Kai and Massoli, 1994) or spheres with a radially varying refractive index (Perelman, 1996), by compounded spheres (see Chapter 8), or by a sphere with arbitrarily located spherical inclusions (Borghese *et al.*, 1994) or irregular inclusions (Videen *et al.*, 1995b)

Rayleigh scattering theory is valid only if the particle is smaller than the wavelength by at least a factor of order 10. If we restrict ourselves to consideration of scattering in the visible and near infrared wavelengths, we may conclude that even for the smallest bacteria, Rayleigh scattering theory is not valid. The size and refractive index ranges of cells therefore limit the theories for describing light scattering of these cells to RGD scattering, AD theory, PO theory (and their improvements), and exact theories.

All these theories have been used with varying success. For instance, Mullaney and Dean (1970) modeled cells as homogeneous spheres with $m = 1.05$ and investigated the near forward scattering using Lorenz–Mie theory. They showed that in this model, the logarithm of the $\Theta = 0.5°$ scattered intensity is proportional to the volume of the cell, for the size parameter range $10 < x < 100$, where Θ is the scattering angle. However, they emphasized that the structure of actual cells is sufficiently complex that the concept of a mean or effective refractive index for all cells in a population may be an oversimplification.

The next logical step is to model cells as two concentric spheres, where the inner sphere models the nucleus and the coating models the cytoplasm. Brunsting and Mullaney (1972) proposed this model and investigated scattering by a concentric sphere in the (x, m) range of biological cells. They found that the principal influence of the cell nucleus occurs at angles outside the main scattering lobe in the forward direction. In an experimental study on the scattering of CHO cells in the M and G_1 phase, which can be modeled by homogeneous and concentric spheres, respectively, these theoretical conclusions were confirmed (Brunsting and Mullaney, 1974). However, Meyer and Brunsting (1975) studied the concentric sphere model in somewhat more detail, and they noted that the small angle scattering is highly dependent on the nucleus to cell diameter ratio. Furthermore, they found that the scattering patterns of nucleated cells exhibit a fine lobe (high frequency) structure dependent on the whole cell size, whereas the envelope lobe (low frequency) structure appears to be dependent on the relative size of the nucleus.

We applied a modified version of RGD scattering (Sloot and Figdor, 1986) to explain the anomalous forward scattering of human lymphocytes that undergo volume changes as a result of a change in the osmolarity of the suspending medium (Sloot *et al.*, 1988a). Contrary to common belief, the forward scattering of these osmotically shocked lymphocytes is inversely proportional to their volume. Using the modified RGD, we could explain this effect by including the change in the refractive index of the cell, caused by the swelling or shrinking. Actually, our light-scattering calculations led us to predict a nuclear volume change in the osmotically stressed lymphocytes, an effect that was absolutely not expected. Later, we showed that the nucleus of the lymphocyte does indeed change its volume as a result of a change in the osmolarity of the suspending medium (Hoekstra *et al.*, 1991). Although we proposed an explanation for this effect, namely, a mechanical link between the cell membrane and the nuclear envelope through the cytoskeleton, it still remains an open issue. The effect of osmolarity on scattering from bacterial cells, yeast, and chloroplasts was studied by Latimer and co-authors (see Latimer, 1982, and references therein).

Soini *et al.* (1997) measured scattering from single, oriented red blood cells and using WKB theory were able to reproduce these experimental results for scattering angles in the range of 15°–35°.

Examples of the use of AD theory in the context of biological cells are provided by Stramski *et al.* (1988), who studied elastic light scattering from phytoplanktonic cells. Streekstra *et al.* (1993, 1994) examined forward scattering from red blood cells deformed into an ellipsoidal shape in an ektacytometer (see also Section III.C) using AD theory. Mazeron *et al.* (1997a, b) studied the same system using the PO approximation.

Recently Videen and Ngo (1998) showed the impact of new developments in exact scattering theory for calculating scattering from biological cells. They modeled a cell as a spherical cytoplasm, containing a nonconcentric spherical nucleus and surrounded by a concentric membrane, and derived exact formulations for this model particle, with organelles modeled using effective medium approximations for the refractive index of the cytoplasm.

B. WHICH THEORY TO CHOOSE?

For cells modeled as near index matching particles, we can apply Rayleigh–Gans–Debye scattering, anomalous diffraction, or the physical optics approach, improvements of these theories, or exact theories. RGD scattering applies when $(m-1) \ll 1$, and the phase shift in the particle $\rho = 2x(m-1) \ll 1$ (van de Hulst, 1957; Bohren and Huffman, 1983). This means that if m is very close to 1, RGD can be applied to large values of x as long as ρ remains small. Anomalous diffraction applies for particles with large x, but is restricted to small scattering angles; the value of ρ is not of importance. Finally, if both x and ρ are large, the domain of physical optics is entered.

In practical applications one wants to know the location of the ranges of applicability of the different theories or, more specifically, the magnitude of the errors that are induced by using a certain approximation as a function of the position in the (x, m) plane. In general, such a question is difficult to answer, as the error depends on the shape and composition of the particles and also on the scattering matrix element and range in scattering angles that is of interest. We can proceed in two ways.

First, one may create a number of benchmark particles that are typical for certain classes of biological cells and compare the RGD, AD, or PO calculations with rigorous numerical methods, such as the discrete dipole approximation (DDA) method (see Chapter 5 of this book and Section II.D). Here we must assume that the results for the small number of benchmark particles are representative of the full class. Second, one can compare, as a function of x and m, exact scattering theories, for example, spheres or ellipsoids, with the RGD, AD, or PO calculations. The resulting error maps are relatively easy to generate. We have, for example, compared RGD scattering from a homogeneous sphere with exact Lorenz–Mie theory (unpublished results; see, however, http://www.wins.uva.nl/~alfons/rdg-vs-mie/rdg.html). We find that the RGD theory is able to cover a large range of the (x, m) plane relevant to biological cells if we accept average errors up to 10% in the F_{11} scattering matrix (see Chapter 1) element. The modified RGD theory (Shimizu, 1983) reduces the average error for F_{11}, but does not improve on the relatively larger errors found for the other scattering matrix elements. We encourage readers to generate their own error maps and, as we have done, make the results available on the World Wide Web.

It is apparent that the question posed by the title of this section has no single answer. The selection of an appropriate approximation theory will always depend on the precise characteristics of the index matching particles and the magnitude of the errors that can be tolerated. Additional studies that systematically investigate the usefulness of the index matching theories for the full (x, m) range of biological particles are certainly warranted.

C. Polarization

Although successful in many circumstances, the approximation theories considered previously all have their limitations, which are most obvious if one wishes to take the polarization properties of the incident and scattered fields into account. It is well known, as illustrated by the examples discussed later, that the polarization state of the scattered light carries an enormous amount of relevant biomedical and biophysical information, which is difficult to obtain by considering only the total scattered intensity. Therefore, there has been substantial interest in measuring the complete scattering matrix for biological cells.

As an example, consider an experiment with cross-polarized elements; that is, the incident light is polarized perpendicular to the scattering plane and the component of the scattered light parallel to the scattering plane is measured. The intensity on the detector is proportional to (see Bohren and Huffman, 1983, Table 13.1) $(F_{11} - F_{12} + F_{21} - F_{22})/4$, where **F** is the scattering matrix as defined in Chapter 1. For both (modified) RGD scattering as well as (concentric) spheres $F_{11} = F_{22}$ and $F_{12} = F_{21}$, so under these circumstances we expect a zero intensity. However, de Grooth *et al.* (1987) measured considerable signals for all types of human white blood cells, even allowing them to distinguish between two types of granulocytes. So the polarization of light scattered by biological cells can reveal properties not evident from measurements of the total scattered intensity alone (see also Section III.B). Deviations of the F_{22} element from the F_{11}

element have also been reported for different kinds of pollen (Bickel and Stafford, 1980) and marine organisms (Voss and Fry, 1984; Lofftus *et al.*, 1992). We may conclude that we cannot model these experimental results with (modified) RGD scattering or with models that view a cell as a (concentric) sphere with equivalent volume and refractive index. Although AD and PO allow the inclusion of polarization effects, they are not suited for the smaller biological cells, which also exhibit strong depolarization signals.

Another scattering matrix element that has received much attention is the F_{34} element. This scattering matrix element has an extreme sensitivity to small morphological changes. For instance, we showed that a small surface roughness on a sphere has a pronounced effect on the F_{34} element (Hoekstra and Sloot, 1993). Measurements of the F_{34} element for many biological cells have been reported (Bronk *et al.*, 1995; van de Merwe *et al.*, 1997; Bickel and Stafford, 1980; Voss and Fry, 1984; Lofftus *et al.*, 1992). All these results show an extreme sensitivity of the F_{34} element to very small morphological changes. The importance of the F_{34} element was recently demonstrated by Bronk *et al.* (1995), who showed that it allows the diameter of rod-shaped bacteria (*Escherichia coli* cells) to be obtained. This parameter used to be very difficult to measure with other techniques such as optical microscopy or electron microscopy. Later, van de Merwe *et al.* (1997) used this technique to measure the effect of growth conditions on the diameter of these rod-shaped bacteria. Because the (modified) RGD scattering theory results in $F_{34} = 0$, it is obviously not suited to F_{34} calculations of biological cells.

Finally we refer to circular intensity differential scattering (CIDS) (Bustamante *et al.*, 1980a, b, 1981, 1982, 1984). In a CIDS experiment the difference between scattered intensities for left and right circularly polarized incident light is measured. In this way, as can be derived from Table 13.1 in Bohren and Huffman (1983), the F_{14} element of the scattering matrix is measured.

RGD scattering for a particle with an isotropic polarizability results in a zero CIDS signal. Anisotropic polarizabilities are required in RGD theory to produce CIDS signals. The so-called "form-CIDS" is an anisotropy caused by the helical structure of a particle such as that demonstrated in the helical sperm head of the octopus *Eledone cirrhosa* (Maestre *et al.*, 1982; Shapiro *et al.*, 1994a). CIDS is also relevant in the study of secondary and ternary structures of macromolecules, such as the supercoiling of DNA (Zietz *et al.*, 1983). As a final example, CIDS can be used to study the polymerization of hemoglobin in sickle cells (Gross *et al.*, 1991).

Approximate theories such as (modified) RGD scattering can be very successful in describing certain scattering properties of biological cells. In particular, near-forward scattering and total cross sections are often treated adequately by these theories. However, if the polarization properties of the scattered light are of interest, more realistic models beyond (modified) RGD, AD, WKB, or PO are required. A number of examples have been reported in the literature (e.g., Acquista, 1976), and many possible theories are available (e.g., Karam, 1997). What is missing is a critical survey of these theories in the range of size parameters and relative refractive indices of biological cells and a comparison with analytical results, brute-force numerical simulations, and, most important, experimental data.

D. Numerical Solutions of Maxwell's Equations

Instead of taking simplified particle models such as concentric spheres or using approximate theories such as RGD or AD, one can choose to rely on more involved methods. One possibility is the use of formal solutions of the scattering problem (e.g., the extended boundary condition method), which require numerical computations to satisfy the boundary conditions. The T-matrix method (see Chapter 6 of this book) is best known and most widely used. Nilsson *et al.* (1998) applied it to calculate scattering from red blood cells. Other formulations exploiting multiple multipoles are now also gaining more interest (Doicu and Wriedt, 1997a, b; Wriedt and Doicu, 1997).

Another possibility is to use methods that numerically solve Maxwell's equations, either in the time domain (the finite difference time domain method, FDTD; see Chapter 7) or in the frequency domain. The discrete dipole approximation (DDA; see Chapter 5) is an example of a method that numerically solves Maxwell's equations in the frequency domain. Many more frequency domain methods are known; see Shafai (1991) and Chapter 2, for example.

The FDTD technique was applied by Dunn and Richards-Kortum (1996) and Dunn *et al.* (1996, 1997) to simulate scattering by a cell model, both in two and three dimensions. The cell is modeled as a slightly ellipsoidal particle with a mean diameter of approximately 11 μm. It contains a nucleus with a diameter of 3 μm and in some cases contains organelles with diameters that are uniformly distributed between 0.4 and 1 μm. The presence of organelles (8% volume fraction), such as mitochondria or melanin, strongly enhances the scattering beyond the forward scattering lobe, suggesting that they play an important role in determining sideward and backward scattering. An interesting question not addressed by Dunn *et al.* is how these organelles influence the polarization characteristics of the scattered light. From this study, Dunn *et al.* concluded that the presence of the nucleus affects the scattered intensity only in the forward direction, with a larger nucleus resulting in a stronger signal. These results for the forward scattering are in agreement with those based on concentric sphere calculations as reported in Section II.A. However, the conclusions seem to contradict the earlier results with respect to larger scattering angles. This issue certainly merits further investigation, because the change of size and shape of the nucleus is an important characteristic of cancer cells.

The experimental results obtained by Bronk *et al.* (1995) on the scattering from rod-shaped bacteria were actually interpreted using the DDA method. Bronk *et al.* succeeded in calculating F_{34} from rod-shaped bacteria, and by fitting the DDA calculations to the experimental results, they were able to deduce the diameter of the bacteria.

Shapiro *et al.* (1994a) used the DDA method to interpret their experimental data on the scattering matrix elements of the octopus sperm *Eledone cirrhosa*. On the basis of a comparison between the DDA model calculations and the experimental data, they proposed a model for the packing of DNA fibers in the sperm head.

Our own interest is in scattering from human red and white blood cells. One of the goals is to find approximate models, beyond the modified RGD, that allow for a rapid calculation of the scattering cross sections, including depolarization effects, of human blood cells. Such models allow tuning of light-scattering experiments to be as sensitive as possible for certain morphological features of the cells. To validate the models, we will compare them with simulations of scattering from white blood cells, using a DDA program that was optimized to execute on parallel computers (Hoekstra and Sloot, 1995; Hoekstra *et al.*, 1998). We have carried out first DDA simulations of small human white blood cells (lymphocytes) (Hoekstra *et al.*, 1998) and red blood cells (unpublished results) and are in the process of simulating larger types of white blood cells (the granulocytes).

The main disadvantage of many full numerical solutions is that they require an enormous amount of computational power, far beyond the capabilities of typical desktop computers. For instance, we estimated that accurate DDA computations of scattering from human white blood cells require models with up to 10 million dipoles (Hoekstra *et al.*, 1998). Fortunately, biological cells have small enough relative refractive indices that allow these very large DDA models to be solved, although it requires supercomputers and vast amounts of memory ($\sim$1 GB). Furthermore, the numerical solutions usually suffer from the fact that orientation averaging of the particles requires repeating the computations, one for each orientation. For larger particles such as human cells, this is a huge task and is not really practical in most applications. However, as we already suggested in Section II.B, large-scale numerical simulations can be used to test the validity of approximate methods by performing simulations on a set of benchmark particles. The approximate methods can then be used for calculating orientation averages and for taking into account the size, shape, and refractive index ranges that biological cell types exhibit.

III. EXPERIMENTAL TECHNIQUES

A. INTRODUCTION

Just as we have many different theories and approximations for light scattering, many different experimental techniques exist for measuring cross sections or the complete scattering matrix. Chapters 2, 12, 13, and 18 cover techniques for measuring scattering matrices of small particles and such techniques have also been applied extensively to measure scattering matrices from biological cells. We will not review all possible techniques for measuring light scattering (see, e.g., Bohren and Huffman, 1983, for a nice overview), but restrict ourselves to two experimental techniques that have found routine use in clinical and biomedical applications, flowcytometry and ektacytometry.

B. FLOWCYTOMETRY

Flowcytometry (van Dilla *et al.*, 1985) is an important technique in the biomedical sciences for identifying and separating various populations of white blood cells, for example. A flowcytometer (shown schematically in Fig. 2) consists of a cuvet, in which the cells are forced to flow into a linear sequence using a technique called hydrofocusing. A laser beam is focused on the cells such that cells enter the focused beam one by one. Each cell produces a pulse of scattered light, which is

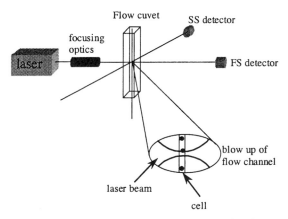

Figure 2 Schematic drawing of a flowcytometer. Cells enter a hydrofocusing cuvet and one by one pass a laser beam, which is focused by two crossed cylindrical lenses. The scattered light is detected in the forward and sideward directions.

usually measured in the two principal directions: the forward scattered (FS; usually $1° < \Theta < 3°$) and the sideward scattered (SS; usually $65° < \Theta < 115°$). The FS and SS signals consist of scattered light integrated over a small solid angle defined by the aperture of the detectors.

In most flowcytometry experiments, cells are stained with fluorescent probes, which bind to specific molecules on the cell surface or in the cell, and several fluorescence signals are measured in combination with light-scattering signals. In this way N independent signals are measured for every cell in the sample and the sample is represented as an N-dimensional histogram. After identifying separate clusters in the histogram, either by visual inspection or by fully automated statistical techniques (see, e.g., Sloot *et al.*, 1988b), different subsets in the original cell sample can be assessed. Many more details of this important technique, used extensively in analytical cytology, can be found in van Dilla *et al.* (1985).

We have argued that in many research and clinical applications, staining of cells is undesirable (Sloot *et al.*, 1989). It is thus preferable that a complete characterization of the sample be obtained solely on the basis of light-scattering measurements. Furthermore, even when fluorescence is the required observable, light-scattering signals are often used as triggers to extract interesting cells from the background. Here we will concentrate on the possibilities of using only light scattering in a flowcytometer to identify cell types. More specifically, we will restrict ourselves to the important case of human white blood cells.

In classical flowcytometry two light-scattering signals (FS and SS) are measured. The incident light is provided by a laser, which usually emits linearly polarized light. If we assume perpendicularly polarized incident light, the classical light-scattering measurements in flowcytometry are of $P_\perp U$ type (Bohren and Huffman, 1983; $P_\perp$ for incident perpendicular polarized light and U for measuring the total intensity of scattered light, with undefined polarization) and therefore FS, SS $= C \int_{\text{detector surface}} (F_{11} - F_{12}) \, dA$, with C a constant. Only a very small portion of the information potentially present in the scattered light is exploited. Yet, this suffices to identify the three main classes of human white blood cells (lymphocytes, monocytes, and granulocytes), as was first demonstrated by Salzman *et al.* (1975). This experiment has become very important and today Salzman's result is used as a standard procedure for discriminating between the main classes of white blood cells in flowcytometry. Figure 3 schematically shows the FS–SS histogram that was measured by Salzman.

Lymphocytes and monocytes can be differentiated on the basis of their FS signal, and by measuring the SS signal at the same time, the granulocytes can also be discriminated. Note that measuring only the FS signal would not be enough, as the granulocytes would completely overlap the lymphocytes and monocytes. The clusters in the histogram are Gaussian distributions (see, e.g., Sloot *et al.*, 1988b) and the width of the histograms is explained by the biological variability of the cells.

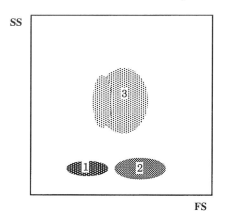

Figure 3 Schematic drawing of the FS–SS histogram of human white blood cells as measured in a flowcytometer by Salzman *et al.* (1975).

A popular, qualitative interpretation of the histogram of Fig. 3 is that FS (small angle scattering) is sensitive to cell size and that SS (side scattering) is sensitive to cell structure or cellular granularity. These arguments are based on the early work of Brunsting, Mullany, and Latimer (as discussed in Section II.A) and are also in agreement with the FDTD simulations of Dunn *et al.* (see Section II.D). Although these arguments help to understand some features of FS–SS histograms, one should be cautious in using them. Light scattering is much more subtle and every experiment should be considered in its own right. Nonetheless, because monocytes are larger than lymphocytes and granulocytes have more internal structure (granules, polymorphological nucleus) than both lymphocytes and monocytes, these arguments allow us to appreciate the structure of the typical histogram illustrated by Fig. 3.

Detection and data analysis techniques in flowcytometry are becoming more and more refined, allowing better and more accurate identification of the (sub) classes of human white blood cells. Furthermore, de Grooth *et al.* (1987) introduced the crossed polarizer experiment (see Section II.C) into flowcytometry of human white blood cells. In this way they could separate two subclasses of granulocytes, the eosinophils and the neutrophils, as illustrated in Fig. 4. This figure is based on data from de Grooth *et al.* (1987) and Terstappen *et al.* (1988a, 1990). The FS–SS histogram contains a number of extra, slightly overlapping distributions. A small cluster containing basophils (a third type of granulocyte) was located between, or overlapping, the lymphocyte and monocyte clusters. Note that the large granulocyte cluster in the FS–SS histogram could be separated into neutrophil and eosinophil clusters by the crossed polarizer experiment. Therefore, by

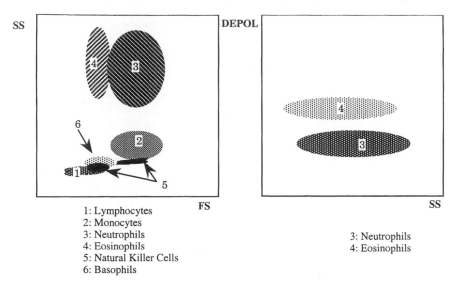

1: Lymphocytes
2: Monocytes
3: Neutrophils
4: Eosinophils
5: Natural Killer Cells
6: Basophils

3: Neutrophils
4: Eosinophils

Figure 4   Schematic drawing of the FS–SS histogram and SS–Depol histogram of human white blood cells, as measured in flowcytometer. This figure is based on data from de Grooth *et al.* (1987) and Terstappen *et al.* (1988a, 1990).

measuring only three light-scattering signals of each cell, a clear distinction between lymphocytes, monocytes, neutrophils, and eosinophils can be made. Furthermore, a subset of lymphocytes, the natural killer (NK) cells, can be found at larger FS and SS signals in the lymphocyte fraction, and some NK cells can be distinguished at low SS signals in the monocyte cluster. A detailed analysis of the SS signal of the lymphocytes even suggests that lymphocyte subclasses can be observed. The B lymphocytes and helper/suppressor cells have a low SS, whereas the cytotoxic cells have a large SS (Terstappen *et al.*, 1986).

The use of light scattering to study pathologies of human white blood cells is not widespread, but some examples exist. We restrict ourselves to chronic lymphocyte leukemia (CLL) (Gale and Foon, 1985), the most common form of leukemia in Europe and the United States. In more than 95% of the cases, CLL develops from a malignant transformation of B lymphocytes or B-lymphocyte precursors. Several forms of this leukemia are described with clear morphological features of the cells and correlated immunological properties (den Ottolander *et al.*, 1985). Examples are

• True B chronic lymphocytic leukemia (B-CLL): small cells, hardly any cytoplasm

- Prolymphocytic leukemia (PLL): larger cells, abundant cytoplasm, nucleus with very prominent central nucleolus
- Hairy cell leukemia: cells of moderate size, eccentric, oval nuclei, and "hairy" cytoplasmic projections

Once more, these morphological differences suggest that light scattering can be used to identify the cells, thus allowing an initial, fast screening of cell samples. Flowcytometric experiments, which measured the SS histograms of these cells, only lift a corner of the veil, and many more theoretical and experimental studies should be devoted to understand the potential of light scattering for such applications.

van Bockstaele *et al.* (1986) showed that hairy cells have a three- to fourfold larger SS than lymphocytes. They were able to count the concentration of hairy cells and follow the effects of treatment with alpha interferon. Terstappen *et al.* (1988b) studied the SS of B-CLL cells. The SS histogram of the lymphocytes of B-CLL is clearly different from those of other CLL such as leukemic follicular non-Hodgkin's lymphoma and a prolymphocytoid transformation of B-CLL (Terstappen *et al.*, 1988c). Finally, the effect of splenic irradiation of a B-CLL patient could be detected in the SS histogram (Terstappen *et al.*, 1988d).

These examples show that light-scattering measurements in a flowcytometer allow identification of many subclasses of human white blood cells. Unfortunately, the histogram regions overlap in many cases, and we should therefore look for alternative light-scattering signals that allow for a better discrimination of the cell subclasses. This is still a barely explored direction of research. However, it is very important, especially if more complicated cell mixtures, such as bone marrow, should be analyzed.

Several groups have gone beyond the classical FS–SS measurements. The importance of the crossed polarization experiment of de Grooth *et al.* (1987) has been mentioned previously. Using their modified RGD theory, Sloot and Figdor (1986) extensively studied differential cross sections of cells and concluded that measurement of backscattered (BS; $160° < \Theta < 174°$) light, combined with FS and SS, provides additional and independent morphological information on the cells. Based on this theoretical work and in the spirit of Brunsting, Sloot *et al.* (1989) have included BS measurements into flowcytometry and also showed that the complete scattering matrix can be measured in a flowcytometer. Many experiments were performed and the data suggest that BS, as well as, for example, the F_{34} element, does provide valuable information. However, also in this case much more experimental data and theoretical support is needed in order to further improve the usability of flowcytometry.

Recently, Chernyshev *et al.* (1995) and Soini *et al.* (1998) introduced a new type of optical design into flowcytometry, allowing them to measure the scattered intensity in a range of scattering angles from 5° to 120°. This new setup has significant potential to further increase the usefulness of flowcytometry.

C. EKTACYTOMETRY

An important physiological parameter is the deformability of red blood cells (RBCs). The diameter of an RBC is larger than that of the smallest blood vessels of microcirculation. This implies that RBCs must be able to deform quite drastically in order to enter the microcirculation and to deliver oxygen. If RBCs are sheared, their resting shape, a biconcave disk, changes to an ellipsoid (Fischer and Schmidt-Schönbein, 1977). The ratio of the major to minor axes depends on the shear rate, and the exact nature of the relationship has important clinical relevance (see, e.g., Hardeman *et al.*, 1988).

An ektacytometer measures the deformation of RBCs under shear stress using forward (small angle) light scattering. Figure 5 is a schematic drawing of an ektacytometer in the configuration used by Streekstra *et al.* (1993). A dilute suspension of RBCs is sheared in a Couette flow, induced by two concentric cylinders, with the inner cylinder rotating at a constant rate. A laser beam is sent through the sheared suspension and the scattered light is measured using a charge-coupled device (CCD) camera. Usually the system is configured in such a way that the CCD camera collects the main diffraction lobe and one or two side rings (i.e., the first few minima in the scattered light pattern are measured).

The measured light is due to single scattering only. The diffraction pattern that is measured by the CCD camera is ellipsoidal in shape. Because all RBCs are oriented in the same way in the Couette flow, and assuming that they are all deformed in the same way, a simple argument based on Fraunhofer diffraction suffices to explain the shape of the measured ellipsoidal diffraction pattern. The aspect ratio of the measured diffraction pattern is equal to the aspect ratio of the

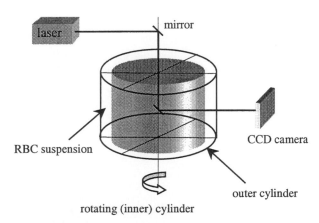

Figure 5   Schematic drawing of an ektacytometer.

deformed RBCs. Therefore, measuring the aspect ratios of the diffraction patterns as a function of the shear rate is sufficient to measure the deformability of the RBCs. Actually, light scattering from RBCs cannot be described by Fraunhofer diffraction. Streekstra *et al.* (1993, 1994) used anomalous diffraction theory to describe the light scattering from deformed RBCs. They showed that within the framework of anomalous diffraction, the basic assumption in ektacytometry, that the aspect ratio of the measured ellipsoidal diffraction pattern is equal to that of the deformed red blood cell, remains valid. Moreover, they showed that even in the presence of a realistic biological variability in the volume of the RBCs, the resulting diffraction pattern is ellipsoidal and reflects the aspect ratio of the deformed RBCs.

Later, Mazeron *et al.* (1997a, b) studied in detail the intensity distributions in the rings around the main diffraction lobe. For true ellipsoidal particles, the intensity should be constant along the ellipsoids. However, they observe variations in the intensity. Using physical optics theory, this intensity reinforcement, as they call it, can be explained by assuming that the deformed RBCs are not exact ellipsoids, but are thinned at the tips.

IV. CONCLUDING REMARKS

The use of single scattering in biomedical and biophysical applications is widespread and has proven to be an invaluable tool in both clinical and research environments. From a theoretical viewpoint, scattering from biological cells can be seen as the problem of scattering from irregularly shaped near index matching particles. This concept opens the potential of a number of scattering theories and many of them have been used with varying success.

We have argued that the state of polarization of the scattered light carries a significant amount of relevant biophysical and biomedical information, as was shown in many experiments. Despite many remarkable results, the theoretical analysis and experimental exploitation of the complete scattering matrix for biological particles still is rather unexplored and we hope and may expect to see many exiting new results in the coming few years.

References

Abdelazeez, M. K. (1983). Wave scattering from a large sphere with rough surface. *IEEE Trans. Antennas Propag.* **31**, 375–377.

Ackerman, T. P., and Toon, O. B. (1981). Absorption of visible radiation in atmospheres containing mixtures of absorbing and nonabsorbing particles. *Appl. Opt.* **20**, 3661–3668.

Acquista, C. (1976). Light scattering by tenuous particles: A generalization of the Rayleigh–Gans–Rocard approach. *Appl. Opt.* **15**, 2932–2936.

Adams, C. N., and Kattawar, G. W. (1970). Solutions of the equations of radiative transfer by an invariant imbedding approach. *J. Quant. Spectrosc. Radiat. Transfer* **10**, 341–366.

Adams, C. N., and Kattawar, G. W. (1993). Effect of volume–scattering function on the errors induced when polarization is neglected in radiance calculations in an atmosphere–ocean system. *Appl. Opt.* **32**, 4610–4617. [Errata: **33**, 453 (1994).]

Aden, A. L., and Kerker, M. (1951). Scattering of electromagnetic waves from two concentric spheres. *J. Appl. Phys.* **22**, 1242–1246.

Agarwal, G. S. (1976). Relation between Waterman's extended boundary condition and the generalized extinction theorem. *Phys. Rev. D* **14**, 1168–1171.

Aitchison, J., and Brown, J. A. C. (1963). "The Lognormal Distribution." Cambridge Univ. Press, Cambridge.

Allan, L. E., and McCormick, G. C. (1978). Measurements of the backscatter matrix of dielectric spheroids. *IEEE Trans. Antennas Propag.* **26**, 579–587.

Allan, L. E., and McCormick, G. C. (1980). Measurements of the backscatter matrix of dielectric bodies. *IEEE Trans. Antennas Propag.* **28**, 166–169.

Al-Rizzo, H. M., and Tranquilla, J. M. (1995a). Electromagnetic scattering from dielectrically coated axisymmetric objects using the generalized point-matching technique. *J. Comput. Phys.* **119**, 342–373.

Al–Rizzo, H. M., and Tranquilla, J. M. (1995b). Electromagnetic wave scattering by highly elongated and geometrically composite objects of large size parameters: The generalized multipole technique. *Appl. Opt.* **34**, 3502–3521.

Anderson, R. (1992). Measurement of Mueller matrices. *Appl. Opt.* **31**, 11–13.

Anderson, D. G. M., and Barakat, R. (1994). Necessary and sufficient conditions for a Mueller matrix to be derivable from a Jones matrix. *J. Opt. Soc. Am. A* **11**, 2305–2319.

Anderson, T. L., Covert, D. S., Marshall, S. F., *et al.* (1996). Performance characteristics of a high-sensitivity, three-wavelength, total scatter/backscatter nephelometer. *J. Atmos. Oceanic Technol.* **13**, 967–986.

Andreae, M. O. (1986). The ocean as a source of atmospheric sulfur compounds. *In* "The Role of Air–Sea Exchange in Geochemical Cycling" (P. Buat-Menard, Ed.), pp. 331–362. Reidel, Dordrecht.

Andrew, W. V., Balanis, C. A., Tirkas, P. A., *et al.* (1997). Finite-difference time-domain analysis of HF antennas on helicopter airframes. *IEEE Trans. Electromagn. Compat.* **39**, 100–113.

Andsager, K., Beard, K. V., and Laird, N. S. (1999). A laboratory study of oscillations and axis ratios for large raindrops. *J. Atmos. Sci.* **56**, 2673–2683.

Ansmann, A., Wandinger, U., Riebesell, M., *et al.* (1992). Independent measurement of extinction and backscatter profiles in cirrus clouds by using a combined Raman elastic-backscatter lidar. *Appl. Opt.* **31**, 7113–7131.

Ansmann, A., Bösenberg, J., Brogniez, G., *et al.* (1993). Lidar network observation of cirrus morphological and scattering properties during the International Cirrus Experiment 1989: The 18 October 1989 case study and statistical analysis. *J. Appl. Meteorol.* **32**, 1608–1622.

Arakawa, E. T., Williams, M. W., and Inagaki, T. (1977). Optical properties of arc-evaporated carbon films between 0.6 and 3.8 eV. *J. Appl. Phys.* **48**, 3176–3177.

Arfken, G. B., and Weber, H. J. (1995). "Mathematical Methods for Physicists." Academic Press, San Diego.

Arnold, S., Liu, C. T., Whitten, W. B., and Ramsey, J. M. (1991). Room-temperature microparticle based persistent spectral hole burning memory. *Opt. Lett.* **16**, 420–422.

Arnold, S., Communale, J., Whitten, W. B., *et al.* (1992). Room-temperature microparticle-based persistent hole-burning memory spectroscopy. *J. Opt. Soc. Am. B* **9**, 4081–4093.

Arnold, S., Ghaemi, A., and Fuller, K. A. (1994). Morphological resonances detected from a cluster of two microspheres. *Opt. Lett.* **19**, 156–158.

Arnott, W. P., and Marston, P. L. (1991). Unfolded optical glory of spheroids: Backscattering of laser light from freely rising spheroidal air bubbles in water. *Appl. Opt.* **30**, 3429–3442.

Arnott, W. P., Dong, Y. Y., and Hallett, J. (1995). Extinction efficiency in the infrared (2–18 μm) of laboratory ice clouds: Observations of scattering minima in the Christiansen bands of ice. *Appl. Opt.* **34**, 541–551.

Asano, S. (1979). Light scattering properties of spheroidal particles. *Appl. Opt.* **18**, 712–723.

Asano, S. (1983). Light scattering by horizontally oriented spheroidal particles. *Appl. Opt.* **22**, 1390–1396.

Asano, S., and Sato, M. (1980). Light scattering by randomly oriented spheroidal particles. *Appl. Opt.* **19**, 962–974.

Asano, S., and Yamamoto, G. (1975). Light scattering by a spheroidal particle. *Appl. Opt.* **14**, 29–49.

Ashkin, A., and Dziedzic, J. (1980). Observation of light scattering from nonspherical particles using optical levitation. *Appl. Opt.* **19**, 660–668.

Aspens, D. E. (1982). Local-field effects and effective-medium theory: A microscopic perspective. *Am. J. Phys.* **50**, 704–709.

Atlas, D. (1964). Advances in radar meteorology. *In* "Advances in Geophysics" (H. E. Landsberg and J. Van Miehem, Eds.), pp. 317–478. Academic Press, New York.

Atlas, D. (1990). "Radar in Meteorology." Am. Meteorol. Soc., Boston.

Atlas, D., and Ulbrich, C. W. (1977). Path- and area-integrated rainfall measurements by microwave attenuation in the 1–3 cm band. *J. Appl. Meteorol.* **16**, 1322–1331.

Atlas, D., Harper, W. G., Ludlam, F. H., and Macklin, W. C. (1960). Radar scatter by large hail. *Quart J. R. Meteorol. Soc.* **86**, 468–482.

Atlas, D., Matrosov, S. Y., Heymsfield, A. J., *et al.* (1995). Radar and radiation properties of ice clouds. *J. Appl. Meteorol.* **34**, 2329–2345.

Awaka, J., and Oguchi, T. (1982). Bistatic radar reflectivities of Pruppacher-and-Pitter form raindrops at 34.8 GHz. *Radio Sci.* **17**, 269–278.

Aydin, K., and Daisley, S. (1998). Effects of raindrop canting and oscillation on rainfall rate estimates from 35 GHz differential attenuation. *Proc. Int. Geosci. Remote Sens. Symp., Seattle, WA, 1998*, 153–155.

Aydin, K., and Giridhar, V. (1991). Polarimetric C-band radar observables in melting hail: A computational study. *Prepr. Vol., Conf. Radar Meteorol., 25th, Paris, 1991*, 733–736.

Aydin, K., and Giridhar, V. (1992). C-band dual-polarization radar observables in rain. *J. Atmos. Oceanic Technol.* **9**, 383–390.

Aydin, K., and Seliga, T. A. (1984). Radar polarimetric backscattering properties of conical graupel. *J. Atmos. Sci.* **41**, 1887–1892.

Aydin, K., and Tang, C. (1997a). Millimeter wave radar scattering from model ice crystal distributions. *IEEE Trans. Geosci. Remote Sens.* **35**, 140–146.

Aydin, K., and Tang, C. (1997b). Relationships between IWC and polarimetric radar measurands at 94 and 220 GHz for hexagonal columns and plates. *J. Atmos. Oceanic Technol.* **14**, 1055–1063.

Aydin, K., and Walsh, T. M. (1996). Computational study of millimeter wave scattering from bullet rosette type ice crystals. *Proc. Int. Geosci. Remote Sens. Symp., Lincoln, NE, 1996*, 563–565.

Aydin, K., and Walsh, T. M. (1998). Separation of millimeter wave radar reflectivities of aggregates and pristine ice crystals in a cloud. *Proc. Int. Geosci. Remote Sens. Symp., Seattle, WA, 1998*, 440–442.

Aydin, K., and Zhao, Y. (1990). A computational study of polarimetric radar observables in hail. *IEEE Trans. Geosci. Remote Sens.* **28**, 412–422.

Aydin, K., Seliga, T. A., and Bringi, V. N. (1984). Differential radar scattering properties of model hail and mixed phase hydrometeors. *Radio Sci.* **19**, 58–66.

Aydin, K., Seliga, T. A., and Balaji, V. (1986). Remote sensing of hail with a dual linear polarization radar. *J. Clim. Appl. Meteorol.* **25**, 1475–1484.

Aydin, K., Direskeneli, H., and Seliga, T. A. (1987). Dual polarization radar estimation of rainfall parameters compared with ground-based disdrometer measurements: 29 October 1982, Central Illinois Experiment. *IEEE Trans. Geosci. Remote Sens.* **28**, 834–844.

Aydin, K., Lure, Y. M., and Seliga, T. A. (1990a). Polarimetric radar measurements of rainfall compared with ground-based rain gauges during MAYPOLE'84. *IEEE Trans. Geosci. Remote Sens.* **28**, 443–449.

Aydin, K., Zhao, Y., and Seliga, T. A. (1990b). A differential reflectivity radar hail measurement technique: Observations during the Denver hailstorm of 13 June 1984. *J. Atmos. Oceanic Technol.* **7**, 104–113.

Aydin, K., Tang, C., Pazmany, A., *et al.* (1994). 95 GHz polarimetric radar measurements in a cloud compared with model computations. *Atmos. Res.* **34**, 135–144.

Aydin, K., Bringi, V. N., and Liu, L. (1995). Rain-rate estimation in the presence of hail using S-band specific differential phase and other radar parameters. *J. Appl. Meteorol.* **34**, 404–410.

Aydin, K., Park, S. H., and Walsh, T. M. (1998). Bistatic dual-polarization scattering from rain and hail at S- and C-band frequencies. *J. Atmos. Oceanic Technol.* **15**, 1110–1121.

Azzam, R. M. A., and Bashara, N. M. (1987). "Ellipsometry and Polarized Light." North-Holland, New York.

Bacon, N. J., Swanson, B. D., Baker, M. B., and Davis, E. J. (1998). Laboratory studies of light scattering by single levitated ice crystals. *Proc. Conf. Cloud Phys., Everett, WA, 1998*, 427–428.

Bahar, E., and Chakrabarti, S. (1985). Scattering and depolarization by large conducting spheres with rough surfaces. *Appl. Opt.* **24**, 1820–1825.

Bahar, E., and Fitzwater, M. A. (1986). Scattering and depolarization by conducting cylinders with rough surfaces. *Appl. Opt.* **25**, 1826–1832.

Balakrishnan, N., and Zrnic, D. S. (1990a). Estimation of rain and hail rates in mixed-phase precipitation. *J. Atmos. Sci.* **47**, 565–583.

Balakrishnan, N., and Zrnic, D. S. (1990b). Use of polarization to characterize precipitation and discriminate large hail. *J. Atmos. Sci.* **47**, 1525–1540.

Balch, W. M., Holligan, P. M., Ackleson, S. G., and Voss, K. J. (1991). Biological and optical properties of mesoscale coccolithophore blooms in the Gulf of Maine. *Limnol. Oceanogr.* **36**, 629–643.

Baltzer, F., Jett, S. D., and Rubahn, H.-G. (1998). Alkali cluster films on insulating substrates: comparison between scanning force microscopy and extinction data. *Chem. Phys. Lett.* **297**, 273–280.

Bantges, R. J., Russell, J. E., and Haigh, J. D. (1998). Cirrus cloud radiance spectra in the thermal infrared. *Prepr. Vol., Conf. Light Scat. Nonspher. Part.: Theory Meas. Appl., New York, 1998*, 305–308.

Barabanenkov, Yu. N., Kravtsov, Yu. A., Ozrin, V. D., and Saichev, A. I. (1991). Enhanced backscattering in optics. *In* "Progress in Optics" (E. Wolf, Ed.), Vol. XXIX, pp. 65–197. Elsevier, Amsterdam.

Barakat, R. (1981). Bilinear constraints between elements of the 4×4 Mueller–Jones transfer matrix of polarization theory. *Opt. Commun.* **38**, 159–161.

Baran, A. J., Foot, J. S., and Mitchell, D. L. (1998). Ice-crystal absorption: A comparison between theory and implications for remote sensing. *Appl. Opt.* **37**, 2207–2215.

Barber, P. W. (1977). Resonance electromagnetic absorption by nonspherical dielectric objects. *IEEE Trans. Microwave Theory Tech.* **25**, 373–381.

Barber, P. W., and Hill, S. C. (1990). "Light Scattering by Particles: Computational Methods." World Scientific, Singapore.

Barber, P. W., and Wang, D.-S. (1978). Rayleigh–Gans–Debye applicability to scattering by nonspherical particles. *Appl. Opt.* **17**, 797–803.

Barber, P., and Yeh, C. (1975). Scattering of electromagnetic waves by arbitrarily shaped dielectric bodies. *Appl. Opt.* **14**, 2864–2872.

Barber, P. W., Miller, E. K., and Sarkar, T. K., Eds. (1994). Feature issue on Scattering by Three-Dimensional Objects. *J. Opt. Soc. Am. A* **11**, 1379–1545.

Barkey, B., Liou, K. N., Takano, Y., *et al.* (1999). An analog light scattering experiment of hexagonal icelike particles. II. Experimental and theoretical results. *J. Atmos. Sci.* **56**, 613–625.

Barrick, D. E. (1970). Rough surfaces. *In* "Radar Cross Section Handbook" (G. T. Ruck, D. E. Barrick, W. D. Stuart, and C. K. Krichbaum, Eds.), Chap. 9. Plenum, New York.

Barton, J. P. (1998). Electromagnetic field calculations for a sphere illuminated by a higher-order Gaussian beam. II. Far-field scattering. *Appl. Opt.* **37**, 3339–3344.

Barton, J. P., Alexander, D. R., and Shaub, S. A. (1989). Internal fields of a spherical particle illuminated by a tightly focused laser beam: Focal point positioning effects at resonance. *J. Appl. Phys.* **65**, 2900–2906.

Bates, R. H. T. (1975). Analytic constraints on electromagnetic field computations. *IEEE Trans. Microwave Theory Tech.* **23**, 605–623.

Bates, R. H. T., and Wall, D. J. N. (1977). Null field approach to scalar diffraction. *Philos. Trans. R. Soc. London, Ser. A* **287**, 45–114.

Bates, T. S., Calhoun, J. A., and Quinn, P. K. (1992). Variations in the methanesulfonate to sulfate molar ratio in submicrometer marine aerosol particles over the South Pacific Ocean. *J. Geophys. Res.* **97**, 9859–9865.

Battaglia, A., Muinonen, K., Nousiainen, T., and Peltoniemi, J. I. (1999). Light scattering by Gaussian particles: Rayleigh-ellipsoid approximation. *J. Quant. Spectrosc. Radiat. Transfer* **63**, in press.

Battan, L. J. (1973). "Radar Observation of the Atmosphere." Univ. of Chicago Press, Chicago.

Batten, C. E. (1985). Spectral optical constants of soots from polarized angular reflectance measurements. *Appl. Opt.* **24**, 1193–1199.

Bauer, P., and Schluessel, P. (1993). Rainfall, total water, ice water, and water vapor over sea from polarized microwave simulations and special sensor microwave/imager data. *J. Geophys. Res.* **98**, 20,737–20,759.

Bauer, P., Schanz, L., Bennartz, R., and Schlüssel, P. (1998). Outlook for combined TMI–VIRS algorithm for TRMM: Lessons from the PIP and AIP projects. *J. Atmos. Sci.* **55**, 1644–1673.

Baum, W. A., Kreidl, T. J., and Schleicher, D. G. (1992). Cometary grains. *Astrophys. J.* **104**, 1216–1225.

Bayliss, A., and Turkel, E. (1980). Radiation boundary conditions for wave-like equations. *Commun. Pure Appl. Math.* **33**, 707–725.

Beard, K. V., and Chuang, C. (1987). A new model for the equilibrium shape of raindrops. *J. Atmos. Sci.* **44**, 1509–1524.

Beard, K. V., and Kubesh, R. J. (1991). Laboratory measurements of small raindrop distortion. 2. Oscillation frequencies and modes. *J. Atmos. Sci.* **48**, 2245–2264.

Beard, K. V., Johnson, D. B., and Jameson, A. R. (1983). Collisional forcing of raindrop oscillations. *J. Atmos. Sci.* **40**, 455–462.

Beard, K. V., Kubesh, R. J., and Ochs, H. T., III (1991). Laboratory measurements of small raindrop distortion. 1. Axis ratios and fall behavior. *J. Atmos. Sci.* **48**, 698–710.

Beardsley, G. F. (1968). Mueller scattering matrix of sea water. *J. Opt. Soc. Am.* **58**, 52–57.

Bebbington, D. H. O., McGuinness, R., and Holt, A. R. (1987). Correction of propagation effects in S-band circular polarisation diversity radars. *IEEE Proc. H: Microwaves, Antennas Propag.* **134**, 431–437.

Beckering, G., Zilker, S. J., and Haarer, D. (1997). Spectral measurements of the emission from highly scattering gain media. *Opt. Lett.* **22**, 1427–1429.

Beckmann, P., and Spizzichino, A. (1963). "The Scattering of Electromagnetic Waves from Rough Surfaces." Macmillan, New York.

Ben-David, A. (1998). Mueller matrix for atmospheric aerosols at CO_2 laser wavelengths from polarized backscattering lidar measurements. *J. Geophys. Res.* **103**, 26,041–26,050.

Benson, M. C., McDougal, D. C., and Coffey, D. S. (1984). The application of perpendicular and forward light scatter to access nuclear and cellular morphology. *Cytometry* **5**, 515–522.

Berenger, J.-P. (1994). A perfectly matched layer for the absorption of electromagnetic waves. *J. Comput. Phys.* **114**, 185–200.

Berenger, J.-P. (1996). Three-dimensional perfectly matched layer for the absorption of electromagnetic waves. *J. Comput. Phys.* **127**, 363–379.

Berkman, R. M., and Wyatt, P. J. (1970). Differential light scattering measurements of heat-treated bacteria. *Appl. Microbiol.* **20**, 510–512.

Berkman, R. M., Wyatt, P. J., and Phillips, D. T. (1970). Rapid detection of penicillin sensitivity in *Staphylococcus aureus*. *Nature (London)* **228**, 458–460.

Berntsen, S., and Hornsleth, S. N. (1994). Retarded time absorbing boundary conditions. *IEEE Trans. Antennas Propag.* **42**, 1059–1064.

Beuthan, J., Minet, O., Helfmann, J., *et al.* (1996). The spatial variation of the refractive index in biological cells. *Phys. Med. Biol.* **41**, 369–382.

Bhandari, R. (1985). Scattering coefficients for a multilayered sphere: Analytic expressions and algorithms. *Appl. Opt.* **24**, 1960–1967.

Bickel, W. S., and Stafford, M. E. (1980). Biological particles as irregularly shaped scatterers. *In* "Light Scattering by Irregularly Shaped Particles" (D. W. Schuerman, Ed.), pp. 299–305. Plenum, New York.

Bickel, W. S., Davidson, J. F., Huffman, D. R., and Kilkson, R. (1976). Application of polarization effects in light scattering: A new biophysical tool. *Proc. Natl. Acad. Sci. U.S.A.* **73**, 486–490.

Blaschak, J. G., and Kriegsmann, G. A. (1988). A comparative study of absorbing boundary conditions. *J. Comput. Phys.* **77**, 109–139.

Bohren, C. F. (1974). Light scattering by an optically active sphere. *Chem. Phys. Lett.* **29**, 458–462.

Bohren, C. F. (1978). Scattering of electromagnetic waves by an optically active cylinder. *J. Colloid Interface Sci.* **66**, 105–109.

Bohren, C. F. (1986). Applicability of effective-medium theories to problems of scattering and absorption by nonhomogeneous atmospheric particles. *J. Atmos. Sci.* **43**, 468–475.

Bohren, C. F., and Battan, L. J. (1980). Radar backscattering by inhomogeneous precipitation particles. *J. Atmos. Sci.* **37**, 1821–1827.

Bohren, C. F., and Battan, L. J. (1982). Radar backscattering of microwaves by spongy ice spheres. *J. Atmos. Sci.* **39**, 2623–2628.

Bohren, C. F., and Huffman, D. R. (1983). "Absorption and Scattering of Light by Small Particles." Wiley, New York.

Bohren, C. F., and Singham, S. B. (1991). Backscattering by nonspherical particles: A review of methods and suggested new approaches. *J. Geophys. Res.* **96**, 5269–5277.

Bolin, F. P., Preuss, L. E., Taylor, R. C., and Ference, R. J. (1989). Refractive index of some mammalian tissues using a fiber optic cladding method. *Appl. Opt.* **28**, 2297–2303.

Borghese, F., Denti, P., Toscano, G., and Sindoni, O. I. (1979). Electromagnetic scattering by a cluster of spheres. *Appl. Opt.* **18**, 116–120.

Borghese, F., Denti, P., Saija, R., et al. (1984). Multiple electromagnetic scattering from a cluster of spheres. I. Theory. *Aerosol Sci. Technol.* **3**, 227–235.

Borghese, F., Denti, P., Saija, R., et al. (1987). Extinction coefficients for a random dispersion of small stratified spheres and a random dispersion of their binary aggregates. *J. Opt. Soc. Am. A* **4**, 1984–1991.

Borghese, F., Denti, P., Saija, R., and Sindoni, O. I. (1992). Optical properties of spheres containing a spherical eccentric inclusion. *J. Opt. Soc. Am. A* **9**, 1327–1335.

Borghese, F., Denti, P., and Saija, R. (1994). Optical properties of spheres containing several spherical inclusions. *Appl. Opt.* **33**, 484–493. [Errata: **34**, 5556 (1995).]

Borghese, F., Denti, P., Saija, R., et al. (1998). Optical resonances of spheres containing an eccentric spherical inclusion. *J. Opt. (Paris)* **29**, 28–34.

Born, M., and Wolf, E. (1970). "Principles of Optics." Pergamon, Oxford.

Bottcher, C. J. F., and Bordewijk, P. (1978). "Theory of Electromagnetic Polarization." Elsevier, Amsterdam.

Bottiger, J. R., Fry, E. S., and Thompson, R. C. (1980). Phase matrix measurements for electromagnetic scattering by sphere aggregates. *In* "Light Scattering by Irregularly Shaped Particles" (D. W. Schuerman, Ed.), pp. 283–290. Plenum, New York.

Bouligand, Y., Soyer, M.-O., and Puiseaux-Dao, S. (1968). La structure fibrillaire et l'orientation des chromosomes chez les dinoflagelles. *Chromosoma (Berlin)* **24**, 251–287.

Bourrely, C., Chiappetta, P., and Torresani, B. (1986). Light scattering by particles of arbitrary shape: A fractal approach. *J. Opt. Soc. Am. A* **3**, 250–255.

Bourrely, C., Chiappetta, P., and Lemaire, T. (1989). Electromagnetic scattering by large rotating particles in the eikonal formalism. *Opt. Commun.* **70**, 173–176.

Bowell, E., Hapke, B., Domingue, D., et al. (1989). Application of photometric models to asteroids. *In* "Asteroids II" (R. Binzel, T. Gehrels, and M. Matthews, Eds.), pp. 524–556. Univ. of Arizona Press, Tucson, AZ.

Bréon, F.-M., Deuzé, J.-L., Tanré, D., and Herman, M. (1997). Validation of spaceborne estimates of aerosol loading from Sun photometer measurements with emphasis on polarization. *J. Geophys. Res.* **102**, 17,187–17,195.

Briggs, R. O., and Hatchett, G. L. (1965). Techniques for improving underwater visibility with video equipment. *Ocean Sci. Ocean Eng.* **1&2**, 1284–1308.

Bringi, V. N., and Hendry, A. (1990). Technology of polarization diversity radars for meteorology. *In* "Radar in Meteorology: Battan Memorial and 40th Anniversary Radar Meteorology Conference" (D. Atlas, Ed.), pp. 153–198. Am. Meteorol. Soc., Boston.

Bringi, V. N., and Seliga, T. A. (1977a). Scattering from axisymmetric dielectrics or perfect conductors imbedded in an axisymmetric dielectric. *IEEE Trans. Antennas Propag.* **25**, 575–580.

Bringi, V. N., and Seliga, T. A. (1977b). Scattering from non-spherical hydrometeors. *Ann. Telecommun.* **32**, 392–397.

Bringi, V. N., Seliga, T. A., and Mueller, E. A. (1982). First comparisons of rainfall rates derived from radar differential reflectivity and disdrometer measurements. *IEEE Trans. Geosci. Remote Sens.* **20**, 201–204.

Bringi, V. N., Seliga, T. A., and Cherry, S. M. (1983). Statistical properties of the dual polarization differential reflectivity (ZDR) radar signal. *IEEE Trans. Geosci. Remote Sens.* **21**, 215–220.

Bringi, V. N., Seliga, T. A., and Aydin, K. (1984). Hail detection with a differential reflectivity radar. *Science* **225**, 1145–1147.

Bringi, V. N., Rasmussen, R. M., and Vivekanandan, J. (1986a). Multiparameter radar measurements in Colorado convective storms. I. Graupel melting studies. *J. Atmos. Sci.* **43**, 2545–2563.

Bringi, V., Vivekanandan, J., and Tuttle, J. D. (1986b). Multiparameter radar measurements in Colorado convective storms. II. Hail detection studies. *J. Atmos. Sci.* **43**, 2564–2577.

Bringi, V. N., Burrows, D. A., and Menon, S. M. (1991). Multiparameter radar and aircraft study of raindrop spectral evolution in warm-based clouds. *J. Appl. Meteorol.* **30**, 853–880.

Bringi, V. N., Chandrasekar, V., and Xiao, R. (1998). Raindrop axis ratios and size distributions in Florida rainshafts: An assessment of multiparameter radar algorithms. *IEEE Trans. Geosci. Remote Sens.* **36**, 703–715.

Britt, C. L. (1989). Solution of electromagnetic scattering problems using time domain techniques. *IEEE Trans. Antennas Propag.* **37**, 1181–1191.

Bronk, B. V., Smith, M. J., and Arnold, S. (1993). Photon-correlation spectroscopy for small spherical inclusions in a micrometer-sized electrodynamically levitated droplet. *Opt. Lett.* **18**, 93–95.

Bronk, B. V., Druger, S. D., Czégé, J., and van de Merwe, W. P. (1995). Measuring diameters of rod-shaped bacteria in vivo with polarized light scattering. *Biophys. J.* **69**, 1170–1177.

Brownlee, D. E. (1985). Collection of cosmic dust: Past and future. *In* "Properties and Interactions of Interplanetary Dust" (R. H. Giese and P. Lamy, Eds.), pp. 143–147. Reidel, Dordrecht.

Bruggeman, D. A. G. (1935). Berechnung vershiedener physikalischer Konstanten von heterogenen Substanzen. 1. Dielektrizitatskonstanten und Leitfahigkeiten der Mischkorper aus isotropen Substanzen. *Ann. Phys. (Leipzig)* **24**, 636–664.

Bruning, J. H., and Lo, Y. T. (1971). Multiple scattering of EM waves by spheres. *IEEE Trans. Antennas Propag.* **19**, 378–400.

Brunsting, A. (1974). Can light scattering techniques be applied to flow-through analysis? *J. Histochem. Cytochem.* **22**, 607–615.

Brunsting, A., and Mullaney, P. F. (1972). Light scattering from coated spheres: Model for biological cells. *Appl. Opt.* **11**, 675–680.

Brunsting, A., and Mullaney, P. F. (1974). Differential light scattering from spherical mammalian cells. *Biophys. J.* **10**, 439–453.

Bryant, H. C., and Cox, A. J. (1966). Mie theory and the glory. *J. Opt. Soc. Am.* **56**, 1529–1532.

Buettner, K. (1963). Regenortnung vom Wettersatelliten mit Hilfe von Zentimeterwellen. *Die Naturwissenschaftler* **50**, 591–592.

Buratti, B. J., Hillier, J. K., and Wang, M. (1996). The lunar opposition surge: Observations by Clementine. *Icarus* **124**, 490–499.

Burns, M. M., Fournier, J.-M., and Golovchenko, J. A. (1990). Optical matter: Crystallization and binding in intense optical fields. *Science* **249**, 749–754.

Burrows, M. L. (1969). Equivalence of the Rayleigh solution and the extended-boundary-condition solution for scattering problems. *Electron. Lett.* **5**, 277–278.

Bustamante, C., Maestre, M. F., and Tinoco, I., Jr. (1980a). Circular intensity differential scattering of light by helical structures. I. Theory. *J. Chem. Phys.* **73**, 4273–4281.

Bustamante, C., Maestre, M. F., and Tinoco, I., Jr. (1980b). Circular intensity differential scattering of light by helical structures. II. Applications. *J. Chem. Phys.* **73**, 6046–6055.

Bustamante, C., Tinoco, I., Jr., and Maestre, M. F. (1981). Circular intensity differential scattering of light by helical structures. III. A general polarizability tensor and anomalous scattering. *J. Chem. Phys.* **74**, 4839–4850.

Bustamante, C., Tinoco, I., Jr., and Maestre, M. F. (1982). Circular intensity differential scattering of light by helical structures. IV. Randomly oriented species. *J. Chem. Phys.* **76**, 3440–3446.

Bustamante, C., Maestre, M. F., Keller, D., and Tinoco, I., Jr. (1984). Differential scattering of circularly polarized light by dense particles. *J. Chem. Phys.* **80**, 4817–4823.

Cai, Q., and Liou, K.-N. (1982). Polarized light scattering by hexagonal ice crystals: Theory. *Appl. Opt.* **21**, 3569–3580.

Cai, J., Lu, N., and Sorensen, C. M. (1993). Comparison of size and morphology of soot aggregates as determined by light scattering and electron microscope analysis. *Langmuir* **9**, 2861–2867.

Cameron, B. D., Raković, M. J., Mehrübeoğlu, M., *et al.* (1998). Measurement and calculation of the two-dimensional backscattering Mueller matrix of a turbid medium. *Opt. Lett.* **23**, 485–487.

Carder, K. L., Beardsley, G. F., and Pak, H. (1971). Particle size distributions in the Eastern Equatorial Pacific. *J. Geophys. Res.* **76**, 5070–5077.

Cariou, J., Le Jeune, B., Lotrian, J., and Guern, Y. (1990). Polarization effects of seawater and underwater targets. *Appl. Opt.* **29**, 1689–1695.

Carslaw, K. S., Wirth, M., Tsias, A., *et al.* (1998). Particle microphysics and chemistry in remotely observed mountain polar stratospheric clouds. *J. Geophys. Res.* **103**, 5785–5796.

Carswell, A. I. (1983). Lidar measurements of the atmosphere. *Can. J. Phys.* **61**, 378–395.

Carswell, A. I., and Pal, S. R. (1980). Polarization anisotropy in lidar multiple scattering from clouds. *Appl. Opt.* **19**, 4123–4126.

Caylor, I. J., and Chandrasekar, V. (1996). Time-varying ice crystal orientation in thunderstorms observed with multiparameter radar. *IEEE Trans. Geosci. Remote Sens.* **34**, 847–858.

Caylor, I. J., and Illingworth, A. J. (1987). Radar observations and modeling of warm rain initiation. *Quart. J. R. Meteorol. Soc.* **113**, 1171–1191.

César, S. L., Vasconcelos, E. F., Freire, V. N., and Farias, G. A. (1994). Scattering coefficients for a spherical particle with stationary stochastic roughness. *Prepr. Vol., Conf. Atmos. Rad., 8th, Nashville, TN, 1994*, 350–352.

Chandrasekar, V., and Bringi, V. N. (1988a). Error structure of multiparameter radar and surface measurements of rainfall. I. Differential reflectivity. *J. Atmos. Oceanic Technol.* **5**, 783–795.

Chandrasekar, V., and Bringi, V. N. (1988b). Error structure of multiparameter radar and surface measurements of rainfall. II. X-band attenuation. *J. Atmos. Oceanic Technol.* **5**, 796–802.

Chandrasekar, V., Cooper, W. A., and Bringi, V. N. (1988). Axis ratios and oscillations of raindrops. *J. Atmos. Sci.* **45**, 1323–1333.

Chandrasekar, V., Bringi, V. N., Balakrishnan, N., and Zrnic, D. S. (1990). Error structure of multiparameter radar and surface measurements of rainfall. III. Specific differential phase. *J. Atmos. Oceanic Technol.* **7**, 621–629.

Chandrasekar, V., Gorgucci, E., and Scarchilli, G. (1993). Optimization of multiparameter radar estimates of rainfall. *J. Appl. Meteorol.* **32**, 1288–1293.

Chandrasekar, V., Hubbert, J., Bringi, V. N., and Meischner, P. F. (1994). Transformation of dual-polarized radar measurements to arbitrary polarization bases. *J. Atmos. Oceanic Technol.* **11**, 937–949.

Chandrasekhar, S. (1950). "Radiative Transfer." Oxford Univ. Press, London (also Dover, New York, 1960).

Charalampopoulos, T. T., and Chang, H. (1991). Effects of soot agglomeration on radiative transfer. *J. Quant. Spectrosc. Radiat. Transfer* **46**, 125–134.

Charlson, R. J., Lovelock, J. E., Andreae, M. O., and Warren, S. G. (1987). Oceanic phytoplankton, atmospheric sulphur, cloud albedo and climate. *Nature (London)* **326**, 655–661.

Chen, H. Y., Iskander, M. G., and Penner, J. E. (1991). Empirical formula for optical absorption by fractal aerosol agglomerates. *Appl. Opt.* **30**, 1547–1551.

Cheng, L., and English, M. (1983). A relationship between hailstone concentration and size. *J. Atmos. Sci.* **40**, 204–213.

Chepfer, H., Brogniez, G., and Fouquart, Y. (1998). Cirrus clouds' microphysical properties deduced from POLDER observations. *J. Quant. Spectrosc. Radiat. Transfer* **60**, 375–390.

Chernyshev, A. V., Prots, V. I., Doroshkin, A. A., and Maltsev, V. P. (1995). Measurement of scattering properties of individual particles with a scanning flowcytometer. *Appl. Opt.* **34**, 6301–6305.

Chew, W. C. (1990). "Waves and Fields in Inhomogeneous Media." Van Nostrand–Reinhold, New York.

Chew, W. C., and Wagner, R. L. (1992). A modified form of Liao's absorbing boundary condition. *IEEE AP-S Int. Symp. Digest, 1992*, 536–539.

Chew, W. C., and Weedon, W. H. (1994). A 3D perfectly matched medium from modified Maxwell's equations with stretched coordinates. *Microwave Opt. Technol. Lett.* **7**, 599–604.

Chew, W. C., Lu, C. C., and Wang, Y. M. (1994). Efficient computation of three-dimensional scattering of vector electromagnetic waves. *J. Opt. Soc. Am. A* **11**, 1528–1537.

Chiappetta, P. (1980a). Multiple scattering approach to light scattering by arbitrarily shaped particles. *J. Phys. A: Math. Gen.* **13**, 2101–2108.

Chiappetta, P. (1980b). A new model for scattering by irregular absorbing particles. *Astron. Astrophys.* **83**, 348–353.

Chuang, C., and Beard, K. V. (1990). A numerical model for the equilibrium shape of electrified raindrops. *J. Atmos. Sci.* **47**, 1374–1389.

Chýlek, P., and Hallett, J. (1992). Enhanced absorption of solar radiation by cloud droplets containing soot particles in their surface. *Quart. J. R. Meteorol. Soc.* **118**, 167–172.

Chýlek, P., and Klett, J. D. (1991a). Extinction cross sections of nonspherical particles in the anomalous diffraction approximation. *J. Opt. Soc. Am. A* **8**, 274–281.

Chýlek, P., and Klett, J. D. (1991b). Absorption and scattering of electromagnetic radiation by prismatic columns: Anomalous diffraction approximation. *J. Opt. Soc. Am. A* **8**, 1713–1720.

Chýlek, P., and Srivastava, V. (1983). Dielectric constant of a composite inhomogeneous medium. *Phys. Rev. B* **27**, 5098–5106.

Chýlek, P., and Videen, G. (1994). Longwave radiative properties of polydispersed hexagonal ice crystals. *J. Atmos. Sci.* **51**, 175–190.

Chýlek, P., and Videen, G. (1998). Scattering by a composite sphere and effective medium approximations. *Opt. Commun.* **146**, 15–20.

Chýlek, P., Grams, G. W., and Pinnick, R. G. (1976). Light scattering by irregular randomly oriented particles. *Science* **193**, 480–482.

Chýlek, P., Ramaswamy, V., Cheng, R., and Pinnick, R. G. (1981). Optical properties and mass concentration of carbonaceous smokes. *Appl. Opt.* **20**, 2980–2985.

Chýlek, P., Ramaswamy, V., and Cheng, R. J. (1984a). Effect of graphitic carbon on the albedo of clouds. *J. Atmos. Sci.* **41**, 3076–3084.

Chýlek, P., Gupta, B. R. D., Knight, N. C., and Knight, C. A. (1984b). Distribution of water in hailstones. *J. Clim. Appl. Meteorol.* **23**, 1469–1472.

Chýlek, P., Srivastava, V., Pinnick, R. G., and Wang, R. T. (1988). Scattering of electromagnetic waves by composite spherical particles: Experiment and effective medium approximations. *Appl. Opt.* **27**, 2396–2404.

Chýlek, P., Pinnick, R. G., and Srivastava, V. (1991). Effect of topology of water–ice mixture on radar backscattering by hailstones. *J. Appl. Meteorol.* **30**, 954–959.

Chýlek, P., Videen, G., Ngo, D., *et al.* (1995). Effect of black carbon on the optical properties and climate forcing of sulfate aerosols. *J. Geophys. Res.* **100**, 16,325–16,332.

Chýlek, P., Lesins, G. B., Videen, G., *et al.* (1996). Black carbon and absorption of solar radiation by clouds. *J. Geophys. Res.* **101**, 23,365–23,371.

Cialdi, Cmdr. A. (1866). Sul moto ondosodel mare e su le correnti de osso specialmente auquelle littorali, pp. 258–288. Cited in *ONI Transl. A-655*, 1. Hydrographic Office, 1955.

Clark, R. E. H. (1978). Intermediate coupling collision strengths from LS coupled R-matrix elements. *Comput. Phys. Commun.* **16**, 119–127.

Clebsch, R. F. A. (1863). Ueber die Reflexion an einer Kugelfläche. *Crelle's J.* **61**, 195–251.

Clothiaux, E. E., Miller, M. A., Albrecht, B. A., *et al.* (1995). An evaluation of a 94-GHz radar for remote sensing of cloud properties. *J. Atmos. Oceanic Technol.* **12**, 201–229.

Cloude, S. R. (1986). Group theory and polarisation algebra. *Optik* **75**, 26–36.

Cloude, S. R. (1989). Conditions for the physical realisability of matrix operators in polarimetry. *Proc. SPIE* **1166**, 177–185.

Cloude, S. R. (1992a). Uniqueness of target decomposition theorems in radar polarimetry. *In* "Direct and Inverse Methods in Radar Polarimetry" (W.-M. Boerner *et al.*, Eds.), pp. 267–296. Kluwer Academic, Dordrecht.

Cloude, S. R. (1992b). Measurement and analysis in radar polarimetry. *In* "Direct and Inverse Methods in Radar Polarimetry" (W.-M. Boerner *et al.*, Eds.), pp. 773–791. Kluwer Academic, Dordrecht.

Coffeen, D. L. (1979). Polarization and scattering characteristics in the atmospheres of Earth, Venus, and Jupiter. *J. Opt. Soc. Am.* **69**, 1051–1064.

Cole, J. B. (1995). A high accuracy FDTD algorithm to solve microwave propagation and scattering problems on a coarse grid. *IEEE Trans. Microwave Theory Tech.* **43**, 2053–2058.

Cole, J. B. (1998). Generalized nonstandard finite differences and applications. *Comput. Phys.* **12**, 82–87.

Coletti, A. (1984). Light scattering by nonspherical particles: A laboratory study. *Aerosol Sci. Technol.* **3**, 39–52.

Cooper, J., Hombach, V., and Schiavoni, A. (1996). Comparison of computational electromagnetic codes applied to a sphere canonical problem. *IEE Proc. Microwave Antennas Propag.* **143**, 309–316.

Cooray, M. F. R. (1990). Electromagnetic scattering by systems of arbitrarily oriented spheroids. Ph.D. Thesis, Univ. of Manitoba, Winnipeg.

Cooray, M. F. R., and Ciric, I. R. (1989a). Rotational–translational addition theorems for vector spheroidal wave functions. *COMPEL* **8**, 151–166.

Cooray, M. F. R., and Ciric, I. R. (1989b). Electromagnetic wave scattering by a system of two spheroids of arbitrary orientation. *IEEE Trans. Antennas Propag.* **37**, 608–618.

Cooray, M. F. R., and Ciric, I. R. (1990). Scattering of electromagnetic waves by a system of two conducting spheroids of arbitrary orientation. *In* "Radar Cross Sections of Complex Objects" (W. R. Stone, Ed.), pp. 343–355. IEEE Press, New York.

Cooray, M. F. R., and Ciric, I. R. (1991a). Scattering of electromagnetic waves by a system of two dielectric spheroids of arbitrary orientation. *IEEE Trans. Antennas Propag.* **39**, 680–684.

Cooray, M. F. R., and Ciric, I. R. (1991b). Scattering by systems of spheroids in arbitrary configurations. *Comput. Phys. Commun.* **68**, 279–305.

Cooray, M. F. R., and Ciric, I. R. (1992). Scattering of electromagnetic waves by a coated dielectric spheroid. *J. Electromagn. Waves Appl.* **6**, 1491–1507.

Cooray, M. F. R., and Ciric, I. R. (1993). Wave scattering by a chiral spheroid. *J. Opt. Soc. Am. A* **10**, 1197–1203.

Cooray, M. F. R., Ciric, I. R., and Sinha, B. P. (1990). Electromagnetic scattering by a system of two parallel dielectric prolate spheroids. *Can. J. Phys.* **68**, 376–384.

Cox, C., and Munk, W. (1954). Measurement of the roughness of the sea surface from photographs of the sun's glitter. *J. Opt. Soc. Am.* **44**, 838–850.

Crépel, O., Gayet, J. F., Fournol, J. F., and Oschepkov, S. (1997). A new airborne polar nephelometer for the measurements of optical and microphysical cloud properties. II. Preliminary tests. *Ann. Geophys.* **15**, 460–470.

Cross, J. D. (1968). Study of the surface of ice with a scanning electron microscope. *In* "Physics of Ice," pp. 81–94. Plenum, New York.

Cruzan, O. R. (1962). Translational addition theorems for spherical vector wave functions. *Quart. Appl. Math.* **20**, 33–40.

Czekala, H. (1998). Effects of ice particle shape and orientation on polarized microwave radiation for off-nadir problems. *Geophys. Res. Lett.* **25**, 1669–1672.

Czekala, H., and Simmer, C. (1998). Microwave radiative transfer with nonspherical precipitating hydrometeors. *J. Quant. Spectrosc. Radiat. Transfer* **60**, 365–374.

Dalmas, J., and Deleuil, R. (1985). Multiple scattering of electromagnetic waves from two infinitely conducting prolate spheroids which are centered in a plane perpendicular to their axes of revolution. *Radio Sci.* **20**, 575–581.

Dalmas, J., Deleuil, R., and MacPhie, R. H. (1989). Rotational–translational addition theorems for spheroidal vector wave functions. *Quart. Appl. Math.* **47**, 351–364.

d'Almeida, G. A., Koepke, P., and Shettle, E. P. (1991). "Atmospheric Aerosols." Deepak, Hampton, VA.

Dalzell, W. H., and Sarofim, A. F. (1969). Optical constants of soot and their application to heat-flux calculations. *J. Heat Transfer* **91**, 100–104.

Das, B. B., Liu, F., and Alfano, R. R. (1997). Time-resolved fluorescence and photon migration studies in biomedical and random media. *Rep. Prog. Phys.* **60**, 227–292.

Debye, P. (1909). Der Lichtdruk auf Kugeln von beliebigem Material. *Ann. Phys.* **30**, 57–136.

de Daran, F., Vignéras-Lefebvre, V., and Parneix, J. P. (1995). Modeling of electromagnetic waves scattered by a system of spherical particles. *IEEE Trans. Magn.* **31**, 1598–1601.

de Grooth, B. G., Terstappen, L. W. M. M., Puppels, G. J., and Greve, J. (1987). Light scattering polarization measurements as a new parameter in cytometry. *Cytometry* **8**, 539–544.

de Haan, J. F. (1987). Effects of aerosols on the brightness and polarization of cloudless planetary atmospheres. Ph.D. Dissertation, Free Univ., Amsterdam.

de Haan, J. F., Bosma, P. B., and Hovenier, J. W. (1987). The adding method for multiple scattering calculations of polarized light. *Astron. Astrophys.* **183**, 371–391.

Deirmendjian, D. (1969). "Electromagnetic Scattering on Spherical Polydispersions." Elsevier, New York.

den Ottolander, G. J., Schuit, H. R. E., Waayer, J. L. M., *et al.* (1985). Chronic B-cell leukemias: Relation between morphological and immunological features. *Clin. Immunol. Immunopathol.* **35**, 92–102.

Derr, V. E., Abshire, N. L., Cupp, R. E., and McNice, G. T. (1976). Depolarization of lidar returns from virga and source cloud. *J. Appl. Meteorol.* **15**, 1200–1203.

de Wolf, D. A., and Ligthart, L. P. (1993). Multipath effects due to rain at 30–50 GHz frequency communication links. *IEEE Trans. Antennas Propag.* **41**, 1132–1138.

Dibbern, J. (1987). Dependence of radar parameters on polarization properties of rain for bistatic CW radar. *Radio Sci.* **22**, 769–779.

Djaloshinski, L., and Orenstein, M. (1998). Coupling of concentric semiconductor microring lasers. *Opt. Lett.* **23**, 364–366.

Dobbins, R. A. and Megaridis, C. M. (1991). Absorption and scattering by polydisperse aggregates. *Appl. Opt.* **30**, 4747–4754.

Dodge, J. D., and Greuet, C. (1987). Dinoflagellate ultrastructure and complex organelles. *In* "The Biology of Dinoflagellates" (F. J. R. Taylor, Ed.), pp. 92–119. Blackwell Sci., Oxford.

Doicu, A., and Wriedt, T. (1997a). Multiple multipole extended boundary condition method. *Optik* **105**, 57–60.

Doicu, A., and Wriedt, T. (1997b). Extended boundary condition method with multipole sources located in the complex plane. *Opt. Commun.* **139**, 85–91.

Doicu, A., and Wriedt, T. (1997c). Computation of the beam-shape coefficients in the generalized Lorenz–Mie theory by using the translational addition theorem for spherical vector wave functions. *Appl. Opt.* **36**, 2971–2978.

Doicu, A., Eremin, Yu. A., and Wriedt, T. (1999). Convergence of the T-matrix method for light scattering from a particle on or near a surface. *Opt. Commun.* **159**, 266–277.

Dolginov, A. Z., Gnedin, Yu. N., and Silant'ev, N. A. (1995). "Propagation and Polarization of Radiation in Cosmic Media." Gordon and Breach, Basel.

Donn, B., and Powell, R. S. (1963). Angular scattering from irregularly shaped particles with application to astronomy. *In* "ICES Electromagnetic Scattering" (M. Kerker, Ed.), pp. 151–158. Pergamon, New York.

Doviak, R. J., and Zrnic, D. S. (1993). "Doppler Radar and Weather Observations." Academic Press, San Diego.

Draine, B. T. (1988). The discrete-dipole approximation and its application to interstellar graphite grains. *Astrophys. J.* **333**, 848–872.

Draine, B. T., and Flatau, P. J. (1994). Discrete-dipole approximation for scattering calculations. *J. Opt. Soc. Am. A* **11**, 1491–1499.

Draine, B. T., and Flatau, P. J. (1997). User guide for the discrete dipole approximation code DDSCAT (Version 5a). Princeton Observatory Preprint POPe-695 (http://www.astro.princeton.edu/~draine/UserGuide/UserGuide.html).

Draine, B. T., and Goodman, J. (1993). Beyond Clausius-Mossotti: Wave propagation on a polarizable point lattice and the discrete dipole approximation. *Astrophys. J.* **405**, 685–697.

Draine, B. T., and Malhotra, S. (1993). On graphite and the 2175 Å extinction profile. *Astrophys. J.* **414**, 632–645.

Draine, B. T., and Weingartner, J. C. (1996). Radiative torques on interstellar grains. I. Superthermal rotation. *Astrophys. J.* **470**, 551–565.

Draine, B. T., and Weingartner, J. C. (1997). Radiative torques on interstellar grains. II. Alignment with the magnetic field. *Astrophys. J.* **480**, 633–646.

Druger, S. D., Kerker, M., Wang, D.-S., and Cooke, D. D. (1979). Light scattering by inhomogeneous particles. *Appl. Opt.* **18**, 3888–3889.

Duce, R. A., and Woodcock, A. H. (1971). Difference in chemical composition of atmospheric sea-salt particles produced in the surf zone and on the open sea in Hawaii. *Tellus* **23**, 427–435.

Duck, F. A. (1990). "Physical Properties of Tissue: A Comprehensive Reference Book." Academic Press, San Diego.

Dugin, V. P., and Mirumyants, S. O. (1976). The light scattering matrices of artificial crystalline clouds. *Izv. Acad. Sci. USSR, Atmos. Oceanic Phys.* (Engl. Transl.) **12**, 606–608.

Dugin, V. P., Volkovitskiy, O. A., Mirumyants, S. O., and Nikiforova, N. K. (1977). Anisotropy of light scattering by artificial crystalline cloud formations. *Izv. Acad. Sci. USSR, Atmos. Oceanic Phys.* (Engl. Transl.) **13**, 22–25.

Dungey, C. (1990). Backscattering by nonspherical particles using the coupled-dipole approximation method: An application in radar meteorology. Ph.D. Thesis, Pennsylvania State Univ., University Park, PA.

Dungey, C. E., and Bohren, C. F. (1991). Light scattering by nonspherical particles: A refinement to the coupled-dipole method. *J. Opt. Soc. Am. A* **8**, 81–87.

Dungey, C. E., and Bohren, C. F. (1993). Backscattering by nonspherical hydrometeors as calculated by the coupled-dipole method: An application in radar meteorology. *J. Atmos. Oceanic Technol.* **10**, 526–532.

Dunn, A., and Richards-Kortum, R. (1996). Three-dimensional computation of light scattering from cells. *IEEE J. Select. Top. Quant. Electron.* **2**, 898–905.

Dunn, A., Smithpeter, C., Welch, A. J., and Richards-Kortum, R. (1996). Light scattering from cells. *OSA Tech. Digest—Biomed. Opt. Spectrosc. Diagn.*, 50–52. Washington, DC.

Dunn, A., Smithpeter, C., Welch, A. J., and Richards-Kortum, R. (1997). Finite-difference time-domain simulation of light scattering from single cells. *J. Biomed. Opt.* **2**, 262–266.

Dyson, R. D. (1974). "Cell Biology: A Molecular Approach." Allyn and Bacon, Boston.

Eberhard, W. L. (1992). Ice-cloud depolarization of backscattering for CO_2 and other infrared lidars. *Appl. Opt.* **31**, 6485–6490.

Edmonds, A. R. (1974). "Angular Momentum in Quantum Mechanics." Princeton Univ. Press, Princeton, NJ.

Eiden, R. (1975). Fourier representation of the energy distribution of an electromagnetic field scattered by spherical particles. *Appl. Opt.* **14**, 2486–2491.

Eloranta, E. W., and Piironen, P. (1994). Depolarization measurements with the high spectral resolution lidar. *Prepr. Vol., Int. Laser Radar Conf., 17th, Sendai, Japan*, 127–128.

Engquist, B., and Majda, A. (1977). Absorbing boundary conditions for the numerical simulation of waves. *Math. Comput.* **31**, 629–651.

Eremin, Yu. A., and Ivakhnenko, V. I. (1998). Modeling of light scattering by non-spherical inhomogeneous particles. *J. Quant. Spectrosc. Radiat. Transfer* **60**, 475–482.

Eremin, Yu., and Orlov, N. (1998). Modeling of light scattering by non-spherical particles based on discrete sources method. *J. Quant. Spectrosc. Radiat. Transfer* **60**, 451–462.

Erma, V. A. (1968a). An exact solution for the scattering of electromagnetic waves from conductors of arbitrary shape. I. Case of cylindrical symmetry. *Phys. Rev.* **173**, 1243–1257.

Erma, V. A. (1968b). Exact solution for the scattering of electromagnetic waves from conductors of arbitrary shape. II. General case. *Phys. Rev.* **176**, 1544–1553.

Erma, V. A. (1969). Exact solution for the scattering of electromagnetic waves from bodies of arbitrary shape. III. Obstacles with arbitrary electromagnetic properties. *Phys. Rev.* **179**, 1238–1246.

Essien, M., Armstrong, R. L., and Pinnick, R. G. (1993). Lasing emission from an evaporating layered microdroplet. *Opt. Lett.* **18**, 762–764.

Evans, B. T. N., and Fournier, G. R. (1994). Analytic approximation to randomly oriented spheroid extinction. *Appl. Opt.* **33**, 5796–5804.

Evans, B. G., and Holt, A. R. (1977). Scattering amplitudes and cross-polarisation of ice particles. *Electron. Lett.* **13**, 342–344.

Evans, K. F., and Stephens, G. L. (1991). A new polarized atmospheric radiative transfer model. *J. Quant. Spectrosc. Radiat. Transfer* **46**, 413–423.

Evans, K. F., and Stephens, G. L. (1995a). Microwave radiative transfer through clouds composed of realistically shaped ice crystals. I. Single scattering properties. *J. Atmos. Sci.* **52**, 2041–2057.

Evans, K. F., and Stephens, G. L. (1995b). Microwave radiative transfer through clouds composed of realistically shaped ice crystals. II. Remote sensing of ice clouds. *J. Atmos. Sci.* **52**, 2058–2072.

Evans, K. F., and Vivekanandan, J. (1990). Multiparameter radar and microwave radiative transfer modeling of nonspherical atmospheric ice particles. *IEEE Trans. Geosci. Remote Sens.* **28**, 423–437.

Evans, K. F., Walter, S. J., Heymsfield, A. J., and Deeter, M. N. (1998). Modeling of submillimeter passive remote sensing of cirrus clouds. *J. Appl. Meteorol.* **37**, 184–205.

Evans, K. F., Evans, A. H., Nolt, I. G., and Marshall, B. T. (1999). The prospect for remote sensing of cirrus clouds with a submillimeter-wave spectrometer. *J. Appl. Meteorol.* **38**, 514–525.

Farafonov, V. G., Voshchinnikov, N. V., and Somsikov, V. V. (1996). Light scattering by a core–mantle spheroidal particle. *Appl. Opt.* **35**, 5412–5426.

Federer, B., and Waldvogel, A. (1975). Hail and raindrop size distributions from a Swiss multicell storm. *J. Appl. Meteorol.* **14**, 91–97.

Fikioris, J. G., and Uzunoglu, N. K. (1979). Scattering from an eccentrically stratified dielectric sphere. *J. Opt. Soc. Am.* **69**, 1359–1366.

Fischer, T., and Schmidt-Schönbein, H. (1977). Tank tread motion of red cell membranes in viscometric flow: Behavior of intracellular and extracellular markers. *Blood Cells* **3**, 351–365.

Fitzgerald, J. W. (1991). Marine aerosols—a review. *Atmos. Environ. A* **25**, 533–545.

Flammer, C. (1957). "Spheroidal Wave Functions." Stanford Univ. Press, Stanford, CA.

Flatau, P. J. (1997). Improvements in the discrete-dipole approximation method of computing scattering and absorption. *Opt. Lett.* **22**, 1205–1207.

Flatau, P. J. (1998). SCATTERLIB: Light Scattering Codes Library. URL: atol.ucsd.edu/~pflatau/scatlib/.

Flatau, P. J., Fuller, K. A., and Mackowski, D. W. (1993). Scattering by two spheres in contact: Comparisons between the discrete dipole approximation and modal analysis. *Appl. Opt.* **32**, 3302–3305.

Flesia, C., Mugnai, A., Emery, Y., *et al.* (1994). Interpretation of lidar depolarization measurements of the Pinatubo stratospheric aerosol layer during EASOE. *Geophys. Res. Lett.* **21**, 1443–1446.

Francis, P. N. (1995). Some aircraft observations of the scattering properties of ice crystals. *J. Atmos. Sci.* **52**, 1142–1154.

Freeman, R. G., Grabar, K. C., Guthrie, A. P., *et al.* (1995). Self-assembled metal colloid monolayers: An approach to SERS substrates. *Science* **267**, 1629–1632.

Fresnel, A. (1866a). "Oevres Completes D'Augustin Fresnel, Vol. II." Imprimerie Impériale, Paris.

Fresnel, A. (1866b). "Oevres Completes D'Augustin Fresnel, Vol. III." Imprimerie Impériale, Paris.

Freudenthaler, V., Homburg, F., and Jäger, H. (1996). Optical parameters of contrails from lidar measurements: Linear depolarization. *Geophys. Res. Lett.* **23**, 3715–3718.

Freund, R. W. (1992). Conjugate gradient-type methods for linear systems with complex symmetric coefficient matrices. *SIAM J. Sci. Statist. Comput.* **13**, 425–448.

Friedman, B., and Russek, J. (1954). Addition theorems for spherical waves. *Quart. Appl. Math.* **12**, 13–23.

Frost, I. R., Goddard, J. W. F., and Illingworth, A. J. (1991). Hydrometeor identification using cross polar radar measurements and aircraft verification. *Prepr. Vol., Conf. Radar Meteorol., 25th, Paris, 1991*, 658–661.

Fry, E. S., and Kattawar, G. W. (1981). Relationships between elements of the Stokes matrix. *Appl. Opt.* **20**, 2811–2814.

Fry, E. S., and Voss, K. J. (1985). Measurement of the Mueller matrix for phytoplankton. *Luminol. Oceanogr.* **30**, 1322–1326.

Fucile, E., Borghese, F., Denti, P., and Saija, R. (1993). Theoretical description of dynamic light scattering from an assembly of large axially symmetric particles. *J. Opt. Soc. Am. A* **10**, 2611–2617.

Fucile, E., Borghese, F., Denti, P., and Saija, R. (1995). Effect of an electrostatic field on the optical properties of a cloud of dielectric particles. *Appl. Opt.* **34**, 4552–4562.

Fujiki, N. M., Geldart, D. J., and Chýlek, P. (1994). Effects of air bubbles on radar backscattering by hailstones. *J. Appl. Meteorol.* **33**, 304–308.

Fuller, K. A. (1987). Cooperative electromagnetic scattering by ensembles of spheres. Ph.D. Dissertation, Texas A&M Univ., College Station, TX.

Fuller, K. A. (1989). Some novel features of morphology dependent resonances of bispheres. *Appl. Opt.* **28**, 3788–3790.

Fuller, K. A. (1991). Optical resonances and two-sphere systems. *Appl. Opt.* **33**, 4716–4731.

Fuller, K. A. (1993a). Scattering and absorption by inhomogeneous spheres and sphere aggregates. *Proc. SPIE* **1862**, 249–257.

Fuller, K. A. (1993b). Scattering of light by coated spheres. *Opt. Lett.* **18**, 257–259.

Fuller, K. A. (1994a). Scattering and absorption cross sections of compounded spheres. I. Theory for external aggregation. *J. Opt. Soc. Am. A* **11**, 3251–3260.

Fuller, K. A. (1994b). Absorption cross sections and Mueller matrices of sulfate/soot composite particles. *Proc. AGU/AWMA Int. Specialty Conf. Aerosols Atmos. Opt., VIP-41, Vol. A*, 497–508. Air and Waste Management Association, Pittsburgh, PA.

Fuller, K. A. (1994c). Morphology dependent resonances in eccentrically stratified spheres. *Opt. Lett.* **19**, 1272–1274.

Fuller, K. A. (1995a). Scattering and absorption cross sections of compounded spheres. II. Calculations for external aggregation. *J. Opt. Soc. Am. A* **12**, 881–892.

Fuller, K. A. (1995b). Scattering and absorption cross sections of compounded spheres. III. Spheres containing arbitrarily located spherical inhomogeneities. *J. Opt. Soc. Am. A* **12**, 893–904.

Fuller, K. A., and Kattawar, G. W. (1988a). Consummate solution to the problem of classical electromagnetic scattering by ensembles of spheres. I. Linear chains. *Opt. Lett.* **13**, 90–92.

Fuller, K. A., and Kattawar, G. W. (1988b). Consummate solution to the problem of classical electromagnetic scattering by ensembles of spheres. II. Clusters of arbitrary configuration. *Opt. Lett.* **13**, 1063–1065.

Fuller, K. A., Kattawar, G. W., and Wang, R. T. (1986). Electromagnetic scattering from two dielectric spheres: Further comparisons between theory and experiment. *Appl. Opt.* **25**, 2521–2529.

Fuller, K. A., Stephens, G. L., and Jersak, B. D. (1994a). Some advances in understanding light scattering by nonspherical particles. *Prepr. Vol., Conf. Atmos. Rad., 8th, Nashville, TN, 1994*, 319–321.

Fuller, K. A., Stephens, G. L., Malm, W. C., *et al.* (1994b). Relationship between IM-PROVE program parameters, aerosol-modified albedo, and climate forcing. *Proc. AGU/AWMA Int. Specialty Conf. Aerosols Atmos. Opt., VIP-41, Vol. A*, 328–338. Air and Waste Management Association, Pittsburgh, PA.

Fuller, K. A., Gonzalez, R. J., and Kochar, M. S. (1998). Light scattering from dimers: Latex–latex and gold–latex. *Proc. SPIE* **3256**, 186–194.

Fuller, K. A., Malm, W. C., and Kreidenweis, S. M. (1999). Effects of mixing on extinction by carbonaceous particles. *J. Geophys. Res.*, in press.

Fung, A. K. (1994). "Microwave Scattering and Emission Models and Their Applications." Artech House, Boston.

Furse, M., Mathur, S. P., and Gandhi, O. P. (1990). Improvements on the finite-difference time-domain method for calculating the radar cross section of a perfectly conducting target. *IEEE Trans. Microwave Theory Tech.* **38**, 919–927.

Fusco, M. A. (1990). FDTD algorithm in curvilinear coordinates. *IEEE Trans. Antennas Propag.* **38**, 76–89.

Fusco, M. A., Smith, M. V., and Gordon, L. W. (1991). A three-dimensional FDTD algorithm in curvilinear coordinates. *IEEE Trans. Antennas Propag.* **39**, 1463–1471.

Gale, R. P., and Foon, K. A. (1985). Chronic lymphocytic leukemia, recent advances in biology and treatment. *Ann. Intern. Med.* **103**, 101–120.

Gao, B.-C., and Kaufman, Y. J. (1995). Selection of the 1.375-ìm MODIS channel for remote sensing of cirrus clouds and stratospheric aerosols from space. *J. Atmos. Sci.* **52**, 4231–4237.

Gasiewski, A. J. (1993). Microwave radiative transfer in hydrometeors. *In* "Atmospheric Remote Sensing by Microwave Radiometry" (M. A. Janssen, Ed.), pp. 91–144. Wiley, New York.

Gasiewski, A. J., and Kunkee, D. B. (1994). Polarized microwave emission from water waves. *Radio Sci.* **29**, 1449–1466.

Gasster, S. D., and Flaming, G. M. (1998). Overview of the Conical Microwave Imager/Sounder development for the NPOESS Program. *IGARSS 1998*, Seattle, WA, Vol. I, 268–270.

Gathman, S. G., and Davidson, K. L. (1993). The Navy oceanic vertical aerosol model. Tech. Rep. 1634, Naval Command, Control and Ocean Surveillance Center, San Diego.

Gautier, A., Salamin, L. M., Couture, E. T., *et al.* (1986). Electron microscopy of the chromosomes of dinoflagellates in situ: Confirmation of Bouligand's liquid crystal hypothesis. *J. Ultra. Mol. Struct. Res.* **97**, 10–30.

Gayet, J. F., Crépel, O., Fournol, J. F., and Oschepkov, S. (1997). A new airborne polar nephelometer for the measurements of optical and microphysical cloud properties. I. Theoretical design. *Ann. Geophys.* **15**, 451–459.

Gayet, J.-F., Auriol, F., Oshchepkov, S., *et al.* (1998). In situ measurements of the scattering phase function of stratocumulus, contrails, and cirrus. *Geophys. Res. Lett.* **25**, 971–974.

Gelfand, I. M., Minlos, R. A., and Shapiro, Z. Ya. (1963). "Representations of the Rotation and Lorentz Groups and Their Applications." Pergamon, Oxford.

Geller, P. E., Tsuei, T. G., and Barber, P. W. (1985). Information content of the scattering matrix for spheroidal particles. *Appl. Opt.* **24**, 2391–2396.

Gérardy, J. M., and Ausloos, M. (1980). Absorption spectrum of clusters of spheres from the general solution of Maxwell's equations: The long wavelength limit. *Phys. Rev. B* **22**, 4950–4959.

Gérardy, J. M., and Ausloos, M. (1982). Absorption spectrum of clusters of spheres from the general solution of Maxwell's equations. II. Optical properties of aggregated metal spheres. *Phys. Rev. B* **25**, 4204–4229.

Germogenova, O. A. (1963). The scattering of a plane electromagnetic wave by two spheres. *Izv. Acad. Sci. USSR, Geophys. Ser.* (Engl. Transl.) **4**, 403–405.

Gibson, E. K., Jr. (1992). Volatiles in interplanetary dust particles: A review. *J. Geophys. Res.* **97**, 3865–3875.

Giese, R. H., Weiss, K., Zerull, R. H., and Ono, T. (1978). Large fluffy particles: A possible explanation of the optical properties of interplanetary dust. *Astron. Astrophys.* **65**, 265–272.

Giese, R. H., Kneissel, B., and Rittich, V. (1986). Three-dimensional models of the zodiacal dust cloud: A comparative study. *Icarus* **68**, 395–411.

Gilbert, G. D., and Pernicka, J. C. (1966). Improvement of underwater visibility by reduction of backscatter with a circular polarization technique. *Proc. SPIE, Underwater Photo-Optics Seminar, Santa Barbara, CA.*

Gillette, D. A., and Walker, T. R. (1977). Characteristics of airborne particles produced by wind erosion of sandy soil, high plains of west Texas. *Soil Sci.* **123**, 97–110.

Giuli, D., Toccafondi, A., Gentili, G. B., and Freni, A. (1991). Tomographic reconstruction of rainfall fields through microwave attenuation measurements. *J. Appl. Meteorol.* **30**, 1323–1340.

Givens, C. R., and Kostinski, A. B. (1993). A simple necessary and sufficient condition on physically realizable Mueller matrices. *J. Mod. Opt.* **40**, 471–481.

Gobbi, G. P. (1998). Polarization lidar returns from aerosols and thin clouds: A framework for the analysis. *Appl. Opt.* **37**, 5505–5508.

Gobbi, G. P., Di Donfrancesco, G., and Adriani, A. (1998). Physical properties of stratospheric clouds during the Antarctic winter of 1995. *J. Geophys. Res.* **103**, 10,859–10,873.

Göbel, G., Lippek, A., and Wriedt, T. (1997). Monte Carlo approach to the light scattering of inhomogeneous nonspherical particles. *Abstr. Vol., Workshop on Light Scattering by Non-Spherical Particles, Univ. of Helsinki, 1997*, 19–20.

Goddard, J. W. F., Cherry, S. M., and Bringi, V. N. (1982). Comparison of dual polarization radar measurements of rain with ground-based disdrometer measurements. *J. Appl. Meteorol.* **21**, 252–256.

Goedecke, G. H., and O'Brien, S. G. (1988). Scattering by irregular inhomogeneous particles via the digitized Green's function algorithm. *Appl. Opt.* **27**, 2431–2438.

Goldsmith, J. E. M., Blair, F. H., Bisson, S. E., and Turner, D. D. (1998). Turn-key Raman lidar for probing atmospheric water vapor, clouds, and aerosols. *Appl. Opt.* **37**, 4979–4990.

Golestani, Y., Chandrasekar, V., and Bringi, V. N. (1989). Intercomparison of multiparameter radar measurements. *Prepr. Vol., Conf. Radar. Meteorol., 24th, Tallahassee, FL, 1989,* 309–314.

Goncharenko, A. V., Semenov, Yu. G., and Venger, E. F. (1999). Effective scattering cross section of small ellipsoidal particles. *J. Opt. Soc. Am. A* **16**, 517–522.

Goodman, S. J., Buechler, D. E., and Wright, P. D. (1988). Lightning and precipitation history of a microburst-producing storm. *Geophys. Res. Lett.* **15**, 1185–1188.

Goodman, J. J., Draine, B. T., and Flatau, P. J. (1991). Application of fast-Fourier-transform techniques to the discrete-dipole approximation. *Opt. Lett.* **16**, 1198–1200.

Goody, R. M., and Yung, Y. L. (1989). "Atmospheric Radiation: Theoretical Basis." Oxford Univ. Press, New York.

Gordon, H. R. (1997). Atmospheric correction of ocean color imagery in the Earth Observing System era. *J. Geophys. Res.* **102**, 17,081–17,106.

Gorman, A. M., Samali, A., McGowan, A. J., and Cotter, T. G. (1997). Use of flow cytometry techniques in studying mechanisms of apoptosis in leukemic cells. *Cytometry* **29**, 97–105.

Gorodetsky, M. L., Savchenkov, A. A., and Ilchenko, V. S. (1996). Ultimate Q of optical microsphere resonators. *Opt. Lett.* **21**, 453–455.

Grabar, K. C., Freeman, R. G., Hommer, M. B., and Natan, M. J. (1995). Preparation and characterization of Au colloid monolayers. *Anal. Chem.* **67**, 735–743.

Gras, J. L., and Ayers, G. P. (1983). Marine aerosol at southern mid-latitudes. *J. Geophys. Res.* **88**, 10,661–10,666.

Greenberg, J. M., and Gustafson, B. Å. S. (1981). A comet fragment model for zodiacal light particles. *Astron. Astrophys.* **93**, 35–42.

Greenberg, J. M., Pedersen, N. E., and Pedersen, J. C. (1961). Microwave analog to the scattering of light by nonspherical particles. *J. Appl. Phys.* **32**, 233–242.

Greenler, R. (1990). "Rainbows, Halos, and Glories." Cambridge Univ. Press, New York.

Grenfell, T. C. (1983). A theoretical model of the optical properties of sea ice in the visible and near infrared. *J. Geophys. Res.* **88**, 9723–9735.

Grenfell, T. C., and Hedrick, D. (1983). Scattering of visible and near infrared radiation by NaCl ice and glacier ice. *Cold Reg. Sci. Technol.* **8**, 119–127.

Gresh, D. L. (1990). Voyager radio occultation by the Uranian rings: Structure, dynamics, and particle sizes. Sci. Rep. D845-1990-1, Center for Radar Astronomy, Stanford Electronics Laboratory, Stanford Univ., Stanford, CA.

Griffin, M. (1983). Complete Stokes parameterization of laser backscattering from artificial clouds. M.S. Thesis, Univ. of Utah, Salt Lake City, UT.

Grimes, C. A. (1991). Electromagnetic properties of random material. *Waves Random Media* **1**, 265–273.

Gross, C. T., Salamon, H., Hunt, A. J., *et al.* (1991). Hemoglobin polymerization in sickle cells studied by circular polarized light scattering. *Biochem. Biophys. Acta* **1079**, 152–160.

Grote, M. J., and Keller, J. B. (1998). Nonreflecting boundary conditions for Maxwell's equations. *J. Comput. Phys.* **139**, 327–342.

Grün, E., Staubach, P., Hamilton, D. P., *et al.* (1997). South–north and radial traverses through the interplanetary dust cloud. *Icarus* **129**, 270–288.

Gustafson, B. Å. S. (1980). "Scattering by Ensembles of Small Particles, Experiment, Theory, and Application." Lund Observatory Press, Lund, Sweden.

Gustafson, B. Å. S. (1983). Dominant particle parameters in side scattering by some aggregates of cylinders. *In* "1982 CSL Scientific Conference on Obscuration and Aerosol Research" (R. H. Kohl, Ed.), pp. 281–292. U.S. Army Chemical Systems Laboratory, Aberdeen, MD.

Gustafson, B. Å. S. (1985). Laboratory studies on polarization properties of elongated particles and comparisons to dust in the tail of comet Ikeya-Seki. *In* "Properties and Interactions of Interplanetary Dust" (R. H. Giese and P. Lamy, Eds.), pp. 227–230. Reidel, Dordrecht.

Gustafson, B. Å. S. (1996). Microwave analog to light scattering measurements: A modern implementation of a proven method to achieve precise control. *J. Quant. Spectrosc. Radiat. Transfer* **55**, 663–672.

Gustafson, B. Å. S. (1999). Scattering by complex systems. I. Methods. *In* "Formation and Evolution of Solids in Space" (J. M. Greenberg and A. Li, Eds.), pp. 535–548. Kluwer Academic, Dordrecht.

Gustafson, B. Å. S., and Kolokolova, L. (1999). A. systematic study of light scattering by aggregate particles using the microwave analog technique: angular and wavelength dependence of intensity and polarization. *J. Goephys. Res.*, in press.

Gustafson, B. Å. S., Kolokolova, L., Thomas-Osip, J. E., *et al.* (1999). Scattering by complex systems. II. Results from microwave measurements. *In* "Formation and Evolution of Solids in Space" (J. M. Greenberg and A. Li, Eds.), pp. 549–564. Kluwer Academic, Dordrecht.

Haferman, J. L., Smith, T. F., and Krajewski, W. F. (1997). A multi-dimensional discrete-ordinates method for polarized radiative transfer. I. Validation for randomly oriented axisymmetric particles. *J. Quant. Spectrosc. Radiat. Transfer* **58**, 379–398. [Errata: **60**, I (1998).]

Hafner, C. (1990). "The Generalized Multipole Technique for Computational Electromagnetics." Artech House, Boston.

Hage, J. I., and Greenberg, J. M. (1990). A model for the optical properties of porous grains. *Astrophys. J.* **361**, 251–259.

Hage, J. I., Greenberg, J. M., and Wang, R. T. (1991). Scattering from arbitrarily shaped particles: Theory and experiment. *Appl. Opt.* **30**, 1141–1152.

Hall, M. P. M., Cherry, S. M., Goddard, J. W. F., and Kennedy, G. R. (1980). Raindrop sizes and rainfall rate measured by dual-polarization radar. *Nature (London)* **285**, 195–198.

Hall, M. P. M., Goddard, J. W. F., and Cherry, S. M. (1984). Identification of hydrometeors and other targets by dual-polarization radar. *Radio Sci.* **19**, 132–140.

Hallock, H. B., and Hallajian, J. (1983). Polarization imaging and mapping. *Appl. Opt.* **22**, 964–966.

Hamermesh, M. (1962). "Group Theory and Its Application to Physical Problems." Addison-Wesley, Reading, MA.

Hamid, A.-K., Ciric, I. R., and Hamid, M. (1990). Electromagnetic scattering by an arbitrary configuration of dielectric spheres. *Can. J. Phys.* **68**, 1419–1428.

Hamid, A.-K., Ciric, I. R., and Hamid, M. (1991). Iterative solution of the scattering by an arbitrary configuration of conducting or dielectric spheres. *IEE Proc. H* **138**, 565–572.

Hamid, A.-K., Ciric, I. R., and Hamid, M. (1992). Analytic solutions of the scattering by two multilayered dielectric spheres. *Can. J. Phys.* **70**, 696–705.

Hänel, G. (1976). The properties of atmospheric aerosol particles as functions of the relative humidity at thermodynamic equilibrium with the surrounding moist air. *In* "Advances in Geophysics" (H. E. Landsberg and J. van Mieghem, Eds.), Vol. 19, pp. 73–188. Academic Press, New York.

Hansen, W. W. (1935). A new type of expansion in radiation problems. *Phys. Rev.* **47**, 139–143.

Hansen, W. W. (1937). Transformations useful in certain antenna calculations. *J. Appl. Phys.* **8**, 282–286.

Hansen, J. E. (1971). Multiple scattering of polarized light in planetary atmospheres. I. The doubling method. *J. Atmos. Sci.* **28**, 120–125.

Hansen, M. Z., and Evans, W. H. (1980). Polar nephelometer for atmospheric particulate studies. *Appl. Opt.* **19**, 3389–3395.

Hansen, J. E., and Hovenier, J. W. (1974). Interpretation of the polarization of Venus. *J. Atmos. Sci.* **27**, 265–281.

Hansen, J. E., and Travis, L. D. (1974). Light scattering in planetary atmospheres. *Space Sci. Rev.* **16**, 527–610.

Hapke, B. (1993). "Theory of Reflectance and Emittance Spectroscopy." Cambridge Univ. Press, New York.

Hapke, B., Cassidy, W., and Wells, E. (1975). Effects of vapor-phase deposition processes on the optical, chemical and magnetic properties of the lunar regolith. *Moon* **13**, 339–353.

Hapke, B., Nelson, R., and Smythe, W. (1998). The opposition effect of the Moon: Coherent backscattering and shadow hiding. *Icarus* **133**, 89–97.

Haracz, R. D., Cohen, L. D., and Cohen, A. (1984). Perturbation theory for scattering from dielectric spheroids and short cylinders. *Appl. Opt.* **23**, 436–441.

Haracz, R. D., Cohen, L. D., and Cohen, A. (1985). Scattering of linearly polarized light from randomly oriented cylinders and spheroids. *J. Appl. Phys.* **58**, 3322–3327.

Haracz, R. D., Cohen, L. D., Cohen, A., and Acquista, C. (1986). Light scattering from dielectric targets composed of a continuous assembly of circular disks. *Appl. Opt.* **25**, 4386–4395.

Hardeman, M. R., Bauersachs, R. M., and Meiselman, H. J. (1988). RBC laser diffractometry and RBC aggregometry with a rotational viscometer: Comparison with rheoscope and myrenne aggregometer. *Clin. Hemorheol.* **8**, 581–593.

Harrington, R. F. (1968). "Field Computation by Moment Methods." Macmillan, New York.

Heine, V. (1960). "Group Theory in Quantum Mechanics: An Introduction to Its Present Usage." Pergamon, New York (also Dover, New York, 1993).

Heintzenberg, J., and Charlson, R. J. (1996). Design and application of the integrating nephelometer: A review. *J. Atmos. Oceanic Technol.* **13**, 987–1000.

Heintzenberg, J., and Wendisch, M. (1996). On the sensitivity of cloud albedo to the partitioning of particulate absorbers in cloudy air. *Beitr. Phys. Atmos.* **69**, 491–499.

Hendry, A., and Antar, Y. M. M. (1984). Precipitation particle identification with centimeter wavelength dual-polarization radars. *Radio Sci.* **19**, 115–122.

Hendry, A., McCormick, G. C., and Barge, B. L. (1976). The degree of common orientation observed by polarization diversity radars. *J. Appl. Meteorol.* **15**, 633–640.

Herman, B. M., and Battan, L. J. (1961). Calculations of Mie back-scattering from melting ice spheres. *J. Meteorol.* **18**, 468–478.

Herman, B. M., Caudill, T. R., Flittner, D. E., *et al.* (1995). Comparison of the Gauss–Seidel spherical polarized radiative transfer code with other radiative transfer codes. *Appl. Opt.* **34**, 4563–4572.

Herman, M., Deuzé, J.-L., Devaux, C., *et al.* (1997). Remote sensing of aerosols over land surfaces including polarization measurements and application to POLDER measurements. *J. Geophys. Res.* **102**, 17,039–17,049.

Hess, M., Koelemeijer, R. B. A., and Stammes, P. (1998). Scattering matrices of imperfect hexagonal ice crystals. *J. Quant. Spectrosc. Radiat. Transfer* **60**, 301–308.

Heymsfield, G. M., and Fulton, R. (1994a). Passive microwave and infrared structure of mesoscale convective systems. *Meteorol. Atmos. Phys.* **54**, 123–140.

Heymsfield, G. M., and Fulton, R. (1994b). Passive microwave structure of severe tornadic storms on 16 November 1987. *Mon. Weather Rev.* **122**, 2587–2595.

Heymsfield, A. J., Miller, K. M., and Spinhirne, J. D. (1990). The 27–28 October 1988 FIRE IFO cirrus case study: Cloud microstructure. *Mon. Weather Rev.* **118**, 2313–2328.

Hiatt, R. E., Senior, T. B. A., and Weston, V. H. (1960). A study of surface roughness and its effects on the backscattering cross section of spheres. *Proc. IRE* **48**, 2008–2016.

Hielscher, A. H., Eick, A. A., Mourant, J. R., *et al.* (1997). Diffuse backscattering Mueller matrices of highly scattering media. *Opt. Express* **1**, 441–453.

Hill, S. C., Hill, A. C., and Barber, P. W. (1984). Light scattering by size/shape distributions of soil particles and spheroids. *Appl. Opt.* **23**, 1025–1031.

Hill, S. C., Saleheen, H. I., and Fuller, K. A. (1995). Light scattering by an inhomogeneous sphere. *J. Opt. Soc. Am. A* **13**, 905–915.

Hillier, J. K. (1997). Scattering of light by composite particles in a planetary surface. *Icarus* **130**, 328–335.

Ho, K. C., and Allen, F. S. (1994). An approach to the inverse obstacle problem from the scattering Mueller matrix. *Inverse Problems* **10**, 387–400.

Hobbs, P. V., Funk, N. T., Weis, R. R., Sr., and Locatelli, J. D. (1985). Evaluation of a 35 GHz radar for cloud physics research. *J. Atmos. Oceanic Technol.* **2**, 35–48.

Hodkinson, J. R. (1963). Light scattering and extinction by irregular particles larger than the wavelength. *In* "ICES Electromagnetic Scattering" (M. Kerker, Ed.), pp. 87–100. Pergamon, New York.

Hodkinson, J. R. (1964). Refractive index and particle-extinction efficiency factors for carbon. *J. Opt. Soc. Am.* **54**, 846.

Hoekstra, A. G., and Sloot, P. M. A. (1993). Dipolar unit size in coupled-dipole calculations of the scattering matrix elements. *Opt. Lett.* **18**, 1211–1213.

Hoekstra, A. G., and Sloot, P. M. A. (1995). Coupled dipole simulations of elastic light scattering on parallel systems. *Int. J. Mod. Phys. C* **6**, 663–679.

Hoekstra, A. G., Aten, J. A., and Sloot, P. M. A. (1991). Effect of aniosmotic media on the T-lymphocyte nucleus. *Biophys. J.* **59**, 765–774.

Hoekstra, A. G., Grimminck, M., and Sloot, P. M. A. (1998). Large scale simulations of elastic light scattering by a fast discrete dipole approximation. *Int. J. Mod. Phys. C* **9**, 87–102.

Hofer, M., and Glatter, O. (1989). Mueller matrix calculations for randomly oriented rotationally symmetric objects with low contrast. *Appl. Opt.* **28**, 2389–2400.

Hohn, D. H. (1969). Depolarization of a laser beam at 6328 Å due to atmospheric transmission. *Appl. Opt.* **8**, 367–370.

Holland, A. C., and Gagne, G. (1970). The scattering of polarized light by polydisperse systems of irregular particles. *Appl. Opt.* **9**, 1113–1121.

Holland, R., Simpson, L., and Kunz, K. (1980). Finite-difference analysis of EMP coupling to lossy dielectric structures. *IEEE Trans. Electromagn. Compat.* **22**, 203–209.

Holland, R., Cable, V. R., and Wilson, L. C. (1991). Finite-volume time-domain (FVTD) techniques for EM scattering. *IEEE Trans. Electromagn. Compat.* **33**, 281–293.

Holler, H., Bringi, V. N., Hubbert, J., *et al.* (1994). Life cycle and precipitation formation in a hybrid-type hailstorm revealed by polarimetric and Doppler radar measurements. *J. Atmos. Sci.* **51**, 2500–2522.

Holler, S., Pan, Y., Chang, R. K., *et al.* (1998). Two-dimensional angular optical scattering for the characterization of airborne microparticles. *Opt. Lett.* **23**, 1489–1491.

Holt, A. R. (1982). The scattering of electromagnetic waves by single hydrometeors. *Radio Sci.* **17**, 929–945.

Holt, A. R. (1984). Some factors affecting the remote sensing of rain by polarization diversity radar in the 3–35 GHz frequency range. *Radio Sci.* **19**, 1399–1412.

Holt, A. R., and Shepherd, J. W. (1979). Electromagnetic scattering by dielectric spheroids in the forward and backward directions. *J. Phys. A: Math. Gen.* **12**, 159–166.

Holt, A. R., Uzunoglu, N. K., and Evans, B. G. (1978). An integral equation solution to the scattering of electromagnetic radiation by dielectric spheroids and ellipsoids. *IEEE Trans. Antennas Propag.* **26**, 706–712.

Hoppel, W. A., Fitzgerald, J. W., Frick, G. M., *et al.* (1990). Aerosol size distributions and optical properties found in the marine boundary layer over the Atlantic Ocean. *J. Geophys. Res.* **95**, 3659–3686.

Horn, R. A., and Johnson, C. R. (1991). "Topics in Matrix Analysis." Cambridge Univ. Press, Cambridge.

Hornbostel, A., and Schroth, A. (1995). Determination of propagation parameters by Olympus Beacon, rain rate, radiometer, and radar measurements. *Frequenz* **49**, 224–231.

Hornbostel, A., Schroth, A., and Kutuza, B. G. (1995). Polarimetric measurements and model calculations of downwelling rain brightness temperatures. *In* "Microwave Radiometry and Remote Sensing of the Environment" (D. Solomini, Ed.), pp. 239–252. VSP, Zeist, The Netherlands.

Hornbostel, A., Schroth, A., Kutuza, B. G., and Evtuchenko, A. (1997). Dual polarisation and multifrequency measurements of rain rate and drop size distribution by ground-based radar and radiometers. *IGARSS 1997*, Singapore.

Houston, J. D., and Carswell, A. I. (1978). Four-component polarization measurements of lidar atmospheric scattering. *Appl. Opt.* **17**, 614–620.

Hovenier, J. W. (1969). Symmetry relations for scattering of polarized light in a slab of randomly oriented particles. *J. Atmos. Sci.* **26**, 488–499.

Hovenier, J. W. (1971). Multiple scattering of polarized light in planetary atmospheres. *Astron. Astrophys.* **13**, 7–29.

Hovenier, J. W. (1994). Structure of a general pure Mueller matrix. *Appl. Opt.* **33**, 8318–8324.

Hovenier, J. W., Ed. (1996). Special issue on Light Scattering by Non-Spherical Particles. *J. Quant. Spectrosc. Radiat. Transfer* **55**, 535–694.

Hovenier, J. W., and Mackowski, D. W. (1998). Symmetry relations for forward and backward scattering by randomly oriented particles. *J. Quant. Spectrosc. Radiat. Transfer* **60**, 483–492.

Hovenier, J. W., and van der Mee, C. V. M. (1983). Fundamental relationships relevant to the transfer of polarized light in a scattering atmosphere. *Astron. Astrophys.* **128**, 1–16.

Hovenier, J. W., and van der Mee, C. V. M. (1988). Scattering of polarized light: Properties of the elements of the phase matrix. *Astron. Astrophys.* **196**, 287–295.

Hovenier, J. W., and van der Mee, C. V. M. (1995). Bounds for the degree of polarization. *Opt. Lett.* **20**, 2454–2456.

Hovenier, J. W., and van der Mee, C. V. M. (1996). Testing scattering matrices: A compendium of recipes. *J. Quant. Spectrosc. Radiat. Transfer* **55**, 649–661.

Hovenier, J. W., and van der Mee, C. V. M. (1997). Basic properties of matrices describing scattering of polarized light in atmospheres and oceans. *In* "IRS 96: Current Problems in Atmospheric Radiation" (W. L. Smith and K. Stamnes, Eds.), pp. 797–800. Deepak, Hampton, VA.

Hovenier, J. W., van de Hulst, H. C., and van der Mee, C. V. M. (1986). Conditions for the elements of the scattering matrix. *Astron. Astrophys.* **157**, 301–310.

Hovenier, J. W., Lumme, K., Mishchenko, M. I., *et al.* (1996). Computations of scattering matrices of four types of non-spherical particles using diverse methods. *J. Quant. Spectrosc. Radiat. Transfer* **55**, 695–705.

Hu, Ch.-R., Kattawar, G. W., Parkin, M. E., and Herb, P. (1987). Symmetry theorems on the forward and backward scattering Müller matrices for light scattering from a nonspherical dielectric scatterer. *Appl. Opt.* **26**, 4159–4173.

Huang, R., and Liou, K.-N. (1983). Polarized microwave radiation transfer in precipitating cloudy atmospheres. *J. Geophys. Res.* **88**, 3885–3893.

Hubbert, J., and Bringi, V. N. (1996). Specular null polarization theory: Applications to radar meteorology. *IEEE Trans. Geosci. Remote Sens.* **34**, 859–873.

Hubbert, J., Chandrasekar, V., Bringi, V. N., and Meischner, P. F. (1993). Processing and interpretation of coherent dual-polarized radar measurements. *J. Atmos. Oceanic Technol.* **10**, 155–164.

Hubbert, J., Bringi, V. N., and Carey, L. D. (1998). CSU–CHILL polarimetric radar measurements from a severe hail storm in eastern Colorado. *J. Appl. Meteorol.* **37**, 749–775.

Huffman, D. R. (1991). Solid C_{60}. *Phys. Today* **44**, 22–29.

Huffman, D. R., and Bohren, C. F. (1980). Infrared absorption spectra of non-spherical particles in the Rayleigh-ellipsoid approximation. *In* "Light Scattering by Irregularly Shaped Particles" (D. W. Schuerman, Ed.), pp. 103–112. Plenum, New York.

Huffman, P. J., and Thursby, W. R. (1969). Light scattering by ice crystals. *J. Atmos. Sci.* **26**, 1073–1077.

Hull, P. G., and Quinby-Hunt, M. S. (1997). A neural network to extract size parameter from light scattering data. *Proc. SPIE* **2963**, 448–454.

Hull, P. G., Hunt, A. J., Quinby-Hunt, M. S., and Shapiro, D. B. (1991). Coupled-dipole approximation: Predicting scattering from non-spherical marine organisms. *Proc. SPIE* **1537**, 21–29.

Hull, P. G., Shaw, F. G., Quinby-Hunt, M. S., *et al.* (1994). Light scattering by marine microorganisms: Comparison of Mueller matrix elements predicted by analytical methods with those obtained by experimental measurements. *Proc. SPIE* **2258**, 613–622.

Humphries, R. G. (1974). Observations and calculations of depolarization effects at 3 GHz due to precipitation. *J. Rech. Atmos.* **8**, 155–161.

Hunt, A. J. (1974). An experimental investigation of the angular dependence of polarization of light scattered from small particles. Ph.D. Thesis, Univ. of Arizona, Tucson, AZ.

Hunt, A. J., and Huffman, D. R. (1973). A new polarization-modulated light scattering instrument. *Rev. Sci. Instrum.* **44**, 1753–1762.

Hunt, A. J., and Huffman, D. R. (1975). A polarization-modulated light scattering instrument for determining liquid aerosol properties. *Jpn. J. Appl. Phys.* **14** (Suppl. 14-1), 435–440.

Hunt, A. J., Quinby-Hunt, M. S., and Shapiro, D. B. (1990). Effects of λ-dependent absorption on the polarization of light scattered from marine *Chlorella*. *Proc. SPIE* **1302**, 269–280.

Hunt, A. J., Quinby-Hunt, M. S., and Shepherd, I. G. (1998). Diesel exhaust particles characterization by polarized light scattering. SAE Paper No. 982629.

Husson, D., and Pointin, Y. (1989). Quantitative estimation of the hail fall intensity with a dual polarization radar and a hail pad network. *Prepr. Vol., Conf. Radar. Meteorol., 24th, Tallahassee, FL, 1989*, 318–321.

Huynen, J. R. (1970). Phenomenological theory of radar targets. Ph.D. Thesis, Delft Technical Univ., Delft, The Netherlands.

Iaquinta, J., Isaka, H., and Personne, P. (1995). Scattering phase function of bullet rosette ice crystals. *J. Atmos. Sci.* **52**, 1401–1413.

Illingworth, A. J., and Caylor, I. J. (1989). Cross polar observations of the bright band. *Prepr. Vol., Conf. Radar. Meteorol., 24th, Tallahassee, FL, 1989*, 323–327.

Illingworth, A. J., Goddard, J. W. F., and Cherry, S. M. (1987). Polarized radar studies of precipitation development in convective storms. *Quart. J. R. Meteorol. Soc.* **113**, 469–489.

Inada, H. (1974). Backscattered short pulse response of surface waves from dielectric spheres. *Appl. Opt.* **13**, 1928–1933.

Inoue, M., and Ohtaka, K. (1985). Enhanced Raman scattering by a two-dimensional array of dielectric spheres. *Phys. Rev. B* **26**, 3487–3490.

Ioannidou, M. P., Skaropoulos, N. C., and Chrissoulidis, D. P. (1995). Study of interactive scattering by clusters of spheres. *J. Opt. Soc. Am. A* **12**, 1782–1789.

Ishimaru, A. (1978). "Wave Propagation and Scattering in Random Media." Academic Press, San Diego.

Iskander, M. F., and Lakhtakia, A. (1984). Extension of the iterative EBCM to calculate scattering by low-loss or lossless elongated dielectric objects. *Appl. Opt.* **23**, 948–953.

Iskander, M. F., Lakhtakia, A., and Durney, C. H. (1983). A new procedure for improving the solution stability and extending the frequency range of EBCM. *IEEE Trans. Antennas Propag.* **31**, 317–324.

Iskander, M. F., Olson, S. C., Benner, R. E., and Yoshida, D. (1986). Optical scattering by metallic and carbon aerosols of high aspect ratio. *Appl. Opt.* **25**, 2514–2520.

Iskander, M. F., Chen, H. Y., and Penner, J. E. (1989a). Optical scattering and absorption by branched chains of aerosols. *Appl. Opt.* **28**, 3083–3091.

Iskander, M. F., Chen, H. Y., and Duong, T. V. (1989b). A new sectioning procedure for calculating scattering and absorption by elongated dielectric objects. *IEEE Trans. Electromagn. Compat.* **31**, 157–163.

Ito, S., and Oguchi, T. (1994). Circular depolarization ratio of radar returns from rain: Validity of the second-order solution of RT equation. *IEE Proc. Microwave Antennas Propag.* **141**, 257–260.

Ito, S., Oguchi, T., Iguchi, T., *et al.* (1995). Depolarization of radar signals due to multiple scattering in rain. *IEEE Trans. Geosci. Remote Sens.* **33**, 1057–1062.

Ivakhnenko, V. I., Stover, J. C., and Eremin, Yu. A. (1998). Modelling of light scattering by non-spherical inhomogeneous particles located on silicon wafer. *In* "Electromagnetic and Light Scattering—Theory and Applications III" (T. Wriedt and Yu. Eremin, Eds.), pp. 125–132. Univ. of Bremen, Bremen, Germany.

Ivezić, Z., and Mengüc, P. (1996). An investigation of dependent/independent scattering regimes using a discrete dipole approximation. *Int. J. Heat Mass Transfer* **39**, 811–822.

Ivezić, Z., Mengüc, P., and Knauer, T. G. (1997). A procedure to determine the onset of soot agglomeration from multi-wavelength experiments. *J. Quant. Spectrosc. Radiat. Transfer* **57**, 859–865.

Iwasaka, Y., and Hayashida, S. (1981). The effects of the volcanic eruption of St. Helens on the polarization properties of stratospheric aerosols: Lidar measurements at Nagoya. *J. Meteorol. Soc. Jpn.* **59**, 611–614.

Jackson, J. D. (1975). "Classical Electrodynamics." Wiley, New York.

Jaggard, D. L., Hill, C., Shorthill, R. W., *et al.* (1981). Light scattering from particles of regular and irregular shape. *Atmos. Environ.* **15**, 2511–2519.

Jalava, J.-P., Taavitsainen, V.-M., Haario, H., and Lamberg, L. (1998). Determination of particle and crystal size distribution from turbidity spectrum of TiO$_2$ pigments by means of T-matrix. *J. Quant. Spectrosc. Radiat. Transfer* **60**, 399–409.

Jameson, A. R. (1983). Microphysical interpretation of multi-parameter radar measurements in rain. I. Interpretation of polarization measurements and estimation of raindrop shapes. *J. Atmos. Sci.* **40**, 1792–1802.

Jameson, A. R. (1985). On deducing the microphysical character of precipitation from multiple-parameter radar polarization measurements. *J. Clim. Appl. Meteorol.* **24**, 1037–1047.

Jameson, A. R. (1987). Relations among linear and circular polarization parameters measured in canted hydrometeors. *J. Atmos. Oceanic Technol.* **4**, 634–645.

Jameson, A. R. (1989). The interpretation and meteorological application of radar backscatter amplitude ratios at linear polarizations. *J. Atmos. Oceanic Technol.* **6**, 908–919.

Jameson, A. R. (1993). Estimating the path-average rainwater content and updraft speed along a microwave link. *J. Atmos. Oceanic Technol.* **10**, 478–485.

Jameson, A. R., and Caylor, I. J. (1994). A new approach to estimating rainwater content by radar using propagation differential phase shift. *J. Atmos. Oceanic Technol.* **11**, 311–322.

Jameson, A. R., and Davie, J. H. (1988). An interpretation of circular polarization measurements affected by propagation differential phase shift. *J. Atmos. Oceanic Technol.* **5**, 405–415.

Jameson, A. R., and Kostinski, A. B. (1999). Non-Rayleigh signal statistics in clustered statistically homogeneous rain. *J. Atmos. Oceanic Technol.* **16**, 575–583.

Jameson, A. R., Murphy, M. J., and Krider, E. P. (1996). Multiple-parameter radar observations of isolated Florida thunderstorms during the onset of electrification. *J. Appl. Meteorol.* **35**, 343–354.

Jensen, H. P., Schellman, J. A., and Troxell, T. (1986). Modulation techniques in polarization spectroscopy. *Appl. Spectrosc.* **32**, 192–200.

Jin, Y.-Q. (1991). An approach to two-dimensional vector thermal radiative transfer for spatially inhomogeneous random media. *J. Appl. Phys.* **69**, 7594–7600.

Jin, Y.-Q. (1998). Some issues on polarized scattering from random nonspherical particles and applications in microwave remote sensing. *Prepr. Vol., Conf. Light Scat. Nonspher. Part.: Theory Meas. Appl., New York, 1998*, 317–320.

Jones, A. R. (1979). Electromagnetic wave scattering by assemblies of particles in the Rayleigh approximation. *Proc. R. Soc. London, Ser. A* **366**, 111–127. [Errata: **375**, 453–454 (1981).]

Jones, A. R. (1999). Light scattering for particle characterization. *Progr. Energy Combust. Sci.* **25**, 1–53.

Joo, K., and Iskander, M. F. (1990). A new procedure of point-matching method for calculating the absorption and scattering of lossy dielectric objects. *IEEE Trans. Antennas Propag.* **38**, 1483–1490.

Junge, C. E. (1972). Our knowledge of the physico-chemistry of aerosols in the undisturbed marine environment. *J. Geophys. Res.* **77**, 5183–5200.

Junge, C. E., Robinson, E., and Ludwig, F. L. (1969). A study of aerosols in the Pacific air masses. *J. Appl. Meteorol.* **8**, 340–347.

Jurgens, T. G., Taflove, A., Umashankar, K., and Moore, T. G. (1992). Finite-difference time-domain modeling of curved surfaces. *IEEE Trans. Antennas Propag.* **40**, 357–366.

Kadyshevich, Ye. A., and Lyubovtseva, Yu. S. (1973). Certain characteristics of ocean hydrosols from the scattering matrices. *Izv. Acad. Sci. USSR, Atmos. Oceanic Phys.* (Engl. Transl.) **9**, 659–663.

Kadyshevich, Ye. A., Lyubovtseva, Yu. S., and Plakina, I. N. (1971). Measurement of matrices for light scattered by sea water. *Izv. Acad. Sci. USSR, Atmos. Oceanic Phys.* (Engl. Transl.) **7**, 367–371.

Kadyshevich, Ye. A., Lyubovtseva, Yu. S., and Rozenberg, G. V. (1976). Light-scattering matrices of Pacific and Atlantic ocean waters. *Izv. Acad. Sci. USSR, Atmos. Oceanic Phys.* (Engl. Transl.) **12**, 106–111.

Kahn, R., West, R., McDonald, D., Rheingans, B., and Mishchenko, M. I. (1997). Sensitivity of multiangle remote sensing observations to aerosol sphericity. *J. Geophys. Res.* **102**, 16,861–16,870.

Kai, L., and Massoli, P. (1994). Scattering of electromagnetic plane waves by radially inhomogeneous spheres: A finely stratified sphere model. *Appl. Opt.* **33**, 501–511.

Kapustin, V. N., and Covert, D. S. (1980). Measurements of the humidification processes of hygroscopic particles. *Appl. Opt.* **19**, 1349–1352.

Karam, M. A. (1997). Electromagnetic wave interactions with dielectric particles. I. Integral equation reformation. *Appl. Opt.* **36**, 5238–5245.

Karam, M. A., and Fung, A. K. (1988). Electromagnetic scattering from a layer of finite length, randomly oriented, dielectric, circular cylinders over a rough interface with application to vegetation. *Int. J. Remote Sens.* **9**, 1109–1134.

Karam, M. A., Amar, F., Fung, A. K., *et al.* (1995). A microwave polarimetric scattering model for forest canopies based on vector radiative transfer theory. *Remote Sens. Environ.* **53**, 16–30.

Kattawar, G. W., and Adams, C. N. (1989). Stokes vector calculations of the submarine light field in an atmosphere–ocean with scattering according to a Rayleigh phase matrix: Effect of interface refractive index on radiance and polarization. *Limnol. Oceanogr.* **34**, 1453–1472.

Kattawar, G. W., and Adams, C. N. (1990). Errors in radiance calculations induced by using scalar rather than Stokes vector theory in a realistic atmosphere–ocean system. *Proc. SPIE* **1302**, 2–12.

Kattawar, G. W., and Dean, C. E. (1983). Electromagnetic scattering from two dielectric spheres: Comparison between theory and experiment. *Opt. Lett.* **8**, 48–50.

Kattawar, G. W., and Eisner, M. (1970). Radiation from a homogeneous isothermal sphere. *Appl. Opt.* **9**, 2685–2690.

Kattawar, G. W., and Fry, E. S. (1982). Inequalities between the elements of the Mueller scattering matrix: Comments. *Appl. Opt.* **21**, 18.

Kattawar, G. W., and Humphreys, T. J. (1980). Electromagnetic scattering from two identical pseudospheres. *In* "Light Scattering by Irregularly Shaped Particles" (D. W. Schuerman, Ed.), pp. 177–190. Plenum, New York.

Kattawar, G. W., Hu, C.-R., Parkin, M. E., and Herb, P. (1987). Mueller matrix calculations for dielectric cubes: Comparison with experiments. *Appl. Opt.* **26**, 4174–4180.

Katz, D. S., Thiele, E. T., and Taflove, A. (1994). Validation and extension to three dimensions of Berenger PML absorbing boundary condition for FD-TD meshes. *IEEE Microwave Guided Wave Lett.* **4**, 268–270.

Kawano, M., Ikuno, H., and Nishimoto, M. (1996). Numerical analysis of 3-D scattering problems using the Yasuura method. *IEICE Trans. Electron.* **E79-C**, 1358–1363.

Kaye, P. H., Hirst, E., Clarke, J. M., and Micheli, F. (1992). Airborne shape and size classification from spatial light scattering profiles. *J. Aerosol Sci.* **23**, 597–611.

Keenan, T., May, P., Zrnic, D., Carey, L., and Rutledge, S. (1997). Tropical rainfall estimation during the maritime continent thunderstorm experiment (MCTEX) using a C-band polarimetric radar. *Proc. 22nd Conf. Hurricanes and Tropical Meteorol.*, 386–387.

Keller, J. B. (1962). Geometrical theory of diffraction. *J. Opt. Soc. Am.* **52**, 116–130.

Keohane, K. W., and Metcalf, W. K. (1959). The cytoplasmic refractive index of lymphocytes, its significance and its changes during active immunisation. *Quart. J. Exper. Physiol. Cognate Med. Sci.* **44**, 343–346.

Kerker, M. (1969). "The Scattering of Light and Other Electromagnetic Radiation." Academic Press, San Diego.

Kerker, M., Ed. (1988). Selected Papers on Light Scattering. SPIE Milestone Series **951**.

Kerker, M. (1991). Founding fathers of light scattering and surface-enhanced Raman scattering. *Appl. Opt.* **30**, 4699–4705.

Khlebtsov, N. G. (1979). Analysis and numerical calculation of the scattering matrix for soft spheroids comparable in size with the wavelength of light. *Opt. Spectrosc.* **46**, 292–295.

Khlebtsov, N. G. (1984). Integral equation for problems of light scattering by particles of the medium. *Opt. Spectrosc.* **57**, 399–401.

Khlebtsov, N. G. (1992). Orientational averaging of light-scattering observables in the *T*-matrix approach. *Appl. Opt.* **31**, 5359–5365.

Khlebtsov, N. G. (1993). Optics of fractal clusters in the anomalous diffraction approximation. *J. Mod. Opt.* **40**, 2221–2235.

Khlebtsov, N. G., and Melnikov, A. G. (1991). Integral equation for light scattering problems: Application to the orientationally induced birefringence of colloidal dispersions. *J. Colloid Interface Sci.* **142**, 396–408.

Khlebtsov, N. G., and Mel'nikov, A. G. (1995). Depolarization of light scattered by fractal smoke clusters: An approximate anisotropic model. *Opt. Spectrosc.* **79**, 605–609.

Khlebtsov, N. G., Melnikov, A. G., and Bogatyrev, V. A. (1991). The linear dichroism and birefringence of colloidal dispersions: Approximate and exact approaches. *J. Colloid Interface Sci.* **146**, 463–478.

Khlebtsov, N. G., Bogatyrev, V. A., Dykman, L. A., and Melnikov, A. G. (1996). Spectral extinction of colloidal gold and its biospecific conjugates. *J. Colloid Interface Sci.* **180**, 436–445.

Khvorostyanov, V. I., and Sassen, K. (1998). Cirrus cloud simulation using explicit microphysics and radiation. II. Microphysics, vapor and mass budgets, and optical and radiative properties. *J. Atmos. Sci.* **55**, 1822–1845.

Kiehl, J. T., Ko, M. W., Mugnai, A., and Chýlek, P. (1980). Perturbation approach to light scattering by nonspherical particles. *In* "Light Scattering by Irregularly Shaped Particles" (D. W. Schuerman, Ed.), pp. 135–140. Plenum, New York.

Kikuchi, S., Mikani, Y., Mukai, T., Mukai, S., and Hough, J. H. (1987). Polarimetry of comet P/Halley. *Astron. Astrophys.* **187**, 689–692.

Kikuchi, S., Mikani, Y., Mukai, T., and Mukai, S. (1989). Polarimetry of comet Bradfield. *Astron. Astrophys.* **214**, 386–388.

Kim, S.-H., and Martin, P. G. (1995). The size distribution of interstellar dust particles as determined from polarization: Spheroids. *Astrophys. J.* **444**, 293–305.

Kim, C. S., and Yeh, C. (1991). Scattering of an obliquely incident wave by a multilayered elliptical lossy dielectric cylinder. *Radio Sci.* **26**, 1165–1176.

Kimura, H., and Mann, I. (1998). Radiation pressure cross section for fluffy aggregates. *J. Quant. Spectrosc. Radiat. Transfer* **60**, 425–438.

King, M. D., Tsay, S. C., Platnick, S. E., *et al.* (1997). Cloud retrieval algorithm for MODIS: Optical thickness, effective particle radius, and thermodynamic phase. MODIS Algorithm Theoretical Basis Document ATBD-MOD-05, MOD06—Cloud Product. NASA Goddard Space Flight Center, Greenbelt, MD.

Kirmaci, I., and Ward, G. (1979). Scattering of 0.627-μm light from spheroidal 2-μm cladosporium and cubical 4-μm NaCl particles. *Appl. Opt.* **18**, 3328–3331.

Kleinman, R. E., and Senior, T. B. A. (1986). Rayleigh scattering. *In* "Low and High Frequency Asymptotics" (V. K. Varadan and V. V. Varadan, Eds.), pp. 1–70. Elsevier, Amsterdam.

Klett, J. D., and Sutherland, R. A. (1992). Approximate methods for modeling the scattering properties of non-spherical particles: Evaluation of the Wentzel–Kramers–Brillouin method. *Appl. Opt.* **31**, 373–386.

Knight, N. C. (1986). Hailstone shape factor and its relation to radar interpretation to hail. *J. Clim. Appl. Meteorol.* **25**, 1956–1958.

Knight, C. A., and Knight, N. C. (1970). The falling behavior of hailstones. *J. Atmos. Sci.* **27**, 672–681.

Knight, J. C., Driver, H. S. T., and Robertson, G. N. (1993). Interference modulation of *Q* values in a cladded-fiber. *Opt. Lett.* **18**, 1296–1298.

Kobayashi, A., Hayashida, S., Okada, K., and Iwasaka, Y. (1985). Measurements of the polarization properties of Kosa (Asian dust-storm) particles by a laser radar in Spring 1983. *J. Meteorol. Soc. Jpn.* **63**, 144–149.

Kobayashi, A., Hayashida, S., Iwasaka, Y., *et al.* (1987). Consideration of depolarization ratio measurements by lidar—in relation to chemical composition of aerosol particles. *J. Meteorol. Soc. Jpn.* **65**, 303–307.

Koepke, P., and Hess, M. (1988). Scattering functions of tropospheric aerosols: The effects of non-spherical particles. *Appl. Opt.* **27**, 2422–2430.

Kokhanovsky, A. (1999). "Optics of Light Scattering Media: Problems and Solutions." Wiley/Praxis, Chichester.

Kolokolova, L., Jockers, K., Chernova, G., and Kiselev, N. (1997). Properties of cometary dust from color and polarization. *Icarus* **126**, 351–361.

Kolokolova, L., Gustafson, B. Å. S., Loesel, J., *et al.* (1998). Polarization and color of aggregate particles from microwave analog experiments: Application to cometary dust. *Prepr. Vol., Conf. Light Scat. Nonspher. Part.: Theory Meas. Appl., New York, 1998*, 180–183.

Können, G., and Tinbergen, J. (1998). Polarization structures in parhelic circles and in 120° parhelia. *Appl. Opt.* **37**, 1457–1464.

Konovalov, N. V. (1985). Polarization matrices corresponding to transformations in the Stokes cone. Preprint 171, Keldysh Inst. Appl. Math., Acad. Sci. USSR, Moscow (in Russian).

Kosarev, A. L., and Mazin, I. P. (1991). An empirical model of the physical structure of upper layer clouds. *Atmos. Res.* **26**, 213–228.

Kostinski, A. B. (1994). Fluctuations of differential phase and radar measurements of precipitation. *J. Appl. Meteorol.* **33**, 1176–1181.

Kostinski, A. B., Givens, C. R., and Kwiatkowski, J. M. (1993). Constraints on Mueller matrices of polarization optics. *Appl. Opt.* **32**, 1646–1651.

Kouzoubov, A., Brennan, M. J., and Thomas, J. C. (1998). Treatment of polarization in laser remote sensing of ocean water. *Appl. Opt.* **37**, 3873–3885.

Köylü, Ü. Ö., and Faeth, G. M. (1992). Structure of overfire soot in buoyant turbulent diffusion flames at long residence times. *Combust. Flame* **89**, 140–156.

Kragh, H. (1991). Ludvig Lorenz and nineteenth century optical theory: The work of a great Danish scientist. *Appl. Opt.* **30**, 4688–4695.

Kraus, E., and Niederman, R. (1990). Changes in neutrophil right-angle light scatter can occur independent of alterations in cytoskeletal actin. *Cytometry* **11**, 272–282.

Kravtsov, Y. A., and Orlov, Y. I. (1998). "Caustics, Catastrophes and Wave Fields." Springer, New York.

Krehbiel, P., Chen, T., McCrary, S., *et al.* (1996). The use of dual circular-polarization radar observations for remotely sensing storm electrification. *Meteorol. Atmos. Phys.* **59**, 65–82.

Kropfli, R. A., Matrosov, S. Y., Uttal, T., *et al.* (1995). Cloud microphysics studies with 8 mm radar. *Atmos. Res.* **35**, 299–313.

Krotkov, N. A., Krueger, A. J., and Bhartia, P. K. (1997). Ultraviolet optical model of volcanic clouds for remote sensing of ash and sulfur dioxide. *J. Geophys. Res.* **102**, 21,891–21,904.

Krotkov, N. A., Flittner, D. E., Krueger, A. J., *et al.* (1999). Effect of particle non-sphericity on monitoring of drifting volcanic ash clouds. *J. Quant. Spectrosc. Radiat. Transfer*, in press.

Kry, P. R., and List, R. (1974). Angular motions of freely falling spheroidal hailstone models. *Phys. Fluids* **17**, 1093–1102.

Ku, J. C. (1993). Comparisons of coupled-dipole solutions and dipole refractive indices for light scattering and absorption by arbitrarily shaped or agglomerated particles. *J. Opt. Soc. Am. A* **10**, 336–342.

Kubesh, R. J., and Beard, K. V. (1993). Laboratory measurements of spontaneous oscillations for moderate-size raindrops. *J. Atmos. Sci.* **50**, 1089–1098.

Kuga, Y. (1991). Third-and fourth-order iterative solutions for the vector radiative transfer equation. *J. Opt. Soc. Am. A* **8**, 1580–1586.

Kuik, F. (1992). Single scattering of light by ensembles of particles with various shapes. Ph.D. Dissertation, Free Univ., Amsterdam.

Kuik, F., Stammes, P., and Hovenier, J. W. (1991). Experimental determination of scattering matrices of water droplets and quartz particles. *Appl. Opt.* **30**, 4872–4881.

Kuik, F., de Haan, J. F., and Hovenier, J. W. (1992). Benchmark results for single scattering by spheroids. *J. Quant. Spectrosc. Radiat. Transfer* **47**, 477–489.

Kuik, F., de Haan, J. F., and Hovenier, J. W. (1994). Single scattering of light by circular cylinders. *Appl. Opt.* **33**, 4906–4918.

Kummerow, C. D., and Weinman, J. A. (1988). Radiative properties of deformed hydrometeors for commonly used passive microwave frequencies. *IEEE Trans. Geosci. Remote Sens.* **26**, 629–638.

Kunz, K. S., and Luebbers, R. J. (1993). "Finite Difference Time Domain Method for Electromagnetics." CRC Press, Boca Raton, FL.

Kunz, K. S., and Simpson, L. (1981). A technique for increasing the resolution of finite-difference solutions of the Maxwell's equation. *IEEE Trans. Electromagn. Compat.* **23**, 419–422.

Kurtz, V., and Salib, S. (1993). Scattering and absorption of electromagnetic radiation by spheroidally shaped particles: Computation of the scattering properties. *J. Imaging Sci. Technol.* **37**, 43–60.

Kuščer, I., and Ribarič, M. (1959). Matrix formalism in the theory of diffusion of light. *Opt. Acta* **6**, 42–51.

Kutuza, B. G., Zagorin, G. K., Hornbostel, A., and Schroth, A. (1998). Physical modeling of passive polarimetric microwave observations of the atmosphere with respect to the third Stokes parameter. *Radio Sci.* **33**, 677–695.

Kuzmin, V. N., and Babenko, V. A. (1981). Light scattering by a weakly anisotropic spherical particle. *Opt. Spectrosc.* **50**, 269–273.

Kwiatkowski, J. M., Kostinski, A. B., and Jameson, A. R. (1995). The use of optimal polarizations for studying the microphysics of precipitation: Nonattenuating wavelengths. *J. Atmos. Oceanic Technol.* **12**, 96–114.

Lacis, A. A., and Mishchenko, M. I. (1995). Climate forcing, climate sensitivity, and climate response: A radiative modeling perspective on atmospheric aerosols. *In* "Aerosol Forcing of Climate" (R. J. Charlson and J. Heintzenberg, Eds.), pp. 11–42. Wiley, New York.

Lacis, A. A., Chowdhary, J., Mishchenko, M. I., and Cairns, B. (1998). Modeling errors in diffuse sky radiation: Vector vs. scalar treatment. *Geophys. Res. Lett.* **25**, 135–138.

Lacoste, D., van Tiggelen, B. A., Rikken, G. L. J. A., and Sparenberg, A. (1998). Optics of a Faraday-active Mie sphere. *J. Opt. Soc. Am. A* **15**, 1636–1642.

Lai, H. M., Lam, C. C., Leung, P. T., and Young, K. (1991). Effect of perturbations on the widths of narrow morphology-dependent resonances in Mie scattering. *J. Opt. Soc. Am. B* **8**, 1962–1973.

Laitinen, H., and Lumme, K. (1998). *T*-matrix method for general star-shaped particles: First results. *J. Quant. Spectrosc. Radiat. Transfer* **60**, 325–334.

Lakhtakia, A. (1991). The extended boundary condition method for scattering by a chiral scatterer in a chiral medium: Formulation and analysis. *Optik* **86**, 155–161.

Lakhtakia, A. (1992). General theory of the Purcell–Pennypacker scattering approach and its extension to bianisotropic scatterers. *Astrophys. J.* **394**, 494–499.

Lakhtakia, A., and Mulholland, G. W. (1993). On two numerical techniques for light scattering by dielectric agglomerated structures. *J. Res. Natl. Inst. Stand. Technol.* **98**, 699–716.

Lakhtakia, A., Iskander, M. F., and Durney, C. H. (1983). An iterative extended boundary condition method for solving the absorption characteristics of lossy dielectric objects of large aspect ratios. *IEEE Trans. Microwave Theory Technol.* **31**, 640–647.

Lakhtakia, A., Varadan, V. K., and Varadan, V. V. (1984). Scattering by highly aspherical targets: EBCM coupled with reinforced orthogonalization. *Appl. Opt.* **23**, 3502–3504.

Lakhtakia, A., Varadan, V. V., and Varadan, V. K. (1985a). Scattering of ultrasonic waves by oblate spheroid voids of high aspect ratio. *J. Appl. Phys.* **58**, 4525–4530.

Lakhtakia, A., Varadan, V. K., and Varadan, V. V. (1985b). Scattering and absorption characteristics of lossy dielectric, chiral, nonspherical objects. *Appl. Opt.* **24**, 4146–4154.

Lamberg, L., Muinonen, K., Lumme, K., and Ylönen, J. (1999). Maximum likelihood estimator for Gaussian random sphere. In preparation.

Landau, L. D., and Lifshitz, E. M. (1960). "Electrodynamics of Continuous Media." Pergamon, Oxford.

Latimer, P. (1975). Light scattering by ellipsoids. *J. Colloid Interface Sci.* **53**, 102–109.

Latimer, P. (1982). Light scattering and absorption as methods of studying cell population parameters. *Ann. Rev. Biophys. Bioeng.* **11**, 129–150.

Latimer, P., and Barber, P. (1978). Scattering by ellipsoids of revolution: A comparison of theoretical methods. *J. Colloid Interface Sci.* **63**, 310–316.

Latimer, P., Moore, D. M., and Dudley Bryant, F. (1968). Changes in total scattering and absorption caused by changes in particle conformation. *J. Theor. Biol.* **21**, 348–367.

Lawson, R. P., Heymsfield, A. J., Aulenbach, S. M., and Jensen, T. J. (1998). Shapes, sizes and light scattering properties of ice crystals in cirrus and a persistent contrail during SUCCESS. *Geophys. Res. Lett.* **25**, 1331–1334.

Lazzi, G., and Gandhi, O. P. (1996). On the optimal design of the PML absorbing boundary condition for the FDTD code. *IEEE Trans. Antennas Propag.* **45**, 914–916.

Lee, J. F. (1993). Obliquely Cartesian finite difference time domain algorithm. *IEE Proc. H* **140**, 23–27.

Leinert, C. (1975). Zodiacal light—a measure of the interplanetary environment. *Space Sci. Rev.* **18**, 281–339.

Leinert, C., and Grün, E. (1990). Interplanetary dust. *In* "Physics and Chemistry in Space—Space and Solar Physics" (R. Schwenn and E. Marsch, Eds.), Vol. 20, pp. 207–275. Springer-Verlag, Berlin.

Leinert, C., Richter, I., Pilz, E., and Planck, B. (1981). The zodiacal light from 1.0 to 0.3 A.U. as observed by the helios space probes. *Astron. Astrophys.* **103**, 177–188.

Leinert, C., Richter, I., Pilz, E., and Hammer, M. (1982). Helios zodiacal light measurements—a tabulated summary. *Astron. Astrophys.* **110**, 355–357.

Leinert, C., Bowyer, S., Haikala, L. K., *et al.* (1998). The 1997 reference of diffuse night sky brightness. *Astron. Astrophys. Suppl. Ser.* **127**, 1–99.

Leitao, M. J., and Watson, P. A. (1984). Application of dual linearly polarized radar data to prediction of microwave path attenuation at 10–30 GHz. *Radio Sci.* **19**, 209–221.

Lemke, H., Okamoto, H., and Quante, M. (1998). Comment on error analysis of backscatter from discrete dipole approximation for different ice particle shapes. *Atmos. Res.* **49**, 189–197.

Lenoble, J., Ed. (1985). "Radiative Transfer in Scattering and Absorbing Atmospheres: Standard Computational Procedures." Deepak, Hampton, VA.

Lenoble, J. (1993). "Atmospheric Radiative Transfer." Deepak, Hampton, VA.

Levasseur-Regourd, A. C. (1991). The zodiacal cloud complex. *In* "Origin and Evolution of Interplanetary Dust" (A. C. Levasseur-Regourd and H. Hasegarva, Eds.), pp. 131–138. Kluwer Academic, Dordrecht.

Levasseur-Regourd, A. C., Hadamcik, E., and Renard, J. B. (1996). Evidence for two classes of comets from their polarimetric properties at large phase angles. *Astron. Astrophys.* **313**, 327–333.

Le Vine, D. M., Schneider, A., Lang, R. H., and Carter, H. G. (1985). Scattering from thin dielectric disks. *IEEE Trans. Antennas Propag.* **33**, 1410–1413.

Lewin, L. (1970). On the restricted validity of point-matching techniques. *IEEE Trans. Microwave Theory Tech.* **18**, 1041–1047.

Lhermitte, R. (1987). A 94-GHz Doppler radar for cloud observations. *J. Atmos. Oceanic Technol.* **4**, 36–48.

Lhermitte, R. (1990). Attenuation and scattering of millimeter wavelength radiation by clouds and precipitation. *J. Atmos. Oceanic Technol.* **7**, 464–479.

Li, L.-W., Kooi, P.-S., Leong, M.-S., *et al.* (1995). Microwave attenuation by realistically distorted raindrops. II. Predictions. *IEEE Trans. Antennas Propag.* **43**, 823–828.

Li, L. W., Yeo, T. S., Kooi, P. S., and Leong, M. S. (1998). Microwave specific attenuation by oblate spheroidal raindrops: An exact analysis of TCS's in terms of spheroidal wave functions. *PIER* **18**, 127–150.

Liang, C., and Lo, Y. T. (1967). Scattering by two spheres. *Radio Sci.* **2**, 1481–1495.

Liang, S., and Mishchenko, M. I. (1997). Calculations of the soil hot-spot effect using the coherent backscattering theory. *Remote Sens. Environ.* **60**, 163–173.

Liao, L., and Sassen, K. (1994). Investigation of relationships between Ka-band radar reflectivity and ice and liquid water contents. *Atmos. Res.* **34**, 231–248.

Liao, Z., Wong, H. L., Yang, B., and Yuan, Y. (1984). A transmitting boundary for transient wave analyses. *Scientia Sinica* **27**, 1063–1076.

Liebe, H. J., Rosenkranz, P. W., and Hufford, G. A. (1992). Atmospheric 60-GHz oxygen spectrum: New laboratory measurements and line parameters. *J. Quant. Spectrosc. Radiat. Transfer* **48**, 629–643.

Liebe, H. J., Hufford, G. A., and Cotton, M. G. (1993). Propagation modeling of moist air and suspended water/ice particles at frequencies below 1000 GHz. Proc. NATO/AGARD Wave Propagation Panel 52nd Meeting, Paper 3/1-10, May 17–20, 1993, Palma de Mallorca, Spain.

Lilienfeld, P. (1991). Gustav Mie: The person. *Appl. Opt.* **30**, 4696–4698.

Lind, A. C., Wang, R. T., and Greenberg, J. M. (1965). Microwave scattering by nonspherical particles. *Appl. Opt.* **4**, 1555–1561.

Liou, K. N. (1980). "An Introduction to Atmospheric Radiation." Academic, San Diego.

Liou, K. N. (1986). The influence of cirrus on weather and climate process: A global perspective. *Mon. Weather Rev.* **114**, 1167–1199.

Liou, K. N. (1992). "Radiation and Cloud Processes in the Atmosphere: Theory, Observation, and Modeling." Oxford Univ. Press, New York.

Liou, K. N., and Coleman, R. F. (1980). Light scattering by hexagonal columns and plates. *In* "Light Scattering by Irregularly Shaped Particles" (D. W. Schuerman, Ed.), pp. 207–218. Plenum, New York.

Liou, K. N., and Hansen, J. E. (1971). Intensity and polarization for single scattering by polydisperse spheres: A comparison of ray optics and Mie theory. *J. Atmos. Sci.* **28**, 995–1004.

Liou, K. N., and Lahore, H. (1974). Laser sensing of cloud composition: A backscattered depolarization technique. *J. Appl. Meteorol.* **13**, 257–263.

Liou, K. N., and Ou, S. C. (1989). The role of cloud microphysical processes in climate: An assessment from a one-dimensional perspective. *J. Geophys. Res.* **94**, 8599–8607.

Liou, K. N., and Takano, Y. (1994). Light scattering by nonspherical particles: Remote sensing and climatic implications. *Atmos. Res.* **31**, 271–298.

Liou, K. N., Cai, Q., Pollack, J. B., and Cuzzi, J. N. (1983). Light scattering by randomly oriented cubes and parallelepipeds. *Appl. Opt.* **22**, 3001–3008.

Liou, K. N., Goody, R. M., West, R., *et al.* (1996). CIRRUS: A low cost cloud/climate mission. Proposal submitted in response to the NASA ESSP Announcement of Opportunity.

Liou, K. N., Yang, P., Takano, Y., *et al.* (1998). On the radiative properties of contrail cirrus. *Geophys. Res. Lett.* **25**, 1161–1164.

Lipkea, W. H., Johnson, J. H., and Vuk, C. T. (1979). The physical and chemical character of diesel particulate emissions—measurement techniques and fundamental considerations. SAE Paper No. 780108.

List, R., Rentsch, U. W., Byram, A. C., and Lozowski, E. P. (1973). On the aerodynamics of spheroidal hailstone models. *J. Atmos. Sci.* **30**, 653–661.

Liu, C.-L., and Illingworth, A. J. (1997). Error analysis of backscatter from discrete dipole approximation for different ice particle shapes. *Atmos. Res.* **44**, 231–241.

Liu, Q., and Simmer, C. (1996). Polarization and intensity in microwave radiative transfer. *Contr. Atmos. Phys.* **69**, 535–545.

Liu, Q., Simmer, C., and Ruprecht, E. (1991). A general analytical expression of the radiation source function for emitting and scattering media within the matrix operator method. *Contr. Atmos. Phys.* **64**, 73–82.

Liu, C., Jonas, P. R., and Saunders, C. P. R. (1996a). Pyramidal ice crystal scattering phase functions and concentric halos. *Ann. Geophys.* **14**, 1192–1197.

Liu, Q., Simmer, C., and Ruprecht, E. (1996b). Three-dimensional radiative transfer effects of clouds in the microwave region. *J. Geophys. Res.* **101**, 4289–4298.

Liu, Y., Arnott, W. P., and Hallett, J. (1998). Anomalous diffraction theory for arbitrarily oriented finite circular cylinders and comparison with exact T-matrix results. *Appl. Opt.* **37**, 5019–5030.

Liu, Y., Arnott, W. P., and Hallett, J. (1999). Particle size distribution retrieval from multispectral optical depth: Influences of particle nonsphericity and refractive index. *J. Geophys. Res.*, in press.

Livolant, F., and Bouligand, Y. (1978). New observations on the twisted arrangement of dinoflagellate chromosomes. *Chromosoma (Berlin)* **68**, 21–44.

Livolant, F., and Bouligand, Y. (1980). Double helical arrangement of spread dinoflagellate chromosomes. *Chromosoma (Berlin)* **80**, 97–118.

Livolant, F., and Maestre, M. (1988). CD microscopy of compact forms of DNA and chromatin in vivo and in vitro: Cholesteric liquid crystalline phases of DNA and single dinoflagellate nuclei. *Biochemistry* **27**, 3056–3068.

Lock, J. A. (1990). Interference enhancement of the internal fields at structural resonances of a coated sphere. *Appl. Opt.* **29**, 3180–3187.

Lock, J. A. (1996). Ray scattering by an arbitrarily oriented spheroid. *Appl. Opt.* **35**, 500–531.

Lofftus, K., Hunt, A. J., Quinby-Hunt, M. S., *et al.* (1988). Immobilization of marine organisms for optical characterization: A new method and results. *Proc. SPIE* **925**, 334–341.

Lofftus, K. D., Quinby-Hunt, M. S., Hunt, A. J., *et al.* (1992). Light scattering by *Prorocentrum micans*: A new method and results. *Appl. Opt.* **31**, 2924–2931.

Logan, N. A. (1965). Survey of some early studies of the scattering of a plane wave by a sphere. *Proc. IEEE* **53**, 773–785.

Logan, N. A. (1990). *Proc. 2nd Int. Conf. Opt. Part. Sizing*, Arizona State University, 7–15.

Lohmeier, S. P., Sekelsky, S. M., Firda, J. M., *et al.* (1997). Classification of particles in stratiform clouds using the 33 and 95 GHz polarimetric cloud profiling radar system (CPRS). *IEEE Trans. Geosci. Remote Sens.* **35**, 256–270.

Longtin, D. R., Bohren, C. F., and Battan, L. J. (1987). Radar backscattering by large, spongy ice oblate spheroids. *J. Atmos. Oceanic Technol.* **4**, 355–358.

Lorentz, H. A. (1880). Ueber die Beziehung zwischen der Fortpflanzungsgeschwindigkeit des Lichtes und der Korperdichte. *Ann. Phys. Chem.* **9**, 641–665.

Lorenz, L. (1880). Ueber die Refractionconstante. *Ann. Phys. Chem.* **11**, 70–103.

Lorenz, L. (1898). "Oeuvres Scientifiques." Lehman and Stage, Copenhagen.

Love, A. E. H. (1899). The scattering of electric waves by a dielectric sphere. *Proc. London Math. Soc.* **30**, 308–321.

Ludwig, A. C. (1991). Scattering by two and three spheres computed by the generalized multipole method. *IEEE Trans. Antennas Propag.* **39**, 703–705.

Lumme, K., Ed. (1998). Special issue on Light Scattering by Non-Spherical Particles. *J. Quant. Spectrosc. Radiat. Transfer* **60**, 301–500.

Lumme, K., and Bowell, E. (1985). Photometric properties of zodiacal light particles. *Icarus* **62**, 54–71.

Lumme, K., and Muinonen, K. (1993). A two-parameter system for linear polarization of some solar system objects. *Proc. Conf. Asteroids, Comets, Meteors, Houston, TX, 1993*, 194.

Lumme, K., and Rahola, J. (1994). Light scattering by porous dust particles in the discrete-dipole approximation. *Astrophys. J.* **425**, 653–667.

Lumme, K., and Rahola, J. (1998). Comparison of light scattering by stochastically rough spheres, best-fit spheroids and spheres. *J. Quant. Spectrosc. Radiat. Transfer* **60**, 439–450.

Lumme, K., Peltoniemi, J. I., and Irvine, W. M. (1990). Diffuse reflection from a stochastically bounded, semi-infinite medium. *Transp. Theory Statist. Phys.* **19**, 317–332.

Lumme, K., Rahola, J., and Hovenier, J. W. (1997). Light scattering by dense clusters of spheres. *Icarus* **126**, 455–469.

Macke, A. (1993). Scattering of light by polyhedral ice crystals. *Appl. Opt.* **32**, 2780–2788.

Macke, A., and Großklaus, M. (1998). Light scattering by nonspherical raindrops: Implications for lidar remote sensing of rainrates. *J. Quant. Spectrosc. Radiat. Transfer* **60**, 355–363.

Macke, A., and Mishchenko, M. I. (1996). Applicability of regular particle shapes in light scattering calculations for atmospheric ice particles. *Appl. Opt.* **35**, 4291–4296.

Macke, A., Mishchenko, M. I., Muinonen, K., and Carlson, B. E. (1995). Scattering of light by large nonspherical particles: Ray tracing approximation versus *T*-matrix method. *Opt. Lett.* **20**, 1934–1936.

Macke, A., Mishchenko, M. I., and Cairns, B. (1996a). The influence of inclusions on light scattering by large ice particles. *J. Geophys. Res.* **101**, 23,311–23,316.

Macke, A., Mueller, J., and Raschke, E. (1996b). Scattering properties of atmospheric ice crystals. *J. Atmos. Sci.* **53**, 2813–2825.

Macke, A., Mishchenko, M. I., and Cairns, B. (1997). The influence of inclusions on light scattering by large hexagonal and spherical ice crystals. *In* "IRS'96: Current Problems in Atmospheric Radiation" (W. L. Smith and K. Stamnes, Eds.), pp. 226–229. Deepak, Hampton, VA.

Mackowski, D. W. (1991). Analysis of radiative scattering for multiple sphere configurations. *Proc. R. Soc. London, Ser. A* **433**, 599–614.

Mackowski, D. W. (1994). Calculation of total cross sections of multiple-sphere clusters. *J. Opt. Soc. Am. A* **11**, 2851–2861.

Mackowski, D. W. (1995). Electrostatic analysis of radiative absorption by sphere clusters in the Rayleigh limit: Application to soot. *Appl. Opt.* **34**, 3535–3545.

Mackowski, D. W., and Jones, P. D. (1995). Theoretical investigation of particles having directionally dependent absorption cross section. *J. Thermophys. Heat Transfer* **9**, 193–201.

Mackowski, D. W., and Mishchenko, M. I. (1996). Calculation of the *T* matrix and the scattering matrix for ensembles of spheres. *J. Opt. Soc. Am. A* **13**, 2266–2278.

Mackowski, D. W., Altenkirch, R. A., and Menguc, M. P. (1990). Internal absorption cross sections in a stratified sphere. *Appl. Opt.* **29**, 1551–1559.

MacLeod, G. C. (1957). The effect of circularly polarized light on the photosynthesis and chlorophyll *a* synthesis of certain marine algae. *Limnol. Oceanogr.* **2**, 360–362.

MacPhie, R. H., Dalmas, J., and Deleuil, R. (1987). Rotational–translational addition theorems for scalar spheroidal wave functions. *Quart. Appl. Math.* **44**, 737–749.

Maestre, M. F., Bustamante, C., Hayes, T. L., *et al.* (1982). Differential scattering of circularly polarized light by the helical sperm head from the octopus Eledone Cirrhosa. *Nature (London)* **298**, 773–774.

Magono, C., and Lee, C. V. (1966). Meteorological classification of natural snow crystals. *J. Fac. Sci. Hokkaido Univ.* **7**, 321–362.

Malm, W. C., Sisler, J. F., Huffman, D., *et al.* (1994). Spatial and seasonal trends in particle concentration and optical extinction in the United States. *J. Geophys. Res.* **99**, 1347–1370.

Malone, T. C. (1971a). The relative importance of nannoplankton and netplankton as primary producers in the California current system. *Fishery Bull.* **69**, 799–820.

Malone, T. C. (1971b). The relative importance of nannoplankton and net plankton as primary producers in tropical oceanic and neritic phytoplankton communities. *Limnol. Oceanogr.* **16**, 633–639.

Mann, I., Okamoto, H., Mukai, T., *et al.* (1994). Fractal aggregate analogues for near solar dust properties. *Astron. Astrophys.* **291**, 1011–1018.

Mannoni, A., Flesia, C., Bruscaglioni, P., and Ismaelli, A. (1996). Multiple scattering from Chebyshev particles: Monte Carlo simulations for backscattering in lidar geometry. *Appl. Opt.* **35**, 7151–7164.

Marchuk, G. I., Mikhailov, G. A., Nazaraliev, M. A., *et al.* (1980). "The Monte Carlo Methods in Atmospheric Optics." Springer-Verlag, Berlin.

Markel, V. A., Shalaev, V. M., and Poliakov, E. Y. (1997). Fluctuations of light scattering by fractal clusters. *J. Opt. Soc. Am. A* **14**, 60–69.

Marshall, J. S., and Palmer, W. McK. (1948). The distribution of raindrops with size. *J. Meteorol.* **5**, 165–166.

Marston, P. L. (1992). Geometrical and catastrophe optics methods in scattering. *In* "Physical Acoustics" (A. D. Pierce and R. N. Thurston, Eds.), Vol. 21, pp. 1–234. Academic Press, San Diego.

Martin, P. G. (1978). "Cosmic Dust." Oxford Univ. Press, Oxford.

Masłowska, A., Flatau, P. J., and Stephens, G. L. (1994). On the validity of the anomalous diffraction theory to light scattering by cubes. *Opt. Commun.* **107**, 35–40.

Massoli, P. (1998). Rainbow refractometry applied to radially inhomogeneous spheres: The critical case of evaporating droplets. *Appl. Opt.* **37**, 3227–3235.

Masuda, K., and Takashima, T. (1992). Feasibility study of derivation of cirrus information using polarimetric measurements from satellite. *Remote Sens. Environ.* **39**, 45–59.

Matrosov, S. Y. (1991). Theoretical study of radar polarization parameters obtained from cirrus clouds. *J. Atmos. Sci.* **48**, 1062–1069.

Matrosov, S. Y. (1992). Radar reflectivity in snowfall. *IEEE Trans. Geosci. Remote Sens.* **30**, 454–461.

Matrosov, S. Y. (1993). Possibilities of cirrus particle sizing from dual-frequency radar measurements. *J. Geophys. Res.* **98**, 20,675–20,683.

Matrosov, S. Y., and Kropfli, R. A. (1993). Cirrus cloud studies with elliptically polarized Ka-band radar signals: A suggested approach. *J. Atmos. Sci.* **10**, 684–692.

Matrosov, S. Y., Reinking, R. F., Kropfli, R. A., and Bartram, B. W. (1996). Estimation of ice hydrometeor types and shapes from radar polarization measurements. *J. Atmos. Oceanic Technol.* **13**, 85–96.

Matson, R. J., and Huggins, A. W. (1980). The direct measurement of the sizes, shapes, and kinematics of falling hailstones. *J. Atmos. Sci.* **37**, 1107–1125.

Matsumura, M., and Seki, M. (1991). Light scattering calculations by the Fredholm integral equation method. *Astrophys. Space Sci.* **176**, 283–295.

Matsumura, M., and Seki, M. (1996). Extinction and polarization by ellipsoidal particles in the infrared. *Astrophys. J.* **456**, 557–565.

Maxwell-Garnett, J. C. (1904). Colours in metal glasses and in metallic films. *Philos. Trans. R. Soc. A* **203**, 385–420.

Mazeron, P., and Muller, S. (1996). Light scattering by ellipsoids in a physical optics approximation. *Appl. Opt.* **35**, 3726–3735.

Mazeron, P., Muller, S., and El Azouzi, H. (1997a). Deformation of erythrocytes under shear: A small-angle light scattering study. *Biorheology* **34**, 99–110.

Mazeron, P., Muller, S., and El Azouzi, H. (1997b). On intensity reinforcements in small-angle light scattering patterns of erythrocytes under shear. *Eur. Biophys. J.* **26**, 247–252.

Mazumder, M. M., Hill, S. C., and Barber, P. W. (1992). Morphology-dependent resonances in inhomogeneous spheres: Comparison of the layered T-matrix method and the time-independent perturbation method. *J. Opt. Soc. Am. A* **9**, 1844–1853.

McClain, W. M., and Ghoul, W. A. (1986). Elastic light scattering by randomly oriented macromolecules: Computation of the complete set of observables. *J. Chem. Phys.* **84**, 6609–6622.

McCormick, G. C. (1985). Optimal polarizations for partially polarized backscatter. *IEEE Trans. Antennas Propag.* **33**, 33–40.

McCormick, G. C., and Hendry, A. (1975). Principles for the radar determination of the polarization properties of precipitation. *Radio Sci.* **10**, 421–434.

McGuire, A. F., and Hapke, B. W. (1995). An experimental study of light scattering by large, irregular particles. *Icarus* **113**, 134–155.

McNeil, W. R., and Carswell, A. I. (1975). Lidar polarization studies of the troposphere. *Appl. Opt.* **14**, 2158–2168.

Mead, J. B., Langlois, P. M., Chang, P. S., and McIntosh, R. E. (1991). Polarimetric scattering from natural surface at 225 GHz. *IEEE Trans. Geosci. Remote Sens.* **39**, 1405–1411.

Mead, J. B., Chang, P. S., Lohmeier, S. P., *et al.* (1993). Polarimetric observations and theory of millimeter-wave backscatter from snow cover. *IEEE Trans. Antennas Propag.* **41**, 38–46.

Medgyesi-Mitschang, L. N., Putnam, J. M., and Gedera, M. B. (1994). Generalized method of moments for three-dimensional penetrable scatterers. *J. Opt. Soc. Am. A* **11**, 1383–1398.

Meeten, G. H. (1982). An anomalous diffraction theory of linear birefringence and dichroism in colloidal dispersions. *J. Colloid Interface Sci.* **87**, 407–415.

Mei, K. K. (1974). Unimoment method of solving antenna and scattering problems. *IEEE Trans. Antennas Propag.* **22**, 760–766.

Meischner, P. F., Bringi, V. N., Heimann, D., and Holler, H. (1991). A squall line in southern Germany: Kinematics and precipitation formation as deduced by advanced polarimetric and Doppler radar measurements. *Mon. Weather Rev.* **119**, 678–701.

Meneghini, R., and Kozu, T. (1990). "Spaceborne Weather Radar." Artech House, Boston.

Meneghini, R., and Kumagai, H. (1994). Characteristics of the vertical profiles of dual-frequency, dual-polarization radar data in stratiform rain. *J. Atmos. Oceanic Technol.* **11**, 701–711.

Meneghini, R., and Liao, L. (1996). Comparisons of cross sections for melting hydrometeors as derived from dielectric mixing formulas and a numerical method. *J. Appl. Meteorol.* **35**, 1658–1670.

Merewether, D. E., Fisher, R., and Smith, F. W. (1980). On implementing a numeric Huygen's source in a finite difference program to illuminate scattering bodies. *IEEE Trans. Nucl. Sci.* **27**, 1829–1833.

Mertens, L. E. (1970). "In-Water Photography." Wiley, New York.

Mészáros, A., and Vissy, K. (1974). Concentration, size distribution, and chemical nature of atmospheric aerosol particles in remote oceanic areas. *J. Aerosol Sci.* **5**, 101–109.

Metcalf, J. I. (1988). A new slant on the distribution and measurement of hydrometeor canting angles. *J. Atmos. Oceanic Technol.* **53**, 1710–1723.

Metcalf, J. (1995). Radar observations of changing orientations of hydrometeors in thunderstorms. *J. Appl. Meteorol.* **34**, 757–772.

Metcalf, J. (1997). Temporal and spatial variations of hydrometeor orientations in thunderstorms. *J. Appl. Meteor.* **36**, 315–321.

Meyer, R. A., and Brunsting, A. (1975). Light scattering from nucleated biological cells. *Biophys. J.* **15**, 191–203.

Michal, F., and Born, G. V. R. (1971). Effect of the rapid change of platelets on the transmission and scattering of light through plasma. *Nature New Biol.* **231**, 220–222.

Michalsky, J. J. (1981). Optical polarimetry of comet West 1976 VI. *Icarus* **47**, 388–396.

Michel, B. (1995). Statistical method to calculate extinction by small irregularly shaped particles. *J. Opt. Soc. Am. A* **12**, 2471–2481.

Mie, G. (1908). Beiträge zur Optik trüber Medien, speziell kolloidaler Metallösungen. *Ann. Phys.* **25**, 377–445.

Mikulski, J. J., and Murphy, E. L. (1963). The computation of electromagnetic scattering from concentric spherical structures. *IEEE Trans. Antennas Propag.* **11**, 169–177.

Millar, R. F. (1973). The Rayleigh hypothesis and a related least-squares solution to scattering problems for periodic surfaces and other scatterers. *Radio Sci.* **8**, 785–796.

Miller, E. K., Medgyesi-Mitschang, L. N., and Newman, E. H. (1992). "Computational Electromagnetics: Frequency-Domain Method of Moments." IEEE Press, New York.

Miller, D., Quinby-Hunt, M. S., and Hunt, A. J. (1994). Polarization dependent measurements of light scattering in sea ice. *Proc. SPIE* **2258**, 908–918.

Miller, D., Quinby-Hunt, M. S., and Hunt, A. J. (1996). A novel bistatic polarization neph-elometer for probing scattering through a planar interface. *Rev. Sci. Instrum.* **67**, 2089–2095.

Miller, D., Quinby-Hunt, M. S., and Hunt, A. J. (1997). Laboratory studies of the angle and polarization dependent light scattering in sea ice. *Appl. Opt.* **36**, 1278–1288.

Mishchenko, M. I. (1990). Multiple scattering of polarized light in anisotropic plane-parallel media. *Transp. Theory Statist. Phys.* **19**, 293–316.

Mishchenko, M. I. (1991a). Light scattering by randomly oriented axially symmetric par-ticles. *J. Opt. Soc. Am. A* **8**, 871–882. [Errata: **9**, 497 (1992).]

Mishchenko, M. I. (1991b). Extinction and polarization of transmitted light by partially aligned nonspherical grains. *Astrophys. J.* **367**, 561–574.

Mishchenko, M. I. (1992a). Coherent propagation of polarized millimeter waves through falling hydrometeors. *J. Electromagn. Waves Appl.* **6**, 1341–1351.

Mishchenko, M. I. (1992b). Enhanced backscattering of polarized light from discrete ran-dom media: Calculations in exactly the backscattering direction. *J. Opt. Soc. Am. A* **9**, 978–982.

Mishchenko, M. I. (1993). Light scattering by size–shape distributions of randomly ori-ented axially symmetric particles of a size comparable to a wavelength. *Appl. Opt.* **32**, 4652–4666.

Mishchenko, M. I. (1996). Coherent backscattering by two-sphere clusters. *Opt. Lett.* **21**, 623–625.

Mishchenko, M. I., and Hovenier, J. W. (1995). Depolarization of light backscattered by randomly oriented nonspherical particles. *Opt. Lett.* **20**, 1356–1358.

Mishchenko, M. I., and Macke, A. (1997). Asymmetry parameters of the phase function for isolated and densely packed spherical particles with multiple internal inclusions in the geometric optics limit. *J. Quant. Spectrosc. Radiat. Transfer* **57**, 767–794.

Mishchenko, M. I., and Macke, A. (1998). Incorporation of physical optics effects and computation of the Legendre expansion for ray-tracing phase functions involving δ-function transmission. *J. Geophys. Res.* **103**, 1799–1805.

Mishchenko, M. I., and Macke, A. (1999). How big should hexagonal ice crystals be to produce halos? *Appl. Opt.* **38**, 1626–1629.

Mishchenko, M. I., and Mackowski, D. W. (1994). Light scattering by randomly oriented bispheres. *Opt. Lett.* **19**, 1604–1606.

Mishchenko, M. I., and Mackowski, D. W. (1996). Electromagnetic scattering by randomly oriented bispheres: Comparison of theory and experiment and benchmark calculations. *J. Quant. Spectrosc. Radiat. Transfer* **55**, 683–694.

Mishchenko, M. I., and Sassen, K. (1998). Depolarization of lidar returns by small ice crystals: An application to contrails. *Geophys. Res. Lett.* **25**, 309–312.

Mishchenko, M. I., and Travis, L. D. (1994a). T-matrix computations of light scattering by large spheroidal particles. *Opt. Commun.* **109**, 16–21.

Mishchenko, M. I., and Travis, L. D. (1994b). Light scattering by polydisperse, rotation-ally symmetric nonspherical particles: Linear polarization. *J. Quant. Spectrosc. Radiat. Transfer* **51**, 759–778.

Mishchenko, M. I., and Travis, L. D. (1994c). Light scattering by polydispersions of ran-domly oriented spheroids with sizes comparable to wavelengths of observation. *Appl. Opt.* **33**, 7206–7225.

Mishchenko, M. I., and Travis, D. L. (1997). Satellite retrieval of aerosol properties over the ocean using polarization as well as intensity of reflected sunlight. *J. Geophys. Res.* **102**, 16,989–17,013.

Mishchenko, M. I., and Travis, L. D. (1998). Capabilities and limitations of a current FORTRAN implementation of the *T*-matrix method for randomly oriented, rotationally symmetric scatterers. *J. Quant. Spectrosc. Radiat. Transfer* **60**, 309–324.

Mishchenko, M. I., Mackowski, D. W., and Travis, L. D. (1995). Scattering of light by bispheres with touching and separated components. *Appl. Opt.* **34**, 4589–4599.

Mishchenko, M. I., Travis, L. D., and Macke, A. (1996a). Scattering of light by polydisperse, randomly oriented, finite circular cylinders. *Appl. Opt.* **35**, 4927–4940.

Mishchenko, M. I., Travis, L. D., and Mackowski, D. W. (1996b). *T*-matrix computations of light scattering by nonspherical particles: A review. *J. Quant. Spectrosc. Radiat. Transfer* **55**, 535–575.

Mishchenko, M. I., Rossow, W. B., Macke, A., and Lacis, A. A. (1996c). Sensitivity of cirrus cloud albedo, bidirectional reflectance and optical thickness retrieval accuracy to ice particle shape. *J. Geophys. Res.* **101**, 16,973–16,985.

Mishchenko, M. I., Travis, L. D., Kahn, R. A., and West, R. A. (1997a). Modeling phase functions for dustlike tropospheric aerosols using a shape mixture of randomly oriented polydisperse spheroids. *J. Geophys. Res.* **102**, 16,831–16,847.

Mishchenko, M. I., Wielaard, D. J., and Carlson, B. E. (1997b). *T*-matrix computations of zenith-enhanced lidar backscatter from horizontally oriented ice plates. *Geophys. Res. Lett.* **24**, 771–774.

Mishchenko, M. I., Travis, L. D., Rossow, W. B., *et al.* (1997c). Retrieving CCN column density from single-channel measurements of reflected sunlight over the ocean: A sensitivity study. *Geophys. Res. Lett.* **24**, 2655–2658.

Mishchenko, M. I., Travis, L. D., and Hovenier, J. W., Eds. (1999). Special issue on Light Scattering by Nonspherical Particles. *J. Quant. Spectrosc. Radiat. Transfer*, in press.

Mitchell, D. L. (1996). Use of mass-and area-dimensional power laws for determining precipitation particle terminal velocities. *J. Atmos. Sci.* **53**, 1710–1723.

Mitchell, D. L., Chai, S. K., Liu, Y., *et al.* (1996). Modeling cirrus clouds. I. Treatment of bimodal size spectra and case study analysis. *J. Atmos. Sci.* **53**, 2952–2966.

Mittra, R., and Ramahi, O. (1990). Absorbing boundary conditions for the direct solution of partial differential equations arising in electromagnetic scattering problems. *In* "Finite Element and Finite Difference Methods in Electromagnetic Scattering" (M. A. Morgan, Ed.), pp. 133–173. Elsevier, New York.

Modest, M. F. (1993). "Radiative Heat Transfer." McGraw–Hill, New York.

Moghaddam, M., and Chew, W. C. (1991). Stabilizing Liao's absorbing boundary conditions using single-precision arithmetic. *IEEE AP-S Int. Symp. Digest, 1991*, 430–433.

Moore, T. G., Blaschak, J. G., Taflove, A., and Kriegsmann, G. A. (1988). Theory and application of radiation boundary operators. *IEEE Trans. Antennas Propag.* **36**, 1797–1812.

Morgan, M. A. (1980). Finite element computation of microwave scattering by raindrops. *Radio Sci.* **15**, 1109–1119.

Morgan, M. A., Ed. (1990). "Finite Element and Finite Difference Methods in Electromagnetic Scattering." Elsevier, New York.

Morgan, M. A., and Mei, K. K. (1979). Finite-element computation of scattering by inhomogeneous penetrable bodies of revolution. *IEEE Trans. Antennas Propag.* **27**, 202–214.

Morita, N. (1979). Another method of extending the boundary condition for the problem of scattering by dielectric cylinders. *IEEE Trans. Antennas Propag.* **27**, 97–99.

Morrison, J. A., and Cross, M.-J. (1974). Scattering of a plane electromagnetic wave by axisymmetric raindrops. *Bell Syst. Tech. J.* **53**, 955–1019.

Morse, P. M., and Feshbach, H. (1953). "Methods of Theoretical Physics." McGraw–Hill, New York.

Moskovits, M. (1985). Surface-enhanced spectroscopy. *Rev. Mod. Phys.* **57**, 783–826.

Motamedi, M., Ed. (1993). Special issue on Photon Migration in Tissue and Biomedical Applications of Lasers. *Appl. Opt.* **32**, 367–434.

Mountain, R. D., and Mulholland, G. W. (1988). Light scattering from simulated smoke agglomerates. *Langmuir* **4**, 1321–1326.

Mugnai, A., and Wiscombe, W. J. (1980). Scattering of radiation by moderately nonspherical particles. *J. Atmos. Sci.* **37**, 1291–1307.

Mugnai, A., and Wiscombe, W. J. (1986). Scattering from nonspherical Chebyshev particles. 1. Cross sections, single-scattering albedo, asymmetry factor, and backscattered fraction. *Appl. Opt.* **25**, 1235–1244.

Mugnai, A., and Wiscombe, W. J. (1989). Scattering from nonspherical Chebyshev particles. 3. Variability in angular scattering patterns. *Appl. Opt.* **28**, 3061–3073.

Muinonen, K. (1989). Scattering of light by crystals: A modified Kirchhoff approximation. *Appl. Opt.* **28**, 3044–3050.

Muinonen, K. (1990). Light scattering by inhomogeneous media: Backward enhancement and reversal of linear polarization. Ph.D. Thesis, Univ. of Helsinki, Helsinki.

Muinonen, K. (1994). Coherent backscattering by solar system dust particles. *In* "Asteroids, Comets, Meteors 1993" (A. Milani, M. Di Martino, and A. Cellino, Eds.), pp. 271–296. Kluwer Academic, Dordrecht.

Muinonen, K. (1996a). Light scattering by Gaussian random particles. *Earth Moon Planets* **72**, 339–342.

Muinonen, K. (1996b). Light scattering by Gaussian random particles: Rayleigh and Rayleigh–Gans approximations. *J. Quant. Spectrosc. Radiat. Transfer* **55**, 603–613.

Muinonen, K. (1998). Introducing the Gaussian shape hypothesis for asteroids and comets. *Astron. Astrophys.* **332**, 1087–1098.

Muinonen, K., and Lagerros, J. (1998). Inversion of shape statistics for small solar system bodies. *Astron. Astrophys.* **333**, 753–761.

Muinonen, K., and Saarinen, K. (1999). Ray optics approximation for Gaussian random cylinders. *J. Quant. Spectrosc. Radiat. Transfer*, in press.

Muinonen, K., Lumme, K., Peltoniemi, J., and Irvine, W. M. (1989). Light scattering by randomly oriented crystals. *Appl. Opt.* **28**, 3051–3060.

Muinonen, K., Nousiainen, T., Fast, P., *et al.* (1996). Light scattering by Gaussian random particles: Ray optics approximation. *J. Quant. Spectrosc. Radiat. Transfer* **55**, 577–601.

Muinonen, K., Lamberg, L., Fast, P., and Lumme, K. (1997). Ray optics regime for Gaussian random spheres. *J. Quant. Spectrosc. Radiat. Transfer* **57**, 197–205.

Mukai, S., Mukai, T., Giese, R. H., *et al.* (1982). Scattering of radiation by a large particle with a random rough surface. *Moon Planets* **26**, 197–208.

Mulholland, G. W., Bohren, C. F., and Fuller, K. A. (1994). Light scattering by agglomerates: Coupled electric and magnetic dipole method. *Langmuir* **10**, 2533–2546.

Mullaney, P. F., and Dean, P. N. (1970). The small angle light scattering of biological cells. *Biophys. J.* **10**, 764–772.

Mur, G. (1981). Absorbing boundary condition for the finite-difference approximation of the time-domain electromagnetic-field equations. *IEEE Trans. Electromagn. Compat.* **23**, 377–382.

Murayama, T., Furushima, M., Oda, A., *et al.* (1996). Depolarization ratio measurements in the atmospheric boundary layer by lidar in Tokyo. *J. Meteorol. Soc. Jpn.* **74**, 571–577.

Murayama, T., Furushima, M., Oda, A., and Iwasaka, N. (1997). Aerosol optical properties in the urban mixing layer studied by polarization lidar with meteorological data. *In* "Advances in Remote Sensing with Lidar" (A. Ansmann *et al.*, Eds.), pp. 19–22. Springer-Verlag, Berlin.

Murayama, T., Sugimoto, N., Matsui, I., *et al.* (1998). Lidar network observation of Asian dust (Kosa) in Japan. *Proc. SPIE* **3504**, 8–15.

Musil, D. J., May, E. L., Smith, P. L., Jr., and Sand, W. R. (1976). Structure of an evolving hailstorm. IV. Internal structure from penetrating aircraft. *Mon. Weather Rev.* **104**, 596–602.

Nag, S., and Sinha, B. P. (1995). Electromagnetic plane wave scattering by a system of two uniformly lossy dielectric prolate spheroids in arbitrary orientation. *IEEE Trans. Antennas Propag.* **43**, 322–327.

Nagirner, D. I. (1993). Constraints on matrices transforming Stokes vectors. *Astron. Astrophys.* **275**, 318–324.

Nakajima, T., Tanaka, M., Yamano, M., *et al.* (1989). Aerosol optical characteristics in the yellow sand events observed in May, 1982 at Nagasaki. II. Models. *J. Meteorol. Soc. Jpn.* **67**, 279–291.

Napper, D. H., and Ottewill, R. H. (1963). Light scattering studies on monodisperse silver bromide sols. *In* "ICES Electromagnetic Scattering" (M. Kerker, Ed.), pp. 377–386. Pergamon, New York.

National Research Council (1996). "Aerosol Radiative Forcing and Climate Change." National Academy Press, Washington, DC.

Nebeker, B. M., Starr, G. W., and Hirleman, E. D. (1998). Evaluation of iteration methods used when modelling scattering from features on surfaces using the discrete-dipole approximation. *J. Quant. Spectrosc. Radiat. Transfer* **60**, 493–500.

Nelson, K. H., and Thompson, T. G. (1954). Deposition of salts from sea water by frigid concentration. *J. Mar. Res.* **13**, 166–182.

Nemarich, J., Wellman, R. J., and Lacombe, J. (1988). Backscatter and attenuation by falling snow and rain at 96, 140, and 225 GHz. *IEEE Trans. Geosci. Remote Sens.* **26**, 319–329.

Nevitt, T. J., and Bohren, C. F. (1984). Infrared backscattering by irregularly shaped particles: A statistical approach. *J. Clim. Appl. Meteorol.* **23**, 1342–1349.

Ngo, D., and Pinnick, R. G. (1994). Suppression of scattering resonances in inhomogeneous microdroplets. *J. Opt. Soc. Am. A* **11**, 1352–1359.

Ngo, D., Videen, G., and Chýlek, P. (1996). A Fortran code for the scattering of EM waves by a sphere with a nonconcentric spherical inclusion. *Comput. Phys. Commun.* **99**, 94–112.

Ngo, D., Videen, G., and Dalling, R. (1997). Chaotic light scattering from a system of osculating, conducting spheres. *Phys. Lett. A* **227**, 197–202.

Nikiforova, N. K., Pavlova, L. N., Petrushin, A. G., *et al.* (1977). Aerodynamic and optical properties of ice crystals. *J. Aerosol Sci.* **8**, 243–250.

Niklasson, G. A., and Granqvist, C. G. (1984). Optical properties and solar selectivity of coevaporated Co–Al$_2$O$_3$ composite films. *J. Appl. Phys.* **55**, 3382–3410.

Niklasson, G. A., Granquist, C. G., and Hunderi, O. (1981). Effective medium models for the optical properties of inhomogeneous materials. *Appl. Opt.* **20**, 26–30.

Nilsson, B. (1979). Meteorological influence on aerosol extinction in the 0.2–40 μm wavelength range. *Appl. Opt.* **18**, 3457–3473.

Nilsson, A. M. K., Alsholm, P., Karlsson, A., and Andersson-Engles, S. (1998). *T*-matrix computations of light scattering by red blood cells. *Appl. Opt.* **37**, 2735–2748.

Nöckel, J. U., and Stone, A. D. (1997). Ray and wave chaos in asymmetric resonant optical cavities. *Nature (London)* **385**, 45–47.

Nousiainen, T., and Muinonen, K. (1999). Light scattering by Gaussian, randomly oscillating raindrops. *J. Quant. Spectrosc. Radiat. Transfer*, in press.

Novakov, T., and Penner, J. E. (1992). The effect of anthropogenic sulfate aerosols on marine cloud droplet concentrations. *Nature (London)* **365**, 823–826.

November, L. J. (1993). Recovery of the matrix operators in the similarity and congruency transformations. *J. Opt. Soc. Am. A* **10**, 719–739.

Nussenzvig, H. M. (1992). "Diffraction Effects in Semiclassical Scattering." Cambridge Univ. Press, Cambridge.

O'Brien, S. G., and Goedecke, G. H. (1988a). Scattering of millimeter waves by snow crystals and equivalent homogeneous symmetric particles. *Appl. Opt.* **27**, 2439–2444.

O'Brien, S. G., and Goedecke, G. H. (1988b). Propagation of polarized millimeter waves through falling snow. *Appl. Opt.* **27**, 2445–2450.

Oguchi, T. (1960). Attenuation of electromagnetic waves due to rain with distorted raindrops. *J. Radio Res. Lab. Jpn.* **7**, 467–485.

Oguchi, T. (1973). Scattering properties of oblate raindrops and cross polarization of radio waves due to rain: Calculations at 19.3 and 34.8 GHz. *J. Radio Res. Lab. Jpn.* **20**, 79–118.

Oguchi, T. (1981). Scattering from hydrometeors: A survey. *Radio Sci.* **16**, 691–730.

Oguchi, T. (1983). Electromagnetic wave propagation and scattering in rain and other hydrometeors. *Proc. IEEE* **71**, 1029–1078.

Oguchi, T., and Hosoya, Y. (1974). Scattering properties of oblate raindrops and cross polarization of radio waves due to rain. II. Calculations at microwave and millimeter wave regions. *J. Radio Res. Lab. Jpn.* **21**, 191–259.

Ogura, H., and Takahashi, N. (1990). Scattering of waves from a random spherical surface—Mie scattering. *J. Math. Phys.* **31**, 61–75.

Ogura, H., Takahashi, N., and Kuwahara, M. (1991). Scattering of an electromagnetic wave from a slightly random cylindrical surface: Horizontal polarization. *Waves Random Media* **1**, 363–389.

Okada, K., Kobayashi, A., Iwasaka, Y., *et al.* (1987). Features of individual Asian duststorm particles collected at Nagoya, Japan. *J. Meteorol. Soc. Jpn.* **65**, 515–521.

Okamoto, H. (1995). Light scattering by clusters: The a_1-term method. *Opt. Rev.* **2**, 407–412.

Okamoto, H., and Xu, Y. (1998). Light scattering by irregular interplanetary dust particles. *Earth Planets Space* **50**, 577–585.

Okamoto, H., Mukai, T., and Kozasa, T. (1994). The 10 micron feature of aggregates in comets. *Planet. Space Sci.* **42**, 643–649.

Okamoto, H., Macke, A., Quante, M., and Raschke, E. (1995). Modeling of backscattering by nonspherical ice particles for the interpretation of cloud radar signals at 94 GHz. An error analysis. *Beitr. Phys. Atmos.* **68**, 319–334.

Olaofe, G. O. (1970). Scattering by two cylinders. *Radio Sci.* **5**, 1351–1360.

Oldenburg, S. J., Averitt, R. D., Westcott, S. L., and Halas, N. J. (1998). Nanoengineering of optical resonances. *Chem. Phys. Lett.* **288**, 243–247.

Oliveira, P. C. de, McGreevy, J. A., and Lawandy, N. M. (1997). External feedback effects in high-gain scattering media. *Opt. Lett.* **22**, 895–897.

Omick, S., and Castillo, S. P. (1993). A new finite-difference time-domain algorithm for the accurate modeling of wide-band electromagnetic phenomena. *IEEE Trans. Electromagn. Compat.* **35**, 315–322.

Onaka, T. (1980). Light scattering by spheroidal grains. *Ann. Tokyo Astron. Observ.* **18**, 1–54.

Onaka, R., and Kawamura, T. (1983). Refractive indices of hexagonal ice. *J. Phys. Soc. Jpn.* **52**, 2947–2953.

O'Neill, E. L. (1963). "Introduction to Statistical Optics." Addison-Wesley, Reading, MA.

Ormerod, M. G., Paul, F., Cheetham, M., and Sun, X.-M. (1995). Discrimination of apoptotic thymocytes by forward light scatter. *Cytometry* **21**, 300–304.

Oseen, C. W. (1915). Uber die Wechselwirkung zwischen zwei elektrichen Dipolen und uber die Drenhung der Polarisationsebene in Kristallen und Flussigkeiten. *Ann. Phys.* **48**, 1–15.

Ossenkopf, V. (1991). Effective-medium theories for cosmic dust grains. *Astron. Astrophys.* **251**, 210–219.

Ovod, V. I., Mackowski, D. W., and Finsy, R. (1998). Modeling the effects of multiple scattering in photon correlation spectroscopy: Plane wave approach. *Langmuir* **14**, 2610–2618.

Pak, H., Zaneveld, J. R. V., and Beardsley, G. F. (1971). Mie scattering by suspended clay particles. *J. Geophys. Res.* **76**, 5065–5069.

Panegrossi, G., Dietrich, S., Marzano, F. S., *et al.* (1998). Use of cloud model microphysics for passive microwave-based precipitation retrieval. *J. Atmos. Sci.* **55**, 1644–1673.

Papadakis, S. N., Uzunoglu, N. K., and Capsalis, C. N. (1990). Scattering of a plane wave by a general anisotropic dielectric ellipsoid. *J. Opt. Soc. Am. A* **7**, 991–997.

Paramonov, L. E. (1995). *T*-matrix approach and the angular momentum theory in light scattering problems by ensembles of arbitrarily shaped particles. *J. Opt. Soc. Am. A* **12**, 2698–2707.

Parungo, F., Kopcewicz, B., Nagamoto, C., *et al.* (1992). Aerosol particles in the Kuwait oil fire plumes: Their morphology, size distribution, chemical composition, transport, and potential effect on climate. *J. Geophys. Res.* **97**, 15,867–15,882.

Paulick, T. C. (1990). Applicability of the Rayleigh hypothesis to real materials. *Phys. Rev. B* **42**, 2801–2824.

Pazmany, A., McIntosh, R. E., Kelly, R. D., and Vali, G. (1994). An airborne 95 GHz dual-polarized radar for cloud studies. *IEEE Trans. Geosci. Remote Sens.* **32**, 731–739.

Peltoniemi, J. I. (1996). Variational volume integral equation method for electromagnetic scattering by irregular grains. *J. Quant. Spectrosc. Radiat. Transfer* **55**, 637–647.

Peltoniemi, J. I., Lumme, K., Muinonen, K., and Irvine, W. M. (1989). Scattering of light by stochastically rough particles. *Appl. Opt.* **28**, 4088–4095.

Peltoniemi, J. I., Nousiainen, T., and Muinonen, K. (1998). Light scattering by Gaussian random particles using various volume-integral-equation techniques. *Prepr. Vol., Conf. Light Scat. Nonspher. Part.: Theory Meas. Appl., New York, 1998*, 195–198.

Perelman, A. Y. (1996). Scattering by particles with radially variable refractive index. *Appl. Opt.* **35**, 5452–5460.

Perovich, D. K., and Grenfell, T. C. (1981). Laboratory studies of the optical properties of young sea ice. *J. Glaciol.* **27**, 331–346.

Perrin, F. (1942). Polarization of light scattered by isotropic opalescent media. *J. Chem. Phys.* **10**, 415–426.

Perrin, J.-M., and Lamy, P. L. (1983). Light scattering by large rough particles. *Opt. Acta* **30**, 1223–1244.

Perrin, J.-M., and Lamy, P. L. (1986). Light scattering by large particles. II. A vectorial description in the eikonal picture. *Opt. Acta* **33**, 1001–1022.

Perrin, J.-M., and Lamy, P. L. (1990). On the validity of effective-medium theories in the case of light extinction by inhomogeneous dust particles. *Astrophys. J.* **364**, 146–151.

Perry, R. J., Hunt, A. J., and Huffman, D. R. (1978). Experimental determinations of Mueller scattering matrices for nonspherical particles. *Appl. Opt.* **17**, 2700–2710.

Peterson, B. (1977). Multiple scattering of waves by an arbitrary lattice. *Phys. Rev. A* **16**, 1363–1370.

Peterson, B., and Ström, S. (1973). T matrix for electromagnetic scattering from an arbitrary number of scatterers and representations of E(3)*. *Phys. Rev. D* **8**, 3661–3678.

Peterson, B., and Ström, S. (1974). T-matrix formulation of electromagnetic scattering from multilayered scatterers. *Phys. Rev. D* **10**, 2670–2684.

Peterson, A. F., Ray, S. L., and Mittra, R. (1998). "Computational Methods for Electromagnetics." IEEE Press, New York.

Petravic, M., and Kuo-Petravic, G. (1979). An ILUCG algorithm which minimizes in the Euclidean norm. *J. Comput. Phys.* **32**, 263–269.

Petrova, E. V. (1999). Mars aerosol optical thickness retrieved from measurements of the polarization inversion angle and the shape of dust particles. *J. Quant. Spectrosc. Radiat. Transfer*, in press.

Petty, G. W., and Turk, J. (1996). Observed multichannel microwave signatures of spatially extensive precipitation in tropical cyclones. *Prepr. Vol., Conf. Sat. Meteorol. Oceanogr., 8th, Atlanta, GA, 1996*, 291–294.

Petzold, T. L. (1972). Volume scattering functions for selected ocean waters. Ref. 72–78, Scripps Inst. Oceanogr. Visibility Lab., Univ. of California, San Diego.

Piironen, J., Muinonen, K., Nousiainen, T., *et al.* (1998). Albedo measurements on meteorite particles. *Planet. Space Sci.* **46**, 937–943.

Pilinis, C., and Li, X. (1998). Particle shape and internal inhomogeneity effects on the optical properties of tropospheric aerosols of relevance to climate forcing. *J. Geophys. Res.* **103**, 3789–3800.

Piller, N. B., and Martin, O. J. F. (1998a). Extension of the generalized multipole technique to three-dimensional anisotropic scatterers. *Opt. Lett.* **23**, 579–581.

Piller, N. B., and Martin, O. J. F. (1998b). Increasing the performance of the coupled-dipole approximation: A spectral approach. *IEEE Trans. Antennas Propag.* **46**, 1126–1137.

Pinnick, R. G., Carroll, D. E., and Hofmann, D. J. (1976). Polarized light scattered from monodisperse randomly oriented nonspherical aerosol particles: Measurements. *Appl. Opt.* **15**, 384–393.

Pinnick, R. G., Fernandez, G., Hinds, B. D., *et al.* (1985). Dust generated by vehicular traffic on unpaved roadways: Sizes and infrared extinction characteristics. *Aerosol Sci. Technol.* **4**, 99–121.

Pitter, M. C., Hopcraft, K. I., Jakeman, E., *et al.* (1998). Polarization structure of Stokes' parameter fluctuations and their relation to particle shape. *Prepr. Vol., Conf. Light Scat. Nonspher. Part.: Theory Meas. Appl., New York, 1998*, 199–202.

Plass, G. N., Kattawar, G. W., and Catchings, F. E. (1973). Matrix operator theory of radiative transfer. 1. Rayleigh scattering. *Appl. Opt.* **12**, 314–329.

Plass, G. N., Humphreys, T. J., and Kattawar, G. W. (1981). Ocean–atmosphere interface: Its influence on radiation. *Appl. Opt.* **20**, 917–930.

Platt, C. M. R. (1977). Lidar observation of a mixed-phase altostratus cloud. *J. Appl. Meteorol.* **16**, 339–345.

Platt, C. M. R. (1978). Lidar backscattering from horizontally oriented ice crystal plates. *J. Appl. Meteorol.* **17**, 482–488.

Platt, C. M. R., and Bartusek, K. (1974). Structure and optical properties of some middle-level clouds. *J. Atmos. Sci.* **31**, 1079–1088.

Platt, C. M. R., Abshire, N. L., and McNice, G. T. (1978). Some microphysical properties of an ice cloud from lidar observations of horizontally oriented crystals. *J. Appl. Meteorol.* **17**, 1220–1224.

Platt, C. M. R., Scott, J. C., and Dilley, A. C. (1987). Remote sounding of high clouds. VI. Optical properties of midlatitude and tropical cirrus. *J. Atmos. Sci.* **44**, 729–747.

Platt, C. M. R., Young, S. A., Manson, P. J., *et al.* (1998). The optical properties of equatorial cirrus from observations in the ARM Pilot Radiation Observation Experiment. *J. Atmos. Sci.* **55**, 1977–1996.

Pluchino, A. (1987). Scattering photometer for measuring single ice crystals and evaporation and condensation rates of liquid droplets. *J. Opt. Soc. Am. A* **4**, 614–620.

Pluchino, A. B., Goldberg, S. S., Dowling, J. M., and Randall, C. M. (1980). Refractive-index measurements of single micron-sized carbon particles. *Appl. Opt.* **19**, 3370–3372.

Pocock, M. D., Bluck, M. J., and Walker, S. P. (1998). Electromagnetic scattering from 3-D curved dielectric bodies using time-domain integral equations. *IEEE Trans. Antennas Propag.* **46**, 1212–1219.

Podzimek, J. (1990). Physical properties of coarse aerosol particles and haze elements in a polluted urban-marine environment. *J. Aerosol. Sci.* **21**, 299–308.

Poggio, A. J., and Miller, E. K. (1973). Integral equation solutions of three-dimensional scattering problems. *In* "Computer Techniques for Electromagnetics" (R. Mittra, Ed.), pp. 159–264. Pergamon, Oxford.

Pollack, J. B., and Cuzzi, J. N. (1980). Scattering by nonspherical particles of size comparable to a wavelength: A new semi-empirical theory and its application to tropospheric aerosols. *J. Atmos. Sci.* **37**, 868–881.

Poole, L. R., Kent, G. S., McCormick, M. P., *et al.* (1990). Dual-polarization lidar observations of polar stratospheric clouds. *Geophys. Res. Lett.* **17**, 389–392.

Pope, S. K., Tomasko, M. G., Williams, M. S., *et al.* (1992). Clouds of ammonia ice: Laboratory measurements of the single-scattering properties. *Icarus* **100**, 203–220.

Prabhakara, C., Kratz, D. P., Yoo, J.-M., *et al.* (1993). Optically thin cirrus clouds: Radiative impact on the warm pool. *J. Quant. Spectrosc. Radiat. Transfer* **49**, 467–483.

Press, W. H., Flannery, B. R., Teukolsky, S. A., and Vetterling, W. T. (1989). "Numerical Recipes (Fortran Version)," p. 523. Cambridge Univ. Press, Cambridge.

Pritchard, B. S., and Elliott, W. G. (1960). Two instruments for atmospheric optics measurements. *J. Opt. Soc. Am.* **50**, 191–202.

Prodi, F., Sturniolo, O., Medini, R., and Battaglia, A. (1998). Simulation of cloud scenarios using monodisperse and polydisperse populations of randomly oriented axially symmetric hydrometeors. *Prepr. Vol., Conf. Light Scat. Nonspher. Part.: Theory Meas. Appl., New York, 1998*, 309–312.

Pruppacher, H. R., and Klett, J. D. (1997). "Microphysics of Clouds and Precipitation." Kluwer Academic, Dordrecht.

Pruppacher, H. R., and Pitter, R. L. (1971). A semi-empirical determination of the shape of cloud and rain drops. *J. Atmos. Sci.* **28**, 86–94.

Purcell, E. M., and Pennypacker, C. R. (1973). Scattering and absorption of light by nonspherical dielectric grains. *Astrophys. J.* **186**, 705–714.

Quinby-Hunt, M. S., and Hunt, A. J. (1988). Effects of structure on scattering from marine organisms: Rayleigh–Debye and Mie predictions. *Proc. SPIE* **925**, 288–295.

Quinby-Hunt, M. S., Hunt, A. J., and Brady, S. A. (1986). Polarization-modulation scattering measurements of well-characterized marine plankton. *Proc. SPIE* **637**, 155–163.

Quinby-Hunt, M. S., Hunt, A. J., Lofftus, K., and Shapiro, D. (1989). Polarized-light scattering studies of marine *Chlorella*. *Luminol. Oceanogr.* **34**, 1587–1600.

Quinby-Hunt, M. S., Hull, P. G., and Hunt, A. J. (1994). Predicting polarization properties of marine aerosols. *Proc. SPIE* **2258**, 735–746.

Quinby-Hunt, M. S., Erskine, L. L., and Hunt, A. J. (1997). Polarized light scattering by aerosols in the marine atmospheric boundary layer. *Appl. Opt.* **36**, 5168–5184.

Quinten, M., and Kreibig, U. (1993). Absorption and elastic scattering of light by particle aggregates. *Appl. Opt.* **32**, 6173–6182.

Quinten, M., Leitner, A., Krenn, J. R., and Aussenegg, F. A. (1998). Electromagnetic energy transport via linear chains of silver nanoparticles. *Opt. Lett.* **23**, 1331–1332.

Quirantes, A. (1999). Light scattering properties of spheroidal coated particles in random orientation. *J. Quant. Spectrosc. Radiat. Transfer*, in press.

Quirantes, A., and Delgado, A. V. (1995). Size–shape determination of nonspherical particles in suspension by means of full and depolarized static light scattering. *Appl. Opt.* **34**, 6256–6262.

Quirantes, A., and Delgado, A. (1998). Experimental size determination of spheroidal particles via the *T*-matrix method. *J. Quant. Spectrosc. Radiat. Transfer* **60**, 463–474.

Raes, F., Wilson, J., and Van Dingenen, R. (1995). Aerosol dynamics and its implication for the global aerosol climatology. *In* "Aerosol Forcing of Climate" (R. J. Charlson and J. Heintzenberg, Eds.), pp. 153–169. Wiley, New York.

Rahola, J. (1996). Efficient solution of dense systems of linear equations in electromagnetic scattering calculations. Ph.D. Thesis, Helsinki Univ. of Technology, Helsinki.

Raković, M. J., and Kattawar, G. W. (1998). Theoretical analysis of polarization patterns from incoherent backscattering of light. *Appl. Opt.* **37**, 3333–3338.

Ramanathan, V., Subasilar, B., Zhang, G. J., *et al.* (1995). Warm pool heat budget and shortwave cloud forcing: A missing physics? *Science* **267**, 499–503.

Rannou, P., Cabane, M., Botet, R., and Chassefière, E. (1997). A new interpretation of scattered light measurements at Titan's limb. *J. Geophys. Res.* **102**, 10,997–11,013.

Rasmussen, R. M., and Heymsfield, A. J. (1987). Melting and shedding of graupel and hail. I. Model physics. *J. Atmos. Sci.* **44**, 2754–2763.

Ravey, J.-C., and Mazeron, P. (1982). Light scattering in the physical optics approximation; application to large spheroids. *J. Opt. (Paris)* **13**, 273–282.

Ravey, J.-C., and Mazeron, P. (1983). Light scattering by large spheroids in the physical optics approximation: Numerical comparison with other approximate and exact results. *J. Opt. (Paris)* **14**, 29–41.

Ray, P. S. (1972). Broadband complex refractive indices of ice and water. *Appl. Opt.* **11**, 1836–1844.

Rayleigh, Lord (1897). On the incidence of aerial and electric waves upon small obstacles in the form of ellipsoids or elliptic cylinders, and on the passage of electric waves through a circular aperture in a conducting screen. *Philos. Mag.* **44**, 28–52.

Rayleigh, Lord (1903). "Scientific Papers, Vol. IV." Cambridge Univ. Press, Cambridge.

Reinking, R. F., Matrosov, S. Y., Bruintjes, R. T., and Martner, B. E. (1997). Identification of hydrometeors with elliptical and linear polarization Ka-band radar. *J. Appl. Meteorol.* **36**, 322–339.

Reitilinger, N. (1957). Scattering of a plane wave incident on a prolate spheroid at an arbitrary angle. Memo 2868-506-M. Radiation Laboratory, Univ. of Michigan, Ann Arbor, MI.

Rietmeijer, F. J. M., and Janeczek, J. (1997). An analytical electron microscope study of airborne industrial particles in Sosnowiec, Poland. *Atmos. Environ.* **31**, 1941–1951.

Rheinstein, J. (1968). Backscatter from spheres: A short pulse view. *IEEE Trans. Antennas Propag.* **16**, 89–97.

Rimmer, J. S., and Saunders, C. P. R. (1997). Radiative scattering by artificially produced clouds of hexagonal plate ice crystals. *Atmos. Res.* **45**, 153–164.

Rimmer, J. S., and Saunders, C. P. R. (1998). Extinction measurements and single scattering albedo in the 0.5 to 1.05 μm range for a laboratory cloud of hexagonal ice crystals. *Atmos. Res.* **49**, 177–188.

Rizzo, P. J. (1987). Biochemistry of the dinoflagellate nucleus. *In* "The Biology of Dinoflagellates" (F. J. R. Taylor, Ed.), pp. 143–173. Blackwell Sci., Oxford.

Ro, P. S., Fahlen, T. S., and Bryant, H. C. (1968). Precision measurement of water droplet evaporation rates. *Appl. Opt.* **7**, 883–890.

Roberti, L., and Kummerow, C. (1999). Monte Carlo calculations of polarized microwave radiation emerging from cloud structures. *J. Geophys. Res.* **104**, 2093–2104.

Roberti, L., Haferman, J., and Kummerow, C. (1994). Microwave radiative transfer through horizontally inhomogeneous precipitating clouds. *J. Geophys. Res.* **99**, 16,707–16,718.

Roberts, S., and von Hippel, A. (1946). A new method for measuring dielectric constant and loss in the range of centimeter waves. *J. Appl. Phys.* **17**, 610–616.

Rockwitz, K.-D. (1989). Scattering properties of horizontally oriented ice crystal columns in cirrus clouds. *Appl. Opt.* **28**, 4103–4110.

Roddy, D. (1996). "Satellite Communications." McGraw–Hill, New York.

Rogers, C., and Martin, P. G. (1979). On the shape of interstellar grains. *Astrophys. J.* **228**, 450–464.

Roll, G., Kaiser, T., Lange, S., and Schweiger, G. (1998). Ray interpretation of multipole fields in spherical dielectric cavities. *J. Opt. Soc. Am. A* **15**, 2879–2891.

Rolland, P., and Liou, K. N. (1998). Remote sensing of optical and microphysical properties of cirrus clouds using MODIS channels. *Proc. Cirrus Topical Meeting, Baltimore, 1998*, 17–19.

Romanov, S. G., Johnson, N. P., and De La Rue, R. M. (1997). Progress in three-dimensional photonic bandgap structures at visible wavelengths. *Opt. Photon. News* **8**, 35–36.

Roos, D. v. d. S., and Carte, A. E. (1973). The falling behavior of oblate and spiky hailstones. *J. Rech. Atmos.* **7**, 39–52.

Rosen, J. M., Pinnick, R. G., and Garvey, D. M. (1997a). Nephelometer optical response model for the interpretation of atmospheric aerosol measurements. *Appl. Opt.* **36**, 2642–2649.

Rosen, J. M., Pinnick, R. G., and Garvey, D. M. (1997b). Measurement of extinction-to-backscatter ratio for near-surface aerosols. *J. Geophys. Res.* **102**, 6017–6024.

Rosenkranz, P. W. (1998). Water vapor microwave continuum absorption: A comparison of measurements and models. *Radio Sci.* **33**, 919–928.

Rother, T. (1998). Generalization of the separation of variables method for non-spherical scattering of dielectric objects. *J. Quant. Spectrosc. Radiat. Transfer* **60**, 335–353.

Rother, T., and Schmidt, K. (1996). The discretized Mie-formalism for plane wave scattering on dielectric objects with non-separable geometries. *J. Quant. Spectrosc. Radiat. Transfer* **55**, 615–625.

Ruf, C. S., Aydin, K., Mathur, S., and Bobak, J. P. (1996). 35-GHz dual-polarization propagation link for rain-rate estimation. *J. Atmos. Oceanic Technol.* **13**, 419–425.

Ruppin, R. (1990). Electromagnetic scattering from finite dielectric cylinders. *J. Phys. D: Appl. Phys.* **23**, 757–763.

Ruppin, R. (1998). Polariton modes of spheroidal microcrystals. *J. Phys.: Condens. Matter* **10**, 7869–7878.

Russchenberg, H. W. J., and Ligthart, L. P. (1996). Backscattering by and propagation through the melting layer of precipitation: A new polarimetric model. *IEEE Trans. Geosci. Remote Sens.* **34**, 3–14.

Ryde, N. P., and Matijević, E. (1994). Color effects of uniform colloidal particles of different morphologies packed into films. *Appl. Opt.* **33**, 7275–7281.

Ryzhkov, A. V., and Zrnic, D. S. (1994). Precipitation observed in Oklahoma mesoscale convective systems with a polarimetric radar. *J. Appl. Meteorol.* **33**, 455–464.

Ryzhkov, A. V., and Zrnic, D. S. (1995). Comparison of dual-polarization radar estimators of rain. *J. Atmos. Oceanic Technol.* **12**, 249–256.

Ryzhkov, A. V., and Zrnic, D. S. (1996). Rain in shallow and deep convection measured with a polarimetric radar. *J. Atmos. Sci.* **53**, 2989–2995.

Ryzhkov, A. V., and Zrnic, D. S. (1998). Beamwidth effects on the differential phase measurements of rain. *J. Atmos. Oceanic Technol.* **15**, 624–634.

Ryzhkov, A. V., Zrnic, D. S., and Gordon, B. A. (1998). Polarimetric method for ice water content determination. *J. Appl. Meteorol.* **37**, 125–134.

Sachidananda, M., and Zrnic, D. S. (1986). Differential propagation phase shift and rainfall rate estimation. *Radio Sci.* **21**, 235–247.

Sachidananda, M., and Zrnic, D. S. (1987). Rainfall rate estimates from differential polarization measurements. *J. Atmos. Oceanic Technol.* **4**, 588–598.

Sadiku, M. N. O. (1985). Refractive index of snow at microwave frequencies. *Appl. Opt.* **24**, 572–575.

Şahin, A., and Miller, E. L. (1998). Recursive T-matrix methods for scattering from multiple dielectric and metallic objects. *IEEE Trans. Antennas Propag.* **46**, 672–678.

Salzman, G. C., Crowell, J. M., Martin, J. C., *et al.* (1975). Cell classification by laser light scattering: Identification and separation of unstained leukocytes. *Acta Cytol.* **19**, 374–377.

Sasse, C., and Peltoniemi, J. I. (1995). Angular scattering measurements and calculations of rough spherically shaped carbon particles. *SPIE Proc.* **2541**, 131–139.

Sasse, C., Muinonen, K., Piironen, J., and Dröse, G. (1996). Albedo measurements on single particles. *J. Quant. Spectrosc. Radiat. Transfer* **55**, 673–681.

Sassen, K. (1974). Depolarization of laser light backscattered by artificial clouds. *J. Appl. Meteorol.* **13**, 923–933.

Sassen, K. (1976). Polarization diversity lidar returns from virga and precipitation: Anomalies and the bright band analogy. *J. Appl. Meteorol.* **15**, 292–300.

Sassen, K. (1977a). Ice crystal habit discrimination with the optical backscatter depolarization technique. *J. Appl. Meteorol.* **16**, 425–431.

Sassen, K. (1977b). Lidar observations of high plains thunderstorm precipitation. *J. Atmos. Sci.* **34**, 1444–1457.

Sassen, K. (1977c). Optical backscattering from near-spherical water, ice, and mixed phase drops. *Appl. Opt.* **16**, 1332–1341.

Sassen, K. (1980). Remote sensing of planar ice crystal fall attitudes. *J. Meteor. Soc. Jpn.* **58**, 422–429.

Sassen, K. (1987a). Polarization and Brewster angle properties of light pillars. *J. Opt. Soc. Am. A* **4**, 570–580.

Sassen, K. (1987b). Ice cloud content from radar reflectivity. *J. Clim. Appl. Meteorol.* **26**, 1050–1053.

Sassen, K. (1991). The polarization lidar technique for cloud research: A review and current assessment. *Bull. Am. Meteorol. Soc.* **72**, 1848–1866.

Sassen, K. (1994). Advances in polarization diversity lidar for cloud remote sensing. *Proc. IEEE* **82**, 1907–1914.

Sassen, K. (1995). Lidar cloud research. *Rev. Laser Eng.* **23**, 148–153.

Sassen, K., and Chen, T. (1995). The lidar dark band: An oddity of the radar bright band analogy. *Geophys. Res. Lett.* **22**, 3505–3508.

Sassen, K., and Horel, J. D. (1990). Polarization lidar and synoptic analyses of an unusual volcanic aerosol cloud. *J. Atmos. Sci.* **47**, 2881–2889.

Sassen, K., and Hsueh, C. (1998). Contrail properties derived from high-resolution polarization lidar studies during SUCCESS. *Geophys. Res. Lett.* **25**, 1165–1168.

Sassen, K., and Liao, L. (1996). Estimation of cloud content by W-band radar. *J. Appl. Meteorol.* **35**, 932–938.

Sassen, K., and Liou, K.-N. (1979a). Scattering of polarized laser light by water droplet, mixed-phase and ice crystal clouds. I. Angular scattering patterns. *J. Atmos. Sci.* **36**, 838–851.

Sassen, K., and Liou, K.-N. (1979b). Scattering of polarized laser light by water droplet, mixed-phase and ice crystal clouds. II. Angular depolarizing and multiple-scattering behavior. *J. Atmos. Sci.* **36**, 852–861.

Sassen, K., and Zhao, H. (1995). Lidar multiple scattering in water droplet clouds: Toward an improved treatment. *Opt. Rev.* **2**, 394–400.

Sassen, K., Liou, K. N., Kinne, S., and Griffin, M. (1985). Highly super-cooled cirrus cloud water: Confirmation and climatic implications. *Science* **227**, 411–413.

Sassen, K., Zhao, H., and Yu, B.-K. (1989a). Backscatter laser depolarization studies of simulated stratospheric aerosols: Crystallized sulfuric acid droplets. *Appl. Opt.* **28**, 3024–3029.

Sassen, K., Starr, D. O'C., and Uttal, T. (1989b). Mesoscale and microscale structure of cirrus clouds: Three case studies. *J. Atmos. Sci.* **46**, 371–396.

Sassen, K., Grund, C. J., Spinhirne, J. D., *et al.* (1990a). The 27–28 October 1986 FIRE IFO cirrus case study: A five lidar overview of cloud structure and evolution. *Mon. Weather Rev.* **118**, 2288–2311.

Sassen, K., Huggins, A. W., Long, A. B., *et al.* (1990b). Investigations of a winter mountain storm in Utah. II. Mesoscale structure, supercooled liquid cloud development, and precipitation processes. *J. Atmos. Sci.* **47**, 1323–1350.

Sassen, K., Zhao, H., and Dodd, G. C. (1992). Simulated polarization diversity lidar returns from water and precipitating mixed phase clouds. *Appl. Opt.* **31**, 2914–2923.

Sassen, K., Peter, T., Luo, B. P., and Crutzin, P. J. (1994a). Volcanic Bishop's ring: Evidence for a sulfuric acid tetrahydrate particle aureole. *Appl. Opt.* **33**, 4602–4606.

Sassen, K., Knight, N. C., Takano, Y., and Heymsfield, A. J. (1994b). Effects of ice crystal structure on halo formation: Cirrus cloud experimental and ray-tracing modeling studies. *Appl. Opt.* **33**, 4590–4601.

Sassen, K., Starr, D. O'C., Mace, G. G., *et al.* (1995). The 5–6 December 1991 FIRE II jet stream cirrus case study: Possible influence of volcanic aerosols. *J. Atmos. Sci.* **52**, 97–123.

Sassen, K., Mace, G. G., Hallett, J., and Poellot, M. R. (1998). Corona-producing ice clouds: A case study of a cold cirrus layer. *Appl. Opt.* **37**, 1477–1585.

Saunders, M. (1980). The effect of a distorting electric field on the backscattered radiance from a single water drop. *In* "Light Scattering by Irregularly Shaped Particles" (D. W. Schuerman, Ed.), pp. 237–242. Plenum, New York.

Saunders, C., Rimmer, J., Jonas, P., Arathoon, J., and Liu, C. (1998). Preliminary laboratory studies of the optical scattering properties of the crystal clouds. *Ann. Geophys.* **16**, 618–627.

Schaefer, R. W. (1980). Calculations of the light scattered by randomly oriented ensembles of spheroids of size comparable to the wavelength. Ph.D. Dissertation, State Univ. of New York, Albany, NY.

Schelkunoff, S. A. (1943). "Electromagnetic Waves." Van Nostrand, New York.

Schertler, D. J., and George, N. (1994). Backscattering cross section of a roughened sphere. *J. Opt. Soc. Am. A* **11**, 2286–2297.

Schiffer, R. (1985). The effect of surface roughness on the spectral reflectance of dielectric particles. Application to the zodiacal light. *Astron. Astrophys.* **148**, 347–358.

Schiffer, R. (1989). Light scattering by perfectly conducting statistically irregular particles. *J. Opt. Soc. Am. A* **6**, 385–402.

Schiffer, R. (1990). Perturbation approach for light scattering by an ensemble of irregular particles of arbitrary material. *Appl. Opt.* **29**, 1536–1550.

Schiffer, R., and Thielheim, K. O. (1979). Light scattering by dielectric needles and disks. *J. Appl. Phys.* **50**, 2476–2483.

Schiffer, R., and Thielheim, K. O. (1982a). Light reflection from randomly oriented convex particles with rough surface. *J. Appl. Phys.* **53**, 2825–2830.

Schiffer, R., and Thielheim, K. O. (1982b). A scattering model for the zodiacal light particles. *Astron. Astrophys.* **116**, 1–9.

Schmehl, R., Nebeker, B. M., and Hirleman, E. H. (1997). Discrete-dipole approximation for scattering by features on surfaces by means of a two-dimensional fast Fourier transform technique. *J. Opt. Soc. Am. A* **14**, 3026–3036.

Schmid, W., and Waldvogel, A. (1986). Radar hail profiles in Switzerland. *J. Clim. Appl. Meteorol.* **25**, 1002–1011.

Schmidt, K., Rother, T., and Wauer, J. (1998). The equivalence of the extended boundary condition and the continuity conditions for solving electromagnetic scattering problems. *Opt. Commun.* **150**, 1–4.

Schmitt, J. M., and Kumar, G. (1996). Turbulent nature of refractive-index variations in biological tissue. *Opt. Lett.* **21**, 1310–1312.

Schneider, J. B., and Peden, I. C. (1988). Differential cross section of a dielectric ellipsoid by the *T*-matrix extended boundary condition method. *IEEE Trans. Antennas Propag.* **36**, 1317–1321.

Schneider, T. L., and Stephens, G. L. (1995). Theoretical aspects of modeling backscattering by cirrus ice particles at millimeter wavelengths. *J. Atmos. Sci.* **52**, 4367–4385.

Schneider, J., Brew, J., and Peden, I. C. (1991). Electromagnetic detection of buried dielectric targets. *IEEE Trans. Geosci. Remote Sens.* **29**, 555–562.

Schols, J., Haferman, J., Weinman, J., *et al.* (1997). Polarized microwave brightness temperature model of melting hydrometeors. *Prepr. Vol., Conf. Atmos. Rad., 9th, Long Beach, CA, 1997*, 270–273.

Schotland, R. M., Sassen, K., and Stone, R. J. (1971). Observations by lidar of linear depolarization ratios by hydrometeors. *J. Appl. Meteorol.* **10**, 1011–1017.

Schroth, A., Hornbostel, A., Kutuza, B. G., and Zagorin, G. K. (1998). Utilization of the first three Stokes parameters for the determination of precipitation characteristics. *IGARSS 1998*, Seattle, WA, Vol. I, 141–143.

Schuerman, D. W., Ed. (1980a). "Light Scattering by Irregularly Shaped Particles." Plenum, New York.

Schuerman, D. W. (1980b). The microwave analog facility at SUNYA: Capabilities and current programs. *In* "Light Scattering by Irregularly Shaped Particles" (D. W. Schuerman, Ed.), pp. 227–232. Plenum, New York.

Schuerman, D. W., Wang, R. T., Gustafson, B. Å. S., and Schaefer, R.W. (1981). Systematic studies of light scattering. 1. Particle shape. *Appl. Opt.* **20**, 4039–4050. [Errata: **21**, 369 (1982).]

Schultz, F. V. (1950). Scattering by a prolate spheroid. Report VMM-42. Willow Run Research Center, Univ. of Michigan, Ann Arbor, MI.

Schulz, F. M., Stamnes, K., and Stamnes, J. J. (1998a). Scattering of electromagnetic waves by spheroidal particles: A novel approach exploiting the *T* matrix computed in spheroidal coordinates. *Appl. Opt.* **37**, 7875–7896.

Schulz, F. M., Stamnes, K., and Stamnes, J. J. (1998b). Modeling the radiative transfer properties of media containing particles of moderately and highly elongated shape. *Geophys. Res. Lett.* **25**, 4481–4484.

Schulz, F. M., Stamnes, K., and Stamnes, J. J. (1999a). Point group symmetries in electromagnetic scattering. *J. Opt. Soc. Am. A* **16**, 853–865.

Schulz, F. M., Stamnes, K., and Weng, F. (1999b). VDISORT: An improved and generalized discrete ordinate method for polarized (vector) radiative transfer. *J. Quant. Spectrosc. Radiat. Transfer* **61**, 105–122.

Schumann, U. (1994). On the effect of emissions from aircraft engines on the state of the atmosphere. *Ann. Geophys.* **12**, 365–384.

Sebak, A. R., and Sinha, B. P. (1992). Scattering by a conducting spheroidal object with dielectric coating at axial incidence. *IEEE Trans. Antennas Propag.* **40**, 268–274.

Seeliger, H. (1887). Zur Theorie Baleuchtung der Grossen Planeten Inbesondere des Saturn. *Abhandl. Bayer. Akad. Wiss. Math. Naturw. Kl. II* **16**, 405–516.

Sekelsky, S. M., and McIntosh, R. E. (1996). Cloud observations with a polarimetric 33 GHz and 95 GHz radar. *Meteorol. Atmos. Phys.* **59**, 123–140.

Seliga, T. A. (1980). Dual polarization radar. *Nature (London)* **285**, 191–192.

Seliga, T. A., and Bringi, V. N. (1976). Potential use of radar differential reflectivity measurements at orthogonal polarizations for measuring precipitation. *J. Appl. Meteorol.* **15**, 69–76.

Seliga, T. A., and Bringi, V. N. (1978). Differential reflectivity and differential phase shift: Applications in radar meteorology. *Radio Sci.* **13**, 271–275.

Seliga, T. A., Bringi, V. N., and Al-Khatib, H. H. (1979). Differential reflectivity measurements in rain: First experiments. *IEEE Trans. Geosci. Electron.* **17**, 240–244.

Seliga, T. A., Bringi, V. N., and Al-Khatib, H. H. (1981). A preliminary study of comparative measurements of rainfall rate using the differential reflectivity technique and a raingage network. *J. Appl. Meteorol.* **20**, 1362–1368.

Seliga, T. A., Aydin, K., and Bringi, V. N. (1982a). Behavior of the differential reflectivity and circular depolarization ratio radar signals and related propagation effects in rainfall. *Proc. URSI Comm. F Open Symp., Bournemouth, U.K.*, 35–42.

Seliga, T. A., Aydin, K., Cato, C. P., and Bringi, V. N. (1982b). Use of the radar differential reflectivity technique for observing convective systems. *In* "Cloud Dynamics" (E. M. Agee and T. Asai, Eds.), pp. 285–300. Reidel, Dordrecht.

Seliga, T. A., Aydin, K., Direskeneli, H., and Bringi, V. N. (1983). Possible evidence for strong vertical electric fields in thunderstorms from differential reflectivity measurements. *Prepr. Vol., Conf. Radar Meteorol., 21st, 1983*, 500–502.

Seliga, T. A., Aydin, K., and Bringi, V. N. (1984). Differential reflectivity and circular depolarization ratio radar signals and related drop oscillation and propagation effects in rainfall. *Radio Sci.* **19**, 81–89.

Seliga, T. A., Aydin, K., and Direskeneli, H. (1986). Disdrometer measurements during an intense rainfall event in central Illinois: Implications for differential reflectivity observations. *J. Clim. Appl. Meteorol.* **25**, 835–846.

Senior, T. B. A., and Weil, H. (1977). Electromagnetic scattering and absorption by thin walled dielectric cylinders with application to ice crystals. *Appl. Opt.* **16**, 2979–2985.

Seow, Y.-L., Li, L.-W., Leong, M.-S., *et al.* (1998). An efficient TCS formula for rainfall microwave attenuation: T-matrix approach and 3-D fitting for oblate spheroidal raindrops. *IEEE Trans. Antennas Propag.* **46**, 1176–1181.

Shafai, L., Ed. (1991). Thematic issue on Computational Electromagnetics. *Comput. Phys. Commun.* **68**, 1–498.

Shao, X. M., and Krehbiel, P. R. (1996). The spatial and temporal development of intracloud lightning. *J. Geophys. Res.* **101**, 26,641–26,668.

Shapiro, D. B., Quinby-Hunt, M. S., and Hunt, A. J. (1990). Origin of the induced circular polarization in the light scattering from a dinoflagellate. *Proc. SPIE* **1302**, 281–289.

Shapiro, D. B., Quinby-Hunt, M. S., Hunt, A. J., and Hull, P. G. (1991). Circular-polarization effects in the light scattering from single and suspensions of dinoflagellates. *Proc. SPIE* **1537**, 30–41.

Shapiro, D. B., Hunt, A. J., Quinby-Hunt, M. S., *et al.* (1992). A theoretical and experimental study of polarized light scattering by helices. *Proc. SPIE* **1750**, 56–72.

Shapiro, D. B., Maestre, M. F., McClain, W. M., *et al.* (1994a). Determination of the average orientation of DNA in the octopus sperm *Eledone cirrhossa* through polarized light scattering. *Appl. Opt.* **33**, 5733–5744.

Shapiro, D. B., Hull, P. G., Shi, Y., *et al.* (1994b). Towards a working theory of polarized light scattering from helices. *J. Chem. Phys.* **100**, 146–157.

Shapiro, D. B., Hull, P. G., Hunt, A. J., and Hearst, J. E. (1994c). Calculations of the Mueller scattering matrix for a DNA plectonemic helix. *J. Chem. Phys.* **101**, 4214–4222.

Sheen, D. M., Ali, S. M., Abouzahra, M. D., and Kong, J. A. (1990). Application of the three-dimensional finite-difference time-domain method to the analysis of planar microstrip circuits. *IEEE Trans. Microwave Theory Tech.* **38**, 849–857.

Sheng, X.-Q., Jin, J.-M., Song, J., *et al.* (1998). On the formulation of hybrid finite-element and boundary-integral methods for 3-D scattering. *IEEE Trans. Antennas Propag.* **46**, 303–311.

Shepherd, J. W., and Holt, A. R. (1983). The scattering of electromagnetic radiation from finite dielectric circular cylinders. *J. Phys. A: Math. Gen.* **16**, 651–662.

Sheridan, P. J. (1989a). Analytical electron microscope studies of size-segregated particles collected during AGASP-II, flights 201–203. *J. Atmos. Chem.* **9**, 267–282.

Sheridan, P. J. (1989b). Characterization of size segregated particles collected over Alaska and the Canadian high Arctic, AGASP-II flights 204–206. *Atmos. Environ.* **11**, 2371–2386.

Sheridan, P. J., and Musselman, I. H. (1985). Characterization of aircraft-collected particles present in the Arctic aerosol; Alaskan Arctic, spring 1983. *Atmos. Environ.* **19**, 2159–2166.

Sheridan, P. J., Schnell, R. C., Kah, J. D., *et al.* (1993). Microanalysis of the aerosol collected over South-Central New Mexico during the ALIVE field experiment, May–Dec. 1989. *Atmos. Environ.* **8**, 1169–1183.

Shettle, E. P., and Fenn, R. W. (1979). Models for the aerosols of the lower atmosphere and the effects of humidity variations on their optical properties. Report AFGL-TR-79-0214, Air Force Geophysics Laboratory, Project 7670, Hanscomb AFB, MA.

Shifrin, K. S. (1951). "Scattering of Light in a Turbid Medium." Gostehteorizdat, Moscow. [English translation: *NASA Tech. Transl.* **NASA TT F-477** (1968).]

Shifrin, K. S., and Mikulinsky, I. A. (1987). Ensemble approach to the problem of light scattering by a system of tenuous particles. *Appl. Opt.* **26**, 3012–3017.

Shifrin, K. S., and Zolotov, I. G. (1995). Nonstationary scattering of electromagnetic pulses by spherical particles. *Appl. Opt.* **34**, 552–558.

Shimizu, K. (1983). Modification of the Rayleigh–Debye approximation. *J. Opt. Soc. Am.* **73**, 504–507.

Shlager, K. L., Maloney, J. G., Ray, S. L., and Peterson, A. F. (1993). Relative accuracy of several finite-difference time-domain methods in two and three dimensions. *IEEE Trans. Antennas Propag.* **41**, 1732–1737.

Shurcliff, W. A. (1962). "Polarized Light: Production and Use." Harvard Univ. Press, Cambridge, MA.

Shvalov, A. N., Soini, J. T., Chernyshev, A. V., *et al.* (1999). Light-scattering properties of individual erythrocytes. *Appl. Opt.* **38**, 230–235.

Sid'ko, F. Ya., Lopatin, V. N., and Paramonov, L. E. (1990). "Polarization Characteristics of Suspensions of Biological Particles." Nauka, Novosibirsk (in Russian).

Siegel, K. M., Schultz, F. V., Gere, B. H., and Sleator, F. B. (1956). The theoretical and numerical determination of the radar cross section of a prolate spheroid. *IRE Trans. Antennas Propag.* **4**, 266–275.

Silvester, P. P., and Ferrari, R. L. (1996). "Finite Elements for Electrical Engineers." Cambridge Univ. Press, New York.

Simon, R. (1982). The connection between Mueller and Jones matrices of polarization optics. *Opt. Commun.* **42**, 293–297.

Simon, R. (1987). Mueller matrices and depolarization criteria. *J. Mod. Opt.* **34**, 569–575.

Simpson, J., Kummerow, C., Tao, W.-K., and Adler, R. F. (1996). On the Tropical Rainfall Measuring Mission (TRMM). *Meteorol. Atmos. Phys.* **60**, 19–36.

Singham, S. B. (1988). Coupled dipoles in light scattering by randomly oriented chiral particles. *J. Chem. Phys.* **88**, 1522–1527.

Singham, S. B. (1989). Theoretical factors in modeling polarized light scattering by arbitrary particles. *Appl. Opt.* **28**, 5058–5064.

Singham, S. B., and Bohren, C. F. (1987). Light scattering by an arbitrary particle: A physical reformulation of the coupled dipole method. *Opt. Lett.* **12**, 10–12.

Singham, S. B., and Bohren, C. F. (1988). Light scattering by an arbitrary particle: The scattering-order formulation of the coupled-dipole method. *J. Opt. Soc. Am. A* **5**, 1867–1872.

Singham, S. B., and Bohren, C. F. (1993). Scattering of unpolarized and polarized light by particle aggregates of different size and fractal dimension. *Langmuir* **9**, 1431–1435.

Singham, S. B., and Salzman, G. C. (1986). Evaluation of the scattering matrix of an arbitrary particle using the coupled-dipole approximation. *J. Chem. Phys.* **84**, 2658–2667.

Singham, M. K., Singham, S. B., and Salzman, G. C. (1986). The scattering matrix for randomly oriented particles. *J. Chem. Phys.* **85**, 3807–3815.

Sinha, B. P. (1974). Electromagnetic scattering from conducting prolate spheroids in the resonance region. Ph.D. Thesis, Univ. of Waterloo, Waterloo.

Sinha, B. P., and MacPhie, R. H. (1975). On the computation of the spheroidal vector wave functions of the second kind. *J. Math. Phys.* **16**, 2378–2381.

Sinha, B. P., and MacPhie, R. H. (1977). Electromagnetic scattering by prolate spheroids for plane waves with arbitrary polarization and angle of incidence. *Radio Sci.* **12**, 171–184.

Sinha, B. P., and MacPhie, R. H. (1983). Electromagnetic plane wave scattering by a system of two parallel conducting prolate spheroids. *IEEE Trans. Antennas Propag.* **31**, 294–304.

Sinha, B. P., MacPhie, R. H., and Prasad, T. (1973). Accurate computation of eigenvalues for prolate spheroidal wave functions. *IEEE Trans. Antennas Propag.* **21**, 406–407.

Skaropoulos, N. C., Ioannidou, M. P., and Chrissoulidis, D. P. (1994). Indirect mode-matching solution to scattering from a dielectric sphere with an eccentric inclusion. *J. Opt. Soc. Am. A* **11**, 1859–1866.

Skaropoulos, N. C., Ioannidou, M. P., and Chrissoulidis, D. P. (1996). Induced EM field in a layered eccentric spheres model of the head: Plane-wave and localized source exposure. *IEEE Trans. Microwave Theory Tech.* **44**, 1963–1973.

Skolnik, M. (1990). "Radar Handbook." McGraw–Hill, New York.

Sloot, P. M. A., and Figdor, C. G. (1986). Elastic light scattering from nucleated blood cells: Rapid numerical analysis. *Appl. Opt.* **25**, 3559–3565.

Sloot, P. M. A., Hoekstra, A. G., and Figdor, C. G. (1988a). Osmotic response of lymphocytes measured by means of forward light scattering: Theoretical considerations. *Cytometry* **9**, 636–641.

Sloot, P. M. A., van der Donk, E. H. M., and Figdor, C. G. (1988b). Computer assisted centrifugal elutriation. II. Multiparametric statistical analysis. *Comput. Meth. Prog. Biomed.* **27**, 37–46.

Sloot, P. M. A., Hoekstra, A. G., van der Liet, H., and Figdor, C. G. (1989). Scattering matrix elements of biological particles measured in a flow through system: Theory and practice. *Appl. Opt.* **28**, 1752–1762.

Smith, P. L. (1984). Equivalent radar reflectivity factors for snow and ice particles. *J. Clim. Appl. Meteorol.* **23**, 1258–1260.

Smith, E. A., and Mugnai, A. (1989). Radiative transfer to space through a precipitating cloud at multiple microwave frequencies. III. Influence of large ice particles. *J. Meteorol. Soc. Jpn.* **67**, 739–755.

Smith, W. L., Ackerman, S., Revercomb, H., *et al.* (1998). Infrared spectral absorption of nearly invisible cirrus clouds. *Geophys. Res. Lett.* **25**, 1137–1140.

Sobolev, V. V. (1974). "Light Scattering in Planetary Atmospheres." Pergamon, Oxford.

Soini, J. T., Chernyshev, A. V., Shvalov, A. N., and Maltsev, V. P. (1997). Measurement of scattering patterns from individual non-spherical particles using scanning flow cytometer. *Abstr. Vol., Workshop on Light Scattering by Non-Spherical Particles, Univ. of Helsinki, 1997*, 37.

Soini, J. T., Chernyshev, A. V., Hänninen, P. E., *et al.* (1998). A new design of the flow cuvette and optical set-up for the scanning flow cytometer. *Cytometry* **31**, 78–84.

Sokolik, I., Toon, O. B., and Bergstrom, R. W. (1997). Direct radiative forcing by airborne mineral aerosols: Regional heating or cooling. *AAAR'97 Abstr.*, *1997*, 330.

Solomon, J. E. (1981). Polarization mapping. *Appl. Opt.* **20**, 1537–1544.

Sommerfeld, A. (1952). "Electrodynamics." Academic Press, New York.

Somsikov, V. V. (1996). Optical properties of two-layered spheroidal dust grains. *Astron. Lett.* **22**, 696–703.

Sorensen, C. M., and Roberts, G. C. (1997). The prefactor of fractal aggregates. *J. Colloid Interface Sci.* **186**, 447–452.

Sorensen, C. M., Cai, J., and Lu, N. (1992). Light-scattering measurements of monomer size, monomers per aggregate, and fractal dimension for soot aggregates in flames. *Appl. Opt.* **31**, 6547–6557.

Soyer, M.-O., and Haapala, O. K. (1974). Structural changes of dinoflagellate chromosomes by pronase and ribonuclease. *Chromosoma (Berlin)* **47**, 179–192.

Spencer, R. W., Goodman, H. M., and Hood, R. E. (1989). Precipitation retrieval over land and ocean with the SSM/I: Identification and characteristics of the scattering signal. *J. Atmos. Oceanic Technol.* **6**, 254–273.

Spinhirne, J. D. (1993). Micro pulse lidar. *IEEE Trans. Geosci. Remote Sens.* **31**, 48–55.

Spinhirne, J. D., Hansen, M. Z., and Simpson, J. (1983). The structure and phase of cloud tops as observed by polarization lidar. *J. Clim. Appl. Meteorol.* **22**, 1319–1331.

Spinhirne, J. D., Hart, W. D., and Hlavka, D. L. (1996). Cirrus infrared parameters and shortwave reflectance relations from observations. *J. Atmos. Sci.* **53**, 1438–1458.

Spinrad, R. W., and Brown, J. (1993). Effects of asphericity on single-particle polarized light scattering. *Appl. Opt.* **32**, 6151–6158.

Stachnik, W. J. (1988). Simulation of marine light fields for underwater research. *Proc. SPIE* **925**, 415–423.

Stackhouse, P. W., Jr., and Stephens, G. L. (1991). A theoretical and observational study of the radiative properties of cirrus: Results from FIRE 1986. *J. Atmos. Sci.* **48**, 2044–2056.

Stamatakos, G. S., and Uzunoglu, N. K. (1997). An integral equation solution to the scattering of electromagnetic radiation by a linear chain of interacting triaxial dielectric ellipsoids. The case of a red blood cell rouleau. *J. Electromagn. Waves Appl.* **11**, 949–980.

Stamatakos, G. S., Yova, D., and Uzunoglu, N. K. (1997). Integral equation model of light scattering by an oriented monodisperse system of triaxial dielectric ellipsoids: Application in ectacytometry. *Appl. Opt.* **36**, 6503–6512.

Stammes, P. (1989). Light scattering properties of aerosols and the radiation inside a planetary atmosphere. Ph.D. Dissertation, Free Univ., Amsterdam.

Steich, D., Luebbers, R., and Kunz, K. (1993). Absorbing boundary condition convergence comparisons. *IEEE AP-S Int. Symp. Digest, 1993*, 6–9.

Stein, S. (1961). Addition theorems for spherical wave functions. *Quart. Appl. Math.* **19**, 15–24.

Steinhorn, I., and Zrnic, D. S. (1988). Potential uses of the differential propagation phase constant to estimate raindrop and hailstone size distributions. *IEEE Trans. Geosci. Remote Sens.* **26**, 639–648.

Stephens, G. L. (1996). How much solar radiation do clouds absorb. *Science* **271**, 1131–1133.

Stephens, G. L., and Tsay, S.-C. (1990). On the cloud absorption anomaly. *Quart. J. R. Meteorol. Soc.* **116**, 671–704.

Stevenson, A. F. (1953). Solution of electromagnetic scattering problems as power series in the ratio (dimension of scatterer/wavelength). *J. Appl. Phys.* **24**, 1134–1142.

Stevenson, A. F. (1968). Light scattering by spheroidal particles oriented by streaming. *J. Chem. Phys.* **49**, 4545–4550.

Stokes, G. G. (1852). On the composition and resolution of streams of polarized light from different sources. *Trans. Cambridge Philos. Soc.* **9**, 399–423.

Stokes, G. G. (1966a). "Mathematical and Physical Papers, Vol. 2." Johnson Reprint Corporation.

Stokes, G. G. (1966b). "Mathematical and Physical Papers, Vol. 3." Johnson Reprint Corporation.

Stokes, G. G. (1966c). "Mathematical and Physical Papers, Vol. 4." Johnson Reprint Corporation.

Stoyan, D., and Stoyan, H. (1994). "Fractals, Random Shapes and Point Fields: Methods of Geometrical Statistics." Wiley, New York.

Straka, J. M., Zrnic, D. S., Ryzhkov, A. V., and Askelson, M. V. (1999). Bulk hydrometeor classification and quantification using multiparameter data: Synthesis of relations. *J. Appl. Meteorol.*, submitted.

Stramski, D., Morel, A., and Bricaud, A. (1988). Modelling the light attenuation and scattering by spherical phytoplanktonic cells: A retrieval of the bulk refractive index. *Appl. Opt.* **27**, 3954–3956.

Stratton, J. A. (1941). "Electromagnetic Theory." McGraw–Hill, New York.

Stratton, J. A., and Chu, L. J. (1939). Diffraction theory of electromagnetic waves. *Phys. Rev.* **56**, 99–107.

Streekstra, G. J., Hoekstra, A. G., Nijhof, E. J., and Heethaar, R. M. (1993). Light scattering by red blood cells in ektacytometry: Fraunhofer versus anomalous diffraction. *Appl. Opt.* **32**, 2266–2272.

Streekstra, G. J., Hoekstra, A. G., and Heethaar, R. M. (1994). Anomalous diffraction by arbitrarily oriented ellipsoids: Applications in ektacytometry. *Appl. Opt.* **33**, 7288–7296.

Ström, S. (1975). On the integral equations for electromagnetic scattering. *Am. J. Phys.* **43**, 1060–1069.

Ström, S., and Zheng, W. (1987). Basic features of the null field method for dielectric scatterers. *Radio Sci.* **22**, 1273–1281.

Ström, S., and Zheng, W. (1988). The null field approach to electromagnetic scattering from composite objects. *IEEE Trans. Antennas Propag.* **36**, 376–382.

Stroud, D., and Pan, F. P. (1978). Self-consistent approach to electromagnetic wave propagation in composite media: Application to model granular metals. *Phys. Rev. B* **17**, 1602–1610.

Struik, D. J. (1961). "Lectures on Classical Differential Geometry." Dover, New York.

Sturniolo, O., Mugnai, A., and Prodi, F. (1995). A numerical sensitivity study on the backscattering at 35.8 GHz from precipitation-sized hydrometeors. *Radio Sci.* **30**, 903–919.

Su, C.-C. (1989). Electromagnetic scattering by a dielectric body with arbitrary inhomogeneity and anisotropy. *IEEE Trans. Antennas Propag.* **37**, 384–389.

Sullivan, D. M., Borup, D. T., and Gandhi, O. P. (1987). Use of the finite-difference time-domain method in calculating EM absorption in human tissues. *IEEE Trans. Biomed. Eng.* **34**, 148–157.

Swarner, W. G., and Peters, L., Jr. (1963). Radar cross sections of dielectric or plasma coated conducting spheres and circular cylinders. *IEEE Trans. Antennas Propag.* **11**, 558–569.

Taflove, A. (1980). Application of the finite-difference time-domain method to sinusoidal steady-state electromagnetic-penetration problems. *IEEE Trans. Electromagn. Compat.* **22**, 191–202.

Taflove, A. (1995). "Computational Electrodynamics: The Finite-Difference Time-Domain Method." Artech House, Boston.

Taflove, A., Ed. (1998). "Advances in Computational Electrodynamics: The Finite-Difference Time-Domain Method." Artech House, Boston.

Taflove, A., and Brodwin, M. E. (1975). Numerical solution of steady-state electromagnetic scattering problems using the time-dependent Maxwell's equations. *IEEE Trans. Microwave Theory Tech.* **23**, 623–630.

Taflove, A., and Umashankar, K. R. (1990). The finite-difference time-domain method for numerical modeling of electromagnetic wave interactions with arbitrary structures. *In* "Finite Element and Finite Difference Methods in Electromagnetic Scattering" (M. A. Morgan, Ed.), pp. 287–333. Elsevier, New York.

Tai, C. T. (1971). "Dyadic Green's Functions in Electromagnetic Theory." International Textbook Company, Scranton, PA.

Takano, Y. (1987). Multiple scattering of polarized light in cirrus clouds. Ph.D. Thesis, Univ. of Utah, Salt Lake City, UT.

Takano, Y., and Asano, S. (1983). Fraunhofer diffraction by ice crystals suspended in the atmosphere. *J. Meteorol. Soc. Jpn.* **61**, 289–300.

Takano, Y., and Jayaweera, K. (1985). Scattering phase matrix for hexagonal ice crystals computed from ray optics. *Appl. Opt.* **24**, 3254–3263.

Takano, Y., and Liou, K. N. (1989a). Solar radiative transfer in cirrus clouds. I. Single-scattering and optical properties of hexagonal ice crystals. *J. Atmos. Sci.* **46**, 3–19.

Takano, Y., and Liou, K. N. (1989b). Radiative transfer in cirrus clouds. II. Theory and computation of multiple scattering in an anisotropic medium. *J. Atmos. Sci.* **46**, 20–36.

Takano, Y., and Liou, K. N. (1995). Solar radiative transfer in cirrus clouds. III. Light scattering by irregular ice crystals. *J. Atmos. Sci.* **52**, 818–837.

Tang, C., and Aydin, K. (1995). Scattering from ice crystals at 94 and 220 GHz millimeter wave frequencies. *IEEE Trans. Geosci. Remote Sens.* **33**, 93–99.

Taniguchi, H., Nishiya, M., Tanosaki, S., and Inaba, H. (1996). Lasing behavior in a liquid spherical dye laser containing highly scattering nanoparticles. *Opt. Lett.* **21**, 263–265.

Temperton, C. (1992). A generalized prime factor FFT algorithm for any $N = 2^p 3^q 5^r$. *SIAM J. Sci. Statist. Comput.* **13**, 676–686.

Terstappen, L. W. M. M., de Grooth, B. G., ten Napel, C. H. H., *et al.* (1986). Discrimination of human cytotoxic lymphocytes from regulatory and B-lymphocytes by orthogonal light scattering. *J. Immunol. Meth.* **95**, 211–216.

Terstappen, L. W. M. M., de Grooth, B. G., Visscher, K., *et al.* (1988a). Four-parameter white blood cell differential counting based on light scattering measurements. *Cytometry* **9**, 39–43.

Terstappen, L. W. M. M., de Grooth, B. G., van Berkel, W., *et al.* (1988b). Abnormal distribution of CD8 subpopulation in B-chronic lymphocytic leukemia identified by flow cytometry. *Leuk. Res.* **12**, 551–557.

Terstappen, L. W. M. M., de Grooth, B. G., van Berkel, W., *et al.* (1988c). Flow cytometric characterization of chronic lymphocytic leukemias using orthogonal light scattering and quantitative immunofluorescence. *Blut* **56**, 201–208.

Terstappen, L. W. M. M., de Grooth, B. G., van Berkel, W., *et al.* (1988d). Application of orthogonal light scattering for routine screening of lymphocyte samples. *Cytometry* **9**, 220–225.

Terstappen, L. W. M. M., Mickaels, R. A., Dost, R., and Loken, M. R. (1990). Increased light scattering resolution facilitates multidimensional flow cytometric analysis. *Cytometry* **11**, 506–512.

Testud, J., Amayenc, P., Dou, X., and Tani, T. (1996). Tests of rain profiling algorithms for a spaceborne radar using raincell models and real data precipitation fields. *J. Atmos. Oceanic Technol.* **13**, 426–453.

Thomas-Osip, J. E. (2000). Ph.D. Thesis, Univ. of Florida, Gainesville, FL.

Thompson, R. C. (1978). An electro-optic light scattering photometric polarimeter. Ph.D. Dissertation, Texas A&M Univ., College Station, TX.

Thompson, R. C., Bottiger, J. R., and Fry, E. S. (1978). Scattering matrix measurements of oceanic hydrosols. *Proc. SPIE* **160**, 43–48.

Thompson, R. C., Bottiger, J. R., and Fry, E. S. (1980). Measurement of polarized light interactions via the Mueller matrix. *Appl. Opt.* **19**, 1323–1332. [Errata: **19**, 2657 (1980).]

Tishkovets, V. P., and Litvinov, P. V. (1996). Coefficients of light extinction by randomly oriented clusters of spherical particles in the double scattering approximation. *Opt. Spectrosc.* **81**, 288–291.

Tokay, A., and Beard, K. V. (1993). The influence of raindrop oscillations on dual-polarization radar measurements of precipitation. *Prepr. Vol., Conf. Radar Meteorol., 26th, Norman, OK, 1993*, 100–102.

Tokay, A., and Short, D. A. (1996). Evidence from tropical raindrop spectra of the origin of rain from stratiform versus convective clouds. *J. Appl. Meteorol.* **35**, 355–371.

Toon, O. B., and Ackerman, T. P. (1981). Algorithms for the calculation of scattering by stratified spheres. *Appl. Opt.* **20**, 3657–3660.

Toon, O. B., and Pollack, J. B. (1976). A global average model of atmospheric aerosols for radiative transfer calculations. *J. Appl. Meteorol.* **15**, 225–246.

Toon, O. B., Pollack, J. B., and Khare, B. N. (1976). The optical constants of several atmospheric aerosol species: Ammonium sulfate, aluminum oxide, and sodium chloride. *J. Geophys. Res.* **81**, 5733–5748.

Toon, O. B., Browell, E. V., Kinne, S., and Jordan, J. (1990). An analysis of lidar observations of polar stratospheric clouds. *Geophys. Res. Lett.* **17**, 393–396.

Torlaschi, E., and Holt, A. R. (1993). Separation of propagation and backscattering effects in rain for circular polarization diversity S-band radars. *J. Atmos. Oceanic Technol.* **10**, 465–477.

Tosun, H. (1994). Boundary condition transfer method for the solution of electromagnetic scattering by rotationally symmetric penetrable bodies. *Radio Sci.* **29**, 723–738.

Trautman, T., Luo, B. P., Tsias, A., *et al.* (1998). Application of light scattering theory to lidar measurements of polar stratospheric clouds. *Prepr. Vol., Conf. Light Scat. Nonspher. Part.: Theory Meas. Appl., New York, 1998*, 112–114.

Trinks, W. (1935). Zur Vielfachstreuung an kleinen Kugeln. *Ann. Phys. Dtsch.* **22**, 561–590.

Tsang, L. (1991). Polarimetric passive microwave remote sensing of random discrete scatterers and rough surfaces. *J. Electromagn. Waves Appl.* **5**, 41–57.

Tsang, L., Kong, J. A., and Shin, R. T. (1984). Radiative transfer theory for active remote sensing of a layer of nonspherical particles. *Radio Sci.* **19**, 629–642.

Tsang, L., Kong, J. A., and Shin, R. T. (1985). "Theory of Microwave Remote Sensing." Wiley, New York.

Tsang, L., Mandt, C. E., and Ding, K. H. (1991). Monte Carlo simulations of the extinction rate of dense media with randomly distributed dielectric spheres based on solution of Maxwell's equations. *Opt. Lett.* **17**, 314–316.

Tsang, L., Ding, K. H., Shih, S. E., and Kong, J. A. (1998). Scattering of electromagnetic waves from dense distributions of spheroidal particles based on Monte Carlo simulations. *J. Opt. Soc. Am. A* **15**, 2660–2669.

Tseng, Y. C., and Fung, A. K. (1994). T-matrix approach to multiple scattering of EM waves from N-spheres. *J. Electromagn. Waves Appl.* **8**, 61–84.

Tsias, A., Carslaw, K. S., Luo, B. P., *et al.* (1998). Combining T-matrix calculations with a microphysical model to unravel polar stratospheric cloud evolution. *Prepr. Vol., Conf. Light Scat. Nonspher. Part.: Theory Meas. Appl., New York, 1998*, 70–72.

Tuchin, V. V. (1998). Coherence-domain methods in tissue and cell optics. *Laser Phys.* **8**, 1–43.

Tuminello, P. S., Arakawa, E. T., Khare, B. N., *et al.* (1997). Optical properties of Bacillus subtilis spores from 0.2 to 2.5 mm. *Appl. Opt.* **36**, 2818–2824.

Turk, J., and Vivekanandan, J. (1995). Effects of hydrometeor shape and orientation upon passive microwave brightness temperature measurements. *In* "Microwave Radiometry and Remote Sensing of the Environment" (D. Solomini, Ed.), pp. 187–196. VSP, Zeist, The Netherlands.

Twersky, V. (1952). Multiple scattering of radiation by an arbitrary configuration of parallel cylinders. *J. Acoust. Soc. Am.* **24**, 42–46.

Tyler, J. E. (1968). The Secchi disc. *Limnol. Oceanogr.* **13**, 1–6.

Ulaby, F. T., and Elachi, C. (1990). "Radar Polarimetry for Geoscience Applications." Artech House, Boston.

Ulaby, F. T., Moore, R. K., and Fung, A. K. (1981). "Microwave Remote Sensing: Active and Passive," Vol. 1. Artech House, Boston.

Ulaby, F. T., Moore, R. K., and Fung, A. K. (1986). "Microwave Remote Sensing: Active and Passive," Vol. 3. Artech House, Boston.

Ulanowski, Z., Wang, Z., Kaye, P. H., and Ludlow, I. K. (1998). Application of neural networks to the inverse light scattering problem for spheres. *Appl. Opt.* **37**, 4027–4033.

Ulbrich, C. W. (1978). Relationships of equivalent reflectivity factor to the vertical fluxes of mass and kinetic energy of hail. *J. Appl. Meteorol.* **17**, 1803–1808.

Ulbrich, C. W. (1983). Natural variations in the analytical form of the raindrop size distribution. *J. Clim. Appl. Meteorol.* **22**, 1764–1775.

Ulbrich, C. W., and Atlas, D. (1982). Hail parameter relations: A comprehensive digest. *J. Appl. Meteorol.* **21**, 22–43.

Umashankar, K., and Taflove, A. (1982). A novel method to analyze electromagnetic scattering of complex objects. *IEEE Trans. Electromagn. Compat.* **24**, 397–405.

Umashankar, K., Taflove, A., and Rao, S. M. (1986). Electromagnetic scattering by arbitrarily shaped three-dimensional homogeneous lossy dielectric objects. *IEEE Trans. Antennas Propag.* **34**, 758–766.

Uno, K., Uozumi, J., and Asakura, T. (1995). Texture analysis of speckles due to random Koch fractals. *Waves Random Media* **5**, 253–266.

Uozumi, J., Kimura, H., and Asakura, T. (1991). Laser diffraction by randomized Koch fractals. *Waves Random Media* **1**, 73–80.

Uthe, E. E. (1978). Remote sensing of aerosol properties using laser radar techniques. *Proc. SPIE* **142**, 67–77.

Uzunoglu, N. K., Alexopoulos, N. G., and Fikioris, J. G. (1978). Scattering from thin and finite dielectric fibers. *J. Opt. Soc. Am.* **68**, 194–197.

Van Bladel, J. (1961). Some remarks on Green's dyadic for infinite space. *IRE Trans. Antennas Propag.* **9**, 563–566.

van Bockstaele, D. R., Berneman, Z. N., and Peetermans, M. E. (1986). Flow cytometric analysis of hairy cell leukemia using right angle light scatter. *Cytometry* **7**, 217–220.

van de Hulst, H. C. (1957). "Light Scattering by Small Particles." Wiley, New York (also Dover, New York, 1981).

van de Hulst, H. C. (1980). "Multiple Light Scattering. Tables, Formulas, and Applications." Academic Press, San Diego.

van der Mee, C. V. M. (1993). An eigenvalue criterion for matrices transforming Stokes parameters. *J. Math. Phys.* **34**, 5072–5088.

van der Mee, C. V. M., and Hovenier, J. W. (1990). Expansion coefficients in polarized light transfer. *Astron. Astrophys.* **228**, 559–568.

van der Mee, C. V. M., and Hovenier, J. W. (1992). Structure of matrices transforming Stokes parameters. *J. Math. Phys.* **33**, 3574–3584.

van de Merwe, W. P., Li, Z.-Z., Bronk, B. V., and Czégé, J. (1997). Polarized light scattering for rapid observation of bacterial size changes. *Biophys. J.* **73**, 500–506.

van der Vorst, H. A. (1992). Bi-CGSTAB: A fast and smoothly converging variant of Bi-CG for the solution of nonsymmetric linear systems. *SIAM J. Sci. Statist. Comput.* **13**, 631–644.

van Dilla, M. A., Dean, P. N., Laerum, O. D., and Melamed, M. R. (1985). "Flowcytometry: Instrumentation and Data Analysis." Academic Press, San Diego.

Vanmarcke, E. (1983). "Random Fields: Analysis and Synthesis." MIT Press, Cambridge, MA.

Varadan, V. K. (1980). Multiple scattering of acoustic, electromagnetic and elastic waves. *In* "Acoustic, Electromagnetic and Elastic Wave Scattering—Focus on the T-Matrix Approach" (V. K. Varadan and V. V. Varadan, Eds.), pp. 103–134. Pergamon, New York.

Varadan, V. K., and Varadan, V. V., Eds. (1980). "Acoustic, Electromagnetic and Elastic Wave Scattering—Focus on the T-Matrix Approach." Pergamon, New York.

Varadan, V. K., Bringi, V. N., Varadan, V. V., and Ishimaru, A. (1983). Multiple scattering theory for waves in discrete random media and comparison with experiment. *Radio Sci.* **18**, 321–327.

Varadan, V. V., Lakhtakia, A., and Varadan, V. K. (1988). Comments on recent criticism of the *T*-matrix method. *J. Acoust. Soc. Am.* **84**, 2280–2284.

Varadan, V. V., Lakhtakia, A., and Varadan, V. K. (1989). Scattering by three-dimensional anisotropic scatterers. *IEEE Trans. Antennas Propag.* **37**, 800–802.

Varshalovich, D. A., Moskalev, A. N., and Khersonskii, V. K. (1988). "Quantum Theory of Angular Momentum." World Scientific, Singapore.

Vechinski, D. A., Rao, S. M., and Sarkar, T. K. (1994). Transient scattering from three-dimensional arbitrarily shaped dielectric bodies. *J. Opt. Soc. Am. A* **11**, 1458–1470.

Veretennikov, V. V., Kozlov, V. S., Naats, I. E., and Fadeev, V. Ya. (1979). Optical studies of smoke aerosols: An inversion method and its applications. *Opt. Lett.* **4**, 411–413.

Vernooy, D. W., Ilchenko, V. S., Mabuchi, H., *et al.* (1998). High-*Q* measurements of fused-silica microspheres in the infrared. *Opt. Lett.* **23**, 247–249.

Videen, G., and Bickel, W. S. (1992). Light-scattering Mueller matrix for a rough fiber. *Appl. Opt.* **31**, 3488–3492.

Videen, G., and Ngo, D. (1998). Light scattering multipole solution for a cell. *J. Biomed. Opt.* **3**, 212–220.

Videen, G., Ngo, D., and Chýlek, P. (1995a). Effective-medium predictions of absorption by graphitic carbon in water droplets. *Opt. Lett.* **21**, 1675–1677.

Videen, G., Ngo, D., Chýlek, P., and Pinnick, R. G. (1995b). Light scattering from a sphere with an irregular inclusion. *J. Opt. Soc. Am. A* **12**, 922–928.

Videen, G., Ngo, D., and Hart, M. B. (1996). Light scattering from a pair of conducting, osculating spheres. *Opt. Commun.* **125**, 275–287.

Videen, G., Pellegrino, P., Ngo, D., *et al.* (1997). Qualitative light-scattering angular correlations of conglomerate particles. *Appl. Opt.* **36**, 3532–3537.

Videen, G., Pinnick, R. G., Ngo, D., *et al.* (1998). Asymmetry parameter and aggregate particles. *Appl. Opt.* **37**, 1104–1109.

Vinh, H., Duger, H., and Van Dam, C. P. (1992). Finite-difference methods for computational electromagnetics (CEM). *IEEE AP-S Int. Symp. Digest, 1992,* 1682–1683.

Vivekanandan, J., Bringi, V. N., and Raghavan, R. (1990). Multiparameter radar modeling and observations of melting ice. *J. Atmos. Sci.* **47**, 549–564.

Vivekanandan, J., Adams, W. M., and Bringi, V. N. (1991). Rigorous approach to polarimetric radar modeling of hydrometeor orientation distributions. *J. Appl. Meteorol.* **30**, 1053–1063.

Vivekanandan, J., Raghavan, R., and Bringi, V. N. (1993a). Polarimetric radar modeling of mixtures of precipitation particles. *IEEE Trans. Geosci. Remote Sens.* **31**, 1017–1030.

Vivekanandan, J., Turk, J., and Bringi, V. N. (1993b). Comparisons of precipitation measurements by the advanced microwave precipitation radiometer and multiparameter radar. *IEEE Trans. Geosci. Remote Sens.* **31**, 860–870.

Vivekanandan, J., Bringi, V. N., Hagen, M., and Meischner, P. (1994). Polarimetric radar studies of atmospheric ice particles. *IEEE Trans. Geosci. Remote Sens.* **32**, 1–10.

Vivekanandan, J., Zrnic, D. S., Ellis, S. M., *et al.* (1999). Cloud microphysics retrieval using S-band dual polarization radar measurements. *Bull. Am. Meteorol. Soc.* **80**, 381–388.

Volakis, J. L., Chatterjee, A., and Kempel, L. C. (1998). "Finite Element Method for Electromagnetics." IEEE Press, New York.

Volkovitskiy, O. A., Pavlova, L. N., and Petrushin, A. G. (1980). Scattering of light by ice crystals. *Izv. Acad. Sci. USSR, Atmos. Oceanic Phys.* (Engl. Transl.) **16**, 98–102.

Volten, H., de Haan, J. F., Hovenier, J.W., *et al.* (1998). Laboratory measurements of angular distributions of light scattered by phytoplankton and silt. *Limnol. Oceanogr.* **43**, 1180–1197.

Volten, H., Jalava, J. P., Lumme, K., *et al.* (1999). Laboratory measurements and T-matrix calculations of the scattering matrix of rutile particles in water, *Appl. Opt.*, in press.

von Hoyningen-Huene, W. (1998). Sky radiance observations and their implications for the scattering behavior of atmospheric aerosol. *Prepr. Vol., Conf. Light Scat. Nonspher. Part.: Theory Meas. Appl., New York, 1998,* 119–122.

von Ross, O. (1971). Method for the solution of electromagnetic scattering problems for inhomogeneous dielectrics as a power series in the ratio (dimension of scatterer)/wavelength. *J. Appl. Phys.* **42**, 4197–4201.

Voshchinnikov, N. V. (1996). Electromagnetic scattering by homogeneous and coated spheroids: Calculations using the separation of variables method. *J. Quant. Spectrosc. Radiat. Transfer* **55**, 627–636.

Voshchinnikov, N. V., and Farafonov, V. G. (1993). Optical properties of spheroidal particles. *Astrophys. Space Sci.* **204**, 19–86.

Voss, K. J., and Fry, E. S. (1984). Measurement of the Mueller matrix for ocean water. *Appl. Opt.* **23**, 4427–4439.

Wachniewski, A., and McClung, H. B. (1986). New approach to effective medium for composite materials: Application to electromagnetic properties. *Phys. Rev. B* **33**, 8053–8059.

Wait, J. R. (1955). Scattering of a plane wave from a circular dielectric cylinder at oblique incidence. *Can. J. Phys.* **33**, 189–195.

Wait, J. R. (1963). Electromagnetic scattering from a radially inhomogeneous sphere. *Appl. Sci. Res. Sect. B* **10**, 441–450.

Waldvogel, A. (1974). The N_0 jump of raindrop spectra. *J. Atmos. Sci.* **31**, 1067–1077.

Waldvogel, A., Schmid, W., and Federer, B. (1978a). The kinetic energy of hailfalls. I. Hailstone spectra. *J. Appl. Meteorol.* **17**, 515–520.

Waldvogel, A., Schmid, W., and Federer, B. (1978b). The kinetic energy of hailfalls. II. Radar and hailpads. *J. Appl. Meteorol.* **17**, 1680–1693.

Wallace, H. B. (1988). Millimeter-wave propagation measurements at the Ballistic Research Laboratory. *IEEE Trans. Geosci. Remote Sens.* **26**, 253–258.

Walsh, T. M. (1998). Polarimetric scattering characteristics of planar and spatial ice crystals at millimeter wave frequencies. Ph.D. Thesis, Pennsylvania State Univ., University Park, PA.

Walsh, T. M., and Aydin, K. (1997). Millimeter wave scattering from capped columns and bullet rosettes. *Prepr. Vol., Conf. Radar Meteorol., 28th, Austin, TX, 1997,* 292–293.

Wang, R. T. (1980). Extinction signatures of non-spherical/non-isotropic particles. *In* "Light Scattering by Irregularly Shaped Particles" (D. W. Schuerman, Ed.), pp. 255–272. Plenum, New York.

Wang, P.-K. (1982). Mathematical description of the shape of conical hydrometeors. *J. Atmos. Sci.* **39**, 2615–2622.

Wang, J. J. H. (1991). "Generalized Moment Methods in Electromagnetics: Formulation and Computer Solution of Integral Equations." Wiley, New York.

Wang, P.-K. (1997). Characterization of ice crystals in clouds by simple mathematical expressions based on successive modification of simple shapes. *J. Atmos. Sci.* **54**, 2035–2041.

Wang, D.-S., and Barber, P. W. (1979). Scattering by inhomogeneous nonspherical objects. *Appl. Opt.* **18**, 1190–1197.

Wang, Y. M., and Chew, W. C. (1993). A recursive T-matrix approach for the solution of electromagnetic scattering by many spheres. *IEEE Trans. Antennas Propag.* **41**, 1633–1639.

Wang, C. S., and Street, R. L. (1978). Measurements of spray at an air–water interface. *Dyn. Atmos. Oceans* **2**, 141–152.

Wang, R. T., and van de Hulst, H. C. (1995). Application of the exact solution for scattering by an infinite cylinder to the estimation of scattering by a finite cylinder. *Appl. Opt.* **34**, 2811–2821.

Wang, D.-S., Chen, H. C. H., Barber, P. W., and Wyatt, P. J. (1979). Light scattering by polydisperse suspensions of inhomogeneous nonspherical particles. *Appl. Opt.* **18**, 2672–2678.

Wang, R. T., Greenberg, J. M., and Schuerman, D. W. (1981). Experimental results of dependent light scattering by two spheres. *Opt. Lett.* **6**, 543–545.

Warner, C., and Hizal, A. (1976). Scattering and depolarization of microwaves by spheroidal raindrops. *Radio Sci.* **11**, 921–930.

Warren, S. G. (1984). Optical constants of ice from the ultraviolet to the microwave. *Appl. Opt.* **23**, 1206–1225.

Warren, J. L., Barrett, R. A., Dodson, A. L., *et al.* (1994). "Cosmic Dust Catalog, Vol. 14." NASA Johnson Space Flight Center, Houston, TX.

Waterman, T. H. (1954). Polarization patterns in submarine illumination. *Science* **120**, 927–932.

Waterman, P. C. (1965). Matrix formulation of electromagnetic scattering. *Proc. IEEE* **53**, 805–812.

Waterman, P. C. (1971). Symmetry, unitarity, and geometry in electromagnetic scattering. *Phys. Rev. D* **3**, 825–839.

Waterman, P. C. (1973). Numerical solution of electromagnetic scattering problems. *In* "Computer Techniques for Electromagnetics" (R. Mittra, Ed.), Vol. 7, pp. 97–157. Pergamon, Oxford.

Waterman, P. C. (1979). Matrix methods in potential theory and electromagnetic scattering. *J. Appl. Phys.* **50**, 4550–4566.

Waterman, T. H. (1988). Polarization of marine light fields and animal orientation. *Proc. SPIE* **925**, 431–437.

Watson, G. N. (1962). "A Treatise on the Theory of Bessel Functions." Cambridge Univ. Press, Cambridge.

Weeks, W. F., and Ackley, S. F. (1986). The growth, structure and properties of sea ice. *In* "Geophysics of Sea Ice" (N. Untersteiner, Ed.), pp. 9–164. Plenum, New York.

Weeks, W. F., and Gow, A. J. (1978). Preferred crystal orientations in the fast ice along the margins of the Arctic Ocean. *J. Geophys. Res.* **83**, 5105–5121.

Weil, H., and Chu, C. M. (1980). Scattering and absorption by thin flat aerosols. *Appl. Opt.* **19**, 2066–2071.

Weiss-Wrana, K. (1983). Optical properties of interplanetary dust: Comparison with light scattering by larger meteoritic and terrestrial grains. *Astron. Astrophys.* **126**, 240–250.

Welch, A. J., and van Gemert, M. C. J. (1992). "Tissue Optics." Academic Press, San Diego.

Wells, K. S., Beach, D. A., Keller, D., and Bustamante, C. (1986). An analysis of circular intensity differential scattering measurements: Studies on the sperm cell of *Eledone cirrhosa. Biopolymers* **25**, 2043–2064.

Wendling, P., Wendling, R., and Weickmann, H. K. (1979). Scattering of solar radiation by hexagonal ice crystals. *Appl. Opt.* **18**, 2663–2671.

West, R. A. (1991). Optical properties of aggregate particles whose outer diameter is comparable to the wavelength. *Appl. Opt.* **30**, 5316–5324.

West, R. A., and Smith, P. H. (1991). Evidence for aggregate particles in the atmospheres of Titan and Jupiter. *Icarus* **90**, 330–333.

West, R. A., Doose, L. R., Eibl, A. M., *et al.* (1997). Laboratory measurements of mineral dust scattering phase function and linear polarization. *J. Geophys. Res.* **102**, 16,871–16,881.

Westcott, S. L., Oldenberg, S. J., Lee, T. R., and Halas, N. J. (1998). Formation and adsorption of clusters of gold nanoparticles onto functionalized silica nanoparticle surfaces. *Langmuir* **14**, 5396–5401.

Wielaard, D. J., Mishchenko, M. I., Macke, A., and Carlson, B. E. (1997). Improved **T**-matrix computations for large, nonabsorbing and weakly absorbing nonspherical particles and comparison with geometrical-optics approximation. *Appl. Opt.* **36**, 4305–4313.

Wiersma, D. S., Bartolini, P., Lagendijk, A., and Righini, R. (1997). Localization of light in a disordered medium. *Nature (London)* **390**, 671–673.

Wiltse, J. C. (1981). Introduction and overview of millimeter waves. *In* "Infrared and Millimeter Waves" (K. J. Barton and J. C. Wiltse, Eds.), Vol. 4, Chap. 1. Academic Press, New York.

Winker, D. M., and Trepte, C. R. (1998). Laminar cirrus observed near the tropical tropopause by LITE. *Geophys. Res. Lett.* **25**, 3351–3354.

Wiscombe, W. J. (1980). Improved Mie scattering algorithms. *Appl. Opt.* **19**, 1505–1509.

Wiscombe, W. J., and Mugnai, A. (1986). Single scattering from nonspherical Chebyshev particles: A compendium of calculations. *NASA Ref. Publ.* **NASA RP-1157**.

Wiscombe, W. J., and Mugnai, A. (1988). Scattering from nonspherical Chebyshev particles. 2. Means of angular scattering patterns. *Appl. Opt.* **27**, 2405–2421.

Witkowski, K., Król, T., Zieliński, A., and Kuteń, E. (1998). A light-scattering matrix for unicellular marine phytoplankton. *Limnol. Oceanogr.* **43**, 859–869.

Wolff, M. J., Clayton, G. C., and Gibson, S. J. (1998). Modeling composite and fluffy grains. II. Porosity and phase functions. *Astrophys. J.* **503**, 815–830.

Woodard, R., Collins, R. L., Disselkamp, R. S., *et al.* (1998). Circular depolarization lidar measurements of cirrus clouds. *Proc. 19th Int. Laser Radar Conf., NASA Conf. Publ.* **NASA/CP-1998–207671/PT1**, 47–50.

Woolf, D. K., Monahan, E. C., and Spiel, D. E. (1988). Quantification of the marine aerosol produced by whitecaps. *Prepr. Vol., Conf. Ocean–Atmos. Inter., Am. Meteorol. Soc., Boston*, 182–185.

Wriedt, T. (1998). List of Electromagnetic Scattering Codes. URL: imperator.cip-iw1.unibremen.de/~fg01/codes2.html.

Wriedt, T., and Comberg, U. (1998). Comparison of computational scattering methods. *J. Quant. Spectrosc. Radiat. Transfer* **60**, 411–423.

Wriedt, T., and Doicu, A. (1997). Comparison between various formulations of the extended boundary condition method. *Opt. Commun.* **142**, 91–98.

Wriedt, T., and Doicu, A. (1998). Formulations of the extended boundary condition method for three-dimensional scattering using the method of discrete sources. *J. Mod. Opt.* **45**, 199–213.

Wu, R., and Weinman, J. A. (1984). Microwave radiances from precipitating clouds containing aspherical ice, combined phase, and liquid hydrometeors. *J. Geophys. Res.* **89**, 7170–7178.

Wu, Z. S., Guo, L. X., Ren, K. F., *et al.* (1997). Improved algorithm for electromagnetic scattering of plane waves and shaped beams by multilayered spheres. *Appl. Opt.* **36**, 5188–5198.

Wurm, G., and Blum, J. (1998). Experiments on preplanetary dust aggregation. *Icarus* **132**, 125–136.

Wurman, J. (1994). Vector winds from a single-transmitter bistatic dual Doppler radar network. *Bull. Am. Meteorol. Soc.* **75**, 983–994.

Wyatt, P. J. (1962). Scattering of electromagnetic plane waves from inhomogeneous spherically symmetric objects. *Phys. Rev.* **127**, 1837–1843.

Wyatt, P. J. (1968). Differential light scattering: A physical method for identifying living bacterial cells. *Appl. Opt.* **7**, 1879–1896.

Wyatt, P. J. (1969). Identification of bacteria by differential light scattering. *Nature (London)* **221**, 1257–1258.

Xiao, R., Chandrasekar, V., and Liu, H. (1998). Development of a neural network based algorithm for radar snowfall estimation. *IEEE Trans. Antennas Propag.* **36**, 716–724.

Xing, Z.-f., and Greenberg, J. M. (1994). Efficient method for the calculation of mean extinction. II. Analyticity of the complex extinction efficiency of homogeneous spheroids and finite cylinders. *Appl. Opt.* **33**, 5783–5795.

Xing, Z., and Hanner, M. S. (1997). Light scattering by aggregate particles. *Astron. Astrophys.* **324**, 805–820.

Xu, Y.-l. (1995). Electromagnetic scattering by an aggregate of spheres. *Appl. Opt.* **34**, 4573–4588.

Xu, Y.-l. (1996). Calculation of the addition coefficients in electromagnetic multisphere scattering theory. *J. Comput. Phys.* **127**, 285–298. [Errata: **134**, 200 (1997).]

Xu, Y.-l. (1997). Electromagnetic scattering by an aggregate of spheres: Far field. *Appl. Opt.* **36**, 9496–9508.

Xu, Y.-l., and Gustafson, B. Å. S. (1996). A complete and efficient multisphere scattering theory for modeling the optical properties of interplanetary dust. *In* "Physics, Chemistry, and Dynamics of Interplanetary Dust" (B. Å. S. Gustafson and M. S. Hanner, Eds.), pp. 417–420. Astron. Soc. of the Pacific Press, Provo, UT.

Xu, Y.-l., and Gustafson, B. Å. S. (1997). Experimental and theoretical results of light scattering by aggregates of spheres. *Appl. Opt.* **36**, 8026–8030.

Xu, Y.-l., and Gustafson, B. Å. S. (1999). Comparison between multisphere light-scattering calculations: (I) Rigorous solution and (II) DDA. *Astrophys. J.* **513**, 896–911.

Xu, L., Zhang, G., Ding, J., and Chen, H. (1997). Light scattering by polydispersions of randomly oriented hexagonal ice crystals: Phase function analysis. *Optik* **106**, 103–114.

Yanamandra-Fisher, P. A., and Hanner, M. S. (1999). Optical properties of non-spherical particles of size comparable to the wavelength of light: Application to comet dust. *Icarus* **138**, 107–128.

Yang, P., and Cai, Q. (1991). Light scattering phase matrix for spheroidal and cylindric large particles. *Chin. J. Atmos. Sci.* **14**, 345–358.

Yang, P., and Liou, K. N. (1995). Light scattering by hexagonal ice crystals: Comparison of finite-difference time domain and geometric optics models. *J. Opt. Soc. Am. A* **12**, 162–176.

Yang, P., and Liou, K. N. (1996a). Finite-difference time domain method for light scattering by small ice crystals in three-dimensional space. *J. Opt. Soc. Am. A* **13**, 2072–2085.

Yang, P., and Liou, K. N. (1996b). Geometric-optics–integral-equation method for light scattering by nonspherical ice crystals. *Appl. Opt.* **35**, 6568–6584.

Yang, P., and Liou, K. N. (1997). Light scattering by hexagonal ice crystals: Solution by a ray-by-ray integration algorithm. *J. Opt. Soc. Am. A* **14**, 2278–2288.

Yang, P., and Liou, K. N. (1998a). Single-scattering properties of complex ice crystals in terrestrial atmosphere. *Contr. Atmos. Phys.* **71**, 223–248.

Yang, P., and Liou, K. N. (1998b). An efficient algorithm for truncating spatial domain in modeling light scattering by finite-difference technique. *J. Comput. Phys.* **140**, 346–369.

Yang, P., Liou, K. N., and Arnott, W. P. (1997). Extinction efficiency and single-scattering albedo for laboratory and natural cirrus clouds. *J. Geophys. Res.* **102**, 21,825–21,835.

Yanovitskij, E. G. (1997). "Light Scattering in Inhomogeneous Atmospheres." Springer-Verlag, Berlin.

Yee, S. K. (1966). Numerical solution of initial boundary value problems involving Maxwell's equations in isotropic media. *IEEE Trans. Antennas Propag.* **14**, 302–307.

Yee, S. K., Chen, J. S., and Chang, A. H. (1992). Conformal finite difference time domain (FDTD) with overlapping grids. *IEEE Trans. Antennas Propag.* **40**, 1068–1075.

Yeh, C. (1964). Perturbation approach to the diffraction of electromagnetic waves by arbitrarily shaped dielectric obstacles. *Phys. Rev. A* **135**, 1193–1201.

Yeh, C., Woo, R., Armstrong, J. W., and Ishimaru, A. (1982). Scattering by Pruppacher–Pitter raindrops at 30 GHz. *Radio Sci.* **17**, 757–765.

Yousif, H. A., and Köhler, S. (1988). Scattering by two penetrable cylinders at oblique incidence. *J. Opt. Soc. Am. A* **5**, 1085–1104.

Yueh, S. H. (1997). Modeling of wind direction signals in polarimetric sea surface brightness temperatures. *IEEE Trans. Geosci. Remote Sens.* **35**, 1400–1418.

Zerull, R. H. (1976). Scattering measurements of dielectric and absorbing nonspherical particles. *Beitr. Phys. Atmos.* **49**, 168–188.

Zerull, R. H. (1985). Laboratory investigations and optical properties of grains. *In* "Properties and Interactions of Interplanetary Dust" (R. H. Giese and P. Lamy, Eds.), pp. 197–206.

Zerull, R., and Giese, R. H. (1974). Microwave analog studies. *In* "Planets, Stars and Nebulae Studied with Photopolarimetry" (T. Gehrels, Ed.), pp. 901–915. Univ. of Arizona Press, Tucson, AZ.

Zerull, R. H., Giese, R. H., and Weiss, K. (1977). Scattering functions of nonspherical dielectric and absorbing particles vs Mie theory. *Appl. Opt.* **16**, 777–778.

Zerull, R. H., Gustafson, B. Å. S., Schulz, K., and Thiele-Corbach, E. (1993). Scattering by aggregates with and without an absorbing mantle: Microwave analog experiments. *Appl. Opt.* **32**, 4088–4100.

Zhang, J., and Xu, L. (1995). Light scattering by absorbing hexagonal ice crystals in cirrus clouds. *Appl. Opt.* **34**, 5867–5874.

Zheng, W. (1988). The null field approach to electromagnetic scattering from composite objects: The case with three or more constituents. *IEEE Trans. Antennas Propag.* **36**, 1396–1400.

Zheng, W., and Ström, S. (1989). The null field approach to electromagnetic scattering from composite objects: The case of concavo-convex constituents. *IEEE Trans. Antennas Propag.* **37**, 373–383.

Zheng, W., and Ström, S. (1991). The null-field approach to electromagnetic resonance of composite objects. *Comput. Phys. Commun.* **68**, 157–174.

Zietz, S., Belmont, A., and Nicolini, C. (1983). Differential scattering of circularly polarized light as a unique probe of polynucleosome superstructures. *Cell Biophys.* **5**, 163–187.

Zrnic, D. S., and Doviak, R. J. (1989). Effect of drop oscillations on spectral moments and differential reflectivity measurements. *J. Atmos. Oceanic Technol.* **6**, 532–536.

Zrnic, D. S., and Ryzhkov, A. (1996). Advantages of rain measurements using specific differential phase. *J. Atmos. Oceanic Technol.* **13**, 454–464.

Zrnic, D. S., Balakrishnan, N., Ziegler, C. L., *et al.* (1993a). Polarimetric signatures in the stratiform region of a mesoscale convective system. *J. Appl. Meteorol.* **32**, 678–693.

Zrnic, D. S., Bringi, V. N., Balakrishnan, N., *et al.* (1993b). Polarimetric measurements in a severe hailstorm. *Mon. Weather Rev.* **121**, 2223–2238.

Zrnic, D. S., Raghavan, R., and Chandrasekar, V. (1994). Observations of copolar correlation coefficient through a bright band at vertical incidence. *J. Appl. Meteorol.* **33**, 45–52.

Zuffada, C., and Crisp, D. (1997). Particle scattering in the resonance regime: Full-wave solution for axisymmetric particles with large aspect ratios. *J. Opt. Soc. Am. A* **14**, 459–469.

Zurk, L. M., Tsang, L., Ding, K. H., and Winebrenner, D. P. (1995). Monte Carlo simulations of the extinction rate of densely packed spheres with clustered and nonclustered geometries. *J. Opt. Soc. Am. A* **12**, 1772–1781.

Index